An Advanced Treatise on *Meloidogyne*

Volume I: Biology and Control

edited by

J. N. Sasser

and

C. C. Carter

A Cooperative Publication of the
DEPARTMENT of PLANT PATHOLOGY
and the
UNITED STATES AGENCY for
INTERNATIONAL DEVELOPMENT

Printed by North Carolina State University Graphics.
1985

For information on the availability of
this publication, write to

Dr. J. N. Sasser
Department of Plant Pathology
North Carolina State University
Box 7616
Raleigh, N.C. 27695-7616
U.S.A.

Library of Congress Catalog Card Number
84-61978
ISBN 0-931901-01-4 (v.1)

North Carolina State University Graphics
Box 7226
Raleigh, North Carolina 27695-7226
U.S.A.

PREFACE

The world community has focused great attention over the past two decades upon the question of how we can feed our burgeoning human population. Even with the present famine in Africa, it is generally recognized that our world's natural resource endowments can support a population substantially larger than the present one. However for this increased support to occur, substantial areas of new land must be brought into crop production, and productivity must be increased by at least 3% annually on lands presently under cultivation. The proposed gain in productivity is a great challenge to crop-production specialists! If this rate of growth in productivity is to be realized, significant advances must be made in all components of the production systems of the developing world, including those of protecting crops against pests such as insects, pathogens, weeds and plant-parasitic nematodes.

Plant-parasitic nematodes are recognized in the United States and other developed countries as potentially serious constraints to crop productivity. This recognition did not occur generally, however, until after World War II when the study of plant nematodes began to receive attention concomitantly with the development of effective and economical management practices (including chemical control agents). The recognition of nematodes as serious deterrents to crop productivity did not occur simultaneously in the developing countries because there were very few trained nematologists to confirm the presence of damaging population levels and to initiate work toward development of appropriate management strategies and tactics.

The United States Agency for International Development (USAID) recognized the lack of attention to plant-parasitic nematodes in the developing nations in 1975 when it funded a project at North Carolina State University entitled "Research on Integrated Crop Protection Systems with Emphasis on the Root-Knot Nematodes (*Meloidogyne* spp.) Affecting Economic Food Crops in Developing Nations." The project was given a world-wide mandate and focused on the *Meloidogyne* species because they are: (a) widely distributed geographically, (b) adapted to a very wide host range, and (c) generally recognized as causing more economic damage to food crops than any other single group of plant-parasitic nematodes. Since problems caused by species of *Meloidogyne* are common to much of the developing world, it was reasoned that this concentration would attract great interest and commitment by scientists in those nations.

The activity, which came to be known as the International *Meloidogyne* Project (IMP), did draw the attention of scientists throughout the world as evidenced by the many collaborators primarily from more than 60 developing nations. A truly peer-type relationship was established between the collaborating country scientists and the IMP staff. Even though the IMP was conceptualized and funded before the Collaborative Research Support Programs (CRSP's) were established under Title XII, it embodies most of the CRSP characteristics. The collaborating scientists jointly planned the work that was to be done in the various regions of the world, and all of them worked on problems that were of common interest. In a few short years and with modest funding inputs through the IMP, the science of nematology was greatly advanced throughout the developing world. Scientists were recognized within their respective countries for their interest and expertise in nematology. Special in-service training was provided through workshops and conferences to provide a common focus and knowledge base for the projects. Work was initiated collaboratively within each participating country which demonstrated the need to manage plant-parasitic nematode populations below the damage threshold levels. Finally, various management strategies were evaluated as the basis for the identification of appropriate technology for each of the countries/regions.

In the process of conducting the project on a world-wide basis, scientists at North Carolina State University had the unique opportunity of assembling a vast amount of observational and research data on *Meloidogyne* spp., such as (a) geographic distribution patterns of species and races and the ecological factors that determine these patterns, (b) cytogenetic basis of species and their evolution, (c) morphology and taxonomy, (d) comparative biochemistry of species and races, and (e) sources of genetic resistance within food crops. The IMP provided the scientific base for the development of economical and ecologically sound management strategies for root-knot nematodes on a worldwide basis. Few groups of plant pathogens have been treated in such a comprehensive manner through a world network of collaborating scientists.

Perhaps, the most important product of the IMP is this publication *An Advanced Treatise on Meloidogyne* with Volume I devoted to *Biology and Control* and Volume II devoted to *Methodology*. These books will be of great value to crop-protection specialists, both researchers and extension specialists, throughout the world. They present state-of-the-art knowledge of this important group of plant pathogens, and a significant component of that knowledge is a product of the IMP. Both volumes represent the proceedings of the final conference conducted under the IMP at Raleigh, North Carolina in April, 1983. This Conference was attended by 95 scientists representing 39 countries.

North Carolina State University is proud of the achievements realized under the IMP. Appreciation is expressed to all of those who have collaborated on the project thereby making it an outstanding effort which will contribute much to enhanced food production throughout the world. We also express appreciation to those representatives of USAID who had the vision and determination to sustain the funding for the IMP during a 9-year period.

J. Lawrence Apple, Coordinator
International Programs
North Carolina State University

TABLE OF CONTENTS

Part One

Introduction to the Importance of Nematodes

CHAPTER 1

AGRICULTURAL RESEARCH AND FOOD PRODUCTION

N. C. Brady

United States Agency for International Development
Washington, D. C., USA

It is a priviledge for me today to be invited to open this three-week "Advanced Course on Root-Knot Nematodes." I am indeed impressed with the broad global representation of specialists from 43 nations with the common goal of managing and controlling root-knot nematodes to help ameliorate food and fiber production in your respective countries. Your efforts in basic and applied research and the subsequent transfer of the developed technology to growers through effective extension programs are to be commended. You are members of a global cadre of agricultural researchers dedicated to applying science and technology to solve perhaps humankind's most pressing problem—that of meeting the food and fiber demands of a growing global populace.

Daily world leaders of both developing and more developed countries are confronted with the question: can the world produce enough food to feed the human family as it adds another 70 million members annually? Can it simultaneously enable its poorest members to eat more and better food than their parents and grandparents did? Concomitantly, can we meet the increased expectation of a human population with more money to spend not only for more food but also for food of better quality such as meat and poultry?

Population increases in developing nations are due to traditionally high birth rates and to the fact that modern medical science is helping to save the lives of people who without such services could not survive. Infant and child mortality has decreased markedly in the past three decades, and longevity of the populace has increased accordingly. World population increased from 2.8 billion in 1950 to 4.4 billion in 1980. Even though some countries, especially those in East Asia, have had remarkable success in reducing their rates of population increase, overall increases have been staggering. They have, in many cases, surpassed increases in food production. Consequently, per capita food production has improved only slightly worldwide, and in some areas, such as Africa south of the Sahara, it has actually declined by about 15 percent over the past 12 years.

Today it is estimated that 75 percent of the people in sub-Sahara Africa are eating less than the recommended dietary allowance (3). The poorest of the populace in the low-income countries are, in some cases, worse off today than they were in the 1970's.

THE GREEN REVOLUTION

While the production of some cash crops such as rubber, tea, coffee, and others was commercially successful, increases in food crop production in most developing countries was slow and uneven until the 1950's. It was scarcely high enough to meet the basic needs for food and fiber of the growing populations. In the 1960's, leaders of both developing and more developed countries began to recognize the seriousness of the world food problem. In the meantime, some farsighted visionaries concerned with this global challenge were persuaded that agricultural research, properly focused, could help stimulate food production for a large part of the world's population. Research jointly sponsored by the Government of Mexico and the Rockefeller Foundation yielded fertilizer-responsive wheats that possessed the capacity to adapt to widely different ecological conditions. These wheats were tested in other countries, adopted by the farmer, and found to be markedly successful in increasing, and in some cases doubling (Turkey, China), the yield of this cereal grain.

The initial successes of the wheat and maize research in Mexico prompted the Rockefeller and the Ford Foundations to join the Government of the Philippines to establish in 1960 the first of a series of international centers of excellence dedicated to agricultural research on specific crops and commodities. The initial goal of this new center, The International Rice Research Institute (IRRI), was to improve rice production, and four years after the dedication of its buildings, IRRI released 'IR8', an improved rice variety that became the miracle rice. Together with the wheats from the research program in Mexico, IR8 helped trigger the Green Revolution.

Progress, however, was not confined to the work of these two international centers. Research networks of cooperating scientists in developing countries were quickly established. Not only did this accelerate the spread of the new grain varieties, but also more importantly, it encouraged the strengthening of national research institutions and the development of locally adapted varieties. It also changed the attitude of leaders worldwide toward science and research. National leaders from both developing and more developed countries began to give more than lip service to research. They provided funds for the development of national institutions and for the training of scientists and educators. The Consultative Group for International Agricultural Reasearch (CGIAR) was born, and ultimately 13 international agricultural research centers (IARCs) were established and supported.

As a result of such international and national cooperation, agricultural growth has been remarkable during the past three decades. Much of this growth can be attributed to developing countries' increased capacity to produce more food. In turn, this capacity is due, at least in part, to improved crop varieties and new technology, along with national policies that made it profitable for farmers to accept the new technologies. Global grain supplies have about doubled, and although the growth rates of agricultural output declined in the industrial nations, rapid growth occurred in the developing countries as a group (8). India tripled its wheat production in 20 years. Indonesia doubled its rice production in 10 years, and Colombia doubled its rice production in 5 years with the new varieties.

The success of the Green Revolution initially was achieved where environmental constraints were not too great. The new varieties flourished where there was an adequate supply of water and nutrients, without excessive water and flooding, and where the climate was favorable to their growth and yield production.

While successes with wheat and rice were commendable, overall worldwide increases in food production per capita have changed little in the past 30 years. In fact, for some of the lower income countries, per capita food production has actually decreased during this period. Food production increases have simply not kept up with population growth.

CHALLENGE OF THE FUTURE

Prospects for Future Food Production Increases in the Developing World

An encouraging potential exists for boosting food production in the developing world. Additional land can be brought into agricultural production, principally in Africa and Latin America. However, this is a costly process in time and finances, and involves surmounting many technical obstacles. For most of Asia and even in many countries of Africa and Latin America, the major means of improving food and fiber production in developing countries is to increase yields and intensify the cropping on land already under cultivation. This is very nearly the sole option for Asia where 75 percent of the world's undernourished live and where the majority of population surges up to the 21st Century are expected to occur. On that continent most of the arable land is already being cultivated, and extra food must come from increased per hectare yields.

To achieve these increased yields, farmers in the developing countries must continue to increase the yields of cereal grains and other crops in those areas where the environmental constraints are minimal, where the water and nutrient needs can be easily supplied. Additionally, the farmers must constantly upgrade their pest management systems and adopt cropping systems that permit two crops where only one is now grown and three where only two crops are now realized.

Areas with marked environmental constraints—too little water, adverse soils, and heavy pest infestation—must receive much greater attention in the future. Presently, in large areas and regions with such serious constraints, no applicable technology is available for improving food production in economically feasible ways. Such areas include semi-arid areas like the Sahel, central India, and some humid tropical areas notorious for flooding and for frequent droughts.

New Technology Needs

For the desired goal of improved indigenous food and fiber production in developing countries to be achieved, new technology will be needed. In the next generation of agricultural research, we must go beyond the Green Revolution to develop seeds that can produce crops under adverse climatic conditions and thrive with a minimum of costly inputs. We need to remove obstacles to the control of plant diseases and insect pests, and to overcome problem soil conditions, nutrient deficiencies, and excesses of water. We must help farmers break other production constraints that continue to prevent millions of third-world farmers from realizing more than a meager existence from their land and labor.

The need for new technology to modify and improve indigenous agricultural production is perhaps most acute in Africa. The prevalence of the deadly tsetse fly in Africa restricts animal production and thus precludes the effective use of forages and crop residues for animal feed in the humid regions of central Africa. This pest must be controlled before Africa's food production potential can be realized. Also population pressures are threatening the stability of the traditional slash-and-burn system of crop cultivation. Yields are being reduced and soil erosion increased because population pressures are forcing a shortening of the traditional 6 to 15 years of fallow time that is required in order for plant nutrients to be cycled adequately to crop root zone areas. Thus, ways must be found to improve or replace slash-and-burn systems of rural land use in much of the tropics.

In all developing countries every effort must be made to increase crop yields by economically sound practices. Analysis of the economics of the expansion of tillable land areas reveals that the cost per unit of additional food production is more than likely to rise. This expected increase is due to the fact that there will be expansion into land of poorer quality. Furthermore, there will likely be greater reliance upon agricultural inputs (fertilizers, pesticides, and others), and they are costly. Likewise, we can expect resource problems such as deteriorating soil and water quality, desertification and erosion. Appropriate technology has the potential to reduce unit costs while concomitantly improving the overall volume of agricultural production.

Pest Management—A Major Requirement For Increased Food Production

Future agricultural research must focus upon the integral parts of production, and the problem pests and pest management are crucial. Worldwide, pests are destroying about 35% of all food crops in the field. After the crops are harvested, up to 20% of the crops are destroyed by insects, microorganisms, rodents and birds. Thus, nearly 50% of the potential world food supply is being destroyed annually by pests (5). This situation occurs despite all efforts to manage and control pests with pesticides and alternate nonchemical controls. These enormous direct losses of food by pests cannot continue indefinitely without seriously constraining our ability to produce the food demanded by a growing world population. Adkisson (1) has pointed out that "improved pest control technology offers one of the best and quickest ways for increasing food production on a global scale."

The management of insect pests, plant and animal diseases, weeds and vertebrate pests (rats, birds) will continue to be heavily dependent upon nonchemical methods of control. Chemicals, however, will remain a helpful component of the overall strategy for pest management, particularly in crop production. For many years before and during World War II, simple inorganic chemicals, such as sulfur, arsenicals and fluorides, as well as soap and miscible oils had been used to control plant pests and diseases. Then around 1950, a new era in pest control began with the introduction of synthetic organic pesticides such as the insecticide DDT and the herbicide 2,4-D. Subsequently, numerous synthetic organic fungicides, nematicides, rodenticides, and other chemicals were developed and found to be effective and low in cost. They rapidly became the principal means of pest control in productive agricultural regions.

As the target organisms developed resistance to some of the widely used pesticides, a plethora of organochlorine, organophosphate, and carbamate pesticides was developed and marketed. Meanwhile, scientists and citizens of countries around the world were voicing grave concern about the hazards of these compounds to human health and the environment. Concomitantly, the continued use of organic chemical pesticides was found to be challenged by a variety of such factors as (1) the development of genetic resistance to pesticides by crop pests, (2) increased cost of these chemicals, and (3) lack of overall reduction in annual crop losses due to pests.

Rangaswami (6) cautioned that although the industrial nations have good cause for alarm over the ill effects of chemical pollution on humans and their environment, some developing countries have not reached such critical levels of pollution. He further observed that some developing countries have made progress in making farmers aware of the damage caused by pests and the benefits of chemical control. Rangaswami reported that a recent survey of farmers in India showed that only about 40 percent were aware of the benefits of plant protection measures and less than half of these adopted the measures scientifically to realize the maximum benefit. Obviously, there is a real need for educating farmers on the proper and careful use of plant protection chemicals and making them aware of the potential damages that can ensue with continued use of excessive doses of such agrichemicals. This lack of farmer education was reported by Kumar (4) when he noted that in west Africa poor education of farmers markedly influenced crop protection practices. He further observed that government-initiated extension services are not very effective in most west African countries.

In addition to the lack of education and training of farmers in developing countries regarding pest management practices, another problem is the fact that the vast majority of the world's small farmers have almost no ready access to pesticides. Davies (2) noted that the infrastructure for the acquisition and delivery of pesticides simply does not exist, and where it does exist, it is generally rudimentary. He noted that the problems of ensuring that the chemicals remain potent and available in sufficient quantity are formidable in many of the remote areas where they are needed. Additionally, the costs involved for farmers and government are large.

Many existing traditional agricultural systems do include some type of pest control practice. Kumar (4) noted that west African farmers practice traditional methods of pest control by manipulating crops and land. On the other hand, sociocultural considerations preclude the use of even "safe" insecticides on food crops in some developing countries because farmers are not educated on their safety, nonhazard characteristics to human reproduction, or efficacy. Thus, the application of an integrated pest management program in many areas of the developing world is difficult and requires remedial action, but the lack of trained personnel and available extension services must be overcome before this is possible. For example, in 1976, it was estimated that there were no more than 10 entomologists per 4.5 million people in Nigeria, and the ratio elsewhere in west Africa is even more lamentable (7).

Agency For International Development Programs

The U.S. Agency for International Development (AID) works with national agricultural programs in developing countries as a supporting partner in their research, education, and extension activities and projects. These national programs have two major foci: (1) building and strengthening the indigenous capacity to develop some improved technologies appropriate for local and regional conditions and (2) the capability to test and validate, under individual country conditions, the improved technology developed by regional and international research organizations. These AID programs are planned and designed to assist developing countries in forming, developing and/or strengthening self-reliant, self-perpetuating research capabilities that improve indigenous food and fiber production.

The agency's most extensive support for agricultural research is for that conducted at the thirteen international agricultural research centers around the world. The majority of the research programs of the international centers provide continuity in agricultural research.

These international agricultural centers provide training experience in research and extension, particularly for scientists and extension workers from national agricultural programs in developing countries. They also serve as a resource pool of agricultural specialists who can collaborate with counterparts in national programs in the design, planning, and evaluation of national and cooperative programs undertaken by developing countries.

Pest management and control are among the highest of priorities for these centers. While chemical control methods are being evaluated, perhaps greatest attention is being given to host tolerance and (or) resistance. Each plant-oriented center has collected genetic resources especially from the center of origin of the crop land in question. For example, the International Rice Research Institute has about 70,000 accessions in its gene bank. These accessions are being routinely evaluated for sources of resistance to major insect pests. Sources of resistance are quickly made available to cooperating scientists around the world.

The Agency further provides some support for Collaborative Research Support Programs (CRSPs) between national agricultural programs in developing nations and cooperating American universities, colleges and other institutions of learning and research. The primary focus of each of these CRSPs is research to increase the production of specific crops or animals, including fish. The expertise of 4 to 9 institutions in the United States is joined with that of an even greater number of cooperating institutions in the developing countries in each of CRSP's subject areas. Through this collaborative mechanism significant results are already being obtained on crops such as bean, cowpea, sorghum and millet, and on small ruminants. Future research efforts will be focused as well on peanut culture, soil management, and fishery assessments. Even though the CRSPs have not been underway nearly as long as the international centers, some significant benefits already have accrued from them.

Through the CRSP mechanism, both universities and vocational schools in developing countries are being strengthened through cooperative arrangements with counterpart institutions in developing countries and the United States. When feasible, both the training and associated institution-building are done in the United States, and the thesis research is carried out in developing countries. The participating institutions can be involved in the research networks mentioned previously.

Pest Management Strategies

The strategies used by AID and the programs it supports in developing nations are designed to minimize human health hazards and other adverse environmental impacts. Each Mission Bureau has an Environmental Officer who is concerned with pesticide residues and human health in the region served by the respective mission. Also, agency environmental coordinators are responsible for interacting with the pesticide industry.

AID has adopted regulations which require all pesticides recommended for the projects to which it provides support to be reviewed for their potential effects on humans and the environment. The agency recognized that pesticides are now an integral part of the modern agricultural technology used by developing countries to increase and maintain high crop productivity and also to save human lives.

Among the specialized programs dedicated to pest management strategies, research and other methods for increasing developing countries' indigenous capabilities for food and fiber production is the partial support AID provides for the Consortium for International Crop Protection (CICP). This non-profit organization was formed in 1978 by a group of U.S. universities for the principal purpose of cooperating with developing countries to reduce food crop losses caused by pests while concomitantly safeguarding the environment. CICP's basic goal is to advance economically efficient and environmentally sound crop protection practices in developing countries.

CICP-member institutions offer large and diverse technical resources in the areas of crop protection, integrated pest management and agromedicine. Training is a high priority; each year in collaboration with national institutions, CICP conducts training for developing country extension workers, researchers, agricultural and health officials, pest control practitioners and related workers. Upon request by the developing countries, CICP will provide technical specialists to assist in the design, implementation, and evaluation of projects in crop protection and pesticide management.

The Agency's Central Bureau of Science and Technology funds four other pest management projects: (1) weeds, (2) rodents, (3) pest management and related environmental protection, and (4) nematodes. I shall comment on only the last two of these efforts. The purpose of the more general pest management project is to furnish training and technical assistance, upon request from developing nations, to specialists in pest and pesticide management. This training is accomplished through a variety of activities designed to assist these scientists, extension specialists, and others working in pest management in planning, developing, and implementing plant pest and disease control programs which maximize food and fiber production while minimizing human health hazards and other adverse environmental impacts. Examples of the activities and training projects carried out under this pest management and related environmental protection project are:

—In-country pesticide management workshop/seminars for mid-level managers, researchers, and related workers from government, universities, and other local institutions;

—In-country pesticide residue sampling and analysis familiarization course for chemical technicians to receive an overview of this subject as it relates to their location conditions;

—Short courses in integrated crop protection in developing countries where pest control problems on crops grown by small farmers are most serious;

—Training courses on aerial and ground application techniques for aerial applicators, supporting ground crews, ground applicators and other key personnel of the requesting developing country.

The International *Meloidogyne* Project is the one that brings you all here today for this three-week advanced and intensive workshop on root-knot nematode methodology, biology and control. Since its inception in 1975, this Project, which is based at North Carolina State University in Raleigh, has achieved considerable progress in the study of these destructive plant parasites—root-knot nematodes. These pests are widespread in developing nations but often go unnoticed. Farmers simply assume that their yields should be low.

Through the concerted efforts of you and your colleagues who are working with the Project and are dedicated to this area of scientific speciality, integrated crop protection systems designed for efficacy in regions around the world are being proven successful. This achievement is something of which you and others involved in project conduct and management can be justifiably proud.

This workshop builds on the findings from the first five years of the Project and can now expand into those crucial areas of the development of root-knot resistant cultivars, cropping systems research for root-knot control, and the transfer of technology to growers by researchers and extension workers in developing countries. These areas bring us beyond the initial successes of the Green Revolution and direct agricultural production to ever-expanding supplies in developing countries.

There are other ways through which AID is encouraging cooperation among the scientists in the developing countries and the U.S. We are establishing Memoranda of Understanding with a small number of committed U.S. universities to give greater continuity to their involvement in international agriculture. We have a small grants program which encourages scientists in the developing countries to initiate small, but significant, research efforts. Some of the research efforts are being supported through the U.S. National Academy of Sciences and some directly through the Agency.

We have also developed the concept of a Joint Career Corp which initially will permit 30 U.S. university scientists to serve one third of their professional career with AID and remaining two-thirds with their respective universities. The AID service will be spent overseas as part of AID country missions. This program will help develop further opportunities for U.S. university involvement in overseas AID programs.

We are also reevaluating the agency's research priorities in both the field and in this country. We hope to increase the agency's research portfolio overall, although this will likely be a slow process. The enhanced research initiative will be focused on strengthening the capacities of the developing countries.

Our job ahead is not an easy one. This year, there is a surplus of food in the United States and in some of the other more developed countries, but worldwide, there are serious food shortages. Our friends in the developing countries want to be as self-sufficient as possible, and we should help them achieve that goal. They want our technical support and collaboration, not our charity. Projects such as the one you are now implementing will help achieve that objective.

I wish each of you in this workshop and in the Project continued excitement, perserverance, and success in your endeavors to manage and control root-knot nematodes. Thank you.

Literature Cited

1. Adkisson, P. L. 1981. Keynote address: essential role of pest control in human health and food production, pp. 21-25 *In* Proc. of Seminar and Workshop on Pests and Pesticide Management in the Caribbean, 3-7 November 1980. Consortium for International Crop Protection and the U.S. Agency for International Development.
2. Davies, J. C. 1983. Integrated approaches to pest management—principal pests of food, pp. 97-107 *In* Chemistry and world food supplies: the new frontiers, CHEMRAWN II, ed. L. W. Shemilt. Pergamon Press, Oxford. 664 pp.
3. International Development Conference. 1981. Executive summary, p.10 *In* Proc. of the 1981 Int. Dev. Conf.: The U. S. Stake in World Development.
4. Kumar, R. 1982. Status of crop protection in West Africa, pp. 33-36 *In* Proceedings of a course: concepts, techniques, and application of integrated pest management in rice in

West Africa, 10-28 January 1982, ed. E. A. Akinsola, B. Ouayogode, and I. Akintayo. West Africa Rice Development Association (WARDA) and Consortium for International Crop Protection (CICP).

5. Pimentel, D. 1983. Environmental aspects of pest management, pp. 185-201 *In* Chemistry and world food supplies: the new frontiers, CHEMRAWN II, ed. L. W. Shemilt. Pergamon Press, Oxford. 664 pp.
6. Rangaswami, G. 1983. Principal diseases of food crops, pp. 109-119 *In* Chemistry and world food supplies: the new frontiers, CHEMRAWN II, ed. L. W. Shemilt. Pergamon Press, Oxford. 664 pp.
7. Toye, S. A. 1976. The role of entomology in the economic development of Nigeria, pp. 3-10 *In* Proc. Ninth Annu. Conf. and Tenth Celebration of Entomol. Soc. Nigeria.
8. Walters, H. E. 1982. Agriculture and development. Finance and Development 19(3):6-11.

CHAPTER 2

PLANT-PARASITIC NEMATODES: THEIR THREAT TO AGRICULTURE

W. F. Mai

Department of Plant Pathology
Cornell University
Ithaca, New York, USA

According to my first nematology teacher, Dr. B. G. Chitwood, "The most severe nematology problems occur where good host crops are grown too frequently for too long on the same land." When our keynote speaker, Dr. Nyle Brady, was Director of Agricultural Research at Cornell, I went with him to observe damage to red beets by the sugar-beet cyst nematode. On this trip, Dr. Brady said that any time monoculture is practiced for too long a nematode problem is likely to appear. He added that if the nematode problem is not recognized the 19th year, it may appear the 20th year. Each year there is more evidence to support these observations.

Nematode damage often is overlooked. For example, in microplots, a 50% reduction in potato yields caused by *Globodera rostochiensis* cannot be predicted by inspection of the upper plant. Also, a 50% reduction in cabbage yields caused by *Heterodera schachtii* in microplots cannot be assessed accurately by inspection alone. The cabbage heads in the nematode-infested plots are much less firm and therefore, weigh less. In a large orchard experiment, we have been unable to detect a 50% yield reduction in apple yields by visual observation alone. For this type of damage to be evident to future observers, yield in plots with different levels of nematodes must be measured.

Nematode damage goes unnoticed all around us, and the yield losses may equal those associated with more severe and easily identifiable plant damage. Waste of increasingly more expensive water and fertilizer are other important indirect losses associated with nematode infestation. Nematode-damaged roots do not utilize water and fertilizers as effectively as uninjured roots do. In competitive situations, uncontrolled and often unrecognized root damage is the difference between profit and loss. Although monetary losses are considerable, I do not plan to give estimates of those due to nematodes. Because much of the damage goes undetected, and for other reasons, I do not think we have estimates which warrant quoting.

GEOGRAPHICAL DISTRIBUTION OF LOSSES

Although nematode problems occur in all areas of the world where crops are grown, the most evident damage occurs in warm areas because: 1) higher temperatures and longer growing seasons result in more generations/yr and thus higher nematode populations and more crop damage; 2) the greater number of susceptible crops per year in warm areas results in higher nematode build-up; 3) some of the more damaging species, such as *Meloidogyne incognita*, occur in warmer areas; and 4) more severe disease complexes occur in warmer areas. There are, however, important exceptions to this general statement. In cooler areas, severe crop damage is caused by *G. rostochiensis* to potato, *H. schachtii* to sugar beet, *H. avenae* to oat, *Anguina tritici* to wheat, and *Ditylenchus dipsaci* to a number of crops, including alfalfa or lucerne. Perennials with nematode-damaged roots have been shown to be more susceptible to cold injury than plants with undamaged roots. Often this kind of nematode damage is attributed to cold injury.

In general, nematodes cause more damage to plants growing in "old tired soils," that is, those which have been cultivated for a long period of time. Again, there are notable exceptions. For example, in areas recently cleared of forest trees for banana production, the nematode *Radopholus similis* causes severe damage within a few years when nematode-infected vegetative sets are planted. Likewise in areas recently cleared for onion production, *D. dipsaci* causes severe damage in a relatively short time if infected vegetative sets are planted.

HISTORY AND MILESTONES IN THE DEVELOPMENT OF PLANT NEMATOLOGY

Nematode parasites of man and animals have been known for a long time. The oldest reference to nematode parasites of man is from China, approximately 2700 B.C. Also in 4500 B.C., nematode parasites of animals were mentioned in early Egyptian records. Although world-wide recognition of nematodes as important causal agents of plant diseases did not occur until the middle of this century, plant-parasitic nematodes were studied in both the British Isles and Europe more than 100 years earlier.

I would like to mention and discuss briefly what I consider to be some of the milestones in the development of plant nematology. The first observation of a plant-parasitic nematode, the wheat gall nematode (*A. tritici*), occurred in 1743. In the 1850's, it was discovered that a root-knot nematode (*Meloidogyne* sp.) caused galls on cucumber roots. Later it was discovered that the sugar-beet nematode *H. schachtii* damaged sugar beet. Damage to sugar beet was widespread and, in some instances in Europe and in the USA, resulted in the closing of sugar factories. This was some of the first evidence of the widespread destructiveness of nematodes.

The introduction of practical field nematicides did not occur until the middle of this century—about 1950. These treatments were important not only in providing practical field control, but also in demonstrating just how well crops could grow and yield when nematode populations were reduced. It is difficult to overestimate the importance of this milestone. Today much of the crop damage caused by nematodes is still unrecognized. Such damage can be recognized only when crops growing on nematicide-treated areas are available for comparison in the same field. Likewise, much of the damage caused by complex diseases can be recognized only when crops growing on strips treated with a general biocide are available for comparison. In other words, much of the unrecognized nematode damage is potential yield which is not now being realized.

The discovery of the destructive potato cyst nematode *Globodera rostochiensis* on Long Island in the USA and the demonstration that burrowing nematodes cause spreading decline of citrus gave considerable publicity and visibility to the nematode problems of crop plants which, I believe, increased interest and support in plant nematology. The recognition of the serious damage caused by migratory endoparasitic nematodes and the nematodes feeding at root surfaces (ectoparasites) greatly increased the number of nematodes known to be pathogenic to plant roots and the number of diseases known to be caused by nematodes. These discoveries resulted in an entirely new area of plant pathology.

DISEASE COMPLEXES

Another important milestone was the realization that some of the most important root diseases are caused by nematodes in combination with other soil organisms (disease complexes). This general understanding was the result of the research results of a number of nematologists and plant pathologists. Some of the important research was conducted here in North Carolina by Powell and associates (1-9).

Although some progress has been made in recent years, we know very little about complex interactions among plant-parasitic nematodes, soil microorganisms (usually more than one) and host roots. The fungus-nematode and the bacterium-nematode interactions are so numerous and varied that they provide a wide-open field for profitable research. For example, weakly parasitic fungal and bacterial parasites can cause considerable damage once they gain entry into plant roots in the presence of feeding nematodes. One might say that nematodes create new plant-pathogenic fungi and bacteria and increase the pathogenicity of others.

Nematode attacks sometimes lower the resistance of plants to diseases caused by other organisms, such as vascular wilt diseases caused by fungi and bacteria. Thus the possibility of nematode attack must be considered by plant breeders when developing cultivars with resistance to fungi and bacteria. The basis for this so-called breakdown of resistance is not understood. Also the biochemical or physiological roles of nematodes in disease complexes are not known. In fact, in most instances it is not known if the presence of two or more pathogens on a single host has an additive or synergistic effect on disease severity.

VIRUS TRANSMISSION BY NEMATODES

The discovery that some Dorylaim plant-parasitic nematodes transmit viruses revealed a new way in which nematodes damage plants.

Because one or a few nematodes may transmit a virus, the level of control must be increased. Although nematodes are now known to transmit a number of soilborne viruses, the nature of virus transmission by nematodes is essentially unknown. Also the importance of nematodes in the damage caused by these virus diseases is understood incompletely. It appears likely that estimates of the economic importance of nematodes as virus transmitters will be increased as new information accumulates.

NON-PHYTOTOXIC NEMATICIDES

Another step forward was the development of new nematicides that are almost nonvolatile and considerably less phytotoxic to crop plants than most fumigants are. These nematicides are less difficult and expensive to apply than fumigants. Most importantly, some of these new nematicides can be applied at planting time thus making nematicide application a much more practical cultural practice.

It has been shown also that some non-phytotoxic nematicides have systemic action. Most effective movement in plants is downward. Thus nematodes in roots can be killed by spraying, which has a number of advantages over soil application. Although this practice is not now widely used, I am sure that for some diseases it will become economically important in the future.

PLANT RESISTANCE

One of the most economical and effective ways to control plant-parasitic nematodes is the growing of nematode-resistant plant cultivars. The importance of discoveries over the years showing that resistance to nematodes is inherited genetically and that resistance occurs in both cultivated and wild species cannot be overemphasized. Most of you are thinking that certainly resistance is inherited and that it occurs in cultivated and wild cultivars, but when I first started to search for resistance to the golden nematode, a well-known plant breeder said to me, "Do you think you can get resistance to those animals?"

We now have a sizeable number of good commercial cultivars with nematode resistance. These nematode-resistant varieties decrease yield losses, increase profits and result in more food and fiber. Still, we do not have anywhere near as many as we need or could have.

Information and techniques are available to produce many more resistant varieties. A sufficient number of nematologists and plant breeders have been trained to do the job. Funds are all that is needed to employ these scientists.

CROP ROTATION

Crop rotation is still the most widely used control measure and, when used properly, a very effective one. In recent years, increased information on nematode taxonomy, population dynamics and host ranges have increased the effectiveness of this control measure. With more information on these and related topics, rotation could become an even more effective control measure, particularly when used in pest management strategies. Recently, races of several important plant-parasitic nematodes which differ markedly in host range and pathogenicity have been found. Undoubtedly, there are other unrecognized races among plant pathogens. Knowledge of the existence and behavior of these races will greatly improve control by means of crop rotation schemes.

BIOLOGICAL CONTROL

Under natural conditions, plant-parasitic nematodes are attacked by numerous and varied soil organisms. These parasites and predators include fungi, bacteria, viruses, protozoans, other nematodes and invertebrates. Because the activities of these organisms and their effects on nematodes are little understood, very few of them have been used effectively in biological control. Recently there have been several noteworthy exceptions including: 1) a bacterial parasite *Bacillus penetrans* effective against the root-knot nematode, *Meloidogyne incognita*, and 2) a fungus *Verticillium chlamydosporium* effective against the oat cyst nematode. Although at present neither of these organisms is widely used as a biological control agent, both show great promise.

Now I would like to mention two setbacks in our progress. The first is the loss or restricted use of several good nematicides, particularly DBCP, which is a very effective nematicide for control of some diseases and which will be difficult to replace. Also, it appears likely that others might be lost or their use restricted. A second setback is the finding of Temik, and later other nematicides, in soil water. Such occurrences are relatively infrequent, but predictive models indicate that they will become a widespread phenomenon. In some soils, it will take a long time for nematicides to reach soil water, but they will get there in time.

These setbacks emphasize the need to learn to use nematicides properly and to find alternate control measures. Research on alternate control strategies must be strongly emphasized. In fact, we must put our "heart and soul" into this effort.

NEMATODE PROBLEMS OF MAJOR CROPS

One test of the importance of nematodes in world agriculture is whether or not they are important pathogens of leading food or fiber crops. Some of the major food crops do have important nematode diseases.

Soybean

Soybean is now not only a leading crop but also one of the most rapidly increasing crops in the world. Unless the soybean cyst nematode is controlled, however, it is impossible to grow soybean economically. This nematode is spreading rapidly throughout the soybean-growing areas of the world. Thus, there is every indication that it will become an even more important pest of this crop in the future.

Potato

This crop is the main food source in a number of areas of the world, particularly colder areas. *Globodera rostochiensis*, the potato root eelworm or golden nematode, is a limiting factor for growing potatoes in many of these areas. Over the years this nematode has been transported to most major potato-growing areas.

Sugar Beet

The sugar-beet cyst nematode is present in most sugar-beet growing areas, and if control measures are not practiced, profitable crops cannot be grown. The primary control measure is crop rotation. Thus, the nematode causes less sugar production per area of land.

Rice

Although rice is important as a world crop, little research has been done on nematode pathogens. One reason is that it is difficult to conduct research on root problems of rice. Most nematologists believe that nematodes are responsible for considerable unrecognized damage to this crop. Some of the nematode pathogens considered to be economically important are the white tip nematode, *Aphelenchoides besseyi*; the rice stem nematode, *Ditylenchus angustus*; the rice root nematodes, *Hirschmaniella* spp.; and the rice cyst nematode, *Heterodera oryzae*.

One way to evaluate the importance of nematode diseases to world agriculture is to rank them relative to damage caused by other diseases. World-wide crop damage caused by root-knot nematodes ranks very high when compared to damage by other disease agents. Some reasons for this high ranking are: 1) the world-wide distribution of root-knot nematodes, 2) their broad host range, 3) the destructive nature of diseases caused by root-knot nematodes, and 4) their role in many destructive disease complexes. Inestimable crop losses are caused by root-knot nematodes, and there is every indication that these losses will increase considerably in future years.

THE IMPORTANCE OF NEMATODES IN WORLD AGRICULTURE

The importance of nematodes in world agriculture can be judged by whether or not their damage is catastrophic to major crops of limited areas. Several plant nematodes are responsible for this kind of damage.

Rhadinaphelenchus on Coconut Palm

On some islands and other areas of the Caribbean, the coconut palm is a major source of oil for cooking, and copra is an important export crop. In addition to coconut being an important source of oil and export funds, fiber and matting are obtained from this crop. Red ring of coconut caused by *Rhadinaphelenchus cocophilus* is a very destructive disease and, on some islands, threatens the existance of this crop. The extensive losses caused by this nematode are a major catastrophe to natives of these areas.

Radopholus Similis on Banana and Plantain

Banana and plantain are important food sources and cash crops for families of many small growers in the tropics. The burrowing nematode *Radopholus similis* is probably the most important pathogen of these crops. It causes a devastating disease, often referred to as "Blackhead Toppling Disease," which essentially can eliminate a banana planting in a relatively short time. This is one of the most spectacular nematode diseases I have seen. The disease is particularly difficult for small growers to control because 1) they have insufficient land to rotate crops, often less than 1 acre; 2) they lack the necessary training to apply some soil treatments; 3) nematicides are too expensive; and 4) often fuel is unavailable for a hot-water treatment. As a result, this disease often results in a scarcity of food and cash for these poor families.

Radopholus Similis on Black Pepper

A black-pepper-yellows pathotype or race of *R. similis* is largely responsible for a very serious disease of black pepper. This disease is thought by many to be caused by a nematode-fungus complex. It has reduced greatly and is possibly killing off the black-pepper industry in parts of Indonesia.

Globodera Rostochiensis and Nacobbus Spp. on Potato

In the Puna area of Peru, two nematodes are combining to reduce potato yields greatly. These nematodes are the potato root eelworm and the false root-knot nematode, *Nacobbus* spp. The Puna is a poor area, and land holdings by individual families are extremely small. Thus, these families can ill afford these drastic yield reductions caused by the two nematodes.

Root-knot Nematodes on Vegetables in Tropics and Semi-tropics

Vegetables are much prized and badly needed for balanced nutrition in areas of the world where very little food is imported. Because of root-knot nematodes, particularly *Meloidogyne incognita*, it is very difficult and sometimes impossible to grow important vegetables, such as tomato, in the tropics and semi-tropics. The disease complex caused by root-knot nematodes and the bacterial wilt organisms is one of the most lethal that is known currently.

LOSSES DUE TO NEMATODES

Crop losses caused by nematode diseases are more serious in areas where food and fiber are scarce than they are in affluent countries. In general, nematodes cause greatest losses where soils have been cropped for a long time and scarcity of land has forced limited or no rotation. Often these are poorer areas of the world where food and fiber are needed urgently.

The most serious nematode damage is root damage which is difficult to diagnose by foliar symptoms. Nematodes spend most of their lives in roots or soil, and soil-borne root-feeding organisms are more difficult to control than foliar parasites. Plant-pathogenic nematodes will become even more important to world agriculture in the future because of their spread by man, and because of more intensive agriculture on better soils to satisfy the food and fiber needs of increasing world populations.

FUTURE CHALLENGES IN PLANT NEMATOLOGY

Disease Complexes

Many root diseases are caused by nematodes in combination with other soil organisms. Therefore, for the benefit of growers and consumers alike, we must control not just the nematode disease but the entire complex. In order for these disease complexes to be controlled better, the causal agents must be identified and their relative importance determined. Also, overlooked peripheral areas need to be investigated further, including changes induced in plant constituents by parasitism, immunology, and enzyme action, as well as physiology, biochemistry, and genetics of nematodes and other organisms involved. There is a great need for new and imaginative techniques to carry out such experiments. Because of the very considerable damage caused by disease complexes, significant "breakthroughs" in this type of research could be very rewarding in increasing crop yields.

Taxonomy

Correct species identification is basic to efficient nematode control. Nematologists need these identifications to carry out research, teaching, extension, and other activities. Thus, in the future, nematode taxonomy at all levels must be emphasized. Research must be conducted to develop more accurate nematode identification methods through the combined use of morphology, biology, genetics, and biochemical systematics.

Taxonomists in cooperation with other nematologists must be urged to study the taxonomic relations among closely related species and, particularly, the influence of environmental factors on morphologic variation. Research workers must be encouraged to preserve nematodes used in an experiment or series of experiments. Otherwise, considerable research results will become useless when new taxonomic data necessitates re-identification.

Taxonomists must become more familiar with the role of taxonomy in other areas of nematology. Including a taxonomist as an advisor or co-worker in appropriate nematological investigations should be encouraged. Last but not least, results of taxonomic research should be more readily available. Computers make this a practical possibility by aiding 1) data storage and retrieval, 2) rapid communication, and 3) perhaps more rapid identification.

Efforts should be made to use computers to make nematological information more readily available, e.g. host ranges, geographical distribution of species, and influence of factors such as temperature and soil type on population dynamics.

New Control Strategies

Although new approaches to control are needed for all growers, they are absolutely essential for small growers. Control strategies for small growers may be very different than those for large growers. Actually, very little has been done to develop nematode control strategies for small growers.

Plant Resistance and Tolerance

Because plant resistance and tolerance are safe, economical, and easy to use, a much greater research effort must be carried out to produce many more cultivars with resistance and tolerance to nematodes and disease complexes. Molecular genetics is a potentially promising tool for developing new cultivars and should receive considerable research effort. The value of conventional plant-breeding techniques is well documented. Therefore, these techniques should also receive major research emphasis.

Biological Control

In biological control, a major research effort should be to identify promising parasites and predators and develop practical ways to use them for nematode control. One of the most successful strategies has been to find soils in which plant-parasitic nematode populations are considerably lower than predicted and conduct research to find out why there are fewer nematodes present. The role of mycorrhizae in nematode diseases and disease complexes should be investigated also. Preliminary results indicate that mycorrhizae may play an important role in reducing severity of these root diseases when plant nutrient levels, particularly those of phosphorus, are low.

Nematicides

Research to produce effective, inexpensive, environmentally-safe, specific nematicides should be accelerated. The search should not just be for better nematicides similar to the ones now available, but attempts should be made to develop treatments such as 1) nematicides that biodegrade in soil, 2) chemicals that attract nematodes to lethal materials or specific areas of the soil, 3) chemicals that neutralize host attractants or hatching agents, 4) chemicals that prevent nematodes from maturing, or juvenile hormones which prevent reproduction, 5) chemicals that hasten maturity when juvenile stages are most damaging (anti-juvenile hormones), and 6) chemicals that reduce or stop nematode feeding.

CONCLUSIONS

In comparison to that of above-ground parts of plants, the protection of roots has been grossly neglected. I have often told grower groups that they would not tolerate such badly damaged roots if the roots grew in the air where they could be easily observed. If roots grew in the air, I do not believe that research administrators would tolerate the problem. They would insist that more funds be given to find ways to increase protection from root pests.

Seriously, everything possible must be done to convince farmers, scientists, and administrators of the importance of controlling root diseases. It must never be forgotten that this is an important research area. Control of root diseases on major food crops could well become one of the most dramatic developments in world agriculture in the future.

Literature Cited

1. Batten, C. K. and N. T. Powell. 1971. The *Rhizoctonia-Meloidogyne* disease complex in flue-cured tobacco. J. Nematol. 3:164-169.
2. Johnson, H. A. and N. T. Powell. 1969. Influence of root knot nematodes on bacterial wilt development in flue-cured tobacco. Phytopathology 59:486-491.
3. Melendéz, P. L. and N. T. Powell. 1967. Histological aspects of the *Fusarium* wilt-root knot complex in flue-cured tobacco. Phytopathology 57:286-292.
4. de Moura, R. M., E. Echandi, and N. T. Powell. 1975. Interaction of *Corynebacterium michiganense* and *Meloidogyne incognita* on tomato. Phytopathology 65:1332-1335.
5. Porter, D. M. and N. T. Powell. 1967. Influence of certain *Meloidogyne* species on

Fusarium wilt development in flue-cured tobacco. Phytopathology 57:282-285.

6. Powell, N. T. 1963. The role of plant-parasitic nematodes in fungus diseases. Phytopathology 53:28-35.
7. Powell, N. T. 1970. Disease complexes in tobacco involving nematode and fungal pathogens, pp. 85-89 *In* Proc. Vth Int. Tobacco Sci. Congr., Hamburg 1970.
8. Powell, N. T., P. L. Melendez, and C. K. Batten. 1971. Disease complexes in tobacco involving *Meloidogyne incognita* and certain soil-borne fungi. Phytopathology 61:1332-1337.
9. Yang, H., N. T. Powell, and K. R. Barker. 1976. The influence of *Trichoderma harzianum* on the root-knot *Fusarium* wilt complex in cotton. J. Nematol. 8:81-86.

OVERVIEW OF THE INTERNATIONAL *MELOIDOGYNE* PROJECT 1975-1984

J. N. Sasser and C. C. Carter
Principal Investigator, IMP, and Research Assistant, respectively
North Carolina State University
Raleigh, North Carolina, USA

One of the greatest challenges facing today's world is the feeding of an ever-increasing population. Low agricultural productivity in developing nations and inadequate means of food distribution contribute greatly to this problem. Factors responsible for minimal productivity include low-yielding varieties, poor soil fertility, and inadequate pest management practices, to name a few. Clearly, if world hunger is to be alleviated, such obstacles must be overcome.

One of the most widespread pests limiting world agricultural productivity is the root-knot nematode *Meloidogyne* species (5). Almost all of the plants that account for the majority of the world's food supply are susceptible to infection by this pest (3,4,5). Although average crop yield losses are thought to be about 5%, small farmers in developing nations commonly experience much larger losses (4). In some developing countries, root-knot nematode infection is so common and widespread that galled roots have been considered "normal." In such cases, poor stands or yields have been attributed to vague agricultural ailments such as "worn-out land" or "soil exhaustion."

Because root-knot nematodes occur throughout most of the world (Fig. 1), infect all major crop plants, and cause substantial reduction in crop yield and quality, scientists at North Carolina State University (NCSU) decided to initiate a worldwide investigation of the problem. In July of 1975, a grant from the United States Agency for International Development made this endeavor possible and led to the formation of the International *Meloidogyne* Project (IMP) headquartered at NCSU. The main goal of the IMP was to assist developing nations in increasing yields of economic food crops. The strategy for attaining this goal was to promote knowledge about root-knot nematodes and to protect crops from the damaging effects of these pests.

ORGANIZATION OF THE IMP

Although headquartered in the Department of Plant Pathology at NCSU, the IMP also established a cooperative or consultant liaison with the NCSU Departments of Genetics, Statistics, and Soil Science as well as the N.C. Agronomic Division. To broaden the scope of the project even further, six prominent foreign nematologists were invited to Raleigh along with six prominent U.S. nematologists for a general research and planning conference. On the basis of this conference, and future meetings with foreign scientists, the eight geographic regions of the IMP were established: Region I, Mexico, Central America and Caribbean; Region II, South America (excluding Brazil); Region III, Brazil; Region IV, West Africa; Region V, East Africa; Region VI, Asia; Region VII, Mediterranean and the Middle East; and Region VIII, India. Over 100 scientists from universities and research institutes around the world agreed to collaborate with IMP nematologists. Regional conferences have been conducted periodically for the planning of objectives and the discussion of research findings.

RESEARCH EMPHASIS

To facilitate IMP research in the areas of *Meloidogyne* biology, taxonomy, ecology, management and technology transfer, a live root-knot nematode collection has been established and is maintained at North Carolina State University. This collection of about 850 populations, the result of extensive sampling by Project cooperators, includes at least 15 species

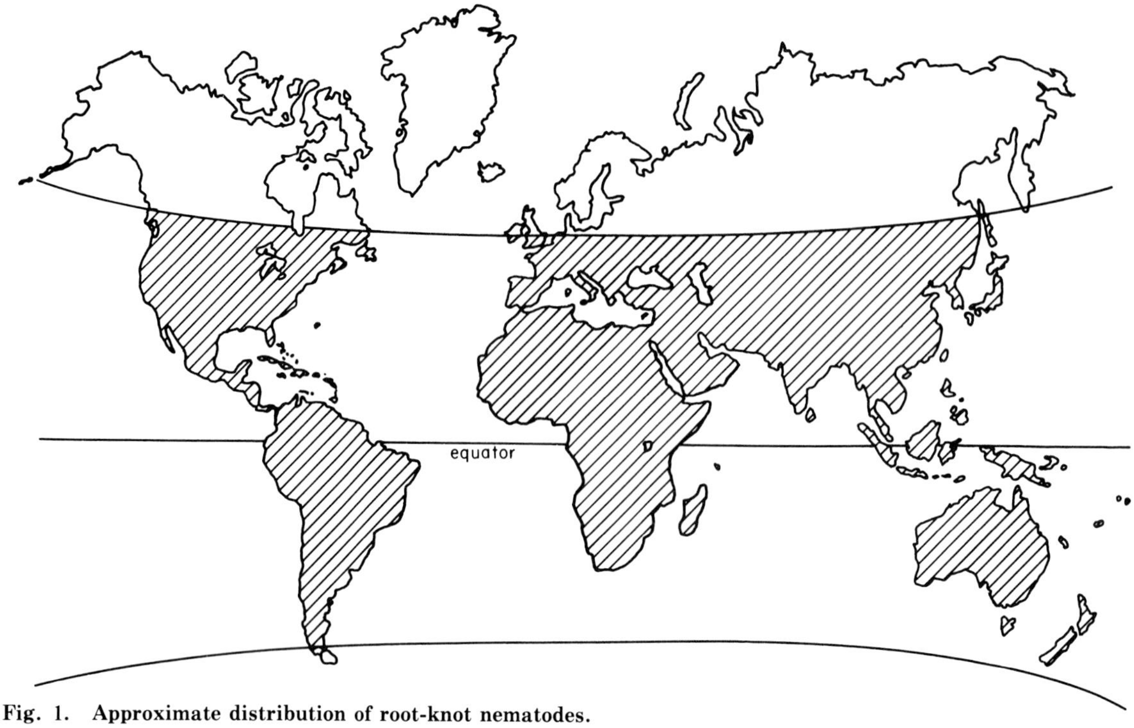

Fig. 1. Approximate distribution of root-knot nematodes.

and several host and cytogenetic races from over 70 countries.

Root-knot nematode egg masses, collected from infected plants all over the world, are sent in saline solution to the Project Center. There, seedlings of 'Rutgers' tomato, a highly susceptible host of root-knot nematodes, are inoculated with the egg masses. The nematode population increases rapidly on this host and can be maintained indefinitely by periodic transfer of eggs to new 'Rutgers' seedlings. Egg masses that have been collected from woody hosts (e.g. coffee, blueberry, azalea) or from grasses (e.g. rice, barley) are also inoculated onto the known host, in addition to tomato, if possible. Populations infecting such hosts are very likely to be uncommon, host-specific species.

Biology and Taxonomy

The early years of the Project were devoted to basic biological studies that provided a foundation for development of reliable identification techniques and management strategies. Because the more than 50 *Meloidogyne* species have significantly different host ranges, accurate identification of the species and race in question is essential to the selection of appropriate management strategies. In developing countries especially, where crop rotation is the primary and most practical management approach, simple and reliable identification techniques are necessary. To address this need, the IMP has explored differences in *Meloidogyne* host preferences, morphology, cytology, mode of reproduction, and biochemistry in order to discover more reliable and expedient characters for nematode identification.

NORTH CAROLINA DIFFERENTIAL HOST TEST Each population of nematodes received at the Project Center is subjected to the North Carolina differential host test. This test involves the inoculation of six standard host plants: cotton 'Deltapine 61,' tobacco 'NC 95,' pepper 'California Wonder,' watermelon 'Charleston Gray,' peanut 'Florunner,' and tomato 'Rutgers' (4). On the basis of host susceptibility (indicated by +) or resistance (indicated by —), pure populations of the four major species—*M. incognita*, *M. javanica*, *M. arenaria*, and *M. hapla*—can be easily distinguished (Table 1). Susceptibility and resistance are based on the average egg-mass and root-gall indices of three replications of the host differential.

The four major species account for more than 95% of the root-knot nematodes received from agricultural soils. Therefore, most pure sample populations encountered by the Project can be reliably identified by means of the North Carolina differential host test. Four host races of *M. incognita* and two of *M. arenaria* have

Table 1. North Carolina differential host test reaction chart.

Meloidogyne Species & Race	Differential Host Plants: Cotton 'Deltapine 61'	Tobacco 'NC 95'	Pepper 'California Wonder'	Watermelon 'Charleston Gray'	Peanut 'Florunner'	Tomato 'Rutgers'
M. incognita						
race 1	−[a]	−	+	+	−	+
race 2	−	+	+	+	−	+
race 3	+	−	+	+	−	+
race 4	+	+	+	+	−	+
M. javanica	−	+	−	+	−	+
M. arenaria						
race 1	−	+	+	+	+	+
race 2	−	+	−	+	−	+
M. hapla	−	+	+	−	+	+

[a] Box indicates key differential host plants.

been established on the basis of consistent responses of these nematodes to the differential hosts.

MORPHOLOGY Identification techniques, other than the North Carolina differential host test, are necessary if mixed populations and/or rare species are to be detected and verified. For this reason, much taxonomic research on *Meloidogyne* species is focused on distinctive morphological characters (1). Such characters include: the perineal pattern, which is an external series of markings in the perineal region of the adult female; male head shape; and stylet morphology.

Under a scanning electron microscope (SEM), minute morphological details can be examined thoroughly. The SEM gives a three-dimensional image with high resolution and large depth of focus of the external surface morphology. Since the SEM is expensive and not readily available to many nematologists, its primary value is the clarification of morphological details that often can be seen subsequently by light microscopy (LM). Only those features readily visible by LM will prove useful in the routine process of nematode identification. Identification of *Meloidogyne* species by means of perineal patterns, male head shape, and male morphology is much more rapid than by the differential host test. However, races cannot be distinguished by this method and mixed populations can easily escape detection.

REPRODUCTIVE AND CYTOLOGICAL CHARACTERS Over 500 root-knot nematode populations have been evaluated with regard to mode of reproduction and cytogenetics. The purpose of this work has been not only to discover reliable taxonomic characters but also to elucidate more fully the phyletic relationships and evolution of root-knot nematodes. Mode of reproduction and chromosome number show much promise as taxonomic characters (6). Some species reproduce only by cross-fertilization (amphimixis); other species reproduce without fertilization (mitotic parthenogenesis); still other species can reproduce with or without fertilization (facultative, meiotic parthenogenesis).

Races of some of the more common *Meloidogyne* species have been identified on the basis of chromosome number and mode of reproduction (Table 2). These races are called cytogenetic races and are designated by letters such as "A" or "B," instead of by numbers, so they can be distinguished from host races. No correlation has been found between host and cytogenetic races.

BIOCHEMISTRY Biochemical differences in protein and enzyme content among species are also being investigated. Major bands of esterase activity have been found to be noticeably different among the four major species (1). Other distinct isozyme patterns are being studied and show much promise for use as taxonomic characters.

Ecology

In addition to taxonomic research, the Project also conducts ecological studies. Once nematode populations are identified at the Project Center in Raleigh, the soil samples and ecological data that accompany each sample are analyzed. Based on this information, the geographical distribution of *Meloidogyne* species is characterized according to host range, temperature,

Table 2. Chromosome numbers and modes of reproduction of the four major *Meloidogyne* species.

***Meloidogyne* Species & Cytological Race**	**Range of Chromosome Numbers**	**Mode of Reproduction**
M. incognita		
race A	40-46	mitotic parthenogenesis
race B	32-36	mitotic parthenogenesis
M. javanica	42-48	mitotic parthenogenesis
M. arenaria		
race A	54 (50-56)	mitotic parthenogenesis
race B	34-37	mitotic parthenogenesis
M. hapla		
race A	14-17	facultative meiotic parthenogenesis
race B	30-32, 43, 45, 48	mitotic parthenogenesis

precipitation, soil composition, and other ecological characters.

Preliminary findings indicate that *M. incognita*, *M. javanica*, and *M. arenaria* inhabit areas with average annual temperatures between 15 and 33° C (5). *M. hapla* occurs more frequently in areas with average annual temperatures of less than 15° C (5). In areas of low rainfall (500 mm or less/yr), *M. javanica* has accounted for about two-thirds and *M. incognita* for one-third of the *Meloidogyne* species encountered in the survey (based on data in 5). In general, root-knot nematodes occur most frequently in soils with less than 10% clay, less than 30% silt, and at least 60% sand. Basically, however, root-knot nematodes can occur anywhere host plants grow.

Management

The current emphasis in IMP research is management of root-knot nematodes. Earlier studies concerning general biology and identification of *Meloidogyne* species have proven to be valuable prerequisites for research into management practices. Research efforts have been concentrated on five aspects of management: host resistance, cropping systems, chemical control, biological control, and integrated crop protection systems.

HOST RESISTANCE Much of the screening and evaluation of cultivars for resistance is being done in cooperation with International Agriculture Research Centers. Some of these centers and the crops on which cooperative work is being done appear in Table 3. Individual cooperators are also involved in screening for resistance. Some crops being studied include: cowpea in Nigeria, tomato in Nepal and Panama, bean and potato in Peru, and a wide range of vegetable crops in the Philippines.

CROPPING SYSTEMS Cropping systems research helps pinpoint crop plants that prevent increases in *Meloidogyne* populations. These plants can then be incorporated into cropping sequences or crop rotation schemes. Cropping sequences for each region are analyzed so those

Table 3. Cooperating research centers & the major crops being tested for root-knot nematode resistance

Research Center	**Crops**
Centro Internacional de la Papa (CIP)	potato
Asian Vegetable Research & Development Center (AVRDC)	Chinese cabbage, tomato
International Crops Research Institute for the Semi-Arid Tropics (ICRISAT)	pigeon pea, chickpea
International Institute of Tropical Agriculture (IITA)	cowpea
Centro Internacional de Agricultura Tropical (CIAT)	bean, *Desmodium*
Centro Internacional de Mejoramiento de Maíz y Trigo (CIMMYT)	maize, wheat, barley, triticale

that most effectively control root-knot nematodes and increase yields can be selected.

CHEMICAL CONTROL Investigations into chemical control include tests for nematicide efficacy and determination of application rates. In developing countries, however, nematicides often cannot be incorporated into control programs. The chemicals may be too expensive or application equipment may not be available. Even so, nematicide tests can serve a very useful purpose by demonstrating to farmers the yield losses attributable to nematodes and the benefits possible when the nematodes are controlled effectively.

BIOLOGICAL CONTROL Research into biological control of root-knot nematodes is being conducted in Peru and at the Project Center at North Carolina State University. The fungus *Paecilomyces lilacinus* occurs naturally in many soils. However, a particularly pathogenic isolate from soils in the Andes mountains has been found to control *Meloidogyne* populations. The fungus penetrates the nematode eggs, destroys the embryo and thus prevents hatch. Under experimental conditions, introduced fungal inoculum has persisted for at least a year in certain soils (2). Additional tests on its performance are being conducted under a wide range of environmental and edaphic conditions.

INTEGRATED CROP PROTECTION SYSTEMS The ultimate goal of root-knot nematode management studies is increased crop production, particularly food crop production, and development of integrated crop protection systems is a step in this direction. These systems will utilize the best combinations of resistant cultivars, crop rotation, nematicides, biological control agents, and sanitary and cultural practices. In this way, a broader spectrum of pests can be managed simultaneously.

TECHNOLOGY TRANSFER

Through the process of technology transfer, current research has already begun to focus on development of management strategies for root-knot nematodes. Basic information on *Meloidogyne* biology, ecology, and methodology is being made available to Project cooperators in developing countries so they can adapt or develop management strategies suited to their particular agricultural situations. The publication of books, conference proceedings, journal articles, and posters is all part of this effort. Certain publications have been made available in Spanish or in English with Spanish or Portuguese summaries.

Many IMP cooperators throughout the world have expressed interest in participating in the process of technology transfer by such methods as training technicians and farmers and setting up mobile agricultural libraries. These cooperators are encouraged to stay up to date through conferences and field and laboratory training sessions. As a result of the IMP's guidance and encouragement, nematological research capabilities in developing countries are continually advancing.

SUMMARY

The International *Meloidogyne* Project has made several achievements in keeping with its original goals. These include: 1) the promotion of awareness concerning the importance of root-knot nematodes in developing nations; 2) survey of the distribution, frequency and relative importance of *Meloidogyne* species and races infesting agricultural soils; 3) the discovery and description of new species; 4) the discovery of more reliable characters for identification; 5) clarification of phyletic relationships on the basis of cytogenetic and biochemical characteristics; 6) the elucidation of ecological factors affecting survival, distribution, and pathogenicity; and 7) the enhancement of research capabilities in developing nations through conferences, publications, and field and laboratory training sessions.

These accomplishments, along with increased communication and cooperation among nematologists throughout the world, are bringing the ultimate goal of increased food production in developing countries closer and closer to reality.

Literature Cited

1. Eisenback, J. D., H. Hirschmann, J. N. Sasser, and A. C. Triantaphyllou. 1981. A guide to the four most common species of root-knot nematodes (*Meloidogyne* species) with a pictorial key. Coop. Publ. Deps. Plant Pathol. and Genet., North Carolina State Univ., and U.S. Agency Int. Dev. Raleigh, N.C. 48 pp.

2. International Potato Center. 1980. A fungus biocontrol for root-knot nematode. CIP Circ. 8(10):1-3.
3. Sasser, J. N., C. C. Carter, and A. L. Taylor. 1982. A guide to the development of a plant nematology program. Coop. Publ. Dep. Plant Pathol., North Carolina State Univ., and the U.S. Agency Int. Dev. Raleigh, N.C. 21 pp.
4. Taylor, A. L. and J. N. Sasser. 1978. Biology, identification and control of root-knot nematodes (*Meloidogyne* species). Coop. Publ. Dep. Plant Pathol., North Carolina State Univ., and U.S. Agency Int. Dev. Raleigh, N.C. 111 pp.
5. Taylor, A. L., J. N. Sasser, and L. A. Nelson. 1982. Relationship of climate and soil characteristics to geographical distribution of *Meloidogyne* species in agricultural soils. Coop. Publ. Dep. Plant Pathol., North Carolina State Univ., and U.S. Agency Int. Dev. Raleigh, N.C. 65 pp.
6. Triantaphyllou, A. C. 1982. Cytogenetics and sexuality of root-knot and cyst nematodes, pp. 71-76 *In* Nematology in the southern region of the United States, ed. R. D. Riggs. South. Coop. Ser. Bull. 276. 206 pp.

Part Two

Taxonomic Approaches

SOME TAXONOMIC PRINCIPLES

Hedwig Hirschmann
Department of Plant Pathology
N. C. State University
Raleigh, North Carolina, USA

Taxonomy is the theory and practice by which kinds of organisms are delimited and classified (24). It is built upon the basic fields of morphology, physiology, ecology, biogeography, and genetics. The first step in the resolution of any kind of biological problem is the classification of the phenomena in an orderly system. This means ultimately the naming, description and classification of plants and animals.

MICROTAXONOMY AND MACROTAXONOMY

Microtaxonomy and macrotaxonomy may be recognized as the two subfields of taxonomy (25). Microtaxonomy deals with the methods and principles by which the various kinds of organisms are recognized and delimited. The basic entity is the species. Individuals must be identified before proper placement in the established classification. Identification deals with relatively few characters and is based on deductive reasoning. Macrotaxonomy, in contrast, is concerned with the methods and principles by which organisms are arranged in the form of classifications. Populations and taxa are assembled into groups and these in turn into larger and larger groups. Genera, families, orders, classes and phyla must be arranged in order to prevent a chaotic accumulation of species descriptions. More speculation and theorizing are involved. Classification evaluates many characters and is an inductive procedure.

Several theories of classification have been formulated over the years. Some of them originated in pre-Darwinian times.

Essentialism or Typology

This philosophy attempts to assign the variability of nature to a fixed number of basic types at various levels: i.e., all members of a given taxon conform to the same type. Variation is considered as trivial or irrelevant.

Nominalism

According to this philosophy, only individuals exist. All groupings, including species, as well as higher taxa, are considered artifacts of the human mind. Emphasis is placed on names coined by man and conferred to groups of similar individuals. Taxa are proposed on the basis of the degree of observed similarity, rather than the fact that they share a common heritage.

Numerical Phenetics

The main objective in this approach is the development of methods to allow quantification of degrees of similarity and thus convert a more or less subjective taxonomy into an objective, numerical taxonomy (28). The relationship of taxa is determined by calculating an overall, unweighted similarity value. All characters are equivalent: i.e., they have the same importance (weight) in classification. In addition, all phyletic information is ignored in order to be strictly objective.

Cladistics

According to this theory (17), classification should be based exclusively on genealogy: i.e., on the branching pattern of phylogeny. Phylogeny consists of a sequence of dichotomies, each representing the splitting of a parental species into two sister (=daughter) species. The ancestral species is assumed to cease to exist at the time of dichotomy, and the sister groups are given the same categorial rank. Classification is, therefore, according to "recency of common decent," and categorial status depends on the position of the branching points on the phylogenetic tree.

Traditional or Evolutionary Methodology (Eclectic Taxonomy)

This philosophy attempts to represent in the classification not only the branching of phyletic lines but also their subsequent divergence (24). Due weight is given to any unequal divergence of the descendent lines. The ranking of a taxon is based on the relative weight of the derived characters acquired by one sister group but not by the other. In addition to these traditional approaches, some of the more newly developed methods are also utilized in the construction of an evolutionary classification including certain numerical phenetic methods and the ancestral-derived partitioning of characters typical of cladistics.

Now that this brief introduction about taxonomy in general has been presented, the remainder of this chapter will deal with the more practical aspects of microtaxonomy only.

TAXONOMIC PROCEDURE

After a problem has been selected, the taxonomic procedure should follow a number of successive steps.

Collection and Care of Material

Ideally, the determination of the geographic variation of a species should be done in two steps: a) all the material already available in collections should be assembled and analyzed, and b) new material should be collected to fill the critical gaps shown in the preserved material of collections. Although preserved specimens are the indispensable basis of all taxonomic research, the information that can be derived from dead, preserved specimens is limited. The modern taxonomist needs much additional information which can only be derived from living or at least freshly preserved organisms. Due to the great variability of most natural populations, adequate samples should be collected throughout the range of the species in order to provide an unbiased population data set. If the species shows seasonal variation, collections should be made in various seasons. If various stages of the life cycle are of taxonomic importance, an effort must be made to have them all represented (in nematodes, e.g. infective juveniles, eggs).

Live cultures of organisms should be propagated in the laboratory or greenhouse (in case of plant-parasitic forms including nematodes) to provide specimens for histological, cytological or biochemical research in addition to the morphological studies.

Specimens should be preserved properly to be least subject to any subsequent deterioration. Satisfactory preservation of important taxonomic traits is necessary for later examination by other taxonomists.

The labeling of the specimen is also important. The following information should be recorded on the label: locality, date collected, habitat, host, number and sex of specimen, and other useful data.

Identification and Taxonomic Discrimination

Identification is the fundamental work upon which all progress in taxonomy is based. Every species name is based on a published description and also on type specimens. Identification or determination is the association of other specimens with the appropriate description or type specimen. In spite of inherent difficulties, correct identifications are possible for a majority of specimens in most groups of animals.

Before the actual taxonomic work begins, pertinent literature must be gathered, and the investigator should proceed according to the following successive steps.

1. Consult keys to orders and families.
2. Consult keys to genera and species if recent monographs or revisions are available.
3. If there is no recent monograph or revision, refer to the most recent catalog for literature citations leading to the description of all species known up to the time of completion.
4. Refer to current bibliographies for literature published subsequent to the most recent catalogs: *Zoological Record*, *Biological Abstracts*, *Helminthological Abstracts* (for nematological papers).
5. Although the secondary literature is often of great help, consider always the original or more recent authoritative descriptions.
6. If it is impossible to make a satisfactory species determination from the literature alone, use "authentic material" for comparison. Paratypes may be sufficient in most cases. Type specimens are the most authentic but are too valuable to be used for routine identifications. They should be used, however, in case of monographic studies.

When identifying material, particularly in less known groups, the taxonomist frequently comes across specimens that defy classification. They do not key out properly, or they do not agree with the specimens of the species with which they key out. The question arises: do these specimens belong to an undescribed species? Well-defined species are usually characterized by a) morphological and physiological differences, b) reproductive isolation, and c) ecological differences. If a taxonomist has accurate information on these three points, he is rarely in doubt, whether or not to classify a population as a new species. Presence or absence of morphological differences as well as geographic relationships can usually be determined. It is the lack of information on reproductive isolation which causes most of the difficulties. A high percentage of taxonomic errors can be avoided if the taxonomist has a thorough understanding of the amount of intraspecific, as well as geographic variation. Sibling species which did not acquire conspicuous morphological differences during speciation and are almost identical, may differ in their pathogenicity or biochemistry, and correct identification will be possible after detailed examination. The occurrence of polyploid populations raises particular difficulties since they are usually parthenogenetic (no gene exchange) and may also be morphologically very similar.

Preparation of Publication

After a taxonomic study is completed, there remains the important task of recording the findings and presenting them in a useful manner. Taxonomic publications range from short descriptions of a new taxon to lengthy monographs, revisions and reviews.

The main objective of a description is to aid in the subsequent recognition of a particular taxon. There are two kinds of descriptions which are not identical: the *diagnosis* and the *general description.*

DIAGNOSIS The diagnosis serves to distinguish the species from other known similar or closely related ones. It is a brief listing of the most important characters or character combinations that are peculiar to a given taxon and by which that taxon can be differentiated from other closely related ones (differential diagnosis). The characters should be presented in order of their diagnostic importance.

GENERAL DESCRIPTION The general description has a broader function. It gives general information and presents details not only on characters that are diagnostic in respect to already described species but also characteristics that may differentiate the species from yet undescribed species. Descriptions thus include all characters that are known to be useful or potentially useful in distinguishing other taxa at the same categorial level. The description should be presented in such a manner that subsequent workers will be able to identify specimens without having to refer to type specimens.

Within a particular group of organisms, it is preferable to standardize descriptions with respect to form and order of presentation, and thus increase their effectiveness and utility. In nematodes, e.g. the sequence of presentation would be to describe the various body parts, organs and structures from anterior to posterior. Available behavioral, ecological and other biological data should follow.

The description should be based on the entire variability of the species. Precise numerical data should be added and can be included in tables. The type specimens should be described briefly and their special features pointed out. Additional information on type specimens, prepared and deposited at a collection at the time of the original description, should also be included.

The style of the description should be telegraphic and concise, and is usually characterized by elimination of articles and verbs and by the selection of adjectives and nouns of precise meaning. How much subject matter should be included in a description depends on the group concerned and the state of knowledge in that group. Excessively long descriptions obscure the essential points; very short descriptions may omit pertinent data.

The description should be supplemented with illustrations (photographs, drawings, etc.) which in most instances are much superior to a verbal description. Preparation of keys is an integral part of the taxonomic procedure as well as a means of presenting findings. The title of a taxonomic article should be long enough to be specific as to the contents of the paper, but brief enough to be easily indexed. Every taxonomic paper should include an introductory part stating the scope of the paper. The essential methods used and material studied should be included. A summary is usually not necessary.

TAXONOMIC CHARACTERS

In microtaxonomy, as well as in macrotaxonomy, we are dealing with certain attributes of organisms which are known as taxonomic characters. "A taxonomic character is any attribute of a member of a taxon by which it differs or may differ from a member of a different taxon" (24). Taxonomic characters have a dual function: a) in microtaxonomy (= identification), they have a diagnostic aspect as indicators of difference at the level of lower categories (species or lower), and b) in macrotaxonomy (=

classification), they function as indicators of relationship to help in the correct delimitation and ranking of higher taxa (genera, families, orders, etc.).

Identification as well as classification are based on the proper analysis and evaluation (weighting) of taxonomic characters. The usefulness of a character depends on its information content: i.e., each character is assigned a certain weight according to its ability to reveal relationship (in the traditional evolutionary taxonomy; 24). Characters have also a differential significance with respect to categorial rank: i.e., different kinds of characters are useful at different categorial levels, such as subspecies, species, genera and higher taxa. To be truly diagnostic, a taxonomic character must be constant for all members of a given category and for no others. The use of taxonomic characters involves a study of their reliability. The same character does not always indicate the same amount of taxonomic difference. Highly variable characters have low weight and are poor indicators of relationship. Such characters include regressive traits, i.e. loss of certain structures.

Early taxonomists used morphological characters almost exclusively to distinguish taxonomic categories. Although morphological characters are still more useful than others, they are being supplemented to an ever-increasing extent by other kinds of characters, such as physiological, ecological, ethological and geographical characters (Table 1). This great broadening of the use of new kinds of taxonomic characters has increased the reliability of classifications.

Most of the taxonomic characters are of a qualitative nature (Table 1). The use of quantitative data (measurements and countable characters), however, is important because related species often differ not by the presence or absence of a structure, but rather by its size, or proportions, or number.

Meristic (countable) characters permit greater accuracy than measurements and are therefore favored wherever possible. Most important for measurements is that they be standarized and accurate. The particular distance measured should be specified exactly. It is also important to indicate how the specimens were prepared for measurement. Widely deviating measurements should be checked for possible errors in the measuring technique or in recording. Although overall body length may be very variable in a population, especially in species in which the adults continue to grow, the proportions of the various body parts to each other may remain rather constant. For this reason, taxonomists often use ratios or percentages rather than linear measurements.

Quantitative data should be evaluated through statistical methods. Two essential statistics are the arithmetic mean M and the standard deviation SD which measures variability of a sample. For evaluation of variability of a given character, the coefficient of variability CV can be calculated. In highly heterogeneous samples, more complicated methods of analysis (multivariate analysis) have to be employed (See Vol. II, Chapter 15). Computer programs have only recently been initiated for the identification of plant-parasitic nematodes (R. Fortuner, personal communication).

Table 1. Kinds of taxonomic characters [from Mayr (24)].

Morphological characters

a. General external morphology
b. Special structures (e.g. genitalia)
c. Internal morphology (anatomy)
d. Embryology
e. Karyology (and other cytological differences)

Physiological characters

a. Metabolic factors
b. Serological, protein, and other biochemical differences
c. Body secretions
d. Genetic sterility factors

Ecological characters

a. Habitats and hosts
b. Food
c. Seasonal variations
d. Parasites
e. Host reactions

Ethological characters

a. Courtship and other ethological isolating mechanisms
b. Other behavior patterns

Geographical characters

a. General biogeographical distribution patterns
b. Sympatric-allopatric relationship of populations

TAXONOMIC CHARACTERS USEFUL IN NEMATOLOGY

A wealth of different kinds of taxonomic characters have been employed in nematology (Table 1), and only a few examples can be presented here to highlight the various categories.

Morphological Characters

All morphological characters listed in Table 1 have been used in nematode taxonomy.

The light microscope has been employed extensively in morphological studies. More recently, however, the application of the scanning electron microscope (SEM) has revealed a number of reliable external characters, especially in the anterior and posterior body regions (2,18; see also Chapters 6,7,8). The SEM has been useful as a taxonomic tool particularly in animal and plant-parasitic nematodes including *Meloidogyne* spp., and it was found that all life stages of nematodes possess characters of taxonomic value. SEM characteristics help to support and complement our present classification of lower as well as higher categories and are especially useful in supplementing species descriptions.

Transmission electron microscope (TEM) studies can demonstrate differences in the structures of certain organs better than light microscope investigations, especially in regard to higher categories (1).

Quantitative morphological characters, such as certain measurements and morphometric relations (a, b, c, d, V, etc.), and countable characters (number of gonads, body annules, incisures in lateral field, head sensory structures, etc.) are utilized in the elucidation of the taxonomic position of taxa. Body width, esophagus length and stylet length are usually correlated with body length, and these relations are useful in the clarification of family relationships (15,16). Measurements and ratios, however, are subject to considerable variation, and some of the ratios are more reliable than others: e.g., V is very constant and useful, while c is rather variable and thus unreliable (14). Changes in morphometrics of certain organs and structures occur also during postembryonic growth and differences in this respect have been found in various taxa (11, 12, 13, 14).

Morphology of juvenile stages and eggs can also provide good taxonomic information (19).

Cytogenetic differences involving chromosome numbers, chromosome morphology, and mode of reproduction are very important. They should always be considered, if available, for characterizing nematode genera, species and subspecific categories (See Chapter 9).

Physiological Characters

Several aspects of molecular taxonomy have been employed in nematology. In most cases, soluble proteins have potential as taxonomic criteria, and studies on DNA are still limited to few nematode species and need greater emphasis in the future.

PROTEIN ELECTROPHORESIS This technique produces specific protein and enzyme profiles and has been successfully tested in a considerable number of different free-living, plant-parasitic and animal-parasitic nematodes (27), including *Meloidogyne* species (4, 5, 6, 23).

SEROLOGICAL STUDIES Such studies on the proteins of free-living and plant-parasitic nematodes are sparse (22). Although some studies with plant-parasitic forms were successful in species separation (21, 26), other results were highly variable, and reliable differentiation of races could not be achieved by immunoelectrophoresis (30). In general, attempts at serological identification of nematodes are preliminary but encouraging.

DNA STUDIES This work has been initiated only recently in nematodes and is being extended to the determination of cellular DNA content, base composition, renaturation behavior, and restriction endonuclease fragmentation (27).

Ecological Characters

Most plant-parasitic nematodes are ectoparasites and are characterized by considerable ecological and biological variability. These forms can attack a wide range of different host plants. A few groups, such as cyst and root-knot nematodes, however, have become highly specialized sedentary plant parasites feeding on one site and may have coevolved with their hosts (29). The highly specialized life habits of such sedentary forms usually have resulted in a limitation of their host range to a single genus, family or group of related plant families, and some of these species can be differentiated on the basis of their host preferences. In addition, specific host reactions can also be useful for the separation of taxa.

Ethological Characters

Behavior patterns are important isolating mechanisms in animals since new adaptations are often initiated by changes in behavior.

There are a number of behaviors that might be important in nematode taxonomy, since differences can occur among related taxa. These behaviors include feeding and attraction to host plants in plant parasites, as well as attraction to the opposite sex (3). Most of these behaviors are widespread, but some are highly specialized, such as the nictating in "dauer larvae" of Rhabditids and the swarming ("nest" formation) in certain Tylenchids, and thus can be useful for the differentiation of taxa. Differences among taxa can occur in feeding habits, duration of feeding and host-plant parts preferred. Behavioral differences due to sex pheromones may indicate reproductive incompatibility and thus can lead to recognition of new species (20).

Geographical Characters

Studies of the distribution of organisms have revealed well-defined geographic patterns. Such patterns can be of taxonomic value, especially for the interpretation of evolution of higher taxonomic categories. Present-day biogeographical distribution patterns have hardly been studied in nematodes. Ferris et al., 1981 (8), investigated the phylogeny of the family Leptonchidae and proposed a biogeographic hypothesis for this group based on cladistic analyses. These authors have also used vicariance (=allopatry) interpretation, as an alternative to the classical dispersal model, to explain present-day distributions in several other nematode groups (7, 9, 10).

Literature Cited

1. Baldwin, J. G. 1983. Fine structure of body wall cuticle of females of *Meloidodera charis*, *Atalodera lonicerae*, and *Sarisodera hydrophila* (Heteroderidae). J. Nematol. 15:370-381.
2. Corbett, D. C. M. and S. A. Clark. 1983. Surface features in the taxonomy of *Pratylenchus* species. Rev. Nématol. 6:85-98.
3. Croll, N. A. and M. V. K. Sukhdeo. 1981. Hierarchies in nematode behavior, pp. 227-251 *In* Plant parasitic nematodes, Vol. III, ed. B. M. Zuckerman and R. A. Rohde. Academic Press, New York. 508 pp.
4. Dalmasso, A. and J. B. Bergé. 1978. Molecular polymorphism and phylogenetic relationship in some *Meloidogyne* spp.: Application to the taxonomy of *Meloidogyne*. J. Nematol. 10:323-332.
5. Dickson, D. W., D. Huisingh, and J. N. Sasser. 1971. Dehydrogenases, acid and alkaline phosphatases, and esterases for chemotaxonomy of selected *Meloidogyne*, *Ditylenchus*, *Heterodera* and *Aphelenchus* spp. J. Nematol. 3:1-16.
6. Dickson, D. W., J. N. Sasser, and D. Huisingh. 1970. Comparative disc-electrophoretic protein analyses of selected *Meloidogyne*, *Ditylenchus*, *Heterodera*, and *Aphelenchus* spp. J. Nematol. 2:286-293.
7. Ferris, V. R. 1983. Phylogeny, historical biogeography and the species concept in soil nematodes, pp. 143-161 *In* Concepts in nematode systematics, ed. A. R. Stone, H. M. Platt, and L. F. Khalil. Academic Press, New York. 388 pp.
8. Ferris, V. R., J. M. Ferris, and C. G. Goseco. 1981. Phylogenetic and biogeographic hypotheses in Leptonchidae (Nematoda: Dorylaimida) and a new classification. Proc. Helminthol. Soc. Wash. 48:163-171.
9. Ferris, V. R., C. G. Goseco, and J. M. Ferris. 1976. Biogeography of free-living soil nematodes from the perspective of plate tectonics. Science 193:508-510.
10. Ferris, V. R., C. G. Goseco, and J. M. Ferris. 1980. Revisions of *Oxydirus* and *Tarjanius* n. gen. in Oxydiridae, Belondiroidea (Nematoda: Dorylaimida); and *Oxydiroides* in Prodorylaimidae, Dorylaimoidea. Purdue Univ. Res. Bull. 965:1-29.
11. Geraert, E. 1978. On growth and form in nematoda: Oesophagus and body width in Tylenchida. Nematologica 24:127-158.
12. Geraert, E. 1978. On growth and form in nematodes: II. Oesophagus and body width in Dorylaimida. Nematologica 24:347-360.
13. Geraert, E. 1979. Growth and form in nematodes: III. Comparison of oesophagus and body shape. Nematologica 25:1-21.
14. Geraert, E. 1979. Growth and form in nematodes: IV. Tail length and vulva position. Nematologica 25:439-444.
15. Geraert, E. and D. Crappe. 1981. Morphometric comparison of the Tylenchida species (Nematoda). Meded. Fac. Landbouwwet. Rijksuniv. Gent. 46:713-720.
16. Geraert, E. and L. Jacobs. 1981. Morphometric comparison of the Dorylaimida species (Nematoda). Meded. Fac. Landbouwwet. Rijksuniv. Gent. 46:721-727.
17. Hennig, W. 1966. Phylogenetic systematics. University of Illinois Press, Urbana, Ill. 263 pp.

18. Hirschmann, H. 1983. Scanning electron microscopy as a tool in nematode taxonomy, pp. 95-111 *In* Concepts in nematode systematics, ed. A. R. Stone, H. M. Platt, and L. F. Khalil. Academic Press, New York. 388 pp.
19. Hirschmann, H. and A. C. Triantaphyllou. 1979. Morphological comparison of members of the *Heterodera trifolii* species complex. Nematologica 25:458-481.
20. Huettel, R. N., D. W. Dickson, and D. T. Kaplan. 1984. *Radopholus citrophilus* sp. n. (Nematoda), a sibling species of *Radopholus similis*. Proc. Helminthol. Soc. Wash. 51:32-35.
21. Hussey, R. S. 1972. Serological relationships of *Meloidogyne incognita* and *M. arenaria*. J. Nematol. 4:101-104.
22. Hussey, R. S. 1979. Biochemical systematics of nematodes—a review. Helminthol. Abstr., Series B, Plant Nematol. 48(4):141-148.
23. Janati, A., J. B. Bergé, A. C. Triantaphyllou, and A. Dalmasso. 1982. Nouvelles données sur l'utilisation des isoestérases pour l'identification des *Meloidogyne*. Rev. Nématol. 5:147-154.
24. Mayr, E. 1969. Principles of systematic zoology. McGraw-Hill Book Co., New York. 428. pp.
25. Mayr, E. 1982. The growth of biological thought. Diversity, evolution, and inhertance. The Belknap Press of Harvard University Press, Cambridge, Mass. 974 pp.
26. Misaghi, I. and M. A. McClure. 1974. Antigenic relationship of *Meloidogyne incognita, M. javanica*, and *M. arenaria*. Phytopathology 64:698-701.
27. Platzer, E. G. 1981. Potential use of protein patterns and DNA nucleotide sequences in nematode taxonomy, pp. 3-21 *In* Plant parasitic nematodes, Vol. III, ed. B. M. Zuckerman and R. A. Rohde. Academic Press, New York. 508 pp.
28. Sokal, R. R. and P. H. A. Sneath. 1963. Principles of numerical taxonomy. W. H. Freeman, London (Revised ed., 1973).
29. Stone, A. R. 1979. Co-evolution of nematodes and plants. Symb. Bot. Ups. XXII (4):46-61.
30. Webster, J. M. and D. J. Hooper. 1968. Serological and morphological studies on the inter- and intraspecific differences of plant-parasitic nematodes *Heterodera* and *Ditylenchus*. Parasitology 58:879-891.

CHAPTER 5

THE CLASSIFICATION OF THE FAMILY MELOIDOGYNIDAE

Hedwig Hirschmann
Department of Plant Pathology
N. C. State University
Raleigh, North Carolina, USA

HISTORIC BACKGROUND

Over the years the genus *Meloidogyne* has been classified in various ways regarding its assignment to higher categories. Earlier, the root-knot nematodes were included in the family Heteroderidae which had been established by Filipjev, 1934 (6), for both root-knot and cyst nematodes. Skarbilovich, 1947 (14), included in the family the two subfamilies Heteroderinae and Tylenchulinae, but Thorne, 1949 (17), redefined the family and recognized Heteroderinae as the only subfamily with the genera *Heterodera, Meloidogyne* and *Meloidodera*. According to Chitwood and Chitwood, 1950 (1), the family comprised three subfamilies: Heteroderinae, Hoplolaiminae and Nacobbinae. In 1959, Skarbilovich (15) redefined Heteroderidae by its sexual dimorphism and included in it five subfamilies: Heteroderinae, Tylenchulinae, Sphaeronematinae, Nacobbinae and Meloidogyninae. The latter was proposed as a new subfamily to contain *Meloidogyne* and *Meloidodera*. Paramonov, 1967 (13), and Wouts and Sher, 1971 (20), recognized only the two subfamilies Heteroderinae and Meloidogyninae. Golden, 1971 (7), raised the family Heteroderidae to superfamily Heteroderoidea with the families Heteroderidae and Nacobbidae. In addition to Heteroderinae and Meloidogyninae, he recognized the new subfamily Meloidoderinae. Wouts and Sher, 1971 (20), amended Thorne's definition of the family Heteroderidae and redefined the two subfamilies Meloidogyninae and Heteroderinae. These authors acknowledged *Meloidogyne* as the only genus in Meloidogyninae. In 1973, Wouts (18) added the subfamily Ataloderinae to the family, and since Meloidogyninae seemed distinctly different, he removed the latter from the Heteroderidae and established an independent family Meloidogynidae, whose diagnosis was amended in 1979 (19). In addition to *Meloidogyne, Meloidoderita* was transferred as a valid genus to the family Meloidogynidae by Wouts in 1973. In the same year, however, *Meloidoderita* was re-examined by Kirjanova and Poghossian (12) and placed in the new family Meloidoderitidae which was rightfully included in Criconematoidea.

Meloinema was described by Choi and Geraert in 1974 (3) and was placed in the Meloidogynidae at that time, but Husain in 1976 (10) proposed to transfer it to the new subfamily Meloineminae, based on the fact that the young adult female is vermiform. The same year that Choi and Geraert published their description of *Meloinema*, Golden and Jensen (8) described the new genus *Nacobbodera* and placed it in a new subfamily, Nacobboderinae. However, according to Stone, 1977 (16), *Nacobbodera* is very similar to *Meloinema* with respect to morphology, female development and life habits. He, therefore, considered it as a junior synonym of *Meloinema*. The subfamily Meloineminae Husain, 1976, thus became a junior synonym of Nacobboderinae Golden and Jensen, 1974.

Meloidoderella indica Khan and Husain, 1972 (11), a cyst-forming species with predominantly *Meloidogyne*-like characters was placed by Husain, 1976 (10), in a new subfamily Meloidoderellinae.

Presently, therefore, there are in the family Meloidogynidae three monotypic subfamilies with the genera *Meloidogyne, Meloinema* and *Meloidoderella* (Table 1).

DIFFERENTIATING CHARACTERS OF THE GENERA OF MELOIDOGYNIDAE

Of the three genera of Meloidogynidae, *Meloidogyne* and *Meloinema* (2,3) are well characterized, but *Meloidoderella* (11) is inadequately described. Little information is available on juveniles and males, no illustrations are presented in the original description, and thus the validity of this genus is greatly doubted. According to the original description, *Meloidoderella* has some relationship to *Heterodera*, apparently showing characters morphologically intermediate between *Heterodera* and *Meloidogyne*.

Females (Table 2; Figs. 1,2,5,6)

A cystoid stage with lace-like cuticle pattern is present only in *Meloidoderella*. *Meloidogyne*

Table 1. Present classification of the family Meloidogynidae.

Subfamily	Genus
Meloidogyninae Skarbilovich, 1959	*Meloidogyne* Goeldi, 1887
Nacobboderinae Golden & Jensen, 1974 (Syn. Meloineminae Husain, 1976)	*Meloinema* Choi & Geraert 1974 (Syn. *Nacobbodera* Golden & Jensen, 1974)
Meloidoderellinae Husain, 1976	*Meloidoderella* Khan & Husain, 1972

Table 2. Differentiating characters of females of the genera of Meloidogynidae.

	Meloidogyne	*Meloinema*	*Meloidoderella*
Cyst stage	Absent	Absent	Present
Body shape	Pear-shaped; in some species oval with posterior protuberance	Young, preparasitic female vermiform; posterior end of parasitic female swollen; neck long	Oval-shaped
Cuticle pattern	Entire body annulated; perineal pattern in posterior body region	Entire body annulated; no perineal pattern in parasitic female	Females with obscure perineal pattern; cysts with lace-like pattern of ridges and punctations
Stylet	10-25 μm long; in majority of species 11-17 μm long	23-25 μm long	13-16 μm long
Excretory pore	Opposite or anterior to median bulb	Slightly posterior to stylet knobs in young female; generally anterior to median bulb in parasitic female	At level of median bulb
Vulva	Terminal	Subterminal, 167-200 μm from tail end in parasitic female	Terminal
Anus	Terminal, close to vulva	Subterminal; 60-85 μm from tail end in parasitic female	Located at edge of posterior vulval lip

and *Meloinema* remain soft-bodied with regular body annulations. The mature adult females are oval to pear-shaped, but in *Meloinema*, the young preparasitic female is vermiform. Vulva and anus of *Meloinema* remain also subterminal in the parasitic female, whereas in the other two genera, these body openings are terminal in position. A perineal pattern is formed in *Meloidogyne* and *Meloidoderella*; such a pattern is absent in the parasitic adult female of *Meloinema*. The cephalic framework of *Meloinema* is more sclerotized, and the stylet is larger than those of *Meloidogyne*. *Meloidogyne* and *Meloinema* have the excretory pore shifted anterior to the median bulb, whereas *Meloidoderella* apparently has it located further posterior.

Males (Table 3; Figs. 3,4,7)

Males of the three genera are similar in general body shape, although those of *Meloidoderella* appear to be much smaller in size. Stylet and spicule size is largest in *Meloinema*, and the tail is characteristically shaped; it is irregular conical with nipple-like terminus. The cephalic framework is heavily sclerotized in *Meloinema*, and the stylet is more robust as compared to *Meloidogyne*. The life history of *Meloinema* is not known, but it may be assumed that juvenile males feed since males are large in size with well-developed anatomical features. Little information is available about males of *Meloidoderella*. They seem to have stylets and spicules of moderate size in comparison.

Juveniles (Table 4; Figs. 3,4,7)

Except for *Meloidogyne spartinae*, which has exceptionally long juveniles (up to 912 μm), *Meloidogyne* juveniles usually are short and slender with mean body lengths of 320-543 μm. Juveniles of *Meloinema* are larger and more robust with an average body length of 650 μm. They have a more heavily sclerotized cephalic framework and a larger stylet as compared to species of *Meloidogyne*. The hemizonid is located posterior to the excretory pore, whereas in most species of *Meloidogyne*, it is anterior to the excretory pore. *Meloidoderella* juveniles are short (380-470 μm) with small stylets and resemble in these characters juveniles of *Meloidogyne*.

Table 3. Differentiating characters of males of the genera of Meloidogynidae.

	Meloidogyne	***Meloinema***	***Meloidoderella***
Body length	Up to 2.0 mm long	Up to 2.5 mm long	0.98 mm long
Stylet	13-33 μm long; in majority of species 16-25 μm long	34-40 μm long; stylet knobs strongly developed	21 μm long
Tail	Usually short	Short, tapering, irregular conical; ventral side flattened	No information
Spicules	19-44 μm long; in majority of species up to 37 μm long; spicule head small	50-64 μm long, narrow; slightly curved with finely rounded tips	28 μm long

Table 4. Differentiating characters of juveniles of the genera of Meloidogynidae.

	Meloidogyne	***Meloinema***	***Meloidoderella***
Cephalic framework	Weakly developed	Well developed	No information
Stylet	9-23 μm long; in majority of species 10-15 μm long; stylet knobs generally small	23-29 μm long; stylet knobs large	14-16 μm long

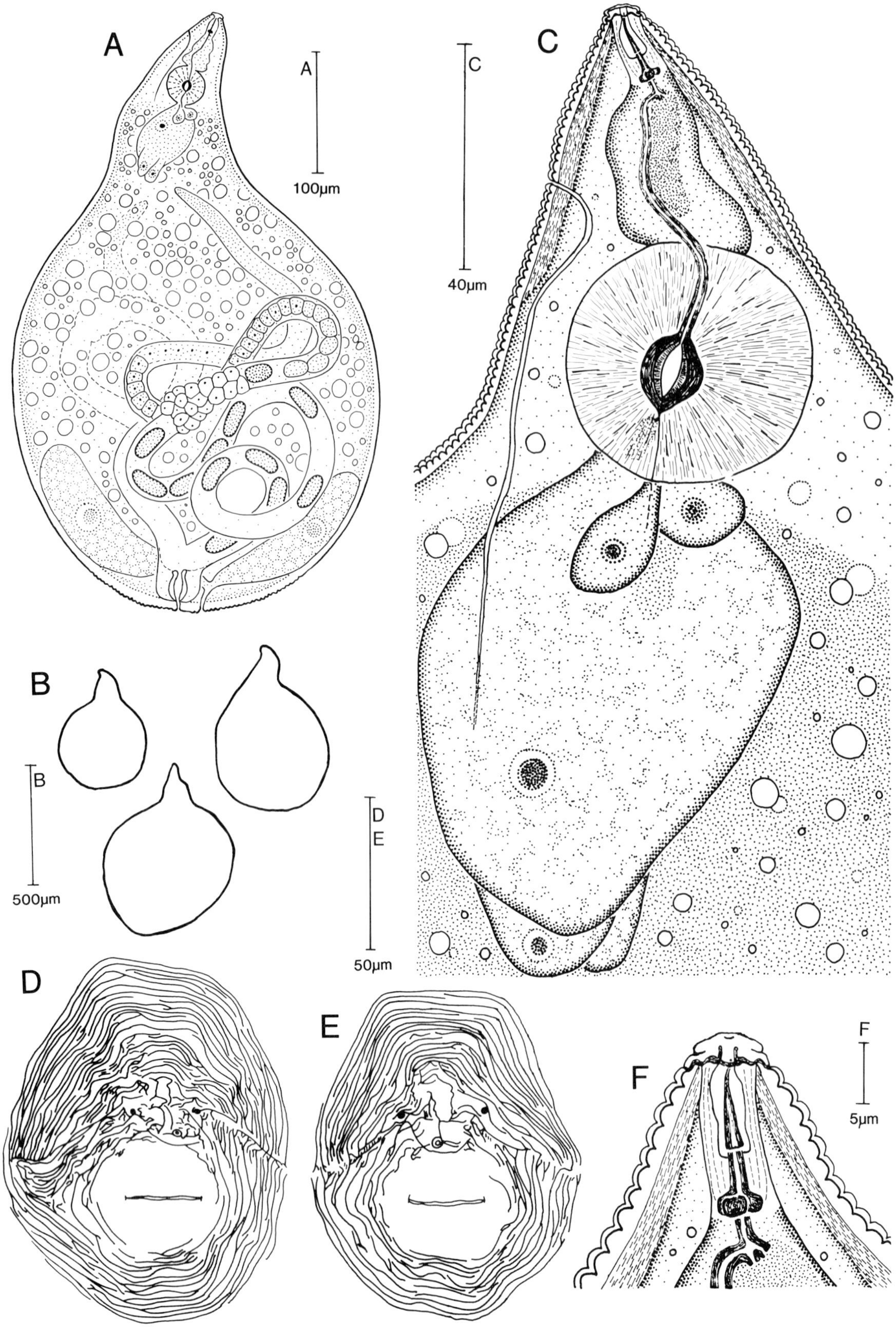
A
A
100µm
C
C
40µm
B
B
500µm
D
E
50µm
D
E
F
F
5µm

Fig. 1. (left). Line drawings of females of *Meloidogyne microcephala* Cliff & Hirschmann. A) Gross morphology. B) Outlines of entire females. C) Esophageal region (lateral). D,E) Perineal patterns showing variation typical for *M. microcephala*. F) Cephalic region (lateral). [B-F from Cliff & Hirschmann (5)].

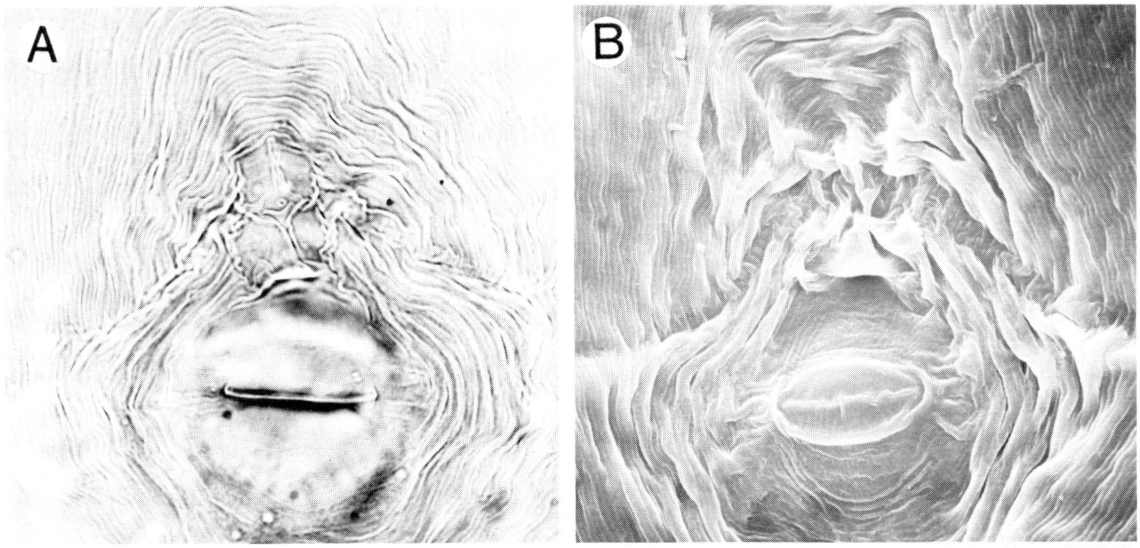

Fig. 2. LM and SEM photographs, respectively, of perineal pattern of *Meloidogyne microcephala*. [from Cliff & Hirschmann (5)].

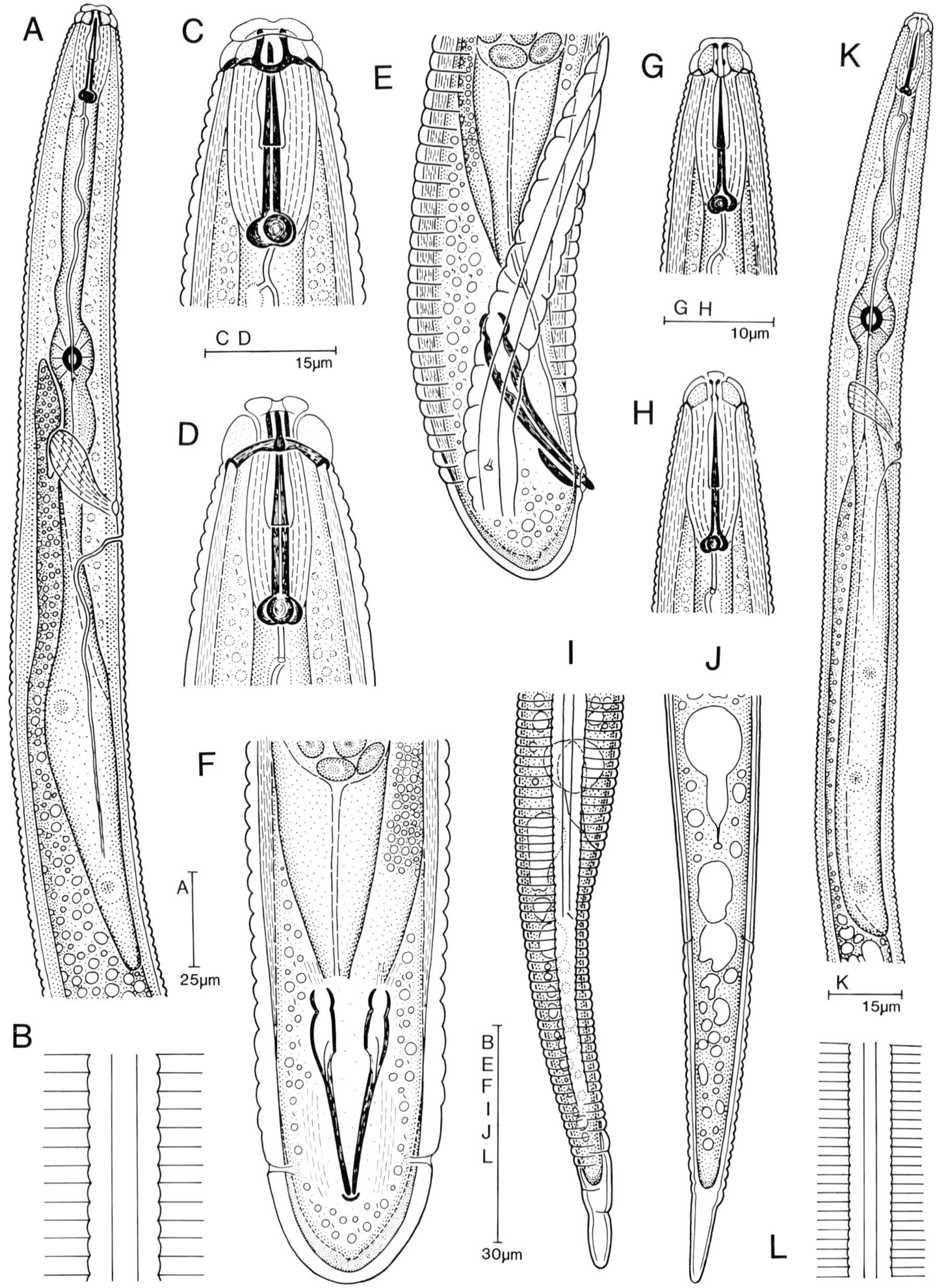
A
C
E
G
K
C D
15μm
G H
10μm
D
H
I
J
F
A
25μm
K
15μm
B
B
E
F
I
J
L
30μm
L

Fig. 3. (left) Line drawings of males (A-F) and second-stage juveniles (G-L) of *Meloidogyne platani* Hirschmann. A) Esophageal region (lateral). B) Lateral field near midbody. C) Cephalic region (lateral). D) Cephalic region (dorsal). E) Tail (lateral). F) Tail (ventral). G) Cephalic region (lateral). H) Cephalic region (dorsal). I) Tail (lateral). J) Tail (ventral). K) Esophageal region (lateral). L) Lateral field near midbody. [from Hirschmann (9)].

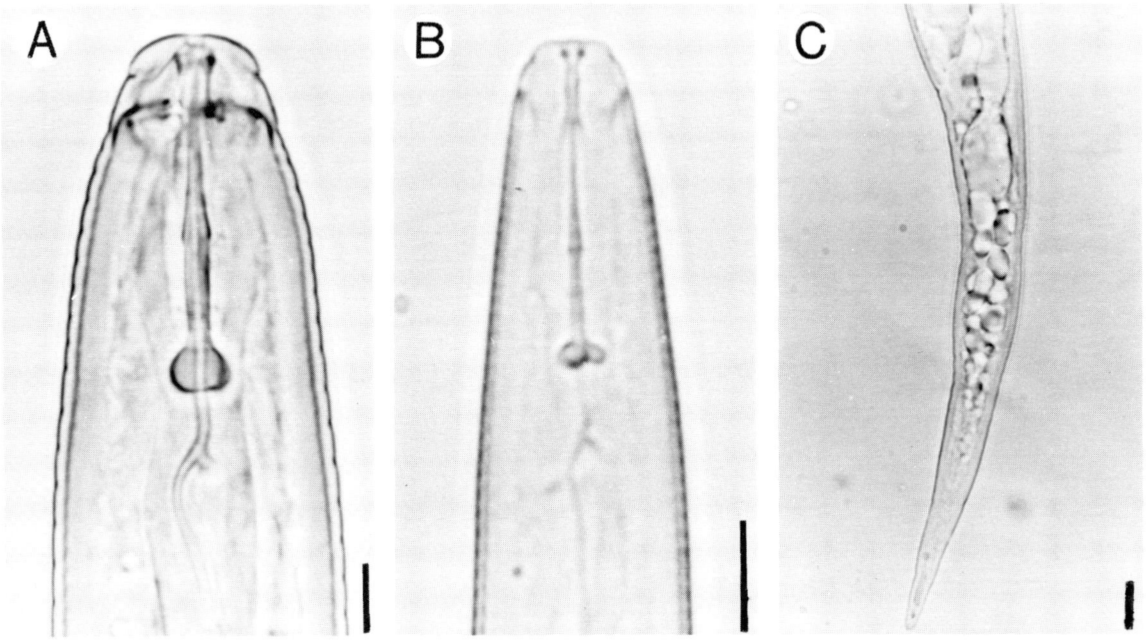

Fig. 4. LM photographs of male and second-stage juveniles of *Meloidogyne arenaria* (Neal) Chitwood. A) Anterior portion of male (lateral). B) Anterior portion of second-stage juvenile (lateral). C) Tail of second-stage juvenile (lateral). [from Cliff (4)]. Each bar = 5 μm.

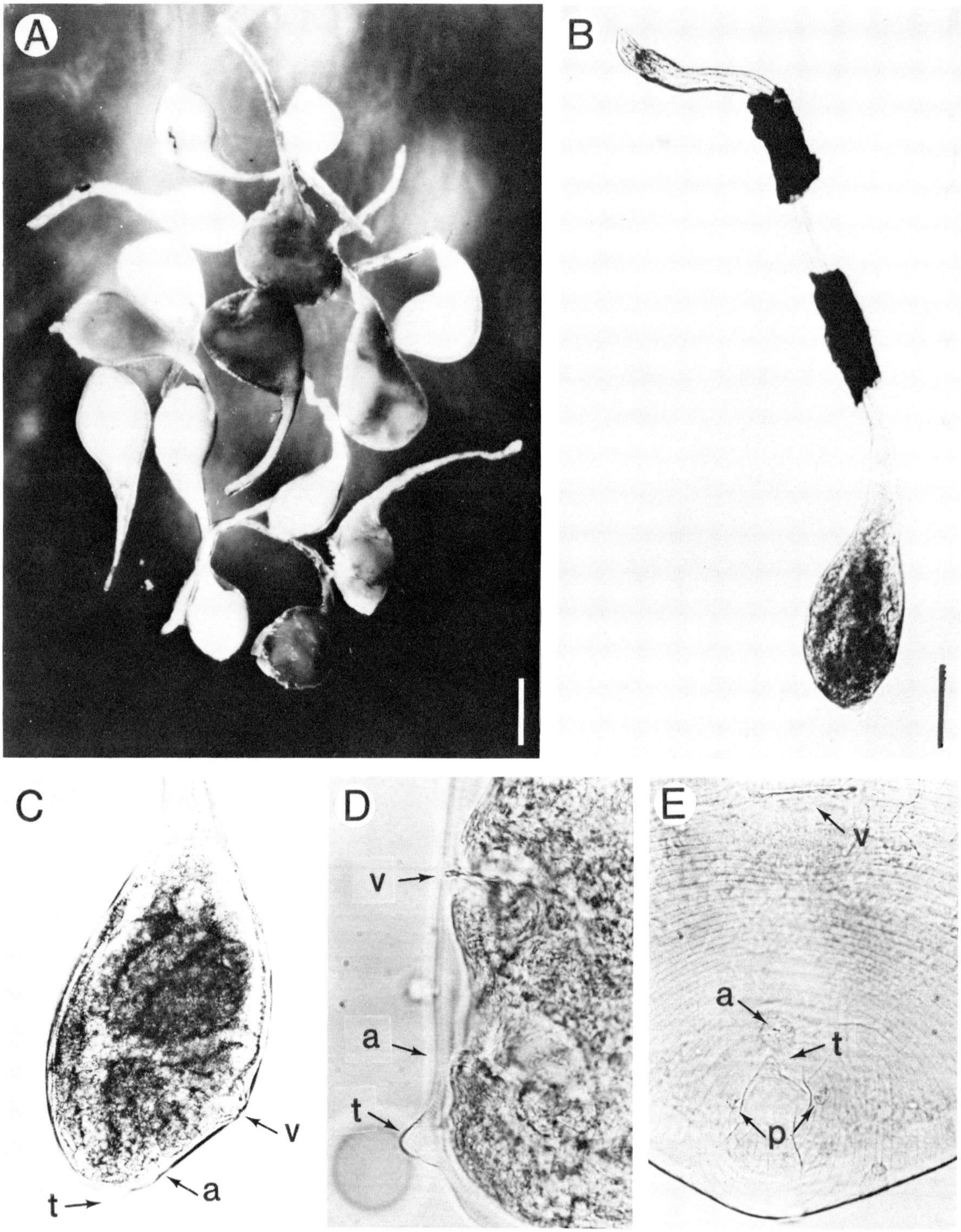

Fig. 5. LM photographs of females of *Meloinema* Choi & Geraert. A) Gross shape of females of *Meloinema chitwoodi* Golden & Jensen. Bar = 500 μm. B) Entire female of *M. kerongense* Choi & Geraert with remnants of root tissue attached in neck region. Bar = 200 μm. C) Posterior portion of egg-laying parasitic female of *M. kerongense* (lateral). D,E) Posterior portions of parasitic females of *M. chitwoodi* enlarged (lateral and ventral, respectively). a = anus; p = phasmid; t = tail; v = vulva. [B,C courtesy of Y. E. Choi; A,D,E from Golden & Jensen (8)].

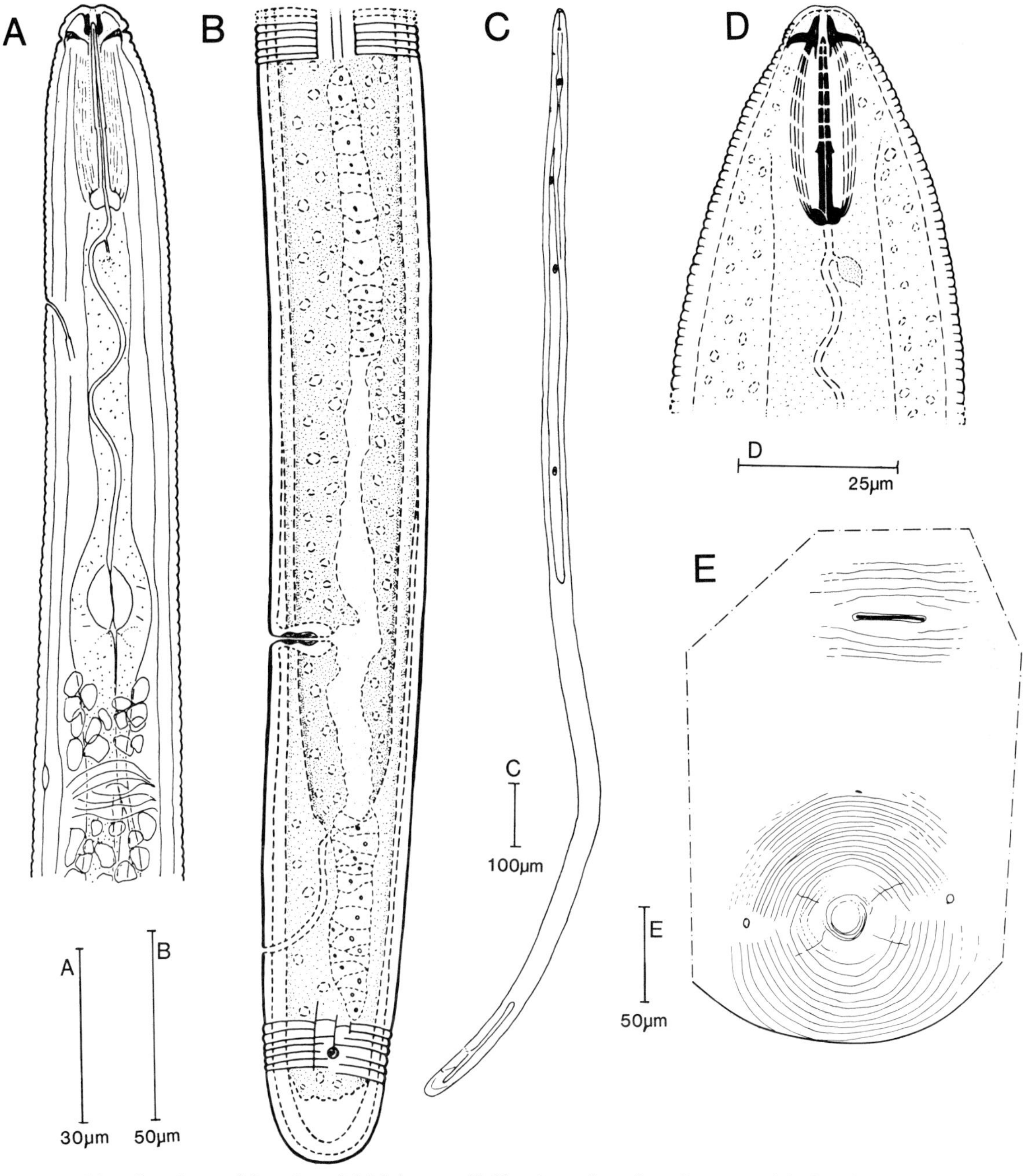

Fig. 6. Line drawings of females of *Meloinema*. A) Esophageal region of preparasitic female of *M. kerongense* (lateral). B) Posterior region of preparasitic female of *M. chitwoodi* (lateral). C) Outline of entire preparasitic female of *M. kerongense*. D) Cephalic region of parasitic female of *M. chitwoodi*. E) Perineal area of parasitic female of *M. kerongense*. [A,C,E from Choi & Geraert (3); B,D from Golden & Jensen (8)].

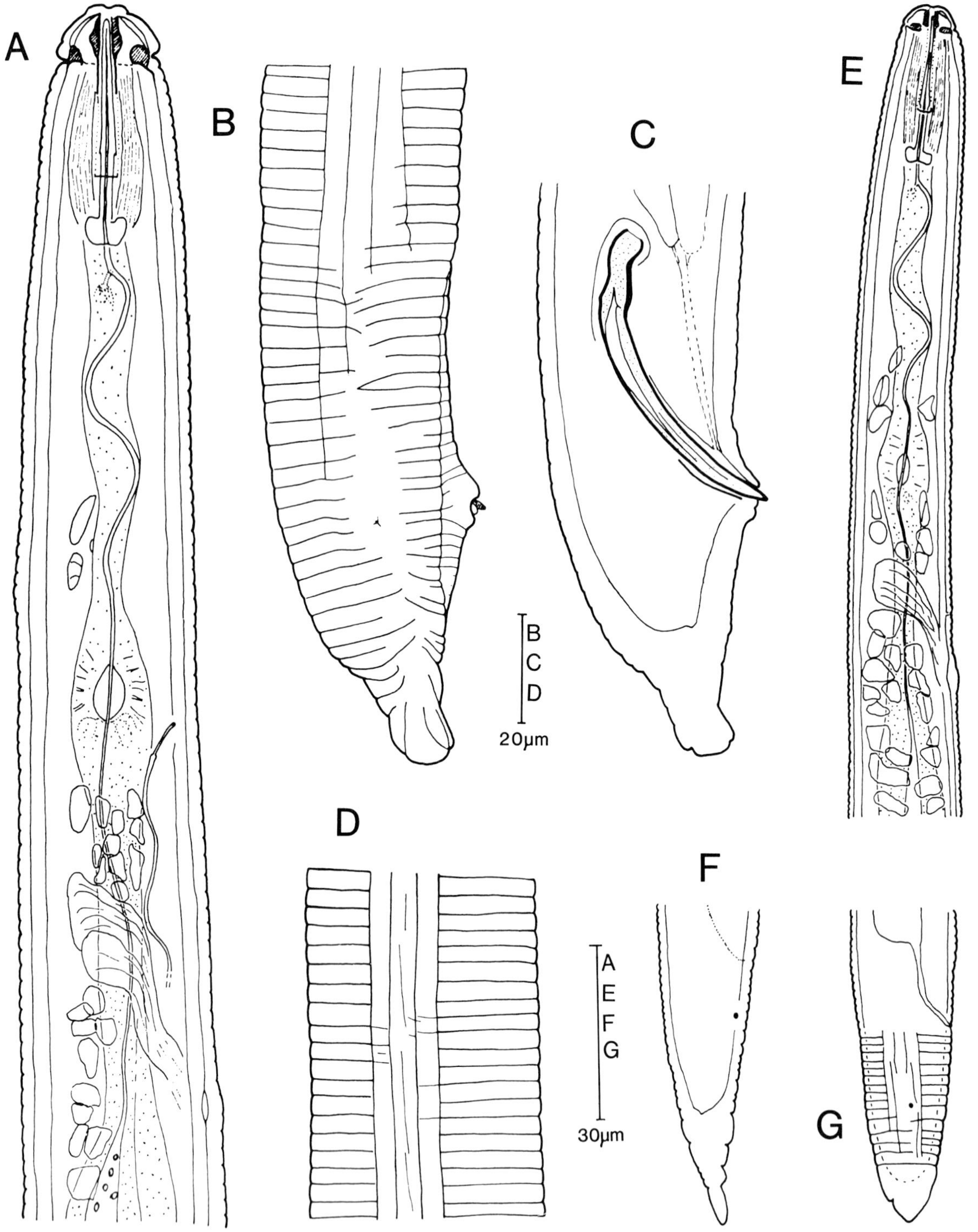

Fig. 7. Line drawings of males and second-stage juveniles of ***Meloinema kerongense***. A) Esophageal region of male. B,C) Male tails (lateral). D) Lateral field of male near midbody. E) Esophageal region of second-stage juvenile. F,G) Tails of second-stage juveniles showing variation in shape (lateral). [from Choi & Geraert (3)].

Literature Cited

1. Chitwood, B. G. and M. B. Chitwood. 1950. An introduction to nematology. Monumental Printing Co., Baltimore, Md. 213 pp.
2. Choi, Y. E. 1981. Additional information on *Meloinema kerongense* (Nematoda: Tylenchida). Nematologica 27:125-127.
3. Choi, Y. E. and E. Geraert. 1974. Description of *Meloinema kerongense* n. g., n. sp. (Nematoda: Meloidogynidae) from Korea. Nematologica 19:334-341.
4. Cliff, G. M. 1983. Morphological and serological comparisons of members of the *Meloidogyne arenaria* species complex, with characterizations of two new forms. Ph.D. thesis, North Carolina State Univ., Raleigh N.C. 153 pp.
5. Cliff, G. M. and H. Hirschmann. 1984. *Meloidogyne microcephala* n. sp. (Meloidogynidae), a root-knot nematode from Thailand. J. Nematol. 16:183-193.
6. Filipjev, I. N. 1934. The classification of the free-living nematodes and their relation to the parasitic nematodes. Smithson. Misc. Collect. 89:1-63.
7. Golden, A. M. 1971. Classification of the genera and the higher categories of the order Tylenchida (Nematoda), pp. 191-232 *In* Plant parasitic nematodes, Vol. I, ed. B. M. Zuckerman, W. F. Mai, and R. A. Rohde. Academic Press, New York. 345 pp.
8. Golden, A. M. and H. J. Jensen. 1974. *Nacobbodera chitwoodi*, n. gen., n. sp. (Nacobbidae: Nematoda) on Douglas fir in Oregon. J. Nematol. 6:30-37.
9. Hirschmann, H. 1982. *Meloidogyne platani* n. sp. (Meloidogynidae), a root-knot nematode parasitizing American sycamore. J. Nematol. 14:84-95.
10. Husain, S. I. 1976. Phylogeny and interrelationship of the superfamily Heteroderoidea (Skarbilovich, 1947) Golden, 1971. Geobios (Jodhpur) 3:9-12.
11. Khan, A. M. and S. I. Husain. 1972. *Meloidoderella indica* n. gen., n. sp., parasitizing *Lycopersicon lycopersicum* at Mahraval, U.P., India. Proc. 59th Indian Sci. Congr. Part III:592-593 (Abstr.).
12. Kirjanova, E. S. and E. E. Poghossian. 1973. [A redescription of *Meloidoderita kirjanovae* Poghossian, 1966 (Nematoda; Meloidoderitidae, fam. n.)]. Parazitologiya 7:280-285. (In Russian)
13. Paramanov, A. A. 1967. [A critical survey of the suborder Tylenchina (Filipjev, 1934) Nematoda-Secernentea]. Proc. Acad. Sci. Georgian SSR Biol. Ser. 18:78-101. (In Russian)
14. Skarbilovich, T. S. 1947. [On the reorganization of the systematics of the nematodes belonging to the family Anguillulinidae Baylis and Daubney, 1926]. Dokl. Akad. Nauk. SSSR 57:307-308. (In Russian)
15. Skarbilovich, T. S. 1959. On the structure of the systematics of the nematode order Tylenchida Thorne, 1949. Acta Parasitol. Pol. 7:117-132.
16. Stone, A. R. 1977. Recent developments and some problems in the taxonomy of cyst-nematodes, with a classification of the Heteroderoidea. Nematologica 23:273-288.
17. Thorne, G. 1949. On the classification of the Tylenchida, new order (Nematoda, Phasmidia). Proc. Helminthol. Soc. Wash. 16:37-73.
18. Wouts, W. M. 1973. A revision of the family Heteroderidae (Nematoda: Tylenchoidea). III. The subfamily Ataloderinae. Nematologica 19:279-284.
19. Wouts, W. M. 1979. Characterization of the family Meloidogynidae with a discussion on its relationship to other families of the suborder Tylenchina based on gonad morphology, pp. 21-35 *In* Root-knot nematodes (*Meloidogyne* species) systematics, biology and control, ed. F. Lamberti and C. E. Taylor. Academic Press, New York. 477 pp.
20. Wouts, W. M. and S. A. Sher. 1971. The genera of the subfamily Heteroderinae (Nematoda: Tylenchoidea) with a description of two new genera. J. Nematol. 3:129-144.

CHAPTER 6

DETAILED MORPHOLOGY AND ANATOMY OF SECOND-STAGE JUVENILES, MALES, AND FEMALES OF THE GENUS *MELOIDOGYNE* (ROOT-KNOT NEMATODES)

J. D. Eisenback

Department of Plant Pathology
North Carolina State University
Raleigh, North Carolina, USA

Morphology and anatomy are important in taxonomic studies and in understanding physiological function. Species identification within the genus *Meloidogyne* is based primarily upon morphological characters (29,31,34,36,54). Also, phylogenetic relationships and classification into higher categories are interpreted from the form of certain body systems (35,55).

Morphological observations are often important in formulating functional roles of certain tissues and structures because of the small size of these complex organisms (51,52). Detailed information about nematode morphology and anatomy is necessary for understanding the interaction of the nematode with its environment, and also provides insight into the intricate relationship between the nematode and the host (7,8,9,10,18). Furthermore, the host-parasite interaction is thought to have had a direct influence on the morphology of the nematode (11).

The morphology of root-knot nematodes changes during their life cycle (22,46). The first-stage juvenile, formed at the end of embryogenesis, immediately molts while still in the egg, becoming a second-stage juvenile (21,44). After the establishment of a host-parasite relationship and an initial feeding period of 3-8 weeks, the second-stage juvenile rapidly molts three times and develops into an adult male or female (6,37,50). This chapter describes the morphology that is considered characteristic for these three longest lived stages—second-stage juveniles, males, and females—of the genus *Meloidogyne*. Body shapes and overall dimensions of the three stages are compared in Fig. 1.

Second-stage juveniles (Figs. 2-15) are initially motile and vermiform, but later become sedentary and saccate during the parasitic mode of their life cycle. Although the juveniles sometimes remain in the egg and overwinter under a diapause (22), they generally hatch and migrate through the soil to find a suitable host.

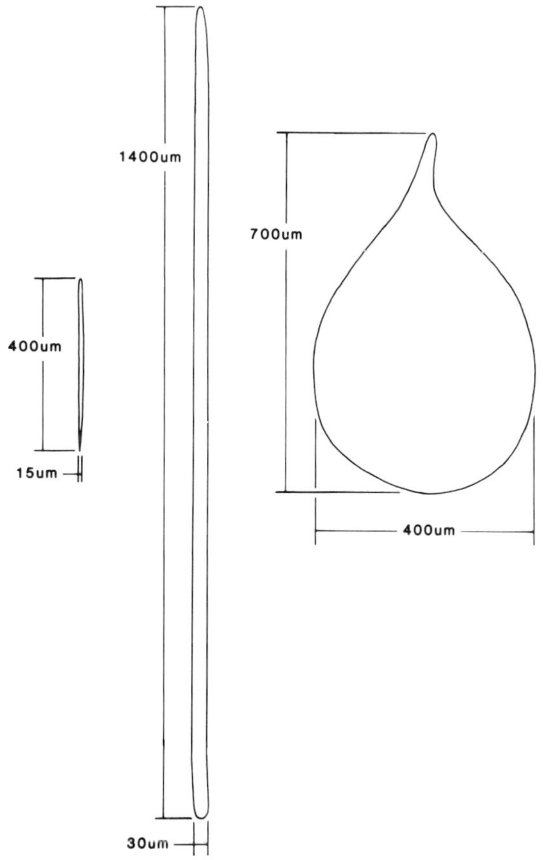

Fig. 1. Diagram of the general body shapes and dimensions of second-stage juveniles, males, and females, respectively, of the genus *Meloidogyne*.

The preparasitic juveniles penetrate the host, move through the tissue to a preferred feeding site, and establish a complex host-parasite relationship with the plant. Several body systems, the body wall, nervous, digestive, and excretory systems, must work together to allow such a tiny organism to accomplish these difficult tasks. The reproductive system remains small in the preparasitic juvenile but begins to develop shortly after feeding is initiated (42).

Males (Figs. 16-29) are vermiform and motile and do not feed. Shortly after the fourth molt, they leave the root and move freely through the soil. Even though males are not necessary for reproduction in the four most common species, many *Meloidogyne* species reproduce by cross-fertilization or facultative meiotic parthenogenesis (49). In populations which reproduce by amphimixis, the ability of the male to find a suitable mate is critical for survival of the species. Also, males within meiotically parthenogenetic populations assure genetic diversity through recombination. The production of males in mitotically parthenogenetic populations is a survival mechanism as well. In plants that are under stress, juveniles develop into males rather than females. Because males do not feed or produce additional offspring that require nutrition, the pressure on a heavily stressed plant would be reduced (47).

The body wall and the nervous and reproductive systems are important to the male for mobility, finding a mate, and reproduction. The digestive and excretory systems, in comparison, are probably less important because feeding may not occur. All of the energy available to the male for movement and reproduction is incorporated and stored within the digestive system during the second juvenile stage.

The saccate and sedentary females (Figs. 30-38) resume feeding shortly after the fourth molt and continue feeding for the remainder of their life. The digestive system is specialized for the maintenance of the host-parasite relationship, extraction of nutrients from the plant, and enhancement of the reproductive capacity of the nematode. Most of the nutrients obtained from the plant are utilized by the reproductive system.

BODY WALL

Movement of preparasitic juveniles and males through the soil is primarily dependent upon the body wall which is composed of cuticle, hypodermis, and somatic muscles.

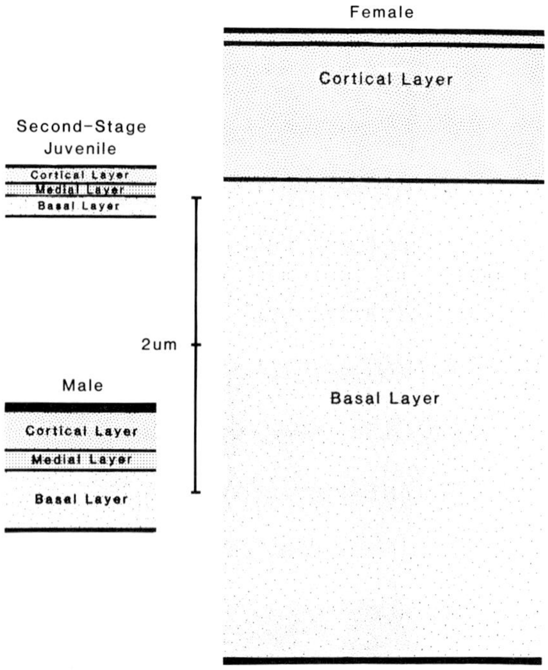

Fig. 2. Diagram of cross-sections of the cuticle of second-stage juveniles, males, and females of the genus *Meloidogyne* [After Bird (14)].

Cuticle

The cuticle is the outermost layer of the body wall which covers the entire nematode as well as invaginations of all of the body systems (Figs. 2,3,16,30). It is noncellular, elastic, and triple-layered and interfaces the nematode with its environment. The cuticle protects the organism from many brutal physical, chemical, and biological forces, and selectively regulates the flow of fluids through the body wall. The rigidity of the cuticle in opposition to the turgor pressure inside the body acts as an exoskeleton; yet the flexibility of the cuticle allows freedom of movement.

Externally, the markings of the cuticle in the head region of second-stage juveniles, males, and females are similar in occurrence, but the expression of the characters is considerably different (Figs. 4,5,17,18,31,32). The anteriormost portion of the cuticle is marked by a slit-like stomatal opening that is located in an ovoid depression on the prestoma and surrounded by the small pore-like openings of the inner labial sensilla. The prestoma of juveniles is oval shaped, and in the adult stages it is either oval shaped or hexagonal, depending on the location of the inner labial sensilla. If the sensilla open out on the labial disc, the prestoma is oval

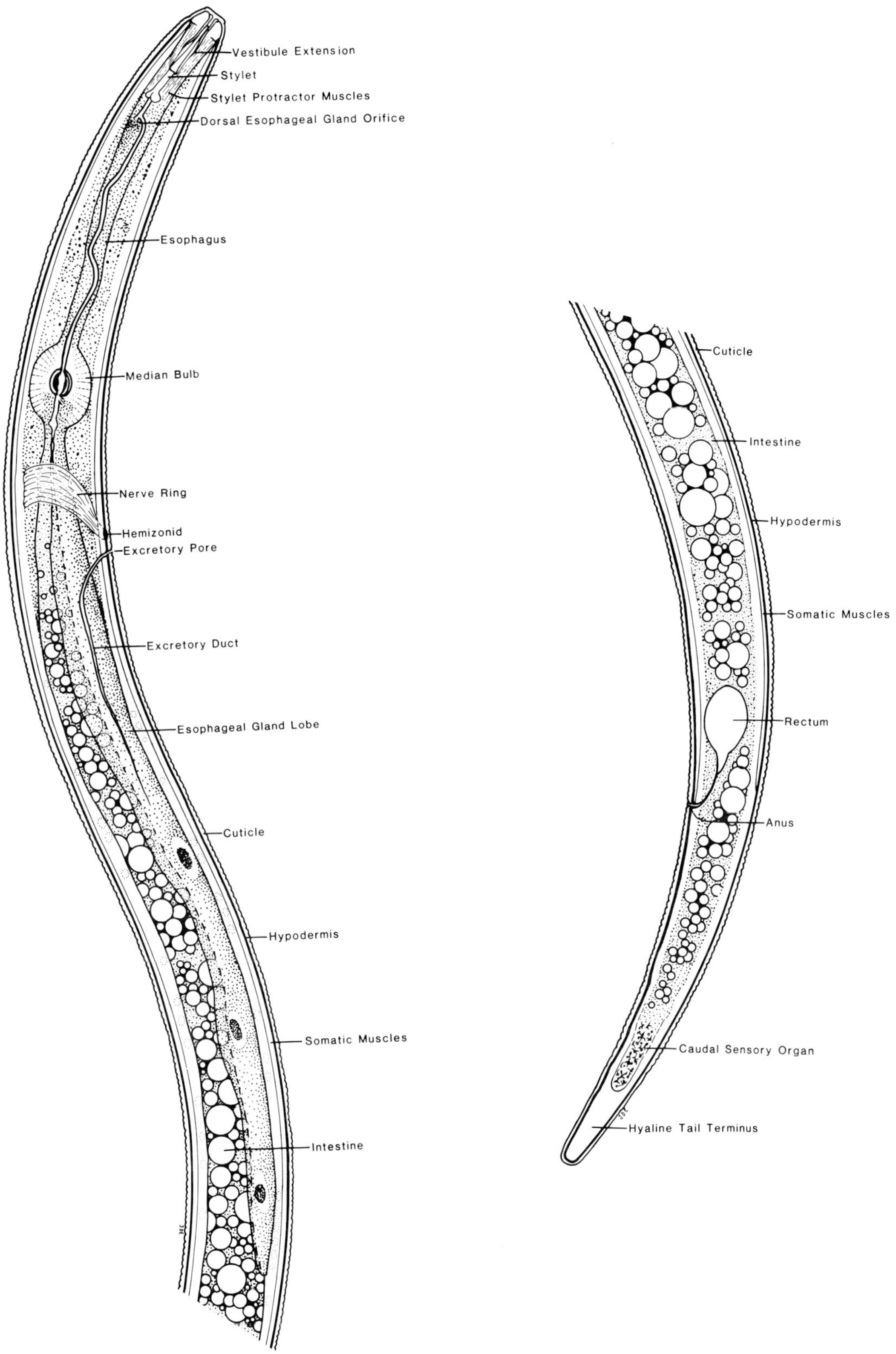

Fig. 3. **Gross morphology and anatomy of anterior and posterior body regions of a second-stage juvenile root-knot nematode as revealed by light microscopy.**

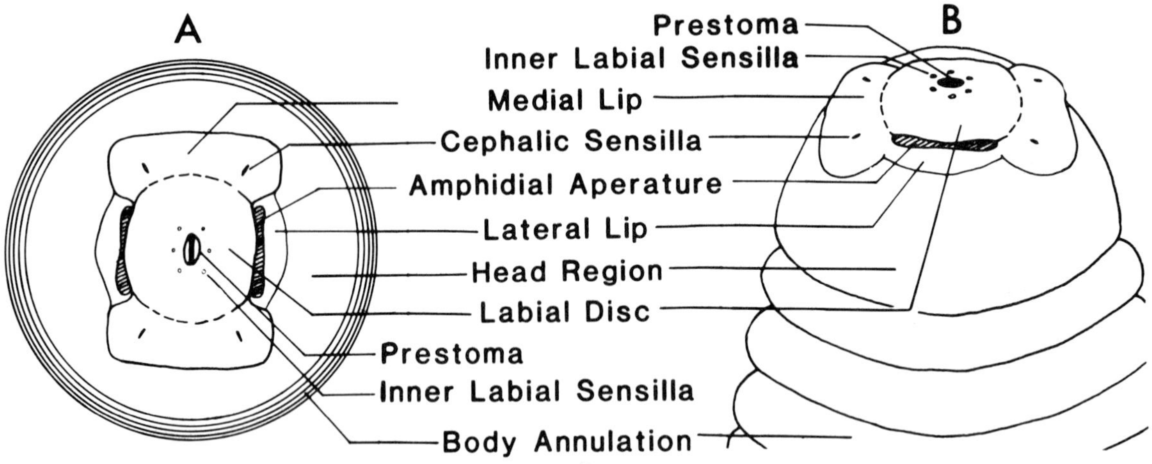

Fig. 4. External head morphology of a second-stage juvenile root-knot nematode in A) face and B) lateral views, as revealed by scanning electron microscopy [After Eisenback & Hirschmann (27)].

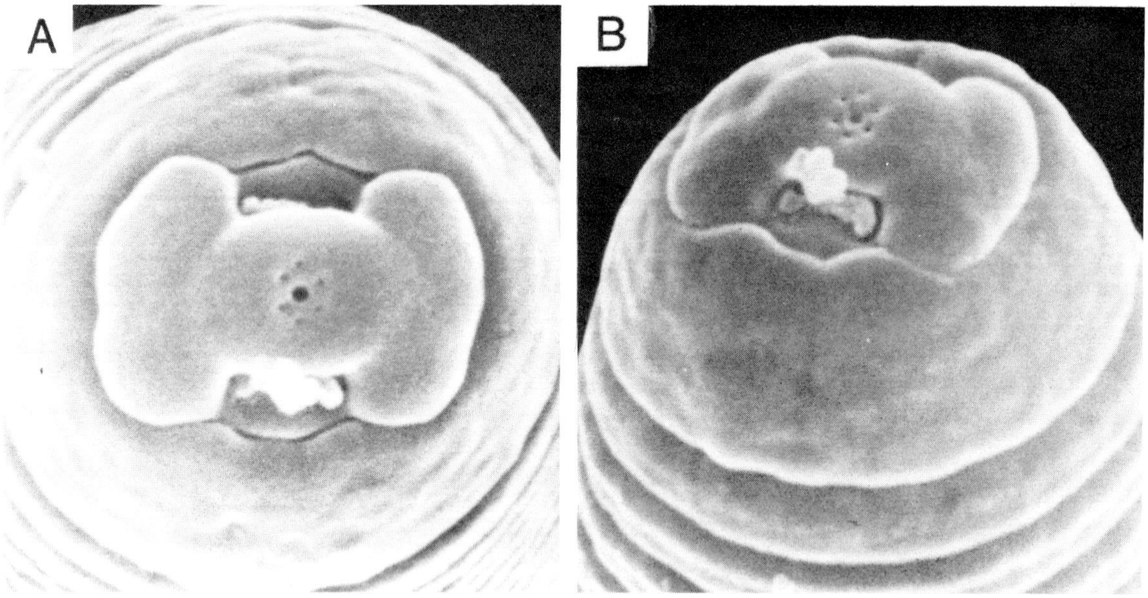

Fig. 5. Scanning electron micrographs of the head of a second-stage juvenile of *Meloidogyne hapla* in A) face and B) lateral views. [After Eisenback & Hirschmann (27)].

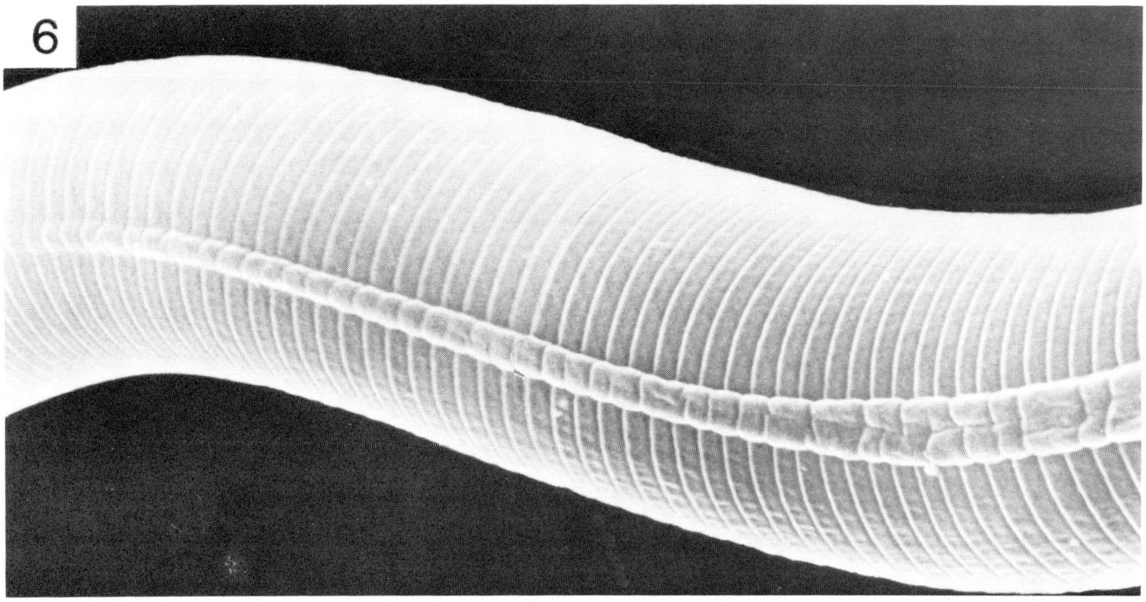

Fig. 6. Scanning electron micrographs of the anterior portion of the lateral field and body of a second-stage root-knot nematode [After Eisenback & Hirschmann (27)].

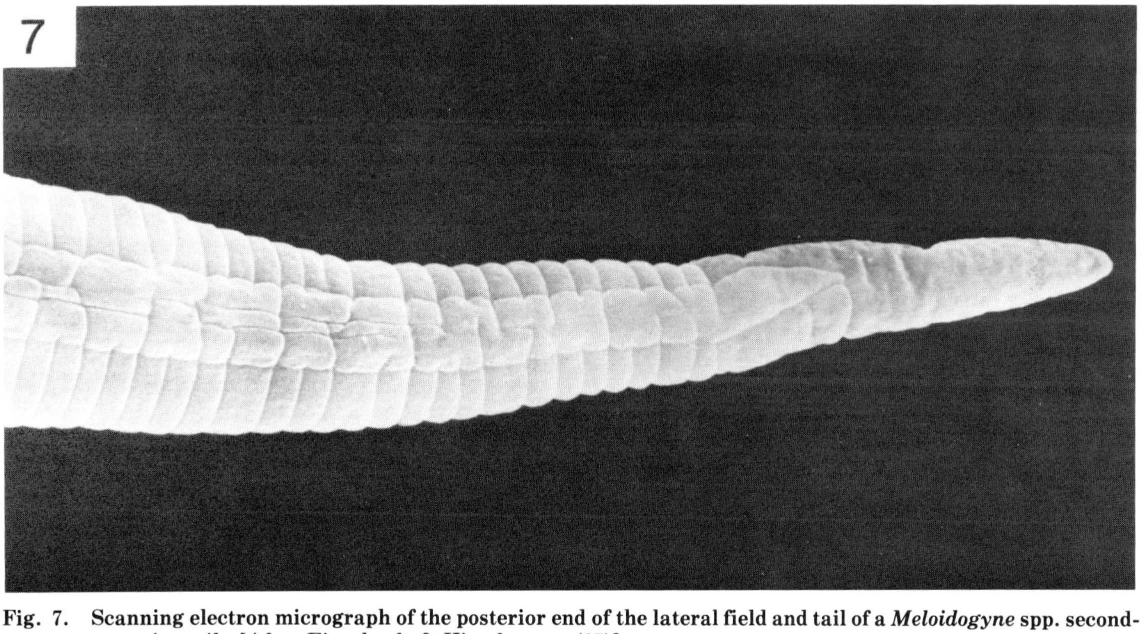

Fig. 7. Scanning electron micrograph of the posterior end of the lateral field and tail of a *Meloidogyne* spp. second-stage juvenile [After Eisenback & Hirschmann (27)].

shaped, but if they open into the prestoma, it is hexagonal. The labial disc is small and rounded in second-stage juveniles, large and rounded in males, and small and rectangular in females. In a face view, the labial disc slightly covers the amphidial apertures in juveniles and females, but in males, the openings are completely obscured by the large, rounded labial disc. The labial disc may be slightly elevated above the medial lips, or fused with them to form one smooth, continuous structure—the head cap. The subdorsal and subventral lip pairs are generally fused and form one dorsal and one ventral lip. They are termed medial lips because dorsal and ventral orientations are difficult to establish in scanning electron microscopy (SEM) (26,27,28,30).

Each medial lip is larger than the labial disc in second-stage juveniles and females, but in males, each medial lip is less than half the size of the labial disc. Each medial lip contains a pair of cephalic sensilla which may be expressed externally in some species as small, rounded cuticular depressions (27,28,30).

The lateral lips are usually smaller than the medial lips. In juveniles and males, they are lower than the head cap, in contour with the head region, but in females, they are raised above the head region, level with the medial lips and labial disc. The lateral lips often fuse with the medial lips in juveniles and females and with the head region of males and juveniles; however, some juveniles and most males do not have lateral lips. When present, they are marked off on the head region by deep grooves in the cuticle, beginning near the lateral edges of the medial lips, and extending posteriorly onto the head region. Sometimes the two grooves completely encircle the lateral lip, but generally they disappear and the lateral lip fuses with the head region posteriorly. In juveniles and females, the lateral lips can be long and narrow, triangular, or partially fused with the head region (27,28,30).

The amphidial apertures are located between the labial disc and medial lips. The amphid openings in juveniles and females appear as elongate ovals, but in males they are more slit-like. They are often obscured by an amphidial exudate in SEM. Posterior to the head cap and lateral lips, the head region can be completely smooth or marked by 1-4 complete or incomplete and irregular head annulations. The head region is generally large and easily distinguished from regular body annules in juveniles and males, but in females, it is smaller and sometimes resembles regular body annulations (27,28,30).

Regular body annulations mark the remainder of the body posterior to the head region in juveniles and males. They are approximately 0.8-1.0 μm apart in second-stage juveniles and 2.0-2.5 μm apart in males. Annulations near the head region may be closer together, whereas annulations in the tail region may be farther apart. Body annulations are interrupted laterally by the ridge-like projection of the lateral fields and ventrally by the insertions of the excretory pore and anus or cloacal opening (27,28).

The lateral field begins as a single ridge near the base of the stylet, 8-15 annules from the head region where body annulations may become disrupted (Figs. 6,19). At a short distance along the lateral field posteriorly, two incisures develop separately or from a single incisure, and subdivide the ridge longitudinally into three ridges which run the entire length of the nematode (27,28).

The lateral field narrows in the tail region of juveniles (Fig. 7), and the two inner incisures fuse into one near the middle of the tail. Shortly thereafter, this single incisure disappears and the lateral field ends near the tail terminus, where the two outer incisures merge. In comparison, the lateral field of males either wraps completely around the tail or spreads out and merges with the tail tip (Fig. 21). The two inner incisures extend nearly to the tail tip or disappear near the cloacal opening. The entire lateral field is areolated, and the areolation often corresponds to body annulations but sometimes becomes irregular and incomplete (27,28).

Body annulations in females mark the anterior portion of the body, including all of the neck. As they proceed posteriorly, the annulations become shallower and farther apart until they disappear completely. These annulations gradually reappear on the posterior portion of the body and become closer together and deeper (unpublished, J. D. Eisenback). The annulations form a cuticular fingerprint-like perineal pattern around the anus and vulva (Figs. 37,38).

Regular body annulations are interrupted laterally by remnants of the lateral fields, which begin 10-15 annulations from the head region near the base of the stylet (Fig. 33). The number of lines in the lateral field is difficult to discern because they are irregular and incomplete. The lateral field usually disappears before the regular body annules become shallow and reappears posteriorly after the annulations become distinct. The lateral field generally does not form a ridge, except posteriorly within the perineal patterns of a few species (unpublished, J. D. Eisenback).

The openings of the phasmids of second-stage juveniles and females are not visible externally but may be hidden by annulations or areolations within the lateral field. The phasmidial apertures in males are located within the inner ridge of the lateral field and appear as small slit-like openings near the level of the cloacal opening (Fig. 21) (27,28,30).

The excretory pore is a simple round to oval-shaped opening located in a depression within the cuticle; however, sometimes in the male, a depression may not ocur (Fig. 20). The body annulations usually are interrupted or deviate slightly from their course as they pass around the pore; this disruption is most severe in females. The anal opening in juveniles and females is a small, round hole in the cuticle and may be covered by a cuticular flap in the female. The hole is so small in the juvenile that normal body annules are not affected. The cloacal opening in the male is transversely elongate, and annulations on the ventral side of the tail anterior and posterior to the opening are generally indistinct or absent (27,28,30).

The cuticle is approximately 0.3-0.4 μm thick in second-stage juveniles, 1.5 μm in males, and 4-6μm in females. Cuticle thickness of females varies according to age and body region, and the cuticle is generally thickest in the perineal region (5).

Internally, the cuticle consists of cortical, median, and basal layers or zones common to most nematodes (Fig. 2) (12,14). The basic layers can be subdivided into sublayers, or some of the layers may be missing. Thus, the cortical layer is divided into the external and internal cortical layers. The external cortical layer is made up of a thin electron-translucent layer sandwiched between two thin osmiophilic layers and is approximately 35-40 nm thick in second-stage juveniles, 100 nm in males, and 100-125 nm in females (2,14,39).

The internal cortical layer is fibrillar and varies in thickness because of indentations of the body annulations. This layer varies from 50-100 nm in juveniles, from 400-500 nm in males, and from 700-900 nm in females (2,14,39).

In the living nematode, the medial layer may be filled with fluids. It contains electron-dense globules within a less dense substrate that may be continuous with the internal cortical layer. The medial layer in juveniles is approximately 100 nm thick under the annules, but disappears completely under the annulations. The medial layer in males occurs under the annules as well as the annulations but varies from 300-400 nm thick, and the globules appear more dense beneath the annules and the anterior portion of the lateral fields. The medial layer is missing in the cuticle of females (2,14,39).

The basal layer is approximately 100-125 nm thick in juveniles, 400-500 nm in males, and 3000-4000 nm in females. In preparasitic juveniles and males, the basal layer is striated

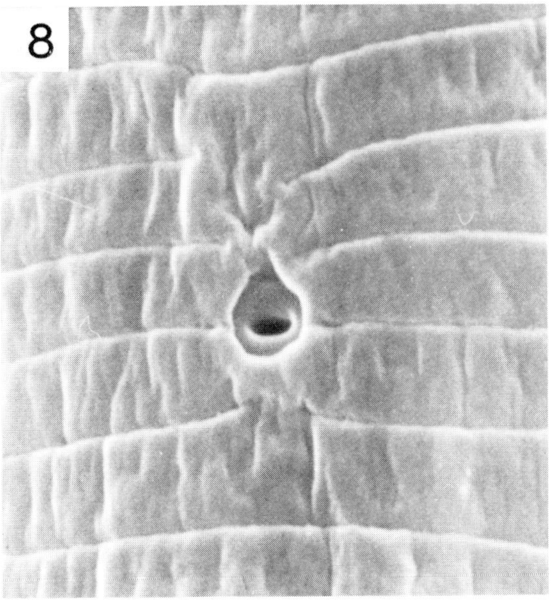

Fig. 8. Scanning electron micrograph of the external morphology of the excretory pore of a second-stage juvenile of a *Meloidogyne* spp. [After Eisenback & Hirschmann (27)].

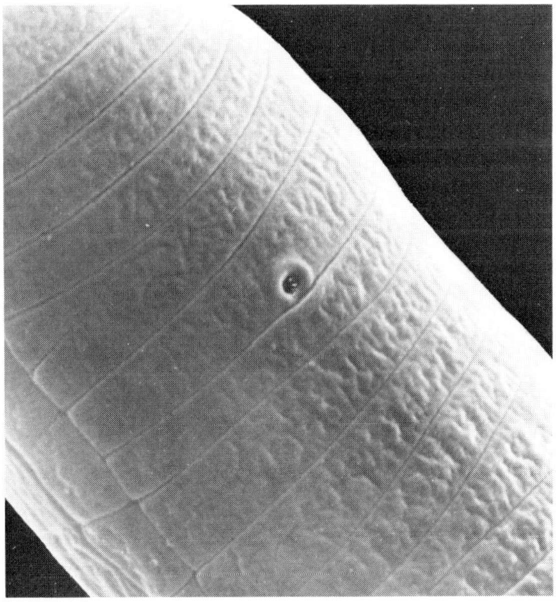

Fig. 9. Scanning electron micrograph of the external morphology of the anus of a root-knot nematode second-stage juvenile.

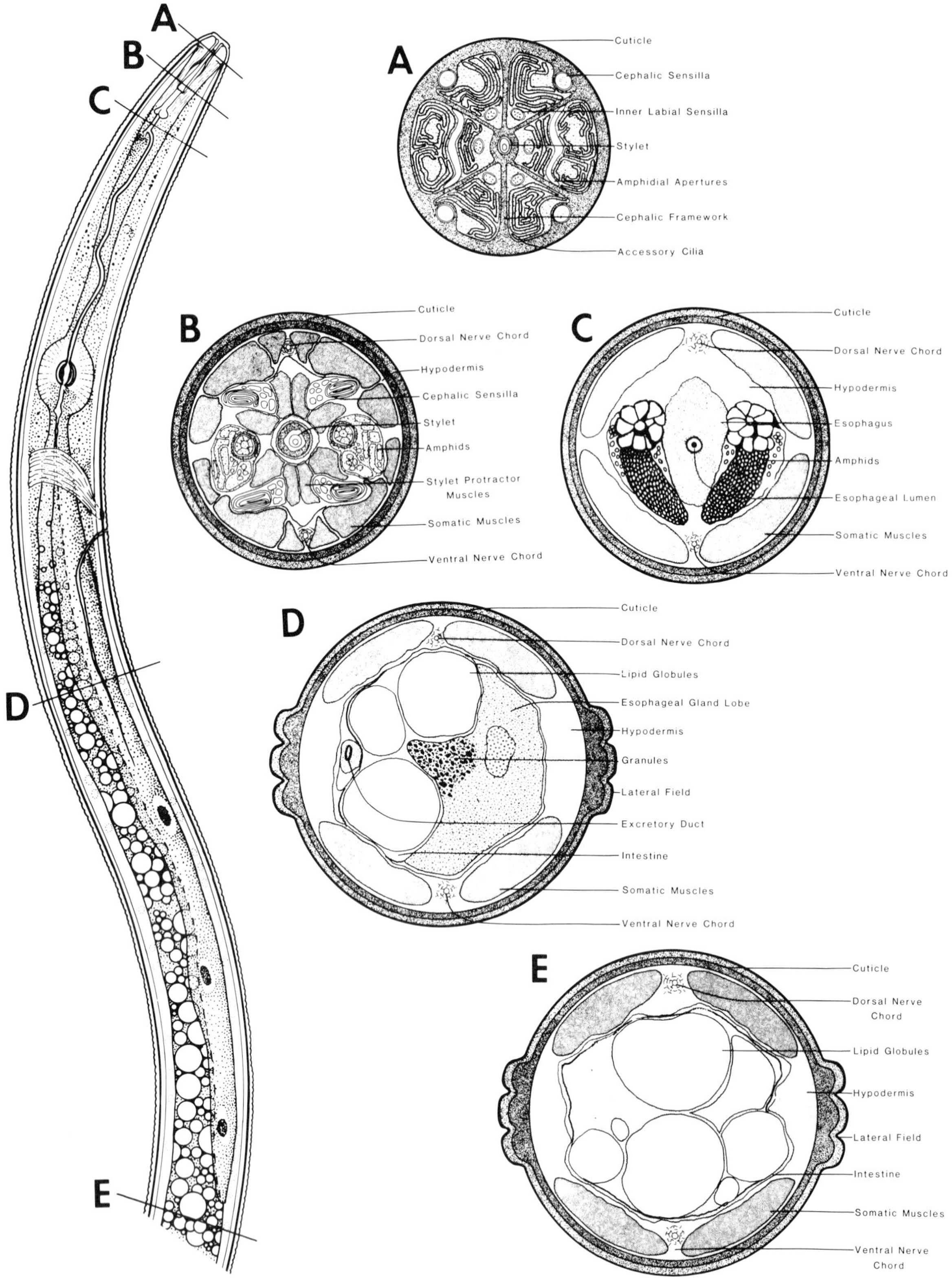

Fig. 10. Diagrams of cross-sections of the anterior portion of a second-stage juvenile of a root-knot nematode as revealed by transmission electron microscopy [A-C after Wergin & Endo (52); D-E after Bird (14)].

by two sets of laminae oriented at right angles to each other (43). The parallel striae are regularly spaced approximately 22 nm apart, and the crystalline nature of these laminae indicates a possible importance in protection of the nematodes from the environment or in service as a skeletal component. Interestingly, these laminae degenerate a couple of days after penetration by second-stage juveniles. They are not present in the basal layer of sedentary females, but reappear in motile, free-living males (11). These laminae are not present in the cuticle anterior to the body annulations, and they alter their form beneath the lateral fields. At the edge of the lateral field, the striae fork and are replaced by obliquely oriented fibrillar layers (2,14,16,39). They become quite disorganized near the tail tip of second-stage juveniles (15).

Hypodermis

The plasma-membrane-bound hypodermis is a living syncytium that lies beneath and secretes the cuticle (Figs. 10,11,36). Anteriorly, somatic hypodermis extends between the elements of the stylet protractor muscles and is continuous with the hypodermis that surrounds the stomatal cavity, cephalic framework, sensory structures, and stylet shaft (2,32,39,53). Posterior to the cephalic framework, the hypodermis expands in between the rows of somatic muscles and forms four hypodermal

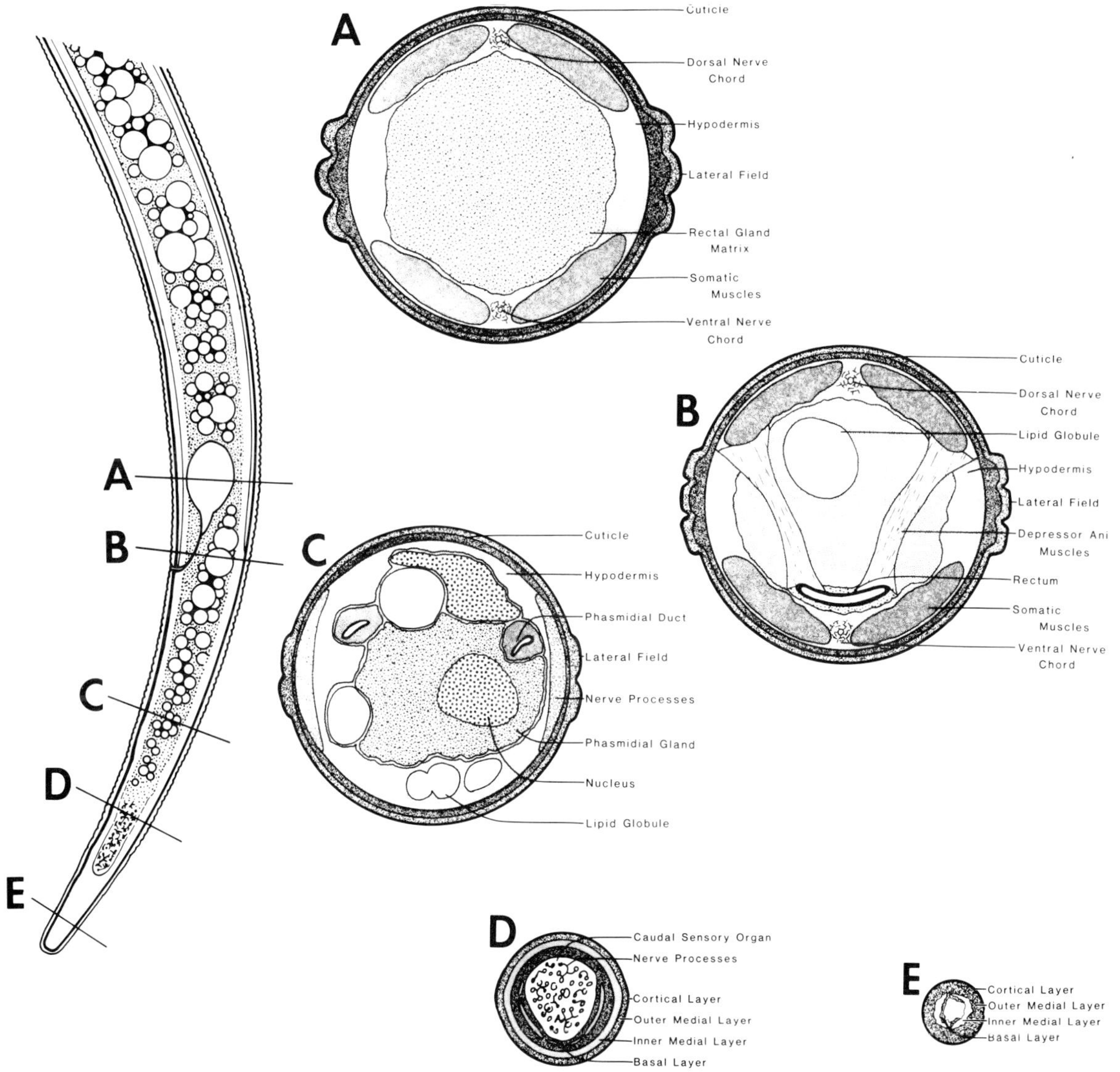

Fig. 11. Diagrams of cross-sections of the posterior portion of a root-knot second-stage juvenile as revealed by transmission electron microscopy [After Bird (15)].

chords. The lateral chords are thick and wide, correspond to the lateral fields, and contain nerves, nerve cells, and the excretory duct. The dorsal and ventral chords are smaller and contain nerves and nerve cells. The chordal hypodermis possesses many nuclei, mitochondria, ribosomes, Golgi bodies, smooth and rough endoplasmic reticulum, lipid globules, and various other organelles (2,23,32,39,53). Interchordal hypodermis lacks cell organelles and contains only nerve processes and numerous hemidesmosomes. The hemidesmosomes are modifications of the plasma membrane and function as attachment points for the cuticle and somatic muscles (2,14). The interchordal hypodermis is approximately 100 nm thick in juveniles and males, and of varying thickness in females (2,14,32).

Somatic Muscles

The somatic musculature is necessary to second-stage juveniles for hatching (8), movement through the soil, and penetration and migration within the host plant. All of the muscles, except those in the anterior end that move the head from one giant cell to another, quickly degenerate after the establishment of a host-parasite relationship (7,11).

The somatic muscles enable males to leave the root tissue, progress freely through the soil, and move toward a suitable mate. The restoration of mobility in adult males is necessary for successful reproduction in the amphimictic species. Females remain saccate and sedentary, and somatic muscles that allow the female to turn her head to feed on several different giant cells are limited to the anterior portion of the neck (14,32).

The anterior portion of second-stage juveniles and males contains eight longitudinally oriented, spindle-shaped muscles that are grouped into four rows between the hypodermal chords in cross-section. The number of cells between the hypodermal chords increases along the body posteriorly. There are two cells between the chords in the head region, 3-4 at the base of the stylet and 4-5 in the esophageal region and throughout the remainder of the body (2). The total number of somatic muscle cells in juveniles has been estimated to be around 50-100. They begin at the cephalic framework of juveniles and males, running the entire length of the nematode, and end at the level of the anus and cloacal opening (15), but in females, they end near the posterior portion of the median bulb.

In cross-section, the obliquely striated contractile portion of the muscle lies next to the hypodermis, and the noncontractile portion projects into the pseudocoelom. The tips of the spindle-shaped muscle cells contain only fine filaments, but the medial portion contains 5-6 cycles of I, A, H, A, and I bands. I bands possess only fine filaments; A bands are made up of thick filaments surrounded by 10-15 fine filaments arranged in a hexagonal pattern; and H bands contain only thick filaments. The thick filaments of myosin are 22-24 nm in diameter, and the thin filaments of actin are 6 nm in diameter. Contraction of the muscle is powered by the myosin-splitting high-energy phosphate bonds, and the muscle is shortened by the action of the actin filaments sliding past the myosin filaments (2,14).

The noncontractile portion of a muscle cell contains the nucleus, gycogen, lipid globules, smooth endoplasmic reticulum, and numerous mitochondria and is usually distinctly marked off from the contractile portion of the muscle. The muscles are innervated by processes located near the broadest area of the noncontractile portion of muscle. The innervation processes merge with the nerves extending from major bundles of longitudinal nerves (1).

NERVOUS SYSTEM

The function of the nervous system is twofold: motor control and sensory capacity. It controls the precise coordination of the somatic muscles that enable second-stage juveniles to move through the soil to find a suitable host, the males to find mates, and females to move their heads to feed on different giant cells.

The sensory function of the nervous system interfaces the nematode with the environment to give the movement direction. A sense of direction is needed by juveniles to find a root, recognize it as a host, and penetrate and migrate to a preferred feeding site. Males need a sense of direction to locate and copulate with females, and females need some sensory function to assist them in feeding.

Most of the available specific information about the nervous system of root-knot nematodes is confined to the anterior and posterior ends of second-stage juveniles and the anterior end of males. Very little is known about the nerves of the male tail, even though numerous nerves and sensory structures have been described. Also, detailed information on the nervous system of the female is missing. Generally, however, it is probably very similar to that described for juveniles and males. The nervous system has been estimated to contain approximately 300 cells (14).

In the light microscope, the most obvious parts of the nervous system are the nerve ring and hemizonid (Figs. 3,16), although some features of the amphids and phasmids may also be visible. The nerve ring, the central coordinating center, encircles the esophagus just posterior to the median bulb and consists mainly of nerve processes. The cell bodies and nuclei of the nerve processes lie anteriorly and posteriorly to the nerve ring (12,18). The hemizonid appears as a transparent semicircle on the ventral side of juveniles and males and is probably a major lateroventral commissure (14).

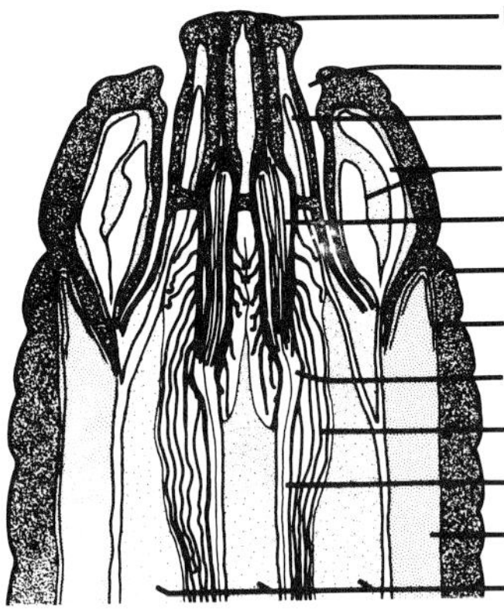

Fig. 12. Diagram of a longitudinal-section through the head of a second-stage juvenile of a *Meloidogyne* spp. illustrating the morphology of the sense organs located in the anterior portion of the nematode as revealed by transmission electron microscopy [After Endo & Wergin (33)].

A number of longitudinal nerves extend both anteriorly and posteriorly from the nerve ring. Four nerves have been detected to proceed posteriorly in juveniles and males, but more may occur. They run obliquely away from the esophagus and migrate into the hypodermal chords, one nerve in each chord. All four nerves are interconnected by transverse commissures. The ventral and dorsal nerves probably function primarily for motor control, whereas the lateral nerves are probably related to sensory function (14). Six nerves, primarily sensory in function, proceed anteriorly from the nerve ring: two subventrals, two subdorsals, and two laterals (2,33).

Cephalic Sensory Structures

Cephalic sensory structures of root-knot nematodes include inner labial sensilla, cephalic sensilla, accessory sense structures, and amphids (1,3,33,53). Each of the sensory structures is innervated by axons that proceed anteriorly from the nerve ring.

Externally, the pore-like openings of the six inner labial sensilla surround the slit-like stomatal opening on the labial disc (Figs. 4,5,17,18,31,32). Internally, each pore is lined with cuticle and surrounded by hypodermal tissue, and contains two receptor cilia of different lengths (Figs. 10A, 12). Because these sensilla have direct contact with the environment, they are probably chemoreceptors and may have some secretory function (33).

The four cephalic sensilla appear as pairs of small ovoid depressions on the medial lips (Figs.

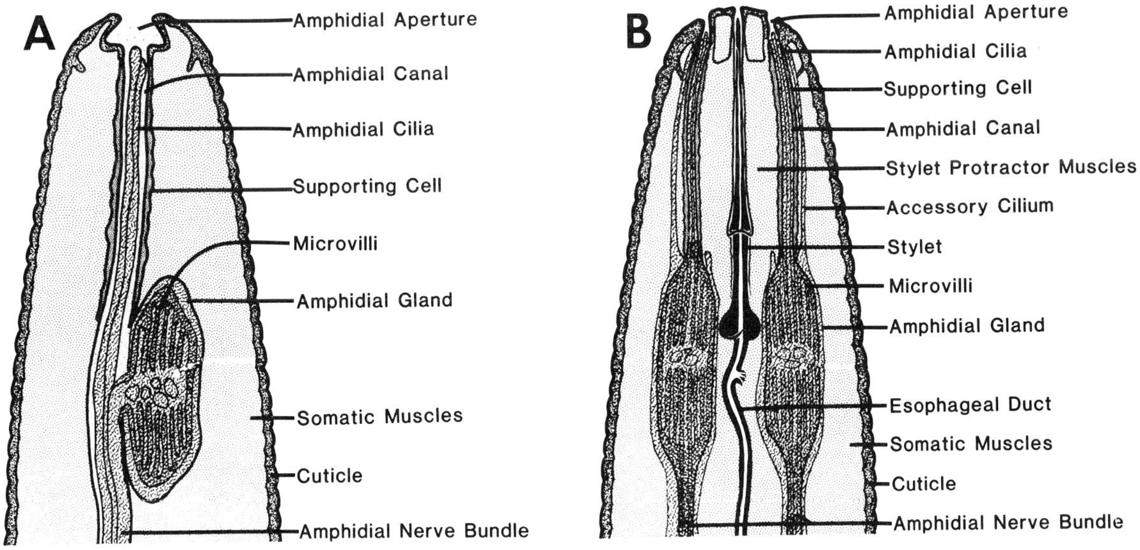

Fig. 13. Diagrams of A) dorsal and B) sublateral longitudinal sections of the anterior portion of a second-stage juvenile of *Meloidogyne incognita* as revealed by transmission electron microscopy [After Wergin & Endo (52)].

4,17,18,31). They are similar to the inner labial sensilla in many respects, except each is innervated by only a single ciliary receptor which ends beneath the cuticle (Figs. 10A,10B,12). The absence of a cuticular opening indicates that the cephalic sensilla are probably tactoreceptors (1,33).

Several other accessory sensilla that are not clearly defined are contained in the anterior region of the head. These sensilla are not surrounded by cuticle, end beneath the cuticle, and are probably tactoreceptors. Accessory sensilla occur in the two subdorsal and subventral sectors of the head region, are associated with the amphidial canals, and are closely related to the two lateral inner labial sensilla (Figs. 10A,12,13) (1,33).

The amphids are paired organs that are marked externally on the nematode by long slit-like openings between the labial disc and lateral lips (Figs. 4,5,17,18,31,32). An exudate secreted by the gland and plugging the opening is often visible by SEM (Fig. 5). The amphids are made up of support cells, nerve cells, and secretory cells (Figs. 10,13,22). The cuticular amphidial canals are surrounded by a support cell (socket cell) that extends from the anterior portion of the canal to the amphidial gland. The canal is filled with seven modified cilia which innervate the amphid and end near the amphidial aperture. Several accessory cilia lie outside the canal and terminate beneath the cuticle in the lateral sectors of the cephalic framework. The amphidial gland is a large, irregularly shaped cell (sheath cell) composed of numerous mitochondria, microvilli, and intercellular space filled with granules and ducts. Each of the amphids are innervated by axons that proceed anteriorly from the nerve ring. Morphological evidence suggests that the amphids are secretory and chemoreceptors have a possible tactoreceptor function (1,53).

Caudal Sensory Structures

Phasmids are located within the lateral fields of second-stage juveniles, males, and females. They occur midway between the anal opening and tail tip of juveniles; in males, they are located near the level of the cloacal opening; and in females, they are anterior to the tail terminus. Morphological details of phasmids of males and females are not known, but they are probably similar to those of second-stage juveniles. Each phasmid of juveniles is composed of a gland cell, nerve process, and a duct (Fig. 11C). The hypodermis that surrounds the phasmidial gland cell is filled with nerve tissue, and nerve processes extend from the nerve tissue into the basal layer of the cuticle in the lateral field. A single cilium extends into the phasmidial duct. Because the cilia have direct contact with the external environment, they probably function as secretory chemoreceptors and may have some tactile function (15).

The juvenile also has a specialized caudal sensory organ located posterior to the phasmids, directly anterior to the hyaline tail terminus (Figs. 3,11D). The caudal sensory organ is 5-10 μm long and consists of many branched nerve processes that originate from axons of the median caudal nerve. Since the caudal sensory organ is completely contained within the hypodermal tissue, it probably functions as a tactoreceptor (15).

The male has some specialized sensory structures associated with the male accessory organs. Two pore-like openings of spicular sensory structures are located at the tip of each spicule (Fig. 28) (unpublished, Eisenback). Details of their anatomy and function are not known, but perhaps they play a role in mate finding or copulation and insemination. They are probably chemoreceptors since they open into the environment.

DIGESTIVE SYSTEM

The digestive system of second-stage juveniles aids in hatching from the egg shell, penetration of the plant root, establishment and maintenance of the host-parasite relationship, and nutrition of the nematode (7,8,9,10,18). If the juvenile develops into a female, it must store enough energy for the three molts without further feeding. The digestive system of the adult female maintains the host-parasite relationship and obtains nutrients for the benefit and operation of the reproductive system. If the juvenile develops into a male, it must store enough energy for three molts, as well as the entire life span of the adult, because males apparently do not feed (50).

The digestive system is composed of stoma, esophagus, intestine, and rectum. The esophagus is the main driving unit of the digestive system; the intestine serves as a storage organ; and the rectum is modified by specialized rectal glands that are related to reproductive capacity.

Stoma and Esophagus

The slit-like stomatal opening is located within an ovoid prestomatal depression on the labial disc (Figs. 4,5,17,18,31,32). The hollow, protrusible stylet lies within the stomatal cavity surrounded by the heavily cuticularized hexaradiate cephalic framework (Figs. 10A,14,23,34). The cephalic framework includes the vestibule, six framework blades, and a basal ring. The vestibule extends posterior to the cephalic framework as the vestibule extension. The lining of the vestibule and extension is continuous with the external layer of cuticle of the body wall. The blades of the cephalic framework are slightly arched posteriorly and extend from the vestibule to the body wall. One blade is located dorsally, two subdorsally, two subventrally, and one ventrally, and the lateral sectors are larger than the subdorsal or subventral sectors. The basal ring of the cephalic framework and the basal layer of the cuticle of the body wall are continuous (Fig. 23) (3,33,53).

The stylet is held in place by the vestibule extension which attaches near the junction of the cone and shaft, and by three stylet protractor muscles. The muscles attach to the cephalic framework and body wall anteriorly and the stylet knobs posteriorly (13). The stylet consists of a cone, shaft, and three knobs. The cone is pointed and gradually widens posteriorly. In second-stage juveniles and females, the opening of the stylet is located near the tip on the ventral side of the cone (25,29), but in males it is 2-3 μm posterior to the tip. The opening is often marked by a slight posterior protuberance (25,29,30), and in some males the opening may be absent (25). The nearly cylindrical stylet shaft extends anteriorly into the cone. One dorsal and two subventral knobs are either set off from the shaft or merge with it in varying degrees. The shaft and knobs are homogeneous but less electron dense than the cone (3).

The stylet functions much like a two-way hypodermic needle. Contraction of the stylet protractor muscles forces the tip of the stylet out through the stomatal opening and into the substrate. The hollow stylet is then used to inject substances into the plant cells and to ingest nutrients from the giant cells (7,8,9,10,18). During protraction of the stylet, the vestibule extension folds backward, and the hypodermal tissues surrounding the shaft are flexible and allow the stylet to move. The elasticity of the folded vestibule extension, hypodermal tissues surrounding the shaft, and the esophagus are probably responsible for the retraction of the stylet, because retractor muscles are absent (3).

Immediately posterior to the stylet, the procorpus contains the noncontractile portions of the stylet protractor muscles (3,4). The cuticular lumen lining of the procorpus is round in cross-section, and the dorsal esophageal gland orifice branches into several channels a short distance (2-8 μm) from the base of the stylet (Figs. 14,15,24,25,33,35) (4,25,29,30).

At the base of the procorpus in second-stage juveniles and females, the enlarged metacorpus functions as the pumping mechanism of the digestive system (Figs. 3,15,34,35). The lumen lining changes from round in the anterior portion of the bulb into a heavily sclerotized, triradiate structure. The outside edges of the enlarged lumen lining are thick and rigid, and the inner portions are thin and flexible (25,30). The suction that is needed for feeding is produced by the muscles pulling apart the thin inner portions of the lumen lining. The thick outer edges of the lining support the metacorporeal pump. Muscle cells within the

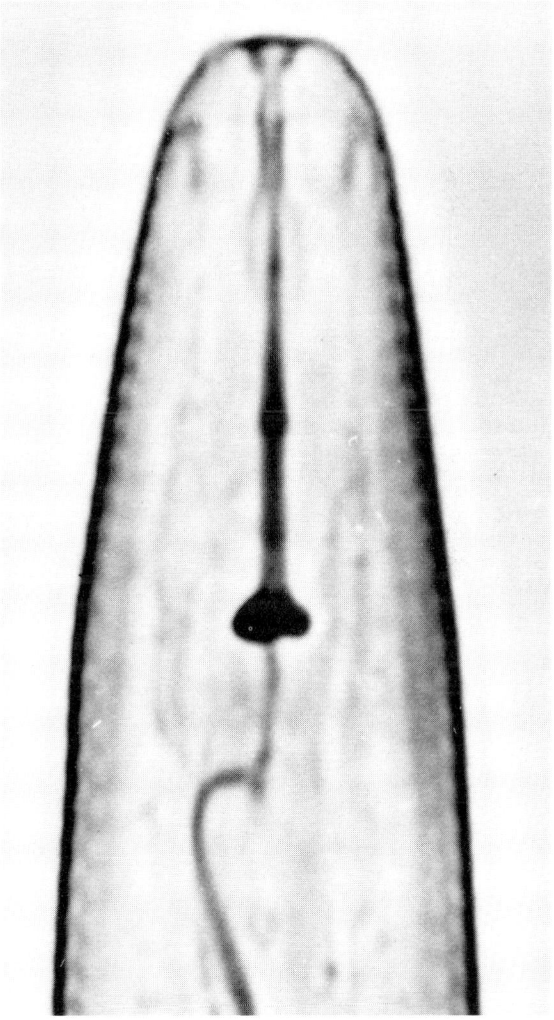

Fig. 14. Light micrograph of the anterior portion of a second-stage juvenile of a *Meloidogyne* spp. [After Eisenback (25)].

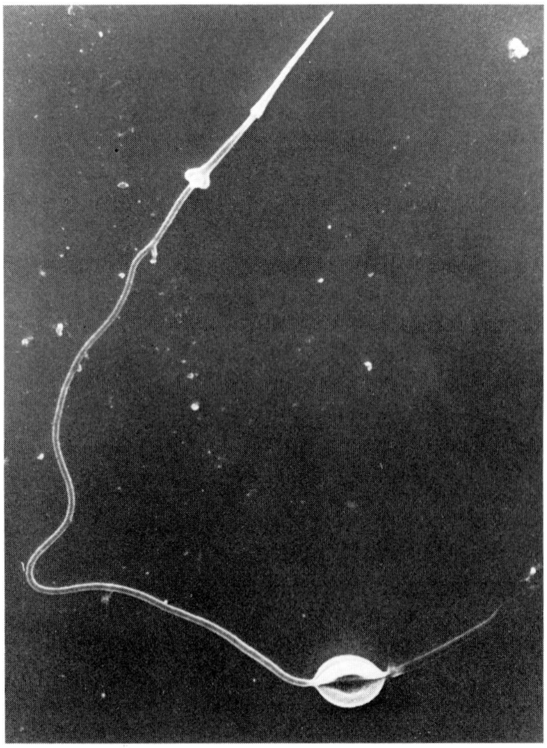

Fig. 15. Scanning electron micrograph of an excised stylet and attached cuticular lumen lining of the esophagus of a second-stage juvenile of a *Meloidogyne* spp. [After Eisenback (25)].

metacorpus are attached to the basement membrane of the lumen lining, and the myofilaments are dispersed radially (14). The branched orifices of the subventral esophageal glands are located ventrally, immediately posterior to the triradiate lumen lining. Posterior to the median bulb, the lumen lining of the isthmus is very thin (25,30).

The basic morphology of the procorpus, metacorpus, and isthmus of males is very similar to that of second-stage juveniles and females (Figs. 16,25). The metacorpus of males, however, appears to be functionally degenerate. The triradiate lumen lining is thin throughout (29), and the radial muscles of the metacorpus are sparse, poorly developed, and disorganized (4). Posterior to the median bulb, the lumen lining of the isthmus is very thin, and the lumen ends as a series of folded membranes. The membranes are surrounded by 2 or 3 esophago-intestinal cells which lead to the intestine (4).

The digestive glands are located at the base of the isthmus and overlap the intestine ventrally (Figs. 3,16,30). In second-stage juveniles and females, the dorsal gland cell is uninucleate and lies anterior to the two uninucleate subventral gland cells. A process of the dorsal esophageal gland extends anteriorly where it forms a distinct ampulla which empties into the esophageal lumen near the base of the stylet. The subventral glands form distinct ampullae in the posterior portion of the metacorpus adjacent to the orifices that open into the lumen of the metacorpus.

The three gland cells are located in a series on the ventral side of the body and overlap and wrap around the intestine in second-stage juveniles. The esophageal glands in females consist of one large dorsal gland lobe and two smaller subventral gland lobes that are variable in position, shape, and size (37). In males, the dorsal esophageal gland is often reduced to an indistinct ampulla and a small gland process near the dorsal esophageal gland orifice. Only processes of the subventral glands are located in the metacorpus, isthmus, and gland lobe. The two uninucleated gland lobes are placed in a series. One gland lobe becomes dorsal to the other, and the ventral gland extends posteriorly. The gland lobes overlap and wrap around the intestine ventrally (4). The morphology and chemical nature of the esophageal glands change as the parasitic status of the nematode changes (7,8,9,10,11,13,18).

Two specialized esophago-intestinal cells are located in the isthmus at the junction of the esophagus and intestine. Although they are not visible by light microscopy in the preparasitic juvenile, they enlarge enough to be seen after 10-12 days of active feeding. They appear as two small, rounded uninucleated cells at the dorsal base of the isthmus, and remain visible in all of the stages following the second-stage juvenile (32,37) (Fig.30). These cells are completely contained within the isthmus of the male (4).

Intestine

The intestinal cells of *Meloidogyne* species are not distinct, and few details of their fine structure have been resolved. Most of the morphology is obscured by numerous lipid globules that pack the intestine. A clearly defined lumen with microvilli is apparently absent in second-stage juveniles (14,15). Males and young adult females may have a partially defined lumen (32). Males have a distinct intestinal caecum, which is syncytial like the intestine and has no lumen. The caecum is located dorsally at the base of the isthmus and extends anteriorly, sometimes nearly to the base of the metacorpus (Fig. 16) (4). In mature females, the

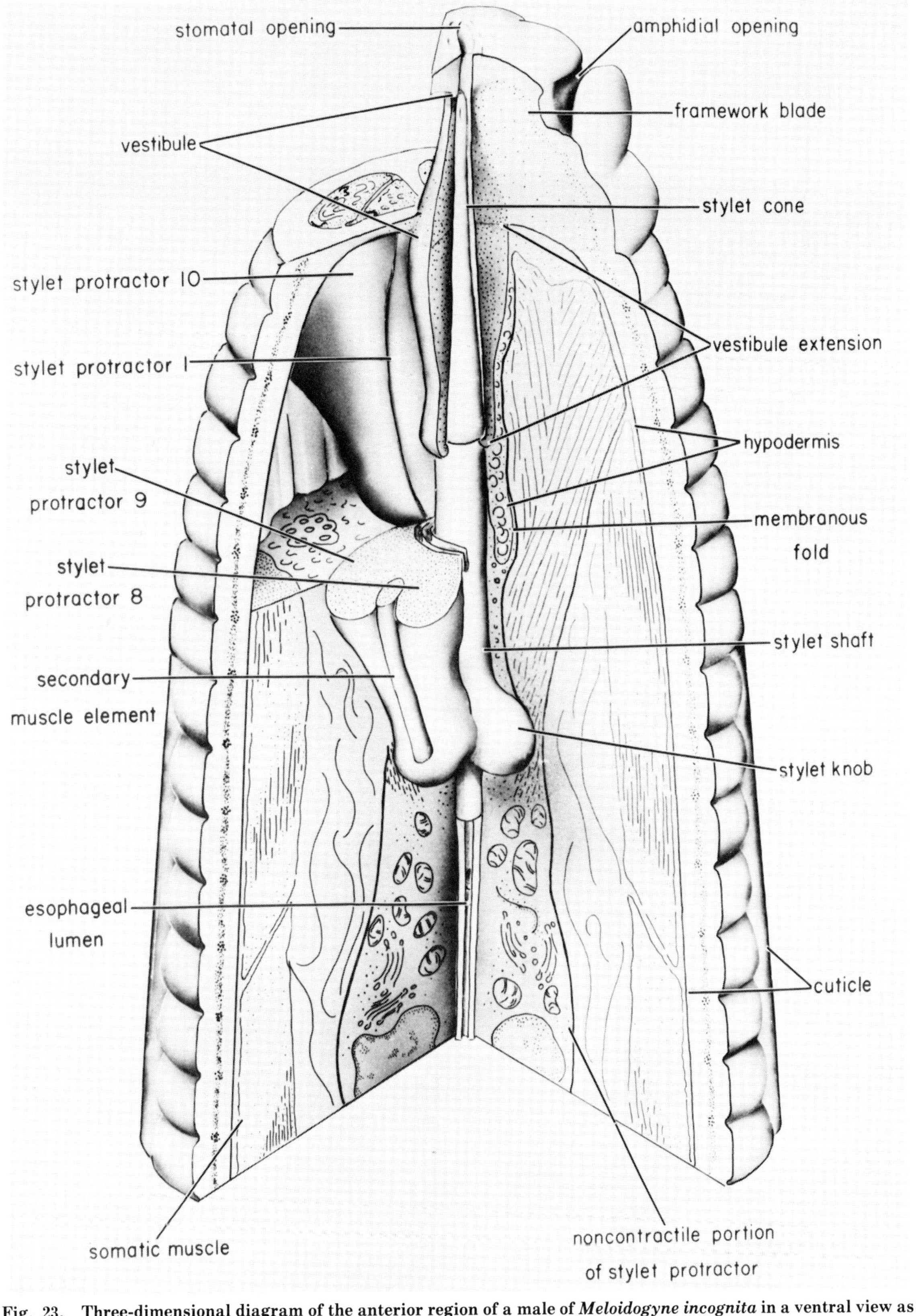

Fig. 23. Three-dimensional diagram of the anterior region of a male of *Meloidogyne incognita* in a ventral view as revealed by transmission electron microscopy [After Baldwin & Hirschmann (3)].

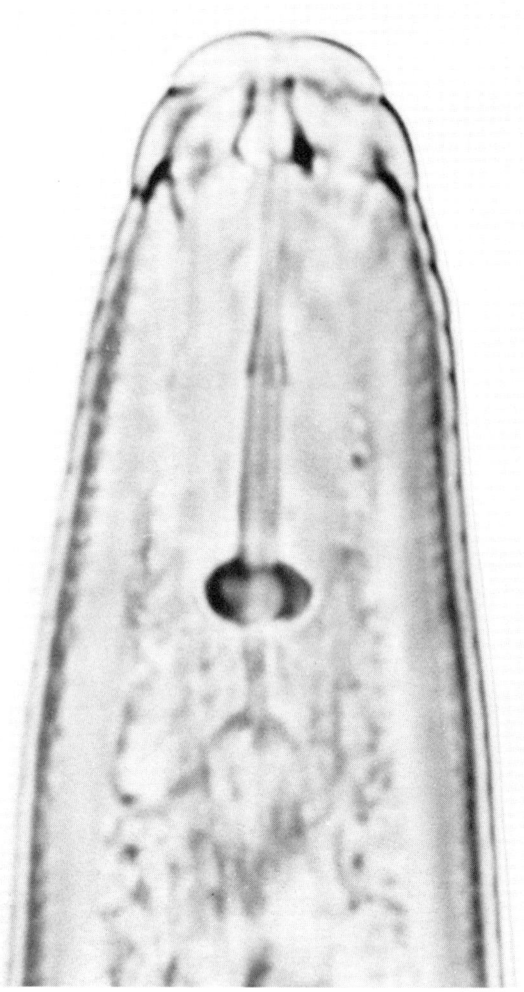

Fig. 24. Light micrograph of the anterior portion of a root-knot nematode male [After Eisenback & Hirschmann (29)].

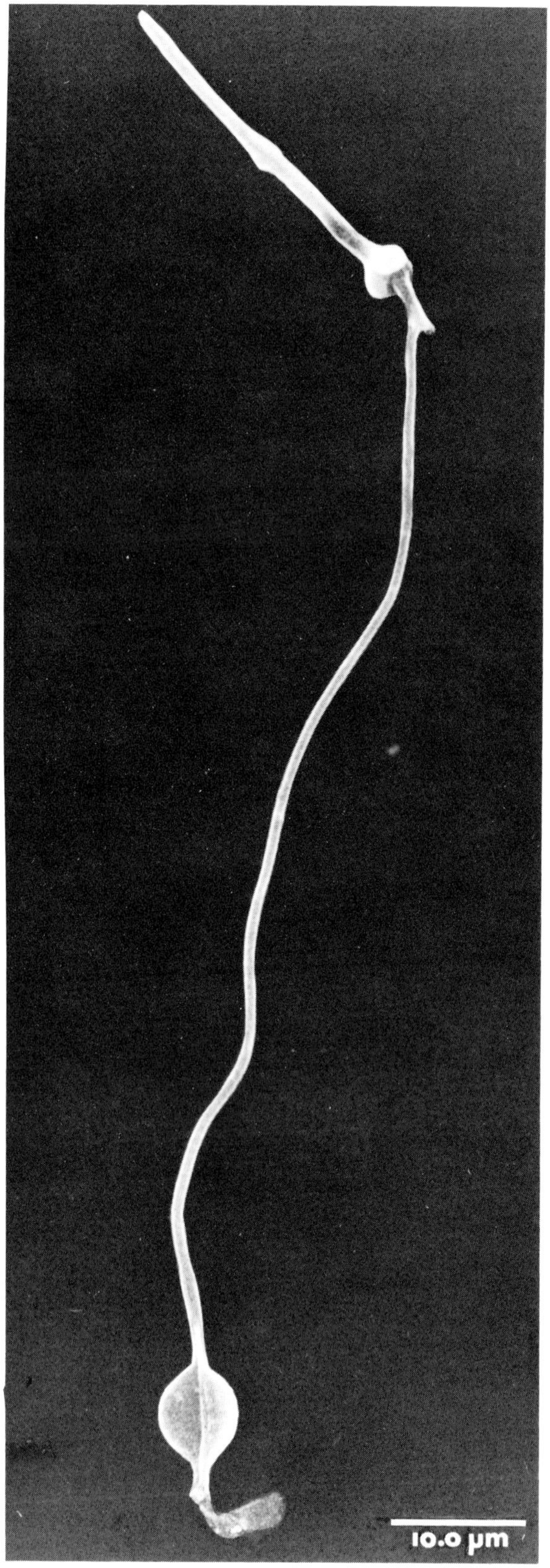

Fig. 25. (right) Scanning electron micrograph of an excised stylet and attached cuticular lumen lining of the esophagus of a male of a *Meloidogyne* spp. [After Eisenback & Hirschmann (29)].

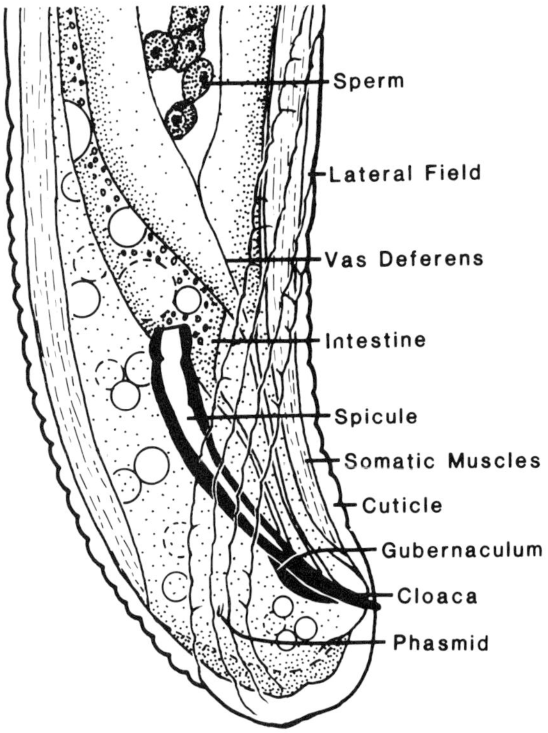

Fig. 26. (above) Diagram of the tail of a root-knot nematode male as revealed by light microscopy.

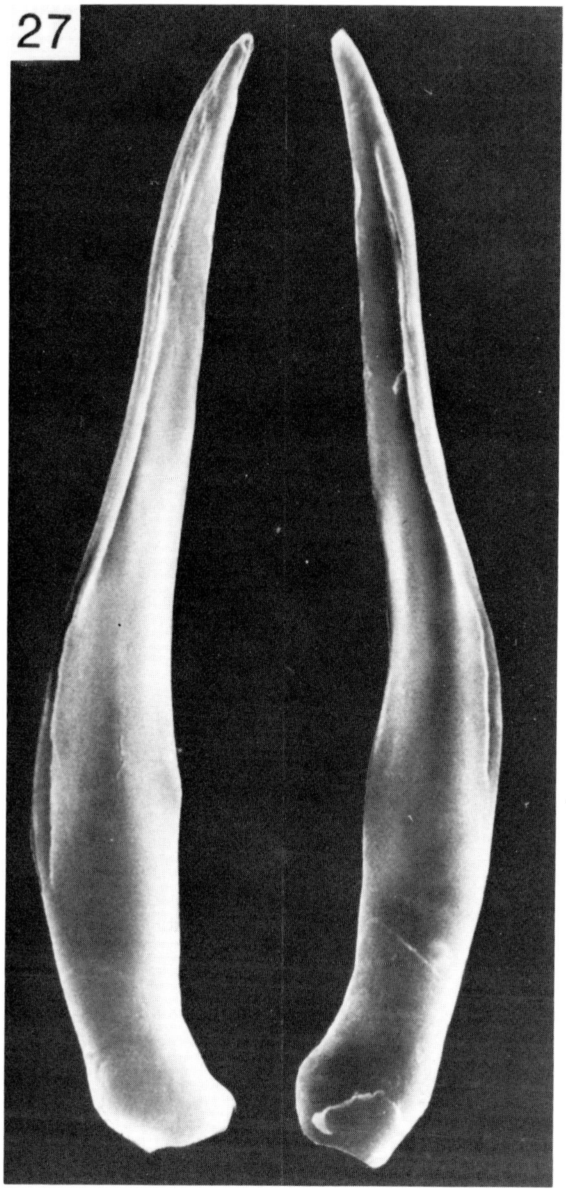

Fig. 27. (above) Scanning electron micrograph of excised spicules of a *Meloidogyne incognita* male.

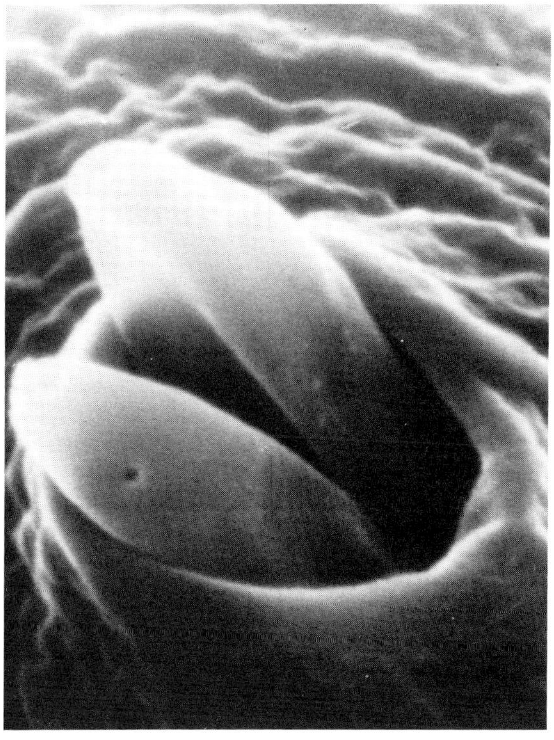

Fig. 28. (left) Scanning electron micrograph of the tips of spicules protruding through the cloacal opening of a male of a root-knot nematode.

morphology of the tail of *Meloidogyne* males are not known (14).

The anal opening of the female is located between the tail terminus and the vulva (Figs. 37,38). The rectal lumen is large in the young adult, but becomes smaller with maturity (Fig. 36). There are six rectal glands that are connected to the rectum: two subdorsal, two lateral, and two subventral. The two subventral glands are separated by the uterus (Fig. 36B), and generally the six glands are equally distributed around the body. They are closely oppressed to the hypodermis and nearly touch each other laterally (40).

Each of the gland cells is pyriform and approximately 100 μm long and 25μm in diameter. A large nucleus is located in the anterior portion of each gland. The nuclei are the largest found in the adult female, approximately 25 μm long with nucleoli about 15 μm in diameter. The cytoplasm of the rectal gland cells is dense and contains many Golgi bodies, mitochondria, anastomosing endoplasmic reticulum with many ribosomes, and multivesicular lamellar bodies. The lamellae of the multivesicular lamellar bodies repeat with a periodicity of 75 Å. Apparently, these organelles and the Golgi bodies play a major role in the secretion of the gelatinous matrix. Sinus canals are visible throughout each gland. They merge and form a single long duct that opens into the anterior end of the rectal lumen. Depressor ani muscles are attached to the rectum and aid in the deposition of the gelatinous matrix through the anus (17,40).

The rectal gland cells produce a very large amount of gelatinous matrix. The matrix is an irregular mesh-like substrate containing proteins, a mucopolysaccharide, and various enzymes. Gelatinous matrix material is produced before and during egg production, and the eggs are laid within and surrounded by it (5,17). The matrix protects the eggs from predators and dehydration and is an important survival mechanism to the sedentary *Meloidogyne* nematodes (11,19,24).

EXCRETORY SYSTEM

The excretory pore is an oval-shaped opening located ventrally in a depression in the cuticle (Figs. 8,20,33). Usually, the body annules deviate slightly as they pass the opening or are interrupted. The excretory pore is located posterior to the metacorpus in second-stage juveniles and males, usually 2-8 μm below the hemizonid (Figs. 3,16). In females, however, it is located anterior to the median bulb, often near

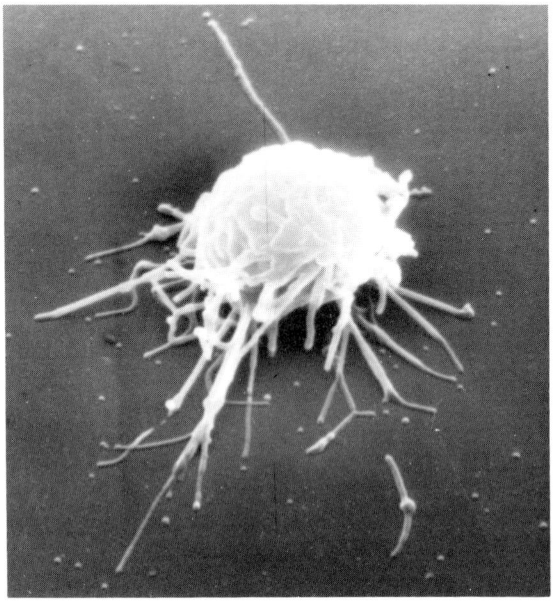

Fig. 29. Scanning electron micrograph of an excised sperm from a *Meloidogyne* male nematode.

the base of the stylet (Fig. 30). The cuticular excretory duct of second-stage juveniles is approximately 0.3 μm in diameter with a lumen less than 0.2 μm in diameter, and it is 1 μm in diameter with a lumen of approximately 0.6 μm in diameter in females (14). The duct extends posteriorly through the hypodermis and changes into a noncuticular canal. The anterior end of the duct of the female is surrounded by numerous vesicles (14), and the posterior end of the excretory canal in all three stages is closely associated with a large sinus gland cell (14,32,37). The function of the excretory system is not clearly understood, but it is assumed that it removes toxic wastes from the body of the nematode (12,14).

REPRODUCTIVE SYSTEM

The reproductive system is directly responsible for the production of new individuals to ensure the survival of the species. In preparasitic second-stage juveniles, it is small, but after feeding begins, it grows rapidly and differentiates into either testes or ovaries (42). The pyriform female shape in *Meloidogyne* species allows for the development of an enlarged female reproductive system that greatly increases the reproductive capacity of the species; however, mobility is lost. The male, in comparison, stays vermiform and motile, and the reproductive system remains of normal size (32).

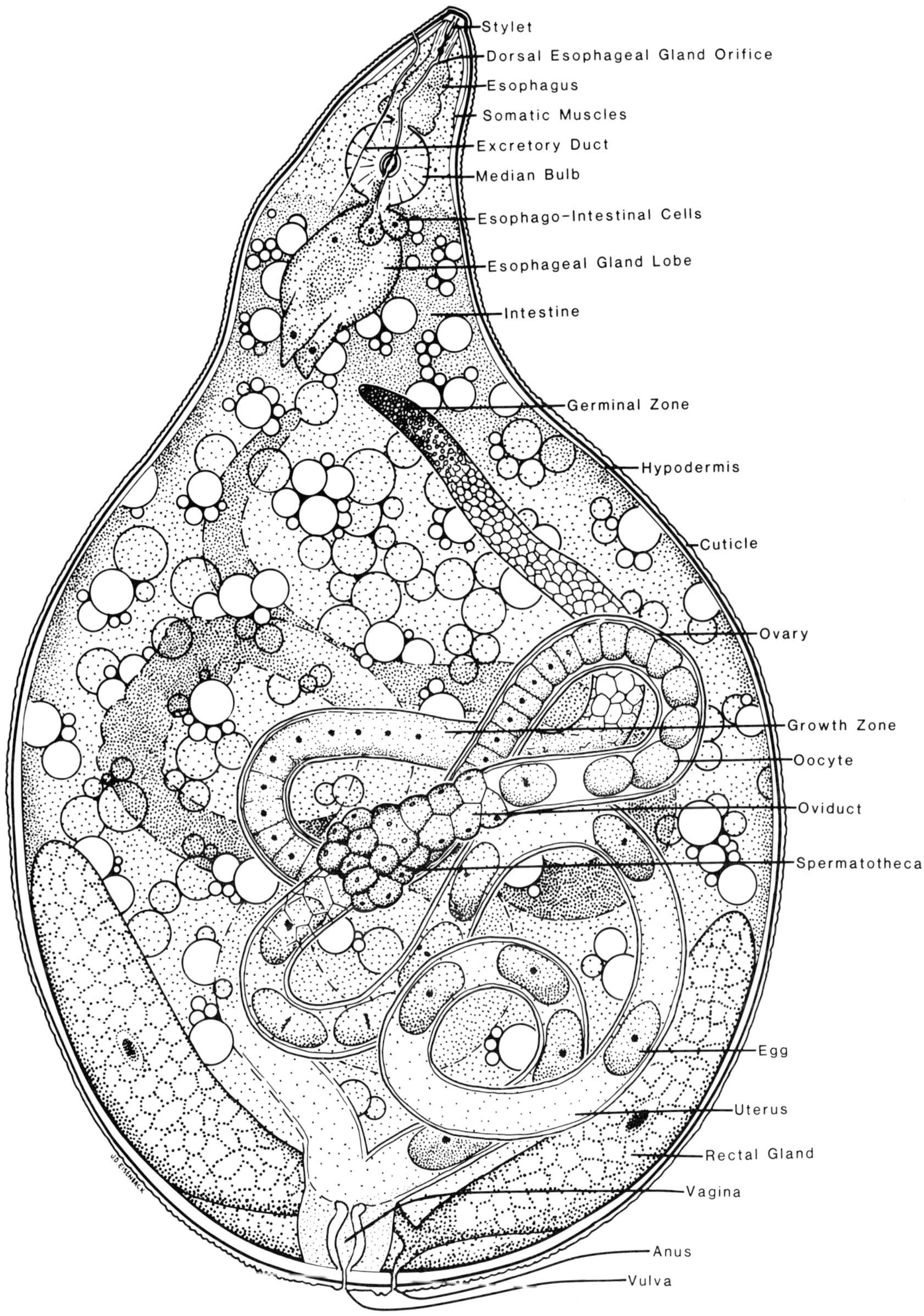

Fig. 30. **Gross morphology and anatomy of a female root-knot nematode as revealed by light microscopy [Ovary after Triantaphyllou (47)].**

Second-Stage Juveniles

The genital primordium of the preparasitic juvenile is parallel with the body wall and located about 65% of the body length from the head. It is made up of four cells: two small, flat, somatic cells surrounding two large, spherical, germinal cells. Several days after penetration and feeding, the primordium begins to develop into ovaries or testes, depending mainly upon the influence of the environment (42).

Males

Normally, *Meloidogyne* males have one testis. Many root-knot males, however, are sex-reversed females and have two testes. The number of testes varies according to the period of gonadal development at which the reversal occurs. If it is early, the male will have only one testis, and if reversal occurs later, two testes will develop (42).

The male gonad is an elongate, tubular structure located in the posterior two-thirds to one-half of the body (Fig. 16). Approximately one-half of the anterior portion of the gonad is testis. One of the original somatic cells, present in the undifferentiated genital primordium, forms the cap cell on the anterior portion of the testis. The germinal zone occupies a short anterior region of the testis, and the remaining posterior portion forms the growth zone. Spermatozoa are stored in the seminal vesicle immediately posterior to the testis. The vas deferens is glandular (45) and empties into the cloaca ventrally near the cloacal opening. A single layer of thin epithelial cells covers the entire gonad (32).

Spicules are located within a cuticular invagination of the cloaca. They are paired, cuticularized structures that are inserted into the vagina during copulation and form a tube to assist in the efficient transfer of spermatozoa into the female (Figs. 16,27,28). Each spicule consists of a head, shaft, and blade. The head and shaft have a cylindrical cytoplasmic core, and the blade is flattened with incurved edges. Nerve endings enter through the base of each spicule and open out through two small pores on the spicule tip (Fig. 28). The gubernaculum is located within an invagination of the cloaca, dorsally to the spicules. Several spicule protractor and retractor muscles attach to the base of the spicules and the body wall (14).

Mature spermatozoa that are stored in the seminal vesicle are approximately 6 μm in diameter and 12 μm long. One part of the spermatozoon contains the nucleus which is surrounded by most of the mitochondria and fibrillar bodies. The other part contains most of the cytoplasm and forms a large pseudopodium (Fig. 29). Microtubule-like structures are numerous in both regions covering the surface. Very long, sometimes branched, and abundant filopodia emerge from the region of the spermatozoon that contains the nucleus (38,45; unpublished, J. D. Eisenback).

Females

In the female, a large portion of the body cavity is filled by a pair of tubular, highly convoluted gonads (Fig. 30). Approximately 60% of the gonad is occupied by the ovaries. A cap cell marks the anterior end of each ovary. The germinal zone occupies a short anterior portion of the ovary. The developing oogonia are attached radially to a rachis by cytoplasmic bridges. The remaining posterior portion of the ovary is the growth zone. The rachis disappears, and the cell boundaries between oocytes become more distinct. The oocytes gradually accumulate lipid globules, refringent bodies, and glycogen as they pass through the growth zone and into the oviduct in single file (32,41,48).

The oviduct consists of two staggered rows of cells. Each row contains four tightly packed cells with large, irregularly shaped nuclei with large cytoplasmic invaginations. The lumen in the valve is very narrow and oocytes stretch and flatten these cells as they pass through. The oocytes become ovoid in this part of the gonad (32,41,48).

The spermatotheca is located immediately posterior to the oviduct and consists of 14 rounded, lobe-like cells with deeply invaginated plasmalemma. Numerous microtubules extend from densely vesiculated bodies that are located between the lobes of the spermatothecal wall. The tubules project into the lumen, contact the developing egg, and aid in the formation of the protein membrane of the egg shell (41).

Posterior to the spermatotheca, the uterus can be differentiated into distinct regions. Cells in the ovarial end of the uterus possess large intracytoplasmic spaces lined with endoplasmic reticulum. They function in the formation of the chitin layer of the egg shell. Cells in the middle of the uterus form the glycolipid layer. Cells in the posterior end of the uterus consist of dense cytoplasm with large areas of compact endoplasmic reticulum (41).

The two uteri of the female reproductive tract fuse and form one common duct. In the normal position, the lumen is flattened and folded. Muscle fibers radiate outward from the vagina which aid in the dilation of the duct during egg

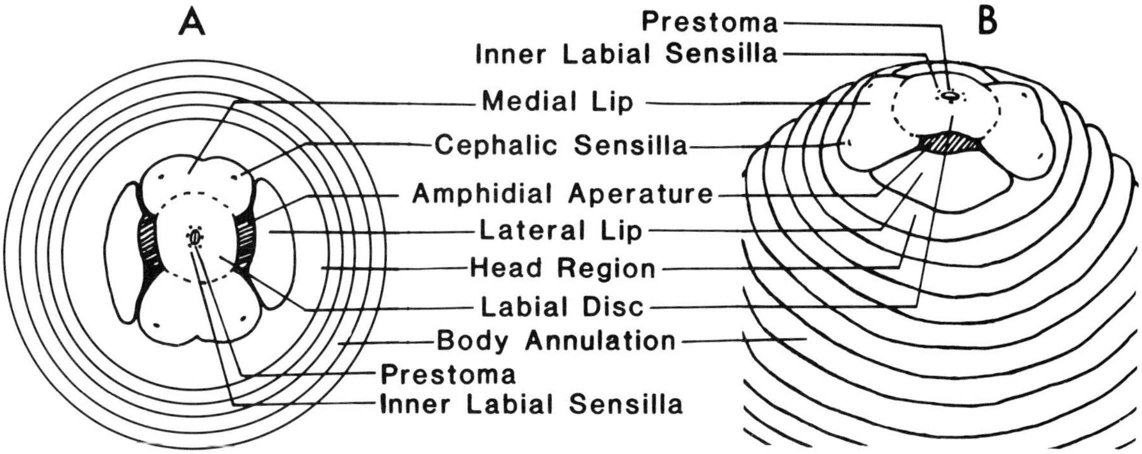

Fig. 31. External head morphology of a female root-knot nematode in A) face and B) lateral views as revealed by scanning electron microscopy [After Eisenback et al. (30)].

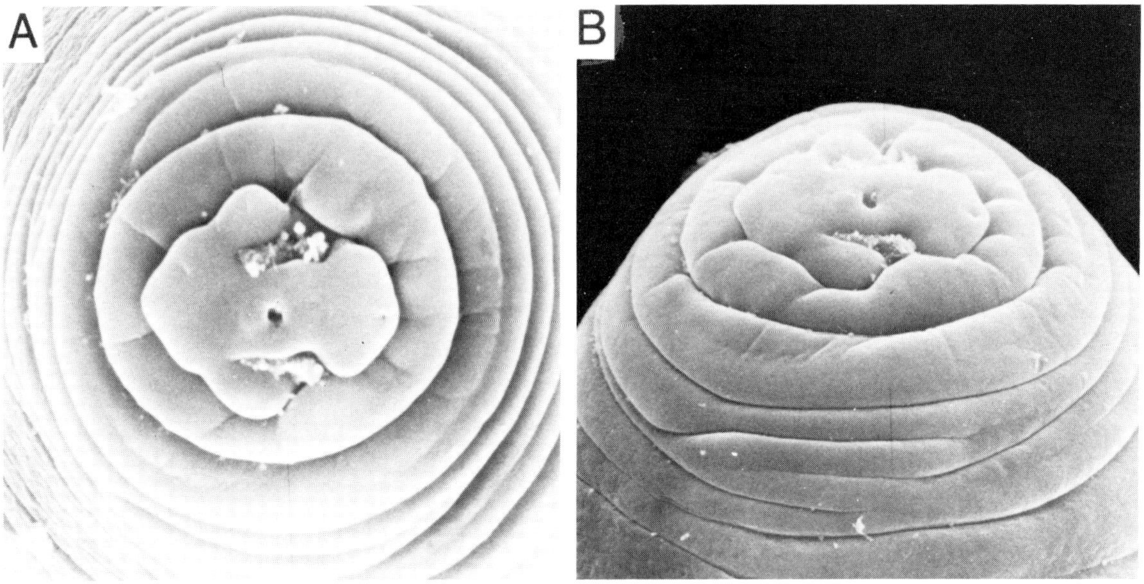

Fig. 32. Scanning electron micrographs of the head of a female of *Meloidogyne hapla* in A) face and B) lateral views [After Eisenback et al. (30)].

laying. The vagina is lined with thick cuticle. Muscles are attached to the vagina and the body wall. The vulva is transversely located on the posterior portion of the body wall and is surrounded by two cuticular lips that are slightly elevated (32).

The eggs of *Meloidogyne* species are approximately 95 μm long and 40 μm wide. The egg shell consists of an outer vitelline layer, a middle chitinous layer, and an inner glycolipid layer. The vitelline layer is approximately 30 nm thick. The chitinous layer is about 400 nm thick, and the glycolipid layer is thickest at the poles. The glycolipid layer makes the eggs of root-knot nematodes very resistant to chemicals (20).

CONCLUSIONS

The overall morphology and anatomy of *Meloidogyne* species has been adequately described with a few exceptions. Detailed morphological studies utilizing transmission

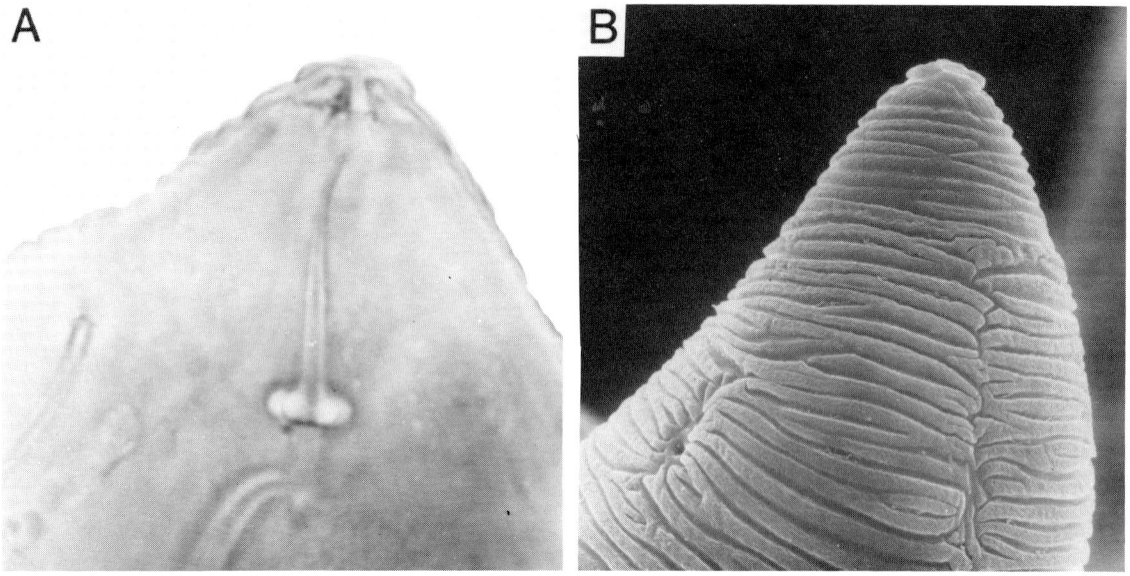

Fig. 33. A) Light and B) scanning electron micrographs of the anterior portion of a female root-knot nematode [After Eisenback et al. (30)].

electron microscopy (TEM) are needed to visualize the structure of the female head, male tail, and esophagus of second-stage juveniles. Also, most of the ultrastructural studies have utilized one of the four most common species of root-knot nematodes, and the available morphological details of the remaining described species are generally limited to the original description. All of the species need to be examined in detail by light microscopy and SEM and, in some cases, by TEM so that the taxonomic position and phylogenetic relationships of members of the genus can be clarified.

Literature Cited

1. Baldwin, J. G. and H. Hirschmann. 1973. Fine structure of the cephalic sense organs in *Meloidogyne incognita* males. J. Nematol. 5:285-302.
2. Baldwin, J. G. and H. Hirschmann. 1975. Body wall fine structure of the anterior region of *Meloidogyne incognita* and *Heterodera glycines* males. J. Nematol. 7:175-193.
3. Baldwin, J. G. and H. Hirschmann. 1976. Comparative fine structure of the stomatal region of males of *Meloidogyne incognita* and *Heterodera glycines*. J. Nematol. 8:1-17.
4. Baldwin, J. G., H. Hirschmann and A. C. Triantaphyllou. 1977. Comparative fine structure of the esophagus of males of *Heterodera glycines* and *Meloidogyne incognita*. Nematologica 23:239-252.
5. Bird, A. F. 1958. The adult female cuticle and egg sac of the genus *Meloidogyne* Goeldi, 1887. Nematologica 3:205-212.
6. Bird, A. F. 1959. Development of the root-knot nematodes *Meloidogyne javanica* (Treub) and *Meloidogyne hapla* Chitwood in the tomato. Nematologica 4:31-42.
7. Bird, A. F. 1967. Changes associated with parasitism in nematodes. I. Morphology and physiology of preparasitic larvae of *Meloidogyne javanica*. J. Parasitol. 53:768-776.
8. Bird, A. F. 1968. Changes associated with parasitism in nematodes. III. Ultrastructure of the egg shell, larval cuticle, and contents of the subventral esophageal glands in *Meloidogyne javanica*, with some observations on hatching. J. Parasitol. 54:475-489.
9. Bird, A. F. 1968. Changes associated with parasitism in nematodes. IV. Cytochemical

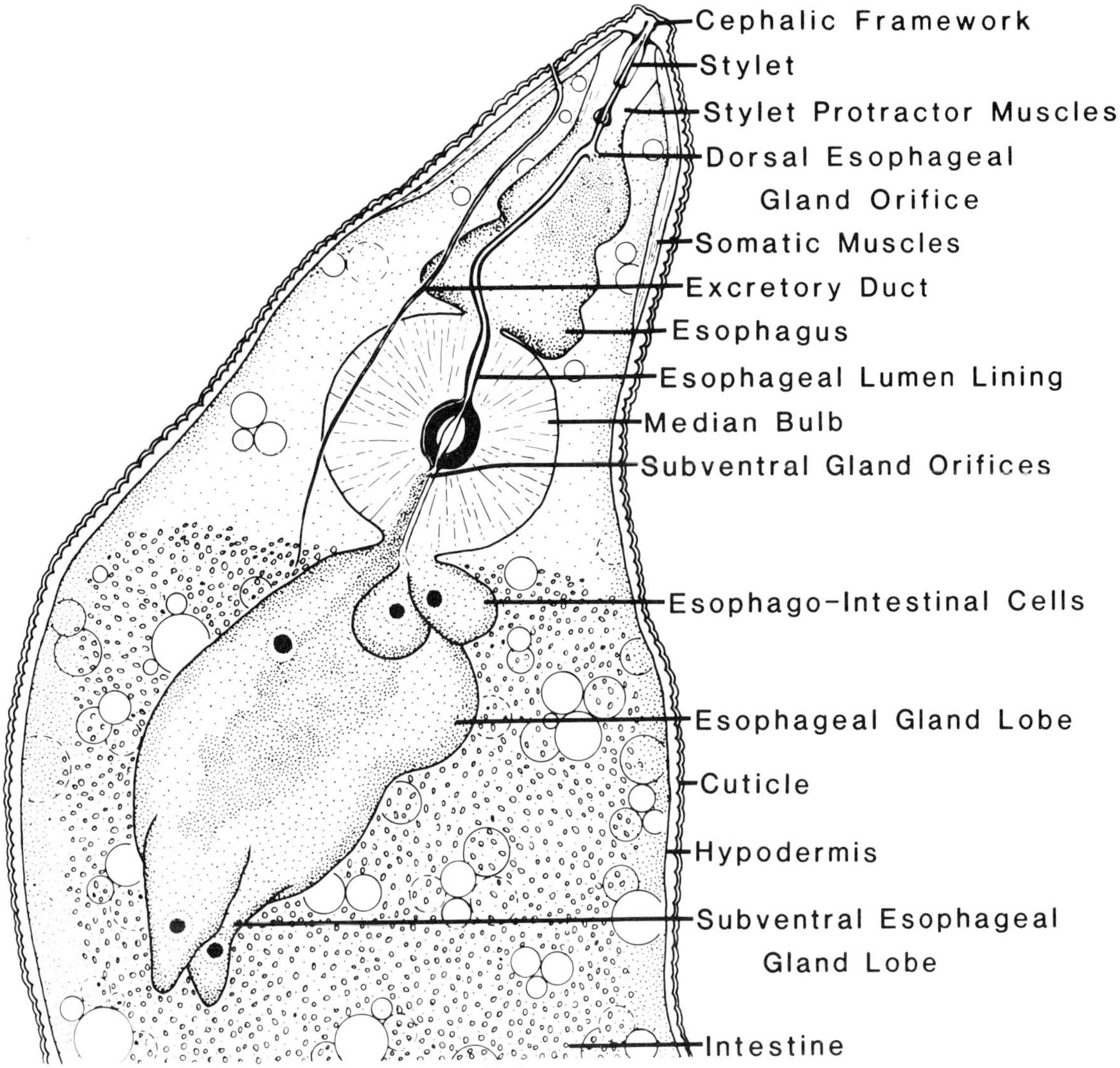

Fig. 34. Diagram of the anterior portion of a female of a *Meloidogyne* species as revealed by light microscopy.

studies on the ampulla of the dorsal esophageal gland of *Meloidogyne javanica* and on exudations of the buccal stylet. J. Parasitol. 54:879-890.

10. Bird, A. F. 1969. Changes associated with parasitism in nematodes. V. Ultrastructure of the stylet exudation and dorsal esophageal gland contents of female *Meloidogyne javanica*. J. Parasitol. 55:337-345.
11. Bird, A. F. 1971. Specialized adaptations of nematodes to parasitism, pp. 35-49 *In* Plant parasitic nematodes, Vol. II., ed. B. M. Zuckerman, W. F. Mai, and R. A. Rohde. Academic Press, New York. 347 pp.
12. Bird, A. F. 1971. The structure of nematodes. Academic Press, New York. 318 pp.
13. Bird, A. F. 1979. Histopathology and physiology of syncytia, pp. 155-171 *In* Root-knot nematodes (*Meloidogyne* species) systematics, biology, and control, ed. F. Lamberti and C. E. Taylor. Academic Press, New York. 477 pp.
14. Bird, A. F. 1979. Morphology and ultrastructure, pp. 59-84 *In* Root-knot nematodes (*Meloidogyne* species) systematics, biology and control, ed. F. Lamberti and C. E. Taylor. Academic Press, New York. 477 pp.
15. Bird, A. F. 1979. Ultrastructure of the tail

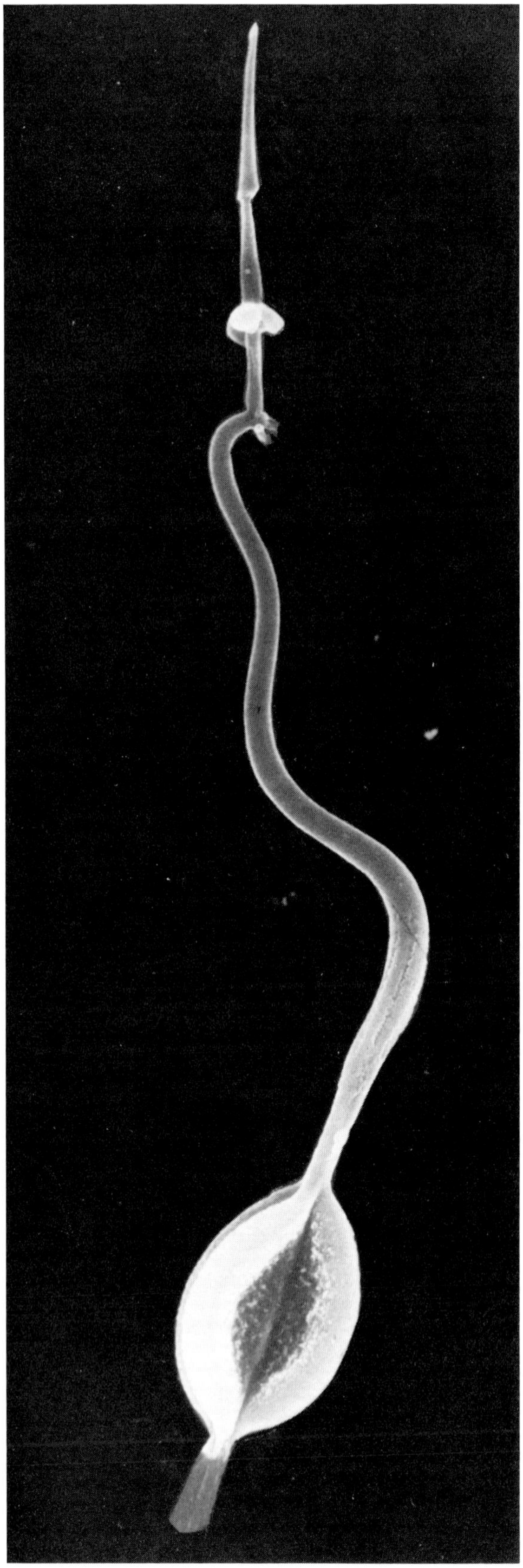

Fig. 35. (left) Scanning electron micrograph of an excised stylet and attached cuticular lumen lining of the esophagus of a female of a *Meloidogyne* spp. [After Eisenback et al. (30)].

region of the second-stage preparasitic larva of the root-knot nematode. Int. J. Parasitol. 9:357-370.

16. Bird, A. F. and G. E. Rogers. 1965. The ultrastructure of the cuticle and its formation *Meloidogyne javanica*. Nematologica 11:224-230.
17. Bird, A. F. and G. E. Rogers. 1965. Ultrastructural and histochemical studies of the cells producing the gelatinous matrix in *Meloidogyne*. Nematologica 11:231-238.
18. Bird, A. F. and W. Saurer. 1967. Changes associated with parasitism in nematodes. II. Histochemical and microspectrophotometric analyses of preparasitic and parasitic larvae of *Meloidogyne javanica*. J. Parasitol. 53:1262-1269.
19. Bird, A. F. and A. Soeffky. 1972. Changes in the ultrastructure of the gelatinous matrix of *Meloidogyne javanica* during dehydration. J. Nematol. 4:166-169.
20. Bird, A. F. and M.A. McClure. 1976. The tylenchoid (Nematoda) egg shell: structure, composition and permeability. Parasitology 72:19-28.
21. Christie, J. R. and G. S. Cobb. 1941. Notes on the life history of the root-knot nematode *Heterodera marioni*. Proc. Helminthol. Soc. Wash. 8:24-26.
22. de Guiran, G. and M. Ritter. 1979. Life cycle of *Meloidogyne* species and factors influencing their development, pp. 172-191 *In* Root-knot nematodes (*Meloidogyne* species) systematics, biology and control, ed. F. Lamberti and C. E. Taylor. Academic Press, New York. 477 pp.
23. Dropkin, V. H. and J. Acedo. 1974. An electron microscopic study of the glycogen and lipid in female *Meloidogyne incognita* (root-knot nematode). J. Parasitol. 60:1013-1021.
24. Dropkin, V. H. and A. F. Bird. 1978. Physiological and morphological studies on secretion of a protein-carbohydrate complex by a nematode. Int. J. Parasitol. 8:225-232.
25. Eisenback, J. D. 1982. Morphological comparison of head shape and stylet morphology of second-stage juveniles of *Meloidogyne* species. J. Nematol. 14:339-343.

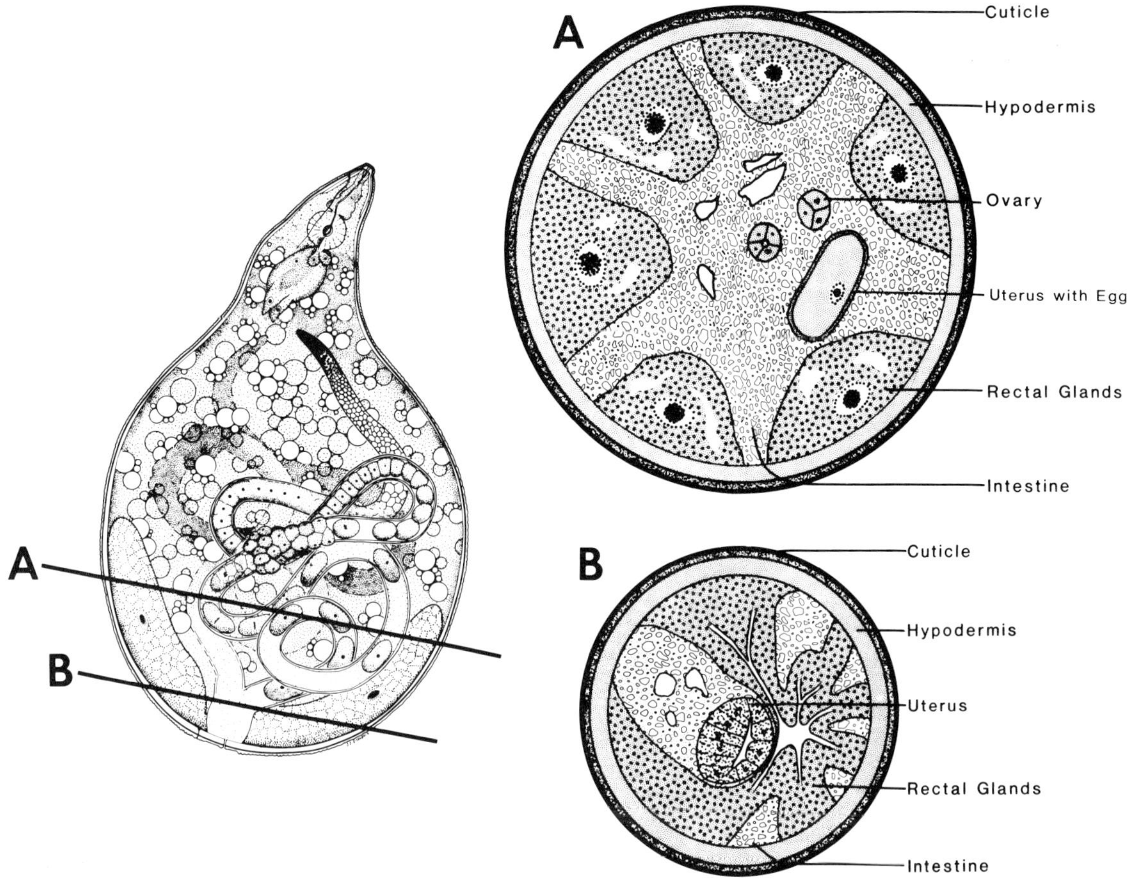

Fig. 36. Diagrams of cross-sections of the posterior end of a female root-knot nematode as revealed by light microscopy [After Maggenti & Allen (40)].

26. Eisenback, J. D. and H. Hirschmann. 1979. Morphological comparison of second-stage juveniles of several *Meloidogyne* species (root-knot nematodes) by scanning electron microscopy. Scanning Electron Microsc. III:223-230.
27. Eisenback, J. D. and H. Hirschmann. 1979. Morphological comparison of second-stage juveniles of six populations of *Meloidogyne hapla* by SEM. J. Nematol. 11:5-16.
28. Eisenback, J. D. and H. Hirschmann. 1980. Morphological comparison of *Meloidogyne* males by scanning electron microscopy. J. Nematol. 12:23-32.
29. Eisenback, J. D. and H. Hirschmann. 1981. Identification of *Meloidogyne* species on the basis of head shape and stylet morphology. J. Nematol. 13:513-521.
30. Eisenback, J. D., H. Hirschmann, and A. C. Triantaphyllou. 1980. Morphological comparison of *Meloidogyne* female head structures, perineal patterns, and stylets. J. Nematol. 12:300-313.
31. Eisenback, J. D., H. Hirschmann, J. N. Sasser, and A. C. Triantaphyllou. 1981. A guide to the four most common species of root-knot nematodes (*Meloidogyne* species), with a pictorial key. Coop. Publ. Deps. Plant Pathol. and Genet., North Carolina State Univ., and U.S. Agency Int. Dev. Raleigh, N.C. 48 pp.
32. Elsea, J. R. 1951.The histological anatomy

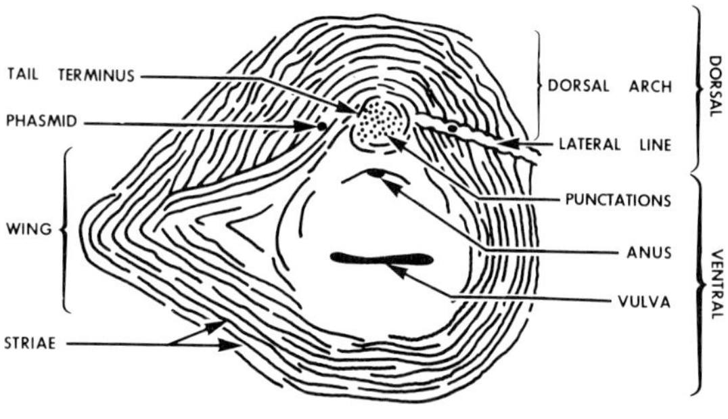

Fig. 37. Diagram of a perineal pattern of a *Meloidogyne* female.

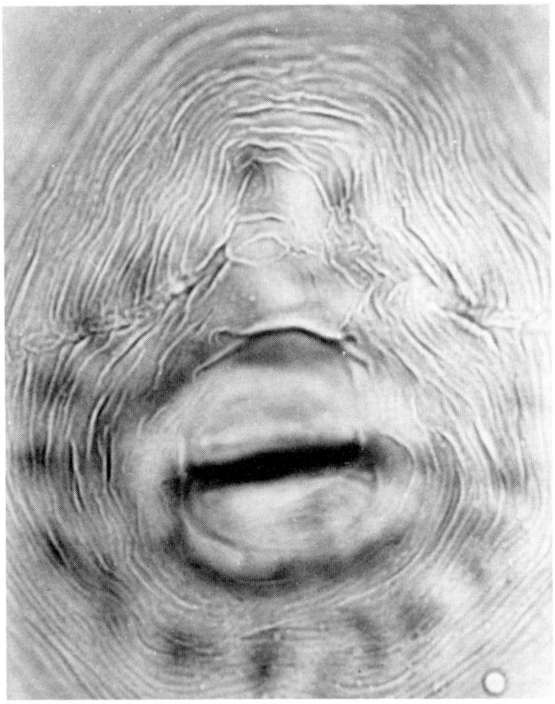

Fig. 38. Light micrograph of a perineal pattern of *Meloidogyne javanica*.

of the nematode *Meloidogyne hapla* (Heteroderidae). Proc. Helminthol. Soc. Wash. 18:53-63.

33. Endo, B. Y. and W. P. Wergin. 1977. Ultrastructure of anterior sensory organs of the root-knot nematode, *Meloidogyne incognita*. J. Ultrastruc. Res. 59:231-249.
34. Esser, R. P., V. G. Perry, and A. L. Taylor. 1976. A diagnostic compendium of the genus *Meloidogyne* (Nematoda: Heteroderidae).Proc. Helminthol.Soc. Wash. 43:138-150.
35. Franklin, M. T. 1971. Taxonomy of Heteroderidae, pp. 139-162 *In* Plant parasitic nematodes, Vol. I., ed. B. M. Zuckermann, W. F. Mai, and R. A. Rohde. Academic Press, New York. 345 pp.
36. Franklin, M. T. 1978. *Meloidogyne*—root-knot eelworms, pp. 98-124 *In* Plant nematology, ed. J. F. Southey. Minist. Agric., Fish., and Food Tech. Bull. 7. Her Majesty's Stationery Office, London. 440 pp.
37. Garcia-Martinez, R. 1982. Post-infection development and morphology of *Meloidogyne cruciani*. J. Nematol. 14:332-338.
38. Goldstein, P. and A. C. Triantaphyllou. 1980. The ultrastructure of sperm development in the plant-parasitic nematode *Meloidogyne hapla*. J. Ultrastruc. Res. 71:143-153.
39. Johnson, P. W. and W. G. Graham. 1976. Ultrastructural studies on the cuticle of the second-stage larvae of four root-knot nematodes: *Meloidogyne hapla*, *M. javanica*, *M. incognita*, and *M. arenaria*. Can. J. Zool. 54:96-100.
40. Maggenti, A. R. and M. W. Allen. 1960. The origin of the gelatinous matrix in *Meloidogyne*. Proc. Helminthol. Soc. Wash. 27:4-10.
41. McClure, M. A. and A. F. Bird. 1976. The tylenchoid (Nematoda) egg shell: formation of the egg shell in *Meloidogyne javanica*. Parasitology 72:29-39.
42. Papadopoulou, J. and A. C. Triantaphyllou. 1982. Sex differentiation in *Meloidogyne incognita* and anatomical evidence of sex reversal. J. Nematol. 14:549-566.
43. Popham, J. D. and J. M. Webster. 1978. An alternative interpretation of the fine structure of the basal zone of the cuticle of the dauerlarva of the nematode *Caenorhabditis elegans* (Nematoda). Can. J. Zool. 56:1556-1563.
44. Saigusa, T. 1957. On the egg development and its morphological observations of the root knot nematode, *Meloidogyne* spp. Jpn. J. Appl. Entomol. Zool. 1:238-243.
45. Shepherd, A. M. and S. A. Clark. 1983. Spermatogenesis and sperm structure in some *Meloidogyne* species (Heteroderoidea, Meloidogynidae) and a comparison with those in some cyst nematodes (Heteroderoidea, Heteroderidae). Rev. Nematol. 6:17-32.
46. Taylor, A. L. and J. N. Sasser. 1978. Biology, identification and control of root-knot nematodes (*Meloidogyne* species). Coop. Publ. Dep. Plant Pathol., North Carolina State Univ., and the U.S. Agency Int. Dev. Raleigh, N.C. 111 pp.
47. Triantaphyllou, A. C. 1960. Sex determina-

tion in *Meloidogyne incognita* Chitwood, 1949 and intersexuality in *M. javanica* (Treub, 1885) Chitwood, 1949. Ann. Inst. Phytopathol., Benaki 3:12-31.

48. Triantaphyllou, A. C. 1962. Oogenesis in the root-knot nematode *Meloidogyne javanica*. Nematologica 7:105-113.
49. Triantaphyllou, A. C. 1979. Cytogenetics of root-knot nematodes, pp. 85-109 *In* Root-knot nematodes (*Meloidogyne* species) systematics, biology and control, ed. F. Lamberti and C. E. Taylor. Academic Press, New York. 477 pp.
50. Triantaphyllou, A. C. and H. Hirschmann. 1960. Post infection development of *Meloidogyne incognita* Chitwood, 1949. Ann. Inst. Phytopathol., Benaki 3:1-11.
51. Viglierchio, D. R. 1979. Selected aspects of root-knot nematode physiology, pp. 115-154 *In* Root-knot nematodes (*Meloidogyne* species) systematics, biology, and control, ed. F. Lamberti and C. E. Taylor. Academic Press, New York. 477 pp.
52. Wergin, W. P. 1981. Scanning electron microscopic techniques and applications for use in nematology, pp. 175-204 *In* Plant parasitic nematodes, Vol. III, ed. B. M. Zuckerman and R. A. Rhode. Academic Press, New York 508 pp.
53. Wergin, W. P. and B. Y. Endo. 1976. Ultrastructure of a neurosensory organ in a root-knot nematode. J. Ultrastruc. Res. 56:258-276.
54. Whitehead, A. G. 1968. Taxonomy of *Meloidogyne* (Nematodea: Heteroderidae) with descriptions of four new species. Trans. Zool. Soc., London 31:263-401.
55. Wouts, W. M. 1979. Characterization of the family Meloidogynidae with a discussion on its relationship to other families of the sub-order Tylenchina based on gonad morphology, pp. 21-35 *In* Root-knot nematodes (*Meloidogyne* species) systematics, biology, and control, ed. F. Lamberti and C. E. Taylor. Academic Press, New York. 477 pp.

THE GENUS *MELOIDOGYNE* AND MORPHOLOGICAL CHARACTERS DIFFERENTIATING ITS SPECIES

Hedwig Hirschmann
Department of Plant Pathology
North Carolina State University
Raleigh, North Carolina, USA

THE GENUS *MELOIDOGYNE*

Root-knot nematodes were first named by Cornu, 1879 (3), in France. He found nematodes in galls on sainfoin (*Onobrychis* sp.) and believed them to be similar to *Heterodera schachtii* but named them *Anguillula marioni* to avoid subsequent misidentification. In 1884, Müller (24) found worms in galls on the roots of *Dodartia orientalis* and recognized that they were related to the genus *Heterodera*. He mistakenly took them to be *Anguillulina radicicola* which had been described from the same host by Greeff in 1872 (19), and, therefore, called them *Heterodera radicicola*. During that time, other investigators described additional root-knot nematodes either under the genus *Heterodera* or *Anguillula* (e.g. *H. javanica*, *A. arenaria*). None of these workers mentioned a paper by Göldi, first published in 1887 and reprinted in 1892 (17), in which this author described a root-knot nematode from coffee naming it *Meloidogyne exigua*. Although Cobb, 1924 (2), claimed that there were morphological differences between root-knot nematodes and the cyst-forming *Heterodera* species and erected the genus or subgenus *Caconema* for the root-knot nematodes, the name *H. radicicola* continued to be used during the early part of the 20th century. It was not until 1932 that Goodey (18) reviewed the nomenclature of these nematodes and resolved Müller's earlier error. Goodey pointed out that Greeff's *Anguillulina radicicola* was not related to either *Heterodera* or *Meloidogyne* but was what is now known as *Subanguina radicicola*. With respect to the root-knot nematodes, he used Cornu's earlier specific name and renamed the root-knot nematodes *Heterodera marioni*. Goodey felt it unjustified to erect a new genus to separate root-knot nematodes from the existing *Heterodera* species. In 1949, Chitwood (1) re-erected the genus *Meloidogyne* Göldi, 1887, to receive all root-knot nematodes and synonymized Cobb's genus *Caconema*. He redescribed the type species *M. exigua* Göldi, 1887, and the three other species that had been named earlier: *M. javanica* (Treub, 1885), *M. incognita* (Kofoid & White, 1919) and *M. arenaria* (Neal, 1889). Chitwood also described the new species *M. hapla* and a new variety, *M. incognita* var. *acrita*. He gave a generic diagnosis for *Meloidogyne* and differentiated it from *Heterodera* A. Schmidt, 1871. Whitehead, 1968 (25), in his comprehensive review redefined the genus and confirmed the morphological distinctness of 23 species of *Meloidogyne*. In her 1972 review, Franklin (15) included 32 species and treated critically the most recent developments of the systematics of *Meloidogyne* with emphasis on the distinguishing characters of the various life stages. Esser, Perry and Taylor, 1976 (14), in their compendium of *Meloidogyne* presented morphometric and morphological data and illustrations to facilitate the identification of 32 species of the genus. Franklin, 1979 (16), listed 36 valid *Meloidogyne* species and summarized the historical background of the genus and the means of identification based on females, juveniles and males. By June 1984, 54 species and 2 subspecies were included in the genus (Table 1).

DIFFERENTIATING CHARACTERS OF SPECIES OF *MELOIDOGYNE*

Females (Table 2)

Aside from slight differences in general body shape, good differentiating characters of females are present either in the anterior or posterior body regions.

Table 1. Species of *Meloidogyne*.

1. *Meloidogyne acronea* Coetzee, 1956
2. *M. africana* Whitehead, 1960
3. *M. aquatilis* Ebsary and Eveleigh, 1983
4. *M. ardenensis* Santos, 1968
5. *M. arenaria* (Neal, 1889) Chitwood, 1949
6. *M. artiellia* Franklin, 1961
7. *M. bauruensis* (Lordello, 1956) Esser, Perry and Taylor, 1976
8. *M. brevicauda* Loos, 1953
9. *M. camelliae* Golden, 1979
10. *M. carolinensis* Eisenback, 1982
11. *M. chitwoodi* Golden, O'Bannon, Santo and Finley, 1980
12. *M. coffeicola* Lordello and Zamith, 1960
13. *M. cruciani* Garcia-Martinez, 1982
14. *M. decalineata* Whitehead, 1968
15. *M. deconincki* Elmiligy, 1968
16. *M. elegans* da Ponte, 1977
17. *M. enterolobii* Yang and Eisenback, 1983
18. *M. ethiopica* Whitehead, 1968
19. *M. exigua* Göldi, 1887
20. *M. grahami* Golden and Slana, 1978
21. *M. graminicola* Golden and Birchfield, 1965
22. *M. graminis* (Sledge and Golden, 1964) Whitehead, 1968
23. *M. hapla* Chitwood, 1949
24. *M. incognita* (Kofoid and White, 1919) Chitwood, 1949
25. *M. incognita acrita* Chitwood, 1949
26. *M. incognita wartellei* Golden and Birchfield, 1978
27. *M. indica* Whitehead, 1968
28. *M. inornata* Lordello, 1956
29. *M. javanica* (Treub, 1885) Chitwood, 1949
30. *M. kikuyensis* de Grisse, 1960
31. *M. kirjanovae* Terentyeva, 1965
32. *M. kralli* Jepson, 1983
33. *M. litoralis* Elmiligy, 1968
34. *M. lordelloi* da Ponte, 1969
35. *M. lucknowica* Singh, 1969
36. *M. mali* Itoh, Ohshima and Ichinohe, 1969
37. *M. megadora* Whitehead, 1968
38. *M. megatyla* Baldwin and Sasser, 1979
39. *M. megriensis* (Poghossian, 1971) Esser, Perry and Taylor, 1976
40. *M. microcephala* Cliff and Hirschmann, 1984
41. *M. microtyla* Mulvey, Townshend and Potter, 1975
42. *M. naasi* Franklin, 1965
43. *M. nataliei* Golden, Rose and Bird, 1981
44. *M. oryzae* Maas, Sanders and Dede, 1978
45. *M. oteifae* Elmiligy, 1968
46. *M. ottersoni* (Thorne, 1969) Franklin, 1971
47. *M. ovalis* Riffle, 1963
48. *M. platani* Hirschmann, 1982
49. *M. poghossianae* Kirjanova, 1963 = species inquirenda
50. *M. propora* Spaull, 1977
51. *M. querciana* Golden, 1979
52. *M. sewelli* Mulvey and Anderson, 1980
53. *M. spartinae* (Rau and Fassuliotis, 1965) Whitehead, 1968
54. *M. subarctica* Bernard, 1981
55. *M. tadshikistanica* Kirjanova and Ivanova, 1965
56. *M. thamesi* (Chitwood in Chitwood, Specht and Havis, 1952) Goodey, 1963

BODY SHAPE The pyriform body shape is typical for most species, and the neck and perineal region are in line with the longitudinal axis. In some species, the females become more oval; a pronounced posterior protuberance is present; and the neck is at an angle to the longitudinal body axis (14).

PERINEAL PATTERN The most characteristic feature of females of most *Meloidogyne* species is the perineal pattern in the posterior body region. This area comprises the tail terminus, phasmids, lateral lines, anus and vulva surrounded by cuticular folds or striae (see Eisenback, Chapter 6, Fig. 37). The perineal patterns of individuals and populations within species vary, but basic species characteristics do not change significantly over an extended period of culturing, and in no case do pattern characteristics change from one species to those of another (25). These characters include overall shape of the perineal pattern (circular, oval, pear-shaped, hourglass-shaped); the presence or absence of markings in the lateral-line areas and of punctations on the tail terminus; the form of the striae (smooth, unbroken, broken, wavy, zigzag); or whether or not the striae form wings on one or both sides of the pattern. Significant dif-

Table 2. Differentiating characters of females of *Meloidogyne*.

Morphometric (quantitative)	Morphologic (qualitative)
Perineal pattern[a]	Body shape
Stylet length[a]	Head structures in SEM
Stylet knob width/height	Shape of stylet cone, shaft and knobs[a]
Distance of DEGO to base of stylet	Perineal pattern[a]
Size of metacorpus "valve plates"	
Position of excretory pore	

[a]Most reliable features.

ferences between species and populations may also occur with respect to the relative positions of the vulva, anus, tail terminus, and phasmids. These features can be expressed morphometrically as interphasmidial distance, distance of anus to tail terminus, and anus to center of vulva. The characteristic features of the perineal patterns are well demonstrated in light micrographs (Fig. 1). In the scanning electron microscope (SEM), the cuticular features of the patterns are emphasized and a three-dimensional impression results (Fig. 2). Aside from the considerable overall variation, it is often difficult to describe the various features of perineal patterns precisely. In order to analyze the pattern characteristics better, Esser et al., 1976 (14), found it useful to divide the perineal pattern into four definite zones. Improved methods for assessing pattern morphology, however, will be necessary in order for it to continue to be a good differentiating character in the future.

EXCRETORY PORE The position of the excretory pore may be helpful in differentiating some species, although this character is quite variable within populations and species. The number of annules between the excretory pore and head end has frequently been used, but annule number is often difficult to determine precisely, and this feature appears unreliable. The position of the excretory pore from the head end as related to stylet length (14) may be of use in some species and subspecies, although it is variable in others.

METACORPUS Some differences occur between species in length and width of the metacorpus and length of the metacorpus "valve plates" (=feeding pump lining). From our studies it appears that "valve plate" size may be a useful character, but bulb size seems to be too variable with too much overlap between species to be helpful diagnostically.

HEAD CAP AND HEAD REGION Some differences are present in the head morphology of females. These differences are usually too obscure to be detected with the light microscope (LM) and can be observed more clearly in the SEM (Eisenback, Chapter 6, Figs. 31, 32). Differences between populations and species are present in the symmetry of the labial disc and medial lips, size and shape of medial and lateral lips, the expression of sensilla and head annulations (Fig. 3). However, for practical identification, the female head structures are not very useful because the head morphology can be influenced by age, environmental factors, and the method of specimen preparation (12).

STYLET The morphology of the stylet including the shape of cone, shaft and knobs is reliable and appears to be species specific in many cases (12, 22; Eisenback, Chapter 8, Fig. 4). This distinction can be demonstrated in the LM, but differences become more obvious in the SEM (Fig. 4). Intraspecific variation of stylet length was found to be low among 16 species studied (22). The stylet length and also the cone length in relation to stylet length are, therefore, important morphometric features. Stylet knob height and width and the ratio width/height may also be useful as they determine the shape of the stylet knobs. In general, female stylet morphology can be used as a good supplemental character for species identification.

DORSAL ESOPHAGEAL GLAND ORIFICE The range of the distance between the stylet knob base and dorsal esophageal gland orifice (DEGO) is broad and seems variable within populations, but this character may have limited use in identification of species groups (22) when preparations are made in a standard manner.

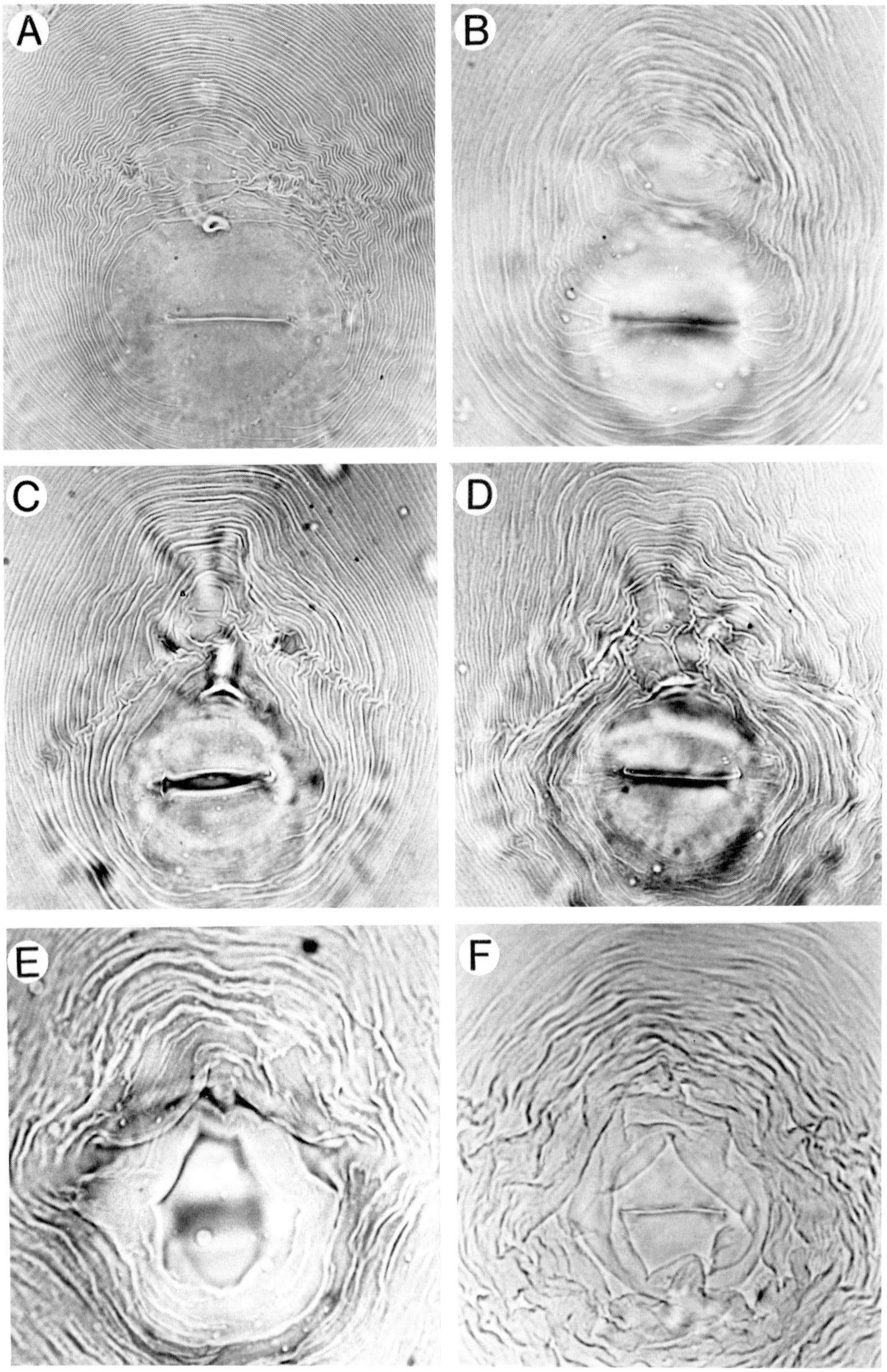

Fig. 1. LM photographs of perineal patterns of females of six different species of *Meloidogyne* showing variation within the genus. A) *M. platani* [from Hirschmann (20)]. B) *Meloidogyne* sp. (courtesy of A. Rammah). C) *M. arenaria* variant [from Cliff (4)]. D) *M. microcephala* [from Cliff & Hirschmann (5)]. E) *M. pini* [from Eisenback et al. (13)]. F) *M. carolinensis* [from Eisenback (6)].

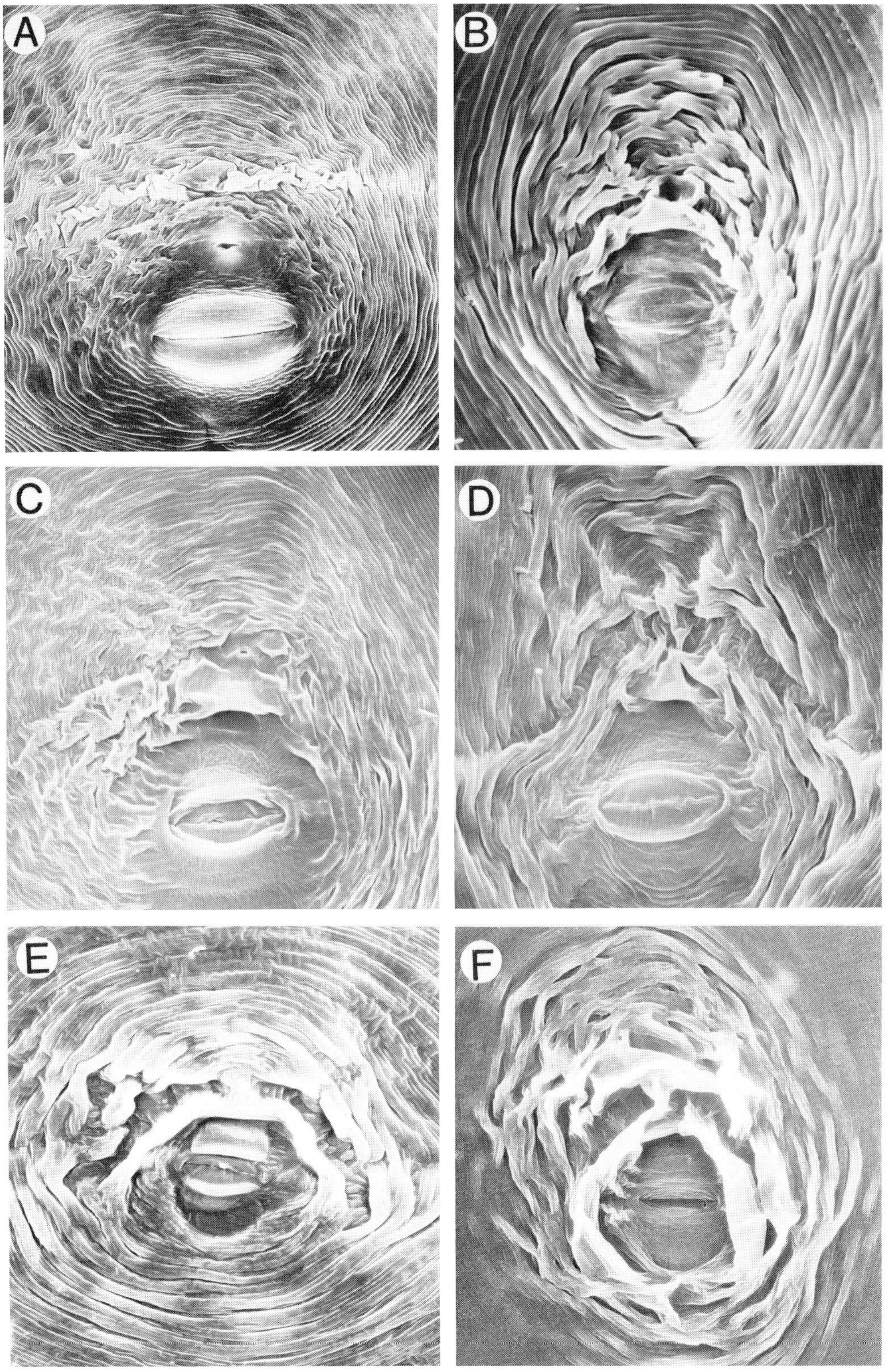

Fig. 2. SEM photographs of perineal patterns of females of six different species of *Meloidogyne* showing variation within the genus. A) *M. platani* [from Hirschmann (20)]. B) *M. chitwoodi* (courtesy of J. D. Eisenback). C) *M. arenaria* variant [from Cliff (4)]. D) *M. microcephala* [from Cliff & Hirschmann (5)]. E) *M. pini* [from Eisenback et al. (13)]. F) *M. carolinensis* [from Eisenback (6)].

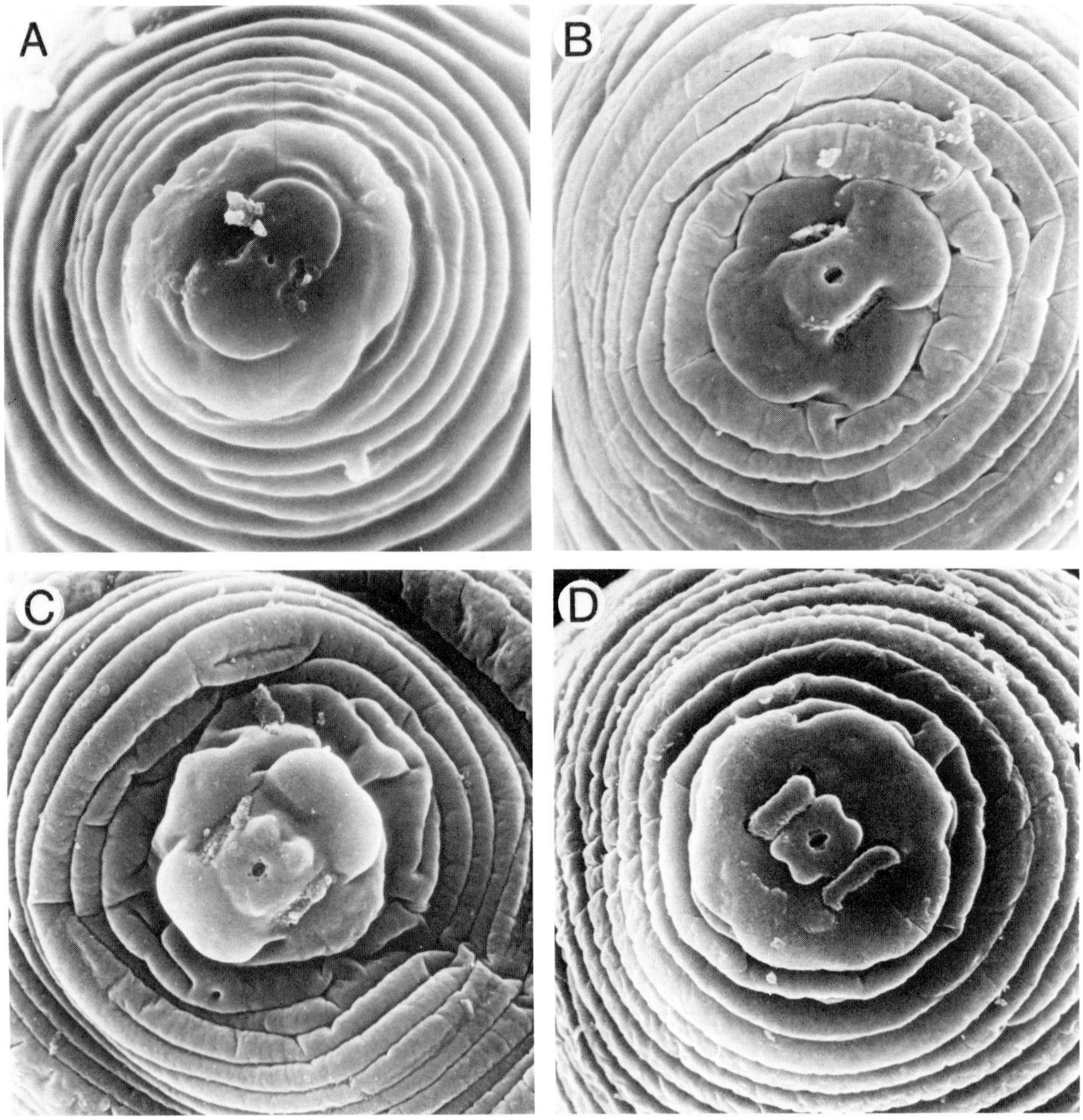

Fig. 3. SEM photographs of head regions of females of *Meloidogyne*. A) *M. platani* [from Hirschmann (20)]. B) *Meloidogyne* sp. C) *M. javanica* [from Eisenback et al. (12)]. D) *M. carolinensis* [from Eisenback (6)].

Males (Table 3)

BODY LENGTH Males vary greatly in body size due to varying conditions existing during their development, and dwarf specimens are often present in heavily infected plant roots. Morphometric characters, therefore, usually have large coefficients of variability.

EXCRETORY PORE The position of the excretory pore expressed variously as a linear measurement, in relation to body length, or in number of annules behind the head region, exhibits large intraspecific variation and is of limited value as a differential character. There may be differences in the surface structure of the excretory pore as seen in SEM (9) that might be useful in some species.

LATERAL FIELD Some investigators hold that lateral lines and areolation can be distinctly seen in the LM and are relatively reliable differentiating features (14). However, the number of incisures varies within the same specimen in different body regions, and the same is true for areolation. Areolation and number of incisures may be occasionally of diagnostic value. Slight irregularities in body annulation are present near the beginning of the lateral field in

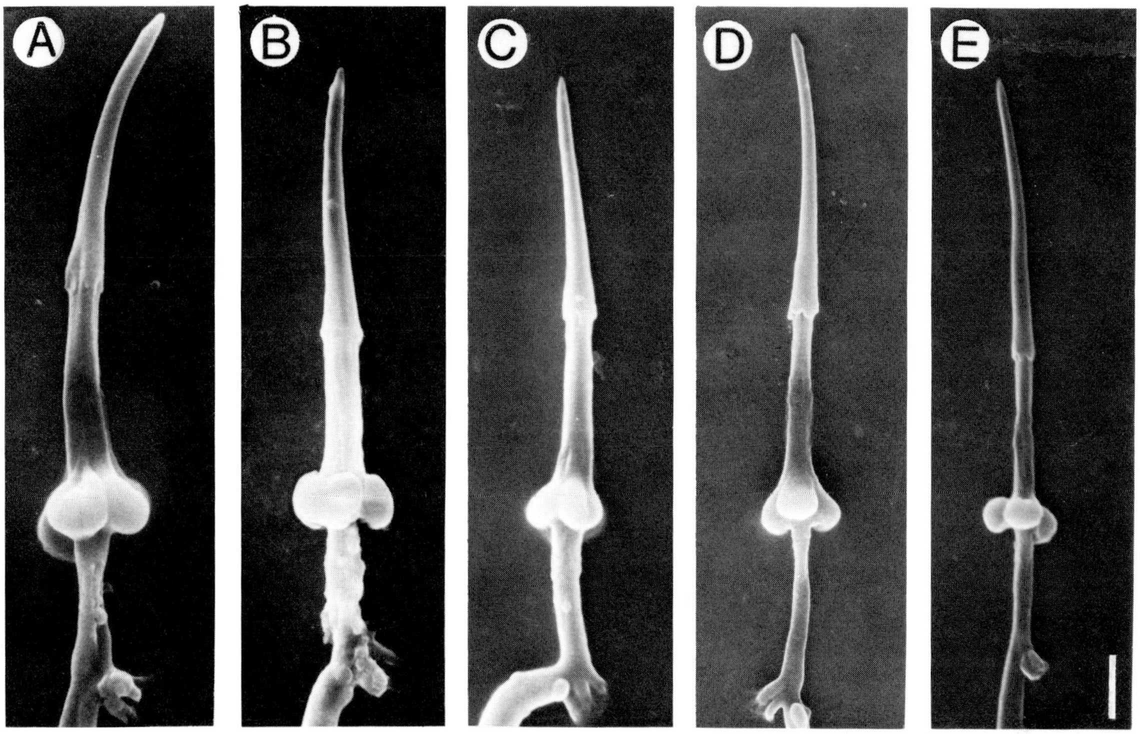

Fig. 4. **SEM photographs of excised stylets of females of different species of *Meloidogyne*. A) *M. megatyla* [from Eisenback et al. (13)]. B) *Meloidogyne* sp. (courtesy of A. Rammah). C) *Meloidogyne* sp. (courtesy of A. Rammah). D) *M. carolinensis* [from Eisenback (6)]. E) *M. pini* [from Eisenback et al. (13)]. Bar = 2 μm.**

specimens of some populations that show a distinct bulge in this region.

CEPHALIDS AND HEMIZONID These are structures difficult to observe in the LM. Differences were reported in the location of the cephalids in some species (25). The hemizonid is located anterior to the excretory pore in most species, but in a few, it is posterior to it and may thus be helpful in identification.

METACORPUS As in the female, there seem to be differences in the length of the "valve plates" of the metacorpus (23).

HEAD CAP AND HEAD REGION As in the other life stages, qualitative morphological differences are expressed mainly in the head characters of males. In the SEM, males have the same basic head characters as second-stage juveniles and females, only the morphological expression of these characters differs slightly (Eisenback, Chapter 6, Figs. 17, 18, 19). The expression of labial and cephalic sensilla, shape and proportion of labial disc and lips, and presence or absence of annulations in the head region can be used to distinguish species and populations within species (Fig. 5). Some of these cephalic features can also be seen clearly in the LM. The head morphology differs in size, height, shape, and slope of the head cap, the number of head annulations and width of the head region relative to the first body annule, or the way the head region fuses with the body region, i.e., whether it is set off or not set off (Fig. 6).

STYLET In addition to head morphology, stylet morphology can be considered an excellent supporting character that is distinct enough in the LM to be useful in species identification. Differences in stylets occur in size and shape of the cone, shaft and knobs (10). These features can be seen clearly in the LM (Fig. 6) but become more distinct in the SEM (Fig. 7). Stylet length was found to be the best differentiating character because of its broad range within the genus (16-27 μm) and low coefficient of variability (4%) (23). Stylet cone length, which is correlated with stylet length and usually is approximately half of the stylet length, can be helpful in some species which have a much shorter cone. Stylet knob width and length are variable and only useful in those species with larger knobs.

DORSAL ESOPHAGEAL GLAND ORIFICE The distance between the stylet knob base and the dorsal esophageal gland orifice (DEGO) exhibits much

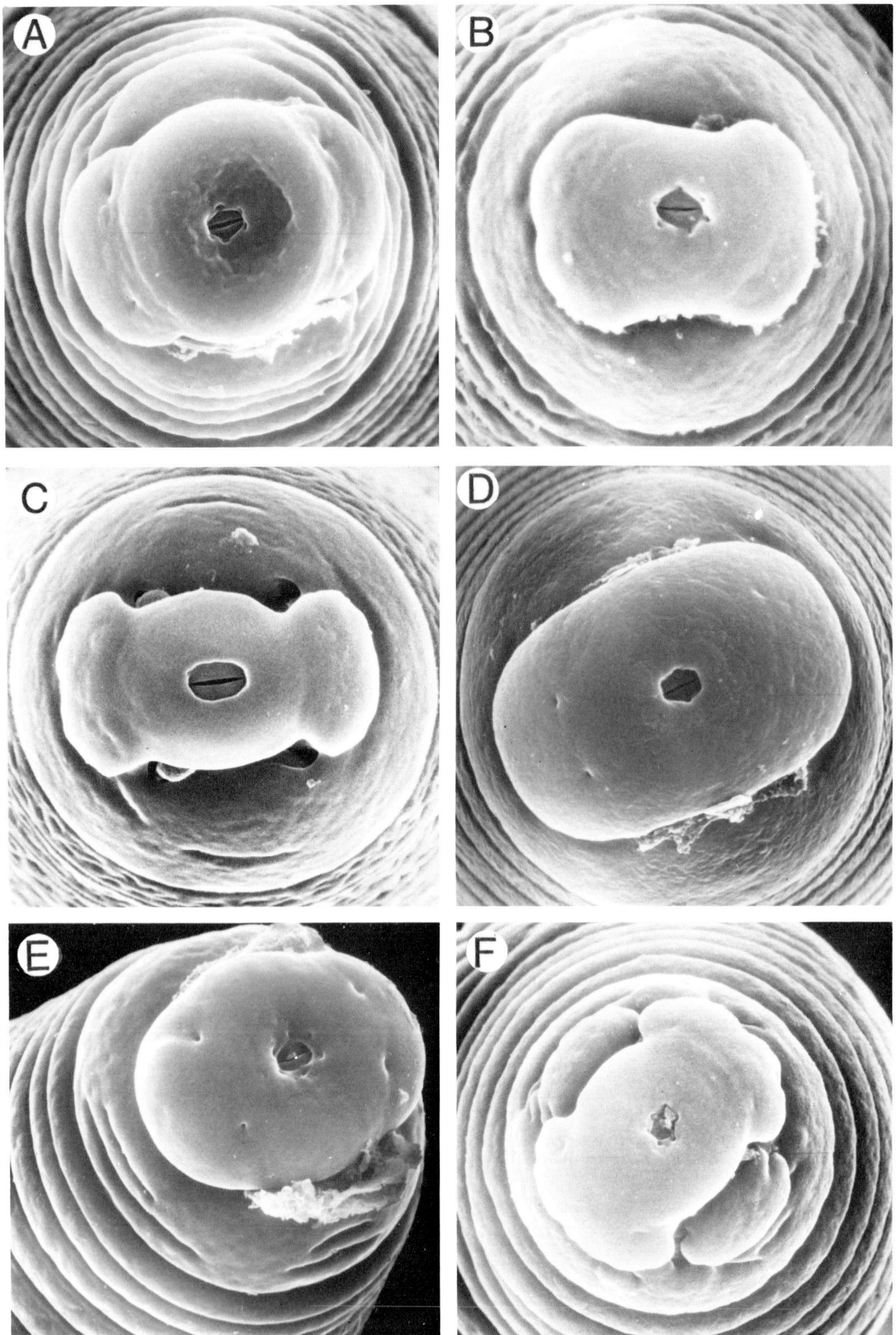

Fig. 5. SEM photographs of head regions of males of different species of *Meloidogyne*. A) *M. incognita*. B) *M. megatyla* [from Eisenback et al. (13)]. C) *Meloidogyne* sp. D) *Meloidogyne* sp. E) *Meloidogyne* sp. F) *M. chitwoodi* (Courtesy of J. D. Eisenback).

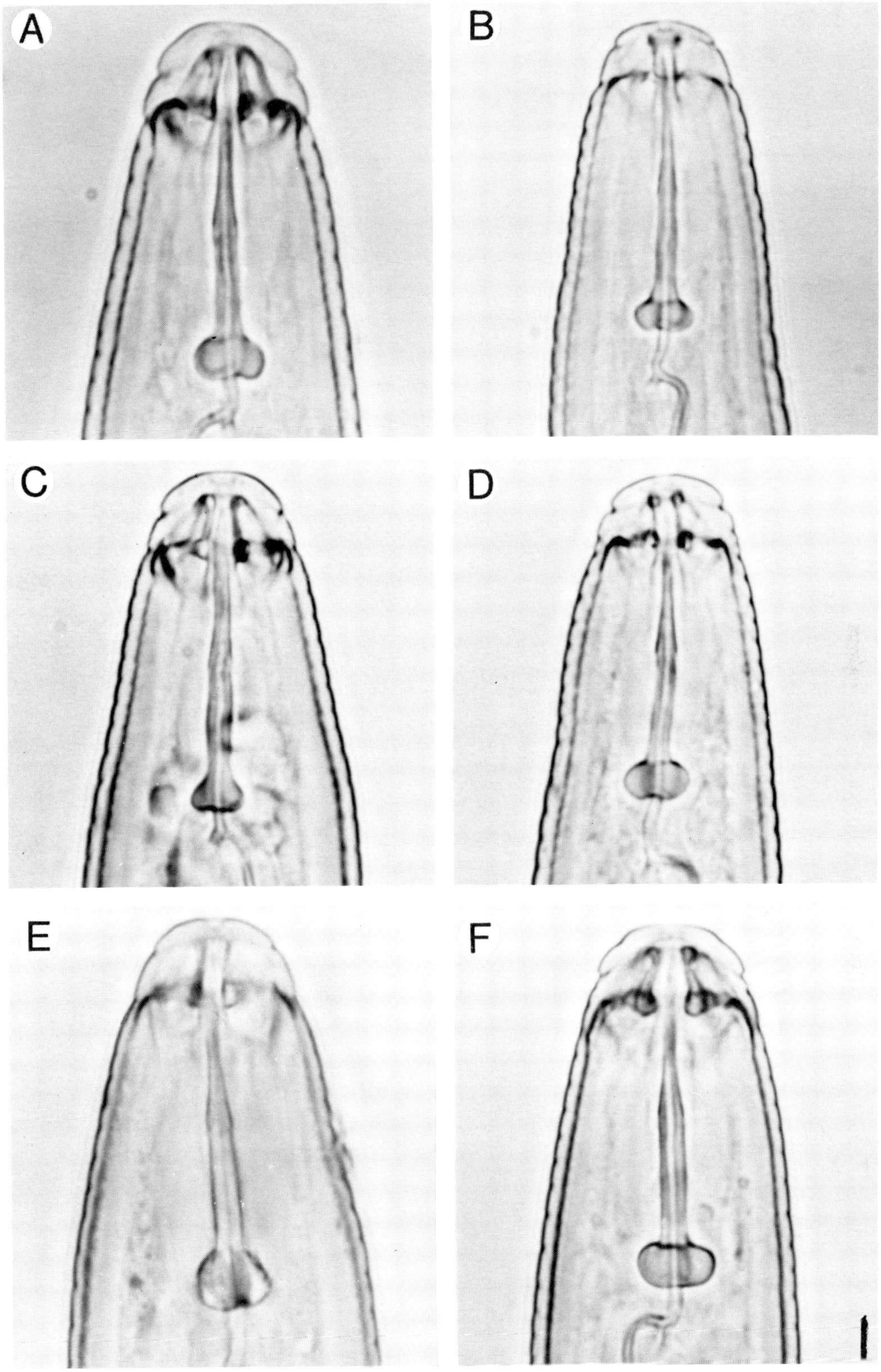

Fig. 6. LM photographs of head regions of males of different species of *Meloidogyne* showing differences in head shape and stylet morphology. A) *Meloidogyne* sp. B) *M. microcephala* [from Cliff & Hirschmann (5)]. C) *M. arenaria* variant [from Cliff (4)]. D) *M. arenaria* variant [from Cliff (4)]. E) *M. megatyla* [from Eisenback et al. (13)]. F) *M. incognita*. Bar = 3 μm.

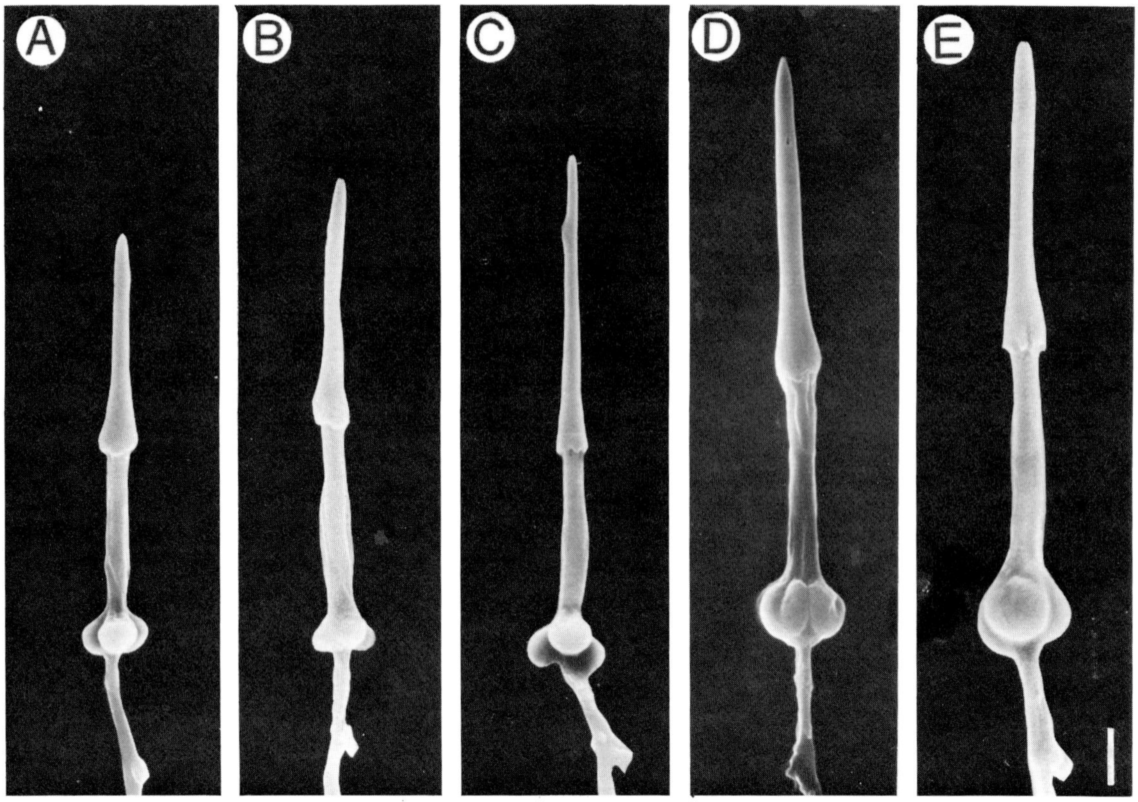

Fig. 7. **SEM photographs of excised stylets of males of different species of *Meloidogyne*. A) *M. exigua* [from Eisenback & Hirschmann (11)]. B) *M. microtyla* [from Eisenback & Hirschmann (11)]. C) *M. carolinensis* [from Eisenback & Hirschmann (11)]. D) *Meloidogyne* sp. (Courtesy of A. Rammah). E) *M. megatyla* [from Eisenback & Hirschmann (11)]. Bar = 2 μm.**

variation in general, although some species can be distinguished on the basis of DEGO distance (23).

SPICULES AND GUBERNACULUM There seem to be slight differences in structure of the spicules between some of the species but no detailed comparative data are available. With respect to size of these structures, specimens have to be in exact lateral view for accurate measurement which is not always possible because male tails are twisted through 90°. Morphometric differences may occur between some species, but there is also considerable overlap.

Second-Stage Juveniles (Table 4)

BODY LENGTH There is considerable variation among species with respect to body length. The range of the genus is narrow (except for *M. spartinae*), and the measurements obviously overlap in many species, although the extreme values can be different, and differences in body length may occur also between populations of the same species. Body length of second-stage juveniles is, therefore, inadequate for species differentiation.

LATERAL FIELD The lateral fields are different in some species with respect to number of incisures and type of areolation. In the four most common *Meloidogyne* species, however, the lateral fields are very similar and not useful for identification (8).

CEPHALIDS AND HEMIZONID As in the males, the location of the hemizonid is considered a fairly useful character. Although in most species it is located near or just anterior to the excretory pore, in some species (e.g., *M. ardenensis*, *M. graminis*), it is posterior to the excretory pore.

The cephalids are so small in second-stage juveniles that they can hardly be observed in the LM. They are not useful as diagnostic characters.

HEAD CAP AND HEAD REGION Qualitative morphological differences are expressed mainly in the head characters which can be observed best with the SEM. Juveniles possess the same basic head characters as females , only their ex-

Table 3. Differentiating characters of males of *Meloidogyne*.

Morphometric (quantitative)	Morphologic (qualitative)
Number of head annulations	Head structures in SEM[a]
Stylet length[a]	Shape of head cap in LM[a]
Distance of DEGO[a]	Shape of stylet cone, shaft and knobs[a]
Position of cephalids	Lateral field
Spicule and gubernaculum length	Excretory pore
Size of metacorpus "valve plates"	Spicule and gubernaculum structure

[a]Most reliable features.

Table 4. Differentiating characters of second-stage juveniles of *Meloidogyne*.

Morphometric (quantitative)	Morphologic (qualitative)
Number of head annulations	Head structures in SEM[a]
Stylet length[a]	Shape of head cap in LM[a]
Distance of DEGO[a]	Shape of stylet knobs[a]
Position of hemizonid	Lateral field
Tail length[a]	Tail shape[a]
Hyaline tail terminus length	
c and *d* ratios	

[a]Most reliable features.

pression is different morphologically (Eisenback, Chapter 6, Figs. 4, 5). Head morphology including shape and proportion of the labial disc, lateral and medial lips, expression of labial and cephalic sensilla, and annulations on the head region is distinctly different among species (8) (Fig. 8). It is more difficult for the various external head characters seen in the SEM to be observed in the LM because the juveniles are very small (Eisenback, Chapter 8, Figs. 11, 13). Slight differences may occur in general head shape. The head may appear squarish or more rounded, and the head cap may vary in size. Annulations in the head region may be present in some species and absent in others, but the number of annules cannot be easily determined since they are usually broken and cannot be counted precisely.

STYLET Stylet morphology is quite distinctive. Differences between species and populations occur especially in size and shape of the stylet knobs and in the way the knobs join the stylet shaft (set off, gradually merging). The stylet knobs may be rounded, back-sloped, flat anteriorly or drawn out laterally (Eisenback, Chapter 8, Figs. 11, 13). These features of stylet morphology are more clearly defined on excised stylets as seen in the SEM (7; Eisenback, Chapter 8, Fig. 12). Although the diagnostic value of stylet length is limited among the four most common species, it may be quite valuable in characterizing exotic and less common species since, in general, its variability is low (23). However, stylet length in juveniles is difficult to measure in fixed specimens because the very fine distal end of the cone cannot be seen easily. Measurements from the stylet knob base to the head end are more reliable when the specimens are properly prepared for measuring.

DORSAL ESOPHAGEAL GLAND ORIFICE The distance of the DEGO to the stylet base is an important distinguishing character, and groups of species may be differentiated by this measurement (23).

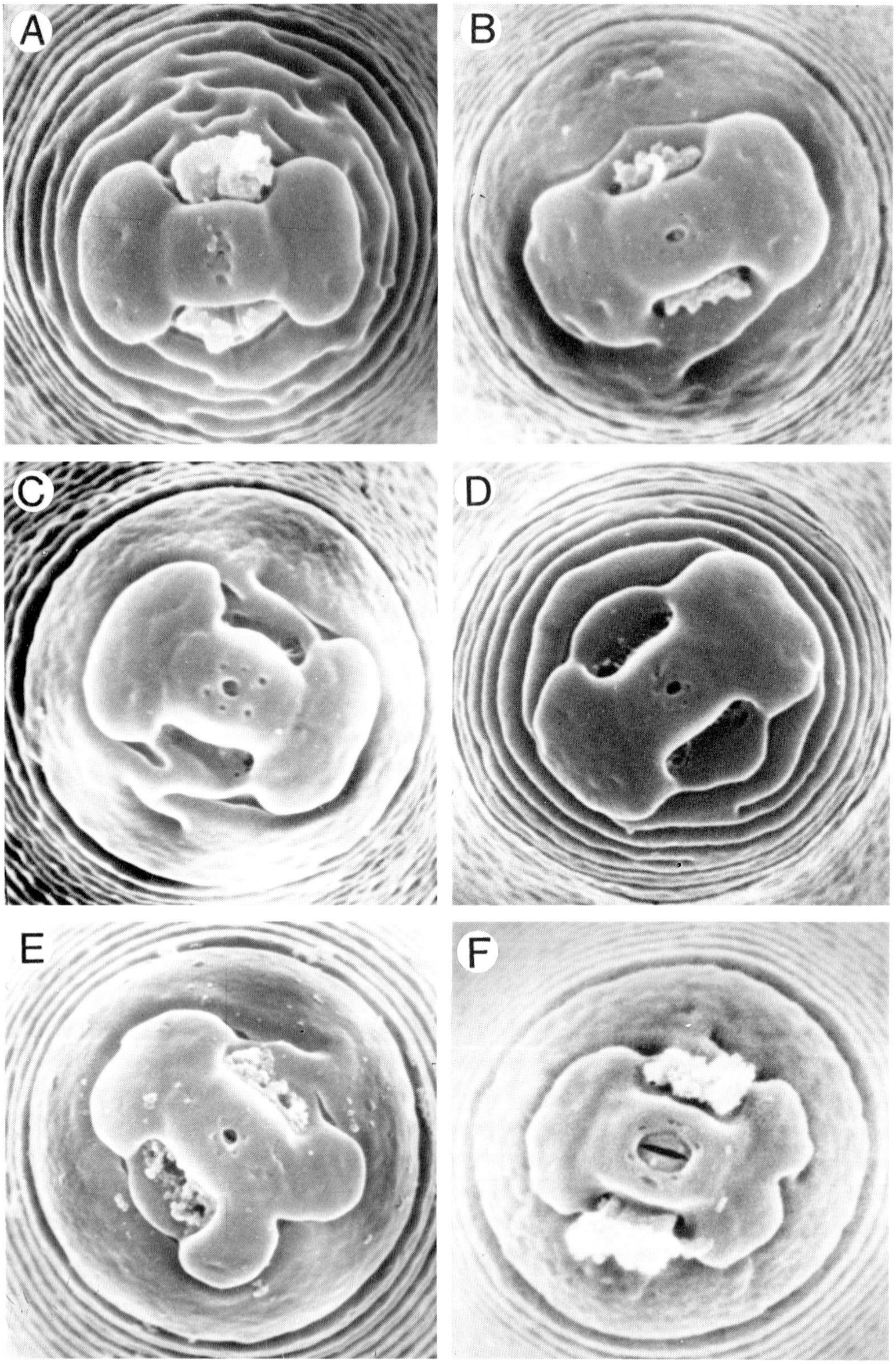

Fig. 8. SEM photographs of head regions of second-stage juveniles of different species of *Meloidogyne*. A) *M. incognita*. B) *Meloidogyne* sp. C) *Meloidogyne* sp. D) *M. incognita*. E) *M. platani* [from Hirschmann (20)]. F) *M. megatyla* [from Eisenback et al. (13)].

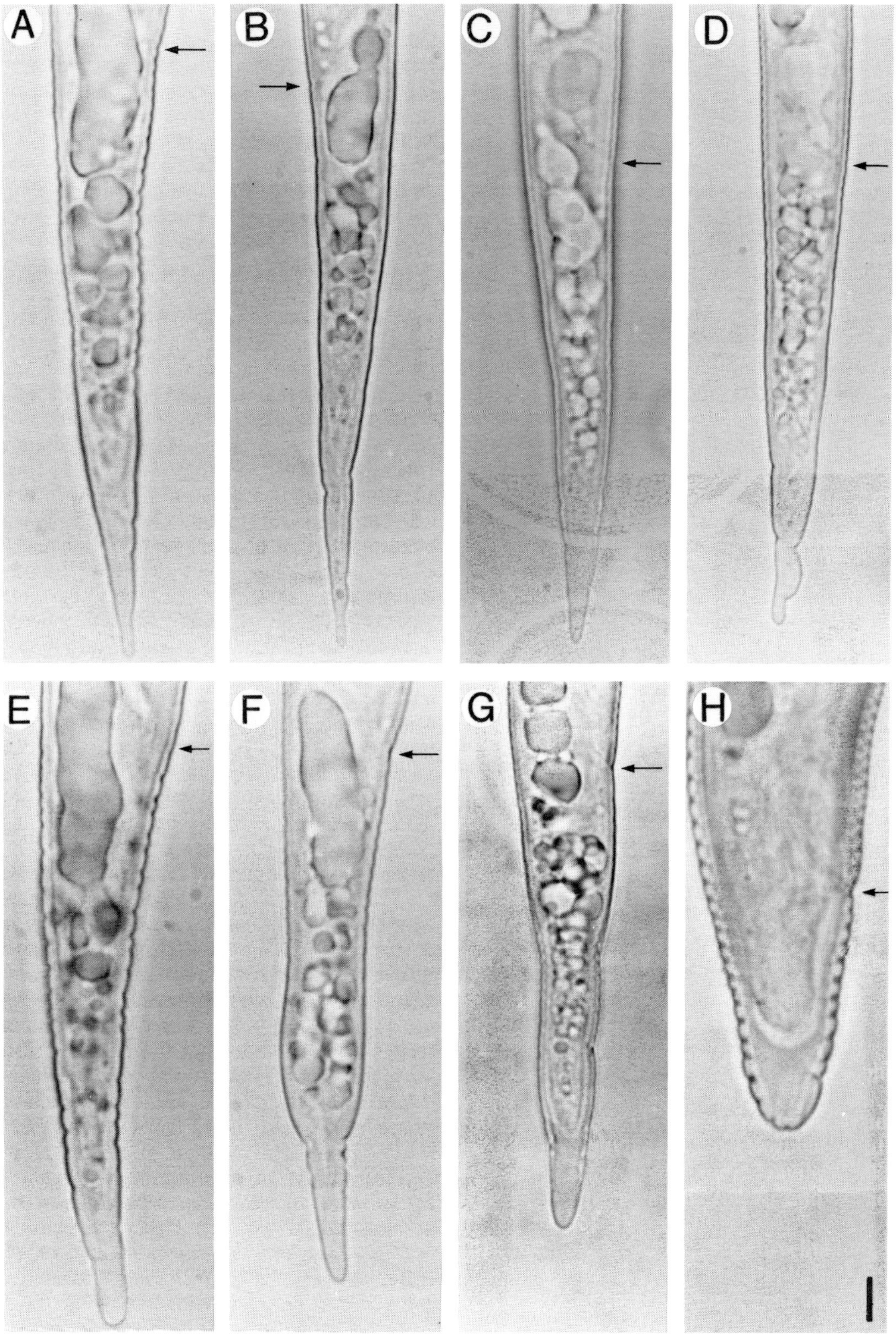

Fig. 9. LM photographs of tails of second-stage juveniles of different species of *Meloidogyne* to show variation in tail length and shape. A) *M. incognita.* B) *M. arenaria* variant [from Cliff (4)]. C) *M. arenaria* [from Cliff (4)]. D) *Meloidogyne* sp. E) *M. incognita.* F) *M. microcephala* [from Cliff & Hirschmann (5)]. G) *M. pini* [from Eisenback et al. (13)]. H) *M. brevicauda* (Courtesy J. Eisenback). Arrow, anal opening. Bar = 4 μm.

TAIL AND HYALINE TAIL TERMINUS Tail length is one of the most useful measurements because it varies considerably among species and has little intraspecific variation (21). Some species are clearly distinct from each other in overall range (e.g., *M. graminicola* with long tail vs. *M. acronea* with short tail). Length of the hyaline tail terminus is often very variable, although in some species it is clearly short and in others always long. *C* ratio (body length/tail length), body width at anus, and the ratio of tail length/body width at anus (*d*) are also important characters. Whitehead, 1968 (25), divided the juveniles into groups of species by tail length: a very short-tailed group; a short-tailed group; an intermediate group; and a longer-tailed group.

Tail shape is the most useful qualitative character, and intraspecific variation is limited (Fig. 9). Jepson, 1983 (21), classified 33 species of *Meloidogyne* into 12 groups on the basis of similarities in tail shape. The combined use of tail shape and tail measurements were considered good differentiating characters to aid in species differentiation.

CONCLUSIONS

Each life stage of *Meloidogyne* has a number of helpful morphologic and morphometric characters. Good differentiating features are present mainly in the anterior and posterior body regions. In females, the most characteristic and useful characters include the general appearance of the perineal pattern and the morphology of the stylet (shape of cone, shaft and knobs), as well as the length of the stylet. In males, the structures in the head region, including particularly the size, height and shape of the head cap, the number of annulations, and the width of the head region as compared to the first body annule, are diagnostically important. Stylet morphology and stylet length are stable and thus can be excellent supporting characters. In second-stage juveniles, head characters are less useful because of the small size of the juveniles, but tail length and tail shape are good characters exhibiting small intraspecific variation.

It should be emphasized here again that reliable identification of taxa should include all the life stages and should be based on as many characters as possible. In addition, the accuracy of identification increases when morphological features are supplemented by other kinds of taxonomic characters such as cytological, biochemical and physiological data.

Literature Cited

1. Chitwood, B. G. 1949. Root-knot nematodes—part I. A revision of the genus *Meloidogyne* Göldi, 1887. Proc. Helminthol. Soc. Wash. 16:90-104.
2. Cobb, N. A. 1924. The amphids of *Caconema* (nom. nov.) and other nemas. J. Parasitol. 11:118-120.
3. Cornu, M. 1879. Etudes sur le *Phylloxera vastatrix*. Mém. Acad. Sci., Paris 26:163-175, 328, 339-341.
4. Cliff, G. M. 1983. Morphological and serological comparisons of members of the *Meloidogyne arenaria* species complex, with characterizations of two new forms. Ph.D. thesis. North Carolina State Univ., Raleigh, N. C. 153 pp.
5. Cliff, G. M. and H. Hirschmann. 1984. *Meloidogyne microcephala* n. sp. (Meloidogynidae), a root-knot nematode from Thailand. J. Nematol. 16:183-193.
6. Eisenback, J. D. 1982. Description of the blueberry root-knot nematode, *Meloidogyne carolinensis* n. sp. J. Nematol. 14:303-317.
7. Eisenback, J. D. 1982. Morphological comparison of head shape and stylet morphology of second-stage juveniles of *Meloidogyne* species. J. Nematol. 14:339-343.
8. Eisenback, J. D. and H. Hirschmann. 1979. Morphological comparison of second-stage juveniles of six populations of *Meloidogyne hapla* by SEM. J. Nematol. 11:5-16.
9. Eisenback, J. D. and H. Hirschmann. 1980. Morphological comparison of *Meloidogyne* males by scanning electron microscopy. J. Nematol. 12:23-32.
10. Eisenback, J. D. and H. Hirschmann. 1981. Identification of *Meloidogyne* species on the basis of head shape and stylet morphology of the male. J. Nematol. 13:513-521.
11. Eisenback, J. D. and H. Hirschmann. 1982. Morphological comparison of stylets of male root-knot nematodes (*Meloidogyne* spp.). Scanning Electron Microscopy II:837-843.
12. Eisenback, J. D., H. Hirschmann, and A. C. Triantaphyllou. 1980. Morphological com-

parison of *Meloidogyne* female head structures, perineal patterns, and stylets. J. Nematol. 12:300-313.
13. Eisenback, J. D., B. Yang, and K. M. Hartman. 1985. Description of *Meloidogyne pini* n. sp., a root-knot nematode parasitic on sand pine (*Pinus clausa*), with additional notes on the morphology of *M. megatyla*. J. Nematol. 17:206-219.
14. Esser, R. P., V. G. Perry, and A. L. Taylor. 1976. A diagnostic compendium of the genus *Meloidogyne* (Nematoda: Heteroderidae). Proc. Helminthol. Soc. Wash. 43:138-150.
15. Franklin, M. T. 1972. The present position in the systematics of *Meloidogyne*. Eur. Mediterr. Plant Prot. Organ. (OEPP/EPPO) Bull. 6:5-15.
16. Franklin, M. T. 1979. Taxonomy of the genus *Meloidogyne*, pp. 37-54 *In* Root-knot nematodes (*Meloidogyne* species), systematics, biology and control, ed. F. Lamberti and C. E. Taylor. Academic Press, New York. 477 pp.
17. Göldi, E. A. 1887. Relatoria sôbre a molestia do cafeiro na provincia do Rio de Janeiro. Arch. Mus. Nac., Rio de Janeiro 8:7-121 (1892).
18. Goodey, T. 1932. On the nomenclature of the root-gall nematodes. J. Helminthol. 10:21-28.
19. Greeff, R. 1872. Ueber Nematoden in Wurzelanschwellungen (Gallen) verschiedener Pflanzen. Sitzber. Ges. Naturwiss. Marburg 11:169-174.
20. Hirschmann, H. 1982. *Meloidogyne platani* n. sp. (Meloidogynidae), a root-knot nematode parasitizing American sycamore. J. Nematol. 14:84-95.
21. Jepson, S. B. 1983. The use of second-stage juvenile tails as an aid in the identification of *Meloidogyne* species. Nematologica 29:11-28.
22. Jepson, S. B. 1983. Identification of *Meloidogyne* species; a comparison of stylets of females. Nematologica 29:132-143.
23. Jepson, S. B. 1983. Identification of *Meloidogyne*: a general assessment and a comparison of male morphology using light microscopy with a key to 24 species. Rev. Nématol. 6:291-309.
24. Müller, C. 1884. Mitteilungen über die unseren Kulturpflanzen schädlichen das Geschlecht *Heterodera* bildenden Würmer. Landwirtsch. Jahrb. Schweiz 13:1-42.
25. Whitehead, A. G. 1968. Taxonomy of *Meloidogyne* (Nematodea: Heteroderidae) with descriptions of four new species. Trans. Zool. Soc. Lond. 31:263-401.

DIAGNOSTIC CHARACTERS USEFUL IN THE IDENTIFICATION OF THE FOUR MOST COMMON SPECIES OF ROOT-KNOT NEMATODES (*MELOIDOGYNE* SPP.)

J. D. Eisenback

Department of Plant Pathology
North Carolina State University
Raleigh, North Carolina, USA

Accurate identification of populations of root-knot nematodes to species and host races is the underlying basis for their effective control and much precise research. Many of the *Meloidogyne* species are easily identified, either by their restricted host ranges, or by certain distinct morphological characters. The majority of these species are also geographically isolated and of minor economic importance (27). Certain other species are very difficult to identify: some because they are poorly described, and others because they are very similar to the four most common root-knot nematode species, *Meloidogyne incognita* (Kofoid and White) Chitwood, *M. javanica* (Treub) Chitwood, *M. arenaria* (Neal) Chitwood, and *M. hapla* Chitwood (17,20). In general, most of the unidentified populations eventually will be shown to belong to one of these four most common species (28).

Difficulty in identification of the four most common species depends upon the particular population. Some populations are easily identified because they possess many characters that are typical for the species. Identification of other populations is difficult due to variant forms or aberrant individuals. Sometimes an identification is further complicated by the widespread occurrence of mixtures of two or more species in the same field, especially when some of the characters overlap among species as often is the case with perineal patterns (24). Because of the present difficulty and confusion about the taxonomy of this group (20,25), the present chapter discusses in detail the diagnosis of the four most common species of root-knot nematodes based on morphological characters.

CHITWOOD'S DIAGNOSIS

When the root-knot nematodes were considered as a single species, the literature contained many conflicting reports on the susceptibility or resistance of particular plant species to the root-knot nematodes (26,30). Christie and Albin (3) showed that populations of root-knot nematodes isolated from various plants had different host ranges. By studying the morphology of several populations with different host responses, Chitwood (2) recognized and clearly described the four most widespread and common *Meloidogyne* species.

Chitwood's descriptions were outstanding and worthy of minute scrutinization. He described and illustrated many morphological differences among the species utilizing details of all major life stages. Initially, the perineal pattern of the adult female became the most important morphological feature used for species identifications, and most of the other taxonomic characters pointed out by Chitwood remained unused until recently (16).

The perineal pattern became the dominant diagnostic character for several reasons. For many populations, it is quite stable and useful for species identifications (6,26). Also, shortly after the description of the four common *Meloidogyne* species, the vulval cyst cones of *Heterodera* species were emphasized as useful taxonomic characters (23). Since the two genera seemed closely related, perineal patterns were thought to be analogous to vulval cyst cones, and thus important taxonomic features.

Compared to other useful characters, perineal patterns are relatively easy to prepare for light microscopic examination because they are relatively large and two-dimensional. Males and second-stage juveniles, however, are smaller, vermiform, and difficult to prepare for observation. Females are used more for identifications, also, because they are more numerous than males and easier to find in infected tissues.

Probably an equally important reason for perineal patterns being the most predominant character used for routine identifications is the nature of the illustrations in the original description (2). Although large, excellent light micrographs of perineal patterns were presented, all other characters were illustrated with simple line drawings. The drawings were well done with unbelievable precision, but they were too small in the description to point out effectively the differences among the species.

In the diagnosis that follows, Chitwood's original illustrations have been copied from the publication, enlarged many times, and reproduced here for a comparison of characters that are useful in the identification of the four most common *Meloidogyne* species.

CURRENT DIAGNOSIS

Various approaches to the taxonomy of *Meloidogyne* include host response (16,24,26,27), cytology (16,29), biochemistry (4,5,16), and morphology (16,17,18,22,28,30). Host-response tests require adequate quantities of nematode inoculum and much greenhouse space and time for culture growth. Cytological and biochemical tests, in contrast, are fast but require sophisticated equipment and technical expertise. Identifications based on morphology, however, are rapid and practical. A light microscope (LM) of moderate quality and a little experience allows an accurate and confident identification of most populations.

All of the taxonomic approaches have supported the validity of the four most common species, even though no single approach can account for the variability within the genus. Within each species there are cytological races (29), host races (27), biochemical enzymatic phenotypes (4), and morphological variants (1,12). The populations that are extremely variable need to be examined in more detail rather than lumped into a species complex and forgotten. Perhaps some of the variable populations are new species or mixtures of species which behave differently together than they would separately (8).

Although it is possible for a population to be identified on the basis of one or two morphological characters of a few specimens, confidence in the accuracy of an identification increases as more kinds of taxonomic characters are examined. Therefore, the more critical the identification, the more kinds of information and numbers of specimens should be used. Within certain species, widespread host races have been described (27), and because these races are apparently morphologically similar, they can be identified only with a differential host test (16,27).

The diagnosis that follows is based entirely upon morphological features. Characters that are valuable in the identification of the four common *Meloidogyne* species include perineal patterns; female stylets; male heads and stylets; and heads and stylets of second-stage juveniles. Although a scanning electron microscope (SEM) is not available to many investigators, scanning electron micrographs are included because they help interpret the images seen in the LM.

Perineal Patterns

The morphology of perineal patterns will probably remain the most important morphological character used for tentative species identifications. Because a certain amount of variability occurs within a particular population or species, many identifications are inconclusive. Characters that overlap among species make the identification of mixed populations especially difficult. For these populations, additional taxonomic characters must be utilized for the final identification. Preparing perineal patterns for observation is not difficult, but it is tedious and time consuming. Important diagnostic characters of perineal patterns of the four most common species are summarized in Table 1.

MELOIDOGYNE INCOGNITA The perineal patterns of *M. incognita* (Figs. 1A,2A) are characterized by the presence of a high, squarish dorsal arch that often contains a distinct whorl in the tail terminal area. The striae are smooth to wavy, sometimes zigzagged. Distinct lateral lines are absent, but the lateral field may be marked by breaks and forks in the striae. Often some striae bend toward the vulval edges.

MELOIDOGYNE JAVANICA The perineal patterns of *M. javanica* (Figs. 1B,2B) are unique because they contain lateral ridges that divide the dorsal and ventral striae. Generally, the ridges run the entire width of the pattern, but gradually disappear near the tail terminus. The dorsal arch is low and rounded to high and squarish and often contains a whorl in the tail terminal area. The striae are smooth to slightly wavy, and some striae may bend toward the vulval edges.

Table 1. Summary of important diagnostic characters of perineal patterns of the four most common *Meloidogyne* species.

	Dorsal arch	Lateral field	Striae	Tail terminus
M. incognita	high, squarish	lateral ridges absent, marked by breaks and forks in striae	coarse, smooth to wavy, sometimes zigzaggy	often with distinct whorl
M. javanica	low, rounded	distinct lateral ridges	coarse, smooth to slightly wavy	often with distinct whorl
M. arenaria	low, rounded, indented near lateral fields	lateral ridges absent, marked by short, irregular and forked striae	coarse, smooth to slightly wavy	usually without distinct whorl
M. hapla	low, rounded	lateral ridges absent	fine, smooth to slightly wavy	whorl absent, marked by subcuticular punctations

MELOIDOGYNE ARENARIA The perineal patterns of *M. arenaria* (Figs. 1C,2C) are distinguished by a low dorsal arch that is slightly indented near the lateral fields to form rounded shoulders. The dorsal and ventral striae often meet at an angle. Distinct lateral lines are absent, but short, irregular, and forked striae mark the lateral fields. The striae are smooth to slightly wavy, and some may bend toward the vulva. Sometimes the patterns are extended laterally and form one or two wings.

MELOIDOGYNE HAPLA Perineal patterns of *M. hapla* (Figs. 1D,2D) are characterized by their overall rounded hexagonal to flattened ovoidal shape, very fine striae, and subcuticular punctations in the smooth tail terminal area. The dorsal arch is usually low and rounded, but may be high and squarish. Lateral ridges are absent but the lateral fields are marked by irregularities in the striae. The dorsal and ventral striae often meet at an angle, and the striae are smooth to slightly wavy. Some patterns may form wings on one or both lateral sides.

Stylets of Females

The morphology of the stylet of females is a species-specific character (14,21). Chitwood (2) also considered it a useful feature (Fig. 3), and recent studies of excised stylets observed with the SEM confirmed and clarified the differences that occur among the four most common species (Fig. 4) (14). Most of the details elucidated by the SEM are clearly visible by the LM (Fig. 5).

The morphology of stylets of females is very stable among populations. In a recent study of 11 populations of *M. hapla* with different chromosome numbers and geographical origins, there were no qualitative differences in stylets of females, whereas all of the other characters examined showed a wide range of variability (15). The usefulness of stylets of females for routine identifications is limited only by the difficulty of specimen preparation for examination with the LM. The specimens must be lateral and level to be of value. Important diagnostic characters of stylets of females of the four most common species are summarized in Table 2.

MELOIDOGYNE INCOGNITA The stylets of females of *M. incognita* (Figs. 3A,4A,5A) characteristically are curved dorsally. The anterior half of the cone is cylindrical and the posterior half is conical. The shaft is slightly wider posteriorly. The knobs are set off from the shaft, anteriorly indented, and transversely elongate. In some specimens, the knobs are deeply indented, and each knob appears as if it were two. The average stylet length is 16 μm, and the range is 15-17 μm.

MELOIDOGYNE JAVANICA The stylets of females of *M. javanica* (Figs. 3B,4B,5B) are similar to those of *M. incognita*, except the cone is only slightly curved dorsally and the shaft is more cylindrical. The knobs are also transversely elongate, but they are not deeply indented anteriorly. The stylet length varies from 14-18 μm, with a mean of 16 μm.

MELOIDOGYNE ARENARIA Female stylets of *M. arenaria* populations are very characteristic (Figs. 3C,4C,5C). The whole stylet is broad and

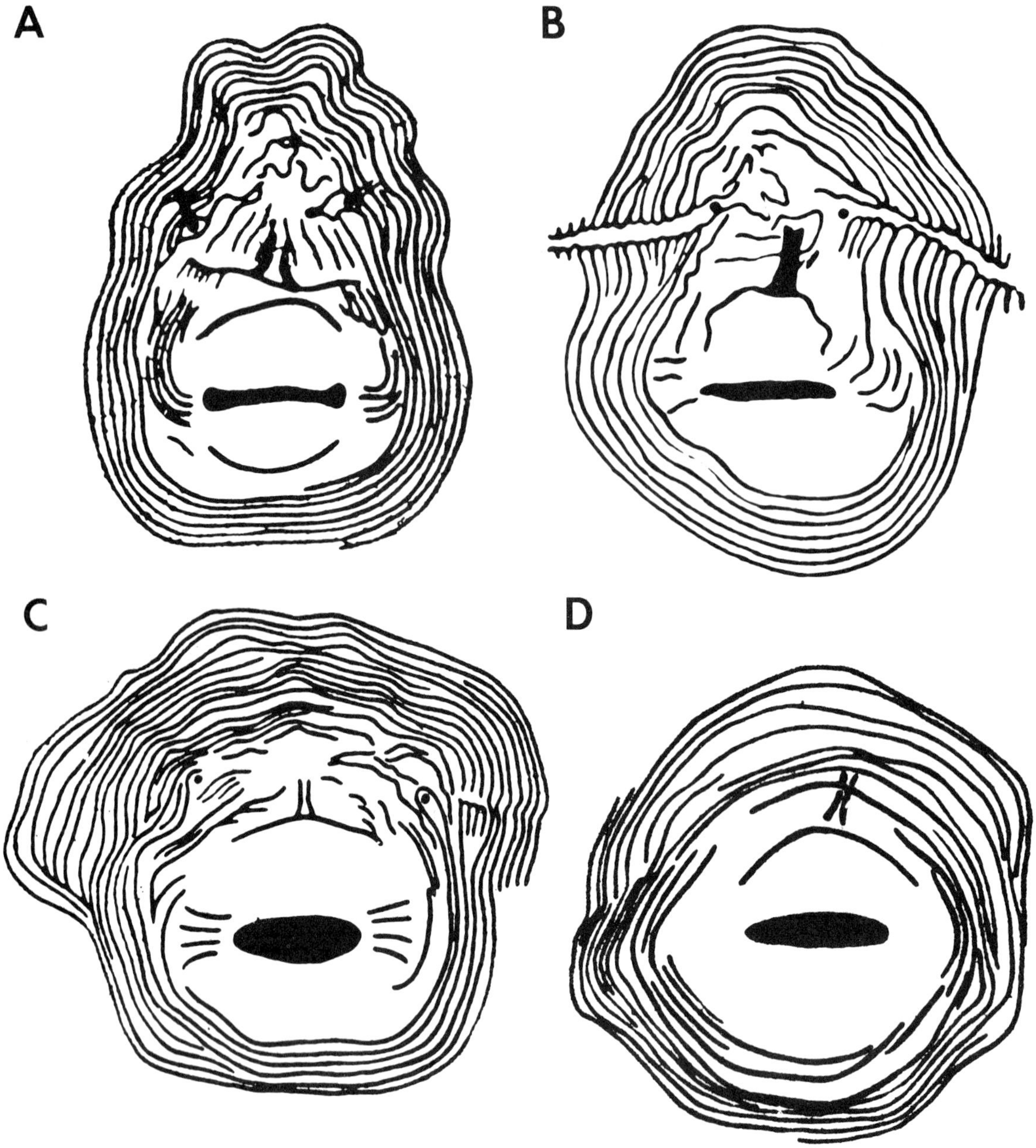

Fig. 1 A-D. Drawings of perineal patterns taken from the original description. A) *Meloidogyne incognita*. B) *M. javanica*. C) *M. arenaria*. D) *M. hapla*. [After B. G. Chitwood (2)].

robust, and the wide, rounded, backward-sloping knobs gradually merge with the shaft. Stylet lengths range from 13-17 μm, with a mean of 15.5 μm.

MELOIDOGYNE HAPLA The stylets of females of *M. hapla* (Figs. 3D,4D,5D) are narrow and delicate compared to the other three species. The cone is only slightly curved dorsally, and the shaft is a little broader posteriorly. The knobs are small, round, and set-off from the shaft. Stylet lengths vary from 13-17 μm with a mean of 15.5 μm.

Heads and Stylets of Males

The shapes of the male head and stylet are extremely useful characters in the identification of *Meloidogyne* species (11,12,13,16). Chitwood placed much emphasis on the morphology of the male (Fig. 6) (2), and Whitehead (30) also considered heads of males to be good taxonomic features too. Unfortunately, the shape of the

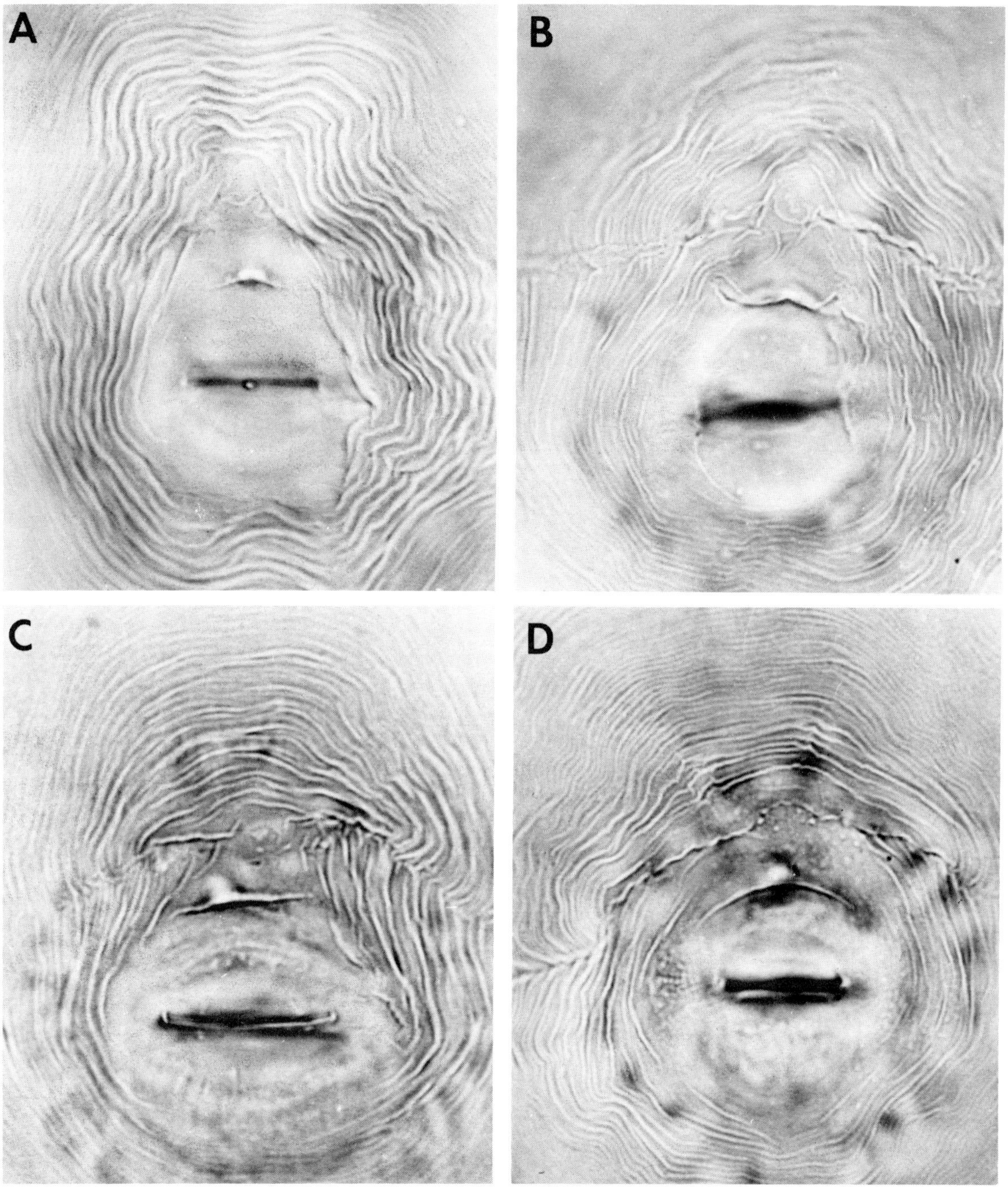

Fig. 2 A-D. Light micrographs of perineal patterns of species of *Meloidogyne*. A) *M. incognita*. B) *M. javanica*. C) *M. arenaria*. D) *M. hapla*. [After J. D. Eisenback et al. (16)].

male head and stylet has not been widely used for routine identifications. Most workers considered males too variable and rare in occurrence to be useful (17,18,19,25).

Recently, the SEM has facilitated clarification of the morphology of the heads and stylets of males (Figs. 7,8) (11,12,13,16). When both of these characters are used together, an accurate, rapid identification based on morphology of the male alone is possible. Furthermore, usually each individual is identifiable with some certainty to species, and the problem of the identification of mixed populations is greatly simplified.

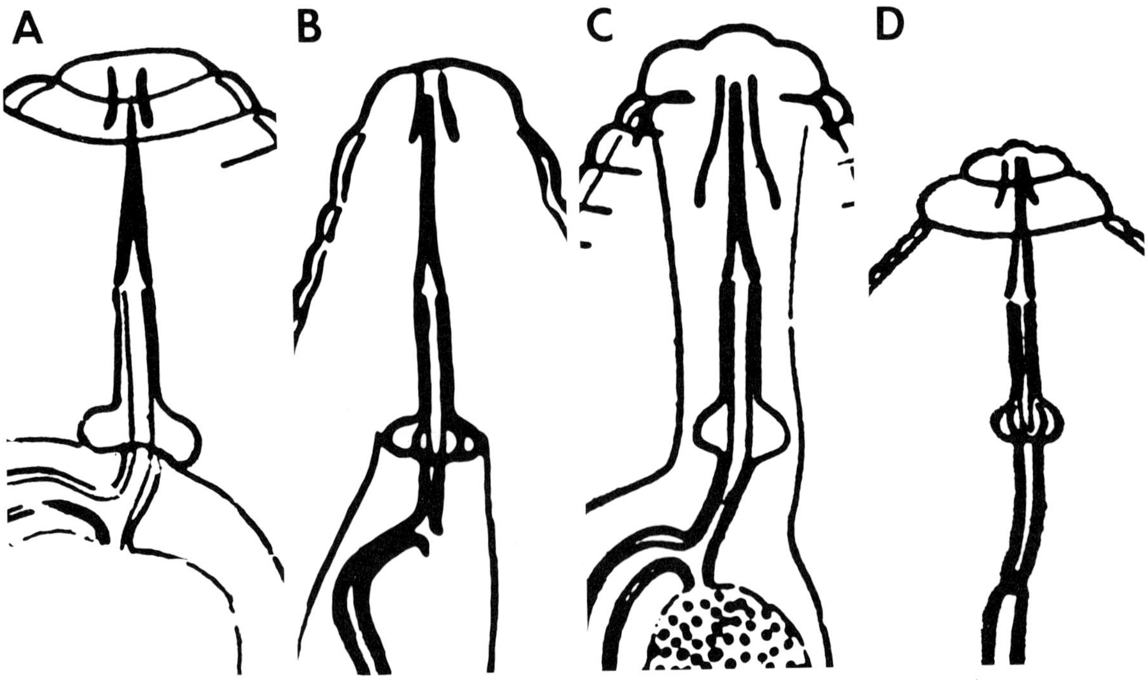

Fig. 3 A-D. Drawings of female heads and stylets taken from the original description. A) *Meloidogyne incognita*. B) *M. javanica*. C) *M. arenaria*. D) *M. hapla*. [After B. G. Chitwood (2)].

Table 2. Summary of important diagnostic characters of stylets of females of the four most common *Meloidogyne* species.

	Stylet cone	Stylet shaft	Stylet knobs	Stylet length
M. incognita	anterior half cylindrical, dorsally curved	slightly wider posteriorly	set off, rounded to transversely elongate, sometimes indented anteriorly	16 μm, mean 15-17 μm, range
M. javanica	slightly curved dorsally	cylindrical	set off, transversely elongate	16 μm, mean 14-18 μm, range
M. arenaria	straight, broad and robust	wider posteriorly	not set off, backward sloping, merging with shaft	15.5 μm, mean 13-17 μm, range
M. hapla	slightly curved dorsally, narrow and delicate	slightly wider posteriorly	set off, small and round	15.5 μm, mean 13-17 μm, range

The factors that limit the usefulness of male heads and stylets are the availability of specimens and adequate specimen preparation. Generally, many males can be obtained when infected plant tissue is placed in a mist chamber or moist chamber. Specimens can be collected in the runoff water each day for several days. Sometimes, however, populations do not produce males and other characters must be used. Also, dwarf males are extremely variable and different in their morphology and should not be considered in a species identification.

Specimens that are mounted in an exact lateral and level position are the easiest to be identified. Much of the reported variability among specimens is probably due to differences

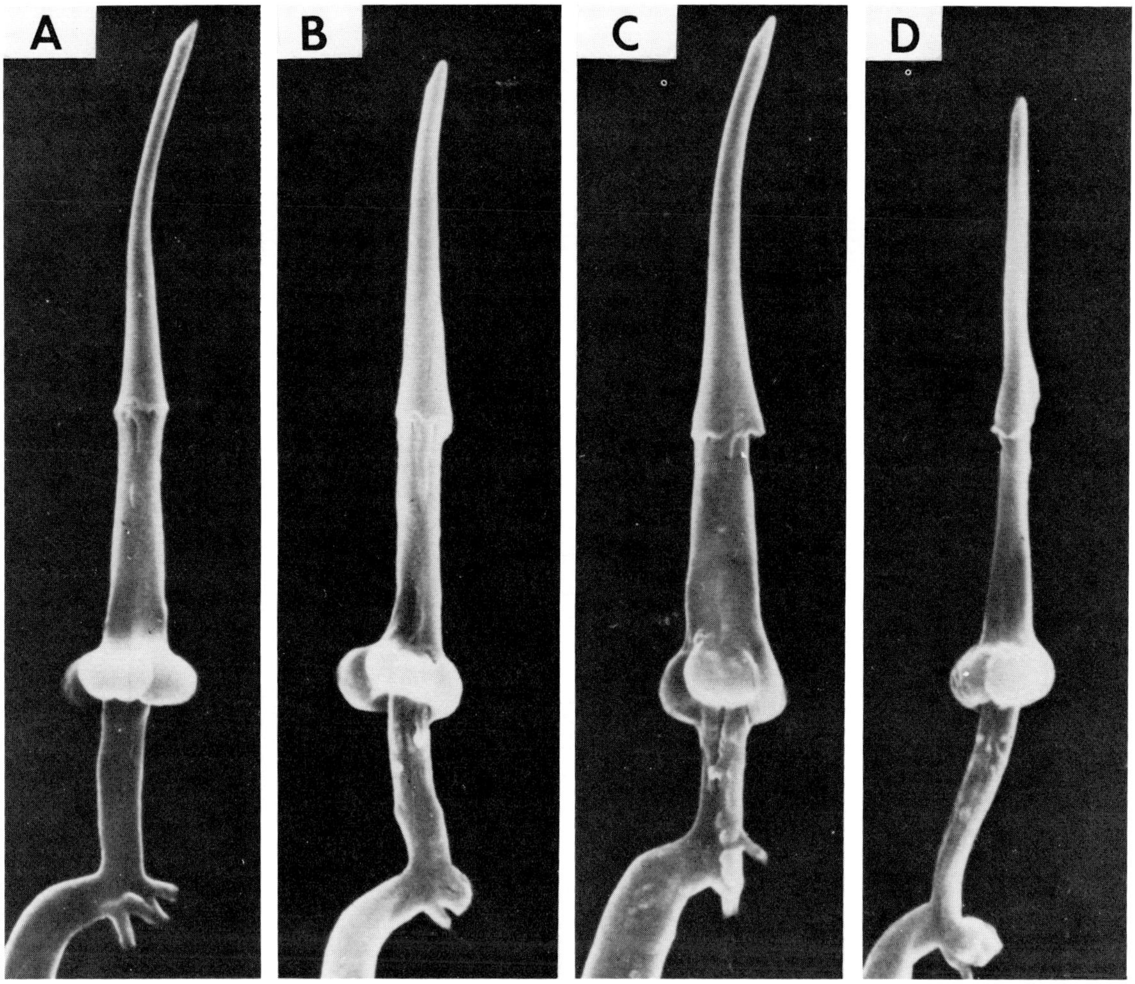

Fig. 4 A-D. Scanning electron micrographs of excised stylets of root-knot females. A) *Meloidogyne incognita.* B) *M. javanica.* C) *M. arenaria.* D) *M. hapla.* [After J. D. Eisenback et al. (16)].

in the orientation of specimens. With experience, however, the position of the specimen becomes less important. It is important that the extent of variation be understood and the least variable characters determined. Generally, for instance, stylet morphology is more stable than head shape. Important diagnostic characters of male head shape and stylet morphology of the four most common species are summarized in Table 3.

MELOIDOGYNE INCOGNITA The shape of the male head of *M. incognita* (Figs. 6A,8A,9A) is characterized by a large, round labial disc that is raised above the medial lips. The labial disc is concave to flat, and the high head cap is nearly as wide as the head region in lateral view. The head region is usually marked by 2-3 incomplete annulations but may be completely smooth. The head region is not distinctly set off from the rest of the body.

The stylet of males of *M. incognita* (Figs. 6A,7A,9A) is distinguished by a blunt, blade-like tip which is wider than the medial portion of the cone. The opening is often marked posteriorly by a pronounced projection. The shaft is usually cylindrical and often narrows near the knobs. The knobs are large, rounded to transversely ovoid, sometimes anteriorly indented, and set off from the shaft. Stylet lengths vary from 23-25 μm, with a mean of 24 μm. Generally, the distance from the dorsal esophageal gland orifice (D.E.G.O.) to the stylet base is relatively short, 2-4 μm, with a mean of 3 μm.

MELOIDOGYNE JAVANICA Male heads of *M. javanica* (Figs. 6B,8B,9B) are distinguished by a

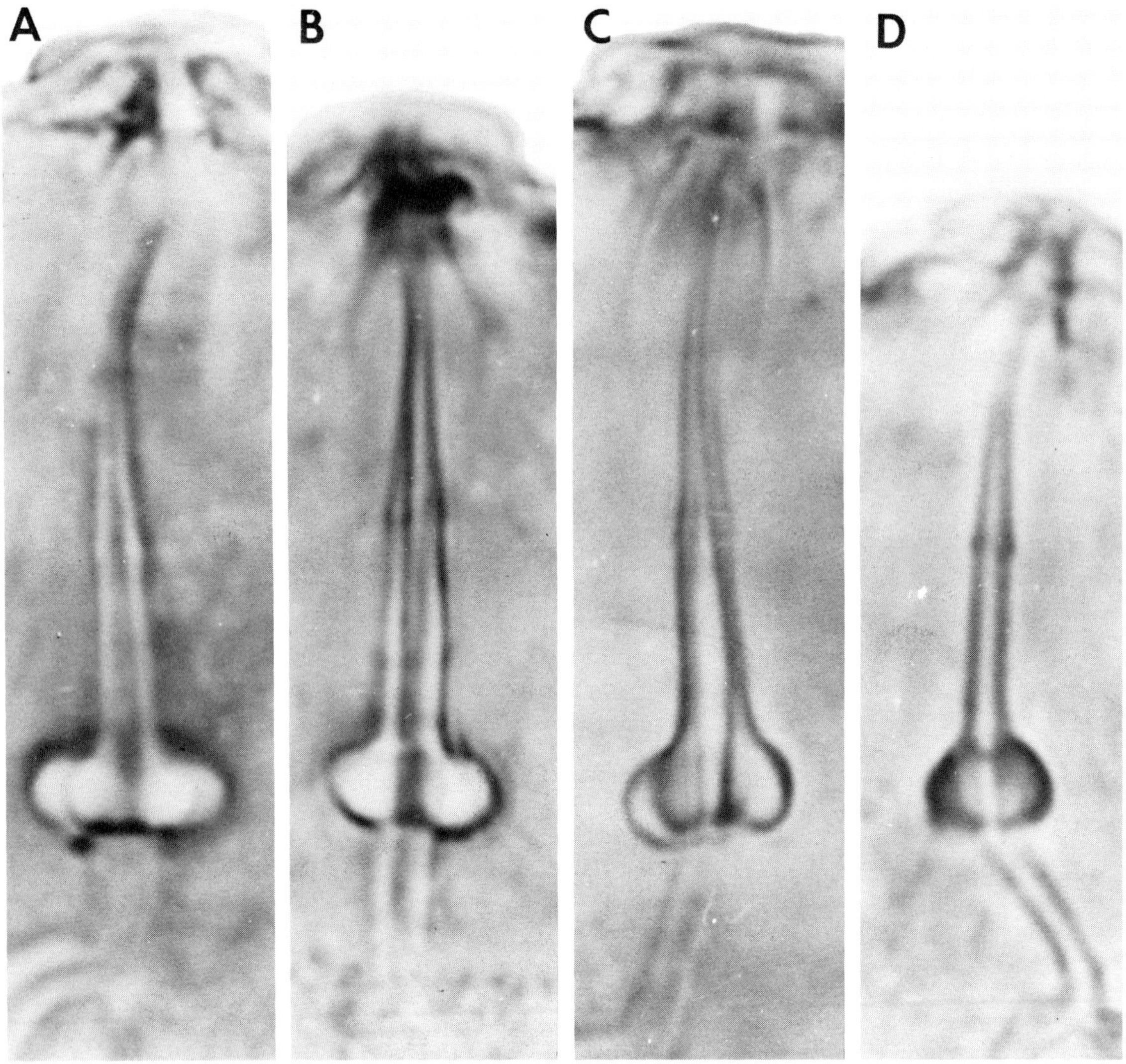

Fig. 5 A-D. Light micrographs of heads and stylets of females of *Meloidogyne* species. A) *M. incognita.* B) *M. javanica.* C) *M. arenaria.* D) *M. hapla.* [After J. D. Eisenback et al. (16)].

high, rounded head cap that is distinctly set off from the head region. The labial disc and medial lips are fused and form one smooth, continuous structure which is almost as wide as the head region in a lateral view. The head region may be smooth or marked by 2-3 incomplete head annulations. The head region is not distinctly set off from the rest of the body.

The stylet of *M. javanica* males is characterized by very broadly elongate stylet knobs (Figs. 6B,7B,9B). The anterior two-thirds of the cone gradually increases in width, while the posterior one-third rapidly widens. The shaft is cylindrical. The knobs are low and wide, sometimes anteriorly indented, and always set off from the shaft. Stylet lengths range from 18-22 μm and average 20 μm. The distance from the D.E.G.O. to the base of the stylet is relatively short, 2-4 μm, with a mean of 3 μm.

Some populations of *M. javanica* produce male intersexes that are very characteristic for the species (Fig. 10) (2,16,18). Male intersexes possess different degrees of secondary female characteristics ranging from a small ventral protuberance to a large protuberance with a rudimentary vagina. The protuberance usually is slightly anterior to the base of the spicules. Many populations produce a large proportion of intersexes, whereas others have only normal males.

MELOIDOGYNE ARENARIA Male heads of *M. arenaria* (Figs. 6C,8C,9C) are characterized by a low head cap that slopes posteriorly. The labial

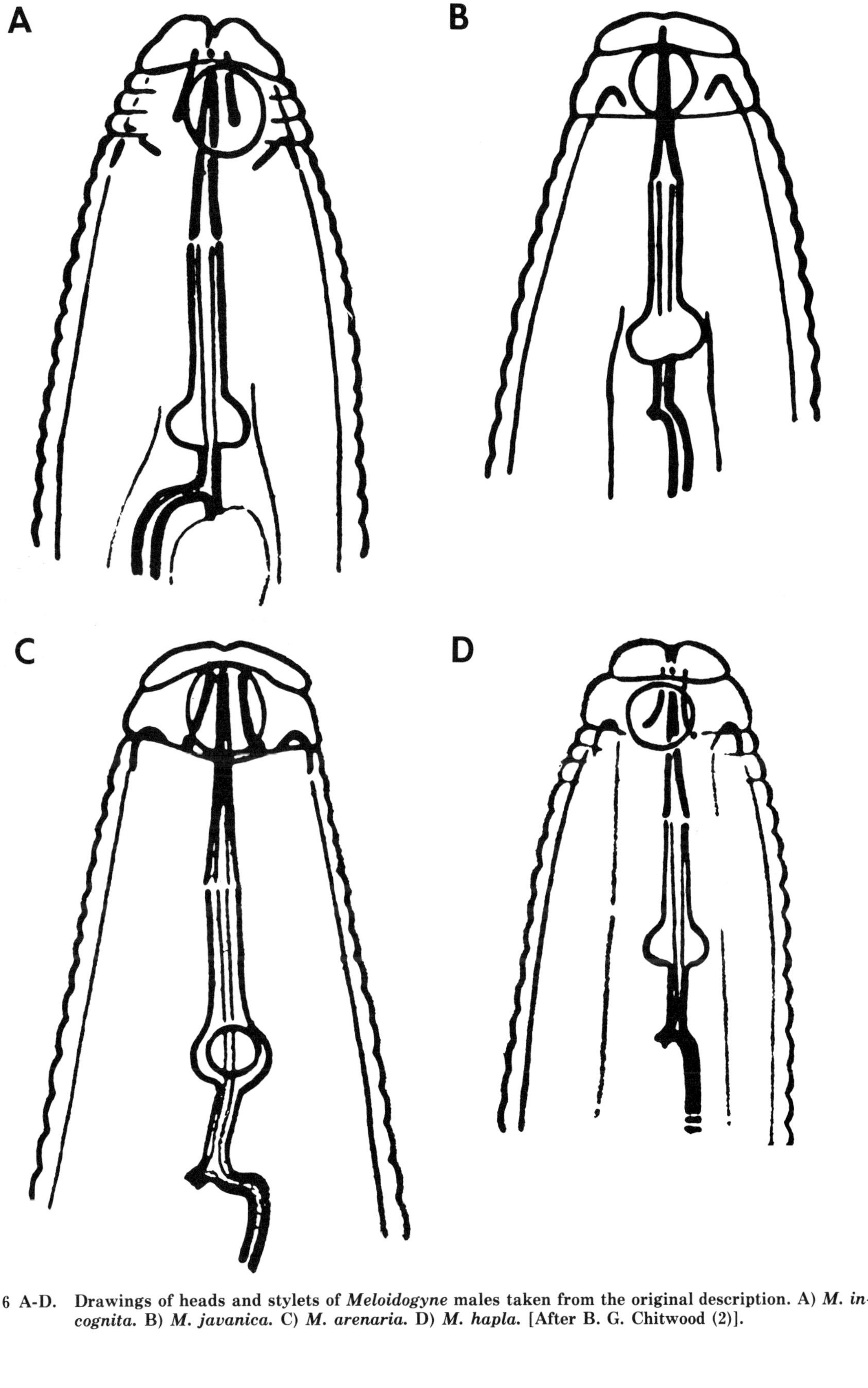

Fig. 6 A-D. Drawings of heads and stylets of *Meloidogyne* males taken from the original description. A) *M. incognita.* B) *M. javanica.* C) *M. arenaria.* D) *M. hapla.* [After B. G. Chitwood (2)].

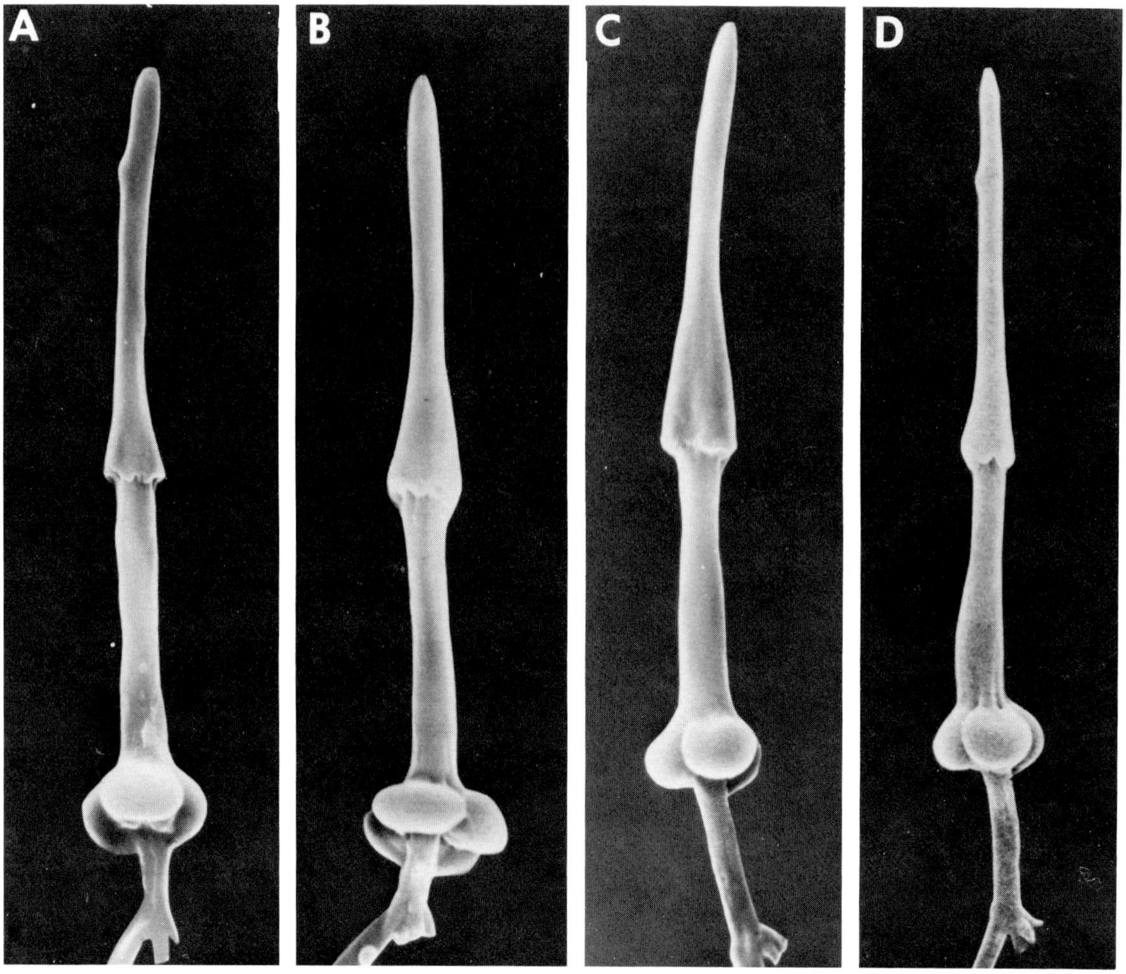

Fig. 7 A-D. Scanning electron micrographs of excised stylets of males of root-knot species. A) *Meloidogyne incognita.* B) *M. javanica.* C) *M. arenaria.* D) *M. hapla.* [After J. D. Eisenback et al. (16)].

disc and medial lips form one smooth, continuous structure that is nearly as wide as the head region in a lateral view. The head region may be smooth or marked by 1-3 incomplete annulations. The head region is not distinctly set off from the rest of the body.

Stylets of males of *M. arenaria* (Figs. 6C,7C,9C) are relatively straight, broad, and robust. The posterior two-thirds of the cone gradually increase in width and the posterior one-third widens rapidly. The posterior portion of the cone is much broader than the anterior portion of the shaft. The shaft is cylindrical, but sometimes widens near its junction with the knobs. The knobs are large, posteriorly rounded, and gradually merge with the shaft. Stylet lengths vary from 20-25 μm, but average around 22 μm. The distance from the D.E.G.O. to the base of the stylet is relatively long, 4-7 μm, and averages around 5.5 μm.

MELOIDOGYNE HAPLA Heads of males of *M. hapla* (Figs. 6D,8D,9D) are characterized by a high and narrow head cap and a set-off head region. The labial disc and medial lips are fused and form one smooth, continuous structure that is much narrower than the head region. The head region is without annulations and distinctly set off from the body annulations. The body annules increase in width and diameter posteriorly. One variant of *M. hapla* with a unique head morphology has been described (11).

The stylets of males of *M. hapla* (Figs. 6D,7D,9D) are very narrow and delicate compared to the other three common species. The cone gradually increases in width, and the base of the cone is only slightly wider than the anterior portion of the shaft. In populations of cytological race A, the shaft tends to be wider at its base, and in those of cytological race B populations, the shaft is generally narrower at its

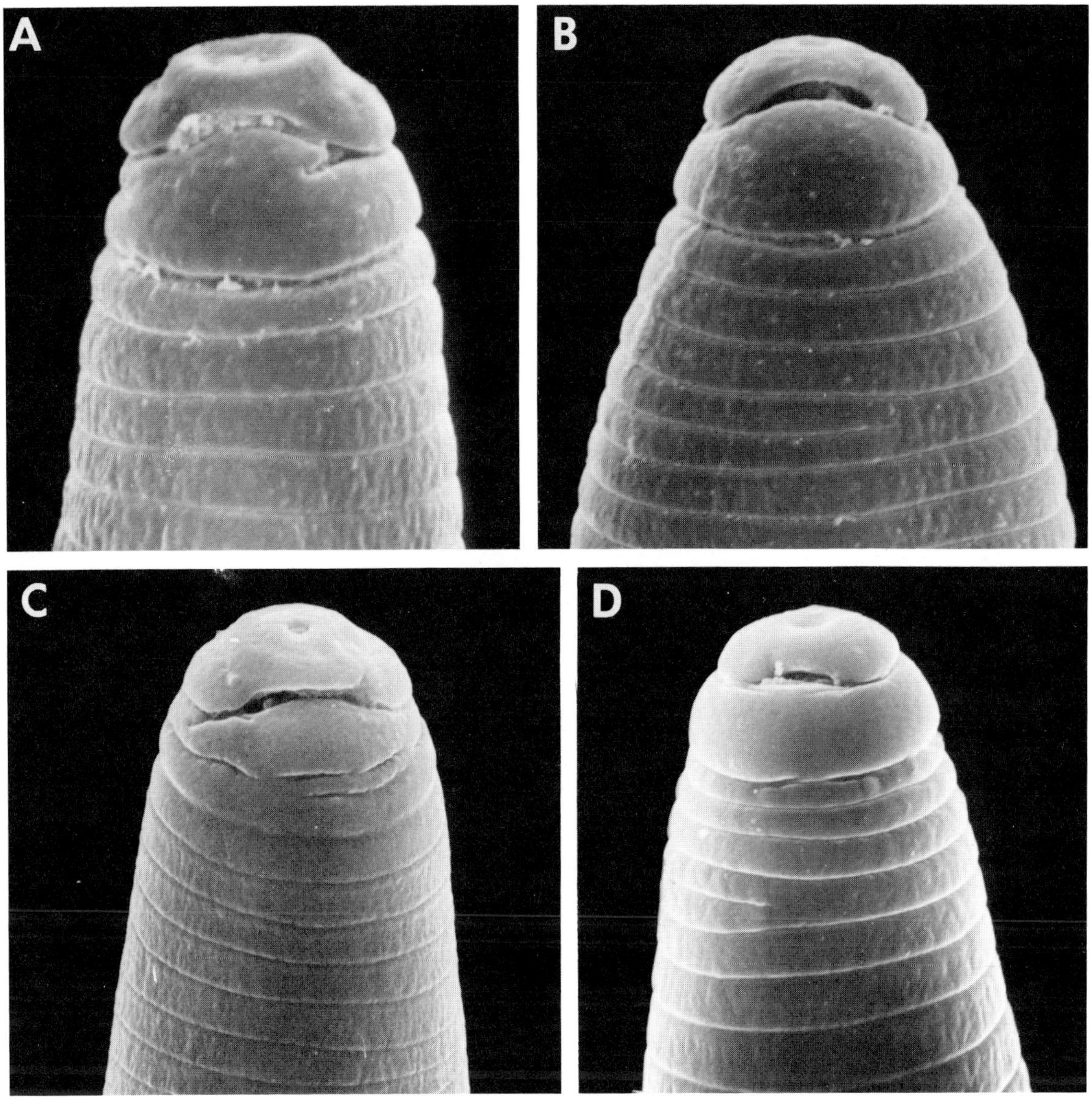

Fig. 8 A-D. Scanning electron micrographs of the anterior portions of males of the four most common species of *Meloidogyne*. A) *M. incognita*. B) M. javanica. C) *M. arenaria*. D) *M. hapla*.

base. The stylet knobs are small, rounded, and set off from the shaft. Stylets of race A are 17-23 µm long with a mean of 20 µm, whereas stylets of race B are 19-23 µm long with a mean of 21 µm. The distance from the D.E.G.O. to the base of the stylet is relatively long, 4-6 µm, with a mean of 5 µm.

Heads and Stylets of Second-stage Juveniles

Although the shape of the head and stylets of second-stage juveniles are useful diagnostic characters, the differences between the four most common species are slight (Figs. 11,12,13) (2,7,9,10,16). These characters are extremely small and difficult to resolve clearly in the LM and will probably be useful only as supplemental characters. Second-stage juveniles are particularly difficult to prepare for light microscopy, and only specimens that are lateral and level are of value for observation and identification. Important diagnostic characters of second-stage juvenile head shape and stylet morphology of the four most common species are summarized in Table 4.

MELOIDOGYNE INCOGNITA The head shape of second-stage juveniles of *M. incognita* as ob-

Table 3. Summary of important diagnostic characters of male head shape and stylet morphology of the four most common *Meloidogyne* species.

	Head cap	Head region	Stylet cone	Stylet shaft	Stylet knobs	Stylet length	D.E.G.O.[a]
M. incognita	flat to concave, labial disc raised above the medial lips	not set off, usually marked by 2-3 incomplete annulations	tip blunt, blade-like	usually cylindrical, often narrows near knobs	set off, rounded to transversely elongate, sometimes indented anteriorly	24 μm mean, 23-25 μm range	short, 3 μm mean, 2-4 μm range
M. javanica	high, rounded, set off from head region	not set off, smooth or marked by 2-3 incomplete annulations	tip pointed, cone straight	usually cylindrical	set off, low & very wide	20 μm mean, 18-22 μm range	short, 3 μm mean, 2-4 μm range
M. arenaria	low, sloping posteriorly	not set off, smooth or marked by 2-3 incomplete annulations	tip pointed, cone broad and robust	usually cylindrical, often broadens near knobs	not set off, backward sloping, merging with shaft	22 μm mean, 20-25 μm range	long, 5.5 μm mean, 4-7 μm range
M. hapla	high and narrow	set off, smooth, larger diameter than first body annule	tip pointed, cone narrow and delicate	cylindrical, often wider or narrower at its base	set off, small and round	20 μm mean, 18-22 μm range	moderately long, 5 μm mean, 4-6 μm range

[a] Distance from the dorsal esophageal gland orifice (D.E.G.O.) to the base of the stylet.

served in the scanning electron microscope has been described (9). The labial disc and medial lips are fused to form one smooth, continuous, elongate structure. The head region may be smooth or marked by 1-3 incomplete to complete annulations. In the LM, the differential details cannot be seen clearly (Figs. 11A,13A) (7). The head cap appears anteriorly flattened, and the head region is rounded posteriorly.

The stylet (Figs. 11A,12A,13A) is characterized by prominent knobs that are posteriorly rounded and distinctly set off from the shaft. The cone and shaft gradually increase in width posteriorly. The distance from the D.E.G.O. to the base of the stylet is relatively short, 2-3 μm, with a mean of 2.5 μm.

MELOIDOGYNE JAVANICA The shape of the head of *M. javanica* second-stage juveniles (Figs. 11B,13B) is very similar to that of *M. incognita*, and is not diagnostic for the species. The head cap is anteriorly flattened, but posteriorly rounded. The labial disc and medial lips are fused and form one smooth, continuous, elongate structure. The head region is usually smooth, but it may be marked with 1-3 incomplete annulations.

The stylet of *M. javanica* second-stage juveniles (Figs. 11B,12B,13B) is characterized by transversely elongate knobs. Otherwise, the stylet is very similar to that of *M. incognita*. The distance from the D.E.G.O. to the base of the stylet is relatively moderate, 3-4 μm, with a mean of 3.5 μm.

MELOIDOGYNE ARENARIA The shape of the head of second-stage juveniles of *M. arenaria* in the LM (Figs. 11C,13C) is very similar to that of *M. incognita* and *M. javanica*. The head cap is anteriorly flattened but posteriorly rounded. The labial disc and medial lips are fused to form one smooth, continuous, elongate structure. The head region is generally smooth, but it may be marked by 1-3 incomplete annulations.

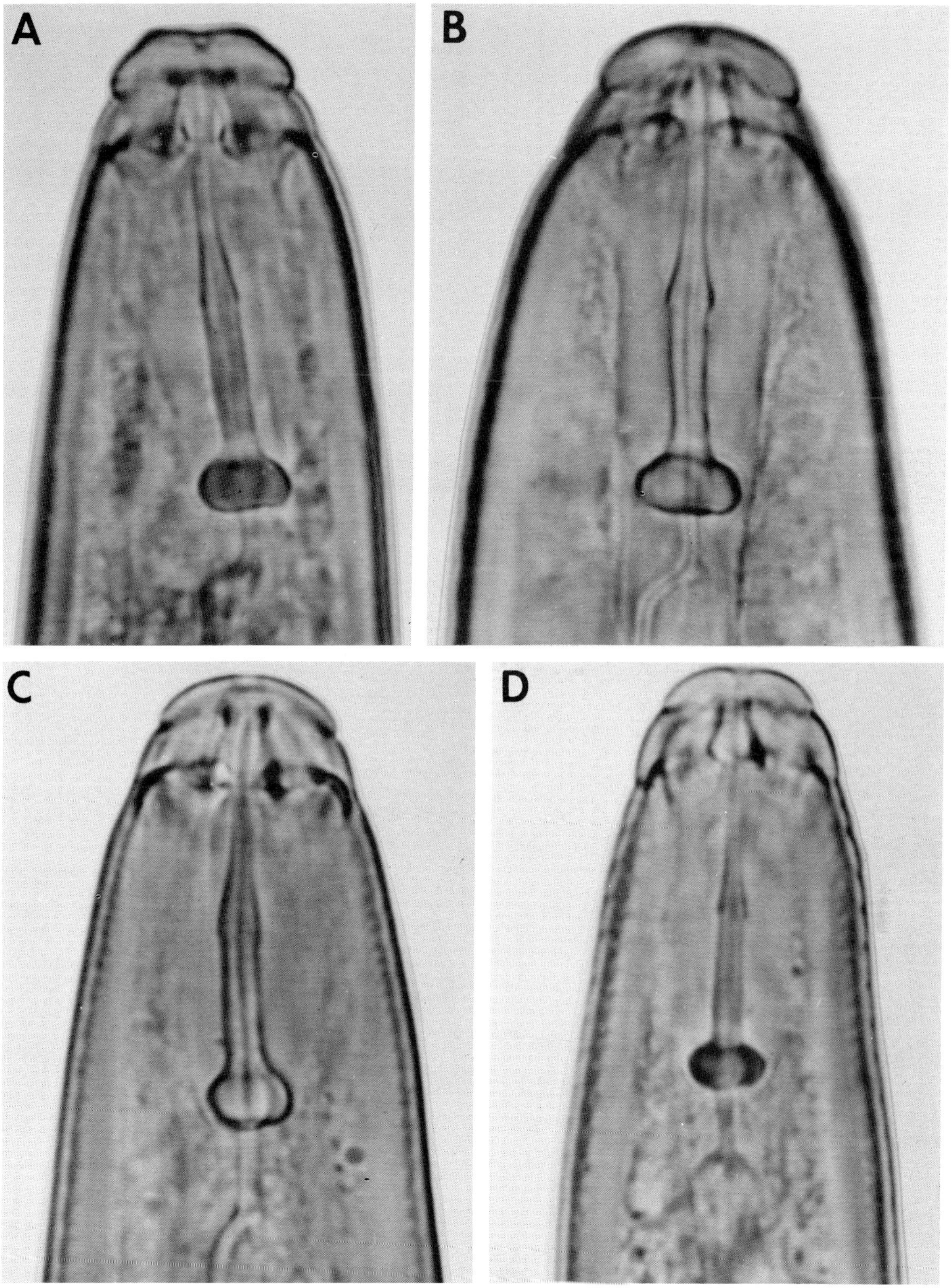

Fig. 9 A-D. Light micrographs of the heads and stylets of males of the four most common *Meloidogyne* species. A) *M. incognita.* B) *M. javanica.* C) *M. arenaria.* D) *M. hapla.* [After J. D. Eisenback et al. (16)].

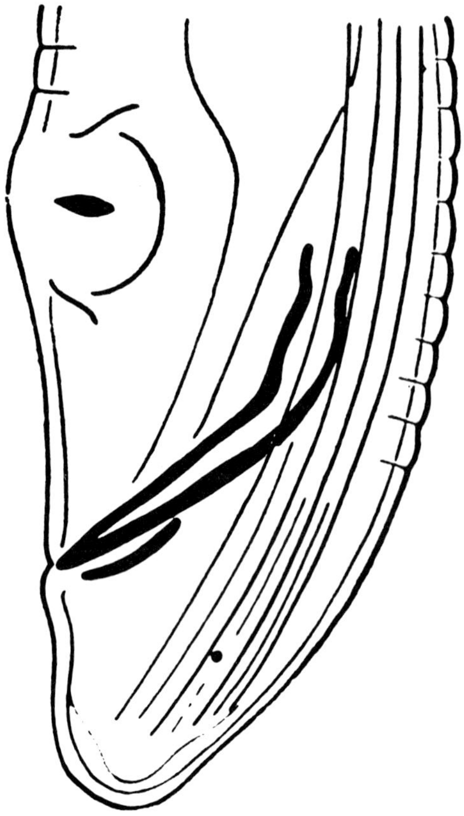

Fig. 10. **Drawing of the tail of a male intersex of *Meloidogyne javanica* taken from the original description of the four most common species of root-knot nematodes. [After B. G. Chitwood (2)].**

The stylet of *M. arenaria* second-stage juveniles (Figs. 11C,12C,13C) is characteristic because the cone and shaft are very wide and robust. The knobs are very broad, posteriorly rounded, sloping backward, and gradually merge with the shaft. The distance from the D.E.G.O. to the base of the shaft is moderately long, 3-4 μm, with a mean of 3.5 μm.

MELOIDOGYNE HAPLA The shape of the head of second-stage juveniles of *M. hapla* (Figs. 11D,13D) is characterized by a narrow head cap and a rounded head region. It is quite different from the other three common species. The labial disc and medial lips are fused and form one smooth, continuous, narrow head cap. The head region is anteriorly rounded and free of annulations.

The stylet of *M. hapla* second-stage juveniles (Figs. 11D,12D,13D) is distinguished by a narrow cone and shaft; and minute, rounded knobs that are set off from the shaft. The distance from the D.E.G.O. to the base of the stylet is relatively long, 4-5 μm, with a mean of 4.5 μm.

CONCLUSIONS

Morphological characters useful in the identification of the four most common species of root-knot nematodes include perineal patterns and stylet morphology of females, head shape and stylet morphology of males, and head shape and stylet morphology of second-stage

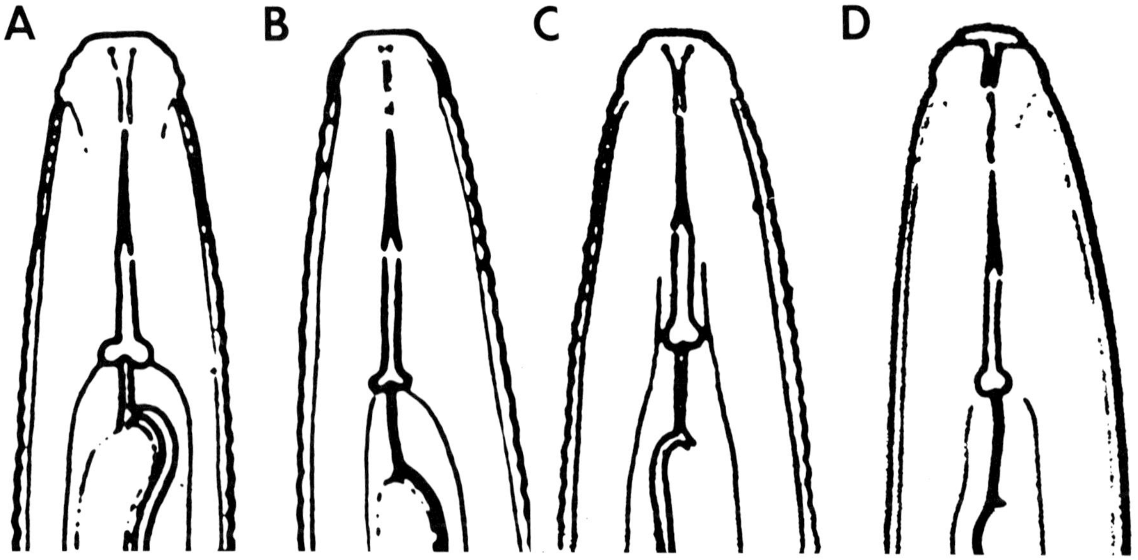

Fig. 11 A-D. **Drawings of heads and stylets of second-stage juveniles of *Meloidogyne* species taken from the original description. A) *M. incognita*. B) *M. javanica*. C) *M. arenaria*. D) *M. hapla*. [After B. G. Chitwood (2)].**

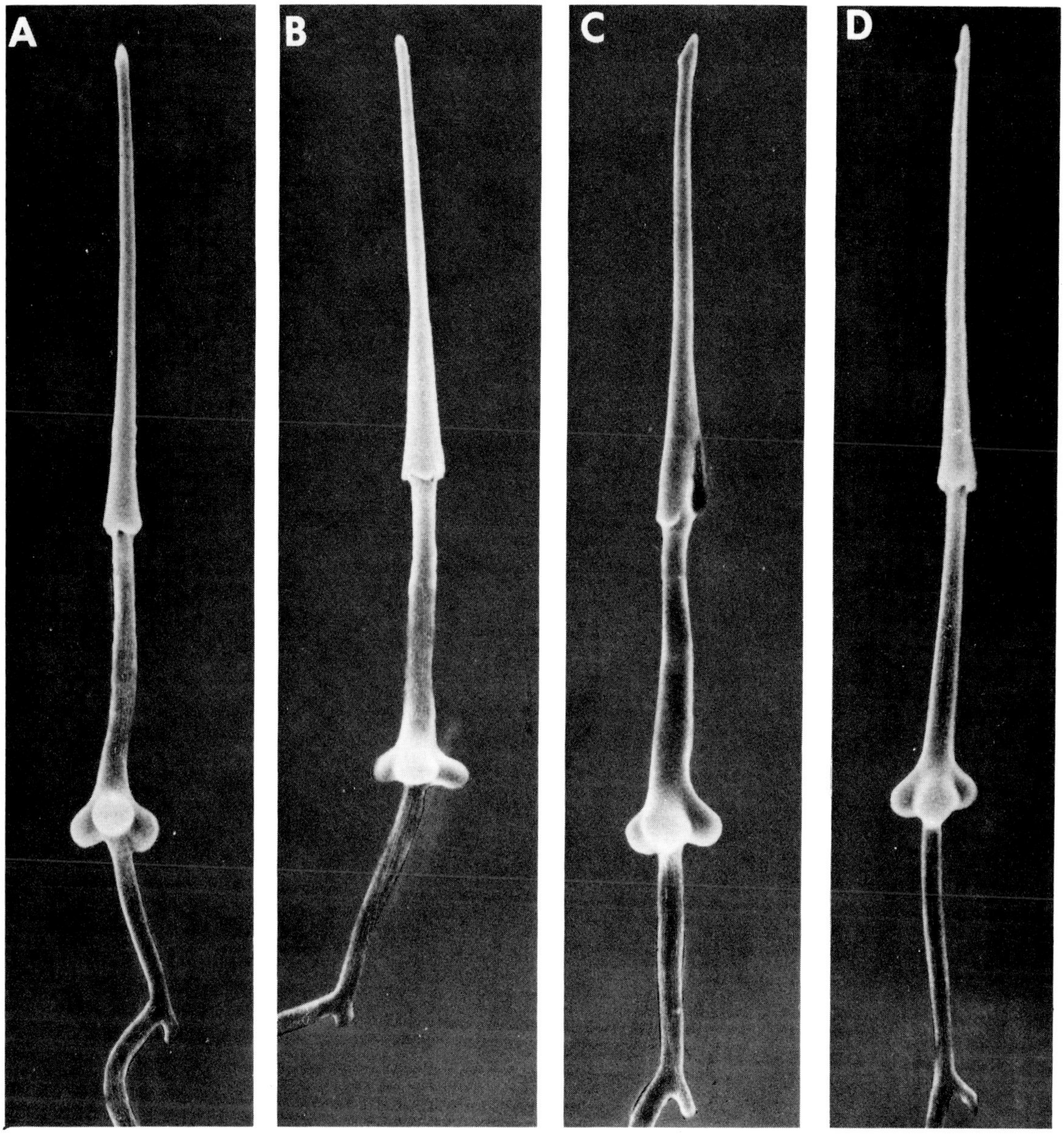

Fig. 12 A-D. Scanning electron micrographs of excised stylets of second-stage juvenile root-knot nematodes. A) *Meloidogyne incognita.* B) *M. javanica.* C) *M. arenaria.* D) *M. hapla.* [After J. D. Eisenback (7)].

juveniles. When available, male heads and stylets provide a rapid and accurate method for identification of populations. If males are absent from the infected root tissue, perineal patterns of females are also useful, particularly when stylet morphology is examined as well, because stylet morphology appears to be quite stable among populations (21). Characters of second-stage juveniles are useful supplemental characters, but cannot be used alone in routine species identifications. The more characters examined, the more precise and accurate the identification. Populations that are used for research purposes should be observed for all useful morphological characters, tested on the standard host differentials (27), and, if possible, examined cytologically and biochemically (16).

In the future, new species descriptions should include accurate, detailed drawings and high quality micrographs of the most important morphological features of preparasitic juveniles, males, and females. Where possible, the SEM and other sophisticated techniques should be utilized as well.

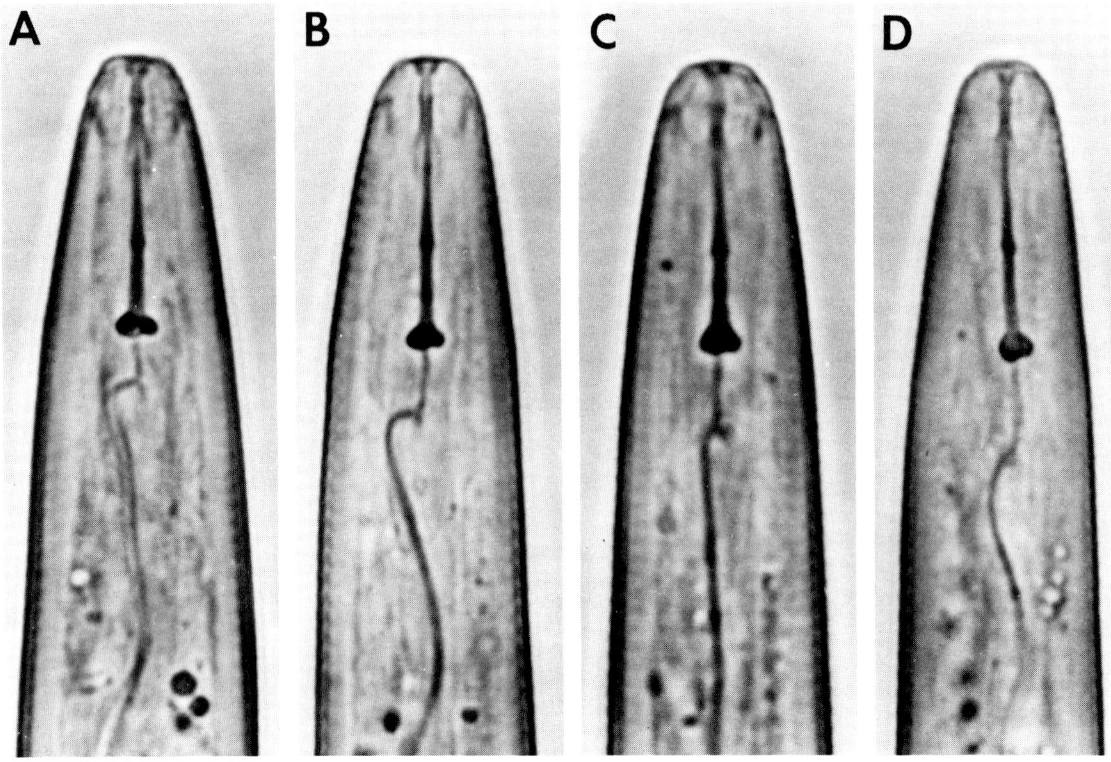

Fig. 13 A-D. Light micrographs of the head and stylets of second-stage juveniles of *Meloidogyne* species. A) *M. incognita.* B) *M. javanica.* C) *M. arenaria.* D) *M. hapla.* [After J. D. Eisenback (7)].

Table 4. Summary of important diagnostic characters of second-stage juvenile head shape and stylet morphology of the four most common *Meloidogyne* species.

	Head cap	Head region	Stylet width	Stylet knobs	D.E.G.O.[a]
M. incognita	anteriorly flattened, elongate	usually marked by 1-3 incomplete annulations	moderately sized cone and shaft	set off, posteriorly rounded sloping backward	short, 3 μm mean, 2-3 μm range
M. javanica	anteriorly flattened, elongate	smooth or marked by 1-3 incomplete annulations	moderately sized cone and shaft	set off, posteriorly, rounded, sloping backward, transversely elongate	moderately long, 3.5 μm mean, 3-4 μm range
M. arenaria	anteriorly flattened, elongate	smooth or marked by 1-3 incomplete annulations	broad cone and shaft	not set off, posteriorly rounded, merging with shaft	moderately long, 3.5 μm mean, 3-4 μm range
M. hapla	rounded and narrow	rounded, usually smooth	narrow cone and shaft	set off small and rounded	long, 4.5 μm mean, 4-5 μm range

[a] Distance from the dorsal esophageal gland orifice (D.E.G.O.) to the base of the stylet.

Literature Cited

1. Allen, M. W. 1952. Observations on the genus *Meloidogyne* Goeldi, 1887. Proc. Helminthol. Soc. Wash. 19:44-51.
2. Chitwood, B. G. 1949. "Root-knot nematodes." I. A revision of the genus *Meloidogyne* Goeldi, 1887. Proc. Helminthol. Soc. Wash. 16:90-104.
3. Christie, J. R. and F. E. Albin. 1944. Host-parasite relationships of the root-knot nematode, *Heterodera marioni*. Proc. Helminthol. Soc. Wash. 11:31-37.
4. Dalmasso, A. and J. B. Berge. 1978. Molecular polymorphism and phylogenetic relationship in some *Meloidogyne* spp. Application to the taxonomy of *Meloidogyne*. J. Nematol. 10:323-332.
5. Dalmasso, A. and J. B. Berge. 1983. Enzyme polymorphism and the concept of parthenogenetic species, exemplified by *Meloidogyne*, pp. 187-195 *In* Concepts in nematode systematics, ed. A. R. Stone, H. M. Platt and L. F. Khalil. Academic Press, New York. 308 pp.
6. Dropkin, V. H. 1953. Studies on the variability of the anal plate patterns in pure lines of *Meloidogyne* spp., the root-knot nematode. Proc. Helminthol. Soc. Wash. 20:32-39.
7. Eisenback, J. D. 1982. Morphological comparison of head shape and stylet morphology of second-stage juveniles of *Meloidogyne* species. J. Nematol. 14:339-343.
8. Eisenback, J. D. 1983. Loss of resistance in tobacco cultivar 'NC 95' by infection of *Meloidogyne arenaria* or *M. hapla*. J. Nematol. 15:478.
9. Eisenback, J. D. and H. Hirschmann. 1979. Morphological comparison of second-stage juveniles of six populations of *Meloidogyne hapla* by SEM. J. Nematol. 11:5-16.
10. Eisenback, J. D. and H. Hirschmann. 1979. Morphological comparison of second-stage juveniles of several *Meloidogyne* species (root-knot nematodes) by scanning electron microscopy. Scanning Electron Microsc. III:223-230.
11. Eisenback, J. D. and H. Hirschmann. 1980. Morphological comparison of *Meloidogyne* males by scanning electron microscopy. J. Nematol. 12:23-32.
12. Eisenback, J. D. and H. Hirschmann. 1981. Identification of *Meloidogyne* species on the basis of head shape and stylet morphology of the male. J. Nematol. 13:513-521.
13. Eisenback, J. D. and H. Hirschmann. 1982. Morphological comparison of stylets of male root-knot nematodes (*Meloidogyne* spp.) Scanning Electron Microsc. II:837-843.
14. Eisenback, J. D., H. Hirschmann, and A. C. Triantaphyllou. 1980. Morphological comparison of *Meloidogyne* female head structures, perineal patterns, and stylets. J. Nematol. 12:300-313.
15. Eisenback, J. D., H. Hirschmann, and A. C. Triantaphyllou. 1985. A more complete morphological characterization of the northern root-knot nematode, *Meloidogyne hapla*. J. Nematol. In press.
16. Eisenback, J. D., H. Hirschmann, J. N. Sasser, and A. C. Triantaphyllou. 1981. A guide to the four most common species of root-knot nematodes (*Meloidogyne* spp.), with a pictorial key. Coop. Publ. Deps. Plant Pathol. and Genet., North Carolina State Univ., and U.S. Agency Int. Dev., Raleigh, North Carolina. 48 pp.
17. Esser, R. P., V. G. Perry, and A. L. Taylor. 1976. A diagnostic compendium of the genus *Meloidogyne* (Nematoda: Heteroderidae). Proc. Helminthol. Soc. Wash. 43:138-150.
18. Franklin, M. T. 1979. Taxonomy of the genus *Meloidogyne*, pp. 37-54 *In* Root-knot nematodes (*Meloidogyne* species) systematics, biology and control, ed. F. Lamberti and C. E. Taylor. Academic Press, New York. 477 pp.
19. Golden, A. M. 1976. *Meloidogyne* taxonomy—current status, some problems, and needs, pp. 13-21 *In* Proc. Res. Plann. Conf. on Root-knot Nematodes, *Meloidogyne* spp., Jan. 12-16, 1976, North Carolina State Univ., Raleigh, NC. 106 pp.
20. Hewlett, T. E. and A. C. Tarjan. 1983. Synopsis of the genus *Meloidogyne* Goeldi, 1887. Nematropica 13:79-102.
21. Jepson, S. B. 1983. The use of stylets in the identification of *Meloidogyne* species females. Nematologica: In press.
22. Jepson, S. B. 1983. Identification of *Meloidogyne*: a general assessment and a comparison of male morphology using light microscopy, with a key to 24 species. Rev. Nematol. 6:291-309.
23. Mulvey, R. H. 1960. The value of cone top and underbridge structures in the separation of some cyst-forming nematodes, pp. 212-215 *In* Nematology, fundamentals and recent advances with emphasis on plant parasitic and soil forms, ed. J. N. Sasser and

W. R. Jenkins. Univ. North Carolina Press, Chapel Hill. 480 pp.

24. Netscher, C. 1978. Morphological and physiological variability of species of *Meloidogyne* in West Africa and implications for their control. Meded. Landbouwhogesch. Wageningen, Nederland 78:1-46.
25. Netscher, C. 1983. Problems in the classification of *Meloidogyne* reproducing by mitotic parthenogenesis, pp. 197-206 *In* Concepts in nematode systematics, ed. A. R. Stone, H. M. Platt, and K. F. Khalil. Academic Press, New York. 308 pp.
26. Sasser, J. N. 1954. Identification and host-parasite relationships of certain root-knot nematodes (*Meloidogyne* spp.). Md. Agric. Exp. Stn. Bull. A-77. 31 pp.
27. Sasser, J. N. 1979. Pathogenicity, host ranges and variability in *Meloidogyne* spp., pp. 257-268 *In* Root-knot nematodes (*Meloidogyne* species) systematics, biology and control, ed. F. Lamberti and C. E. Taylor. Academic Press, New York. 477 pp.
28. Sasser, J. N. 1980. Root-knot nematodes: a global menace to crop production. Plant Dis. 64:36-41.
29. Triantaphyllou, A. C. 1979. Cytogenetics of root-knot nematodes, pp. 85-114 *In* Root-knot nematodes (*Meloidogyne* species) systematics, biology and control, ed. F. Lamberti and C. E. Taylor. Academic Press, New York. 477 pp.
30. Whitehead, A. G. 1968. Taxonomy of *Meloidogyne* (Nematodea: Heteroderidae) with descriptions of four new species. Trans. Zool. Soc. Lond. 31:263-401.

CYTOGENETICS, CYTOTAXONOMY AND PHYLOGENY OF ROOT-KNOT NEMATODES

A. C. Triantaphyllou
Department of Genetics
North Carolina State University
Raleigh, North Carolina, USA

About 600 populations of *Meloidogyne* from 85 countries have been studied cytogenetically during the last 8 years in collaboration with the International *Meloidogyne* Project. These populations represent 24 of the 54 described species of *Meloidogyne*. Investigations on such a large number of populations have demonstrated that root-knot nematodes have undergone extensive cytogenetic diversification, probably unparalleled by that of any other animal group. Characteristic features are the establishment of meiotic and mitotic parthenogenesis in association with various degrees of polyploidy and aneuploidy. Obligatory cross-fertilization also occurs in some diploid and polyploid forms, whereas facultative meiotic and obligatory mitotic parthenogenesis prevails in most polyploid and aneuploid forms. This extensive cytogenetic diversification must have played an important role in the overall biological and genetic evolution of these nematodes. The cytogenetic status of each species may be instructive in the interpretation of its phyletic and, consequently, its taxonomic relationship to the other species of root-knot nematodes.

Cytogenetic features that can be used as taxonomic characters for species identification and can be useful in the interpretation of phyletic relationships of root-knot nematodes include the following:

1) Method of reproduction (cross-fertilization, meiotic parthenogenesis, mitotic parthenogenesis, combination of various methods);

2) Chromosome number (haploid number in forms undergoing meiosis, diploid or somatic number of mitotic forms, variations due to polyploidy or aneuploidy existing within a given species);

3) Chromosome size and general morphology;

4) Chromosome behavior during maturation of oocytes; and

5) Amount of DNA per haploid or diploid nucleus.

The first two features are the most useful for taxonomic purposes and have been studied most extensively. A limited cytological study of adult, egg-laying females is sufficient for observation of these features. The cytological techniques required for such a study are described elsewhere (Vol. II, Chapter 7). In order to facilitate cytological investigations and to help in their interpretation, a brief account of the process of oogenesis and mode of reproduction of root-knot nematodes is given prior to that of cytogenetic characterization of the major species.

Studies of spermatogenesis have also been conducted with some *Meloidogyne* species, but they have been more difficult and less fruitful. Invariably, the chromosomes observed during spermatogenesis in males are similar to those observed during oogenesis (3).

OOGENESIS

The reproductive system of females of all *Meloidogyne* species studied thus far consists of two ovaries (gonads), each connected to the vagina via separate gonoducts (Fig. 1). Each gonoduct consists of a short, glandular oviduct; a spherical, glandular spermatotheca; and a long, cylindrical uterus.

Oogenesis is a two-step process, involving:

1) multiplication of oogonia and,

2) growth and maturation of oocytes.

The oogonial cells are direct descendents of the two germinal (propagation) cells of the genital primordium seen in second stage juveniles. Oogonia occupy the apical portion (germinal zone) of the ovary (Figs. 1-3). In

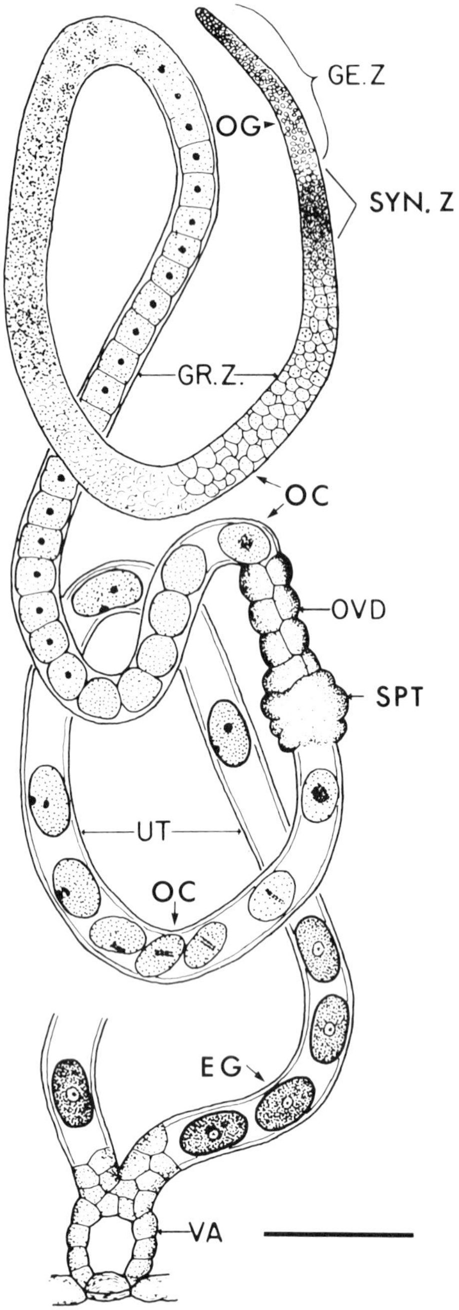

Fig. 1. Female reproductive system of *M. javanica* showing the various regions of one gonad and the process of oogenesis (diagrammatic). EG = egg; GE.Z = germinal zone; GR.Z = growth zone; OC = oocyte; OG = oogonium; OVD = oviduct; SPT = spermatotheca; SYN.Z = synapsis zone; UT = uterus; Bar = 0.2mm.

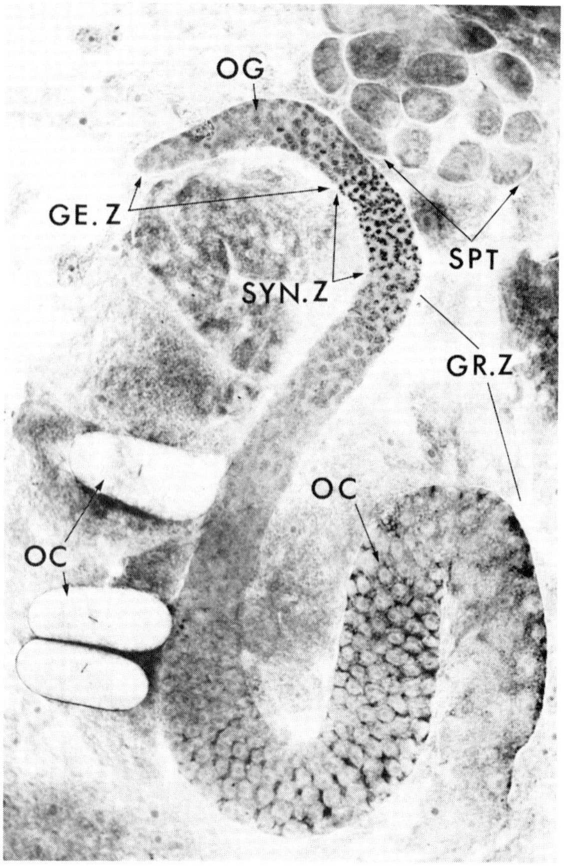

Fig. 2. General view at low magnification (10× objective) of a cytological preparation, showing the anterior half of one ovary and part of the gonoducts of the female reproductive system of *M. arenaria* (propionic orcein stain). GE.Z = germinal zone; GR.Z = growth zone; OC = oocyte; OG = oogonium; SPT = spermatotheca; SYN.Z = synapsis zone.

young females, the oogonia undergo several mitotic divisions which result in the production of a larger number of oogonia, all of which have the somatic (2n) chromosome number. At late prophase or early metaphase of such oogonial divisions, the chromosomes are discrete, spherical or, most often, rod-shaped, but they are small and can be counted precisely only in exceptionally favorable cytological preparations (Fig. 4). Although differences in chromosome size and morphology have been detected among *Meloidogyne* species, differences in chromosome numbers are the only characters which have a practical use for taxonomic purposes.

At the end of the germinal zone, the oogonia stop dividing and enter a new developmental stage. The chromatin inside their nuclei condenses and becomes heavily stained with nuclear dyes (Figs. 2,5). Gonial cells at this stage are called oocytes. They occupy a short,

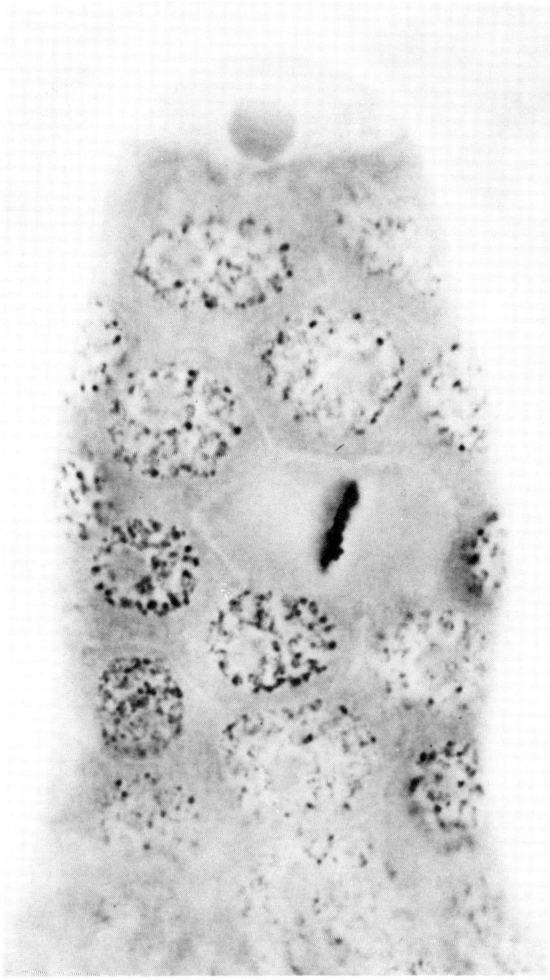

Fig. 3. Apical part of the ovary (germinal zone) of *M. incognita* with a distinct apical, epithelial cell (cap cell), many oogonia at interphase, and one at metaphase (side view).

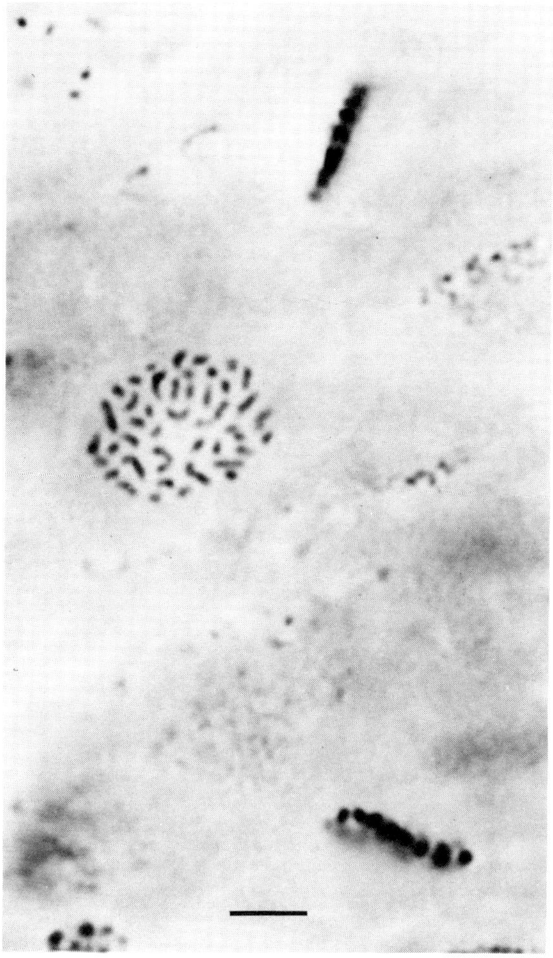

Fig. 4. Dividing oogonia in the germinal zone of the ovary of *M. incognita* at metaphase (two in side view and one in polar view). Bar = 3μm.

distinct zone (synapsis zone) which is stained heavier than the rest of the ovary (Fig. 2). This chromatin condensation and the increased stainability of the nuclei of the young oocytes probably are related to pairing of homologous chromosomes and their subsequent contraction (pachytene stage), which normally occurs in *Meloidogyne* species that undergo meiosis. However, a similar distinct zone is observed also in ameiotic (mitotic) parthenogenetic species. The presence of a zone with condensed chromatin, therefore, does not reveal the type of maturation of oocytes of a particular species and, for this reason, has no taxonomic value.

As the oocytes migrate from the zone of synapsis toward the posterior part of the ovary, they increase progressively in size. Their chromosomes are not clearly visible during this period (diffuse stage). They become sufficiently condensed and can be seen again in more advanced oocytes approaching the oviduct-spermatotheca region (Fig. 6). Such oocytes are at diplotene or diakinesis stage. Fully grown oocytes migrate into the spermatotheca and the uterus, and advance to prophase or metaphase of the first maturation division. Their chromosomes can be recognized as univalent or bivalent.

Univalent chromosomes appear as dyads (Fig. 7); each one consists of two chromatids lying close and parallel to each other.

Bivalent chromosomes appear as tetrads (Fig. 8); each one is formed by the close association of two homologous univalent chromosomes which usually are connected end-to-end.

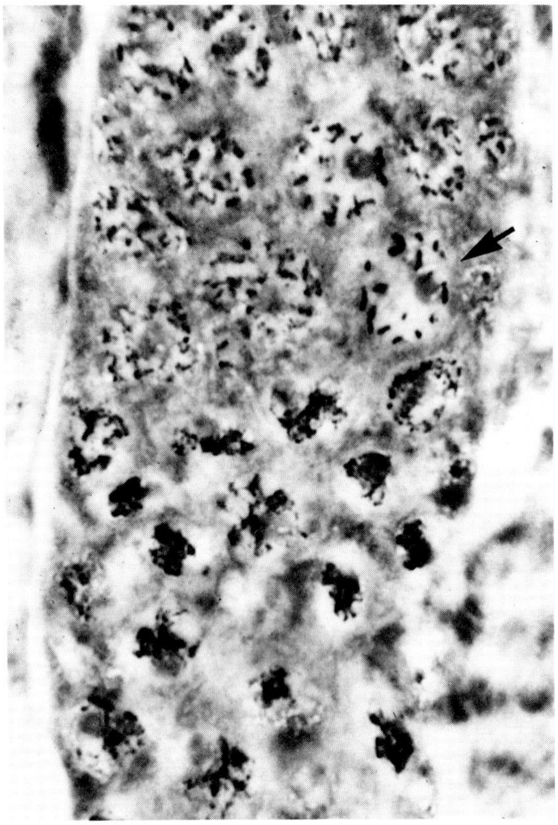

Fig. 5. End part of the germinal zone (upper half of the picture) and synapsis zone (lower half) of the ovary of *M. hapla*. In the germinal zone some oogonia are at interphase and others at mitotic prophase (arrow). In the zone of synapsis, the chromatin of the young oocytes is highly condensed and heavily stained.

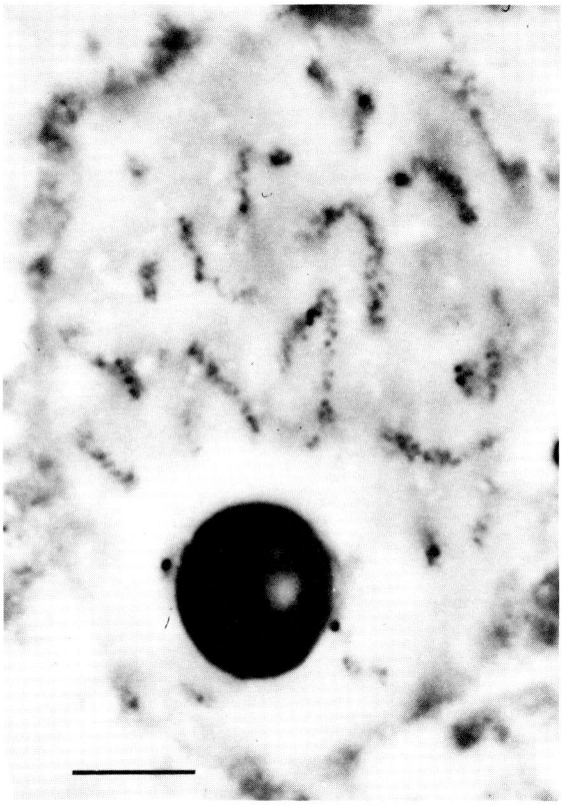

Fig. 6. The nucleus of an oocyte of *M. hapla* after the diffuse stage. The chromosomes (diplotene stage) have started condensing and have become visible again. A large, heavily stained nucleolus is also visible. Bar = 5μm.

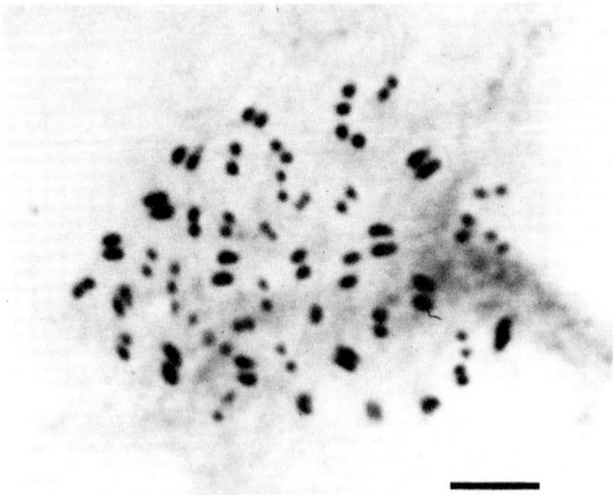

Fig. 7. The 46 univalent chromosomes of an oocyte of *M. javanica* shortly before metaphase I (prometaphase). Each chromosome comprises two chromatids that lie close to each other and form a dyad. Dyads are typical of univalent chromosomes. Bar = 3μm.

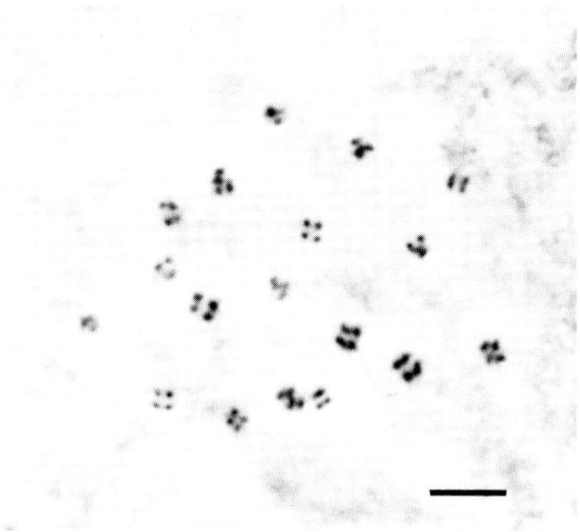

Fig. 8. The 18 bivalent chromosomes of an oocyte of *M. graminis* at prometaphase of the first maturation division. Each bivalent consists of 4 chromatids (2 per univalent) in close association, forming a tetrad. Tetrads are characteristic of bivalent chromosomes and indicate that meiosis is taking place. Bar = 3μm.

The Cytological Pathway Toward Mitotic Parthenogenesis

The presence of univalent chromosomes at, or shortly before metaphase I, indicates that pairing of homologous chromosomes did not take place when the oocytes were in the zone of synapsis. Also, the number of univalent chromosomes at this stage is the same as the number observed in the oogonia of the germinal zone of the ovary. Therefore, there has been no reduction in the chromosome number (no meiosis) during maturation of the oocytes. In more advanced oocytes, the univalent chromosomes divide mitotically, i.e. the two chromatids migrate toward the opposite poles. At the completion of this division, a polar nucleus and the egg pronucleus are formed, each one with the somatic chromosome number. The oocyte has become a mature egg, ready to undergo the first cleavage division. Oogenesis is completed, and embryogenesis is initiated at this stage. This pattern of oogenesis is associated with parthenogenetic reproduction of the mitotic type (mitotic parthenogenesis or apomixis) and is common in many root-knot nematode species, including most of the major species, i.e. *M. incognita*, *M. javanica*, and *M. arenaria*.

The Cytological Pathway Toward Cross-Fertilization and Meiotic Parthenogenesis

The presence of bivalent chromosomes in oocytes located in the spematotheca or the anterior part of the uterus indicates that synapsis of homologous chromosomes occurred while the oocytes were in the zone of synapsis. The remaining stages of meiosis (pachytene, diplotene, diakinesis) and chiasmata terminalization have occurred progressively during migration of the oocytes toward the end of the ovary. The bivalent chromosomes are one-half the number of the somatic chromosomes observed in oogonia, and this reduced number is maintained in mature eggs. Two maturation divisions are completed while the oocytes migrate down the uterus or shortly after egg laying. They result in the formation of two polar nuclei and an egg pronucleus with the haploid (n) chromosomal complement. The somatic chromosome number will have to be re-established before such an oocyte proceeds with cleavage. This is done in two different ways leading to two different methods of reproduction.

1. A sperm nucleus fuses with the egg pronucleus to form the zygote nucleus with the diploid chromosomal complement (Fig. 9).

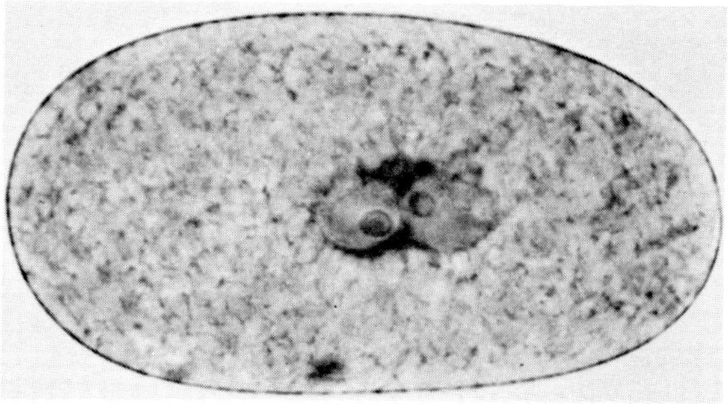

Fig. 9. Mature oocyte of *M. hapla* (race A) with the sperm and egg pronuclei about to fuse and form the zygote nucleus of a fertilized egg.

Reproduction in this case is by cross-fertilization (amphimixis). It occurs obligatorily in some *Meloidogyne* species (*M. carolinensis*, *M. megatyla*, *M. microtyla*), which usually have a large proportion of males. In such species, sperm is stored in the spermatothecae of inseminated females, and subsequently, single spermatozoa enter the oocytes as the latter pass through the spermatothecae.

2. The second polar nucleus fuses with the egg pronucleus to form the egg nucleus with the diploid (2n) chromosomal complement. The egg, then, proceeds with cleavage. Reproduction is by meiotic parthenogenesis (automictic parthenogenesis).

Meiotic parthenogenesis in *Meloidogyne* is facultative, in that species which reproduce by meiotic parthenogenesis can also reproduce by cross-fertilization. In fact, a single population can reproduce by cross-fertilization when males are present or by meiotic parthenogenesis when males are absent. Furthermore, an inseminated female with a limited number of spermatozoa may produce some eggs as the result of cross-fertilization and still continue producing eggs by meiotic parthenogenesis when the sperm in its spermatothecae is used up (3).

CYTOGENETIC FEATURES OF MAJOR MELOIDOGYNE SPECIES

Meloidogyne incognita

This species reproduces exclusively by mitotic parthenogenesis and has two chromosomal forms: a rare diploid with about 36 chromosomes and a prevalent triploid, or hypotriploid, with approximately 42-44 chromosomes (Table 1).

An extensive study of at least 220 populations from 60 countries has revealed that *M. incognita* is cytogenetically distinct and can be easily differentiated from other species by various cytological characteristics. Many oogonial divi-

Table 1. Summary of cytogenetic information related to root-knot nematodes.

***Meloidogyne* species**	**Populations studied (Number)**	**Countries of origin (Number)**	**Chromosome Number n**	**2n**	**Mode of Reproduction**
M. carolinensis	2	1	18		
M. megatyla	1	1	18		Amphimixis
M. microtyla	2	1	18-19		
M. subarctica	1	1	18		
M. exigua	6	1	18		
M. graminicola	1	1	18		
M. graminis	10	5	18		Facultative
M. naasi	1	1	18		
M. ottersoni	1	1	18		Meiotic
M. chitwoodi	6	3	14-18		
M. hapla (Race A)	48	24	13-17		Parthenogenesis
(Polyploid)	1	1	28		
(Polyploid)	2	2	34		
M. arenaria	18	13		30-38	
	34	21		40-48	
	68	32		51-56	
M. cruciani	1	1		42-44	
M. enterolobii	1	1		~46	
M. hapla	6	3		30-32	
	11	8		43-48	Obligatory
M. hispanica[a]	4	4		33-36	
M. incognita	6	6		32-38	
	215	64		41-46	Mitotic
M. javanica	126	45		42-48	
M. microcephala	3	2		36-38	
M. oryzae	2	1		51-55	Parthenogenesis
M. platani	1	1		42-44	
M. querciana	1	1		30-32	
M. spp. (from coffee)	6	3		50-56	
(from rice)	2	2		36	
Unidentified	5	3		36	
	11	7		40-46	
Total	602				

[a] To be described by H. Hirschmann.

sions take place in the germinal zone of the ovary of young, egg-laying females (Fig. 4). This feature is consistent with the relatively high reproductive potential of *M. incognita.* Metaphase of oogonial divisions is the best stage for obtaining an acceptable estimate of the somatic chromosome number of *M. incognita.* Populations with 32 to 38 chromosomes are considered to be representing a diploid form (n=16 to 18), and those with more than 41 chromosomes are assumed to be triploid or to have been derived from triploid forms through loss of chromosomes (hypotriploids). The distinction between the two chromosomal forms, however,is not very clear because of the absence of a distinct gap between them. More than 95% of the populations of *M.incognita* studied have 2n=42-44 chromosomes, suggesting that this chromosomal form is, by far, the most important agronomically.

The most characteristic cytogenetic feature of *M. incognita,* and one of high taxonomic value, is provided by the behavior of its chromosomes at prophase of the first maturation division (5). Thus, contrary to the prophase figures of other *Meloidogyne* species in which the chromosomes are discrete, the prophase chromosomes of *M. incognita* are crowded together in a small spherical area and are not discrete (Fig. 10A). Furthermore, the prophase in *M.incognita* is much more extended in time, compared to that of the other species. As a result of this behavior, most of the oocytes in the uterus of *M. incognita* females are at prophase and their chromosome number cannot be determined. Only a few oocytes, located at the end of the uterus, close to the vagina, may be at metaphase I. In these oocytes, the chromosomes are discrete and can be counted precisely only when the metaphase

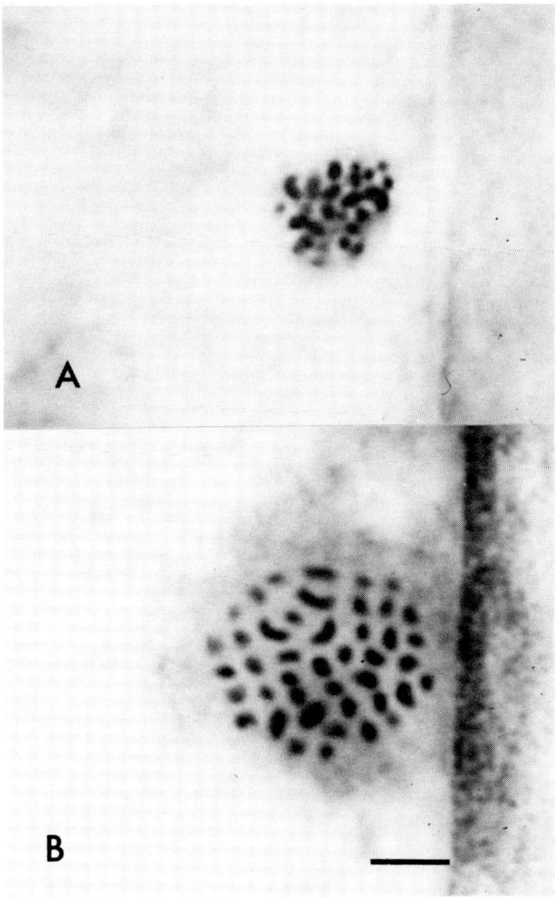

Fig. 10. The chromosomes of *M. incognita* during maturation of the oocytes. A) Prophase chromosomes in an oocyte located in the anterior part of the uterus. The chromosomes are crowded together in a small area and cannot be counted. B) Metaphase chromosomes (polar view) in an oocyte located in the posterior part of the uterus. Bar = 3μm.

plates are in perfect polar view (Fig. 10B). The chromosomes in such metaphase figures are univalent and occur in the same number (somatic) as in oogonial cells. The eggs usually are deposited in the egg mass while they are at metaphase I or telophase I and complete the first maturation division subsequently. They do not undergo a second maturation division but proceed directly with the first cleavage division. In very old females, some eggs may remain in the uterus for a longer period of time and undergo one or more cleavage divisions before they are deposited in the egg mass.

Meloidogyne javanica

Like *M. incognita*, *M. javanica* reproduces exclusively by mitotic parthenogenesis. Among the 126 populations studied thus far, the chromosome number varied from 2n=42 to 2n = 48 (Table 1), but most populations had 2n = 45-46. Chromosomal races have not been recognized in *M. javanica*, and the behavior of the chromosomes during oogenesis is uniform throughout the species.

Oogonial divisions are identical with those of *M. incognita* in frequency of occurrence and behavior of the chromosomes. The chromosome number cannot be determined easily in oogonial divisions, but an acceptable estimate can be obtained from many metaphase figures.

Oocytes are at late prophase or early metaphase I as they pass from the spermatotheca into the uterus. Their chromosomes are quite discrete and spread over a large area. They are univalent, appearing as dyads (Fig. 11A) and can be counted precisely, especially during early metaphase I (Fig. 11B). The oocytes advance to anaphase and telophase I as they migrate toward the posterior part of the uterus. Counting the chromosomes during these stages becomes more difficult. Occasionally, determination of the chromosome number can be made in one or both of the anaphase or telophase plates (Fig. 11C).

Meloidogyne javanica is the most favorable species of *Meloidogyne* for cytological studies. Its chromosomes stain well in young oocytes at prophase or metaphase I and are widely separated from one another. Also, the chromosomes of *M. javanica* appear to be slightly larger than those of *M. arenaria* (diploid and hypotriploid forms) and *M. hapla* (race B). However, this size difference cannot be used in differentiating these species. A wide variation in size is usually observed, even when standard staining procedures are employed.

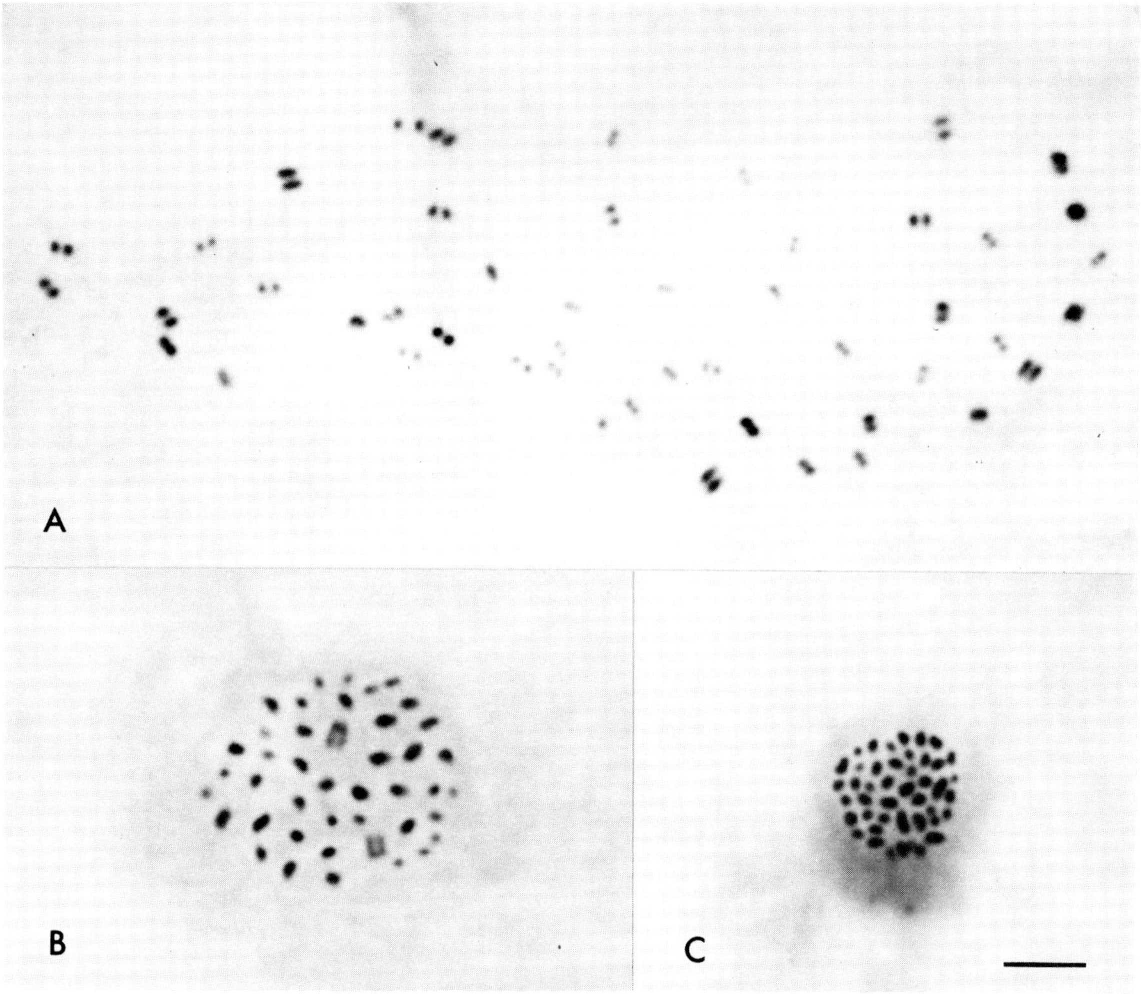

Fig. 11. **The chromosomes of *M. javanica* during maturation of the oocytes. A) Prometaphase figure with 46 univalents (dyads) in an oocyte that has just entered the uterus. B) Metaphase or anaphase figure (polar view). C) Telophase figure in an advanced oocyte located in the posterior part of the uterus. Bar = 3μm.**

Meloidogyne arenaria

Meloidogyne arenaria reproduces exclusively by mitotic parthenogenesis. Cytologically, it is a very complex species in that it includes populations with chromosome numbers varying from 2n = 30 to 2n = 56 (Table 1). A triploid form with somatic chromosome numbers larger than 50 is the most common and, consequently, typical of the species (Fig. 12A). However, diploid populations with 2n = 30 - 38 (Fig. 12B) and hypotriploid ones with 2n = 40 - 48 chromosomes (Fig. 12C) have been found in many parts of the world and are provisionally identified as *M. arenaria*, or variants of it.

More than 100 populations of *M. arenaria* originating from 40 countries have been studied recently (Table 1). Multiplication of oogonia and maturation of oocytes follow approximately the same pattern, as in *M. incognita* and *M. javanica*. The triploid form of *M. arenaria* can be differentiated from the other major species of *Meloidogyne* on the basis of its high chromosome numbers (more than 50, versus smaller numbers for the other species). The diploid form, which is very rare, cannot be differentiated from some diploid populations of *M. hapla*, and the hypotriploid form, which is more common than the diploid, cannot be differentiated from *M. javanica* and from the hypotriploid form of *M. hapla*. However, identification of these forms of *M. arenaria* (diploid and hypotriploid) is also difficult even on the basis of morphological and other criteria, and their taxonomic status is presently rather uncertain.

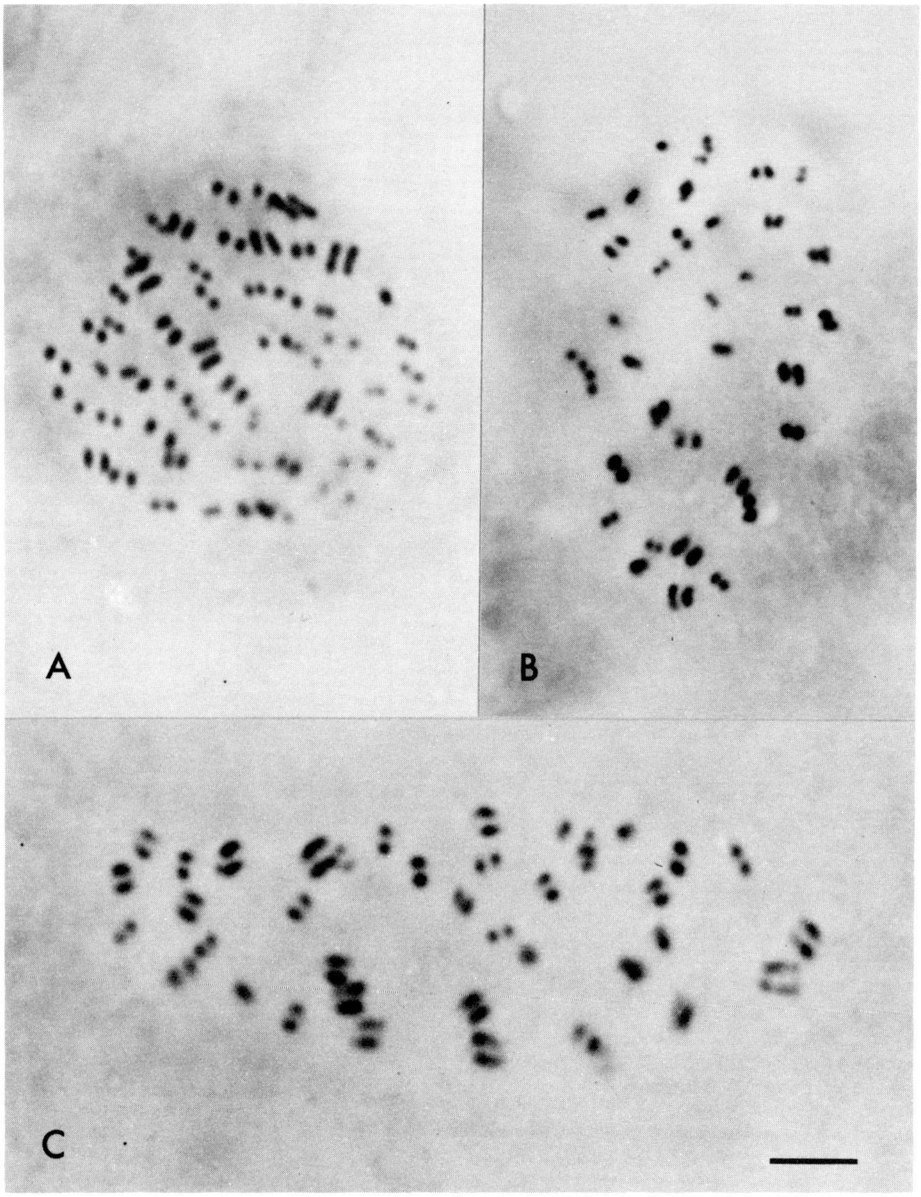

Fig. 12. The chromosomes of *M. arenaria* in oocytes at prometaphase. A) Triploid form with 54 chromosomes. B) Diploid form with 36 chromosomes. C) Hypotriploid form with 42 chromosomes. Bar = 3μm.

Meloidogyne hapla

Cytogenetically, *M. hapla* is the most diversified species of root-knot nematodes (2,3). Most populations reproduce by facultative meiotic parthenogenesis (race A), i.e. usually by meiotic parthenogenesis and less frequently by cross-fertilization. Most have a haploid chromosomal complement of 17; some have n = 16; and a few have n = 15, 14 or 13 (Tables 1, 2; Fig. 13). Some polyploid females with n = 28 - 34 also occur in low frequency within diploid populations which have one half as many chromosomes (6). Other populations of *M. hapla* reproduce obligatorily by mitotic parthenogenesis (race B). Most of them have a somatic chromosome number varying from 43 to 48, but some have only 30 to 32 chromosomes (Tables 1, 2; Fig. 14).

Race A of *M. hapla* is easily distinguished cytogenetically from the other three major species of *Meloidogyne* by the presence of bivalent chromosomes (tetrads) in oocytes located in the spermatotheca or in the anterior part of the uterus. This feature cannot be used in the differentiation of *M. hapla* from minor

Table 2. Distribution of various cytogenetic forms of Meloidogyne hapla.

	Mode of Reproduction and Chromosome Number[a]						
Continent	**Meiotic parthenogenesis**				**Mitotic parthenogenesis**		
	n=13-14	**n=15**	**n=16**	**n=17**	**2n=30-32**	**2n=43-45**	**2n=48**
North America	USA-Minn	USA-NC Canada (2)	USA-NC (3) Canada (3)	USA-Va USA-Tenn USA-Ohio USA-NC (3) USA-SD USA-Mich USA-Minn Canada	USA-Iowa	USA-NC USA-MD	USA-FL
Central and South America		Mexico Colombia	Equador	Costa Rica (2) Equador Chile			Argentina Chile (3)
Europe	Holland		France Italy	France (2) Germany Portugal Holland		England France	
Africa			Libya Kenya	Kenya (2)			
Asia	Taiwan	Korea	Korea (3) China	Korea (5) China	China	Korea (2)	
Total	3	6	15	27	2	6	5

[a] Numbers in parentheses indicate the number of populations studied.

species (see next section) that also undergo meiosis and have bivalent chromosomes in primary oocytes at prometaphase or metaphase I. However, with the exception of some populations of *M. chitwoodi*, which have chromosome numbers similar to those of *M. hapla*, all other meiotic species have n = 18 and, therefore, can be distinguished from *M. hapla* on this basis.

Race B of *M. hapla* with 30 to 32 univalent chromosomes (dyads) in oocytes at metaphase I is similar to the diploid form of *M. arenaria* which most often has a somatic chromosome number of 34 to 36. Other race B populations of *M. hapla* with 2n = 43-48 are indistinguishable from the hypotriploid form of *M. arenaria* with the same number of chromosomes. Similarly, they cannot be differentiated from *M. javanica* which shows the same range of chromosome numbers in spite of the fact that the chromosomes of *M. javanica* are usually slightly larger than those of *M. hapla*, race B.

CYTOGENETIC FEATURES OF MINOR *MELOIDOGYNE* SPECIES

Aside from the four major species discussed in the previous section, the remaining species of *Meloidogyne* are referred to as minor species because they are encountered much less frequently and are less important as plant pathogens on a worldwide basis.

Two of the minor species studied thus far, *M. carolinensis* and *M. megatyla*, are similar cytogenetically in that they reproduce obligatorily by cross-fertilization and have a haploid chromosome complement of 18.

M. microtyla is cytogenetically similar to the previous species, but only a few females of the population studied had n=18 chromosomes. The majority of the females had n = 19 chromosomes. This species is the only *Meloidogyne* species with a chromosome number larger than 18, not including the cases of reported polyploidy in *M. hapla* (6).

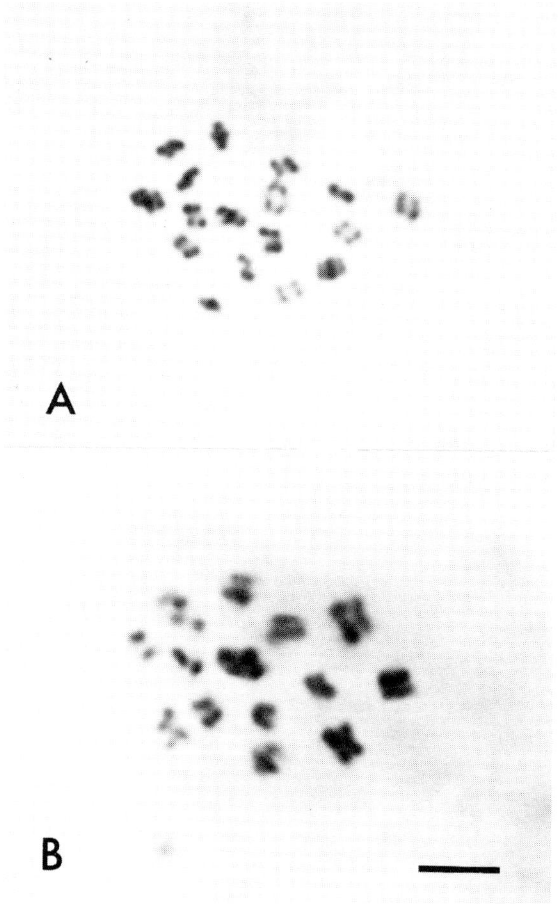

Fig. 13. **Prometaphase figures in primary oocytes of *M. hapla* (race A). A) With 17, and B) with 14 bivalent chromosomes. Bar = 3μm.**

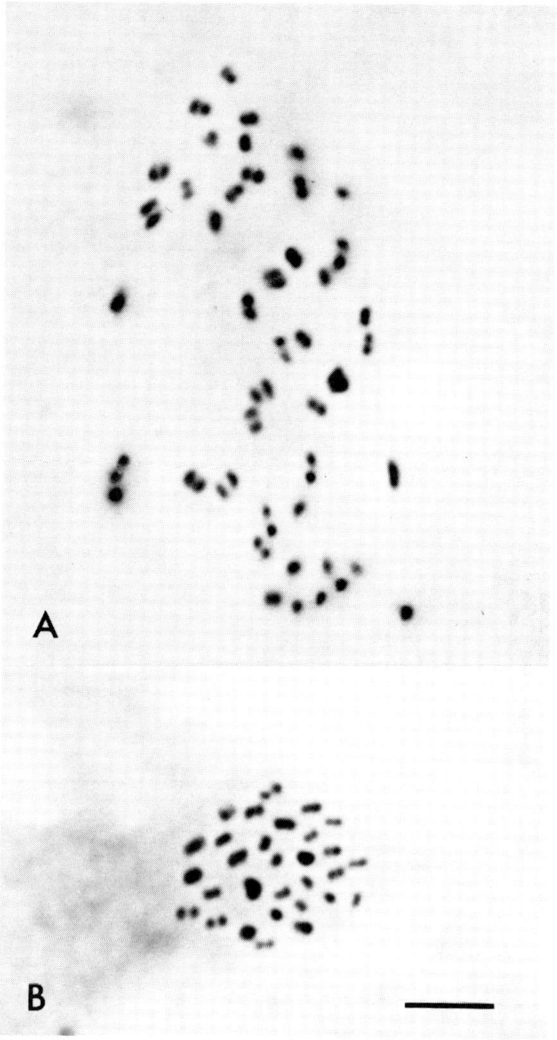

Fig. 14. **Prometaphase figures in oocytes of *M. hapla* (race B). A) With 45, and B) with 32 univalent chromosomes. Bar = 3μm.**

M. subarctica also has n = 18 chromosomes (1), but its mode of reproduction is not known and is only suspected to be cross-fertilization.

Another group of minor species includes *M. exigua*, *M. graminis*, *M. graminicola*, *M. naasi*, and *M. ottersoni* which have n = 18 chromosomes but reproduce by facultative meiotic parthenogenesis (7).

The remainder of the minor species that have been studied cytogenetically do not undergo meiosis and reproduce exclusively by mitotic parthenogenesis. (Table 1). Some of them are diploid with about 30 to 36 chromosomes [*M. hispanica* (to be described by H. Hirschmann), *M. microcephala*, *M. querciana*, and several undescribed forms], some are triploid with 50 to 56 chromosomes (*M. oryzae* and an undescribed species from coffee), and most are hypotriploid with intermediate chromosome numbers (*M. cruciani*, *M. enterolobii*, *M. platani*, and 8 undescribed forms).

EVOLUTIONARY RELATIONSHIPS OF ROOT-KNOT NEMATODES

Cytogenetic information about root-knot nematodes has been very useful in elucidating biological relationships and in suggesting a number of possible pathways of evolution of the present forms of these nematodes (6). Still, interpretation of the interrelationships of the various species is difficult. The fact that most root-knot nematode species reproduce by mitotic parthenogenesis and have variable chromosome numbers complicates the concept of what constitutes a biological species in these organisms. With the exception of *M. incognita* and *M. javanica* which appear to represent uniform biological units with distinct

Table 3. Cytogenetic relationships of root-knot nematodes.

M. carolinensis + 3 minor species		n = 18		AMPHIMIXIS
M. exigua + 3 minor species		n = 18		MEIOTIC PARTHENOGENESIS
M. hapla (Race A) M. chitwoodi	55	14-17 14-18		
M. hapla (Race B) M. arenaria M. incognita + 6 minor species		(Diploid) 36 (30-38)		
M. hapla (Race B) M. arenaria M. incognita M. javanica + 7 minor species	 24 215 131	(Hypotriploid) 45 (41-48)	Hybridization	MITOTIC PARTHENOGENESIS
M. arenaria + 2 minor species	77	(Triploid) (51-56)		

morphological, cytological and biochemical features (see Chapter 11), the biological status of all other obligatorily parthenogenetic (mitotic) species is not clear. How they are related to one another or to the facultatively parthenogenetic (meiotic) species from which supposedly they have been derived is a matter of speculation.

The status of the facultatively parthenogenetic species is also doubtful, although, for all practical purposes, such species could be regarded as equivalent to amphimictic species. Conceptually they can be considered as true biological species if they are really reproductively isolated from each other. However, tests for demonstrating reproductive isolation are technically very difficult because of the facultative nature of parthenogenetic reproduction and the lack of genetic markers that would verify intercrossing.

Only the obligatorily amphimictic species could be regarded as true biological species, but even these have not been tested for reproductive isolation and their high degree of host specialization renders such tests quite difficult.

Based on available cytogenetic information, the following assumptions can be made regarding the cytogenetic evolution of root-knot nematodes:

a) The ancestral root-knot nematodes were amphimictic animals, like most other animals we know today;

b) Parthenogenetic forms evolved from amphimictic ones;

c) Since the presently amphimictic forms have a haploid chromosome number of 18, this number may represent the basic number from which all other chromosomal forms have evolved;

d) High chromosome numbers represent states of polyploidy;

e) Chromosome numbers which are not multiples of 18 may represent various states of aneuploidy resulting through loss of individual chromosomes (in polyploid forms) or through chromosomal fusions.

These assumptions lead to the following interpretation regarding the pathway of evolution of root-knot nematodes (Table 3). The obligatorily amphimictic species (*M. megatyla*, *M. carolinensis*, and *M. microtyla*) are the closest relatives of the assumed ancestral root-knot nematode. Facultatively parthenogenetic species, like *M. exigua*, *M. naasi*, *M. graminis*, etc., with n=18, probably have evolved from an amphimictic ancestor with the same chromosome number, following evolution toward meiotic parthenogenesis. Some forms within this group, like *M. hapla* (race A) and *M. chitwoodi*, may have evolved further by additional modifications of their chromosomal complement that resulted in the reduction of the haploid chromosome number from 18 to 17, 16, 14, and 13.

All the mitotic parthenogenetic forms probably have evolved from meiotic parthenogenetic ancestors, or less likely, from amphimictic ones, following suppression of the meiotic process during maturation of the oocytes. The variation in chromosome numbers observed among the mitotic parthenogenetic forms suggests several pathways of derivation. Species with about 36 chromosomes, apparently are diploid and may have evolved from diploid meiotic forms without any change in the degree of ploidy. Species with about 54 chromosomes could be considered as triploids, and most likely they have been derived following hybridization of meiotic parthenogenetic forms involving, e.g., fertilization of an unreduced egg with 36 (18+18) chromosomes with a normal sperm having 18 chromosomes. Species with intermediate numbers of chromosomes (hypotriploid) may have been derived from the triploid forms through actual loss or fusions of a number of chromosomes. Most likely, however, they may have evolved from meiotic diploid forms with reduced chromosome numbers following fusion of unequal gametes. Thus, a hypotriploid form with 45 chromosomes may have been derived from the fusion of an unreduced egg with 30 chromosomes with a normal sperm with 15 chromosomes.

This pathway of evolution of root-knot nematodes is similar to the pathways of evolution proposed for other parthenogenetic animals and represents the prevailing thought on evolution of parthenogenesis, polyploidy and aneuploidy (8). However, recent studies of the behavior of tetraploid forms of *M. hapla* (6) and further consideration of the chromosomal complement of nematodes in general have offered alternative explanations about the possible pathways of evolution of root-knot nematodes. Since all nematodes, with the exception of some ascarids and a few other polyploid forms, have small chromosome numbers, ranging between 5 and 9 (n), the n=18 chromosomes of the genus *Meloidogyne* may represent a state of polyploidy (tetraploidy ?). The existence of *Hypsoperine* (*Meloidogyne*?) *spartinae*, a species very closely related to root-knot nematodes, with only 7 chromosomes as the haploid number, provides additional support to this assumption. Furthermore, polyploidy is known to be very common only in parthenogenetic animals, whereas it is very rare in amphimictic animals. Therefore, if the present-day forms of *Meloidogyne* are truly polyploid, one may have to assume that such polyploid forms have evolved during a parthenogenetic phase in the evolution of the genus. Consequently, the meiotic parthenogenetic forms, such as *M. graminis*, *M. naasi* and others, with n=18, as well as those with reduced chromosome numbers (*M. hapla* and *M. chitwoodi*) may hold a position closer to the ancestral forms of root-knot nematodes and may have played a significant role in the evolution of all the other forms. The amphimictic forms, *M. carolinensis*, *M. megatyla* and *M. microtyla*, may have evolved to polyploidy while they were meiotically parthenogenetic and converted to obligatory amphimixis later on. These assumptions are very tentative at the present time. Further cytogenetic analysis of additional rare *Meloidogyne* species and other forms related to the root-knot nematodes, as well as consideration of other sources of information (morphology, anatomy, physiology, biochemistry), will be needed for a more realistic evaluation of the pathways of evolution of these nematodes.

Acknowledgement

This research was funded in part by the National Science Foundation grant BSR-8314908.

Literature Cited

1. Bernard, E. C. 1981. Three new species of Heteroderoidea (Nematoda) from the Aleutian Islands. J. Nematol. 13:499-513.
2. Dalmasso, A. and J. Bergé. 1975. Variabilité génétique chez les *Meloidogyne* et plus particulièrement chez *M. hapla*. Cah. ORSTOM, sér. Biol. 10:233-238.
3. Triantaphyllou, A. C. 1966. Polyploidy and reproductive patterns in the root-knot nematode *Meloidogyne hapla*. J. Morphol. 118:403-413.
4. Triantaphyllou, A. C. 1979. Cytogenetics of root-knot nematodes, pp. 85-109 *In* Root-knot nematodes (*Meloidogyne* species)

systematics, biology and control, ed. F. Lamberti and C. E. Taylor. Academic Press, London. 477 pp.

5. Triantaphyllou, A. C. 1981. Oogenesis and the chromosomes of the parthenogenetic root-knot nematode *Meloidogyne incognita*. J. Nematol. 13:95-104.
6. Triantaphyllou, A. C. 1984. Polyploidy in meiotic parthenogenetic populations of *Meloidogyne hapla* and a mechanism of conversion to diploidy. Rev. Nematol. 7:65-72.
7. Triantaphyllou, A. C. and H. Hirschmann. 1980. Cytogenetics and morphology in relation to evolution and speciation of plant-parasitic nematodes. Annu. Rev. Phytopathol. 18:333-359.
8. White, M. J. D. 1978. Modes of speciation. Freeman, San Francisco, Calif. 455 pp.

CHAPTER 10

BIOCHEMISTRY AS A TOOL IN IDENTIFICATION AND ITS PROBABLE USEFULNESS IN UNDERSTANDING THE NATURE OF PARASITISM

Richard S. Hussey

Department of Plant Pathology,
University of Georgia
Athens, Georgia, USA

Biochemical approaches to characterizing *Meloidogyne* species are providing new and useful information for taxonomic and phylogenetic investigations of these nematodes. Data obtained by these new approaches will complement and extend the information provided by classical nematode systematics. Through the application of technology currently being developed in biochemistry, biochemical techniques may soon become a valuable and routine approach to characterizing and identifying *Meloidogyne* species.

The problems associated with taxonomy of *Meloidogyne* species include the variability of morphological characters (e.g. the posterior cuticular patterns of adult females) and the occurrence of physiological races (e.g. *M. incognita* and *M. arenaria*) differentiable only by their reproduction on certain plants (33,39). These problems have caused investigators to search for means other than morphological and anatomical characters to assist in the identification and classification of species and races of this nematode genus.

In the past decade, because of the worldwide importance of *Meloidogyne* species as plant pathogens (38) and the difficulty frequently encountered in identifying species, attention has been focused on cytogenetic and biochemical analyses as tools for identification and for investigation of systematics of these nematodes (46). These studies make it apparent that problems encountered in the separation of *Meloidogyne* species and in elucidation of their phylogenetic relationships will be best resolved by combining information obtained from morphological, cytological, and biochemical investigations, thereby characterizing these nematodes more completely than they have been previously. The application of biochemical systematics to nematodes has been reviewed by Hansen & Buecher (20), Hussey (22), and Platzer (37). This paper contains a brief review of the biochemical approaches being used in studying the taxonomy and phylogeny of species of *Meloidogyne*.

BIOCHEMICAL ANALYSES

Biochemical systematics depends on identification of the subtle molecular differences which underlie taxonomic variation (20). A primary problem is determining which chemical characters will be valuable in providing information for taxonomy. A good chemical taxonomic character (i) should remain constant within samples being studied, (ii) should be unresponsive to environmental influence, and (iii) should correlate with existing classifications constructed using other characters. (7).

The ultimate goal in taxonomy should be to classify the genotypes of organisms; however, a practical method for the analysis of the sequence of nucleotides in a gene is unavailable. DNA nucleotide differences in organisms can be estimated by DNA hybridization studies, although this has not been used to any extent with nematodes (37). Proteins, however, are a manifestation of the sequence of nucleotides in a gene, and analysis of these macromolecules provides a reliable approach for comparing the genotypes of organisms (1). Protein sequencing is not presently feasible with nematodes, and biochemical systematics of these organisms has relied on comparisons of the amino acid composition of proteins via gel electrophoresis and serology.

Axenic culture is the ideal method of propagating nematodes for biochemical analyses because the influence of a host plant on the nematode's proteins is removed and culture techniques can be standardized. However, *Meloidogyne* species are obligate parasites and cannot be cultured axenically. Due to their highly specialized host-parasite relationships, these nematodes are most easily cultured xenically on greenhouse-grown plants, although they can be cultured on excised roots growing on a sterile agar substrate (35).

An important problem to overcome once nematodes are in culture is separating the developmental stages. It has been well established that certain chemical characters in nematodes are age dependent. Dickson et al. (12) demonstrated the occurrence of stage-specific electrophoretic enzyme patterns in *M. incognita*. Four isozymes of malate dehydrogenase were detected in the second-stage juvenile and egg stages whereas only three were found in the adult female stage. Esterase profiles also differed among the three stages. Similar differences were found among the life stages of *Panagrellus silusiae* (9). These studies clearly illustrate the importance of utilizing nematodes in a single developmental stage when biochemical systematic investigations are being conducted.

Meloidogyne species are particularly suitable for biochemical analyses because living globose adult females, second-stage juveniles, or eggs can be readily separated from each other (24,25). This permits analyses to be carried out on nematodes all in a single developmental stage, thereby eliminating most age-dependent variation. In fact, pre-parasitic second-stage juveniles of *Globodera rostochiensis* provided a more reliable source of proteins for analyses than did adult females because the juveniles were free of host proteins which can contribute to variation in electrophoretic protein profiles of adult nematodes (19). This absence would be particularly important if the *Meloidogyne* species being studied could not be cultured on a common host (e.g. tomato). Nevertheless, nematodes in the same stage but at different points of development within that stage might still have slightly different chemical characteristics.

GEL ELECTROPHORESIS

Polyacrylamide-gel electrophoresis (10,36) provides a high resolution technique for separating protein molecules on the basis of size and net charge. The technique is relatively simple to use proficiently, and several books have been written that describe the procedure in detail (8,31,40). Comparative studies of nematode proteins by polyacrylamide-gel electrophoresis have provided information that has been helpful in characterizing and elucidating relationships among nematode species (22).

The first biochemical systematic study with *Meloidogyne* species was conducted by Dickson et al. (13), who compared proteins in buffer extracts of adult females by polyacrylamide-gel electrophoresis. Unique nonenzymatic protein profiles were found for females of the four common *Meloidogyne* species (*M. incognita, M. arenaria, M. hapla,* and *M. javanica*). Four populations of *M. javanica*, three populations each of *M. hapla* and *M. incognita* and one population of *M. arenaria* were compared in this study. Two regions of the protein profiles were considered to have taxonomic value. One region of the profile was considered genus specific, and another region contained interspecific differences. Furthermore, populations of *M. javanica* originating in widely separated geographic regions of the world had very similar protein profiles.

The sensitivity of polyacrylamide-gel electrophoresis is greatly enhanced when specific enzymatic proteins are localized in the gels through staining techniques. Staining recipes for identifying 38 different enzymes in gels are available (40,41). Many enzymes are known to exist in multiple molecular forms which differ in electrophoretic mobility. If they share a common substrate, they are called isozymes; if they are products of a single genetic locus, then they are called allozymes (1).

Dickson et al. (12) demonstrated that electrophoretic analysis of enzymes had potential in the identification of *Meloidogyne* species. Lactate dehydrogenase and acid and alkaline phosphatase were similar for 11 geographic isolates of *Meloidogyne* species. However, electrophoretic patterns of α-glycerophosphate dehydrogenase, malate dehydrogenase, glucose-6-phosphate dehydrogenase and non-specific esterases varied among the four species. Glucose-6-phosphate dehydrogenase was monomorphic but differed in electrophoretic mobility among *M. incognita, M. hapla,* and *M. javanica*. Activity for this enzyme was not detected in *M. arenaria*. The three other characterizing enzymes differed in either band number, band mobility or both among the four species, with the exception of malate dehydrogenase which was similar for *M. incognita* and *M. javanica*. Only esterase profiles exhibited intraspecific variation. Esterase profiles varied in one population each of *M. incognita* and *M. hapla*. Interestingly, no differences were detected in enzyme patterns from two cytological races of *M. hapla*, one having a reduced chromosome number of 15 to 17 and the other with a somatic chromosome number of 45. The results obtained from this study were used to compute a phylogenetic classification of these four *Meloidogyne* species based on the number of nonmatching enzyme bands. *M. hapla* had the greatest number of nonmatching bands and was considered the most primitive member.

Hussey et al. (26) continued the studies of proteins of *Meloidogyne* species, and their results supported the earlier work of Dickson et al. (12). The most striking differences between *M. incognita* and *M. arenaria* were in esterase, malate dehydrogenase and α-glycerophosphate dehydrogenase patterns. Although differences were present in the nonenzymatic protein profiles, these differences were not as diagnostic as the enzyme patterns. Several additional esterase bands for both species, and two more α-glycerophosphate dehydrogenase bands for *M. arenaria* were detected than previously reported (12). Very similar protein and enzyme profiles were obtained for two populations of each species even though the populations exist in widely separated geographical regions of the world and have slightly different host ranges.

Results from studies of the influence of a host plant on nematode enzyme patterns have been variable. Differences have been detected in peroxidase and α-glycerophosphate dehydrogenase patterns of extracts from *M. incognita* propagated on differnt hosts (25,27). Since patterns of several other enzymes, including esterases, were not influenced by varying the host, the host plant may influence only certain enzymes. Ishibashi (28) also reported variation in protein and enzyme patterns of *Meloidogyne* species as affected by host plants and growing conditions of the host. He reported variation in esterase profiles as did Bergé & Dalmasso (4). The culturing of *Meloidogyne* species on four hosts did not affect the enzyme patterns studied by Dickson et al. (12). Because the effects of host plants on nematode enzyme patterns have been found variable, *Meloidogyne* species should be propagated on a common host when possible for biochemical studies.

The most significant development in the biochemical systematics of *Meloidogyne* species was the adaptation of polyacrylamide gel electrophoresis on a microscale in order to separate proteins from individual adult female nematodes. Dalmasso & Bergé (10) were the first to use microelectrophoresis to separate proteins from single female specimens. Acrylamide gels are used as thin slabs (0.7 mm thick) or as thin cylinders in microhematocrit tubes (1.1 mm i.d.). These exceedingly thin gels are capable of separating proteins from one adult female specimen of *Meloidogyne* species. The micro-slab technique allows comparison of extracts of 20 individual nematodes on a single gel.

Dalmasso & Bergé (10) used the microtechniques to study enzymes and nonenzymatic proteins from several populations of the four common *Meloidogyne* species and one population each of *M. naasi* and an undescribed species. So the variability at each locus studied could be assessed, 80 individuals from wild populations and 20 specimens from single egg mass cultures were analyzed; a total of 22,000 individual specimens were examined. Although some differences were apparent in the nonenzymatic protein profiles from individual female nematodes, the greatest discrimination was obtained with the enzyme analyses, and nonspecific esterase patterns proved to be the most valuable. The esterases were polymorphic and shown to be the product of multiple genetic loci. Two types of esterase, differentiated by their activity toward α-naphthyl acetate, were identified in *Meloidogyne* females. Strongly stained esterases were designated b esterases and weakly stained ones were called β esterases. The b esterase patterns were the most reliable for identifying the common *Meloidogyne* species. Three b esterases were identified for *M. javanica* and *M. arenaria* and one for *M. hapla* and *M. incognita*.

Janati et al. (30) continued the investigation of esterases and studied esterase patterns in 75 *Meloidogyne* populations. The b esterases were species specific, and this finding confirmed the earlier report (10); however, eleven of the 75 populations had unusual β-esterase profiles which indicated that they might be undescribed species.

Micro-disc electrophoresis of proteins of *Meloidogyne* females provides a new dimension for characterizing and identifying root-knot nematode species. For diagnostic purposes it is now feasible to conduct enzyme analysis on individual females dissected from galled roots of plants and compare the esterase profiles obtained with the known esterase patterns of the four most common species of root-knot nematodes (14). Though 50 or more *Meloidogyne* species have been described, 98% of the populations worldwide that have been identified in the International *Meloidogyne* Project belonged to either *M. incognita* (54%), *M. javanica* (30%), *M. hapla* (7%) or *M. arenaria* (7%) (38). The posterior cuticular pattern, which is the principal character currently being used for the identification of these four root-knot nematode species, is variable, and its features are not always diagnostic for *M. arenaria* and *M. incognita*. Esterase analysis of adult females, when used in conjunction with morphological analysis, facilitates the identification of these four *Meloidogyne* species.

Gel electrophoresis has also contributed to the systematics and identification of other nematodes. Trudgill & Carpenter (47), during a study of nonenzymatic profiles of cyst nematodes, noted that profiles of pathotype A and pathotypes B and E of *Heterodera rostochiensis* consistently differed by three bands. The electrophoretic data supported the decision to separate the pathotypes into two species, *Globodera pallida* and *G. rostochiensis* (2). Electrophoresis also proved to be a reliable procedure for identifying *Caenorhabditis* species accurately. Four out of five strains received as putative *C. briggsae* were identified as *C. elegans* based primarily on malate dehydrogenase and nonenzymatic protein profiles obtained by gel electrophoresis (16). Since these two species are normally separated by bursal ray arrangement in males, which are uncommon in cultures, electrophoretic analysis of proteins has provided a reliable and accurate means of species identification.

The identification of physiological races of *Meloidogyne* species and determination of their distribution is frequently important in order to reduce crop losses to these nematodes. However, identification of biochemical differences among races of *Meloidogyne* species remains to be accomplished. Avise (1) has pointed out that multi-loci studies of conspecific populations of amphimictic organisms have shown that they usually have a high percentage of biochemical similarity which makes it very difficult for subspecies to be identified by this method. Nevertheless, two pathotypes of *G. rostochiensis* have been differentiated by means of two-dimensional electrophoresis of aqueous extracts (2). Pathotype PO_5 of *G. rostochiensis* had one protein that was absent in the protein pattern of pathotype PO_1. The nature of this protein is unknown. Differences among *Meloidogyne* races probably reside in the contents of the secretory granules found in the dorsal and subventral glands (5,6). The compounds present in these granules are thought to be responsible for establishing the specialized host-parasite relationships of root-knot nematodes and might be responsible for the differences in host ranges among races. Characterization of the compound(s) in the secretory granules and their role in the establishment of the highly specialized host-parasite relationship is one of the most challenging and exciting areas for future research with *Meloidogyne* species.

SEROLOGY

The protein constituents of two organisms can also be compared by serology. Because of the relative specificity of the antigen-antibody reaction, serological techniques are regarded as very reliable in determining homologies among proteins found in different species, thus providing insight into the evolutionary relationship of the species in question (17). These techniques have been employed to determine similarities and differences among nematode antigens (22).

Serology failed to demonstrate great differnces in antigens between adult females of *M. incognita* and *M. arenaria* (23). Double diffusion (Ouchterlony) tests showed that one immunoprecipitate was unique for *M. incognita*, whereas another eight, based on band position and coalescence, were common to both species. The large number of common precipitin bands that formed in the double diffusion tests indicates a close serological relationship between the two *Meloidogyne* species. Preliminary results of comparison of *M. hapla* and *M. javanica* with *M. incognita* and *M arenaria* suggest that *M. javanica* is more closely related to *M. incognita* and *M. arenaria* than is *M. hapla* (22). Fewer precipitin bands were obtained with extracts of *M. hapla* than with those of the other three species, and spurs formed on some of the precipitin bands that did develop.

In Ouchterlony tests, immunoprecipitin patterns resulting from multiple antigen systems often are difficult to interpret. Differences in precipitin bands can be obscured due to the complex nature of a precipitin pattern or to diffuse precipitin bands. Gel diffusion tests provide the most reliable information when single kinds of molecules are compared to show their structural similarity (17). More striking serological differences probably would be obtained with purified nematode proteins, e.g. cuticle or muscle proteins. Multiple antigen systems are perhaps best analyzed by immunoelectrophoresis where only the presence or absence of antigens among species is noted. A larger number of precipitin bands developed when antigens of *M. incognita* and *M. arenaria* were compared by immunodisc electrophoresis rather than by Ouchterlony test (27).

Absorption of antisera for removal of antibodies that react with similar antigens from two *Meloidogyne* species produced a species-specific antiserum (23). When *M. incognita* antiserum that had been previously absorbed with *M. arenaria* antigens was reacted with homologous and heterologous antigens, a precipitin band formed only between the homologous antigens and antibodies.

Misaghi & McClure (32) examined the serological relationships of three *Meloidogyne* species using antigens from second-stage juveniles and eggs. Although a number of precipitin bands were common to all species, juveniles and eggs of *M. incognita* possessed one specific antigen not present in *M. javanica* or *M. arenaria*. Two precipitin bands were also unique to *M. javanica*. The species-specific antigens were confirmed by cross-absorption tests. Baron (3) also reported a strong serological relationship for second-stage juveniles *M. javanica*, *M. incognita* and *M. arenaria* from West Africa.

The usefulness of serology in distinguishing populations or races of *Meloidogyne* species is still uncertain. In tests comparing populations of *M. incognita* from Taiwan and Peru, nine distinct precipitin bands formed; however, all precipitin bands coalesced, indicating that these two populations were serologically similar (23). Serology has been shown to be a reliable method for determining similarities in nematode antigens; hence, these techniques have great potential for elucidating the phylogenetic relationships of root-knot nematodes, and in fact, may facilitate the identification of species.

OTHER BIOCHEMICAL ANALYSES

Compounds other than nucleic acids and proteins may have taxonomic value for classification of *Meloidogyne* species although little attention has been given to investigating other chemical characters for this purpose. In a study comparing the lipid and fatty acid composition of adult females and eggs of *M. incognita* and *M. arenaria*, no qualitative differences were detected in the fatty acids, although relative quantities of certain fatty acids did differ between the two species (29).

PROSPECTS

Biochemical approaches to the taxonomy of *Meloidogyne*, although still being evaluated, have considerable potential for assisting in the identification and classification of members of this important genus of phytoparasitic nematodes. The most promising approach seems to be the comparison of enzymes of adult females with micro-disc electrophoresis. Differences found in enzyme profiles, most notably with esterases, already have been shown to be useful in separating *Meloidogyne* species.

Nonchemical control strategies (crop rotation and resistant cultivars) for *Meloidogyne* species are becoming more important in agriculture. Knowledge of species and even race distribution is necessary for management of these nematodes. Control programs using either resistant cultivars or crop rotation are usually directed at specific *Meloidogyne* species. Since different *Meloidogyne* species are important pathogens on different agronomic crops (45), the species present in a field must be accurately and often rapidly identified. Therefore, the development of a rapid and reliable biochemical procedure for identifying the species present in a field would enhance the ability to control these nematodes. The same speed and reliability in the identification of races would be helpful, although as noted above, attempts to separate them biochemically have not been successful. Two-dimensional electrophoresis (34) can separate more than one-thousand proteins and may allow detection of protein differences among races of *Meloidogyne* species.

Serology has been shown to have potential for elucidating phylogenetic relationships among *Meloidogyne* species and for adaptaion for use in species identification. Production of species-specific antisera could provide a means of rapid identification of *Meloidogyne* species. The new technology of monoclonal antibody production (15) should facilitate the production of species-specific antisera, and enzyme-linked immunosorbent assays provide a highly sensitive method for detection of *Meloidogyne* antigens.

Literature Cited

1. Avise, J. C. 1974. Systematic value of electrophoretic data. Syst. Zool. 23: 465-481.
2. Bakker, J. and F. J. Gommers. 1982. Differentiation of the potato cyst nematodes *Globodera rostochiensis* and *G. pallida* and of two *G. rostochiensis* pathotypes by

means of two-dimensional electrophoresis. Proc. K. Ned. Akad. Wet. Ser. C, Biol. Med. Sci. 85: 309-314.

3. Baron, J. C. 1975. Méthode d'obtention et étude immunologique des protéines solubles de juvéniles d'une population naturelle de *Meloidogyne*. Cah. ORSTOM, ser. Biol. 10: 273-282.
4. Bergé, J. B. and A. Dalmasso. 1975. Caracteristique biochimiques de quelques populations de *Meloidogyne hapla* et *Meloidogyne* spp. Cah. ORSTOM, ser. Biol. 10: 263-271.
5. Bird, A. F. 1969. Changes associated with parasitism in nematode. V. Ultrastructure of the stylet exudation and dorsal esophageal gland contents of female *Meloidogyne javanica*. J. Parasitol. 55: 337-345.
6. Bird, A. F. and W. Saureer. 1967. Changes associated with parasitism in nematodes. II. Histochemical and microspectrophotometric analyses of preparasitic and parasitic larvae of *Meloidogyne javanica*. J. Parasitol. 53:1262-1269.
7. Boulter, D. and D. A. Thurman. 1968. Acrylamide gel electrophoresis of proteins in plant systematics, pp. 39-48 *In* Chemotaxonomy and serotaxonomy, ed. J. G. Hawkes. Academic Press, New York. 299 pp.
8. Brewer, G. J. 1970. An introduction to isozyme techniques. Academic Press, New York. 186 pp.
9. Chow, H. H. and J. Pasternak. 1969. Protein changes during maturation of the free-living nematode, *Panagrellus silusiae*. J. Exp. Zool. 170: 77-84.
10. Dalmasso, A. and J. B. Bergé. 1978. Molecular polymorphism and phylogenetic relationship in some *Meloidogyne* spp.: Application to the taxonomy of *Meloidogyne*. J. Nematol. 10: 323-332.
11. Davis, B. J. 1964. Disc electrophoresis. II. Methods and application to human serum proteins. Ann. N. Y. Acad. Sci. 121: 404-427.
12. Dickson, D. W., D. Huisingh, and J. N. Sasser. 1971. Dehydrogenases, acid and alkaline phosphatases and esterases for chemotaxonomy of selected *Meloidogyne*, *Ditylenchus*, *Heterodera*, and *Aphelenchus* spp. J. Nematol. 3: 1-16.
13. Dickson, D. W., J. N. Sasser, and D. Huisingh. 1970. Comparative disc-electrophoretic protein analyses of selected *Meloidogyne*, *Ditylenchus*, *Heterodera*, and *Aphelenchus* spp. J. Nematol. 2:. 286-293.
14. Eisenback, J. D., H. Hirschmann, J. N. Sasser and A. C. Triantaphyllou. 1981. A guide to the four most common species of root-knot nematodes (*Meloidogyne* species) with a pictorial key. A Coop. Publ. Deps. Plant Pathol. and Genet., North Carolina State Univ., and U.S. Agency Int. Dev., Raleigh, N.C. 48 pp.
15. Eisenbarth, G. S. 1981. Application of non-oclonal antibody techniques to biochemical research. Anal. Biochem. 111: 1-16.
16. Friedman, P. A., E. G. Platzer, and J. E. Eby. 1977. Species differentiation in *Caenorhabditis briggsae* and *Caenorhabditis elegans*. J. Nematol. 9: 197-203.
17. Gell, P. G. H. 1968. Serotaxonomy of vertebrate soluble proteins, pp. 67-76 *In* Chemotaxonomy and serotaxonomy, ed. J. G. Hawkes. Academic Press, New York. 299 pp.
18. Gottlieb, L. D. 1971. Gel electrophoresis: new approach to the study of evolution. Bioscience 21: 939-944.
19. Greet, D. N. and J. Firth. 1977. Influence of host plant on electrophoretic protein patterns of some round-cyst nematode females and use of larvae to obtain less ambiguous results. Nematologica 23: 411-415.
20. Hansen, E. L. and E. J. Buecher. 1970. Biochemical approach to systematic studies with axenic nematodes. J. Nematol. 2: 1-6.
21. Hussey, R. S. 1982. Molecular approaches to taxonomy of Heteroderoidea, pp. 50-53 *In* Nematology in the southern region of the United States, ed. R. D. Riggs. South Coop. Ser. Bull. 276, Arkansas Agric. Exp. Stn., Fayetteville, Ark. 206 pp.
22. Hussey, R. S. 1979. Biochemical systematics of nematodes—a review. Helminthol. Abstr. (B) 48: 141-148.
23. Hussey, R. S. 1972. Serological relationship of *Meloidogyne incognita* and *M. arenaria*. J. Nematol. 4: 101-104.
24. Hussey, R. S. 1971. A technique for obtaining quantities of living *Meloidogyne* females. J. Nematol. 3: 99-100.
25. Hussey, R. S. and J. N. Sasser. 1973. Peroxidase from *Meloidogyne incognita*. Physiol. Plant Pathol. 3: 223-229.
26. Hussey, R. S. and K. R. Barker. 1973. A comparison of methods of collecting inocula of *Meloidogyne* spp., including a new technique. Plant Dis. Rep. 57: 1025-1028.
27. Hussey, R. S., J. N. Sasser, and D. Huisingh. 1972. Disc-electrophoretic studies of soluble proteins and enzymes of *Meloidogyne incognita* and *M. arenaria*. J. Nematol. 4: 183-189.
28. Ishibashi, N. 1970. Variations of the elec-

trophoretic protein patterns of Heteroderidae (Nematodea: Tylenchida) depending on the developmental stages of the nematode and on the growing conditions of the host plant. Appl. Entomol. Zool. 5: 23-32.

29. Krusberg, L. R., R. S. Hussey, and C. L. Fletcher. 1973. Lipid and fatty acid composition of females and eggs of *Meloidogyne incognita* and *M. arenaria*. Comp. Biochem. Physiol. 45B: 335-341.
30. Janati, A., J. B. Berge, A. C. Triantaphyllou, and A. Dalmasso. 1981. New data on isoesterase polymorphism in *Meloidogyne* spp. Application to the taxonomy of some species of this genus. Rev. Nematol. 4: 147-154.
31. Maurer, H. R. 1971. Disc electrophoresis and related techniques of polyacrylamide gel electrophoresis. Walter de Gruyter, New York. 222 pp.
32. Misaghi, I. and M. A. McClure. 1974. Antigenic relationship of *Meloidogyne incognita, M. javanica* and *M. arenaria*. Phytopathology 64: 698-701.
33. Netscher, C. and D. P. Taylor. 1979. Physiological variation with the genus *Meloidogyne* and its implications on integrated control, pp. 268-294 *In* Root-knot nematodes (*Meloidogyne* species) systematics, biology, and control, ed. F. Lamberti and C. E. Taylor. Academic Press, Inc., New York. 477 pp.
34. O'Farrell, P. H. 1975. High resolution two-dimensional electrophoresis of proteins. J. Biol. Chem. 250: 4007-4021.
35. Orion, D., W. P. Wergin., and B. Y. Endo. 1980. Inhibition of syncytia formation and root-knot nematode development to cultures of excised tomato roots. J. Nematol. 12: 196-203.
36. Ornstein, L. 1964. Disc electrophoresis. I. Background and theory. Ann. N. Y. Acad. Sci. 121: 321-349.
37. Platzer, E. G. 1981. Potential use of protein patterns and DNA nucleotide sequences in nematode taxonomy, pp. 3-21 *In* Plant parasitic nematodes, Vol. III, ed. B. M. Zuckerman and R. A. Rohde. Academic Press, Inc., New York. 508 pp.
38. Sasser, J. N. 1980. Root-knot nematodes: a global menace to crop production. Plant Dis. Rep. 64: 36-41.
39. Sasser, J. N. 1979. Pathogenicity, host ranges and variability in *Meloidogyne* species, pp. 257-268 *In* Root-knot nematodes (*Meloidogyne* species) systematics, biology, and control, ed. F. Lamberti and C. E. Taylor. Academic Press, Inc., New York. 477 pp.
40. Shaw, C. R., and A. L. Doen. 1968. Starch gel zone electrophoresis of enzymes, pp. 325-364 *In* Chromatographic and electrophoretic techniques. Volume II. Zone electrophoresis, ed. I. Smith. John Wiley and Sons, Inc., New York. 524 pp.
41. Shaw, C. R. and R. Prasad. 1970. Starch gel electrophoresis of enzymes—a compilaton of recipes. Biochem. Genet. 4: 297-320.
42. Starr, J. L. 1979. Peroxidase isozymes from *Meloidogyne* spp. and their origin. J. Nematol. 11: 1-5.
43. Stone, A. R. 1973. *Heterodera pallida* n. sp. (Nematoda: Heteroderidae), a second species of potato cyst nematode. Nematologica 18: 591-606.
44. Stone,, A. R. and R. D. Williams. 1974. The morphology and soluble protein electrophoresis of *Heterodera avenae* pathotypes. Ann. Appl. Biol. 76: 231-236.
45. Taylor, A. L. and J. N. Sasser. 1978. Biology, identification and control of root-knot nematodes (*Meloidogyne* species). A Coop. Publ. Dep. Plant Pathol., N. C. State Univ., and U.S. Agency Int. Dev., Raleigh, N. C. 111 pp.
46. Triantaphyllou, A. C. 1979. Cytogenetics of root-knot nematodes, pp. 84-109 *In* Root-knot nematodes (*Meloidogyne* species) systematics, biology, and control, ed. F. Lamberti and C. E. Taylor. Academic Press, Inc., New York. 477 pp.
47. Triantaphyllou, A. C. and R. S. Hussey. 1973. Modern approaches in the study of relationships in the genus *Meloidogyne*. Eur. Mediterr. Plant Prot. Organ. (OEPP/EPPO) Bull. 9: 61-66.
48. Trudgill, D. L. and J. M. Carpenter. 1971. Disc electrophoresis of proteins of *Heterodera* species and pathotypes of *Heterodera rostochiensis*. Ann. Appl. Biol. 69: 35-41.

CHAPTER 11

IDENTIFICATION OF MAJOR *MELOIDOGYNE* SPECIES EMPLOYING ENZYME PHENOTYPES AS DIFFERENTIATING CHARACTERS

P. R. Esbenshade and A. C. Triantaphyllou
Departments of Plant Pathology and Genetics, respectively
North Carolina State University
Raleigh, North Carolina, USA

The use of enzyme phenotypes for species identification and for the elucidation of the taxonomic and phyletic relationships of organisms has become quite popular during the last 20 years (16, 17, 21). Enzyme studies of root-knot nematodes were initiated in 1971 (8). This early work indicated clearly that phenotypes of some enzymes are characteristic of certain species or certain populations within a species. Since then, about 30 enzymes have been detected in various *Meloidogyne* populations (Table 1), but only a few of them have been investigated extensively and assessed for their usefulness as taxonomic characters (4, 10, 14). Dalmasso and Bergé (4) conducted an extensive survey involving 83 populations of 5 *Meloidogyne* species. These investigators used a miniaturized extraction and electrophoretic procedure that enabled them to determine the enzyme phenotype of individual females of root-knot nematodes, thus increasing their chances of detecting enzymatic variation within populations. In more recent work, the thin-slab electrophoretic procedure provided even more accurate evaluations of enzyme phenotypes of these nematodes (14). Comparisons of 20 or more nematode-protein extracts in the same slab gel, under exactly the same electrophoretic conditions, can reveal even slight isozymic variations that may exist within and between populations of a species or between species (10).

It is believed that, with standardization of the electrophoretic procedures (protein extraction, electrophoresis, and staining), enzyme phenotypes can provide a number of advantageous and objective characters for the identification of certain species or biological forms of root-knot nematodes and for taxonomic characterization of new forms.

Because the early data about certain enzymes of root-knot nematodes proved to be so promising (1, 4, 8, 12, 14), an extensive investigation was undertaken in collaboration with the International *Meloidogyne* Project in an effort to determine the practicality and reliability of the use of enzyme phenotypes in the identification of root-knot nematodes on a worldwide basis (10). About 300 populations were selected from a collection of more than 600 which had been studied cytologically (see Chapter 9) and with regard to host range and perineal patterns of adult females. Tentative identifications to species had been made on the basis of such information and, for some populations, on supplemental studies of male stylets and other morphological features of females, males and juveniles. The 300 populations selected for the biochemical study had originated from 65 countries and 20 states in the U.S.A., and represented 16 *Meloidogyne* species including the four major species (*M. arenaria*, *M. hapla*, *M. incognita*, and *M. javanica*) and most of the known cytogenetic and physiological (host range) races of these species. Studies have been completed on four enzymes (non-specific esterases, malate dehydrogenase, superoxide dismutase, and glutamate-oxaloacetate transaminase) that gave promising results in the early stages of this work. Of those, esterases proved to be the most useful enzyme for the differentiation of the major *Meloidogyne* species. This conclusion confirmed previous findings by other investigators who had studied smaller numbers of populations, not representing all of the geographic, cytogenetic, and physiological variation known to exist within each species (1, 4, 8, 12, 14). Superoxide dismutase was also useful in the characterization of certain major

Table 1. Proteins and enzymes detected in *Meloidogyne* species.

Protein	Citation
Total protein	1, 4, 7, 12, 13, 15
Acid phosphatase	1, 5, 8, 12, 13
N-Acetylglucosaminidase	1
Alkaline phosphatase	1, 8
Aminopeptidases	1
Amylase	22
Catalase	1, 4, 5, 12
Cellulase	3, 9, 18
Esterase	1, 2, 4, 5, 8, 10, 11, 12, 13, 14
α-Fucosidase	1
Galactosidase	1
Glucose-6-phosphate dehydrogenase	4, 8, 12
Glucosidase	1
β-Glucuronidase	1
Glutamate-oxaloacetate transaminase	10
α-Glycerophosphate dehydrogenase	4, 8, 12
β-Glycosidases	20
Invertase	18
Lactate dehydrogenase	8, 12, 13
Lipase	1
Malate dehydrogenase	4, 5, 8, 10, 12
α-Mannosidase	1
Pectinase	18
Peroxidase	1, 12, 19
Phosphoaminidase	1
6-Phosphogluconate dehydrogenase	4, 12
Protease	1, 6
Superoxide dismutase	4, 5, 10

species, but most of its phenotypes were observed in more than one species. Phenotypes of the other two enzymes (Mdh, Got), with only a few exceptions, were less useful for identification of *Meloidogyne* species.

It is believed that enzyme phenotypes, especially esterase phenotypes, can be used for routine identification of major *Meloidogyne* species, provided that a standardized procedure (see Vol. II, Chapter 8) is followed in every laboratory conducting such identifications. Furthermore, it is essential that (a) only bands of enzymatic activity that can be easily recognized are utilized as identification characters, (b) slab gels are used for accurate comparison of phenotypes that are not strikingly different, and (c) the phenotype of an unknown population is compared in the same gel to the phenotype of a pure population of a known species. It is not advisable to rely on the determination of the phenotype of an unknown population merely by consideration of the rate of electrophoretic mobility of its bands.

In the following sections, the most species-specific enzyme phenotypes detected in this study will be summarized and their use as differentiating characters for the identification of major *Meloidogyne* species will be discussed.

ESTERASES

Three of the four major species of *Meloidogyne* have been identified with a high degree of accuracy on the basis of their characteristic esterase (Est) phenotypes. Thus, all 46 populations of *M. javanica* had an esterase phenotype designated as J3, i.e. *javanica* with 3 bands (Table 2, Fig. 1). The same phenotype did not appear in any of the other 245 populations of *Meloidogyne* included in this study. The J3 esterase phenotype is unquestionably highly

Table 2. Enzyme phenotypes of the 4 common *Meloidogyne* species.

Nematode species and populations	Number of populations	Enzyme phenotypes[a] Est	Mdh	Sod	Got	Chromosome number
M. arenaria	7	A1	N1	JA2	N1	36-52
	19	A2	N1	JA2	N1	37-54
	1	A2	N1	A4	N1	~54
	12	A2	N3	A4	N1	~54
	1	A2	N3	A4	H2	~54
	1	A2	N3b	A4	N1	~49
	4	A2	N3	JA2	N3	42-54
	1	A2	N1	JA2	N3	~43
	7	A3	N1	A4	N1	52-54
	1	A3	N1	JA2	N1	52-54
M. arenaria (atypical populations)	2	S1-M1	N1	I2	N1	37-43
	4	S1-M1	N3	I2	N1	36-42
	3	S2-M1	N1	I2	N1	36-42
	1	M3-F1	N1	JA2	N1	~42
M. hapla	32	H1	H1	H1	H2	13-17 (n) 30-48
	1	null	H1	H1	H2	17 (n)
	1	A1	H1	H1	H2	14 (n)
M. incognita	104	I1	N1	I2	N1	32-46
	1	I1	N3b	I2	N1	~42
	1	S1	N1	I2	N1	42-45
M. javanica	44	J3	N1	JA2	N1	41-48
	2	J3	N3	JA2	N1	42-46

[a] Phenotypes considered to be specific for one or more of the major species are designated with a letter or letters indicative of the species they specify (A-arenaria, H-hapla, I-incognita, J-javanica), or indicative of the relative rate of their migration (S-slow, M-medium, F-fast). Phenotypes of Mdh, Sod and Got not characteristic of any particular species are designated with N, for nonspecific. The number following the letter designation indicates the number of major bands of that phenotype.

species-specific and can be used as a very reliable and least subjective character for identifications of *M. javanica*. Similarly, 105 of the 106 populations of *M. incognita* studied exhibited a distinct phenotype designated as I1. Again, the I1 phenotype did not appear in any populations of the other species or unidentified forms, and this feature renders this phenotype a valuable character for definitive identification of *M. incognita*. Of the 34 populations of *M. hapla* examined, 32 had a unique esterase phenotype designated as H1 (Table 2, Fig. 1). Only one population of *M. hapla* would have been misidentified as *M. arenaria* because it exhibited a phenotype (A1) which is suggestive of *M. arenaria*. Another exceptional population of *M. hapla* showed no bands of esterase activity and could not have been identified to species on this basis.

The fourth major species, *M. arenaria*, was not associated with any particular esterase phenotype. However, most of the populations (84%) of *M. arenaria* exhibited one of three phenotypes designated as A1, A2 and A3 (Fig. 1). The A3 phenotype appeared only in triploid populations which are considered to be the most typical cytological form of *M. arenaria*. The A3 phenotype was not observed in any other species of *Meloidogyne*, and therefore, it should be regarded as a highly specific character for *M. arenaria*. The A2 phenotype was observed in 39 populations representing all three cytological forms of *M. arenaria* (diploid, hypotriploid, and triploid), and in one population which could not be identified to species. Consequently, this phenotype is also very reliable for identifications of *M. arenaria*. The A1 phenotype occurred in 7 populations of *M. arenaria*, three populations of *M. microcephala* and one of *M. hapla*. It is less precise than the other phenotypes for identification of *M. arenaria* (64% accuracy). Still, it is a valuable taxonomic character

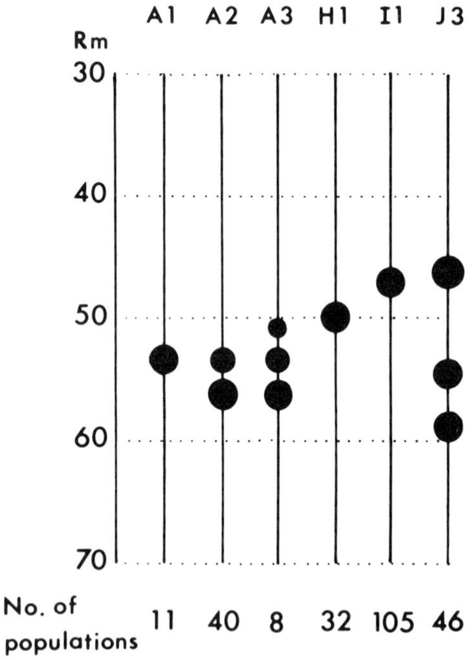

Fig. 1. The six most prevalent esterase phenotypes observed in 238 populations of the four major species of *Meloidogyne*. Phenotypes are designated with a letter suggestive of the species they specify (A=arenaria, H=hapla, I=incognita, J=javanica) and a number indicative of the number of major bands of activity.

because it did not appear in any of the 280 populations of other species studied.

The remaining 16% of the populations of *M. arenaria*, designated in Table 2 as atypical, exhibited three additional, rare esterase phenotypes (S1-M1, S2-M1, M3-F1) that were observed also in some minor species and unidentified populations and, therefore, cannot be considered as phenotypes characteristic of *M. arenaria*. The detection of these rare phenotypes among populations of *M. arenaria* and also the low specificity of the A1 phenotype present difficulties in the identification of about 26% of the populations of *M. arenaria*, if identification is sought solely on the basis of esterase phenotypes. However, 74% of the populations of *M. arenaria* have the A2 or A3 esterase phenotypes and can be identified precisely on this basis.

In addition to the prevalent esterase phenotypes discussed above, another 13 phenotypes were observed (not included in Fig. 1) in about 14% of the populations studied. These phenotypes occurred in less prevalent *Meloidogyne* species and in some populations that could not be identified to species. Some of these phenotypes may prove to be useful for identification of certain minor species (10).

SUPEROXIDE DISMUTASE

The superoxide dismutase (Sod) phenotypes are the next best characters for identification of some *Meloidogyne* species, after esterases. Six phenotypes of Sod were identified (Table 2). Four of them were encountered frequently (Fig. 2), whereas the other two (not illustrated) were observed only rarely.

Phenotype H1 was observed in all 34 populations of *M. hapla* studied, but it was also observed in 4 populations of *M. chitwoodi,* and one population each of *M. carolinensis M. graminicola, M. microtyla, M. naasi, M. oryzae, M. querciana,* and an unidentified population from rice. Thus, the H1 phenotype of Sod can easily differentiate *M. hapla* from the other three major species of *Meloidogyne* but not from many minor species.

Similarly the I2 phenotype (Fig. 2) was observed in all 106 populations of *M. incognita* studied, but it was also encountered in a few atypical populations of *M. arenaria* (Table 2),in some populations of the minor species *M. enterolobii, M. hispanica* (to be described by H. Hirschmann), *M. platani*, and in several populations not identified to species. Still, 81% of the populations with this phenotype were identified as *M. incognita*.

Phenotype A4 was observed only in populations of *M. arenaria*. However, only 22 of the 64 populations of *M. arenaria* exhibited this phenotype. Most of the remaining *M. arenaria* populations (52%) as well as all populations of *M. javanica* gave a phenotype designated as JA2. This phenotype is similar to the A4 phenotype but has only two major bands, whereas the other two bands are much weaker. About 55% of the populations with this phenotype were *M. javanica*, and 40% were *M. arenaria*.

OTHER ENZYMES

Malate dehydrogenase (Mdh) and glutamate-oxaloacetate transaminase (Got) phenotypes were useful for differentiating *M. hapla* from the other major species and many of the less common species (Fig. 2). Thus phenotype H1 of Mdh was observed only in populations of *M. hapla* and none of the other major species. The H2 of Got was also useful in differentiating *M. hapla* from the other major species, with the exception of a single population of *M. arenaria* which also gave the H2 Got phenotype (Table 2).

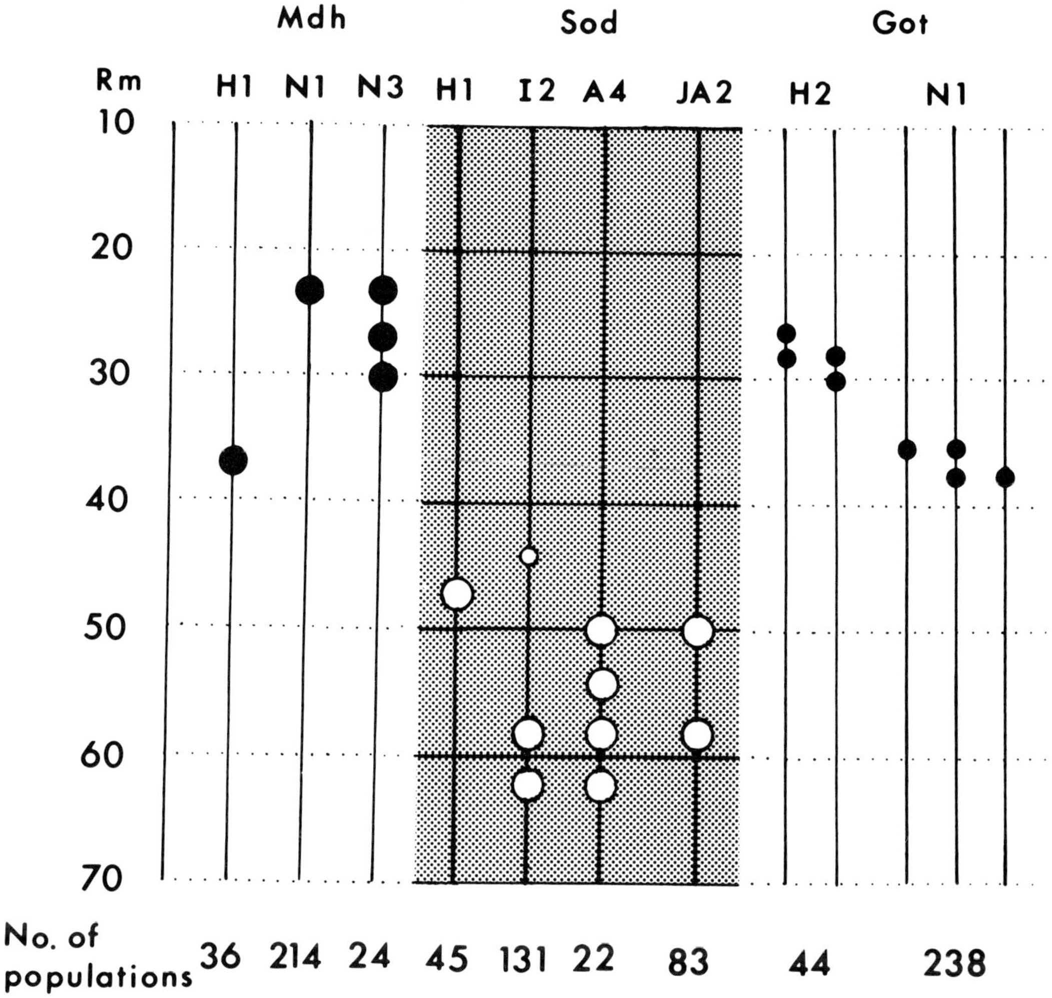

Fig. 2. Prevalent phenotypes of malate dehydrogenase (Mdh), superoxide dismutase (Sod), and glutamate-oxaloacetate transaminase (Got) observed in 245 populations of the four major species of *Meloidogyne*. Phenotypes are designated with a letter indicative of the species that they specify (A=arenaria, H=hapla, I=incognita, J=javanica, N=nonspecific) and a number indicative of the number of major bands of activity.

Most of the populations of *M. arenaria*, *M. incognita*, and *M. javanica* had similar Mdh and Got phenotypes and could not be differentiated from each other on this basis. They exhibited the phenotypes designated with the letter N, for non-species-specific phenotype, followed by a number indicating the number of major bands present (Table 2, Fig. 2).

Many of the 27 enzymes studied by other investigators (Table 1) may prove to be useful in the identification of *Meloidogyne* species. However, extensive studies that will determine the existing enzymatic variation within and between species will be required before the value of these enzymes as taxonomic characters is properly assessed.

Literature Cited

1. Bergé, J. B. and A. Dalmasso. 1975. Characteristiques biochimiques de quelques population de *Meloidogyne hapla* et *Meloidogyne* spp. Cah. ORSTOM ser. Biol. 10:263-271.
2. Bird, A. F. 1966. Esterases in the genus *Meloidogyne*. Nematologica 12:359-361.
3. Bird, A. F., W. J. S. Downton, and J. S. Hawker. 1975. Cellulase secretion by second stage larvae of the root-knot nematode

(*Meloidogyne javanica*). Marcellia 38:165-169.

4. Dalmasso, A. and J. B. Bergé. 1978. Molecular polymorphism and phylogenetic relationship in some *Meloidogyne* spp.: Application to the taxonomy of *Meloidogyne*. J. Nematol. 10:323-332.
5. Dalmasso, A. and J. B. Bergé. 1983. Enzyme polymorphism and the concept of parthenogenetic species, exemplified by *Meloidogyne*, pp. 187-196 *In* Concepts in nematode systematics, ed. A. R. Stone, H. M. Platt, and L. F. Khalil. Academic Press, New York. 388 pp.
6. Dasgupta, D. R. and A. K. Ganguly 1975. Isolation, purification and characterization of a trypsin-like protease from the root-knot nematode *Meloidogyne incognita*. Nematologica 21:370-384.
7. Dickson, D. W., J. N. Sasser, and D. Huisingh. 1970.Comparative disc-electrophoretic protein analyses of selected *Meloidogyne*, *Ditylenchus*, *Heterodera*, and *Aphelenchus* spp. J. Nematol. 2:286-293.
8. Dickson, D. W., D. Huisingh, and J. N. Sasser. 1971. Dehydrogenases, acid and alkaline phosphatases and esterases for chemotaxonomy of selected *Meloidogyne*, *Ditylenchus*, *Heterodera*, and *Aphelenchus* spp. J. Nematol. 1:1-16.
9. Dropkin, V. H. 1963. Cellulase in phytoparasitic nematodes. Nematologica 9:444-454.
10. Esbenshade, P. R. and A. C. Triantaphyllou. 1985. Use of enzyme phenotypes for identification of *Meloidogyne* species. J. Nematol. 17:6-20.
11. Farooqi, T. N. A. and D. R. Dasgupta. 1978. Esterase activity from the root-knot nematode *Meloidogyne incognita*. Indian J. Entomol. 40:197-203.
12. Hussey, R. S., J. N. Sasser, and D. Huisingh. 1972. Disc electrophoretic studies of soluble proteins and enzymes of *Meloidogyne incognita* and *M. arenaria*. J. Nematol. 4:183-189.
13. Ishibashi, N. 1970. Variations of the electrophoretic protein patterns of Heteroderidae (Nematodea: Tylenchida) depending on the developmental stages of the nematode and on the growing conditions of the host plant. Appl. Entomol. Zool. 5:23-32.
14. Janati, A., J. B. Bergé, A. C. Triantaphyllou, and A. Dalmasso. 1982. Nouvelles donnees sur l'utilisation des isoesterases pour l'identification des *Meloidogyne*. Rev. Nematol. 5:147-154.
15. Lawson, E. C. III, G. E. Carter Jr., and S. A. Lewis. 1984. Application of isoelectric focusing to the taxonomic identification of *Meloidogyne* spp. J. Nematol. 16:91-96.
16. Markert, C. L. 1975. Isozymes: Vol. I, Molecular structure. Academic Press, New York. 856 pp.
17. Markert, C. L. 1975. Isozymes: Vol. IV, Genetics and evolution. Academic Press, New York. 965 pp.
18. Meyers, R. F. 1965. Amylase, cellulase, invertase and pectinase in several free-living, mycophagus, and plant parasitic nematodes. Nematologica 11:441-448.
19. Starr, J. L. 1979. Peroxidase isozymes from *Meloidogyne* spp. and their origin. J. Nematol. 11:1-5.
20. Starr, J. L. 1981. Beta-glycosidases from *Meloidogyne incognita* and *M. javanica*. J. Nematol. 13:413-414.
21. Taylor, A. E. R. and R. Muller. 1979. Problems in the identification of parasities and their vectors. Blackwell Scientific Publications, Oxford. 221 pp.
22. Zinoviev, V. G. 1957. Enzymatic activity of nematodes parasitizing plants. Zool. Zh. 36:617-620.

Part Three

Host-Parasite Relationships

CHAPTER PAGE

HOST-PARASITE RELATIONSHIPS AND ASSOCIATED PHYSIOLOGICAL CHANGES

Richard S. Hussey
Department of Plant Pathology,
University of Georgia,
Athens, Georgia, USA

Sedentary endoparasitic nematodes have evolved very specialized and complex relationships with their hosts. Host-parasite relationships involving *Meloidogyne* species typify these specialized associations. The common *Meloidogyne* species (72) differ from most other sedentary endoparasites by having extensive host ranges; parasitizing monocotyledons, dicotyledons and herbaceous and woody plants; and causing the formation of the familiar intercalary knot or gall on roots of susceptible hosts. *Meloidogyne* species parasitize more than 2,000 different plant species, but relatively few host-parasite relationships have been investigated. Nevertheless, a voluminous amount of information is available on certain host-parasite interactions, and different aspects of these relationships have been reviewed often in various articles (12,24,25,29,41,47,82,83).

Although recent literature was emphasized in the preparation of this chapter, many interesting and detailed articles were written prior to 1949 on root-knot nematodes when they were called *Heterodera marioni* (48) and before that *H. radicicola* (35). These early articles provided a solid foundation for research on host-parasite relationships involving this important genus. Even though numerous articles have been published on this subject during the past three decades, little is known about the biochemical basis of this association. Identifying substances of nematode origin that are responsible for initiating the feeding site in host tissue remains one of the most intriguing and challenging areas of future research on *Meloidogyne*.

ATTRACTION AND PENETRATION

Plants can affect the life cycle of *Meloidogyne* prior to egg hatching. Eggs are enclosed in gelatinous egg sacs that are usually deposited on the surface of galled roots, although on certain hosts they may also occur within galls (35). Following embryogenesis, the first molt occurs within the egg giving rise to a second-stage juvenile, which is the first stage that can be affected by plants. Hatching of *Meloidogyne* eggs is usually spontaneous and readily occurs without any stimulus from plant roots; however, in some instances, root diffusates do stimulate egg hatch (79).

Once second-stage juveniles leave the egg masses, they infect nearby galled roots or locate new roots of susceptible plants to continue their life cycles. Second-stage juveniles are the only juveniles of *Meloidogyne* that migrate in the soil free from host tissue and hence constitute the infective stage. Although juveniles may survive in the soil in a quiescent state for an extended period of time, their food reserves are limited and their infectivity decreases with the depletion of these reserves (73). As the greatest loss of body reserves occurs during migration in the soil (73), the ability of second-stage juveniles to perceive stimuli and be attracted to plant roots enables them to conserve food reserves and enhance their pathogenic potential. Research shows that *Meloidogyne* second-stage juveniles do not find roots by random movement, but are attracted to plants in response to stimuli emanating from roots (33,66). These juveniles possess well-developed, innervated cephalic sensory organs that are thought to be chemoreceptive and presumably serve to locate roots (21). The cephalic sensory organs comprise six inner labial sensilla and two amphids. The amphids, because of their large size and structural complexity, are considered to be the principal sensory organs for second-stage juveniles (86). The morphology of the cephalic sensory organs has been thoroughly studied, but the exact role in location of roots is unknown. Chemical agents that specifically affect the different cephalic sensory organs also have not yet been identified.

The attraction of second-stage juveniles to plant roots can be separated into attraction either over long distances (>10 cm) or over short distances. Plant-parasitic nematodes were considered capable of migrating only a few centimeters per year (66). However, recent studies with second-stage juveniles of *Meloidogyne* indicated they could be attracted to and migrate to host roots 25 cm vertically in 10 days (65).

Linford (49) was the first to study the attractiveness of roots to *Meloidogyne* (*Heterodera marioni*) over short distances and found that second-stage juveniles accumulated at the region of cell elongation just behind the root cap of seven plants species, some of which were resistant to root-knot. One plant species failed to attract any second-stage juveniles. When *Pratylenchus pratensis* and *Rotylenchus multicinctus* were mixed with root-knot nematodes, the second-stage juveniles of the root-knot nematode accumulated near the root tip, and the other two species avoided this region, which suggested that different attractants are perceived by different nematode species.

Second-stage juveniles are also attracted to apical meristems, points where lateral roots emerge, penetration sites of other juveniles, and cut surfaces of roots (5,49). Weiser (84) attempted to locate the exudation site of attractant(s) and found the apical 2 mm of an excised tomato root actually repelled *M. hapla* juveniles, and from 3 mm to 8 mm from the tip, the root tissue increased in attractiveness. Beyond 8 mm from the tip the tissue was neutral or somewhat repellent to juveniles. This, however, is the only report of the distal portion of a root being repellent to *Meloidogyne* juveniles. Actively elongating roots were more attractive to juveniles than roots in which extension had slowed. Weiser (85) also studied the attractiveness of other hosts of *M. hapla* and found excised roots of bean somewhat repellent, and those of eggplant and soybean varied in their attractiveness to juveniles. He suggested that an interplay between attractant and repellent factors may occur in the same plant. Viglierchio (76) arrived at similar conclusions in studies on rye and *M. hapla*.

Lownsbery and Viglierchio (52) reported that root attractants were nonspecific and were dialyzable. Juveniles of *M. hapla* aggregated around germinating seeds of tomato, rye, and oats; the latter two plants are both nonhosts for this nematode. They found host selection was not governed by root diffusates *per se*, but the intensity of the attraction was affected by diffusates. Juveniles also were attracted to tomato roots when a dialysis membrane separated the nematodes from roots. In a follow-up study, Viglierchio (76) found young seedlings and 5-week-old plants, growing in either sand or soil, were equally attractive to juveniles. The attractant(s) was able to diffuse through two membranes 10 mm apart and still attract juveniles.

Griffin and Waite (34) compared susceptible and resistant alfalfa cultivars in their attraction to *M. hapla*. They found little difference in attractiveness between individual susceptible and resistant plants; 90% and 70% of the nematode inoculum was attracted to the susceptible and resistant cultivars, respectively. However, when both cultivars were challenged simultaneously, 72% of the juveniles were attracted to the susceptible seedling and 28% migrated to the resistant seedling, which suggests that resistant and susceptible cultivars produce different attractants and/or repellents.

The nature of the stimuli produced by the roots and perceived by second-stage *Meloidogyne* juveniles is not clear. The rhizosphere is greatly modified by growing roots and associated microorganisms (67). Many organic and inorganic compounds excreted by roots and microorganisms form gradients from the root surface out into the soil and may influence nematode attraction over short distances. Klingler (43) regards certain amino acids and CO_2 as the most important root excretions for attracting plant-parasitic nematodes and indeed, some of these compounds act as attractants to *Meloidogyne* juveniles in vitro (4,5). Nevertheless, substances emanating from specific regions of roots attractive to juveniles remain to be identified. The factors influencing long distance migration of juveniles are unknown.

When second-stage juveniles contact plant roots, they often penetrate immediately. Although juveniles primarily enter roots directly behind the root cap, penetration can occur at other sites: points where lateral roots emerge, and galled tissue surrounding adult females. The penetration site of one juvenile frequently becomes an attractive site for other juveniles to enter the root (35). The mechanism of penetration may involve mechanical action by thrusting of the stylet (50). Cellulytic (17) or pectolytic enzymes may also be involved in penetration; however, the roles of specific enzymes in the penetration process still eludes us. Following penetration, especially with multiple infections on the same root, the root tip may enlarge and root growth is often suppressed for a brief period (35).

Juveniles of different *Meloidogyne* species differ in their ability to locate and invade roots. Juveniles of the more aggressive species *M. javanica* are capable of locating and invading tobacco roots earlier and quicker than are *M. incognita* and *M. arenaria* (1). This may explain, in part, why *M. javanica* is more damaging to tobacco than are the latter species.

FEEDING SITE

After juveniles penetrate the root tissue directly behind the root cap, they migrate intercellularly in the cortex to the region of cell differentiation (35,48,50). Endo and Wergin (31) found that this migration activity causes cortical cells to separate along the middle lamella and rarely causes cells to rupture. They noted that cells along the path of migration become distended and compressed by migrating juveniles, but show no signs of nematode feeding. However, Linford (50) observed feeding by migrating juveniles. Also, Endo and Veech (30) reported increases in levels of oxidoreductive enzyme activity in cortical cells along the path of migrating second-stage juveniles in soybean, an indication of increased metabolic activity in these cells. After migrating a short distance in the root, juveniles usually reside in cortical tissue in the zone of differention with their heads in the periphery of the vascular tissue and the remainder of their bodies in the cortex parallel with the long axis of the root.

Although *Meloidogyne* juveniles penetrate roots of most plants, susceptible plants respond to feeding by juveniles and undergo pronounced morphological and physiological changes. Paramount among these changes is the development of elaborate permanent feeding sites, called giant cells (6). Giant cells are highly specialized cellular adaptions induced and maintained in susceptible hosts by the feeding of *Meloidogyne* species. Juveniles fail to develop normally without this unique host response and must feed without killing the host cells. Therefore, formation of giant cells is essential for a successful host-parasite relationship. Tissues preferred for development of giant cells are primary phloem or adjacent parenchyma (18,19,46). Once these tissues are located, each juvenile commences to feed on five to six cells. Feeding transforms these cells into the elaborate nutritive giant cells from which juveniles obtain nourishment to continue their development. If the juvenile fails to elicit a giant cell response in the root it will either starve or, if it has sufficient food reserves, may migrate out of that root and locate another root.

Concurrent with the establishment of giant cells, root tissues around the nematode and its feeding site undergo hyperplasia and hypertrophy giving rise to the characteristic root gall usually associated with *Meloidogyne* infections. Stelar parenchyma cells and pericycle cells undergo hyperplasia and surrounding cortical cells usually only become hypertrophied. Small galls may occasionally be induced by juveniles continuously feeding at the root surface without actually entering the roots (51).

Galls usually develop one to two days after juvenile penetration, and are often the first symptoms observed on an infected plant. On a susceptible plant, gall size is related to the number of nematodes in the tissue (23). However, gall size differs among plant species in response to nematode infection. Galls induced by most *Meloidogyne* species are similar morphologically; however, galls induced by *M. hapla* are usually smaller than those produced by the other common *Meloidogyne* species and characteristically have numerous lateral roots protruding from them. In contrast to giant cell formation, gall development is not essential for normal nematode growth and development. Consequently, gall and giant cell formation are usually considered to be separate, but associated phenomena (82).

Giant cell development and physiology is discussed in detail elsewhere in this book. Giant cell formation is one of the most elaborate responses elicited in plant tissue by any organism, a unique cellular adaptation requiring further investigation.

NEMATODE DEVELOPMENT

Following penetration of host tissue, second-stage juveniles undergo several morphological changes that are associated with parasitism. Juveniles have three esophageal glands, one dorsal and two subventral, that presumably are necessary for the establishment of successful host-parasite relations. The dorsal gland has a long duct that empties into the lumen of the esophagus immediately behind the stylet knobs, and the subventral glands each have a short duct that empties into the lumen of the esophagus behind the cuticular valve in the posterior region of the metacorpus. Each gland has a complex valve that controls the release of glandular substances into the lumen of the esophagus. The esophageal glands produce membrane-bound granules that accumulate around the valves prior to being secreted. Bird and co-workers (8,9,10,11,14) have studied changes that occur in the esophageal glands of *M. javanica* during parasitism. Granules, 0.6 to 0.8 μm in diameter, accumulate in the ducts of the subventral glands of the second-stage juvenile just prior to hatching. Granules in the subventral glands are probably involved in egg hatching and juvenile penetration of root tissue because they disappear 3 days after penetration (8,9). All three esophageal glands and the metacorpus enlarge during early stages of parasitism (8). Bird and Sauer (14) analyzed the contents of the subventral glands histochemically and microspectrographically and reported the ducts of the subventral glands of preparasitic juveniles contained protein but no nucleic acid. Nucleic acid was also absent in these same ducts in parasitic juveniles, but the protein absorption pattern differed from that recorded for preparasitic juveniles. Granules in the subventral glands of parasitic juveniles also stained differently from those produced by preparasitic juveniles (14). These changes in the subventral glands in juveniles following penetration of host tissue suggest that they are responsible for eliciting giant cell formation in a susceptible host.

The dorsal esophageal gland also undergoes noticeable changes during parasitism. The duct of the dorsal gland in preparasitic juveniles rarely contains granules (8). However, soon after penetration, granules appear in the dorsal gland duct. The presence of protein and absence of nucleic acid has been confirmed in parasitic juveniles by UV microspectrographic analysis (8). The esophageal glands become more compact as juveniles mature and are considerably shorter in adult females than they are in second-stage juveniles (11). Since the dorsal esophageal gland in the adult female is the largest of the three esophageal glands, Bird (10) suggests that the granules from this gland provide the stimulus necessary for the maintenance of the giant-cell complex. Although a continuous stimulus must be provided by the adult nematode to maintain the integrity of the giant cells (6), Jones (40) suggests that removal of solutes from giant cells by the adult female nematode may be the stimulus necessary for maintaining the active metabolism of giant cells. Secretions thought to be produced by the dorsal gland of adult females stain positive for basic protein, but are negative for several other enzymes and nucleic acid (10). Peroxidase activity has been detected in secretions of adult female nematodes (39), although recent studies indicate that the peroxidase may have originated from the host (41,71).

Ultrastructural studies revealed differences in the morphology of secretory granules from dorsal and subventral glands. Dorsal gland granules have invaginations, are smaller, and appear more dense than granules from subventral glands (11). Subventral gland granules of preparasitic juveniles have a distinct outer membrane whereas those granules from parasitic juveniles, recovered from host tissue two to three days after penetration, are smaller and have indistinct membranes (9).

MECHANISM OF HOST RESPONSE

Forty-seven years have passed since Christie's classic paper on the development of root-knot nematode galls (19). Although we currently have a better understanding of the formation of giant cells (37,42), the biochemical basis for the induction of these specialized feeding cells remains unknown. Plant growth regulators have been implicated in the development of giant cells and galls. Auxins, promoters of cell growth, and cytokinins, promoters of cell division, have been studied extensively in relation to *Meloidogyne* infections and results of these investigations have been reviewed previously (77).

Auxins have been identified in galls caused by root-knot nematodes, and frequently greater auxin concentrations occur in nematode-infected tissue than in non-galled tissue (2,69). Certain auxins are unique to galled tissue and their concentrations are influenced both by the *Meloidogyne* and host species involved (78,87). Cytokinin levels may also increase in *Meloidogyne*-infected plants (45), and have been reported to be higher in noninfected root-knot susceptible plants than in resistant plants (74). Auxins alone (44), cytokinins alone (27), and combinations of these two (44,68), when exogenously supplied to root-knot resistant plants, reversed the resistance response and made plants susceptible to infection by *Meloidogyne* juveniles. Giant cells and galls usually developed on growth-regulator-treated plants with subsequent normal nematode development frequently occurring. Sawney and Webster (68) found resistance was not always completely reversed with auxin and cytokinin.

Auxins and cytokinins have also been identified in different stages of *Meloidogyne*. Auxins have been detected in juveniles and egg masses of *M. hapla* and in egg masses of *M. javanica* and *M. incognita* (78). Cytokinins have been identified in adult females, juveniles, and egg masses of *M. incognita* (22), and evidence for the secretion of cytokinin-like substances by juveniles of *M. javanica* has been reported (16). In spite of the body of knowledge generated by research on growth regulators associated with *Meloidogyne* infections, our understanding of their role in disease development is incomplete. If growth regulators are involved in giant cell formation, we do not yet know which ones are important or their source, i.e., whether they are of nematode or host origin. Recent evidence (42) indicates that cytokinesis is inhibited (cell plate vesicles fail to fuse) during giant cell formation, suggesting that juveniles probably do not secrete cytokinins into the host tissue.

Another plant growth regulator, ethylene, has been associated with gall formation. Ethylene production was highly correlated with increases in gall weight on *M. javanica*-infected, excised tomato roots, suggesting that it may be involved in the hypertrophy of cortical parenchyma tissue during gall formation (32).

Recent articles (25,40) have eloquently discussed possible mechanisms for giant cell and gall formation. However, the key to understanding the infection process resides in the nematode. Control of the infection process resides in the second-stage juvenile, and particularly in the secretory granules produced by its esophageal glands. Although Bird and coworkers (8,9,10,11,14) have provided significant information about these glands in juveniles and adult female nematodes, information regarding the composition of the contents of the secretory granules is still lacking. The biochemical basis for giant cell formation will be unknown until these substances are identified. Electron cytochemistry (55) of secretory granules *in situ* and biochemical analyses of granules isolated from preparasitic and parasitic juveniles and adult females might be used in identification of the granule contents. We also need to know the relative importance of the subventral glands versus the dorsal gland in the establishment of the host-parasite relationship. Does the location of the subventral gland orifices behind the valve in the metacorpus preclude the secretion of its contents into host tissue? Esophageal gland ablation experiments with a laser microbeam (28) might be used to resolve this question.

PLANT RESPONSE TO DISEASE

That the physiology of *Meloidogyne*-infected plants is drastically altered is reflected by symptoms produced on the shoots and roots. Root alterations, as discussed above, are very characteristic for this host-parasite combination. However, above-ground symptoms exhibited by infected plants are similar to those produced on any plants having a damaged and malfunctioning root system. These include: 1) suppressed shoot growth and accompanying decreased shoot-root ratio (60); 2) nutritional deficiencies showing in the foliage, particularly chlorosis; 3) temporary wilting during periods of mild water stress or during mid-day, when temperatures are high even though adequate soil moisture is available, and 4) suppressed plant yields. The magnitude of these symptoms is often related to the number of juveniles penetrating and becoming established within the root tissue of young plants. Environmental factors influence this plant damage and juvenile density interrelationship. Factors which induce plant stress, e.g. drought or temperature extremes, increase damage to infected plants.

The common explanation for the production of above-ground symptoms on diseased plants is that *Meloidogyne* infection affects water and nutrient absorption and upward translocation by the root system. Reports in the literature, however, are not in complete agreement on this.

Respiration

Respiration of galled tissue varies with host species and physiological age of the host. A study by Myuge (58) found that respiration of *Meloidogyne*-induced galls of cucumber was three times greater than that of noninfected root tissue, whereas in a study on tomato, Bird (13) found respiration of gall tissue was no greater than that in adjacent tissues or in uninfected roots. Owens and Rubinstein (62) reported that galls from 40-day-old tomato plants respired at a lower rate than did noninfected tissue, but when 70-day old plants were examined, the respiration rates were reversed. While the effect of *Meloidogyne* infection on respiration rate varies with nematode-host combinations and plant age, increased enzyme activity (75) and changes in the biochemical components of giant cells (63) show that giant cell metabolism is affected and regulated by the nematode.

Nutritional Deficiencies

Meloidogyne infection markedly affects absorption and translocation of nutrients in plants, although certain reports in the literature do not support this. However, most of the evidence from previous research shows N, P, and K accumulation in above-ground plant parts. Histological studies (18,57) of *Meloidogyne* galls revealed a disruption in the continuity of vessel elements and the presence of abnormal vessel elements. Such alterations of the vascular tissue must disrupt nutrient and water flow in the xylem; however, the magnitude of this dysfunction remains unresolved.

Hunter (38) studied nutrient absorption in tomato infected by *M. incognita*. He found shoot growth was suppressed and foliage was chlorotic on diseased plants with no accompanying alteration of N, P, K, Ca, Mg, and Fe content in the foliage. Roots of infected plants, however, contained larger quantities of N, P, K, and Mg than did uninfected plants. Nutrient accumulation in infected roots may be caused by 1) increased root absorption, 2) impaired translocation to foliage, or 3) mobilization of nutrients from shoots to roots (3). Dropkin and King (26) found that while absorption rates of radioactive phosphorus (32p) in infected tomato roots were retarded near time of harvest only when P was deficient in the soil nutrient solution, the translocation rates of 32p in the infected plants was unaffected. Bergeson (3) studied *M. incognita*-infected tomato plants in split-root experiments and found N and K increased in the nonfertilized root half of a split-root system, indicating that these nutrients were mobilized to the infection site. When he found that N was translocated from infected fertilized roots to noninfected nonfertilized roots, he concluded nematode infection has little effect on mineral translocation. Similar conclusions were drawn in studies with *M. javanica* (70). A different conclusion was reached by Oteifa and Elgindi (61). They found translocation of 32p in *M. incognita*-infected tomato reduced over time, with 32p uptake increasing during the first month following infection and decreasing over the next 2 months. The increases were confined to root tissue, with shoots on infected plants always having less phosphorus than the controls. These data support the idea that nutrients accumulate in the roots, and limited amounts are translocated to the shoots on infected plants.

Translocation of other substances is impaired in *Meloidogyne*-infected plants. Nicotine, which is synthesized in tobacco root tips, increased in root tissue of susceptible and resistant (less damaged) plants and in leaves of only resistant plants following *M. incognita* infection (36). However, leaf content of nicotine decreased by half in the susceptible cultivar where root damage was the greatest. This occurrence indicates that translocation may have been affected.

Supplying nutrients to infected plants will often increase tolerance to *Meloidogyne*. Oteifa (60) found that growth of *M. incognita*-infected lima bean was significantly enhanced with increasing levels of K. Although infected plants had lower concentrations of mineral elements than noninoculated plants had, mineral content increased when K concentration was increased in the nutrient solution. Nematode inoculum level had the greatest effect on the rate of accumulation of K in leaves; K decreased with increasing inoculum level while other elements (Ca, Mg, N, P) increased. Although Oteifa (60) suggested absorption was affected in infected plants, he only measured concentrations in leaf tissue. Therefore, the effect of infection on translocation of these elements is unclear.

Water Relations

Few studies have thoroughly examined the water relations of *Meloidogyne*-infected plants. Water consumption by nematode-infected plants seems to be related to its availability in the soil. Cotton plants infected by *M. incognita* consumed slightly more water than noninfected plants when water was supplied to the plants continuously (59). However, if water was only supplied when soil moisture was depleted to 50 percent field capacity, then infected plants used 184 percent less water than did noninfected plants. Limiting water suppressed growth of infected plants by 78 percent whereas growth of diseased plants receiving continuous water was not affected. These results suggest that water consumption is not affected by nematode infection if soil moisture is not limiting, but under periodic moisture stress, consumption is reduced. Plant development is suppressed by infections that suppress root growth during moisture stress and prevent roots from extending into moist soil.

Meon et al. (57) recently studied the effects of *M. javanica* on water relations in tomato. They found that water flow in infected tomato roots decreased with increasing inoculum densities and with decreasing soil moisture content. They concluded that disruption and development of abnormal vessel elements caused by *M. javanica* infection resulted in restricted water flow. Although there were only slight differences in the transpiration rate between infected and noninfected plants, diffusive resistance of stomata increased and water potential in leaves decreased. They concluded that infected plants respond to moisture stress by closing their stomates.

Photosynthesis

Meloidogyne infection of roots decreases the rate of photosynthesis in leaves. Loveys and

Bird (53) reported that high inoculum levels of *M. javanica* on tomato caused a decline in the net photosynthetic rate within two days after inoculation. This lower rate was maintained throughout subsequent growth of the infected plant. They also found that pruning the apical meristem of tomatoes enhanced the photosynthetic rate of noninoculated plants, but had no effect on infected plants. They concluded that the nematodes interfered with the production and translocation of root-derived factors regulating photosynthesis.

In a similar study, Wallace (7,9) reported that decreases in the photosynthetic rate of young tomato plants varied with inoculum levels and was not greatly influenced by plant age. The incorporation of $^{14}CO_2$ in roots of infected plants was retarded at all inoculum levels whereas ^{14}C-glucose incorporation was affected only at the highest inoculum level (80).

HOW IS PLANT GROWTH AND YIELD SUPPRESSED?

The effect of *Meloidogyne* on plant growth and yield is varied and complex. The damage caused by this nematode on susceptible plants probably involves several mechanisms. Nematodes remove plant nutrients, alter nutrient flow patterns in plant tissue, and retard root growth, all of which may contribute to suppressed plant yield.

Sink Phenomenon

Adult female nematodes require considerable amounts of nutrients for egg production and compete with the host for the pool of nutrients in the roots. It is not clear whether the removal of nutrients by nematodes suppresses plant growth. Experiments conducted by Bird and Loveys (15) and McClure (54) unequivocally show that *Meloiodgyne* functions as a metabolic sink in diseased plants. The increased metabolic activity of giant cells stimulates mobilization of photosynthates from shoots to roots and, in particular, to the giant cells where they are removed and utilized by the feeding nematode. Mobilization and accumulation of substances reaches a maximum when the adult females commence egg laying and declines thereafter (56). Just how much removal of photosynthates by the parasites is needed to suppress plant growth and yield is unknown. Dropkin and King (26), working with galled tomato plants, found that nematodes removed less than ten percent of the total phosphorus in the plant, and concluded that root-knot nematodes were not significant competitors for plant P. However, it has not been determined whether *Meloidogyne* species consume host carbon compounds in sufficient quantitites to affect plant growth.

Nutrient Absorption and Translocation

Although uptake of phosphorus and potassium may be reduced in galled tissue (26), their absorption by nongalled tissue should not be affected since these elements are taken up via the symplastic pathway in roots (20). In a recent study with *Heterodera avenae* in oats (64), the presence of nematodes in seminal roots did not impede absorption or transport of potassium or phosphorus. In fact, uptake of the symplastically absorbed nutrients was enhanced in infected plants, which suggested that roots compensated for the suppressed and altered root growth by increasing rates of nutrient uptake. Nematode blockage of absorption and translocation of nutrients and water may not be as important as originally thought.

Suppressed Root Growth

Root branching and rate of root extension is frequently suppressed by nematode infection. This abnormal root growth results in reduced root surface area and limits the capacity of a root system to explore the soil. Thus the primary cause of poor nutrient uptake and suppressed growth of infected plants could be related to the reduced root system (38). On the other hand, plants that are tolerant to parasitic nematodes might possess the capacity to develop an adequate root system even in the presence of the parasites.

Growth of root tips invaded by *Meloidogyne* juveniles is frequently temporarily inhibited. The resumption of growth of root tips and the formation of galls may create a drain on the plant's resources in shoot tissue, resulting in suppression of shoot growth. The amount of resources diverted, however, would be directly related to the intensity of infection. Low infection levels frequently stimulate plant growth, whereas high nematode densities suppress plant growth and yield (81).

The temporary inhibition of root growth probably also interrupts the synthesis of cytokinins and gibberellins which are produced in root apices. Decreased synthesis of these growth regulators in young plants, when they are most vulnerable to nematode damage, probably has a profound influence on plant growth. However, most studies have concentrated on analysis of physiological response of older infected plants. Greater emphasis should be placed on analyzing physiological changes in diseased plants when they are young and more vulnerable to nematode damage.

It is still not clearly resolved whether specific root alterations caused by *Meloidogyne* are important in suppressing plant yield or if the sum of the effects of the infection are ultimately responsible for retarding plant growth. Disturbance caused by *Meloidogyne* in plant roots obviously have significant impact on the physiology of the whole plant. These nematodes will directly and indirectly have an effect on the translocation patterns of substances between the shoot and roots which may ultimately affect the yield of plants. Nevertheless, as Dropkin suggests (25), under field conditions, the greatest effect of *Meloidogyne* on plant health may be through predisposition of the roots to infection and subsequent damage by other soilborne microorganisms.

Acknowledgment

I would like to thank Dr. John L. Ruehle for a careful review of the manuscript.

Literature Cited

1. Arens, M. L., J. R. Rich, and D. W. Dickson. 1981. Comparative studies on root invasion, root galling, and fecundity of three *Meloidogyne* spp. on a susceptible tobacco cultivar. J. Nematol. 13:201-205.
2. Balasubramanian, M. and G. Rangaswami. 1962. Presence of indole compounds in nematode galls. Nature 194:774-775.
3. Bergeson, G. B. 1966. Mobilization of minerals to the infection site of root-knot nematodes. Phytopathology 56:1287-1289.
4. Bird, A. F. 1959. The attractiveness of roots to the plant parasitic nematodes *Meloidogyne javanica* and *M. hapla*. Nematologica 4:322-335.
5. Bird, A. F. 1960. Additional notes on the attractiveness of roots to plant parasitic nematodes. Nematologica 5:217.
6. Bird, A. F. 1962. The inducement of giant cells by *Meloidogyne javanica*. Nematologica 8:1-10.
7. Bird, A. F. 1962. Orientation of the larvae of *Meloidogyne javanica* relative to roots. Nematologica 8:275-287.
8. Bird, A. F. 1967. Changes associated with parasitism in nematodes. I. Morphology and physiology of preparasitic and parasitic larvae of *Meloidogyne javanica*. J. Parasitol. 53:768-776.
9. Bird, A. F. 1968. Changes associated with parasitism in nematodes. III. Ultrastructure of the egg shell, larval cuticle, and contents of the subventral esophageal glands in *Meloidogyne javanica*, with some observations on hatching. J. Parasitol. 54:475-489.
10. Bird, A. F. 1968. Changes associated with parasitism in nematodes. IV. Cytochemical studies on the ampulla of the dorsal esophageal gland of *Meloidogyne javanica* and on exudations from the buccal stylet. J. Parasitol. 54:879-890.
11. Bird, A. F. 1969. Changes associated with parasitism in nematodes. V. Ultrastructure of the stylet exudation and dorsal esophageal gland contents of female *Meloidogyne javanica*. J. Parasitol. 55:337-345.
12. Bird, A. F. 1974. Plant response to root-knot nematodes. Annu. Rev. Phytopathol. 12:69-85.
13. Bird, A. F. and A. Millerd. 1962. Respiration studies on *Meloidogyne*-induced galls in tomato roots. Nematologica 8:261-266.
14. Bird, A. F. and W. Sauer. 1967. Changes associated with parasitism in nematodes. II. Histochemical and microspectrophotometric analysis of preparasitic and parasitic larvae of *Meloidogyne javanica*. J. Parasitol. 53:1262-1269.
15. Bird, A. F. and B. R. Loveys. 1975. The incorporation of photosynthates by *Meloidogyne javanica*. J. Nematol. 7:111-113.
16. Bird, A. F. and B. R. Loveys. 1980. The involvement of cytokinins in a host-parasite relationship between the tomato (*Lycopersicon esculentum*) and a nematode (*Meloidogyne javanica*). Parasitology 80:497-505.
17. Bird, A. F., W. J. S. Downton, and J. S. Hawker. 1975. Cellulase secretion by second stage larvae of the root-knot nematode (*Meloidogyne javanica*). Marcellia 38:165-169.
18. Byrne, J. M., T. C. Pesacreta, and J. A. Fox. 1977. Vascular pattern change caused by a nematode, *Meloidogyne incognita*, in the lateral roots of *Glycine max* (L.) Merr. Am. J. Bot. 64:960-965.
19. Christie, J. R. 1936. The development of root-knot nematode galls. Phytopathology 26:1-22.

20. Clarkson, D. T. and A. W. Robards. 1975. The endodermis, its structural development and physiological role, pp. 415-436 *In* The development and function of roots, ed. J. G. Torrey and D. T. Clarkson. Academic Press, London. 618 pp.
21. Coomans, A. 1979. The anterior sensilla of nematodes. Rev. Nematol. 2:259-283.
22. Dimalla, G. G. and J. van Staden. 1977. Cytokinins in the root-knot nematode, *Meloidogyne incognita*. Plant Sci. Lett. 10:25-29.
23. Dropkin, V. H. 1954. Infectivity and gall size in tomato and cucumber seedlings infected with *Meloidogyne incognita* var. *acrita* (root-knot nematode). Phytopathology 44:43-49.
24. Dropkin, V. H. 1969. Cellular responses of plants to nematode infection. Annu. Rev. Phytopathol. 7:101-121.
25. Dropkin, V. H. 1972. Pathology of *Meloidogyne*-galling, giant cell formation, effects on host physiology. Eur. Mediterr. Plant Prot. Organ. (OEPP/EPPO) Bull. 6:23-32.
26. Dropkin, V. H. and R. C. King. 1956. Studies on plant parasitic nematodes homogeneously labeled with radiophosphorus. Exp. Parasitol. 5:469-480.
27. Dropkin, V. H., J. P. Helgeson, and C. D. Upper. 1969. The hypersensitivity reaction of tomatoes resistant to *Meloidogyne incognita*: reversal by cytokinins. J. Nematol. 1:55-61.
28. Dusenbery, D. B. 1983. Chemotactic behavior of nematodes. J. Nematol. 15:168-173.
29. Endo, B. Y. 1971. Nematode-induced syncytia (giant cells). Host-parasite relationships of Heteroderidae, pp. 91-117 *In* Plant parasitic nematodes, Vol. II, ed. B. M. Zuckerman, W. F. Mai, and R. A. Rohde, Academic Press, New York. 347 pp.
30. Endo, B. Y. and J. A. Veech. 1969. The histochemical localization of oxidoreductive enzymes of soybeans infected with the root-knot nematode, *Meloidogyne incognita acrita*. Phytopathology 59:418-425.
31. Endo, B. Y. and W. P. Wergin. 1973. Ultrastructural investigation of clover roots during early stages of infection by the root-knot nematode, *Meloidogyne incognita*. Protoplasma 78:365-379.
32. Glazer, I., D. Orion, and A. Apelbaum. 1983. Interrelationships between ethylene production, gall formation and root-knot nematode development in tomato plants infected with *Meloidogyne javanica*. J. Nematol. 15:539-544.
33. Green, C. D. 1971. Mating and host finding behavior of plant nematodes, pp. 247-266 *In* Plant parasitic nematodes, Vol. II, ed. B. M. Zuckerman, W. F. Mai, and R. A. Rohde. Academic Press, New York. 347 pp.
34. Griffin, G. D. and W. W. Waite. 1971. Attraction of *Ditylenchus dipsaci* and *Meloidogyne hapla* by resistant and susceptible alfalfa seedlings. J. Nematol. 3:215-219.
35. Godfrey, G. H. and J. Oliveira. 1932. The development of the root-knot nematode in relation to root tissue of pineapple and cowpea. Phytopathology 22:325-348.
36. Hanowanik, S. B. and W. W. Osborne. 1975. Influence of *Meloidogyne incognita* on the content of amino acids and nicotine in tobacco grown under gnotobiotic conditions. J. Nematol. 7:332-336.
37. Huang, C. S. and A. R. Maggenti. 1969. Mitotic aberrations and nuclear changes in developing giant cells in *Vicia faba* caused by root-knot nematode, *Meloidogyne javanica*. Phytopathology 59:447-455.
38. Hunter, A. H. 1958. Nutrient absorption and translocation of phosphorus as influenced by the root knot nematode (*Meloidogyne incognita acrita*). Soil Sci. 86:245-250.
39. Hussey, R. S. and J. N. Sasser. 1973. Peroxidase from *Meloidogyne incognita*. Physiol. Plant Pathol. 3:223-229.
40. Jones, M. G. K. 1981. Host cell responses to endoparasitic nematode attack: structure and function of giant cells and syncytia. Ann. Appl. Biol. 97:353-372.
41. Jones, M. G. K. 1980. Microgel electrophoretic examination of soluble proteins in giant transfer cells and associated root-knot nematodes (*Meloidogyne javanica*) in balsam roots. Physiol. Plant Pathol. 16:359-367.
42. Jones, M. G. K. and H. L. Payne. 1978. Early stages of nematode-induced giant-cell formation in roots of *Impatiens balsamina*. J. Nematol. 10:70-85.
43. Klingler, J. 1965. On the orientation of plant nematodes and of some other soil animals. Nematologica 11:4-18.
44. Kochba, J. and R. M. Samish. 1971. Effect of kinetin and 1-naphthylacetic acid on root-knot nematodes in resistant and susceptible peach rootstocks. J. Am. Soc. Hortic. Sci. 96:458-461.
45. Krupasagar, V. and K. R. Barker. 1966. Increased cytokinin concentrations in tobacco

infected with the root knot nematode *Meloidogyne incognita*. Phytopathology 56:885 (Abstr.).
46. Krusberg, L. R. and L. W. Nielsen. 1958. Pathogenesis of root-knot nematodes to the Porto Rico variety of sweetpotato. Phytopathology 48:30-39.
47. Lamberti, F. and C. E. Taylor, eds. 1979. Root-knot nematodes (*Meloidogyne* species) systematics, biology and control. Academic Press, New York. 477 pp.
48. Linford, M. B. 1937. The feeding of the root-knot nematode in root tissue and nutrient solution. Phytopathology 27:824-835.
49. Linford, M. B. 1939. Attractiveness of roots and excised shoot tissues to certain nematodes. Proc. Helminthol. Soc, Wash. 6:11-18.
50. Linford, M. B. 1942. The transient feeding of root-knot nematode larvae. Phytopathology 32:580-589.
51. Loewenberg, J. R., T. Sullivan, and M. L. Schuster. 1960. Gall induction by *Meloidogyne incognita* by surface feeding and factors affecting the behavior pattern of the second-stage larvae. Phytopathology 50:322-333.
52. Lownsbery, B. F. and D. R. Viglierchio. 1961. Importance of response of *Meloidogyne hapla* to an agent from germinating tomato seeds. Phytopathology 51:219-221.
53. Loveys, R. R. and A. F. Bird. 1973. The influence of nematodes on photosynthesis in tomato plants. Physiol. Plant Pathol. 3:525-529.
54. McClure, M. A. 1977. *Meloidogyne incognita*: a metabolic sink. J. Nematol. 9:88-90.
55. McClure, M. A. 1981. Electron microscope histochemistry, pp. 89-124 *In* Plant parasitic nematodes, Vol. III, ed. B. M. Zuckerman and R. A. Rohde. Academic Press, New York. 508 pp.
56. Meon, S., J. M. Fisher, and H. R. Wallace. 1978. Changes in free proline following infection of plants with either *Meloidogyne javanica* or *Agrobacterium tumefaciens*. Physiol. Plant Pathol. 12:251-256.
57. Meon, S., H. R. Wallace, and J. M. Fisher. 1978. Water relations of tomato (*Lycopersicon esculentum* Mill. cv. Early Dwarf Red) infected with *Meloidogyne javanica* (Treub), Chitwood. Physiol. Plant Pathol. 13:275-281.
58. Myuge, S. G. 1956. The nutritional physiology of the gall nematodes (*Meloidogyne incognita*). Dokl. Akad. Nauk SSSR 108:164-165.
59. O'Bannon, J. H. and H. W. Reynolds. 1965. Water consumption and growth of root-knot nematode-infected and uninfected cotton plants. Soil Sci. 99:251-255.
60. Oteifa, B. A. 1952. Potassium nutrition of the host in relation to infection by root-knot nematode, *Meloidogyne incognita*. Proc. Helminthol. Soc. Wash. 19:99-104.
61. Oteifa, B. A. and D. M. Elgindi. 1962. Influence of parasitic duration of *Meloidogyne javanica* (Treub) on host nutrient uptake. Nematologica 8:216-220.
62. Owens, R. G. and J. H. Rubinstein. 1966. Metabolic changes induced by root-knot nematodes in host tissues. Contrib. Boyce Thompson Inst. 23:199-214.
63. Owens, R. G. and H. N. Specht. 1966. Biochemical alterations induced in host tissues by root-knot nematodes. Contrib. Boyce Thompson Inst. 23:181-198.
64. Price, N. S., D. T. Clarkson, and N. G. M. Hague. 1982. The uptake of potassium and phosphorus in oats infested with the cereal cyst nematode, *Heterodera avenae* Woll. Rev. Nematol. 5:321-325.
65. Prot, J. C. 1978. Vertical migration of four natural populations of *Meloidogyne*. Rev. Nematol. 1:21-26.
66. Prot, J. C. 1980. Migration of plant-parasitic nematodes towards plant roots. Rev. Nematol. 3:305-318.
67. Russell, R. S. 1977. Plant root systems: their function and interaction with the soil. McGraw-Hill Book Co. Ltd., Berkshire, England. 298 pp.
68. Sawhney, R. and J. M. Webster. 1975. The role of plant growth hormones in determining the resistance of tomato plants to the root-knot nematode, *Meloidogyne incognita*. Nematologica 21:95-103.
69. Setty, K. G. G. and A. W. Wheeler. 1968. Growth substances in roots of tomato infected with root-knot nematodes (*Meloidogyne* spp.). Ann. Appl. Biol. 61:498-501.
70. Spiegel, Y., E. Cohn, U. Kafkafi, and M. Sulami. 1982. Influence of potassium and nitrogen fertilization on parasitism by the root-knot nematode *Meloidogyne javanica*. J. Nematol 14:530-535.
71. Starr, J. L. 1979. Peroxidase isoenzymes from *Meloidogyne* spp. and their origins. J. Nematol. 11:1-5.
72. Taylor, A. L. and J. N. Sasser. 1978. Biology, identification and control of root-knot nematodes (*Meloidogyne* species). Coop. Publ. Dep. Plant Pathol., North Carolina

State Univ., and U.S. Agency Int. Dev., Raleigh, N.C. 111 pp.

73. Van Gundy, S. D., A. F. Bird, and H. R. Wallace. 1967. Aging and starvation in larvae of *Meloidogyne javanica* and *Tylenchulus semipenetrans*. Phytopathology 57:559-571.
74. Van Staden, J. and G. G. Dimalla. 1977. A comparison of the endogenous cytokinins in the roots and xylem exudate of nematode-resistant and susceptible tomato cultivars. J. Exp. Bot. 28:1351-1356.
75. Veech, J. A. and B. Y. Endo. 1969. The histochemical localization of several enzymes of soybeans infected with the root-knot nematode, *Meloidogyne incognita acrita*. J. Nematol. 1:265-276.
76. Viglierchio, D. R. 1961. Attraction of parasitic nematodes by plant root emanations. Phytopathology 51:136-142.
77. Viglierchio, D. R. 1971. Nematodes and other pathogens in auxin-related plant-growth disorders. Bot. Rev. 37:1-21.
78. Viglierchio, D. R. and P. K. Yu. 1968. Plant growth substances and plant parasitic nematodes. II. Host influence on auxin content. Exp. Parasitol. 23:85-95.
79. Viglierchio, D. R. and B. F. Lownsbery. 1960. The hatching response of *Meloidogyne* species to emanations from the roots of germinating tomatoes. Nematologica 5:153-157.
80. Wallace, H. R. 1974. The influence of root-knot nematode, *Meloidogyne javanica*, on photosynthesis and nutrient demand by roots of tomato plants. Nematologica 20:27-33.
81. Wallace, H. R. 1971. The influence of the density of nematode populations on plants. Nematologica 17:154-166.
82. Webster, J. M. 1969. The host-parasite relationships of plant-parasitic nematodes. Adv. Parasitol. 7:1-40.
83. Webster, J. M. 1975. Aspects of the host-parasite relationship of plant-parasitic nematodes. Adv. Parasitol. 13:225-250.
84. Weiser, W. 1955. The attractiveness of plants to larvae of root-knot nematodes. I. The effect of tomato seedlings and excised roots on *Meloidogyne hapla* Chitwood. Proc. Helminthol. Soc. Wash. 22:106-112.
85. Weiser, W. 1956. The attractiveness of plants to larvae of root-knot nematodes. II. The effect of excised bean, eggplant, and soybean roots on *Meloidogyne hapla* Chitwood. Proc. Helminthol. Soc. Wash. 23:59-64.
86. Wergin, W. P. and B. Y. Endo. 1976. Ultrastructure of a neurosensory organ in a root-knot nematode. J. Ultrastruct. Res. 56:258-276.
87. Yu, P. K. and D. R. Viglierchio. 1964. Plant growth substances and parasitic nematodes. I. Root-knot nematodes and tomato. Exp. Parasitol. 15:242-248.

FORMATION, ANATOMY AND PHYSIOLOGY OF GIANT CELLS INDUCED BY ROOT-KNOT NEMATODES

C. S. Huang

Dep. de Biologia Vegetal, Universidade de Brasília
Brasília, DF, Brazil

The second-stage juvenile of *Meloidogyne* penetrates the root and establishes its feeding site on vascular parenchyma. In response to the feeding activity, some parenchyma cells become hypertrophied and multinucleate (Fig. 1) and are known generally as "giant cells." Because of their constant association, there has never been any doubt about the causal relationship between root-knot nematodes and the giant cells located at their feeding sites. Bird (3) conclusively demonstrated that both initiation and continued development of giant cells in tomato roots are dependent on the presence of active *Meloidogyne javanica*. He also showed that functional giant cells are essential for development of the root-knot nematode. Failure of giant cells to develop and the consequent aborted development of the root-knot nematode (or vice versa), which has been observed frequently in resistant cultivars and unsuitable hosts (14,40,51) and in plants treated with antimitotic agents (43), substantiate this view.

Because of their obligate relationship with root-knot nematode infection (3) and perhaps also as a result of their striking morphological features, the large abnormal host cells have incited much curiosity among investigators since their discovery a century ago. Mechanisms of their initiation and subsequent development as well as their physiological and biochemical characteristics have been the subjects of intensive investigations.

The objective of this chapter is to evaluate the conflicting evidence on the mechanisms of the formation of giant cells and to summarize the information on their anatomy and physiology. A comprehensive literature survey is not made here since excellent reviews are available (7,8,13,15,16,18,19,24,36).

MECHANISM OF GIANT CELL FORMATION

For many years there have been opposing views on the mechanism by which giant cells develop. As early as 1898, Beille suggested that giant cell formation involved cell wall dissolution and subsequent cytoplasmic fusion between the neighboring cells (cited in 13). Although experimental evidence was never convincing, this line of thinking prevailed among many investigators (1,6,7,13,36,53). Cell wall fragmentations, which have been claimed to be observed in giant cell cytoplasms in histological sections, were used to support this point of view (2,5,6,47,55). While unequivocal identification of cell wall breakdown with light microscopy (LM) proved to be difficult, transmission (TEM) and scanning (SEM) electron microscopies are powerful systems for the task. Cell-wall dissolutions for giant cells induced by *Meloidogyne* spp. could not be demonstrated with the latter methods despite extensive searches (23,27,29,32).

In sharp contrast to the mechanism induced by *Meloidogyne*, convincing evidence for cell-wall breakdown was obtained, also by means of TEM and SEM of cryofractured specimens, for syncytia induced in host plants by other genera, namely *Heterodera* spp. (26,52), *Nacobbus aberrans* (30), *Rotylenchulus reniformis* (50) and *Longidorus apulus* (11). Jones and Dropkin (25) were able to demonstrate, with LM and epoxy resin embedded semi-thin sections, the striking difference between the absence of cell-wall breakdown in giant cells of soybean roots infected with *M. incognita* and the easily observed wall gaps in syncytia induced by *H. glycines* and *R. reniformis* in the same plant.

Fig. 1. Photomicrograph (10-μm-thick paraffin section) of giant cells in the root of *Vicia faba* 15 days after infected with *Meloidogyne javanica*. Note the uneven thickening of the cell walls.

Fig. 2. Photomicrograph of a portion of a giant cell isolated from the root of *Vicia faba* infected with *Meloidogyne javanica* (Feulgen-stained squash preparation). Note that the interphase nuclei are variable in size and profile. Nuclear connections are also apparent.

Fig. 3. EM photomicrograph (thin section, glutaldehyde-OsO_4 fixation) of a giant cell nucleus in the root of *Vicia faba* infected with *Meloidogyne javanica*. Note the amoeboid form of the nucleus.

Fig. 4. Photomicrograph of a giant cell, undergoing synchronous nuclear division (anaphase), from the root of *Vicia faba* infected with *Meloidogyne javanica* (Feulgen-stained squash preparation). Note that the mitotic apparatus, with different number of chromosome sets, conduct their anaphase migrations along different axes, apparently lacking the normal spindle orientation.

Fig. 5. EM photomicrograph (thin section, glutaldehyde-OsO_4 fixation) of a portion of a giant cell in the root of *Cucumis sativus* 19 days after inoculated with *Meloidogyne javanica*. Note the branching cell wall ingrowths (W_2), delineated by a continuous plasmalemma (Pm).

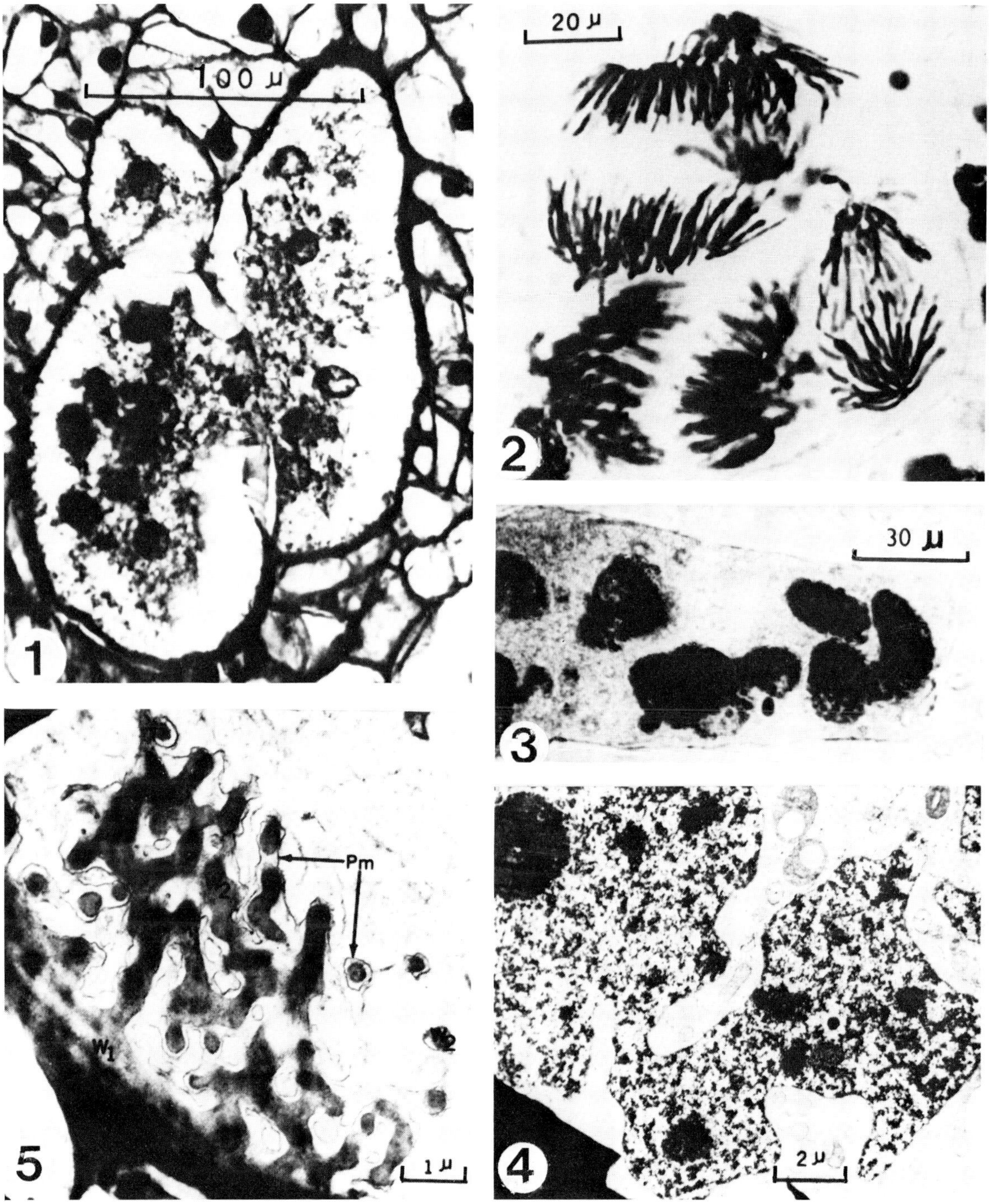
100 µ
1
20 µ
2
30 µ
3
2 µ
4
Pm
W1
W2
1 µ
5

Only twelve years after the suggestion of Beille, Némec in 1910 gave an alternative explanation for the mechanism of formation of giant cells. He observed that nuclei in a giant cell induced by a root-knot nematode divided without forming new cell walls (cited in 13). Although nuclear divisions in giant cells have often been observed since then (2,13,37, 39,47,55), the importance of karyokinesis without subsequent cytokinesis in giant-cell ontogeny attracted little attention for more than half a century.

In 1969, Huang and Maggenti (22) demonstrated that metaphase chromosome numbers of giant cells in *Vicia faba* infected with *M. javanica* followed the geometric progression 24,48,96,192 and 384. They suggested, therefore, that the giant cells were most likely formed through repeated endomitoses without cytokineses. They hypothesized that involvement of cytoplasmic coalescence in the formation of giant cells was unlikely, otherwise one would expect to find chromosome numbers that follow an arithmetic progression. The hypothesis was contested by some investigators. Finding that photomicrometric DNA measurements of giant-cell nuclei were variable and did not fit the ploidy sequence, Bird (4) proposed that cell fusions must also have been involved in the giant cell formation. However, many factors may contribute to the variability of measurements of this kind, including the precision of the method itself, the variation among nuclei of the same giant cells due to irregular fusion of mitotic apparatus and occasional aneuploidy (22), and the "age" of the nuclei when measurements are taken. Rohde & McClure (53) infected cotton seedlings for 24 h with *M. incognita* and then transferred the infected plants to a nutrient solution containing ^{3}H-methyl-thymidine for another 24 h. Using autoradiography, they detected unlabelled binuclear cells in roots 24 h after exposure to the labelled nutrient solution had finished. In addition, an image suggestive of a wall gap near the head of an infective juvenile was recorded as evidence of cell wall dissolution (53). These results led them to conclude that giant cells could also be formed by cytoplasmic fusion between adjacent cells resulting from cell wall dissolution. Presuming that the autoradiographic technique was sufficiently sensitive and precise, the possibility still exists that the unlabelled binucleate giant cells were derived from the nuclei that had remained unlabelled after 24 h of incubation in the "hot" ^{3}H-nutrient solution. The cell-wall dissolution observed is difficult to ascertain with paraffin section because the wax is not firm enough to exclude the possibility that indistinguisable artifacts might have been formed during the sectioning. In addition, the generally thick cell walls of giant cells have thin areas that are close to the limit of LM resolution (23,32), thus giving an impression of "gaps" in sections on some plains.

Bird (6) repeated the chromosome counts of giant cells in *Vicia faba* infected with *M. javanica*. He was unable to obtain the numbers that fit the ploidy sequence obtained by Huang & Maggenti (22), and reported various stages of cell-wall breakdown and cell fusion in the giant cells with phase-contrast microscopy. He, therefore, argued that wall breakdown was indeed involved in the process of giant-cell formation. However, the factors that may have accounted for the discrepancies in the chromosome counts carried out by the aforementioned investigators are not certain. It is difficult to ascertain the cell-wall breakdown and cell fusions claimed to have been observed because the resolving power of the phase contrast LM system is limited.

Advances in instrumentation in the last ten years, such as the cryofracturing technique employed in SEM, have greatly facilitated the studies of giant-cell ontogeny. Experimental evidence with these advanced EM techniques has accumulated against the hypothesis that giant-cell formation involves cell-wall dissolution and subsequent cytoplasmic coalescence. In their EM and LM studies of giant cells of *Impatiens balsamina* induced by *M. javanica* and *M. incognita*, Jones & Payne (32) were not able to detect any evidence of cell-wall dissolution in the first 72 h of their initiation, a period during which thorough observations were made. On the contrary, their studies (32) confirmed the observations of Huang & Maggenti (22) that daughter-cell walls were not formed after nuclear division had occurred in the giant cells of *Vicia faba* infected with *M. javanica*.

After the cytoplasm from cryofractured materials has been digested, considerable portions of the interior surface of a giant cell can be examined with SEM. Such studies have clearly demonstrated the absence of perforations in the giant-cell walls of *I. balsamina* infected with *M. incognita* (27). This absence is considered to be a strong indication of non-involvement of cell-wall lyses in giant-cell formation. This interpretation is probably well founded since similar investigations have clearly shown wall perforations in the syncytia of soybean and tobacco infected with cyst nematodes (26) and in those of tomato infected with *Nacobbus aberrans* (30). These perforations are all known to derive from wall lyses and subsequent cytoplasmic fusions (25).

ANATOMY OF GIANT CELLS

Giant cells vary anatomically with their developmental "age." In the roots of *I. balsamina* 24 hr after being infected with *M. javanica*, nuclear divisions were detected in parenchyma cells up to two cell layers from the head of the nematode (32). Since the vascular parenchyma rarely undergoes mitosis under normal conditions, the nuclear divisions were presumably stimulated by the nematode. These divisions are considered to be the first detectable sign of giant-cell initiation. EM studies by Jones & Payne (32) further showed that, at anaphase, vesicles lined up initially between the two daughter nuclei to form a new cell plate. Golgi bodies, endoplasmic reticulum (ER), mitochondria and small vacuoles were found accumulated around the partially formed cell plate. Contrary to normal mitoses, however, the vesicles aggregated along the presumptive cell plate were posteriorly dispersed, resulting in abortion of the new cell-wall formation. The first karyokinesis without cytokinesis results in the formation of a young binucleate "giant cell," whose size barely differs from that of the neighboring, normal parenchyma cells. After telophase, a partially formed cell plate may become attached to the mother cell wall and develop as a wall "stub" or "flange" with posterior polysaccharide depositions. However, the wall "stubs" should not be interpreted as the remnants of wall decomposition since corresponding wall fragments could not be found anywhere within the giant cell (32). These features of abortive cytokinesis in developing giant cells very closely resemble those of caffeine-induced mitotic aberrations (31).

The binuclei of a young giant cell undergo rapid, synchronous divisions without subsequent cytokineses. Thus, giant cells with up to eight nuclei can be found within 48 hr of nematode infection (32). However, it is not yet known whether in the subsequent nuclear divisions, cell-plate-like structures are also consolidated initially but aborted afterwards. Synchronous nuclear divisions within the same giant cells (Fig. 2) have been observed for many other host plants (2,22,37,47,55), but corresponding daughter cell-wall formations have never been detected.

In a more advanced giant cell, synchrony of the nuclear divisions may not be constant; some occasionally become slightly out of phase as the polyploidy increases (6,22). The polyploidal condition of a giant cell commonly reaches 32 n, and 64 n has been recorded (22).

Nuclei within a giant cell are highly variable both in size and in shape (Fig. 3; see also 22,60). They are generally irregularly lobed and amoeboid in form (Figs. 3 & 4), thus with a tremendously increased surface area. Linkages between neighboring nuclei by the lobes are also common (Fig. 3). The variabilities in nuclear size and profile are attributable to the irregular fusions of neighboring metaphase plates, to the uneven distribution of chromosome sets at anaphase (Fig. 2) from lack of spindle orientations and to the incomplete migration of anaphase chromosome (22). The mitotic aberrations and the resultant nuclear irregularities make nuclear counts and quantitative nuclear DNA determinations practically impossible.

The cytoplasms of developing giant cells are reminiscent of active meristematic cells. Golgi apparatus, mitochrondria, ribosomes, polysomes and ER are abundant (25,28,29,32). As a parenchyma cell is transformed to a developing giant cell, the predominant central vacuole gradually disappears, and small vacuoles increasingly prevail in the cytoplasm (32,59). A crystalline body surrounded by an elaborate smooth membrane system, which is apparently proteinaceous but of unidentified origin, can occasionally be found in giant-cell ground cytoplasm (17,29). Amyloplasts with starch grains were found in the giant cells of soybean roots infected with *M. incognita* (25) and in those of *Coleus blumei* roots infected with *M. arenaria* (29) but not in those of tomato roots infected with *M. javanica* (42). In the cytoplasm of soybean-root giant cells near the head region of *M. incognita*, Jones & Dropkin (25) reported some "clear areas" not stained with toluidine blue except for some membranous inclusions. They suggested that the "clear areas," albeit rare, correspond to packets of saliva originating from the nematode stylet.

The walls of giant cells are apparently continuous and without "gaps" or "perforations" (23,25,27,29,32). There are, however, alternating thick and thin areas on the same wall (Fig. 1). Plasmodesmata are found only in the thin portions of the wall (23,25,27,29).

The thick portions of the wall are associated with secondary ingrowths jutting into the giant-cell cytoplasm (Fig. 5, see also 23,25,27,28,29). The ingrowths vary from simple rods to extensive finger-like branchings and are always followed by a continuous plasmalemma (23,29).

In the giant cells of *I. balsamina*, the branching ingrowths are essentially cylindrical with a fairly constant diameter of 0.37 ± 0.06 μm and can reach a length of 10 μm in the well-developed stage (27). Whether the measurements also apply to giant cells of other plants is not known. The ingrowths are most extensive on the giant-cell walls adjacent to xylem elements and on walls between giant cells, less so on walls next to sieve elements, and least on those separating giant cells and other parenchyma cells (25,29). In young giant cells, numerous small pits are found on walls all around. At later stages however, only the walls between two giant cells have pit fields with plasmodesmata. The walls between a grown giant cell and normal cells are remarkably free of pit fields (27).

PHYSIOLOGY OF GIANT CELLS

A considerable amount of research has been done in the last two decades on the physiological and biochemical aspects of giant cells. The subject was recently reviewed meticulously by Jones (24) who contributed greatly to our knowledge about the physiological functions of these structures.

It is increasingly clear from the studies on the wall ingrowths (28,29,33,34) that giant cells induced by root-knot nematodes are transfer cells similar in function to those found in normal tissues such as companion cells and xylem parenchyma cells (49). The ingrowths greatly increase the surface area of the plasmalemma in a giant cell, with an amplification factor which can reach more than 10× (27). It is assumed that these structures effectively facilitate the transport of solutes. The fact that wall ingrowths proliferate as the nematode develops and degenerate when it ages (27) demonstrates the importance of nematodes in the physiology of the giant cells.

A root-knot nematode requires about a month to reach maturity. An estimate of the biomass of the adult female and the eggs she lays demonstrates that an infective juvenile increases its size by about 1000 times during that period (16). The total biomass is much larger than any of the giant cells on which she feeds. To accomplish this tremendous growth rate, the nematode must draw the bulk of the nutrients required from the half-dozen giant cells surrounding its head and, in so doing, acts as a metabolic sink (9,41). To satisfy the nematode's needs, on the other hand, a large quantity of solutes must obviously be transported from the surrounding tissues to the giant cells. Information gained from the studies of transmembrane potentials (33,34) led Jones (24) to hypothesize that the wall-ingrowth apparatus with its considerably amplified plasmalemma, is equipped with a hydrogen ion efflux pump that enables solutes such as sugars and amino acids to pass efficiently from the cell-wall compartment to the cytoplasmic compartment of the giant cell.

Biochemical analyses of gall-tissue homogenates (12,35,38,45,46,48) do not indicate which metabolites are specific to giant cells, since the preparations also contain other host tissues as well as the nematode, its eggs and the gelatinous matrix. Information about the biochemical characteristics of giant cells comes mainly from histochemical and cytochemical studies. Using specific stains, Dropkin & Nelson (17) found that walls of giant cells contained cellulose and pectin but not lignin, suberin, starch or ninhydrin-positive substances. Bird (2) detected higher carbohydrates, fats, ribonucleic acids and proteins in the cells. Pronounced activities of oxido-reductive enzymes such as NAD and NADP diaphorases, malate, isocitrate, succinate, glucose-6-phosphate and 6-phospho-fluconate dehydrogenases were found in histochemical preparations of them (20). Alkaline and acid phosphotases, non-specific esterases, peroxidases, adenosine triphosphatase and cytochrome oxidase were similarly found (57). Autoradiographies of giant cells supplied with ^{3}H-thymidine showed, as might be expected, heavy DNA synthesis from very early stages of their development (53). In their histochemical work, Orion & Bonner (42) found strong amylase and invertase activities in the cells and their immediate vicinities, with the invertase being more concentrated on the cell walls. Since starch was markedly absent in these cells, they suggested that soluble sugars were being actively produced by starch hydrolysis and diffused to giant cells where they were broken down to monosaccharides and smaller molecules. The histochemical and autoradiographical techniques, however, are not quantitative so conclusions drawn from the data are necessarily speculative.

Gommers & Dropkin (21) were the first to determine quantitatively the metabolites in giant cells. The cells were dissected from lyophilized sections and accurately weighed (5 ng to 50 ng dry weight) on a fishpole balance. The metabolites were then quantitatively converted to equivalent pyridine nucleotides which in turn were subjected to "enzyme cycling" and measured fluorometrically. The results showed that giant cells of soybean roots induced by

Meloidogyne contained 4× and 6× more glucose and free amino acids, respectively, than did the actively growing root tip cells. The ATP, glucose-6-phosphate and protein contents of the two types of cells were similar.

Auxin has long been implicated in root-knot-nematode-induced gall formation, a subject that has been adequately reviewed by Viglierchio (58). Specific evidence for auxin participation in giant-cell physiology is still lacking, perhaps due to the absence of an adequate cytochemical technique for this group of compounds.

The involvement of cytokinins in giant-cell formation has also been suggested. Van Staden & Dimalla (56) demonstrated that susceptible tomato cultivars contained higher endogenous cytokinins than resistant ones. Moreover, root-knot nematode infections elevated these growth regulators in the roots of both cultivars. Bird & Loveys (10) were able to detect cytokinin-like substances in the eggs, juveniles and adults of root-knot nematodes and demonstrated that activity of the growth regulators in a gall was higher when the nematode was in rapid growth. Orion & Wergin (44) reported that *M. incognita* provided redifferentiations of amyloplasts to chloroplast-like plastids in tomato roots, an effect indistinguishable from that of cytokinin treatment. Hence, circumstantial evidence suggests the involvement of cytokinins in root-knot gall formation, as it does with auxins. However, a specific relationship of the growth regulators with giant-cell physiology is not yet known.

The most intriguing questions about the physiology of giant cells are those related to their initiation. There are two distinct, primary events in the initiation of a giant cell induced by a root-knot nematode: the modified nuclear divisions and the abortion of daughter-cell plate formations (32). Those two processes continue to operate also during the subsequent developmental stages of the giant cells. However, little experimental evidence is available to reveal the primary chemical agents or mechanisms responsible for these two fundamental phenomena. The nuclear divisions can be detected as soon as the infective juvenile begins to feed (32) and throughout its development (3), so the substance or substances responsible for triggering the chain of events in the accelerated karyokineses in a giant cell presumably come from the dorso-esophageal glands of all of the feeding stages of the nematode. It is possible that such key substances are highly diffusable small molecules which are also responsible for the hyperplasia observed around giant cells. The inhibitor mechanism for the daughter-cell plate formation, on the other hand, apparently limits its presence to the few modified cells around the head of the nematode, indicating that they probably are either highly labile molecules, not easily diffusable, or are present in extremely low concentrations. Jones & Payne (31) showed that failure of cytokineses in the roots of *I. balsamina* treated with caffeine was practically indistinguishable to that provoked by root-knot nematodes. It is not known, however, whether these nematodes produce any alkylated oxypurines, the group to which caffeine belongs.

CONCLUSIONS

Giant cells induced by root-knot nematodes were discovered about a century ago. However, it is only during the last two decades that detailed studies of the multinucleate cells and their obligate interrelationships with the nematodes have been made. With advanced EM instrumentation, much has been learned in the last ten years about the ultrastructures of these cells. However, histochemical and autoradiographical investigations have accumulated considerable qualitative data concerning their chemical and enzymatic constituents.

Recent cytological evidence suggests that giant cells are formed by repeated endomitoses without subsequent cytokineses (22,25,27,32) and that they are essentially transfer cells (29) passing nutrients to the nematode. However, before more confident generalizations can be made, additional information will be required on two fronts. Firstly, more plant-nematode combinations should be examined with modern EM instrumentation to verify the mode of their giant-cell formation. Current understanding of the giant-cell ontogeny is based on the detailed studies of only half a dozen plant-nematode relationships. There are more than 50 species of *Meloidogyne* (54), all of which apparently stimulate the formation of giant cells in their respective host plants, and we need information on many additional nematode-plant associations. In the area of the physiology of giant cells, direct experimental proof should be provided for the "giant" transfer cell model currently being speculated. For this purpose, we are in great need of ultramicroanalytical techniques with which metabolites in the developing giant cells, as well as in the associated nematode, can be precisely determined.

Acknowledgement

I would like to thank Dr. Anthony Raw, Dep. de Biologia Vegetal, Universidade de Brasília, for the language corrections and valuable criticisms made on the manuscript before submission.

Literature Cited

1. Birchfield, W. 1965. Host-parasite relations and host range studies of a new *Meloidogyne* species in southern USA. Phytopathology 55:1359-1361.
2. Bird, A. F. 1961. The ultrastructure and histochemistry of a nematode-induced giant cell. J. Biophys. Biochem. Cytol. 11:701-715.
3. Bird, A. F. 1962. The inducement of giant cells by *Meloidogyne javanica*. Nematologica 8:1-10.
4. Bird, A. F. 1972. Quantitative studies on the growth of syncytia induced in plants by root-knot nematodes. Int. J. Parasitol. 2:157-170.
5. Bird, A. F. 1972. Cell wall breakdown during the formation of syncytia induced in plants by root-knot nematodes. Int. J. Parasitol. 2:431-432.
6. Bird, A. F. 1973. Observations on chromosomes and nucleoli in syncytia induced by *Meloidogyne javanica*. Physiol. Plant Pathol. 3:387-391.
7. Bird, A. F. 1974. Plant response to root-knot nematode. Annu. Rev. Phytopathol. 12:69-85.
8. Bird, A. F. 1975. Symbiotic relationships between nematodes and plants. Symp. Soc. Exp. Biol. 29:351-371.
9. Bird, A. F. and B. R. Loveys. 1975. The incorporation of phytosynthates by *Meloidogyne javanica*. J. Nematol. 7:111-113.
10. Bird, A. F. and B. R. Loveys. 1980. The involvement of cytokinins in a host-parasite relationship between tomato (*Lycopersicon esculentum*) and a nematode (*Meloidogyne javanica*). Parasitology 80:497-505.
11. Bleve-Zacheo, T., G. Zacheo, F. Lamberti and O. Arrigoni. 1977. Cell wall breakdown and cellular response in developing galls induced by *Longidorus apulus*. Nematol. Mediterr. 5:305-311.
12. Brueske, C. H. and V. H. Dropkin. 1973. Free phenols and root necroses in Nematex tomato infected with the root-knot nematode. Phytopathology 63:329-334.
13. Christie, J. 1936. The development of root knot nematode galls. Phytopathology 26:1-22.
14. Crittenden. H. W. 1958. Histology and cytology of susceptible and resistant soybeans infected with *Meloidogyne incognita acrita*. Phytopathology 48:461.
15. Dropkin. V. H. 1969. Cellular responses of plant cells to nematode infections. Annu. Rev. Phytopathol. 7:101-122.
16. Dropkin, V. H. 1972. Pathology of *Meloidogyne*-galling, giant cell formation, effects on host physiology. Eur. Mediterr. Plant Prot. Organ. (OEPP/EPPO) Bull. 6:23-32.
17. Dropkin, V. H. and P. E. Nelson. 1960. The histopathology of root-knot nematode infections in soybeans. Phytopathology 50:442-447.
18. Endo, B. Y. 1971. Nematode-induced syncytia (giant cells). Host-parasite relationship of Heteroderidae, pp. 91-117 *In* Plant parasitic nematodes, Vol. II, ed. B. M. Zuckerman, W. F. Mai and R. A. Rhode. Academic Press, New York/London. 347 pp.
19. Endo, B. Y. 1975. Pathogenesis of nematode-infected plants. Annu. Rev. Phytopathol. 13:213-238.
20. Endo, B. Y. and J. A. Veech. 1969. The histochemical localization of oxidoreductive enzymes of soybeans infected with the root-knot nematode *Meloidogyne incognita acrita*. Phytopathology 59:418-425.
21. Gommers, F. J. and V. H. Dropkin. 1977. Quantitative histochemistry of nematode-induced transfer cells. Phytopathology 67:869-873.
22. Huang, C. S. and A. R. Maggenti. 1969. Mitotic aberrations and nuclear changes of developing giant cells in *Vicia faba* caused by root knot nematode *Meloidogyne javanica*. Phytopathology 59:447-455.
23. Huang, C. S. and A. R. Maggenti. 1969. Wall modifications in developing giant cells of *Vicia faba* and *Cucumis sativus* induced by root knot nematode, *Meloidogyne javanica*. Phytopathology 59:931-937.
24. Jones, M. G. K. 1981. Host cell responses to endoparasitic nematode attack: structure and function of giant cells and syncytia. Ann. Appl. Biol. 97:353-372.
25. Jones, M. G. K. and V. H. Dropkin. 1975. Cellular alterations induced in soybean roots by three endoparasitic nematodes. Physiol. Plant Pathol. 5:119-124.
26. Jones, M. G. K. and V. H. Dropkin. 1975. Scanning electron microscopy of syncytial transfer cells induced in roots by cyst nematodes. Physiol. Plant Pathol. 7:259-263.
27. Jones, M. G. K. and V. H. Dropkin. 1976. Scanning electron microscopy of nematode-induced giant transfer cells. Cytobios 15:149-161.
28. Jones, M. G. K. and B. E. S. Gunning. 1976. Transfer cells and nematode induced giant

cells in *Helianthemum*. Protoplasma 87:273-279.

29. Jones, M. G. K. and D. H. Northcote. 1972. Multinucleate transfer cells induced in Coleus roots by the root-knot nematode, *Meloidogyne arenaria*. Protoplasma 75:381-395.
30. Jones, M. G. K. and H. L. Payne. 1977. Scanning electron microscopy of syncytia induced by *Nacobbus aberrans* in tomato roots. Nematologica 23:172-176.
31. Jones, M. G. K. and H. L. Payne. 1977. Cytokinesis in *Impatiens balsamina* and the effect of caffeine. Cytobios 20:79-91.
32. Jones, M. G. K. and H. L. Payne. 1978. Early stages of nematode-induced giant-cell formation in roots of *Impatiens balsamina*. J. Nematol. 10:70-84.
33. Jones, M. G. K., A. Novacky and V. H. Dropkin. 1974. "Action potentials" in nematode-induced plant transfer cells. Protoplasma 80:401-405.
34. Jones, M. G. K., A. Novacky and V. H. Dropkin. 1975. Transmembrane potentials of parenchyma cells and nematode-induced transfer cells. Protoplasma 85:15-37.
35. Knypl, J. S. and K. M. Janas. 1975. Synthesis of RNA and protein, with ribonuclease activity, in carrot roots infested with *Meloidogyne hapla* Chitw. Physiol. Plant Pathol. 7:213-220.
36. Krusberg, L. R. 1963. Host response to nematode infection. Annu. Rev. Phytopathol. 1:219-240.
37. Krusberg, L. R. and L. W. Nielsen. 1959. Pathogenesis of root-knot nematodes to the Porto Rico variety of sweetpotato. Phytopathology 48:30-39.
38. Lewis, S. and M. A. McClure. 1975. Free amino acids in roots of infected cotton seedlings resistant and susceptible to *Meloidogyne incognita*. J. Nematol. 7:10-15.
39. Littrell, R. H. 1966. Cellular responses of *Hisbiscus esculentus* to *Meloidogyne incognita acrita*. Phytopathology 56:540-544.
40. Malo, S. E. 1965. Histological studies on the nature of resistance of Okinawa and Nemaguard peach rootstocks to *Meloidogyne javanica* (Treub 1885) Chitwood 1949. Nematologica 11:43.
41. McClure, M. A. 1977. *Meloidogyne incognita*: A metabolic sink. J. Nematol. 9:88-90.
42. Orion, D. and R. Bronner. 1973. The localization of starch, amylase and invertase in *Meloidogyne javanica* galls. Nematologica 19:401-402.
43. Orion, D. and G. Minz. 1971. The influence of morphactin on the root-knot nematode, *Meloidogyne javanica*, and its galls. Nematologica 17:107-112.
44. Orion, D. and W. P. Wergin. 1982. Chloroplast differentiation in tomato root galls induced by the root-knot nematode *Meloidogyne incognita*. J. Nematol. 14:77-83.
45. Owens, R. G. and R. F. Bottino. 1966. Changes in host cell wall composition induced by root-knot nematodes. Contrib. Boyce Thompson Inst. 23:171-180.
46. Owens, R. G. and J. H. Rubinstein. 1966. Metabolic changes induced by root-knot nematodes in host tissues. Contrib. Boyce Thompson Inst. 23:199-214.
47. Owens, R. G. and H. N. Specht. 1964. Root-knot histogenesis. Contrib. Boyce Thompson Inst. 22:471-489.
48. Owens, R. G. and H. N. Specht. 1966. Biochemical alterations induced in host tissues by root-knot nematodes. Contrib. Boyce Thompson Inst. 23:181-198.
49. Pate, J. S. and B. E. S. Gunning. 1972. Transfer cells. Annu. Rev. Plant Physiol. 23:173-196.
50. Rebois, R. V., P. A. Madden and B. Joe Eldridge. 1975. Some ultrastructural changes induced in resistant and susceptible soybean roots following infection by *Rotylenchulus reniformis*. J. Nematol. 7:122-139.
51. Reynolds, H. W., W. W. Carter and J. H. O'Bannon. 1970. Symptomless resistance of alfalfa to *Meloidogyne incognita acrita*. J. Nematol. 2:131-135.
52. Riggs, R. D., K. S. Kim and Ilene Gipson. 1973. Ultrastructural changes in Peking soybean infected with *Heterodera glycines*. Phytopathology 63:76-84.
53. Rohde, R. A. and M. A. McClure. 1975. Autoradiography of developing syncytia in cotton roots infected with *Meloidogyne incognita*. J. Nematol. 7:64-69.
54. Sasser, J. N. 1980. Root-knot nematodes: a global menace to crop production. Plant Dis. 64:36-41.
55. Smith, J. J. and W. F. Mai. 1965. Host-parasite relationships of *Allium cepa* and *Meloidogyne hapla*. Phytopathology 55:693-697.
56. Van Staden, J. and G. G. Dimalla. 1977. A comparison of the endogenous cytokinins in the roots and xylem exudate of nematode-resistant and susceptible tomato cultivars. J. Exp. Bot. 28:1351-1356.
57. Veech, J. A. and B. Y. Endo. 1969. The histochemical localization of several enzymes

of soybeans infected with the root-knot nematode *Meloidogyne incognita acrita*. J. Nematol. 1:265-276.

58. Viglierchio, D. R. 1971. Nematodes and other pathogens in auxin-related plant-growth disorders. Bot. Rev. 37:1-21.
59. Wergin, W. P. and D. Orion. 1981. Scanning electron microscope study of the root-knot nematode (*Meloidogyne incognita*) on tomato root. J. Nematol. 13:358-367.
60. Yousif, G. M. 1979. Histological responses of four leguminous crops infected with *Meloidogyne incognita*. J. Nematol. 11:395-401.

MECHANISMS OF RESISTANCE TO ROOT-KNOT NEMATODES

Jeng-sheng Huang

Department of Plant Pathology
North Carolina State University
Raleigh, North Carolina, USA

Agricultural and horticultural plants grown under field conditions are exposed to a tremendous number of soil microorganisms. One square meter of high-fertility field soil may contain as many as 3×10^{14} (300 g) bacterial cells, 400 g of fungi, 5×10^{8} (38g) protozoa and 1×10^{7} (12 g) nematodes (43). Fortunately, most soils do not harbor this many microorganisms, and many of these are not pathogenic to plants. For example, out of the more than 15,000 species of nematodes described, presently only 2200 species are recognized as plant parasites. Among these, only 50 species have been implicated as pathogens of soybean, 30 species as pathogens of corn, 28 species as pathogens of wheat, and 26 species as pathogens of alfalfa. Less than ten species attack each of these crops severely enough to be considered of economic importance.

The restricted number of plant-parasitic nematodes capable of attacking a given plant species is due to the presence of plant defense mechanisms. Plants possess many types of defense mechanisms. Cowling and Horsfall (9) draw an analogy between the defense in plants and the defense of a medieval castle. Extrusion from plants of chemicals toxic to pathogens is equivalent to the pouring of boiling oil on invaders by castle defenders. Preformed structures and chemicals that ward off pathogens at the plant surface are similar to the castle setup of reinforced walls, trap doors and poisonous baits. These mechanisms constitute the first line of defense. When this defense fails, many plants synthesize toxic substances, such as phytoalexins, and build physical barriers, such as apposition of callose to cell walls to fence off the invading pathogens. These defenses are similar in a way to the throwing of explosives and setting up barricades by the castle defenders to confine and minimize the damage done on the castle by invaders.

The preceding analogy serves to acquaint the reader with the types of resistance mechanisms. However, one must realize that resistance mechanisms are complex and, in many cases, poorly understood. Some plants, for example, show little injury even though the pathogen is allowed to build up to a high population. On the other hand, some plants may exhibit severe injury upon pathogenic infection but at the same time limit the pathogen from reproduction. According to Rohde (45), any reaction that renders the plant the ability to overcome or withstand the attack of a pathogen is a resistance reaction. Mechanisms involved in plant resistance to nematodes have been recently reviewed (31, 54, 55). In this Chapter, discussion will be limited to those mechanisms that are pertinent to the defense against root-knot nematodes before, during, and after their penetration.

PRE-INFECTIONAL RESISTANCE

Root exudates play an important role in both attracting and repelling root-knot nematodes. When an egg mass of *M. hapla* was placed on a water-agar plate equally distant from germinating resistant and susceptible alfalfa seeds, approximately three-fourths of the hatched juveniles were attracted to the susceptible seedlings, and only one-fourth were attracted to the resistant plants (23). Cucumber plants carrying a single dominant bitter gene (*Bi*) produce compounds that repel nematodes. These triterpenoid compounds known as cucurbitacins (Fig. 1) are toxic to a variety of insects. Excised roots from bitter cucumbers attract significantly fewer *M. incognita* than do roots from nonbitter isogenic lines (26).

These examples suggest that root-knot nematodes may be differentially attracted to roots of resistant and susceptible plants. The compounds responsible for repulsion and attrac-

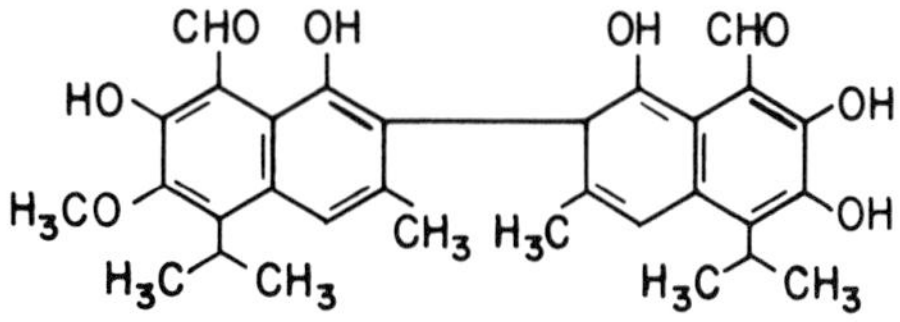

methoxygossypol

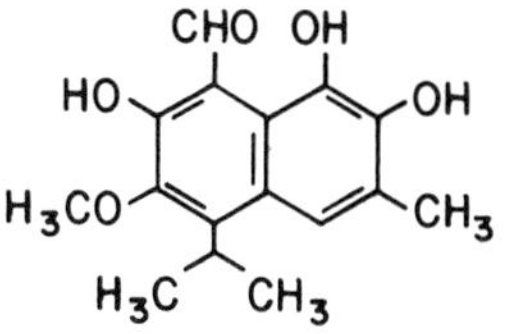

6-methoxyhemigossypol

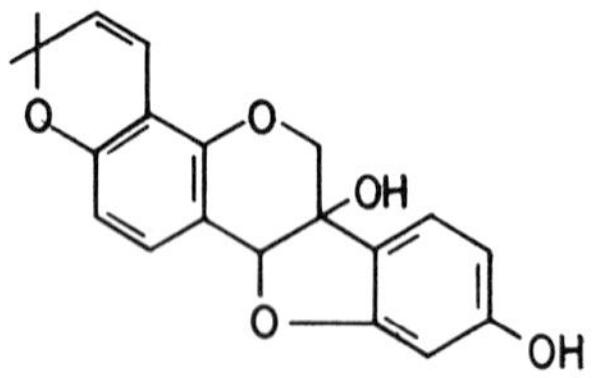

glyceollin I

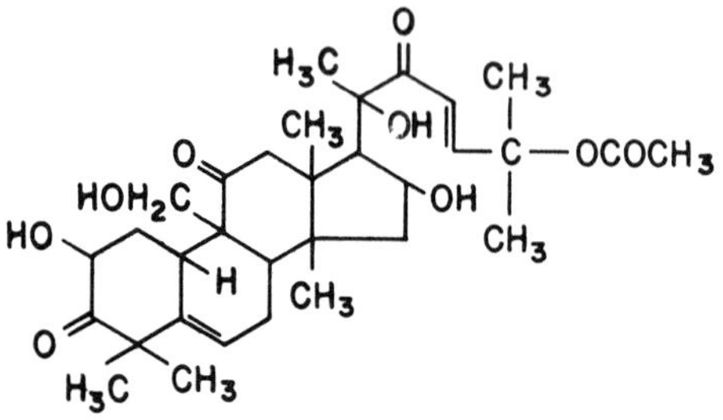

cucurbitacin A

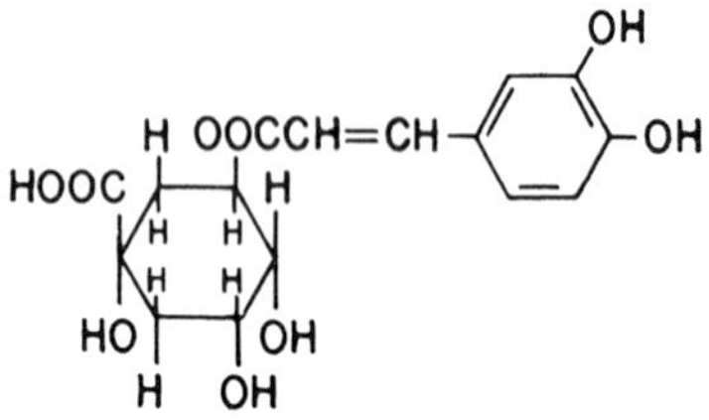

chlorogenic acid

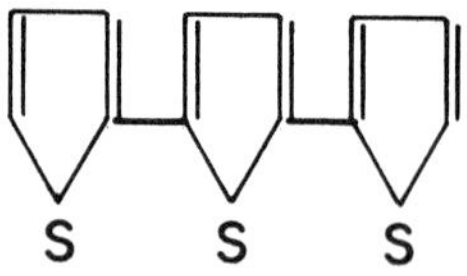

α-terthienyl

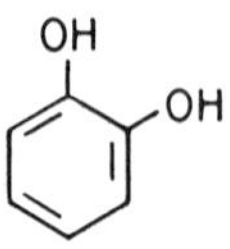

pyrocatechol

Fig. 1. Chemicals involved in resistance to root-knot nematodes.

tion, with few exceptions, have not been identified or characterized.

Root exudates from certain plants are nematicidal. For example, 63% of *M. incognita* juveniles treated with root exudates from 1 month-old marigold (*Tagetes erecta* L.) seedlings were killed in 72 hrs (1). Marigold roots contain a high concentration (~0.1% of dry root wt) of α-terthienyl (Fig. 1) that exhibits high nematicidal activity against several plant-pathogenic nematodes (52). The minimum lethal dosages for *Globodera rostochiensis*, *Anguina tritici*, and *Ditylenchus dipsaci* are 0.2, 0.5, and 5 ppm, respectively. Similar results were obtained with root exudates from margosa (*Azadirachta indica* Juss.). In these roots, nimbidin and thionemone are the active nematicidal compounds (1). Another reported naturally occurring nematicide is pyrocatechol (Fig. 1). This compound accumulates in high concentration in roots of *Eragrostis curvula* and is a potent nematicide against *M. javanica*, *M. hapla*, *M. arenaria thamesi*, and *M. incognita acrita* (48).

Despite its effectiveness in *in vitro* experiments, the role of α-terthienyl as a nematicide in soil has not been unequivocally established in greenhouse or field trials. Hackney and Dickerson (25) noted that populations of *M. incognita* were significantly fewer from marigold than from tomato, and aborted giant cells and dead juveniles and females were frequently observed in roots of marigold. Marigold, however, did not significantly reduce nematode populations in tomato plants when both were cultivated simultaneously. These results suggest that marigold do not reduce the nematode populations in the soil by nematicidal action. Daulton and Curtis (12) reported that several species of *Tagetes* grown in *M. javanica*-infested soil reduced the nematode populations, but addition of α-terthienyl as an aqueous suspension or as a solid at dosages up to 200 ppm had no effect on the nematodes in the soil.

Polyacetylenes extracted from roots of safflower (*Carthamus tinctorius*) and α-chaconine isolated from potato sprouts have also been shown to have potent nematicidal activity against certain nematodes (2, 34). Their effects on *Meloidogyne* spp. have yet to be determined.

RESISTANCE AT ROOT SURFACES

All plant-parasitic nematodes possess a protrusible stylet that functions to rupture their egg shell during hatching, to penetrate plant cell walls during infection, and to feed on the host cells during the feeding process. The pumping action of the stylet (2-24 strokes/sec), combined with cellulases, pectinases and other cell-wall-degrading enzymes released by nematodes (16), is capable of overcoming all superficial mechanical barriers such as plant cell walls and cuticles. Most experimental results indicate little if any difference in penetration of resistant and susceptible cultivars of alfalfa, cotton, tomato and snap bean by *Meloidogyne* spp. (19, 22, 38, 42, 44).

Exceptions to these reports, however, do exist. Sasser (47) tested fifty plant species and varieties commonly used in crop rotation for susceptibility to root-knot nematodes and found that resistant plants were not penetrated as readily as susceptible plants. Similarly, in a histopathological study of soybean-root-knot nematode interaction, Dropkin and Nelson (17) noticed that roots of resistant varieties had fewer nematodes than susceptible plants. Whether this observed difference is due to the root surface acting as a barrier to penetration or other factors has not been investigated.

It is difficult to assess critically the importance of the root surface as a barrier to nematode penetration. Direct observations of penetration using isolated epidermis from resistant and susceptible plants should provide more conclusive evidence for the role of the root surface as a penetration barrier.

POST-INFECTIONAL RESISTANCE

Preformed Chemicals

A number of plants contain compounds that are toxic to root-knot nematodes. Phenolics are involved in the formation of necrotic lesions and have been frequently suggested as the compounds responsible for disease resistance. Positive correlations between the concentrations of phenolics in plants and resistance to root-knot nematodes have been reported. For example, tomato cultivars susceptible to root-knot nematodes contain 40-44 μg phenolic compounds/g root tissue whereas a highly tolerant cultivar contains 79 μg; resistant cultivars, 97-102 μg; and immune cultivars, 107-117 μg (49).

Hung and Rohde (28) compared the chlorogenic acid (Fig. 1) content in two closely related tomato cultivars. They found 'Nemared,' a cultivar resistant to *M. incognita*, to contain 580 μg/g root tissue; and 'B-5,' a susceptible cultivar, 380 μg. Chlorogenic acid is known to affect nematode coordination.

The involvement of phenylalanine ammonia lyase (PAL) in the biosynthesis of phenolic compounds is well documented (11). Therefore, researchers have looked for a correlation between PAL activity and disease resistance. Roots of tomato 'Nematex' grown at 27°C (resistant state) showed a significant increase in PAL activity 12 hr after inoculation with *M. incognita* whereas roots of control and inoculated plants grown at 32°C (susceptible state) showed depressed PAL activity. According to Brueske (8), elevated PAL activity in resistant roots results in the synthesis of more phenolic compounds which in turn prevent and localize the parasitization of the nematode. These results support the importance of phenolics in resistance against root-knot nematodes.

Most phenolic compounds occur in plant tissues in bound forms as glycosides exhibiting low physiological and biochemical activity. Cleavage of these glycosides by β-glycosidases releases physiologically active free phenols. Homogenates of adult females and preparasitic second-stage juveniles of *M. incognita* and *M. javanica* exhibit β-galactosidase and β-glucosidase activities which are able to hydrolyze glycosides (50). However, currently there are no experimental data that clearly demonstrate a positive correlation among β-glycosidases, glycosides and resistance to root-knot nematodes.

Nutrition

Nutrition is important in resistance to root-knot nematodes in two ways: first, absence of certain nutrients may force penetrated nematodes to move out of the infected root; and second, host nutrition may affect sex ratio of root-knot nematodes in infected plants.

A time-course study of penetration, development and migration of *M. incognita acrita* in resistant and susceptible alfalfa cultivars was conducted by Reynolds et al. (42). During the first two days of inoculation, approximately the same number of nematodes penetrated the roots of resistant and susceptible plants; however, this number abruptly decreased in resistant alfalfa 3 to 4 days after inoculation. Examination of the nematode-infected seedlings incubated in water revealed that juveniles migrated out of the resistant roots, with the greatest migration occurring 4 days after inoculation. The abrupt decrease in nematode number inside resistant roots coincided with the massive migration out of resistant plants.

Similar results were obtained by Griffin and Elgin (22) in *M. hapla*-alfalfa interactions. They examined the penetration and development of root-knot nematode in resistant and susceptible cultivars under different temperatures. Significant differences in the number of juveniles in resistant and susceptible hosts were not found during the first 6 days of the experiments. A subsequent decline in nematode number was observed in resistant alfalfa while the number of nematodes in the susceptible remained high. An interpretation of these results is that roots of resistant alfalfa plants are unable to provide essential nutrients for nematode development, thereby, forcing mass exodus of penetrated nematodes. However, the actual mechanisms responsible for movement of juveniles out of roots are still not known (31).

Several reports have shown that sex determination in root-knot nematodes is influenced by a number of environmental factors. Gamma-ray irradiation of host roots increases the male population (30) as does heavy inoculation with nematodes (13). Similarly, continuous incubation of *M. graminis*-infected Bermuda grass at 32°C yielded 74% more males than those kept at 27°C and 80-100% more than those kept at 21 and 16°C (36). Host nutrition also plays an important role in sex determination. Approximately 5% of *M. incognita* juveniles in excised cucumber roots developed into males when the roots were grown on a medium containing 18 g sucrose/liter. This percentage increased as the concentrations of sucrose in the medium decreased. All juveniles developed into males when the sucrose concentration was decreased to 0.67 g/liter (37). Tomato plants deficient in N,P, K, N-P, N-K, P-K, or N-P-K also had higher male populations than those plants receiving complete nutrients (14).

Maleic hydrazide is a plant growth inhibitor that inhibits cell division in the meristematic regions, reduces the rate of respiration, blocks the translocation of photosynthates from leaves to roots and affects plant health in general. Davide and Triantaphyllou (15) reported that more juveniles of *M. javanica* and *M. incognita* developed into males if the tomato plants were treated with maleic hydrazide. Treatment of tomato plants with morphactin, another plant growth regulator affecting plant morphogenesis, produced similar results mainly through its antimitotic activity (39). Since female root-knot nematodes do most of the damage and are responsible for reproduction, a response in the plant which increases the male/female ratio may be considered a resistance reaction. Although high male/female ratios are frequently associated with phytotoxicity and poor growth of plants treated with certain growth inhibitors, manipulation of host nutrition and application of tolerant levels of growth inhibitors may be a feasible mode of control for root-knot nematodes and should not be overlooked.

Phytoalexins

According to Paxton (40), phytoalexins are low molecular-weight, antimicrobial compounds that are both synthesized by and accumulated in plants after exposure to microorganisms. This definition does not relate phytoalexins to disease resistance, and there is no evidence to prove irrevocably that phytoalexins are primary factors in resistance to parasites. Since many higher plants do synthesize and accumulate phytoalexins in response to biotic stress and most phytoalexins have potent antimicrobial activity, the possible role of phytoalexins in inhibiting the invading microorganisms and preventing disease development should not be ignored.

To date, 102 phytoalexins have been isolated and chemically characterized from Leguminosae (29), 25 from Solanaceae (35) and 46 from other families (10). Phytoalexins have diverse chemical structures but the majority of them are isoflavanoids, polyacetylenes, terpenoids and steroids. Production of phytoalexins has been investigated primarily in relation to fungal diseases, but they have been reported also in bacterial, viral and nematode diseases. The most detailed study of accumulation of root-knot-nematode-induced phytoalexin and disease resistance is the one reported by Kaplan et al. (32). They employed two soybean cultivars and two *Meloidogyne* spp. in their study of the association of phytoalexin glyceollin with the incompatible response of soybean roots to root-knot nematodes. Soybean cv. 'Centennial' is resistant to *M. incognita* but susceptible to *M. javanica*. Cultivar 'Pickett 71' is susceptible to both nematodes. Extracts from Centennial roots inoculated with *M. incognita* (a resistant host/nematode combination) contained 40 μg phytoalexin glyceollin/g root tissue tissue 3 days after inoculation and increased to 70 μg/g root after 7 days (Fig. 1). Roots of Centennial inoculated with *M. javanica* and those of Pickett 71 inoculated with either *M. incognita* or *M. javanica* (all three are susceptible host/nematode combinations) contained a very low concentration of glyceollin. These results indicate that glyceollin production is stimulated in a resistant host/nematode combination but not in susceptible interactions. Furthermore, and vascular tissues, the vascular tissue of Centennial inoculated with *M. incognita* had the highest glyceollin concentration. Vascular tissue is the site where the root-knot nematode becomes sedentary after penetration.

In vitro tests have shown that second-stage juveniles of *M. incognita* are extremely sensitive to glyceollin with $ED_{50}=11$ μg/ml. *M. javanica*, however, is not affected by glyceollin even at a concentration of 60 μg/ml.

The mode of action of glyceollin on *M. incognita* is not well understood. Glyceollin is known to inhibit oxygen uptake by second-stage juveniles (33). Work conducted on isolated soybean mitochondria indicates that glyceollin inhibits electron transport at the site between NADH dehydrogenase and coenzyme Q (7). Whether the specific site of inhibition on nematode mitochondria is similar to that on soybean mitochondria remains to be determined.

In this soybean/root-knot-nematode system, glyceollin appears to be involved in disease resistance because: 1) glyceollin is produced in large quantities in resistant host/parasite interactions; 2) glyceollin is accumulated at the feeding site of the nematode; and 3) glyceollin is induced by the nematode that is sensitive to the phytoalexin but not by the one resistant to the phytoalexin.

The accumulation of phytoalexins in resistant plants has also been found in root-knot-nematode-inoculated cotton roots. Veech and McClure (56) were unable to find a direct correlation between preinfectional concentrations of terpenoid aldehydes and resistance of cotton to *M. incognita*. However, upon nematode infection they did detect an increase of terpenoid aldehydes in the resistant cultivar. Veech (53) later identified methoxygossypol and methoxyhemigossypol (Fig. 1) as the compounds that increased in the infected resistant plants. He concluded that the ability to stimulate synthesis of methoxy-substituted terpenoid aldehydes in response to infection was important in host resistance.

Although the ability to produce phytoalexins in response to biotic stress exists in many higher plants, investigations of phytoalexin production in relation to root-knot nematode infection is limited to the above-mentioned reports. Nematologists should take advantage of the established techniques available for phytoalexin isolation and identification to characterize the role of phytoalexin production in resistance to nematode diseases.

Hypersensitive Reaction

Hypersensitivity is a reaction characterized by rapid cell death of infected plant tissues in response to pathogenic attacks. The resultant necrosis, formed as a consequence of cell death, localizes invaded pathogens and prevents the development of the disease. The hypersensitive reaction (HR) occurs when a pathogen attacks an incompatible host or a resistant plant and is considered by plant pathologists as a defense mechanism. Although the HR has been intensively investigated in fungal, viral and bacterial diseases, the occurrence of the HR in resistant plants against nematode infections was first recognized more than a quarter of a century ago (46). To date the HR has been cited as a possible resistance mechanism in many nematode diseases (18, 46, 57).

Dropkin and his coworkers (18) used development of juveniles in infected roots and the induction of necrosis and gall formation as criteria for resistance and susceptibility in their study of tomato/root-knot-nematode interactions. They reported that 73% of the juveniles that entered roots of susceptible plants developed, and 87% of the infected roots formed galls. None of the infected roots showed necrosis. In roots of a resistant variety, only 4% of the juveniles grew, 88% of infected roots showed necrosis and only 29% of the infected roots formed galls. Therefore, they concluded that hypersensitive necrosis of host cells and growth of juveniles are mutually exclusive.

Light and electron microscopy of the hypersensitive reaction in resistant tomato roots inoculated with *M. incognita* revealed that necrotic cells were localized to areas adjacent to the nematode. The cells surrounding the necrotic area showed a marked production of vesicles by the endoplasmic reticulum and had increased deposition of callose (5). The function of callose deposition in this host-parasite interaction has not been investigated but callose has been considered as a barrier against electrolyte loss in tobacco cells undergoing bacterially induced hypersensitive reaction (41).

We do not know the exact cause of hypersensitivity at this time. Some suggest that phytoalexins are involved and others contend that preformed phenolic compounds are responsible. We hope future study in this area will provide more insights to this important phenomenon.

Activation of Cyanide-resistant Respiration

Approximately 95% of the oxygen taken up by higher plants is consumed via cytochrome oxidases which are sensitive to cyanide. The other 5% is utilized by ascorbate oxidase and polyphenol oxidase which are cyanide-resistant. Following wounding, as in the case of nematode penetration, cyanide-resistant respiration is activated.

According to Arrigoni et al. (4), mitochondria isolated from aged slices of Jerusalem artichoke tubers were more resistant to cyanide than those prepared from freshly cut tissues. Proteins from cyanide-resistant mitochondria contain more hydroxyproline than those from cyanide-sensitive ones. The biosynthesis of hydroxyproline-containing proteins during the aging process requires ascorbic acid.

Zacheo et al. (59) was the first to suggest the involvement of cyanide-resistant respiration in resistance to root-knot nematodes. When susceptible tomato plants were treated with ascorbic acid, the population of *M. incognita* juveniles decreased significantly. On the other hand, nematode populations increased in resistant plants treated with lycorine, an inhibitor of ascorbic acid synthesis. These results suggested that ascorbic acid rendered tomato plants resistant to root-knot nematode infection, possibly through the increased synthesis of hydroxyproline-containing proteins and the activation of cyanide-resistant respiration. This contention is further supported by the fact that mitochondrial hydroxyproline-containing proteins increase in resistant tomato varieties but decrease in susceptible plants following inoculation of *M. incognita* (59).

The role of cyanide-resistant respiration in disease resistance was recently suggested by Stelzig et al. (51) and Alves et al. (3). Arachidonic acid, a fatty acid from *Phytophthora infestans*, elicits the accumulation of phytoalexins (rishitin, lubimin and other sesquiterpenes) in potato (6). Potato tuber slices preincubated with salicylhydroxamic acid (SHAM) or tetraethylthiuram disulfide (disulfiram), inhibitors of cyanide-resistant respiration, failed to accumulate the phytoalexins when challenged with arachidonic acid or extracts of *P. infestans* (3, 51). Similarly, accumulation of phytoalexin ipomeamarone was potentiated by aging root slices of sweet potato prior to treatment with $HgCl_2$. Pretreatment of root slices with potassium isothiocyanate or 8-hydroxyquinoline sulfate, other known inhibitors of cyanide-resistant respiration, repressed $HgCl_2$-induced ipomeamarone accumulation (24). These results suggest that cyanide-resistant respiration is involved in biosynthesis of phytoalexins.

The importance of hydroxyproline in disease resistance is also suggested in a study of host/*Heterodera rostochiensis* interactions. Giebel and Stobiecka (21) separated cytoplasmic and cell wall proteins from free amino acids and found that the ratios of proline to hydroxyproline (PRO/HYPRO) in the protein fractions were 1.59-2.11 and 2.20-2.54 in uninfected susceptible and resistant potato varieties, respectively. Upon inoculation with the golden nematode, the ratios increased to 2.30-2.36 in susceptible plants and decreased to 0.93-1.22 in resistant plants. They suggested that in resistant plants higher concentrations of hydroxyproline interfered with IAA-induced extension of plant cells, reduced plasticity of cell walls and inhibited hypertrophy, thus, preventing the nematodes from developing into cysts. Whether the increased hydroxyproline is associated with mitochondrial proteins has not been determined.

Peroxidases and Superoxide Dismutase

Numerous reports have suggested that peroxidase may be involved in disease resistance. Zacheo et al. (58) reported that there was no statistical difference in peroxidase activities between resistant and susceptible tomato cultivars prior to nematode inoculation. Upon infection by *M. incognita*, however, peroxidase activity significantly decreased in susceptible plants but increased in resistant plants. These results are in contrast to those reported by Ganguly and Dasgupta (20) who found that peroxidase activity was generally higher in resistant than in susceptible tomato cultivars prior to nematode infection. Upon inoculation of *M. incognita*, however, peroxidase activity increased over noninoculated susceptible controls. This increase was greater than that observed in resistant plants under identical conditions. Thus, the involvement of peroxidase in disease resistance can not be conclusively verified from the available data. Renewed interest in studying the role of peroxidase in disease resistance has been generated by recent discoveries that superoxide radicals and singlet O_2 can be generated from NADH and hydrogen peroxide through the action of catalase and peroxidase, respectively. Superoxide and singlet O_2 are toxic to living organisms and may play a role in inhibiting pathogen development.

A study of *M. incognita*/tomato interactions has shown that peroxidase activity increases and superoxide dismutase activity decreases in resistant cultivars after nematode inoculation. The reverse is true for susceptible plants (58). An interpretation of these results is that root cells of resistant plants react to the presence of the nematode by developing high peroxidase activity which produces a large quantity of free radicals. These free radicals incur resistance to the plant by inactivating the pathogen. Resistant cultivars maintain a high concentration of free radicals during the defense process by decreasing the level of superoxide dismutase, which would otherwise transform the free superoxide radicals into hydrogen peroxide and eventually into water and oxygen via catalase.

CONCLUSION

A series of biochemical and physiological reactions occur in plants in response to root-knot-nematode infection (27). As a consequence of this plant-parasite interaction, the plant is either overcome by the nematode and disease ensues, or the nematode is localized by the plant and disease development is limited. Detailed characterization of these biochemical and physiological processes is essential to advance our understanding of host-parasite relationship. This information would greatly aid plant breeders in breeding for cultivars resistant to root-knot nematodes.

In this Chapter, mechanisms of resistance to root-knot nematodes are summarized. There are still many gray areas in understanding host defense mechanisms against nematodes and renewed efforts to investigate this host/root-knot-nematode interaction are urgently needed. Among the various types of resistance mechanisms, induced resistance mediated by phytoalexins and hypersensitive reactions are promising areas of investigation because of their frequent occurrence in plant diseases, and they deserve our special attention.

Literature Cited

1. Alam, M. M., A. Masood, and S. I. Husain. 1975. Effect of margosa and marigold root-exudates on mortality and larval hatch of certain nematodes. Indian J. Exp. Biol. 13:412-414.
2. Allen, E. H. and J. Feldmesser. 1971. Nematicidal activity of α-chaconine: effect of hydrogen-ion concentration. J. Nematol. 3:58-61.
3. Alves, L. M., E. G. Heisler, J. C. Kissinger,

J. M. Patterson III, and E. B. Kalan. 1979. Effects of controlled atmospheres on production of sesquiterpenoid stress metabolites by white potato tuber. Possible involvement of cyanide-resistant respiration. Plant Physiol. 63:359-362.

4. Arrigoni, O., A. DeSantis, R. Arrigoni-Liso, and G. Calabrese. 1977. The increase of hydroxyproline-containing proteins in Jerusalem artichoke mitochondria during the development of cyanide-insensitive respiration. Biochem. Biophys. Res. Commun. 74:1637-1641.
5. Bleve-Zacheo, T., G. Zacheo, M. T. Melillo, and F. Lamberti. 1982. Ultrastructural aspects of the hypersensitive reaction in tomato root cells resistant to *Meloidogyne incognita*. Nematol. Mediterr. 10:81-90.
6. Bostock, R. M., J. A. Kuc, and R. A. Laine. 1981. Eicosapentaenoic and arachidonic acids from *Phytophthora infestans* elicit fungitoxic sesquiterpenes in the potato. Science 212:67-69.
7. Boydston, R., J. D. Paxton, and D. E. Koeppe. 1983. Glyceollin: A site-specific inhibitor of electron transport in isolated soybean mitochondria. Plant Physiol. 72:151-155.
8. Brueske, C. H. 1980. Phenylalanine ammonia lyase activity in tomato roots infected and resistant to the root-knot nematode, *Meloidogyne incognita*. Physiol. Plant Pathol. 16:409-414.
9. Cowling, E. B. and J. G. Horsfall. 1980. Prologue: How plants defend themselves, pp. 1-16, *In* Plant disease, an advanced treatise, Vol. V, ed. J. G. Horsfall and E. B. Cowling. Academic Press, New York. 534 pp.
10. Coxon, D. T. 1982. Phytoalexins from other families, pp. 106-132 *In* Phytoalexins, ed. J. A. Bailey and J. W. Mansfield. Halsted Press, London. 334 pp.
11. Creasy, L. L. and M. Zucker. 1974. Phenylalanine ammonialyase and phenolic metabolism. Recent Adv. Phytochem. 8:1-19.
12. Daulton, R. A. C. and R. F. Curtis. 1963. The effects of *Tagetes* spp. on *Meloidogyne javanica* in southern Rhodesia. Nematologica 9:357-362.
13. Davide, R. G. and A. C. Triantaphyllou. 1967. Influence of the environment on development and sex differentiation of root-knot nematodes. I. Effect of infection density, age of the host plant and soil temperture. Nematologica 13:102-110.
14. Davide, R. G. and A. C. Triantaphyllou. 1967. Influence of the environment on development and sex differentiation of root-knot nematodes. II. Effect of host nutrition. Nematologica 13:111-117.
15. Davide, R. G. and A. C. Triantaphyllou. 1968. Influence of the environment on development and sex differentiation of root-knot nematodes. III. Effect of foliar application of maleic hydrazide. Nematologica 14:37-46.
16. Deubert, K. H. and R. A. Rohde. 1971. Nematode enzymes, pp. 73-90 *In* Plant parasitic nematodes, Vol. I, ed. B. M. Zuckerman, W. F. Mai, and R. A. Rohde. Academic Press, New York. 347 pp.
17. Dropkin, V. H. and P. E. Nelson. 1960. The histopathology of root-knot nematode infections in soybeans. Phytopathology 50:442-447.
18. Dropkin, V. H., J. P. Helgeson and C. D. Upper. 1969. The hypersensitivity reaction of tomatoes resistant to *Meloidogyne incognita*: reversal by cytokinins. J. Nematol. 1:55-61.
19. Fassuliotis, G., J. R. Deakin, and J. C. Hoffman. 1970. Root-knot nematode resistance in snap beans: Breeding and nature of resistance. J. Am. Soc. Hortic. Sci. 95:640-645.
20. Ganguly, A. K. and D. R. Dasgupta. 1979. Sequential development of peroxidase (EC 1.11.1.7) and IAA-oxidase activities in relation to resistant and susceptible responses in tomatoes to the root-knot nematode, *Meloidogyne incognita*. Indian J. Nematol. 9:143-151.
21. Giebel, J. and M. Stobiecka. 1974. Role of amino acids in plant tissue response to *Heterodera rostochiensis*. I. Protein-proline and hydroxyproline content in roots of susceptible and resistant solanaceous plants. Nematologica 20:407-414.
22. Griffin, G. D. and J. H. Elgin, Jr. 1977. Penetration and development of *Meloidogyne hapla* in resistant and susceptible alfalfa under differing temperatures. J. Nematol. 9:51-56.
23. Griffin, G. D. and W. W. Waite. 1971. Attraction of *Ditylenchus dipsaci* and *Meloidogyne hapla* by resistant and susceptible alfalfa seedlings. J. Nematol. 3:215-219.
24. Haard, N. F. 1977. Potentiation of wound induced formation of ipomeamarone by cyanide insensitive respiration in sweet potato (*Ipomoea batatas*) root slices. Z. Pflanzenphysiol. 81:364-368.
25. Hackney, R. W. and O. J. Dickerson. 1975.

Marigold, castor bean, and chrysanthemum as controls of *Meloidogyne incognita* and *Pratylenchus alleni*. J. Nematol. 7:84-90.

26. Haynes, R. L. and C. M. Jones. 1976. Effects of the *Bi* locus in cucumber on reproduction, attraction, and response of the plant to infection by the southern root-knot nematode. J. Am. Soc. Hortic. Sci. 101:422-424.
27. Huang, J. S. 1976. Possible mechanisms involved in host specificity, pp. 57-63 *In* Proc. IMP Res. Plann. Conf. on Root-knot Nematodes, *Meloidogyne* spp. January 12-16, 1976, North Carolina State Univ. Raleigh, NC. 106 pp.
28. Hung, C. L. and R. A. Rohde. 1973. Phenol accumulation related to resistance in tomato to infection by root-knot and lesion nematodes. J. Nematol. 5:253-258.
29. Ingham, J. L. 1982. Phytoalexins from the leguminosae, pp. 21-80 *In* Phytoalexins, ed. J. A. Bailey and J. W. Mansfield. Halsted Press, London. 334 pp.
30. Ishibashi, N. 1965. The increase in male adults by gamma-ray irradiation in the root-knot nematode, *Meloidogyne incognita* Chitwood. Nematologica 11:361-369.
31. Kaplan, D. T. and N. T. Keen. 1980. Mechanisms conferring plant incompatibility to nematodes. Rev. Nematol. 3:123-134.
32. Kaplan, D. T., N. T. Keen, and I. J. Thomason. 1980. Association of glyceollin with the incompatible response of roots to *Meloidogyne incognita*. Physiol. Plant Pathol. 16:309-318.
33. Kaplan, D. T., N. T. Keen, and I. J. Thomason. 1980. Studies on the mode of action of glyceollin in soybean incompatibility to the root knot nematode, *Meloidogyne incognita*. Physiol. Plant Pathol. 16:319-325.
34. Kogiso, S., K. Wada, and K. Munakata. 1976. Isolation of nematicidal polycetylenes from *Carthamus tinctorius* L. Agric. Biol. Chem. 40:2085-2089.
35. Kuć, J. 1982. Phytoalexins from the Solanaceae, pp. 81-105 *In* Phytoalexins, ed. J. A. Bailey and J. W. Mansfield. Halsted Press, London. 334 pp.
36. Laughlin, C. W., A. S. Williams, and J. A. Fox. 1969. The influence of temperature on development and sex differentiation of *Meloidogyne graminis*. J. Nematol. 1:212-215.
37. McClure, M. A. and D. R. Viglierchio. 1966. The influence of host nutrition and intensity of infection on the sex ratio and development of *Meloidogyne incognita* in sterile agar cultures of excised cucumber roots. Nematologica 12:248-258.
38. McClure, M. A., K. C. Ellis, and E. L. Nigh. 1974. Resistance of cotton to the root-knot nematode, *Meloidogyne incognita*. J. Nematol. 6:17-20.
39. Orion, D. and G. Minz. 1971. The influence of morphactin on the root-knot nematode, *Meloidogyne javanica*, and its galls. Nematologica 17:107-112.
40. Paxton, J. D. 1980. A new working definition of the term "phytoalexin." Plant Dis. 64:734.
41. Politis, D. J. and R. N. Goodman. 1978. Localized cell wall appositions: incompatibility response of tobacco leaf cells to *Pseudomonas pisi*. Phytopathology 68:309-316.
42. Reynolds, H.W., W. W. Carter, and J. H. O'Bannon. 1970. Symptomless resistance of alfalfa to *Meloidogyne incognita acrita*. J. Nematol. 2:131-134.
43. Richards, B. N. 1976. Introduction to the soil ecosystem. Longman Group Ltd., London. 266 pp.
44. Riggs, R. D. and N. N. Winstead. 1959. Studies on resistance in tomato to root-knot nematodes and on the occurrence of pathogenic biotypes. Phytopathology 49:716-724.
45. Rohde, R. A. 1972. Expression of resistance in plants to nematodes. Annu. Rev. Phytopathol. 10:233-252.
46. Ross, J. P. 1958. Host-parasite relationship of the soybean cyst nematode in resistant soybean roots. Phytopathology 48:578-579.
47. Sasser, J. N. 1954. Identification and host-parasite relationships of certain root-knot nematodes (*Meloidogyne* spp.). Md. Agric. Exp. Stn. Bull. A-77. 31pp.
48. Scheffer, F., R. Kickuth, and J. H. Visser. 1962. Die Wurzelausscheidungen von *Eragostis curvula* (Schrad.) Nees und ihr Einfluss auf Wurzeknoten-nematoden. Z. Pflanzenernaehr. Dueng. Bodenk. D. 98:114-120.
49. Singh, B. and B. Choudhury. 1973. The chemical characteristics of tomato cultivars resistant to root-knot nematodes (*Meloidogyne* spp.). Nematologica 19:443-448.
50. Starr, J. L. 1981. Beta-glycosidases from *Meloidogyne incognita* and *M. javanica*. J. Nematol. 13:413-414.
51. Stelzig, D. A., R. D. Allen, and S. K. Bhatia. 1983. Inhibition of phytoalexin synthesis in arachidonic acid-stressed potato tissue by inhibitors of lipoxygenase and cyanide-

resistant respiration. Plant Physiol. 72:746-749.

52. Uhlenbroek, J. H. and J. D. Bijloo. 1958. Investigations on nematicides. I. Isolation and structure of a nematicidal principle occurring in *Tagetes* roots. Recl. Trav. Chim. Pays-Bas Belg. 77:1004-1009.
53. Veech, J. A. 1978. An apparent relationship between methoxy-substituted terpenoid aldehydes and the resistance of cotton to *Meloidogyne incognita*. Nematologica 24:81-87.
54. Veech, J. A. 1981. Plant resistance to nematodes, pp. 377-403 *In* Plant parasitic nematodes, Vol. III, ed. B. M. Zuckermann and R. A. Rohde. Academic Press, New York. 508 pp.
55. Veech, J. A. 1982. Phytoalexins and their role in the resistance of plants to nematodes. J. Nematol. 14:2-9.
56. Veech, J. A. and M. A. McClure. 1977. Terpenoid aldehydes in cotton roots susceptible and resistant to the root-knot nematode, *Meloidogyne incognita*. J. Nematol. 9:225-229.
57. Wallace, H. R. 1961. The nature of resistance in chrysanthemum varieties to *Aphelenchoides ritzemabosi*. Nematologica 6:49-58.
58. Zacheo, G., T. Bleve-Zacheo, and F. Lamberti. 1982. Role of peroxidase and superoxide dismutase activity in resistant and susceptible tomato cultivars infested by *Meloidogyne incognita*. Nematol. Mediterr. 10:75-80.
59. Zacheo, G., F. Lamberti, R. Arrigoni-Liso, and O. Arrigoni. 1977. Mitochondrial protein-hydroxyproline content of susceptible and resistant tomatoes infected by *Meloidogyne incognita*. Nematologica 23:471-476.

Part Four

Ecology

CHAPTER PAGE

ECOLOGY OF *MELOIDOGYNE* SPP. — EMPHASIS ON ENVIRONMENTAL FACTORS AFFECTING SURVIVAL AND PATHOGENICITY

S. D. Van Gundy

Department of Nematology
University of California
Riverside, California, USA

Ecology of *Meloidogyne* spp. is a complexity of dynamic, interacting environmental factors. In the agroecosystem there is a lack of homogeneity of the soil environment both horizontally and vertically. Superimposed on the complexity of the soil physical environment is the spatial distribution of the interacting biological subsystem. A holistic comprehension of this system is difficult, and Ferris and Van Gundy (9) have made a case for use of derived models through computer simulation to promote a better understanding of this complexity. Most of the current ecological information on *Meloidogyne* spp. has evolved from an examination of single extrinsic and intrinsic factors on individual components of the system. Thus, the interaction of effects among the factors are often overlooked or not investigated because of interpretation and experimentation difficulties. For example nematodes depend on moisture and oxygen; however, in soil it is difficult to separate the interacting effects of soil texture, moisture and aeration on nematode movement and survival.

Often research data on nematode activities is expressed as "optimum conditions" which exhibit the physiological processes functioning at their maximum rate. These optima, however, are not always favorable for longevity and survival of the population or species as a whole nor are they likely to occur in nature for extended periods of time. For example, the optimum temperature for growth and reproduction of *M. javanica* is 25-30°C while the optimum temperature for survival of eggs and juveniles in soil is 10-15°C.

From the holistic point of view the population level of any *Meloidogyne* sp. is the product of the food supply, adaptation to the physical and biological environment and the compatibility of the host and nematode. Depending on host crop and seasonal fluctuations as many as 8×10^4 *Meloidogyne* juveniles can be found in the upper 120 cm of soil (8). In addition to the nematode's ability to reproduce on a host crop is its ability to survive from one crop season to another, to endure long periods of cold and dry and to survive bathing in a variety of chemical solutions. Although the genus *Meloidogyne* is one of the most intensively studied of the plant-parasitic nematodes, there are few generalities that can be made about the ecology of this nematode. The International *Meloidogyne* Project has collected an extensive set of ecological data on 662 populations of seven species from agroecosystems in 76 countries (32). The four most prevalent species were *Meloidogyne incognita* (47%), *M. javanica* (40%), *M. arenaria* (7%) and *M. hapla* (6%). They also found that *M. incognita* and *M. javanica* were widely distributed in tropical, subtropical and warm temperature climates. Though less common, *M. arenaria* possessed similar temperature requirements. *Meloidogyne hapla* was collected in climates with annual temperatures as low as 6°C or about 9 degrees lower than the other three species. There also appeared to be a positive correlation between high pH and low rainfall cropping situations and the collections of *M. incognita* and *M. javanica*. Very few *Meloidogyne* populations were collected from soils with more than 40% clay or more than 50% silt. Clearly, the four main species of *Meloidogyne* have been widely distributed throughout the agricultural regions of the world and appear to be very adaptable to a wide range of agroecosystems.

The ecological relationships of *Meloidogyne* spp. have been reviewed from a holistic systems approach by Ferris and Van Gundy (9) and from a pest-management approach by Van Gundy (40). Additional ecological information on *Meloidogyne* is contained in several sources (17,31,38,41,46). This review will focus on the environmental factors affecting survival and pathogenicity of *Meloidogyne* eggs and juveniles in soil.

DORMANCY

Survival of *Meloidogyne* eggs and juveniles is less persistent (>1 year) than in the closely related genus *Heterodera* where eggs in cysts remain viable for 1-5 years (38). Eggs of *Meloidogyne* are deposited outside the nematode body in a gelatinous matrix. There is some evidence that two kinds of egg masses (white and brown) may be formed, depending on environmental stresses (12,14). The white egg masses are usually formed early in the host growing cycle, and the eggs hatch immediately upon development to ensure more than one generation per growing cycle. As the growing season progresses and/or adverse environmental stresses are imposed on the host, brown egg masses are formed which have a dormancy (diapause) and do not hatch immediately (12). This diapause is present in all strains of the four most common *Meloidogyne* species (13). The development of two kinds of egg masses (white and brown) ensures that at least some eggs will not hatch immediately even if the soil environmental conditions remain optimum for hatching and that there will be some carryover of eggs from one crop season to another. Egg masses of *M. naasi* exhibit a temperature dormancy (diapause) where hatching is stimulated by chilling at 0°C or 10°C for 1 week prior to incubation at 20°C (10). In at least five species there is some physiological dormancy mechanism for delaying hatching and extending the survival of viable eggs in the soil for several months. There is no evidence that survival of eggs is extended in fallow soil beyond one year.

TEMPERATURE

Temperature is the main driving factor in the embryonic development of eggs, in the ability of juveniles to escape from the eggshell, for growth and reproduction, and for survival. The optimum temperature for embryogenesis is between 25°C and 30°C for all the species studied (3,46). In *M. javanica*, hatching was about 2 times as rapid at 25°C as at 20°C and 4 times as rapid as at 15°C (4) while in *M. chitwoodi* and *M. hapla*, hatching at 25°C was only about 2 times as rapid as at 15°C (27).

Within the genus *Meloidogyne* there are two distinct survival groups: thermophils and cryophils, which can be separated on their ability to survive lipid-phase transitions that occur at 10°C. Both *M. hapla* and *M. chitwoodi* and possibly *M. naasi* are cryophils and are able to survive soil temperatures to and below 0°C (27,28). *Meloidogyne incognita, M. javanica, M. arenaria* (33) and possibly *M. exigua* are thermophils and do not have extended survival in soils below 10°C. This finding is further reinforced by hatching data. Both *M. chitwoodi* and *M. hapla* exhibit significant hatching below 10°C (27) while *M. incognita* and *M. javanica* do not exhibit significant hatching below 15°C (3). Hatching requires movement of the nematode and stylet thrusting and presumably a higher metabolic level than minimum maintenance and survival. The susceptibility of *Meloidogyne* juveniles to succumb to temperatures is related to the existence of temperature-induced phase transitions in the lipid position of the biological membranes in the nematode. Lyons et al. (15) found that *M. javanica* exhibited a phase transition at 10°C while *M. hapla* did not. Within a species there may be small differences in temperature minimums between populations. These are called thermotypes. Dao (5) demonstrated that a Netherlands population of *M. incognita* was able to infect plant roots and reproduce at about 5°C lower than a Venezuelan population. This suggests that within the broad temperature range of each species individual populations are able to adapt to local temperature regimes. The temperature optimums for *M. hapla* and *M. chitwoodi* which are generally 5°C lower than that for *M. javanica* are probably the main factor responsible for the worldwide and local distribution of members of this genus. For example in California, *M. hapla* is most common in the cooler coastal areas (30) although it has been reported from the hot Palo Verde Valley of Southern California (24). *Meloidogyne naasi* and *M. thamesi* are found along the coast and in the cooler areas (30). *M. incognita* and *M. javanica* are more commonly found in the hot interior irrigated valleys (30). In East Africa, *M. hapla* occurred in several areas but only flourished above 1800 m altitude. *M. incognita* and *M. javanica* were restricted to areas below 2000 m (47). This pattern is typical of other observations on the worldwide distribution of *Meloidogyne* species as demonstrated in the International *Meloidogyne* Project.

Temperature has a strong influence on the survival of *Meloidogyne* eggs and juveniles in soil. For example, the juveniles of *M. javanica* begin their life in soil with a third of their body weight in lipids as a storage energy reserve. It is used at a rate dependent upon rate of metabolism which is in turn influenced by temperature, moisture, aeration, etc. (39,43,44). The more food reserves the nematode has when the host root enters its environment the greater are its chances of getting to the infection site and entering the root. Extended survival and infectivity in soil must then be related to conservation of food reserves and physiological condition of aged juveniles rather than chronological age. In general the optimum temperatures for survival of eggs and juveniles is from 10-15°C (2,34).

In addition to the effects of temperature on distribution, survival, growth and reproduction, falling or rising temperatures are a critical factor in determination of nematode movement and infection. For example, Prot and Van Gundy (23) demonstrated that about 17% of *M. hapla* juveniles were able to migrate 20 cm at 14°C and infect tomato roots in 10 days while only about 3% of the *M. incognita* were capable of migration at that temperature. Thus, *M. hapla* seems to be able to hatch, migrate, infect roots and begin its life cycle much earlier in the spring than *M. incognita*. The same would also be true for fall plantings where Roberts et al. (26) found that if the planting of winter wheat was delayed until the soil temperatures reached 18°C, there was little infection and egg production by *M. incognita* during the winter growing season. If wheat was planted earlier, there was some infection and egg production by spring which maintained the soil population for infection of a summer crop. Thus, a knowledge of the local temperature requirements of each species can be useful in developing management strategies for control.

In summary, of the *Meloidogyne* species studied, the order of resistance to cold is *M. chitwoodi* > *M. hapla* > *M. incognita* > *M. arenaria* > *M. javanica*.

SOIL TEXTURE, MOISTURE, AERATION AND OSMOTIC POTENTIAL

In field soil it is difficult to separate the interacting effects of soil texture, moisture and aeration. Briefly, soil is a three-phase system: solid, liquid and gas (31). The solid phase is composed of sand, silt and clay particles and organic materials and can be simply analyzed by various techniques. Soil texture is a mixture of solids and is constantly changing both vertically and horizontally in the field. The surfaces of the soil particles and the spaces between particles contain the liquid and gas phases. The liquid phase contains soil solutes and oxygen and is inversely proportioned to the gas phase. The pore spaces contain and distribute oxygen in the soil: the more liquid, the less oxygen. Nematodes are active in soils with moisture levels 40-60% of field capacity and move through the soil in water films. As the soils either dry or increase in moisture, nematode activity decreases. Guiran (11) found that in both a sandy soil and in a clay soil the optimum emergence of juveniles from egg masses was slightly above field capacity (pF 3.0): pF 3.3 in sandy soil and pF 3.6 in clay soil. Emergence decreased as the wilting point (pF 4.2) was approached. Conditions below field capacity inhibited hatch due to reduced oxygenated conditions.

There are many reports and observations that associate the distribution and disease severity of *Meloidogyne* with sandy or high organic soils. Sasser (29) observed that infestations of *M. incognita* and *M. hapla* in Maryland occurred more frequently in sandy loam soils than in clay soils. In Arizona, heaviest infestations of *M. incognita* occurred on coarse-textured soils (18). *Meloidogyne hapla* was prevalent in organic soils in New York (49). In east Africa, Whitehead (48) reported that there was no correlation of *M. javanica* and *M. incognita* with soil texture. Generally, *Meloidogyne* sp. can occur on a wide range of soil types, but their association with crop damage is highly reflective of sandy soils or sandy patches within fields. The presence of the nematode increases the plant stress often found on these soils. Recently, Prot and Van Gundy (22) found that the percentage of juveniles able to migrate 20 cm and penetrate roots decreased as the clay and silt fractions in the soil increased from 14% to 33%. No migration occurred in soils with more than 30% clay plus silt. There seems to be little question that the agricultural importance of *Meloidogyne* is associated with sandy soils.

In drying environments, the gelatinous matrix of the egg sac appears to maintain a high moisture level and provide a barrier to water loss from eggs (44). Embryos and first-stage juveniles are more resistant to water loss than unhatched second-stage juveniles because of changes in egg membrane after the first molt in the egg. Van Gundy et al. (39) found that some *Meloidogyne* juveniles were able to survive in moist soil for periods of a month without much physiological aging as measured by stored food reserves. They suggested that these juveniles were contained in moist soil aggregates at or near anaerobic conditions. Towson and Apt (35) found that the overall survival of *M. javanica* was greatest at a water potential of —1.1 bars. The predominant form of survival was juveniles of which about 90% were coiled and in the state of anhydrobiosis. This report was the first that demonstrated conclusively that *Meloidogyne* can enter the state of anhydrobiosis and prolong their survival in dry soils. In drier soils desiccation effectively reduced numbers of eggs and juveniles. However, the survival of some eggs was due to inhibition of hatching and the survival of juveniles due to anhydrobiosis. Wallace (42) also observed a reduction in hatch in dry sand.

A clear and inverse relationship exists with water potential of soils. At low water potentials the soil pores are filled and oxygen deficiency may result, resulting in decreased hatch (43,44). Exposures of embryonated eggs to anaerobic conditions for one week are lethal, probably due to lactic acid accumulation within the eggshell. At low oxygen levels, the metabolism, movement, and infectivity of juveniles are reduced, but their survival is enhanced; the growth and reproduction of females in roots is also reduced (36,37,39). There is a linear relationship between rate of oxygen diffusion and movement of *M. javanica* juveniles (36). Baxter and Blake (1) found that all activities of *M. javanica* increased as oxygen concentrations increased from 0.2 to 21% and concluded that a favorable environment for development, hatch and mobility would be provided where moist soils drain rapidly and allow oxygen concentrations to increase above 10%. The effect of aeration on invasion, movement and hatch of *M. hapla* in organic soils was adversely affected by CO_2 concentrations above 10% (49).

At high water potentials, there are compounding effects beyond moisture availability. Especially in drying soils, nematodes may also experience high osmotic pressures, particularly after fertilizer applications. The osmotic potential of the soil solution increases and can alter the behavior of *Meloidogyne* sp. Wallace (45) found that hatching of *M. javanica* from egg sacs was stimulated at high nutrient levels (200 mg/pot) and was reduced at osmotic potentials above 2.5 atmospheres (42). Reversat (25) found that *M. javanica* used its food reserves 7.5 times slower under saline conditions than normal conditions. However, infectivity was reduced at solution concentrations of ECe (electrical conductivity of the saturation extract of the soil) 1.5, 2.5, 3.5, and 5.0 mmhos/cm of NaCl or $CaCl_2$ on tomato plants (6). Even small changes in osmotic potential can be important to nematode behavior. Prot (19) has demonstrated "hydrotaxis" in *M. javanica* where juveniles migrate along a moisture gradient from the region of highest salt concentration to regions of lowest salt concentration (0.125×10^{-2} and 0.5×10^{-2} M/liter).

Meloidogyne spp. survive, hatch and reproduce over a wide pH range, 4.0-8.0. Wallace (42) found that the hatch of *M. javanica* eggs was maximum between pH 6.4 and 7.0 and inhibited below pH 5.2. As with other soil solution factors, pH fluctuates with changes in soil moisture and soil salinity. As with soil texture, increased crop damage from *Meloidogyne* sp. is often associated with alkaline soils (7,16). This occurrence appears to be associated with the increased stress placed on the plant by salinity and increased uptake of sodium ions. Another major factor in the soil solutions affecting behavior and distribution of plant-parasitic nematodes in soil are roots and their emanations (21). The spacial distribution of *Meloidogyne* spp. is closely associated with root distribution in grape vineyards (8). Although this is largely a response to food source, there are other factors that influence nematode migration. Prot (20) found that the capacity of juveniles to migrate relatively large distances (25-50 cm) in 5 to 9 days was common within populations of *Meloidogyne* from West Africa. *M. incognita* was most aggressive in migration and penetration of the populations tested.

In summary, the relationship among numbers of *Meloidogyne* in the soil, rate of reproduction and growth of the host plant is a complex ecological situation. Environmental factors may influence both plant and nematode. They may influence the nematode in the soil, in soil water, and in the plant. A good understanding of these interactions is needed for the development of nematode pest-management programs for a specific crop in a specific geographical region.

Literature Cited

1. Baxter, R. I. and C. D. Blake. 1969. Oxygen and the hatch of eggs and migration of larvae of *Meloidogyne javanica*. Ann. Appl. Biol. 63:191-203.
2. Bergeson, G. B. 1959. The influence of temperature on the survival of some species of the genus *Meloidogyne* in the absence of a host. Nematologica 4:344-354.

3. Bird, A. F. and H. R. Wallace. 1965. The influence of temperature on *Meloidogyne hapla* and *M. javanica*. Nematologica 11:581-589.
4. Bird, A. F. 1972. Influence of temperature on embryogenesis in *Meloidogyne javanica*. J. Nematol. 4:206-213.
5. Dao, D. F. 1970. Climatic influence on distribution pattern of plant parasitic and soil inhabiting nematodes. Meded. Landbouwhogesch. Wageningen 70-2. 181 pp.
6. Edongali, E. A., L. Duncan, and H. Ferris. 1982. Influence of salt concentration on infectivity and development of *Meloidogyne incognita* on tomato. Rev. Nematol. 5:111-117.
7. Edongali, E. A. and H. Ferris. 1982. Varietal response of tomato to the interaction of salinity and *Meloidogyne incognita* infection. J. Nematol. 14:57-62.
8. Ferris, H. and M. V. McKenry. 1974. Seasonal fluctuations in the spatial distribution of nematode populations in a California vineyard. J. Nematol. 6:203-210.
9. Ferris, H. and S. D. Van Gundy. 1979. *Meloidogyne* ecology and host interrelationships pp. 205-230 *In* Root-knot nematodes (*Meloidogyne* species) systematics, biology and control, ed. F. Lamberti and C. E. Taylor. Academic Press, New York. 477 pp.
10. Franklin, M. T., S. A. Clark, and J. A. Course. 1971. Population changes and development of *Meloidogyne naasi* in the field. Nematologica 17:575-590.
11. Guiran, G. de and Y. Demeure. 1978. Influence du potential hydrique des soils sur les masses d'oeufes de *Meloidogyne incognita* (Nematoda:Meloidogynidae). Rev. Nematol. 1:119-134.
12. Guiran, G. de. 1980. Facteurs induisant chez *Meloidogyne incognita* un blocage du developpement des ocufs considere comme une diapouse. Rev. Nematol. 3:61-69.
13. Guiran, G. de and M. Villemin. 1980. Spécificité de la diapause embryonnaire des oeufs de *Meloidogyne* (Nematoda). Rev. Nematol. 3:115-121.
14. Ishibashi, N. 1969. Studies on the propagation of the root-knot nematode *Meloidogyne incognita* (Kofoid & White), Chitwood, 1949. Rev. Plant Prot. Res. 2:125-128.
15. Lyons, J. M., A. D. Keith, and I. J. Thomason. 1975. Temperature-induced phase transitions in nematode lipids and their influence on respiration. J. Nematol. 7:98-104.
16. Maggenti. A. R. and A. Hardin. 1973. The effects of soil salinity and *Meloidogyne javanica* on tomato. J. Nematol. 5:231-233.
17. Norton, D. C. 1978. Ecology of plant parasitic nematodes. John Wiley & Sons, N. Y. 268 pp.
18. O'Bannon, J. H. and H. W. Reynolds. 1961. Root-knot nematode damage and cotton yields in relation to certain soil properties. Soil Sci. 92:384-386.
19. Prot, J. C. 1978. Influence of concentration gradients of salts on the movement of second stage juveniles of *Meloidogyne javanica* Rev. Nematol. 1:21-26.
20. Prot, J. C. 1978. Vertical migration of four natural populations of *Meloidogyne*. Rev. Nematol. 1:109-112.
21. Prot, J. C. 1980. Migration of plant-parasitic nematodes toward plant roots. Rev. Nematol. 3:305-318.
22. Prot, J. C. and S. D. Van Gundy. 1981. Effect of soil texture and the clay component on migration of *Meloidogyne incognita* second-stage juveniles. J. Nematol. 13:213-216.
23. Prot, J. C. and S. D. Van Gundy. 1981. Influence of photoperiod and temperature on migration of *Meloidogyne* juveniles. J. Nematol. 13:217-219.
24. Radewald, J. D., P. G. Mowbray, A. O. Paulus, R. Shibuya, and J. M. Rible. 1969. Preplant soil fumigation for California head lettuce. Plant Dis. Rep. 53:385-389.
25. Reversat, G. 1981. Consumption of food reserves by starved second-stage juveniles of *Meloidogyne javanica* under conditions inducing osmobiosis. Nematologica 27:207-214.
26. Roberts, P. A., S. D. Van Gundy, and H. E. McKinney. 1981. Effects of soil temperature and planting date of wheat on *Meloidogyne incognita* reproduction, soil populations and grain yield. J. Nematol. 13:338-344.
27. Santo, G. S. and J. H. O'Bannon. 1981. Effect of soil temperature on the pathogenicity and reproduction of *Meloidogyne chitwoodi* and *M. hapla* on Russet Burbank potato. J. Nematol. 13:483-486.
28. Sayre, R. M. 1964. Cold-hardiness of nematodes. I. Effects of rapid freezing on the eggs and larvae of *Meloidogyne incognita* and *M. hapla*. Nematologica 10:168-179.
29. Sasser, J. N. 1954. Identification and host-parasite relationships of certain root-knot nematodes (*Meloidogyne* sp.). Md. Agric. Exp. Stn. Bull. A-77. 31 pp.
30. Siddiqui, I. A., S. A. Sher, and A. M.

French. 1973. Distribution of plant parasitic nematodes in California. California Dep. Food and Agric. Sacramento, Calif. 324 pp.
31. Stolzy, L. H. and S. D. Van Gundy. 1968. The soil as an environment for microflora and microfauna. Phytopathology 58:889-899.
32. Taylor, A. L., J. N. Sasser, and L. A. Nelson. 1982. Relationship of climate and soil characteristics to geographical distribution of *Meloidogyne* species in agricultural soils. Coop. Publ. Dep. Plant Pathol., North Carolina State Univ., and U.S. Agency Int. Dev. Raleigh, N.C. 65 pp.
33. Thomason, I. J. and B. Lear. 1961. Rate of reproduction of *Meloidogyne* spp. as influenced by soil temperature. Phytopathology 51:520-524.
34. Thomason, I. J., S. D. Van Gundy, and J. D. Kirkpatrick. 1964. Motility and infectivity of *Meloidogyne javanica* as affected by storage time and temperature in water. Phytopathology 54:192-195.
35. Towson, A. J. and W. J. Apt. 1983. Effect of soil water potential on survival of *Meloidogyne javanica* in fallow soil. J. Nematol. 15:110-114.
36. Van Gundy, S. D. and L. H. Stolzy. 1961. Influence of soil oxygen concentrations on the development of *Meloidogyne javanica*. Science 134:665-666.
37. Van Gundy, S. D. and L. H. Stolzy. 1963. Oxygen diffusion rates and nematode movement in cellulose sponges. Nature (London) 200:1187-1189.
38. Van Gundy, S. D. 1965. Factors in survival of nematodes. Annu. Rev. Phytopathol. 3:43-68.
39. Van Gundy, S. D., A. F. Bird, and H. R. Wallace. 1967. Aging and starvation in larvae of *Meloidogyne javanica* and *Tylenchulus semipenetrans*. Phytopathology 57:559-571.
40. Van Gundy, S. D. 1976. Ecology of root-knot nematodes, pp. 64-74 *In* Proc. Res. Plann. Conf. on Root-knot Nematodes, *Meloidogyne* spp., January 12-16, 1976, North Carolina State University, Raleigh, N.C. 106 pp.
41. Wallace, H. R. 1963. The biology of plant parasitic nematodes. Edward Arnold, London. 280 pp.
42. Wallace, H. R. 1966. The influence of moisture stress on the development, hatch and survival of eggs of *Meloidogyne javanica*. Nematologia 12:57-69.
43. Wallace, H. R. 1968. The influence of aeration on survival and hatch of *Meloidogyne javanica*. Nematologica 14:223-230.
44. Wallace, H. R. 1968. The influence of soil moisture on survival and hatch of *Meloidogyne javanica*. Nematologica 14:231-242.
45. Wallace, H. R. 1969. The influence of nematode numbers and of soil particle size, nutrients and temperature on the reproduction of *Meloidogyne javanica*. Nematologica 15:55-64.
46. Wallace, H. R. 1971. Abiotic influences in the soil environment, pp. 257-280 *In* Plant parasitic nematodes, Vol. 1, ed. B. M. Zuckerman, W. F. Mai and R. A. Rohde. Academic Press, N.Y. 345 pp.
47. Wallace, H. R. 1973. Nematode ecology and plant disease. Edward Arnold, London. 228 pp.
48. Whitehead, A. G. 1969. The distribution of root-knot nematodes (*Meloidogyne* spp.) in tropical Africa. Nematologica 15:315-333.
49. Wong, T. K. and W. F. Mai. 1973. *Meloidogyne hapla* in organic soil: effects of environment on hatch, movement and root invasion. J. Nematol. 5:130-138.

INTERACTION OF MELOIDOGYNE WITH FUNGI ON CROP PLANTS

J.M. Webster

Centre for Pest Management
Department of Biological Sciences
Simon Fraser University
Burnaby, Vancouver, Canada

To obtain optimum crop quality and economic production, it is essential to be aware of the array of organisms influencing the crop and of the nature of the various organismic interactions. Under field conditions plants develop close associations with many soil organisms, especially with fungi and nematodes. These organisms independently develop associations with plants that are either beneficial or harmful to the plant, and either facultative or obligative for the organism. Some crops, for example, grow better in the presence than in the absence of a mycorrhizal associate or, on the other hand, may be decimated by fungal wilt and/or root-knot nematodes. In other situations fungi appear to act as biological control agents of *Meloidogyne* spp., but often this is difficult to demonstrate convincingly (1).

SINGLE PARASITE INFECTIONS

In order for multiple associations on parasitism of a crop plant to be understood better, the influence of a single parasite on host plant physiology and its relationship to symptom expression should be examined. If a dwarf variety of bean (*Phaseolus vulgaris*) is exposed to different numbers of *Meloidogyne incognita* juveniles, it can be readily demonstrated that the greater the number of nematodes, the more severe the symptoms (Melakeberhan, personal communication). In particular, there is a greater loss of plant dry weight, a smaller leaf-chlorophyll content and a concomitant decrease in number of flowers, pods and seeds. Associated with these external symptoms, there is less water uptake, an imbalanced nutrient uptake and translocation, diminished photosynthetic rate and, depending on the infection stage, an occasional increased respiration rate compared with nematode-free plants (20) (Fig. 1). This finding parallels, in part, those of Loveys & Bird (18) and of Wallace (36) using *Meloidogyne* spp. on tomato. Somewhat similar visible effects occur on tomato plants infected by *Fusarium oxysporum* f. sp. *lycopersici*. The overall impact of such parasitic infections on the plant and crop is decreased yield.

In contrast to these parasitized bean and tomato plants, some host species are very tolerant to parasites in that they show few symptoms, their yield is not greatly diminished by parasitism and the parasite develops and reproduces so as to maintain a relatively high population. An example of this type of response is that of the lima bean (*Phaseolus lunatus*) to *Meloidogyne javanica*. The lima bean does not show severe symptoms and, because of its tolerance to *Meloidogyne* spp. infections, is frequently used as a stock plant upon which to build up large populations of the nematode. The effect of such a parasite on the physiology of a tolerant host is not known though it seems likely that some balancing compensatory processes occur.

These events, observed either under greenhouse or growth-chamber conditions, are the result of a single nematode or fungal parasite (*Meloidogyne* spp. or *Fusarium* spp.) of a host.

DISEASE COMPLEXES

In the field, multiple-parasite associations and disease complexes of the plants are the norm. Our knowledge of nematode-fungal disease complexes has increased greatly since they were first reported towards the end of the last century. In 1892, George F. Atkinson, at the

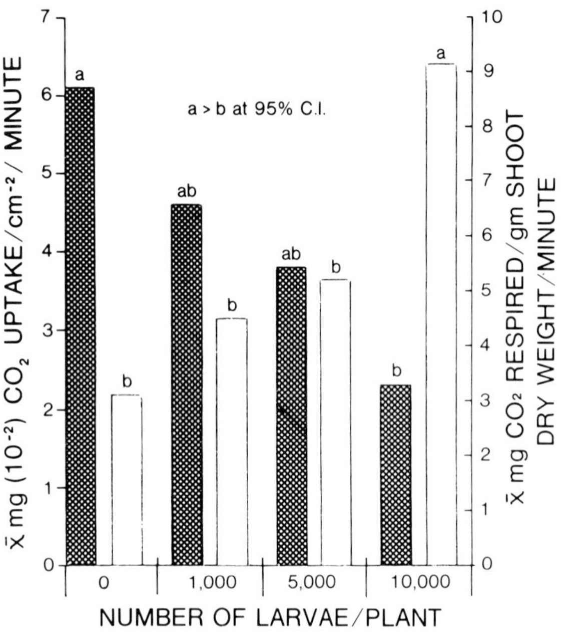

Fig. 1. The effect of *Meloidogyne incognita* on the rates of apparent net photosynthesis and dark respiration (Courtesy of H. Melakeberhan).

Agricultural Experiment Station at Auburn, Alabama, wrote about diseases of cotton and said:

> "Root gall is a disease of the roots caused by minute parasitic worms. These worms live in the roots where they cause abnormal thickenings, resulting in what is known as galls. — I have since found that the organism of 'frenching' was associated with the nematodes in producing that disease, or rather in making it much more serious" (3).

"Frenching" is an old term for *Fusarium* wilt. Hence, for nearly 100 years, scientists have been aware of the concept of disease complexes but have failed to understand fully the mechanism of interaction.

Early observations on nematode-fungal interactions suggested that they were due to the nematode providing a ready means of entry into the host for the fungus. Undoubtedly, this occurs when root-browsing nematodes cause superficial root injury and so enhance fungal access and secondary pathogenicity of the roots. However, the relationship between two or more parasites on a single host is often more sophisticated that this, especially with endogenous nematodes such as *Meloidogyne* spp. (5,9,13, 23). It appears for instance, that when bean (*P. vulgaris*) is infected by both the obligate foliar parasite *Uromyces phaseoli* and the obligate root parasite *M. incognita* that reproduction of each parasite is decreased by the presence of the other (8). Field and plot records suggest that *Meloidogyne* modifies the host's mineral composition and physiology and, consequently, slows host growth, and in many instances breaks the host's resistance to other parasites. Conversely, the plugging of xylem vessels by a wilt fungus prevents nutrient and water uptake, and so decreases host vigour and apparently makes them more vulnerable to challenges by other pathogens, such as *Meloidogyne*.

The *Meloidogyne-Fusarium* interaction has been extensively studied because of its significance on major world crops, such as cotton (28) and tobacco (21). The extent of the disease interaction is greatly influenced by a range of factors. For instance, there is more extensive *Fusarium* colonization of the plant tissues as the number of developing *Meloidogyne* increases. Moreover, symptom expression, as a manifestation of disease interaction, is greater in "young" than in "old" plants (22).

Ultrastructure

Fungal colonization of giant cells and xylem is common in fungus-root-knot interactions. It occurs, for instance, in the serious black-shank disease of tobacco (*Phytophthora parasitica* var. *nicotianae/M. incognita*) (24) as well as in *Fusarium* wilt of tomato.

The ultrastructural response of tomato to *Fusarium*-wilt/root-knot nematode (*F. oxysporum lycopersici/M. javanica*) infection has been described in a temporal sequence (11). For 1 wk there is little evidence of *Fusarium*-wilt or of fungal invasion of the root cells, and the nematode-induced giant cells develop normally. However, after 2 wks fungal hyphae have spread throughout the xylem tissue and occur especially in the xylem vessels adjacent to giant cells. The giant cell cytoplasm becomes less electron dense, and the nuclear structure is greatly modified (Fig. 2). Within 3 wks fungal hyphae and spores ramify throughout the xylem tissue and necrotic giant cells. Ultrastructural changes occur in the giant cells of both *Fusarium*-susceptible and *Fusarium*-resistant cultivars infected by *M. javanica*, but they occur much sooner in the susceptible varieties. Fungal hyphae appear not to enter the giant cells of predisposed resistant varieties until some threshold is reached. Furthermore, the fact that the structure of the giant cell changes prior to fungal entry implies that translocatable fungal secretions rather than the penetrating hyphae initially modify and eventually kill the giant cells (11).

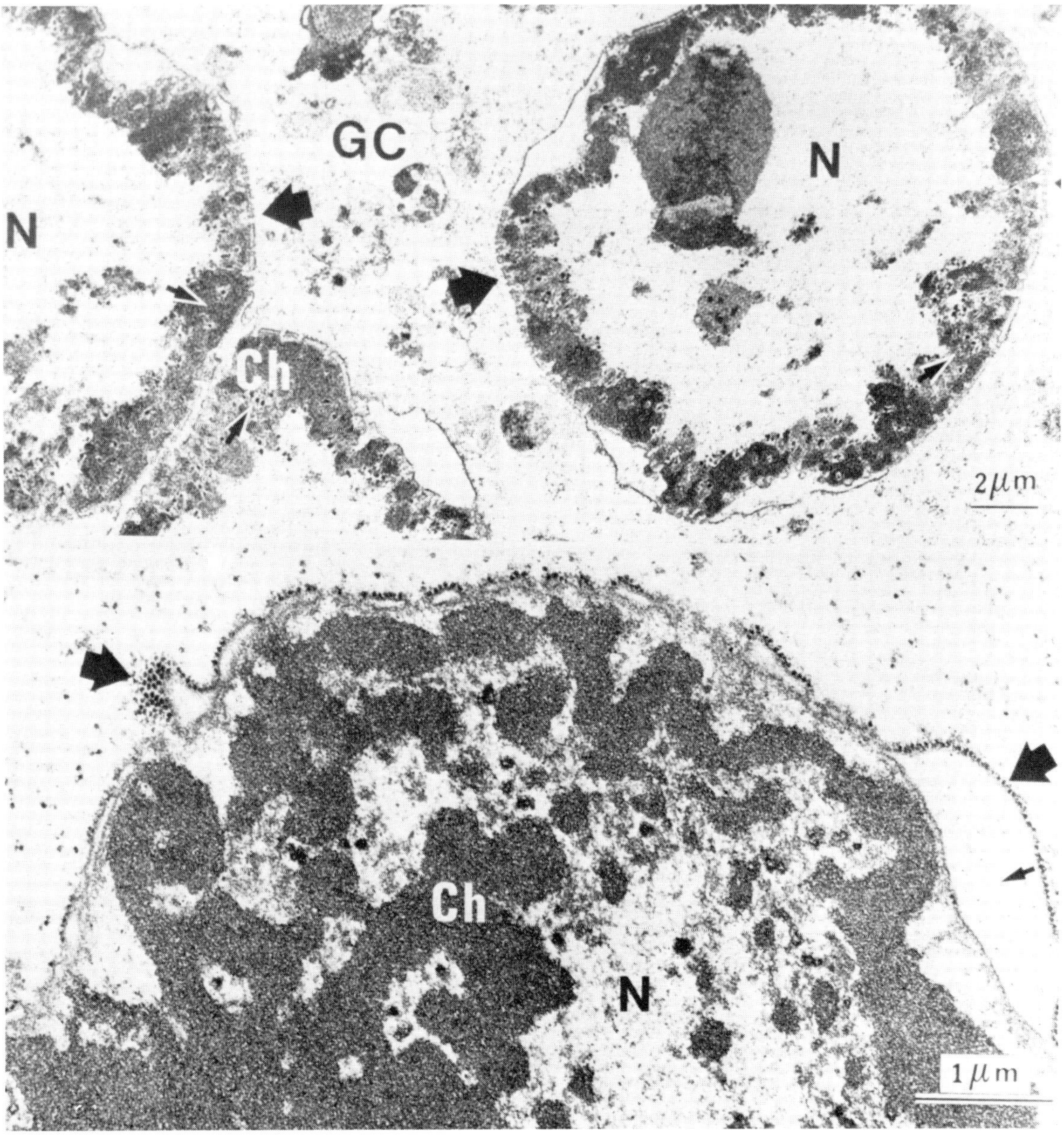

Fig. 2. Electron micrograph of *Meloidogyne javanica*-induced giant cells (GC) in the roots of *Lycopersicon esculentum*, resistant to race 1 *Fusarium oxysporum lycopersici*, 3 wks after fungal inoculation. *Note: (Top)* nuclei (N) with electron-light central area showing fragmental nuclear membrane (large arrows) and dark inclusions (small arrows) within the chromatin material (Ch); (*Bottom*) nucleus with ribosomes attached to the nuclear membrane (large arrows) and nuclear membrane swelling (small arrow) [Fattah & Webster (11)].

This occurrence has been demonstrated by incubation of *M. javanica*-infected tomato roots in a cell-free filtrate of the medium in which *F. oxysporum lycopersici* has been cultured (10). Such filtrate-treated roots undergo ultrastructural changes identical with those of roots from *Fusarium*-infected plants: e.g., the giant cell nuclei become spherical; the nuclear membranes become swollen and ruptured; and nuclear chromatin contains scattered, electron dense spheres each surrounded by an electron-light halo (Fig. 3). In other words, the cell-free filtrate from a fungus culture medium causes many of the same ultrastructural changes to giant cells that the living fungus does in a *Meloidogyne*-infected plant.

In addition to this fungus-induced/produced factor in the *Meloidogyne-Fusarium* disease

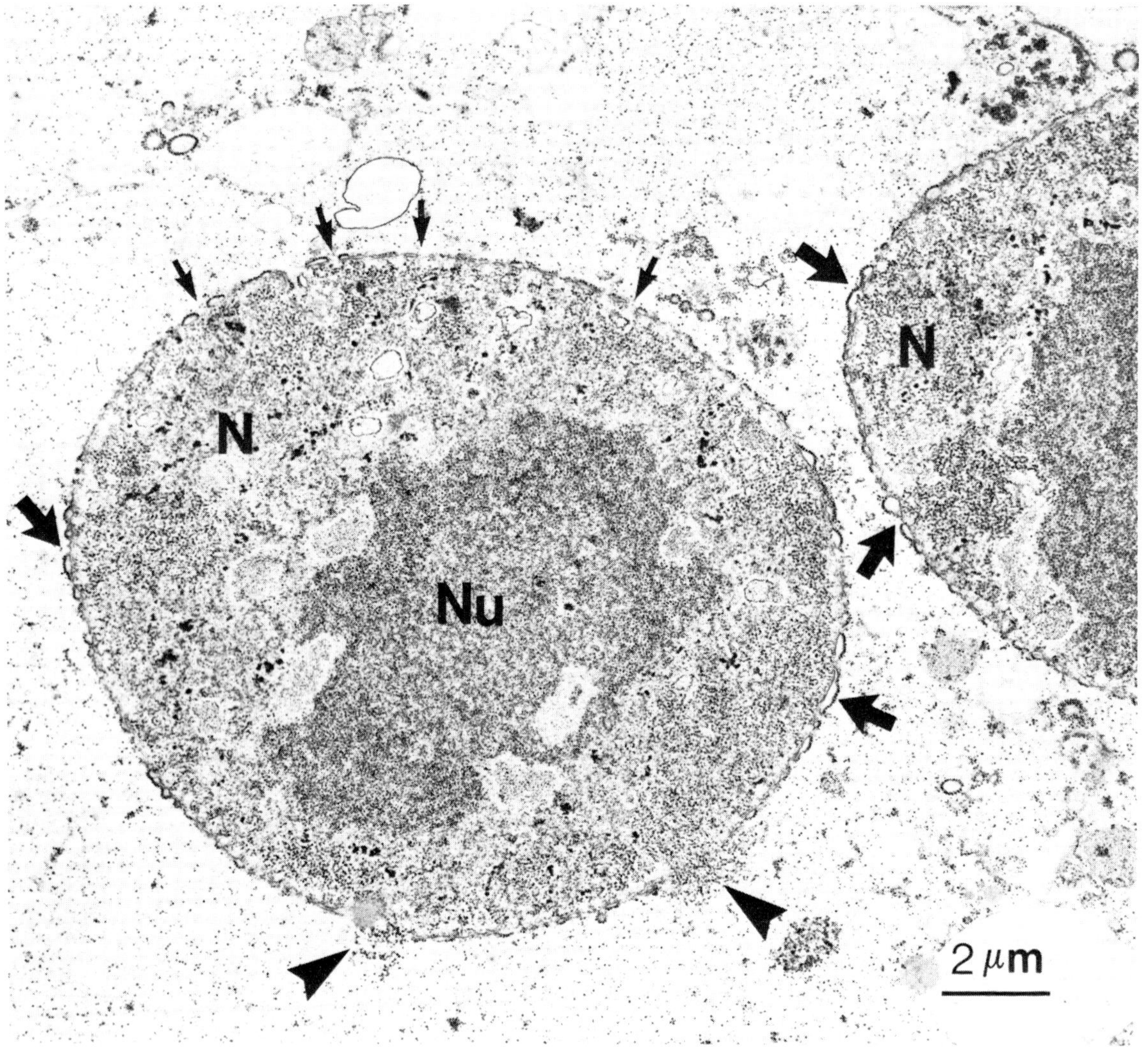

Fig. 3. Electron micrograph of part of a *Meloidogyne javanica*-induced giant cell in the root of a *Fusarium* susceptible cultivar of *Lycopersicon esculentum* culture filtrate for 9 hr. Note the spherical nuclei (N) with fragmented (small arrows), swollen and ribosome covered (large arrows) and ruptured (arrow heads) nuclear membrane; nucleolus (Nu) is less dense and irregularly shaped (Courtesy of F. Fattah).

complex on tomato, the developing nematode produces a factor or induces the plant to produce such a factor that is transmitted upward through the plant and breaks the fungal resistance (i.e. predisposes) of the host to *Fusarium*. The fungus develops and its secretions cause a progressive necrosis of the giant cells which, in turn, causes the nematode to starve to death.

Translocatable Factors

The mobility of the factor that predisposes the plant to the fungus has been demonstrated experimentally using layered tomato plants (30).

When a fungus-resistant scion (cv. 'Chico III') is grafted onto a *F. oxysporum lycopersici*-infected, susceptible plant (cv. 'Bonny Best'), tissues in the resistant scion distal to the fungus infection become susceptible to the fungus. Disease indices of layered tomato plants, based on chlorosis, and corroborated by propagule counts, are larger in the stem exudates distal than in those proximal to the site of fungal inoculation. Stocks of Bonny Best when inoculated with the fungus alone develop wilt symptoms. However, cv. Bonny Best scions, separated from fungal-infected Bonny Best stocks by a fungus-resistant interscion (cv. Chico III) remain relatively resistant. On the other hand, if such interscion and scion-grafted

Table 1. Average levels [a] of free amino acids in xylem exudates from *Meloidogyne incognita* infected (3-week-old galled) (+) and nematode-free (—) tomato cultivars collected 4 weeks after inoculation [based on Sidhu & Webster (31)].

Amino acids	Bonny best		Rutgers		Nematex	
	+	—	+	—	+	—
Aspartic AG	12.71	4.72	13.84	14.22	4.71	5.01
Threonine	3.28	7.53	2.85	8.23	3.07	3.80
Serine	0.06	3.12	0.38	0.53	1.40	1.80
Proline	1.95	9.52	2.05	5.59	2.60	2.10
Glutamic AG	9.05	17.13	16.55	17.17	14.01	14.60
Glycine	0.46	0.74	0.02	0.45	0.18	0.20
Alanine	0.77	2.27	0.59	1.30	1.45	1.80
Valine	2.85	4.88	3.48	5.48	1.40	1.20
Methionine	0.58	4.58	0.12	1.93	0.20	0.22
Isoleucine	3.30	2.96	4.23	3.35	1.01	1.49
Leucine	2.24	1.62	1.82	2.00	1.60	1.88
Tyrosine	1.42	1.27	1.08	1.18	0.20	0.42
Phenylalanine	0.91	0.72	0.89	0.84	0.60	0.68
Lysine	1.50	1.78	1.55	1.93	3.32	3.80
Histidine	0.70	3.10	0.32	2.60	0.12	0.01
Tryptophan	Tr[b]	2.01	Tr	2.40	Tr	Tr
Arginine	0.77	4.63	0.29	5.51	3.30	3.48
Amides	57.45	27.42	49.94	25.29	60.82	57.51
	100.00	100.00	100.00	100.00	100.00	100.00

[a] Expressed as percentage composition of total amino acids present
[b] Tr, trace of amino acid.

plants had the primary root system infected first with the nematode and subsequently with the fungus, typical wilt symptoms (and high propagule counts) would occur throughout the resistant interscion and in the scion (30). This factor makes a resistant host susceptible and a susceptible host even more so.

The qualitative and quantitative amino acid composition of the xylem of tomato cultivars changes during a *Meloidogyne* infection (30) (Table 1). Moreover, root exudates reflect some of these internal changes which, in turn, influence the composition of the rhizosphere. For example, a change in root secretions due to a *Meloidogyne* infection may influence the nutrition-induced dormancy of fungal spores. Hence, nematodes are likely to influence the incidence of mycorrhizae and of root infection by parasitic fungi through this mechanism.

Rhizosphere Environment

When Hiltner (14) introduced the term "rhizosphere" in 1904, it is unlikely that he realized the complexity of the physico-chemical environment surrounding the roots or the number and variety of microorganisms that inhabit that niche (12). As the roots grow, the rhizosphere extends, and as the roots age, the rhizosphere habitat changes significantly due to the changing root exudates. In particular, the exudates are relatively minor in the period immediately following germination and reach a climax at the time of flowering (26). Furthermore, there are major quantitative changes in the amino acid content of the root exudates as the plant ages. The rhizospheric effect, therefore, may change dramatically as the plant matures, and consequently, influence the colonization of this region around the root and the behaviour of the organisms within it. This is likely one of several reasons why young seedlings are more susceptible to parasite attack than older ones are.

The fact that fungi in the rhizosphere of a plant that is susceptible to a particular parasite are different from those in the rhizosphere of cultivars resistant to that parasite is due directly to the influence of root exudates and indirectly to the exudates influencing the composition of the fungal and bacterial flora. The exudates from *M. javanica*-infected tomato roots depress the number of actinomycetes in the rhizosphere and this correspondingly increases the number of propagules of *F. oxysporum lycopersici*. This trend is reversed by addition of a chitin amendment to the soil because the number of actinomycetes increase and there is a corresponding decrease in invasion of root galls by secondary pathogens (6). The competition for the available substances within this rhizosphere niche has an important influence on the plant itself. Hence, Katznelson (16) showed that rhizospheric bacteria can limit the development of a pathogenic fungus on alfalfa roots by competing with the fungus for thiamine and mineral nutrients.

Many of these exudates (e.g., organic acids, vitamins, free amino acids and large protein molecules) accumulate near the root tips where the microflora is abundant and close to the location of entry for *Meloidogyne* juveniles. The bacterial and fungal activity breaks down the larger molecules and so provides a range of nutrients available for the many organisms in the rhizosphere. Increased bacterial and fungal levels in the rhizosphere may in turn lead to larger populations of microbivorous and predatory nematodes. Some of the latter may be predatory on *Meloidogyne* juveniles and adults.

Mycorrhizal Infection

Many of the mycorrhizae play an important role in enabling vigorous plant growth and so facilitating higher crop yields (15). The mycorrhizal mycelial network associated with the roots "filters" the root-exudates and, through differential absorption, causes quantitative and qualitative changes in the exudates. Furthermore, ectomycorrhizae produce a range of substances, some of which may function as antibiotics, and it is these substances plus the filtering effect of the hyphae that result in the change in composition of the rhizosphere community, including fungi and nematodes.

Some of the beneficial effects of mycorrhizae on plants and on the fungus-nematode interactions can be affected by free-living soil nematodes. Hence, *Suillus granulatus*, an ectomycorrhizal associate of *Pinus ponderosa*, is diminished by *Aphelenchoides composticola* feeding on the hyphae. This association decreases the ability of the ponderosa pine to grow well in poor, leached soils. Powell and associates (25) showed that nematicidal applications to the soil around pecan trees enhanced mycorrhizal development, presumably by killing the fungal-feeding nematodes, and subsequently improved tree growth.

The timing of establishment of mycorrhizae in the root in relation to the feeding of parasitic nematodes can be significant especially in those plants where ectomycorrhizae predominate. In those instances where the mycorrhizae are already established, the nematode must penetrate the tightly interwoven network of hyphae and the modified cortical tissue, known as the Hartig net (19), before gaining access to the deeper tissues of the root. The Hartig net mechanically and physiologically slows root invasion by *Meloidogyne* juveniles and other parasites (19).

The vesicular-arbuscular mycorrhizae, such as *Glomus* spp., not only provide the plant with nutrients but often significantly decrease root-knot in several crops. It appears that the root-knot nematode juveniles enter endomycorrhizal roots less readily than they do non-mycorrhizal roots, and those juveniles that do enter the root develop more slowly. Although endomycorrhizal infections appear to stimulate plant growth (2),there are conflicting reports as to their influence on the severity of root-knot infections (17, 27). The range of interactions between mycorrhizae and *Meloidogyne* is wide. In some instances, the vesicles and hyphae of endomycorrhiza are reported to be visably disintegrated within the *Meloidogyne*-induced galls. In other instances apparently fewer galls are produced when the soil population of the mycorrhiza is high. Moreover, care should be taken in interpreting nematode densities on the root system as root ball size is influenced by both the mycorrhiza and the nematode. Supplementary phosphorus is, therefore, often included in the experimental treatments so as to assist root growth and ensure fair comparisons. It is suggested by Thomson Cason and associates (35) that a primary effect of endomycorrhizal infections is to increase tolerance to *Meloidogyne* by improving plant nutrition and growth.

NEMATODE-FUNGAL INTERACTION

Nutritional Basis

Giant cells induced by *Meloidogyne* are maintained in a state of high metabolic activity, especially in terms of protein synthesis and active transport of amino acids through the continuous stimulation by the nematode (39). They contain most DNA, RNA and photosynthates about 3-4 wks after induction (7), which is coincident with the time when *Meloidogyne* has the greatest capability for predisposing plants to attack by *Fusarium* spp. Furthermore, there are greater concentrations of sugars in plants during these late stages of a *Meloidogyne* infection (38). Hence, food availability to a fungus may be a significant factor in fungal-resistant plants that may explain the mechanism of predisposi-

Table 2. Response of tomato cultivars to amino acid mutants of *Fusarium oxysporum f. sp. lycopersici*, following application of the respective amino acids exogenously, recorded as disease indices (mean ± S.E.) 4 weeks after inoculation [based on Sidhu & Webster (31)].

	Disease indices of tomato cultivars[a]		
Fungal type	**Bonny best**	**Rutgers**	**Nematex**
Wild type	3.95±0.06	3.02±0.04	0.20±0.01
Threonine	2.48±0.06	2.90±0.06	2.85±0.04
Proline	2.95±0.01	2.48±0.02	0.15±0.01
Glycine	2.60±0.04	1.49±0.02	0.10±0.01
Methionine	2.05±0.06	2.00±0.02	2.80±0.06
Histidine	2.01±0.01	2.02±0.02	2.60±0.08
Tryptophan	0.95±0.01	0.85±0.02	0.10±0.01
Arginine	1.06±0.02	1.02±0.04	0.15±0.01

[a] Based on degrees of leaf chlorosis and wilting (below 2 = relatively resistant; above 2 = relatively susceptible).

tion. Certainly, a fungus could be expected to grow better on a nutritionally rich than on a nutritionally poor medium. It could be hypothesized that a fungal-resistant plant that is predisposed to fungal attack by prior nematode infection is analagous to a complete synthetic medium for the fungus in that the fungus is now able to develop in the plant because of the introduction or initiation of additional metabolic products in the host by the nematode. This hypothesis is supported by the extensive growth of *Fusarium* hyphae in the nematode-induced galls of tobacco roots, and by the considerable hyphal growth in stem tissue at sites distant from nematode-galled tissue. We have noted above that substances introduced into the plant by the nematode or produced in the roots in response to the nematode are translocated throughout the plant and very likely influence the fungal hyphae as well as overall plant host growth. In the absence of a stimulus from the developing *Meloidogyne* juveniles, there is no translocatable factor, and the hyphae fail to grow. We know, for instance, that free amino acids in a plant are increased up to 700% when a plant is infected with *Meloidogyne* (31). It is possible, therefore, that the genetic resistance of tomato to *Fusarium* may be mediated partially through an amino-acid, nutrient requirement. In general, larger quantities of the free amino acids occur in galled than in the non-galled tomato roots. In experiments with amino-acid-specific auxotrophs it was shown that susceptibility of the tomato cultivars 'Bonny Best' and 'Rutgers' to *Fusarium* was associated with a significant increase of certain amino acids, e.g. threonine, methionine and arginine, in their exudates (31). Interestingly, where amino acids corresponding to their respective fungal auxotrophs were applied exogenously, most cultivars so treated showed greater wilt, and fungal-resistant seedlings showed wilt symptoms (31) (Table 2).

Another possible explanation for the breakdown of resistance to *Fusarium* spp. is the failure of root-knot-infected plants to develop tyloses (4). *Fusarium*-induced tyloses are formed in the xylem vessels by the balloon-like expansion of xylem parenchyma cells through the xylem pits. Xylem parenchyma cells that are modified to form giant cells or are physiologically influenced by adjacent giant cells are unlikely to produce tyloses. Moreover, the flow of substances through the xylem influences the root exudates and/or the uptake of substances from the rhizosphere.

There is an abundance of evidence, therefore, of a dynamic situation in the rhizosphere in which many factors influence the ultimate growth and yield of a plant. However, the underlying genetic basis of some of these factors on the principal organisms in the complex should not be overlooked.

Genetic Basis

The typical response of a crop plant to one or more pathogens in the soil is controlled by one or more genes, and these are countered by virulent genes in the parasite in the classic gene-for-gene pattern. For instance, it has been shown that resistance to the *M. incognita-F. oxysporum lycopersici* disease complex of tomato is controlled by two independently segregating, dominant genes: one effective against the nematode and the other against the fungus (29). Sequential inoculations of F_2 plants, resulting from a resistant susceptible cross, show (as measured by nematode galling, fungal wilt and propagule counts) a classical segregation (9:3:3:1) that is modified to a 9:3:4 ratio reflecting:

resistance to both nematode and fungal parasites,

resistance to the nematode but susceptibility to fungus,

susceptiblity to both nematode and fungus.

This modified ratio reflects that the recessive allele of the gene for nematode resistance modifies the effect of the fungal-resistance gene, and so the 3/16 which genotypically segregate as susceptible to the nematode but resistant to the fungus become phenotypically indistinguishable from those that are susceptible to both pathogens, i.e. the nematode predisposes the plant to the fungus.

These interactions at the genetic level of components of a disease complex have implications for other organisms in the rhizosphere, as we have seen above. They also have important implications for plant breeding. The primary nature of host-parasite interactions is controlled largely by the inherent genetic potential for resistance and susceptibility in the host together with that for virulence in the parasites. However, secondary reaction types occur as a result of an array of epistatic and hypostatic interactions among pathogens in disease complexes (34):

an "epistatic parasite" is one which influences the expression of another parasite, and

a "hypostatic parasite" is the one whose expression is influenced.

These are relative terms in that epistatic and hypostatic parasites may change their relative positions in different disease situations depending on the components of the particular disease complex: e.g., *F. oxysporum lycopersici* behaves as a hypostatic parasite in its relation to *Verticillium albo atrum* on tomato (33).

In a disease-complex situation in the field, the host response is actually a composite response to many parasite-parasite interactions and to many abiotic influences. Consequently, the final reaction type against parasite components of the complex appears to be quantitative rather than qualitative and is expressed as a horizontal type of resistance (34). In fact, examination of field data in the literature on the *Fusarium-Meloidogyne* disease of tomato and on the *Fusarium-Meloidogyne* disease of cotton over several years shows a stable level of resistance to *Fusarium* that is typical of horizontal resistance. Despite our recognition of the occurrence of disease complexes, crop cultivars usually have been bred against specific target parasites and their unique interaction with the host, rather than against a group of parasites in a disease complex. However, there have been several reports (32) that show induced modifications in the host response either in the form of cross protection or predisposition, and some of these should be harnessed for disease control. For instance, it would be worthwhile to establish the epistatic and hypostatic relationships among economically important pathogens of a given crop under controlled field conditions. Now that parasitic epistasis and the genetic basis of resistance against each pathogen have been established, it should be possible for disease management to be improved by:

a) incorporation of tolerance to epistatic parasites that induce resistance to hypostatic parasites and
b) the breeding of cultivars with suitable R-gene combinations against an epistatic parasite so that host resistance is maintained against the hypostatic parasite.

It is preferable, therefore, to achieve long-term disease control of multiple, potentially interacting pathogens rather than of single pathogens. Ideally, R-gene(s) that control the resistance against more than one parasite, including the hypostatic parasites, should be brought together in a common genetic background. Some multigene or multiline cultivars already exist that function as described above, but often this has occurred by chance rather than as a result of the recognition of this need.

There is, therefore, a growing bank of knowledge on plants interacting with more than one parasite, especially with species of *Meloidogyne* and fungi, and with other biotic and abiotic components of the environment. Although caution must be used in the interpretation of the nature of such interaction (37), it behooves us to pursue these studies because of the apparent chronic impact of disease complexes on crop yields.

Literature Cited

1. Al-Hazmi, A.S., D.P. Schmitt, and J.N. Sasser. 1982. The effect of *Arthrobotrys conoides* on *Meloidogyne incognita* population densities in corn as influenced by tem-

perature, fungus inoculum density, and time of fungus introduction in the soil. J. Nematol. 14:168-174.

2. Atilano, R.A., J.A. Menge, and S.D. Van Gundy. 1981. Interaction between *Meloidogyne arenaria* and *Glomus fasciculatus* in grape. J. Nematol. 13: 52-57.
3. Atkinson, G.F. 1982. Some diseases of cotton. Ala. Agric. Exp. Stn. Bull. (Auburn Univ.) 41. 65 pp.
4. Beckman, C.H., D.M. Elgersma, and W.E. MacHardy. 1972. The localization of fusarial infections in the vascular tissue of single-dominant-gene resistant tomatoes. Phytopathology 62: 1256-1260.
5. Bergeson, G.B. 1972. Concepts of nematode-fungus associations in plant disease complexes: a review. Exp. Parasitol. 32: 301-314.
6. Bergeson, G.B., S.D. Van Gundy, and I.J. Thomason. 1970. Effect of *Meloidogyne javanica* on rhizosphere microflora and *Fusarium* wilt of tomato. Phytopathology 60: 1245-1249.
7. Bird, A.F. 1972. Quantitative studies on the growth of syncytia induced in plants by root knot nematodes. Int. J. Parasitol. 2: 157-170.
8. Bookbinder, M.G. and J.R. Bloom. 1980. Interaction of *Uromyces phaseoli* and *Meloidogyne incognita* on bean. J. Nematol. 12: 177-182.
9. Dropkin, V.H. 1980. Introduction to nematology. Wiley & Sons, New York. 293 pp.
10. Fattah, F.A. and J.M. Webster. 1981. Effect of culture filtrate of *Fusarium oxysporum* f. *lycopersici* on the ultrastructure of giant cells induced by *Meloidogyne javanica* in tomato. J. Nematol. 13: 437-438 (Abstr.)
11. Fattah, F.A. and J.M. Webster. 1983. Ultrastructural changes caused by *Fusarium oxysporum* f. sp. *lycopersici* in *Meloidogyne javanica* induced giant cells in *Fusarium* resistant and susceptible tomato cultivars. J. Nematol. 15: 128-135.
12. Foster, R.C., A.D. Rovira, and T.W. Cock. 1983. Ultrastructure of root-soil interface. Am. Phytopathol. Soc., St. Paul, Minn. 157 pp.
13. Garber, R.H., E.C. Jorgenson, S. Smith, and A.H. Hyer. 1979. Interaction of population levels of *Fusarium oxysporum* f. sp. *vasinfectum* and *Meloidogyne incognita* on cotton. J. Nematol. 11: 133-137.
14. Hiltner, L. 1904. Uber Neuere Erfahrungen und Probleme auf dem Gebiet der Bodenbakteriologie und unter besonderer Berücksichtigung der Gründüng und Brache. Arb. Deut. Landwirt.-Ges. 98: 59-78.
15. Hussey, R.S. and R.W. Roncadori. 1982. Vesicular-arbuscular mycorrhizae may limit nematode activity and improve plant growth. Plant Dis. 66: 9-14.
16. Katznelson, H. 1965. Nature and importance of the rhizosphere, pp. 187-209 *In* Ecology of soil-borne plant pathogens, ed. K.F. Baker and W.C. Snyder. Univ. California Press, Berkeley, Calif. 571 pp.
17. Kellam, M.K. and N.C. Schenck. 1977. The effect of a vesicular-arbuscular mycorrhizal fungus, *Glomus macrocarpus* on the amount of galling produced by *Meloidogyne incognita* on soybean. Proc. Am. Phytopathol. Soc. 4: 124 (Abstr.).
18. Loveys, B.R. and A.F. Bird. 1973. The influence of nematodes on photosynthesis in tomato plants. Physiol. Plant Pathol. 3: 525-529.
19. Marx, D.H. 1972. Ectomycorrhizae as biological deterrents to pathogenic root infections. Annu. Rev. Phytopathol. 10: 429-454.
20. Melakeberhan, H., J.M. Webster, and R.C. Brooke. 1984. Improved techniques for measuring the CO_2 exchange rate of *Meloidogyne* infected bean plants. Nematologica In press.
21. Milne, D.L. 1972. Nematodes of tobacco, pp. 159-186 *In* Economic nematology, ed. J.M. Webster. Academic Press, London. 563 pp.
22. Pitcher, J.R. 1974. The reduction of resistance of tomato to *Fusarium* wilt by *Meloidogyne javanica* (Nematoda). MSc thesis. Simon Fraser Univ., Vancouver. 75 pp.
23. Pitcher, R.S. 1965. Interrelationships of nematodes and other pathogens of plants. Helminthol. Abstr. 34: 1-17.
24. Powell, N.T. and C.J. Nusbaum. 1960. The black shank-root-knot complex in flue-cured tobacco. Phytopathology 50:899-906.
25. Powell, W.M., F.F. Hendrix and D.H. Marx. 1968. Chemical control of feeder root necrosis of pecans caused by *Phythium* species and nematodes. Plant Dis. Rep. 52: 577-578.
26. Rambelli, A. 1973. The rhizosphere of mycorrhizae, pp. 299-349 *In* Ectomycorrhizae, their ecology and physiology, ed. G.C. Marks and T.T. Kozlowski. Academic Press, New York. 444 pp.
27. Roncadori, R.W. and R.S. Hussey. 1977. Interaction of the endomycorrhizal fungus *Gigaspora margarita*, and root-knot

nematode on cotton. Phytopathology 67: 1507-1511.

28. Sasser, J.N. 1972. Nematode diseases of cotton, pp. 187-214 *In* Economic nematology, ed. J.M. Webster. Academic Press, London. 563 pp.
29. Sidhu, G. and J.M. Webster. 1974. Genetics of resistance in the tomato to root-knot nematode-wilt-fungus complex. J. Hered. 65: 153-156.
30. Sidhu, G. and J.M. Webster. 1977. Predisposition of tomato to the wilt fungus (*Fusarium oxysporum lycopersici*) by the root-knot nematode (*Meloidogyne incognita*). Nematologica 23: 436-442.
31. Sidhu, G.S. and J.M. Webster. 1977. The use of amino acid fungal auxotrophs to study the predisposition phenomena in the root-knot-wilt fungus disease complex of tomato. Physiol. Plant Pathol. 11: 117-127.
32. Sidhu, G.S. and J.M. Webster. 1977. Genetics of simple and complex host-parasite interactions, pp. 59-79 *In* Induced mutations against plant disease. Int. Atomic Energy Agency, Vienna. 580 pp.
33. Sidhu, G.S. and J.M. Webster. 1979. Genetics of tomato resistance to the *Fusarium-Verticillium* complex. Physiol. Plant Pathol. 15: 93-98.
34. Sidhu, G.S. and J.M. Webster. 1983. Horizontal resistance in tomato against the *Meloidogyne-Fusarium* complex: an artefact of parasitic epistasis. Crop Prot. 2: 205-210.
35. Thomson Cason, K.M., R.S. Hussey and R.W. Roncadori. 1983. Interaction of vesicular-arbuscular mycorrhizal fungi and phosphorus with *Meloidogyne incognita* on tomato. J. Nematol. 15: 410-417.
36. Wallace, H.R. 1974. The influence of root-knot nematode, *Meloidogyne javanica*, on photosynthesis and on nutrient demand by roots of tomato plants. Nematologica 20: 27-33.
37. Wallace, H.R. 1983. Interactions between nematodes and other factors on plants. J. Nematol. 15: 221-227.
38. Wang, L.H. 1973. Biochemical and physiological changes in root exudate, xylem sap and cell permeability of tomato plants infected with *Meloidogyne incognita.* Ph.D. thesis. Purdue Univ., Indiana. 81 pp.
39. Webster, J.M. 1975. Aspects of the host-parasite relationship of plant-parasitic nematodes. Adv. Parasitol. 13: 225-250.

INTERACTIONS AMONG CONCOMITANT POPULATIONS OF NEMATODES

J. D. Eisenback

Department of Plant Pathology
North Carolina State University
Raleigh, North Carolina, USA

Most nematological research involves the effects of a single nematode species on a particular crop plant. In nature, however, plant-parasitic nematodes rarely occur in monospecific communities. Rather, the nematode community is dynamic, and its members are constantly interacting with each other, as well as the plant, the environment, and other organisms.

Interactions among species of plant-parasitic nematodes can be considered ecologically or etiologically, and the two aspects are interrelated. Ecological interactions affect the reproductive capacity of the populations, whereas etiological interactions alter the development of plant disease. The relationship between the ecology and etiology of the interaction is based on the correlation of numbers of nematodes with the amount of disease (53). However, because nematode reproduction is not always related to the susceptibility of the plant to damage, the two aspects of the interaction may behave independently. Thus, it is important to consider nematode-nematode interactions from both approaches.

ECOLOGICAL INTERACTIONS

Plant-parasitic nematodes generally occur in polyspecific communities. The diversity of the community and relative success of the individual species are dependent upon many different factors. Diversity of the species is enhanced by the widespread dispersal, polyphagous nature, persistence, and weak interspecific competition of nematodes (53). The occurrence of a particular species in an environment is related to its dissemination, climate, edaphic factors, cropping history, host suitability, reproductive capacity, persistence, and interactions with other organisms, including nematodes (51,53).

Interactions between species of nematodes may be beneficial to one or both species, have no effect, or most commonly, interfere with the well-being of one or both species. Competition between species occurs when the reproductive capacity of a species is greater by itself than in combination with another species (6). The antagonistic interaction can be caused by a spatial occupation or physical alteration or destruction of feeding sites, or by a physiological alteration of the host that decreases its suitability (51).

According to the principle of competitive exclusion, two closely related species cannot occupy the same ecological niche (18). Even though each species is capable of functioning by itself in a particular niche, the two together interact until one species predominates. The species which predominates may depend on which has a competitive advantage, but competition is also density and time dependent (6).

Application of the competitive exclusion principle to communities of plant-parasitic nematodes is difficult because the soil environment contains many different niches that are dynamic and thus not easily defined or delineated. Furthermore, exclusions may be caused by other factors, and interactions may occur between species in close proximity even though they occupy different niches (6,51). Two species can coinhabit when the competition among individuals in a species is greater than the competition between species (6).

Sometimes the interactions between two nematode species may be beneficial to one or both species. The beneficial effects may result from a mechanical or physiological alteration of the host that makes it a more suitable substrate. Physiological changes can also enhance nutrition or reduce resistance of the host to the parasite (51).

The effects of the nematode population on

each other are generally related to the nature of parasitism. Competition is more severe between species with similar feeding habits, and competitive advantage increases as the host-parasite relationship becomes more complex.

Ectoparasites

The ectoparasites can be divided into several groups according to their parasitic habits. The most primitive parasites feed only on epidermal cells and root hairs; they remain completely outside of the root tissues and pierce the cells with their relatively short stylet. The association with the host is short-term and not very specialized (84). Representative genera include *Tylenchorhynchus*, *Paratylenchus*, *Trichodorus*, and *Paratrichodorus*.

A similar group of ectoparasites may feed externally or the anterior portion of their heads may penetrate into the cortical tissues of some hosts. The host-parasite relationship is usually short-term and not very specialized. Members of this group include *Criconemella*, *Helicotylenchus*, and *Hoplolaimus*.

A more specialized group of ectoparasites has relatively long stylets, feeds on cells below the epidermis, and may incite hypertrophy and hyperplasia. They remain completely outside of the root tissue, but the parasitic relationship with the plant is more complex than with the epidermal feeding ectoparasites (84). Some of the genera that feed below the epidermis include *Xiphinema*, *Longidorus*, *Belonolaimus*, *Dolichodorus* and *Hemicycliophora*.

The mode of parasitism within the ectoparasites affects the degree of competition between species. The shallow epidermal-feeding parasites are more antagonistic to each other than to the deep feeding nematodes. Also, the more specialized host-parasite relationships usually are more competitive than the primitive relationships (33,34,49).

Competition occurred between *Tylenchorhynchus claytoni* and *Paratrichodorus minor* on five different corn varieties in greenhouse tests. *Tylenchorhynchus claytoni* was antagonistic to *P. minor* on one cultivar, and *P. minor* repressed *T. claytoni* on three cultivars after 50 days (34). In similar greenhouse experiments with six varieties of bermuda grass, more competition occurred between *Tylenchorhynchus martini* and *Criconemella ornata* than between *T. martini* and *Belonolaimus longicaudatus* or *C. ornata* and *B. longicaudatus* after 155 days; however, reactions varied according to cultivar (33).

Host suitability affects the interaction between ectoparasites. Populations of *Paratylenchus projectus* were suppressed by *Helicotylenchus pseudorobustus* and *Criconemella similis* on soybean in a greenhouse test after 50 days. Soybean was a good host for *C. similis*, a moderately good host for *H. pseudorobustus*, and a poor host for *P. projectus*. *C. similis* and *H. pseudorobustus* did not interact (42,43).

Competition between ectoparasites is density dependent and can be modified by environmental factors. After 6 weeks in greenhouse tests on plum seedlings, high populations of *Paratylenchus neoamblycephalus* suppressed low numbers of *Criconemella xenoplax* at 20°C, and high numbers of *C. xenoplax* antagonized low population densities of *P. neoamblycephalus*. High numbers of *P. neoamblycephalus* suppressed low populations of *C. xenoplax* at 26°C, and high numbers of *C. xenoplax* severely reduced low populations of *P. neoamblycephalus*. The optimum temperature for *P. neoamblycephalus* was 20°C and 26°C for *C. xenoplax* (5).

Interactions between ectoparasites can be stimulatory. In greenhouse experiments with concomitant populations of *Hoplolaimus columbus* and *Scutellonema brachyurum* on cotton, reproduction of both species was mutually stimulated after 90 days (41). Reproduction by *B. longicaudatus* was increased by the presence of *Hoplolaimus galeatus* in greenhouse experiments after 63 days on cotton (85). The mechanisms involved are unknown. Interactions between ectoparasites are summarized in Table 1.

Ecto- and Migratory Endoparasites

Migratory endoparasitic nematodes move through the root tissues altering root morphology and physiology. Generally, they are antagonistic to the ectoparasitic nematodes (3,7,55,79). Host suitability plays an important role in the interactions. In greenhouse tests with a duration of 267 days with *Tylenchorhynchus martini* and *Pratylenchus penetrans* on red clover, reproduction of *T. martini* was suppressed by 75-90%. Red clover was a good host for both species (7). However, in greenhouse experiments with *T. agri* and *P. penetrans* on creeping bentgrass, no interaction was observed after 10 months. Creeping bentgrass is an excellent host for *T. agri*, but a poor host for *P. penetrans* (66). *P. brachyurus* was suppressed by *P. minor* or *Helicotylenchus dihystera* on most varieties of soybean in greenhouse tests of 50 days duration; however, *H. dihystera* was suppressed by *P. brachyurus* on two varieties (34).

Table 1. Summary of interactions between ectoparasitic nematodes.

Nematode combination	Test host	Dominant species	General response	Comments	References
Paratrichodorus minor + *Tylenchorhynchus claytoni*	Corn	*T. claytoni* on 1 cultivar *P. minor* on 3 cultivars neither on 1 cultivar	Antagonistic Antagonistic Neutral	50 days, greenhouse tests.	34
T. martini + *Criconemella ornata*	Bermuda grass	*T. martini*	Antagonistic		33
T. martini + *Belonolaimus longicaudatus*	Bermuda grass	None	Neutral		
C. ornata + *B. longicaudatus*	Bermuda grass	None	Neutral		
Paratylenchus projectus + *Helicotylenchus pseudorobustus*	Soybean	*H. pseudorobustus*	Antagonistic	50 days, greenhouse tests.	42,43
P. projectus + *C. similis*	Soybean	*C. similis*	Antagonistic	Soybean is a good host for *C. similis*, a moderate host for *H. pseudorobustus*, and a poor host for *P. projectus*.	
C. similis + *H. pseudorobustus*	Soybean	None	Neutral		
High numbers of *P. neoamblyephalus* + low numbers of *C. xenoplax*	Plum	*P. neoamblycephalus*	Antagonistic	Optimum temp. for *P. neoamblycephalus* is 20°C and 26°C for *C. xenoplax*.	5
High numbers of *C. xenoplax* + low numbers of *P. neoamblyephalus*	Plum	*C. xenoplax*	Antagonistic		
Hoplolaimus columbus + *Scutellonema brachyurum*	Cotton	None	Stimulatory	Both species were stimulated 90 days, greenhouse tests.	41
B. longicaudatus + *H. galeatus*	Cotton	None	Stimulatory	Only *B. longicaudatus* was stimulated, 63 days, greenhouse tests.	85

The nematode-nematode interaction may be time dependent. *Tylenchorhynchus agri* suppressed the reproduction of *Pratylenchus penetrans* on red clover after 3 months in a greenhouse experiment but not after 5 or 7 months (3).

Physiological changes in the host induced by *T. claytoni* inhibited penetration by *P. penetrans* by 25-90% on tobacco cultivar 'WS117.' In split-root experiments, translocatable factors altered the attractiveness of the roots to *P. penetrans*. These tests were only several days in duration and long-term interactions were not determined (45).

Positive interactions occurred between *P. minor* and *P. zeae* on several varieties of corn and between *P. minor* and *P. brachyurus* on several soybean cultivars after 50 days in greenhouse experiments. *P. zeae* populations were higher with *P. minor* on two varieties of corn, and populations of *P. minor* were higher with *P. zeae* on all five cultivars. *P. brachyurus* also stimulated reproduction of *P. minor* on three varieties of soybean (34). The mechanisms involved are unknown, but small changes in host suitability may be responsible for an increase in the reproduction rate in certain nematodes. Interactions between ecto- and migratory endoparasites are summarized in Table 2.

Ecto- and Sedentary Endoparasites

The feeding sites of ectoparasitic and sedentary endoparasitic nematodes are quite different, and the two forms can coexist on the same plant without any interaction (27,49). Depending upon the mechanisms involved, the interactions can be suppressive for one nematode, mutually detrimental, or stimulatory.

The ectoparasite can suppress reproduction by the sedentary endoparasite by an indirect or direct competition for feeding sites. If the ectoparasite feeds at the preferred penetration site of the endoparasite, then the number of penetration sites for the endoparasite will become reduced. The ectoparasites can also damage the root system and thus indirectly reduce the number of feeding sites available to the endoparasite. *Paratrichodorus minor* inhibited *Meloidogyne naasi* on creeping bentgrass in the greenhouse after 6 weeks. *P. minor* prefers to feed near the root tips and competes for feeding sites with *M. naasi*. *Tylenchorhynchus agri* also suppressed *M. naasi*, probably by inhibiting root growth (66,67). Likewise, *T. vulgaris* antagonized *M. incognita* on bajra (*Pennisetum typhoides* Stapf. & Hubbard) (80); *Criconemella ornata*, *Hoplolaimus indicus*, and *Tylenchorhynchus nudus* interacted negatively toward *M. incognita* on brinjal (*Solanum melongena* L.) (49); *C. xenoplax* suppressed *M. hapla* on concord grape (63); and *M. javanica* was inhibited by *P. minor* on tomato (81). *H. columbus* antagonized *M. incognita* on cotton probably by altering the physiology of the host rather than by competing for feeding sites (41).

In comparison, sedentary endoparasitic nematodes can suppress the ectoparasites. Because they are separated by host tissue, the mechanisms are probably physiological rather than mechanical. The endoparasite may make the host more attractive or suitable to the ectoparasite. *Meloidogyne incognita* suppressed *Scutellonema brachyurum* on cotton in greenhouse experiments after 60 days (41); *M. incognita* inhibited *Tylenchorhynchus brassicae* on tomato after 30 days (2); and *M. incognita* out-competed *T. agri* on red clover after 3, 5, and 7 months (3). In field tests lasting four years, *M. hapla* antagonized *Xiphinema americanum* on alfalfa. Twice as many *X. americanum* occurred in plots without *M. hapla*. *M. hapla* may limit the number of feeding sites, but most probably changes the physiology of the alfalfa and makes it a less suitable host (50).

The relationship between the ectoparasite and sedentary endoparasitic nematodes can be mutually suppressive. Sometimes either species can predominate according to other environmental or edaphic factors. In a survey of nematodes in a cotton field over 6 years, *H. galeatus* became introduced or increased from undetectable levels and gradually replaced *M. incognita* as the predominant plant-parasitic nematode species. At a different location, *H. galeatus* predominated in one portion of the field, and *M. incognita* was most common in the remaining portion of the field; the presence of either species greatly inhibited the occurrence of both (4). Likewise, *Globodera tabacum* and *T. claytoni* were mutually suppressive in greenhouse experiments after 4 weeks. In the field, however, *G. tabacum* severely inhibited *T. claytoni*, possibly because the cyst nematode was more persistent (46). In other greenhouse tests, *M. incognita* and *H. galeatus* or *B. longicaudatus* on cotton (85), *M. incognita* and *Criconemella ornata* on brinjal (49), and *M. javanica* and *Hemicycliophora arenaria* on tomato (81) were mutually antagonistic.

Finally, reproduction of the ectoparasite may be stimulated by a sedentary endoparasitic nematode or vice versa. After 90 days on cotton in a greenhouse, *M. incognita* increased the

Table 2. Summary of interactions between ectoparasitic and migratory endoparasitic nematodes.

Nematode combination	Test host	Dominant species	General response	Comments	References
Tylenchorhynchus martini + *Pratylenchus penetrans*	Red clover	*P. penetrans*	Antagonistic	267 days, greenhouse tests.	7
T. agri + *P. penetrans*	Creeping bentgrass	None	Neutral	10 months, greenhouse tests, excellent host for *T. agri*, but poor host for *P. penetrans*.	66
P. brachyurus + *Paratrichodorus minor*	Soybean	*P. minor*	Antagonistic	50 days, greenhouse tests.	34
P. brachyurus + *Helicotylenchus dihystera*	Soybean	*H. dihystera* on most cultivars *P. brachyurus* on 2 cultivars	Antagonistic		
T. agri + *P. penetrans*	Red clover	*T. agri*	Antagonistic	Suppression after 3 months but not 5 or 7 months.	3
T. claytoni + *P. penetrans*	Tobacco	*T. claytoni*	Antagonistic	Penetration of *P. penetrans* inhibited by a translocatable factor. Tests were only several days in duration.	45
P. minor + *Pratylenchus zeae* *P. minor* + *Pratylenchus brachyurus*	Corn Soybean	None None	Stimulatory Stimulatory	*P. zeae* populations higher with *P. minor* on 2 corn varieties. *P. minor* populations higher with *P. zeae* on all 5 corn varieties. *P. brachyurus* stimulated *P. minor* on 3 varieties of soybean. 50 days, greenhouse tests.	34

reproduction of *H. columbus* (41), *M. hapla* stimulated the reproduction of *C. xenoplax* on concord grape (63), and *M. incognita* stimulated *Tylenchorhynchus vulgaris* on bajra (80). In comparison, infection of cotton by the ectoparasite *S. brachyurum* stimulated reproduction of the sedentary endoparasite *M. incognita* (41). The mechanisms involved are unknown, but probably are physiological. Interactions between ecto- and sedentary parasites are summarized in Table 3.

Migratory Endoparasites

Migratory endoparasitic nematode species are very competitive with each other because they generally utilize the same feeding sites. Host suitability is an important factor in the interactions. In greenhouse tests lasting three months, *P. penetrans* suppressed *P. alleni* on soybean. The ratio of females to males was higher in *P. penetrans* when it was mixed with *P. alleni* than when it was alone (16). Similar experiments with *Scutellonema bradys* and *P. coffeae* on guinea yam (*Dioscorea rotunda*) supported field observations that *P. coffeae* antagonized populations of *S. bradys* (1).

Other climatic or edaphic factors may be involved in an interaction between two nematode species. *Radopholus similis* and *P. coffeae* were mutually suppressive on citrus after 10-15 months in a greenhouse. Soil texture favored one of the species. *P. coffeae* predominated in fine textured soils, and *R. similis* had the advantage in coarse-textured soils (52). Interactions between migratory endoparasites are summarized in Table 4.

Migratory Endo- and Sedentary Endoparasites

The movement of migratory endoparasitic nematodes generally disrupts root tissues and often disturbs feeding by sedentary endoparasitic nematodes. *Pratylenchus major* greatly suppressed *Meloidogyne* spp. on pineapple (26), *P. brachyurus* inhibited *M. incognita* on cotton (19), and *Pratylenchus* spp. dominated over *M. incognita* on tobacco in field observations (20).

Often the migratory endoparasite penetrates the host faster or inhibits penetration by the sedentary species (19,77,78), but not always (17). In comparison, the sedentary endoparasite may stimulate or suppress penetration by the migratory species depending on suitability of the host. In tomato, which is a better host for *M. incognita* than *P. brachyurus*, prior inoculation of *M. incognita* suppressed penetration by *P. brachyurus*; however in cotton, which is a good host for *P. brachyurus* and a poor host for *M. incognita*, prior infection by *M. incognita* slightly stimulated penetration by *P. brachyurus* (19).

The interaction between species is often time dependent. *P. penetrans* initially inhibited *M. naasi* on creeping bentgrass, but after ten months had no effect (66). Likewise, *M. incognita* had no effect on a population of *P. penetrans* on red clover in greenhouse experiments after 3 months, but suppressed it after 5 months (3).

Sedentary endoparasites in comparison to the migratory species are more advanced parasites and generally establish a very complex relationship with the host that greatly alters the physiology of the plant. Sometimes the changes increase the suitability of the host toward the migratory parasites, and sometimes the host becomes less desirable. *P. penetrans* and *M. incognita* were mutually suppressive in greenhouse tests (14,15). In split-root experiments, translocatable factors induced by *M. incognita* greatly inhibited *P. penetrans*. The antagonism of *P. penetrans* toward *M. incognita* was not translocatable and was probably more closely associated with a disruption of feeding sites (14,15).

Host suitability affects competition between species. On *M. incognita*-susceptible tobacco cultivar 'NC 2326,' *P. brachyurus* and *M. incognita* were mutually suppressive; *P. brachyurus* had no effect on *M. incognita* on susceptible cultivar 'Hicks,' but *M. incognita* inhibited *P. brachyurus*; and *M. incognita* stimulated *P. brachyurus* on root-knot resistant cultivars 'NC 95' and 'NC 2512.' In comparison, *P. brachyurus* suppressed *M. hapla* on Hicks and NC 95, but had not effect on NC 2326 and NC 2512; *M. hapla* inhibited *P. brachyurus* on Hicks and NC 2326, but had no effect on NC 95 and NC 2512 (35).

The effect of the nematode-nematode interaction on host suitability is density dependent. *Pratylenchus penetrans* deposited 37% fewer eggs when low numbers of *M. incognita* were present, and 57% fewer eggs when high numbers of root-knot nematodes were present (9,10). Density may also affect the interactions between species that are mutually suppressive. Low numbers of *P. penetrans* were greatly inhibited by *Globodera tabacum* on tobacco in the greenhouse, and high numbers of *P. penetrans* greatly suppressed *G. tabacum*. In the field, however, *G. tabacum* generally occurred in monospecific communities (44,46).

Several interactions between sedentary and migratory endoparasites were mutually suppressive. Climatic or edaphic factors determine the predominant species. *P. coffeae* and

Tylenchulus semipenetrans were mutually exclusive on citrus, based on field observations (38). Likewise, *Radopholus similis* and *M. incognita* were mutually suppressive on black pepper (*Piper nigrum*) in greenhouse experiments (65).

Interactions between migratory and sedentary endoparasites are not always negative. They may also be neutral (24) or stimulatory. *M. incognita* increased the reproduction of *P. brachyurus* on tobacco (35), and *M. naasi* stimulated *P. penetrans* on creeping bentgrass (66). Prior infection by *Ditylenchus dipsaci* on a root-knot-resistant cultivar of alfalfa reduced the resistance to *M. hapla* (22,23). Interactions between migratory and sedentary endoparasites are summarized in Table 5.

Sedentary Endoparasites

Sedentary endoparasites are highly specialized parasites and have a long-lasting, complex relationship with the host (37). Interactions have been shown to occur among species of *Meloidogyne*, between *Heterodera* and *Meloidogyne* species, between *Rotylenchulus reniformis* and *Meloidogyne* species, and *R. reniformis* and *Tylenchulus semipenetrans*. Competition between two sedentary endoparasitic species is generally mutually suppressive because of the contention for available feeding sites (51) and physiological alterations; however, neutral and stimulatory interactions also occur.

The common, simultaneous occurrence of two or more species of *Meloidogyne* in the same field, root system, or gall (47,72) indicates that competition among species is weak. Other factors are probably more important in the domination of a particular species. Temperature may be important because certain species are more common in cooler climates and others are more common in warmer climates (72).

In greenhouse experiments on tomato, *M. javanica* suppressed *M. incognita* and *M. hapla* under cool temperatures, but *M. incognita* dominated *M. javanica* and *M. hapla* under high temperatures (47). In simultaneous inoculations of *M. incognita* and *M. hapla* at high temperatures, 90% of the females were *M. incognita* and only 10% were *M. hapla*. Only 57% of the females were *M. incognita* at low temperatures (8).

Factors other than temperature may also be important in the domination of a particular *Meloidogyne* species. *M. javanica* dominated *M. hapla* at 20°C even though low temperatures generally favor *M. hapla*. Also competition between species was density dependent, particularly for *M. hapla* (40). *M. incognita* suppressed *M. hapla* on *M. incognita*-resistant tobacco cultivars NC 95 and NC 2512 in greenhouse experiments at moderate temperatures. In simultaneous inoculations of both species, *M. incognita* penetrated the roots much faster than *M. hapla*, and thus all of the infection sites were occupied and destroyed by the hypersensitive reaction (35).

Interactions between *Meloidogyne* and *Heterodera* occur in which *Meloidogyne* inhibits, has no effect, or stimulates *Heterodera* species. In greenhouse tests, *M. graminicola* antagonized *H. oryzicola* on rice after 52 days in simultaneous or spaced inoculations of 7 days (61). No interactions occurred in simultaneous inoculations of *M. incognita* and *H. cajani* on cowpea in the greenhouse (64), or between *M. hapla* and *H. schachtii* on sugar beet (30,32). However, prior inoculation of *M. hapla* on sugar beet, stimulated *H. schachtii* (30,32).

Heterodera species can suppress or have no effect on *Meloidogyne* species. *H. schachtii* inhibited *M. hapla* on tomato (25) and sugar beet (29,30,32). The relationship between cyst and root-knot nematodes is density and time dependent (64). Low populations of *M. incognita* had no effect on *H. glycines* in microplot studies with soybean, yet high populations suppressed cyst nematodes early in the season and stimulated them late in the season (62). In greenhouse experiments, prior inoculation of *M. incognita* was detrimental to *H. cajani* on cowpea, but previous infection by *H. cajani* was suppressive to *M. incognita*. Simultaneous inoculations, in comparison, were density dependent and generally mutually antagonistic (64). Prior inoculation of *H. schachtii* on sugar beet in the greenhouse suppressed *M. hapla*, but previous infection by *M. hapla* stimulated *H. schachtii* (30,32). Double inoculations of *M. hapla* on sugar beet had little effect on each other, whereas *H schachtii* inoculations interacted positively. In double inoculations 10 days apart, there was a 3- to 5-fold increase in the number of infections as compared to a single inoculation of equal numbers of infective juveniles (31).

Interactions between *Meloidogyne* and *Rotylenchulus* species can be suppressive for either or both species. In simultaneous inoculations on soybean in greenhouse experiments, *M. incognita* suppressed *R. reniformis*, but was not affected itself (68). Similar results were obtained with black gram (*Vigna mungo*) (48), but on grape seedlings the two species were mutually antagonistic, although *M. incognita* was a little more competitive over time (60). In comparison, *R. reniformis* generally inhibited *M. incognita* on tomato, except high numbers of *M. incognita* were

Table 3. Summary of interactions between ectoparasitic and sedentary endoparasitic nematodes

Nematode combination	Test host	Dominant species	General response	Comments	References
Paratrichodorus minor + *Meloidogyne naasi*	Creeping bentgrass	*P. minor*	Antagonistic	6 weeks, greenhouse tests. Competition for feeding sites.	66,67
Tylenchorhynchus agri + *M. naasi*	Creeping bentgrass	*T. agri*	Antagonistic	Inhibition of root growth.	66,67
T. vulgaris + *M. incognita*	Bajra	*T. vulgaris*	Antagonistic		80
Criconemella ornata + *M. incognita*	Brinjal	*C. ornata*	Antagonistic		49
Hoplolaimus indicus + *M. incognita*	Brinjal	*H. indicus*	Antagonistic		
T. nudus + *M. incognita*	Brinjal	*T. nudus*	Antagonistic		
C. xenoplax + *M. hapla*	Grape	*C. xenoplax*	Antagonistic		63
P. minor + *M. javanica*	Tomato	*P. minor*	Antagonistic		81
H. columbus + *M. incognita*	Cotton	*H. columbus*	Antagonistic	Physiological alteration of the host.	41
M. incognita + *Scutellonema brachyurum*	Cotton	*M. incognita*	Antagonistic	60 days, greenhouse tests.	41
M. incognita + *T. brassicae*	Tomato	*M. incognita*	Antagonistic	30 days, greenhouse tests.	2
M. incognita + *T. agri*	Red clover	*M. incognita*	Antagonistic	3,5,7 months, greenhouse tests.	3
M. hapla + *Xiphinema americanum*	Alfalfa	*M. hapla*	Antagonistic	4 yrs, field tests. Limit feeding sites & physiological changes.	50
Hoplolaimus galeatus + *M. incognita*	Cotton	*H. galeatus* or *M. incognita*	Antagonistic	Edaphic factors important in the interaction.	4
Globodera tabacum + *T. claytoni*	Tobacco	None	Mutually antagonistic	Mutually suppressive in greenhouse tests, 4 weeks.	46
	Tobacco	*G. tabacum*	Antagonistic	Field tests, cyst nematode is more persistent.	
M. incognita + *Hoplolaimus galeatus*	Cotton	None	Mutually antagonistic		85

M. incognita – Belonolaimus longicaudatus	Cotton	None	Mutually antagonistic		
M. incognita + C. ornata	Brinjal	None	Mutually antagonistic		49
M. javanica + Hemicycliophora arenaria	Tomato	None	Mutually antagonistic		81
M. incognita + H. columbus	Cotton	*H. columbus*	Stimulatory	*H. columbus* reproduction was stimulated by *M. incognita.*	41
M. hapla + C. xenoplax	Grape	*C. xenoplax*	Stimulatory	*C. xenoplax* reproduction increased.	63
M. incognita + T. vulgaris	Bajra	*T. vulgaris*	Stimulatory	*T. vulgaris* reproduction increased.	80
M. incognita + S. brachyurum	Cotton	*M. incognita*	Stimulatory	*M. incognita* reproduction increased.	41

Table 4. Summary of interactions between migratory endoparasitic nematodes.

Nematode combination	Test host	Dominant species	General response	Comments	References
Pratylenchus penetrans + P. alleni	Soybean	*P. penetrans*	Antagonistic	Female to male ratio was higher in *P. penetrans* in the presence of *P. alleni.*	16
Scutellonema bradys + P. coffeae	Guinea yam	*P. coffeae*	Antagonistic	Greenhouse tests supported field observations.	1
Radopholus similis + P. coffeae	Citrus	None	Mutually antagonistic	10-15 months, greenhouse tests. *P. coffeae* dominated in fine-textured soils, *R. similis* dominated in coarse-textured soils.	52

Table 5. Summary of interactions between migratory and sedentary endoparasitic nematodes.

Nematode combination	Test host	Dominant species	General response	Comments	References
Pratylenchus major + *Meloidogyne* spp.	Pineapple	*P. major*	Antagonistic		26
P. brachyurus + *M. incognita*	Cotton	*P. brachyurus*	Antagonistic		19
Pratylenchus spp. + *M. incognita*	Tobacco	*Pratylenchus* spp.	Antagonistic	Field observations.	20
M. incognita + *P. brachyurus*	Tomato	*M. incognita*	Antagonistic	*M. incognita* suppresses penetration by *P. brachyurus*,	19
M. incognita + *P. brachyurus*	Cotton	*P. brachyurus*	Stimulatory	*M. incognita* stimulates penetration by *P. brachyurus*.	
P. penetrans + *M. naasi*	Creeping bentgrass	*P. penetrans*	Antagonistic	Initially antagonistic, but no effect after 10 months.	66
M. incognita + *P. penetrans*	Red clover	*M. incognita*	Antagonistic	Initially neutral, but antagonistic after 3 months.	3
M. incognita + *P. penetrans*	Tomato	None or *M. incognita*	Mutually antagonistic Antagonistic	Concomitant populations. In split-root experiments, translocatable factors antagonistic to *P. penetrans*.	14,15
M. incognita + *P. brachyurus*	Tobacco	None	Mutually antagonistic	Susceptible cv. 'NC2326.'	35
M. incognita + *P. brachyurus*	Tobacco	*M. incognita*	Antagonistic	Susceptible cv. 'Hicks.'	
M. incognita + *P. brachyurus*	Tobacco	*P. brachyurus*	Stimulatory	Resistant cvs. 'NC95' and 'NC2512.'	
M. hapla + *P. brachyurus*	Tobacco	*P. brachyurus*	Antagonistic	Tobacco cvs. 'Hicks' and 'NC95.'	
M. hapla + *P. brachyurus*	Tobacco	None	Neutral	Cvs. 'NC2326' and 'NC2512'.	
M. incognita + *P. penetrans*	Red clover	*M. incognita*	Antagonistic	*P. penetrans* deposited 37% fewer eggs in the presence of low numbers of *M. incognita* and 57% with high numbers.	9,10

Globodera tabacum + low numbers of *P. penetrans*	Tobacco	*G. tabacum*	Antagonistic	In the field, *G. tabacum* generally occurs mono-specifically.	44,46
G. tabacum + high numbers of *P. penetrans*	Tobacco	*P. penetrans*	Antagonistic		
P. coffeae + *Tylenchulus semipenetrans*	Citrus	None	Mutually antagonistic	Field observations.	38
Radopholus similis + *M. incognita*	Black pepper	None	Mutually antagonistic	Greenhouse experiments.	65
M. incognita + *P. brachyurus*	Tobacco	*P. brachyurus*	Stimulatory	*M. incognita* increased the reproduction of *P. penetrans*.	35
M. naasi + *P. penetrans*	Creeping bentgrass	*P. penetrans*	Stimulatory	*M. naasi* increased the *reproduction of P. penetrans*.	66
Ditylenchus dipsaci + *M. hapla*	Alfalfa	None	Stimulatory	Resistance reduced by infection with *D. dipsaci*.	22,23

suppressive to *R. reniformis* (39,83). Likewise, *R. reniformis* initially inhibited *M. javanica*, but was less competitive over time (69,70). The short life cycle of *R. reniformis* permitted it to increase more rapidly and suppress *M. javanica* after 60 days, but no differences occurred after 90 days (59).

Interactions between *Meloidogyne* spp. and *R. reniformis* are also density as well as time dependent. In field studies, low levels of *R. reniformis* inhibited *M. incognita* on sweet potato, whereas *M. incognita* had no affect on *R. reniformis*. High levels of *M. incognita*, however, suppressed *R. reniformis*, and *R. reniformis* had no affect on the root-knot nematode. Each species was capable of suppressing the other to become the dominant species (73,74,75,76). In greenhouse tests, in comparison, both low and high levels of *M. incognita* reduced the reproduction rate of *R. reniformis*, but *R. reniformis* had no effect on the reproduction of *M. incognita*. The effect of *M. incognita* on the reproduction of *R. reniformis* was not translocatable in split-root experiments (73,75).

Interactions between *Tylenchulus semipenetrans* and *R. reniformis* on grape were mutually suppressive in greenhouse experiments. Populations of both nematode species were significantly higher in single infestations than in inoculations of concomitant species (71). Interactions between sedentary endoparasites are summarized in Table 6.

ETIOLOGICAL INTERACTIONS

In most studies involving plant-parasitic nematodes as causes of plant disease, nematodes are considered to be "separate entities solely responsible for symptoms in an infected plant" (51); however, nematodes generally occur in polyspecific communities (53). Consequently two or more species of nematodes may interact with each other and thereby alter the course of the disease. If the amount of disease caused by both nematodes is less than the combined effect of each alone, the interaction is negative; if it is more, the interaction is positive; and if it is the same, there is no interaction (82).

Generally if the amount of disease caused by two or more nematode species is nearly the same as the combined effects of each alone, then there is no interaction (2, 7, 21, 24, 33, 35, 38, 41, 49, 55, 59, 67, 68, 80). However, interactions may have been undetected because single inoculum levels were used in most experiments. Interactions between species of nematodes may be density dependent. In microplot studies with *M. incognita* and *H. glycines* on soybean, yield varied from slightly less to slightly more than additive depending upon nematode inoculum levels (62).

Negative Interactions

Less disease occurs when there is strong competition between nematode species because the less pathogenic species reduce the number of infection sites available for the more pathogenic species (11,12). In all of the experiments where infection by two species caused less suppression in growth than was caused by each species alone, the species involved were complete competitors. Depending on the inoculum level, either of the two species can predominate, but generally they cannot co-inhabit. *M. incognita* suppressed the growth of tomato more by itself than when combined with *P. penetrans* in greenhouse experiments (14). Likewise, the inhibition in growth caused by *M. incognita* and *R. reniformis* on black gram (48) and grapevine seedlings (60) was less in combined inoculations than in single inoculations.

Positive Interactions

Sometimes more disease occurs when two species co-inhabit than when either species is present alone. Mechanisms are not clearly known, but may result from an increase in penetration or an increase in host suitability (58). Alteration of host physiology is probably an important factor because all positive interactions reported thus far involve a *Meloidogyne* species and some other endoparasite.

Combinations of *H. schachtii* and *M. hapla* on tomato suppressed root growth by 65, 64, and 61% below that of controls and single inoculations of *M. hapla* and *H. schachtii*, respectively (25). Likewise, concomitant populations of *M. javanica* and *M. incognita* significantly limited plant size of tobacco cv. Hicks (54). In greenhouse tests with *M. hapla* and *Ditylenchus dipsaci* on alfalfa, growth suppression occurred on cv. 'Ranger,' and infection by *M. hapla* reduced the resistance of cv. 'Vernal 298' to *D. dipsaci* (23). Similarly, *M. arenaria* or *M. hapla* reduced resistance in tobacco cv. NC 95 to *M. incognita* race 1; however, *M. javanica* or *M. incognita* race 4 had no effect (13).

Interactions With Other Pathogens

Nematodes interact with other pathogens to increase the severity of disease (56,57,58); however, the disease may even be more severe when two nematode species interact with another pathogen. Mechanisms of these interactions are unclear, but mechanical damage caused by the nematode and physiological alterations are apparently important. *Meloidogyne incognita* and *B. longicaudatus* and a fungal pathogen of chrysanthemum, *Pythium aphanidermatum*, cause more suppression in plant growth than did any pathogen alone, or any single nematode and fungus combination (36). Likewise, *M. incognita* and *Hoplolaimus indicus* interacted with the fungal pathogen *Macrophomona phaseoli* in jute to produce the most severe disease symptoms and inhibition of plant growth (27). Inoculation of *P. penetrans* and *P. coffeae* with *Fusarium oxysporum* f. sp. *lycopersici* on tomato caused a great inhibition of growth and more severe wilt symptoms than either nematode alone with the fungus. Also the population of *P. coffeae* was greater in roots infected with the fungus than in healthy roots (28). Nematode-nematode interactions and their effects on the cause of disease are summarized in Table 7.

CONCLUSIONS

In nature, plants are usually infected by more than one species of nematode. Concomitant populations of nematodes can interact with each other to affect their reproductive capability, and their interaction can alter the etiology of the plant disease. Most studies on nematode interactions have been qualitative and provide limited information. These preliminary experiments are useful, but more interactions may have been found if different levels of variables had been used (60). As a result, interpretation of the data is not easy and the development of conclusions about nematode-nematode interactions is difficult. However, the following general conclusions may be useful in summarizing this complex subject:

1. Nematode-nematode interactions can be approached from two interrelated aspects of ecology or etiology.
2. Interactions can be simulatory to one or both species, neutral, or detrimental to one or both species.
3. Competition between nematode species generally is weak.
4. Competition between species can restrict the distribution of some species.
5. Competition is more severe between species with similar feeding habits.
6. Competitive advantage increases as the host-parasite relationship becomes more complex. Hence, the endoparasites are more competitive than the ectoparasites, and the sedentary endoparasites are more competitive than the migratory endoparasites.
7. Competition between species may be density and time dependent.
8. Competition can be modified by environmental factors which are difficult to separate from competition (51).
9. Host suitability is a key factor in interactions and often is responsible for one species dominating another.
10. Mechanisms of competition may include mechanical destruction or physical occupation of feeding sites or induced physiological changes in the host's suitability or attractiveness.
11. Generally, the amount of disease caused by two or more nematode species is nearly the same or slightly less than the combined effects of each alone, but sometimes it can be more than additive.
12. Strong competitors cause less disease in combination with another species than singly.
13. Most nematode-nematode interactions which result in more disease in combination than singly involve a sedentary endoparasite.
14. Nematode-nematode interactions can become more complex when other organisms are also involved in the etiology of the disease.
15. Interactions among nematodes are important in nature, and more precise experiments are needed if their effects on the variability of crop growth are to be quantified (60).

Table 6. Summary of interactions between sedentary endoparasitic nematodes.

Nematode combination	Test host	Dominant species	General response	Comments	References
Meloidogyne javanica + *M. incognita*	Tomato	*M. javanica* or *M. incognita*	Antagonistic	Greenhouse tests, *M. javanica* dominated at low temperatures; *M. incognita* at high temperatures.	47
M. javanica + *M. hapla*	Tomato	*M. javanica*	Antagonistic	*M. javanica* dominated at low & high temperatures.	8,40
M. incognita + *M. hapla*	Tomato	*M. incognita*	Antagonistic	Temperature affected level of dominance.	8
M. incognita + *M. hapla*	Tobacco	*M. incognita*	Antagonistic	*M. incognita* resistant cvs. 'NC95' and 'NC2512.' *M. incognita* penetrated faster and "occupied" all of the feeding sites.	35
M. graminicola + *Heterodera oryzicola*	Rice	*M. graminicola*	Antagonistic	52 days, greenhouse tests.	61
M. incognita + *H. cajani*	Cowpea	None	Neutral		64
M. hapla + *H. schachtii*	Sugar beet	None	Neutral or stimulatory	Prior infection by *M. hapla* stimulated infection by *H. schachtii*.	30,32
M. hapla + *H. schachtii*	Tomato	*H. schachtii*	Antagonistic		25
Low numbers of M. incognita + *H. glycines*	Soybean	None	Neutral		62
High numbers of *M. incognita* + *H. glycines*	Soybean	(See comments)	Antagonistic or stimulatory	Antagonistic early in season, stimulatory late in season.	
M. incognita + *H. cajani*	Cowpea	*M. incognita*	Antagonistic or mutually antagonistic	Density dependent.	64

H. schachtii + *M. hapla*	Sugar beet	*H. schachtii*	Antagonistic or stimulatory	Prior inoculation of *H. schachtii* suppressed *M. hapla*, whereas prior inoculation of *M. hapla* stimulated *H. schachtii*.	30,32
M. hapla + *M. hapla*	Sugar beet	(See comments)	Neutral	Double inoculations 10 days apart increase infections 3-5 times.	31
H. schachtii + *H. schachtii*	Sugar beet	(See comments)	Stimulatory		
M. incognita + *Rotylenchulus reniformis*	Soybean	*M. incognita*	Antagonistic		68
M. incognita + *R. reniformis*	Black gram	*M. incognita*	Antagonistic		48
M. incognita + *R. reniformis*	Grape	None	Mutually antagonistic	*M. incognita* more competitive over time.	60
Low numbers of *M. incognita* + *R. reniformis*	Tomato	*R. reniformis*	Antagonistic	Density dependent.	39,83
High numbers of *M. incognita* + *R. reniformis*	Tomato	*M. incognita*	Antagonistic		
M. javanica + *R. reniformis*	Cowpea	None	Neutral	*R. reniformis* is initially more competitive but *M. javanica* is not affected over time.	59,69, 70
Low numbers of *R. reniformis* + *M. incognita*	Sweet potato	*R. reniformis*	Antagonistic	Density dependent.	73,74,75, 76
High numbers of *M. incognita* + *R. reniformis*	Sweet potato	*M. incognita*	Antagonistic	Density dependent.	
Tylenchulus semipenetrans + *R. reniformis*	Grape	None	Mutually antagonistic		71

Table 7. Summary of nematode-nematode interactions that affect disease expression.

Nematode combinations	Additional pathogen	Test host	Plant growth response & comments	References
Negative interactions				
M. incognita + *P. penetrans*	None	Tomato	*M. incognita* suppressed growth more by itself than when combined with *P. penetrans*.	14
M. incognita + *Rotylenchulus reniformis*	None	Black gram	Growth was less in combined inoculations than in single inoculations.	48
M. incognita + *R. reniformis*	None	Grape	Growth was less in combined inoculations than in single inoculations.	60
Positive interactions				
M. hapla + *H. schachtii*	None	Tomato	Combined inoculations suppressed root growth by 65, 64 and 61% below that of controls and single inoculations of *M. hapla* and *H. schachtii*, respectively.	25
M. javanica + *M. incognita*	None	Tobacco	Combined inoculations significantly suppressed growth to cv. 'Hicks' whereas single inoculations did not.	54
M. hapla + *Ditylenchus dipsaci*	None	Alfalfa	Resistance of cv. 'Vernal 298' to *D. dipsaci* was reduced by infection of *M. hapla*.	23
M. arenaria + *M. incognita* *M. hapla* + *M. incognita*	None None	Tobacco Tobacco	Resistance of cv. 'NC95' to *M. incognita* was reduced by prior infection with *M. arenaria* or *M. hapla*.	13

Interactions with other pathogens				
M. incognita + *Belonolaimus longicaudatus*	*Pythium aphanidermatum*	Chrysanthemum	All three pathogens interacted to cause more growth suppression than did any pathogen alone or any single nematode and fungus combination.	36
M. incognita + *Hoplolaimus indicus*	*Macrophomina phaseoli*	Jute	All three pathogens produced the most severe disease symptoms.	27
Pratylenhus penetrans + *P. coffeae*	*Fusarium oxysporium* f. sp. *lycopersici*	Tomato	All three pathogens caused the greatest inhibition of growth and more severe wilt symptoms.	28

Literature Cited

1. Acosta, N. and A. Ayala. 1976. Effects of *Pratylenchus coffeae* and *Scutellonema bradys* alone and in combination on guinea yam (*Dioscorea rotundata*). J. Nematol. 8:315-317.
2. Alam, M. M., N. Hasan, and S. K. Saxena. 1975. Influence of concomitant populations of *Meloidogyne incognita* and *Tylenchorhynchus brassicae* on their development and on the growth of tomato. Indian J. Nematol. 5:247-249.
3. Amosu, J. O. and D. P. Taylor. 1975. Interaction of *Meloidogyne hapla, Pratylenchus penetrans* and *Tylenchorhynchus agri* on Kenland red clover, *Trifolium pratense*. Indian J. Nematol. 4:124-131.
4. Bird, G. W., O. L. Brooks and C. E. Perry. 1974. Dynamics of concomitant field populations of *Hoplolaimus columbus* and *Meloidogyne incognita*. J. Nematol. 6:190-194.
5. Braun, A. L., H. Mojtahedi. and B. F. Lownsbery. 1975. Separate and combined effects of *Paratylenchus neoamblycephalus* and *Criconemoides xenoplax* on 'Myrobalan' plum. Phytopathology 65:328-330.
6. Brewer, R. 1978. Principles of ecology. W. B. Saunders Co., Philadelphia, Penn. 299 pp.
7. Chapman, R. A. 1959. Development of *Pratylenchus penetrans* and *Tylenchorhynchus martini* on red clover and alfalfa. Phytopathology 49:357-359.
8. Chapman, R. A. 1965. Infection of single root systems by larvae of two coincident species of root-knot nematodes. Nematologica 12:89 (Abstr.).
9. Chapman, R. A. and D. R. Turner. 1972. Effect of entrant *Meloidogyne incognita* on reproduction of concomitant *Pratylenchus penetrans* in red clover. J. Nematol. 4:221 (Abstr.).
10. Chapman, R. A. and D. R. Turner. 1975. Effect of *Meloidogyne incognita* on reproduction of *Pratylenchus penetrans* in red clover and alfalfa. J. Nematol. 7:6-10.
11. Duncan, L. and H. Ferris. 1982. Interactions between phytophagous nematodes, pp. 29-51 *In* Nematodes in soil ecosystems, ed. D. Freckman. Univ. of Texas Press, Austin, Texas. 206 pp.
12. Duncan, L. W. and H. Ferris. 1983. Validation of a model for prediction of host damage by two nematode species. J. Nematol. 15:227-234.
13. Eisenback, J. D. 1983. Loss of resistance in tobacco cultivar 'NC 95' by infection of *Meloidogyne arenaria* or *M. hapla*. J. Nematol. 15:478 (Abstr.).
14. Estores, R. A. and T. A. Chen. 1970. Interaction of *Pratylenchus penetrans* and *Meloidogyne incognita acrita* as cohabitants on tomatoes. Phytopathology 60:1291 (Abstr.).
15. Estores, R. A. and T. A. Chen. 1972. Interactions of *Pratylenchus penetrans* and *Meloidogyne incognita* as coinhabitants in tomato. J. Nematol. 4:170-174.
16. Ferris, V. R., J. M. Ferris and R. L. Bernard. 1967. Relative competitiveness of two species of *Pratylenchus* in soybeans. Nematologica 13:143 (Abstr.).
17. Freckman, D. W. and R. A. Chapman. 1972. Infection of red clover seedlings by *Heterodera trifolii* Goffart and *Pratylenchus penetrans* (Cobb). J. Nematol. 4:23-28.
18. Gause, G. F. 1934. The struggle for existence. Williams and Wilkins Co., Baltimore, Md. 163 pp.
19. Gay, C. M. and G. W. Bird. 1973. Influence of concomitant *Pratylenchus brachyurus* and *Meloidogyne* spp. on root penetration and population dynamics. J. Nematol. 5:212-217.
20. Graham, T. W., Z. T. Ford, and R. E. Currin. 1964. Response of root-knot resistant tobacco to the nematode disease complex caused by *Pratylenchus* spp. and *Meloidogyne incognita acrita*. Phytopathology 54:205-210.
21. Green, C. D. and E. B. Dennis. 1981. An analysis of the variability in yield of pea crops attacked by *Heterodera goettingiana, Helicotylenchus vulgaris* and *Pratylenchus thornei*. Plant Pathol. 30:65-72.
22. Griffin, G. D. 1972. Interaction of *Meloidogyne hapla* and *Ditylenchus dipsaci* on root-knot resistant alfalfa. Phytopathology 62:1103 (Abstr.).
23. Griffin, G. D. 1980. Interrelationship of *Meloidogyne hapla* and *Ditylenchus dipsaci* on resistant and susceptible alfalfa. J. Nematol. 12:287-293.
24. Griffin, G. D. 1983. The interrelationship of *Heterodera schachtii* and *Ditylenchus dipsaci* on sugarbeet. J. Nematol. 15:426-432.
25. Griffin, G. D. and W. W. Waite. 1982. Pathological interaction of a combination of *Heterodera schachtii* and *Meloidogyne hapla* on tomato. J. Nematol. 14:182-187.
26. Gueroünt, R. 1968. Competition *Pratylenchus brachyurus-Meloidogyne* n. sp. dans les cultures d'ananas de Cote d'Ivoire, pp. 64-69 *In* Rep. 8th Int. Symp. Nematol., Antibes, Sept. 8-14, 1965, ed. E. J. Brill.

27. Haque, M. S. and M. C. Mukhopadhyaya. 1979. Pathogenicity of *Macrophomina phaseoli* on jute in the presence of *Meloidogyne incognita* and *Hoplolaimus indicus*. J. Nematol. 11:318-320.

28. Hirano, K. and T. Kawamura. 1972. Incidence of complex diseases caused by *Pratylenchus penetrans* or *P. coffeae* and Fusarium wilt fungus in tomato seedlings. Tech. Bull. Soc. Hortic. Chiba Univ. 20:37-43.

29. Jatala, P. and H. J. Jensen. 1976. Histopathology of *Beta vulgaris* to individual and concomitant infections by *Meloidogyne hapla* and *Heterodera schachtii*. J. Nematol. 8:336-341.

30. Jatala, P. and H. J. Jensen. 1976. Parasitism of *Beta vulgaris* by *Meloidogyne hapla* and *Heterodera schachtii* alone and in combination. J. Nematol. 8:200-205.

31. Jatala, P. and H. J. Jensen. 1976. Self-interactions of *Meloidogyne hapla* and of *Heterodera schachtii* on *Beta vulgaris*. J. Nematol. 8:43-48.

32. Jatala, P. and H. J. Jensen. 1983. Influence of *Meloidogyne hapla* Chitwood, 1949 on development and establishment of *Heterodera schachtii* Schmidt, 1871 on *Beta vulgaris* L. J. Nematol. 15:564-566.

33. Johnson, A. W. 1970. Pathogenicity and interaction of three nematode species on six bermudagrasses. J. Nematol. 2:36-41.

34. Johnson, A. W. and C. J. Nusbaum. 1968. The activity of *Tylenchorhynchus claytoni, Trichodorus christiei, Pratylenchus brachyurus, P. zeae,* and *Helicotylenchus dihystera* in single and multiple inoculations on corn and soybean. Nematologica 14:9 (Abstr.).

35. Johnson, A. W. and C. J. Nusbaum. 1970. Interactions between *Meloidogyne incognita, M. hapla,* and *Pratylenchus brachyurus* in tobacco. J. Nematol. 2:334-340.

36. Johnson, A. W. and R. H. Littrell. 1970. Pathogenicity of *Pythium aphanidermatum* to Chrysanthemum in combined inoculations with *Belonolaimus longicaudatus* or *Meloidogyne incognita*. J. Nematol. 2:255-259.

37. Jones, M. G. K. 1981. The development and function of plant cells modified by endoparasitic nematodes, pp. 255-279 *In* Plant parasitic nematodes, Vol. III, ed. B. M. Zuckerman and R. A. Rohde. Academic Press, New York. 508 pp.

38. Kaplan, D. T. and L. W. Timmer. 1982. Effects of *Pratylenchus coffeae-Tylenchulus semipenetrans* interactions on nematode population dynamics in citrus. J. Nematol. 14:368-373.

39. Kheir, A. M. and A. A. Osman. 1977. Interaction of *Meloidogyne incognita* and *Rotylenchulus reniformis* on tomato. Nematol. Mediterr. 5:113-116.

40. Kinloch, R. A. and M. W. Allen. 1972. Interaction of *Meloidogyne hapla* and *M. javanica* infecting tomato. J. Nematol. 4:7-16.

41. Kraus-Schmidt, H. and S. A. Lewis. 1981. Dynamics of concomitant populations of *Hoplolaimus columbus, Scutellonema brachyurum,* and *Meloidogyne incognita* on cotton. J. Nematol. 13:41-48.

42. McGawley, E. C. and R. A. Chapman. 1976. Development of concomitant populations of *Helicotylenchus pseudorobustus, Paratylenchus projectus,* and *Criconemoides simile* on soybean. J. Nematol. 8:296 (Abstr.).

43. McGawley, E. C. and R. A. Chapman. 1983. Reproduction of *Criconemoides simile, Helicotylenchus pseudorobustus,* and *Paratylenchus projectus* on soybean. J. Nematol. 15:87-91.

44. Miller, P. M. 1970. Rate of increase of a low population of *Heterodera tabacum* reduced by *Pratylenchus penetrans* in the soil. Plant Dis. Rep. 54:25-26.

45. Miller, P. M. and J. L. McIntyre. 1975. *Tylenchorhynchus claytoni* feeding on tobacco roots inhibits entry of *Pratylenchus penetrans*. J. Nematol. 17:327 (Abstr.).

46. Miller, P. M. and S. E. Wihrheim. 1968. Mutual antagonism between *Heterodera tabacum* and some other parasitic nematodes. Plant Dis. Rep. 52:57-58.

47. Minz, G. and D. Strich-Harari. 1959. Inoculation experiments with *Meloidogyne* spp. on tomato roots. Ktavim Rec. Agric. Res. Stn. 9:275-279.

48. Mishara, S. D. and H. S. Gaur. 1982. Effect of individual and concomitant inoculation with *Meloidogyne incognita* and *Rotylenchulus reniformis* on the growth of black gram *Vigna mungo*. Indian J. Nematol. 11:25-28.

49. Misra, C. and S. N. Das. 1979. Interaction of some plant parasitic nematodes on root-knot development in brinjal. Indian J. Nematol. 7:46-53.

50. Norton, D. C. 1969. *Meloidogyne hapla* as a factor in alfalfa decline in Iowa. Phytopathology 59:1824-1828.

51. Norton, D. C. 1978. Ecology of plant-parasitic nematodes. John Wiley and Sons, New York. 268 pp.

52. O'Bannon, J. H., J. D. Radewald, A. T. Tomerlin and R. N. Inserra. 1976. Comparative influence of *Radopholus similis* and

Pratylenchus coffeae on citrus. J. Nematol. 8:58-63.

53. Oostenbrink, M. 1966. Major characteristics of the relation between nematodes and plants. Meded. Landbouw. Wageningen No. 66-4. 46 pp.
54. Paez, N., A. Arcia M., and J. A. Meredith. 1976. [Individual and combined effect of *Meloidogyne incognita* and *M. javanica* on four tobacco (*Nicotiana tabacum* L.) cultivars in Venezuela]. Nematropica 6:68-76. (In Spanish)
55. Pinochet, J., D. J. Raski and A. C. Goheen. 1976. Effects of *Pratylenchus vulnus* and *Xiphinema index* singly and combined on vine growth of *Vitis vinifera*. J. Nematol. 8:330-335.
56. Pitcher, R. S. 1978. Interactions of nematodes with other pathogens, pp. 63-77 *In* Plant nematology, ed. J. F. Southey. Minist. Agric., Fish. and Food Tech. Bull. 7. Her Majesty's Stationery Office, London. 440 pp.
57. Powell, N. T. 1971. Interactions between nematodes and fungi in disease complexes. Annu. Rev. Phytopathol. 9:253-274.
58. Powell, N. T. 1979. Internal synergisms among organisms inducing disease, pp. 113-133 *In* Plant disease, an advanced treatise, Vol. IV: How pathogens induce disease, ed. J. G. Horsfall and E. B. Cowling. Academic Press, New York. 466 pp.
59. Ras, B. H. K. and S. K. Prasad. 1971. Population studies of *Meloidogyne javanica* and *Rotylenchulus reniformis* occurring together and separately and their effects on the host. Indian J. Entomol. 32:194-200.
60. Ras, K. T. and A. R. Seshadri. 1981. Studies on interaction between *Meloidogyne incognita* and *Rotylenchulus reniformis* on grapevine seedlings. Indian J. Nematol. 11:101-102.
61. Rao, Y. S. and J. S. Prasad. 1981. Interaction of *Heterodera oryzicola* and *Meloidogyne graminicola* in artificial inoculation in rice. Indian J. Nematol. 11:104 (Abstr.).
62. Ross, J. P. 1964. Interaction of *Heterodera glycines* and *Meloidogyne incognita* on soybeans. Phytopathology 54:304-307.
63. Santo, G. S. and W. J. Bolander. 1977. Separate and concomitant effects of *Macropostonia xenoplax* and *Meloidogyne hapla* on concord grapes. J. Nematol. 9:282-283.
64. Sharma, N. K. and C. L. Sethi. 1978. Interaction between *Meloidogyne incognita* and *Heterodera cajani* on cowpea. Indian J. Nematol. 6:1-12.
65. Sheela, M. S. and T. S. Venkitesan. 1981. Interrelationships of infectivity between the burrowing and root-knot nematode in black pepper, *Piper nigrum*. Indian J. Nematol. 11:105 (Abstr.).
66. Sikora, R. A., D. P. Taylor, R. B. Malek and D. I. Edwards. 1972. Interaction of *Meloidogyne naasi, Pratylenchus penetrans*, and *Tylenchorhynchus agri* on creeping bentgrass. J. Nematol. 4:162-165.
67. Sikora, R. A., R. B. Malek, D. P. Taylor and D. I. Edwards. 1979. Reduction of *Meloidogyne naasi* infection of creeping bentgrass by *Tylenchorhynchus agri* and *Paratrichodorus minor*. Nematologica 25:179-183.
68. Singh, N. D. 1976. Interaction of *Meloidogyne incognita* and *Rotylenchulus reniformis* on soybean. Nematropica 6:76-81.
69. Taha, A. H. Y. and A. S. Kassab. 1979. The Histological reactions of *Vigna sinensis* to separate and concomitant parasitism by *Meloidogyne javanica* and *Rotylenchulus reniformis*. J. Nematol. 11:117-123.
70. Taha, A. H. Y. and A. S. Kassab. 1980. Interrelations between *Meloidogyne incognita, Rotylenchulus reniformis* and *Rhizobium* sp. on *Vigna sinensis*. J. Nematol. 12:57-62.
71. Taha, A. H. Y. and S. A. Sultan. 1977. Populations of *Rotylenchulus reniformis* and *Tylenchulus semipenetrans* on grape seedlings as influenced by coincident infestation. Nematol. Mediterr. 5:253-257.
72. Taylor, A. L., J. N. Sasser and L. A. Nelson. 1982. Relationship of climate and soil characteristics to geographical distribution of *Meloidogyne* species in agricultural soils. Coop. Publ. Dep. Plant Pathol., North Carolina State Univ., and U.S. Agency Int. Dev., Raleigh, N.C. 65 pp.
73. Thomas, J. R. and C. A. Clark. 1980. Interactions between *Meloidogyne incognita* and *Rotylenchulus reniformis* on sweet potato. J. Nematol. 12:239 (Abstr.).
74. Thomas, R. J. and C. A. Clark. 1981. *Meloidogyne incognita* and *Rotylenchulus reniformis* interactions in a sweet potato field. Phytopathology 71:908 (Abstr.).
75. Thomas, R. J. and C. A. Clark. 1983. Effects of concomitant development on reproduction of *Meloidogyne incognita* and *Rotylenchulus reniformis* on sweet potato. J. Nematol. 15:215-221.
76. Thomas, R. J. and C. A. Clark. 1983. Population dynamics of *Meloidogyne incognita* and *Rotylenchulus reniformis* alone and in combination and their effects on sweet potato. J. Nematol. 15:204-211.
77. Turner, D. R. and R. A. Chapman. 1971. Infection of alfalfa and red clover by concomitant populations of *Meloidogyne incognita* and

Pratylenchus penetrans. J. Nematol. 3:332 (Abstr.).
78. Turner, D. R. and R. A. Chapman. 1972. Infection of seedlings of alfalfa and red clover by concomitant populations of *Meloidogyne incognita* and *Pratylenchus penetrans*. J. Nematol. 4:280-286.
79. Upadhyay, K. D. and G. Swarup. 1981. Growth of maize plants in the presence of *Tylenchorhynchus vulgaris* Upadhyay et al., 1973, singly and in combination with *Pratylenchus zeae* Graham, 1951 and *Fusarium moniliforme* Sheld. Indian J. Nematol. 11:29-34.
80. Vaishnav, M. U. and C. L. Sethi. 1978. Pathogenicity of *Meloidogyne incognita* and *Tylenchorhynchus vulgaris* on bajra and their interrelationship. Indian J. Nematol. 8:1-8.
81. Van Gundy, S. D. and J. D. Kirkpatrick. 1975. Nematode-nematode interactions on tomato. J. Nematol. 7:330 (Abstr.).
82. Wallace, H. R. 1983. Interactions between nematodes and other factors on plants. J. Nematol. 15:221-227.
83. Winoto, S. and T. K. Lim. 1972. Interaction of *Meloidogyne incognita* and *Rotylenchulus reniformis* on tomato. Malaysian Agric. Res. 1:6-13. Helminthol. Abstr. 42:102.
84. Wyss, U. 1981. Ectoparasitic root nematodes: feeding behavior and plant cell responses, pp. 325-351 *In* Plant parasitic nematodes, Vol. III, ed. B. M. Zuckerman and R. A. Rohde. Academic Press, New York. 508 pp.
85. Yang, H., N. T. Powell, and K. R. Barker. 1976. Interactions of concomitant species of nematodes and *Fusarium oxysporum* f. sp. *vasinfectum* on cotton. J. Nematol. 8:74-80.

Part Five

Nematode Management

CHAPTER PAGE

BREEDING POTATOES FOR RESISTANCE TO THE ROOT-KNOT NEMATODE *MELOIDOGYNE* SPECIES

H. A. Mendoza and P. Jatala
Heads, Departments of Breeding & Genetics and Nematology & Entomology, respectively
International Potato Center
Lima, Peru

An important feature of the breeding work carried out at the International Potato Center (CIP) is that it is environmentally oriented. Germ plasm has been developed for utilization in the highland tropics, lowland tropics and countries in the subtropical zone. These lines have a combination of genes for adaptation to the climatic conditions of these zones as well as genes controlling tolerance or resistance to their main biotic constraints.

Since the root-knot nematodes are very widespread principally in habitats where the soil temperature is 20°C or more, resistance to these nematodes is an important component in the genetic makeup of the lowland tropic populations. The importance of *Meloidogyne* is not only restricted to its direct damage to roots and tubers, but also its interaction with other pathogens. Jatala (2) demonstrated that cultivars carrying resistance to bacterial wilt, *Pseudomonas solanacearum*, lose their resistance when planted in soils infested by *Meloidogyne*. This interaction requires that factors of resistance to both pathogens be combined in the same genotype so potatoes can be grown successfully in the lowland tropics.

Mendoza and Jatala (5) presented the progress made by CIP in breeding and genetic research for resistance to *Meloidogyne*. The present paper presents further information on the breeding work as well as the confirmation of partial genetic research results reported previously.

BREEDING PHILOSOPHY AND STRATEGY

CIP's breeding program maximizes the utilization of the genetic resources contained in the genus *Solanum*. Commercial cultivars and breeding lines, primitive cultivated species, and a set of selected wild species containing valuable traits of resistance or tolerance to stresses, pests, and diseases are all included. Figure 1 represents the initial gene pool with which CIP started its breeding work in the main research program and the research contracts. All genetic information contained in the genus *Solanum* and related to adaptation, yield per se, and disease and pest resistance is distributed among the large group of clones forming the highly heterozygous and heterogeneous gene pool. However, the genes controlling specific attributes are temporarily dispersed and held in segments or subsets of this initial population. Also, the relative frequency of certain genes is extremely low if one considers the large size of the population. As a consequence of these considerations, the basic goals in CIP's management of this extremely large gene pool have been identified:

1. Maintenance of a wide genetic diversity to secure a high yield and stability of performance,
2. An increase in the frequency of genes controlling desirable attributes, such as adaptation, yield per se, and resistance or tolerance to biotic and abiotic stresses, and
3. Stimulation of the recombination of desirable attributes in the same genotypes and populations.

Several alternative procedures for achievement of these goals were considered and breeding strategies studied. Experiments were set up by which the best selection procedure could be chosen after comparison of several selection methodologies. Individual, family, and combined selection schemes were compared, and it was decided to use recurrent selection with progeny testing for the population improvement. When selection was started on the gene pool, the genes were isolated within species

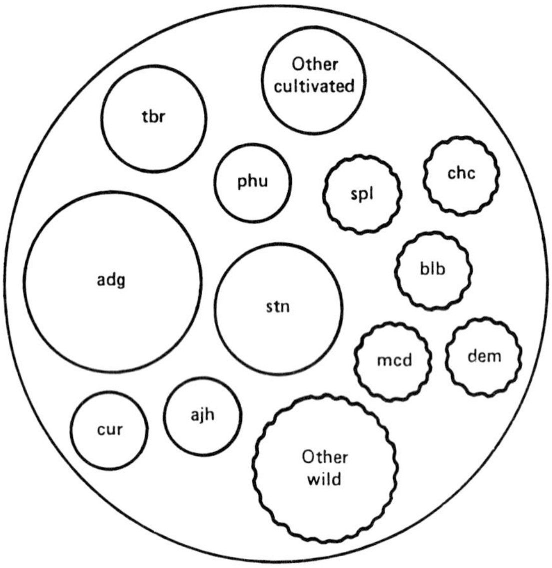

Fig. 1. The most frequently used cultivated and wild *Solanum* species. Cultivated: tbr = *tuberosum*, adg = *andigena*, phu = *phureja*, stn = *stenotomum*, ajh = *ajanhuiri*, cur = *curtilobum*.
Wild: spl = *sparsipilum*, chc = *chacoense*, blb = *bulbocastanum*, mcd = *microdontum*, dem = *demissum* [After Mendoza and Sawyer (7)].

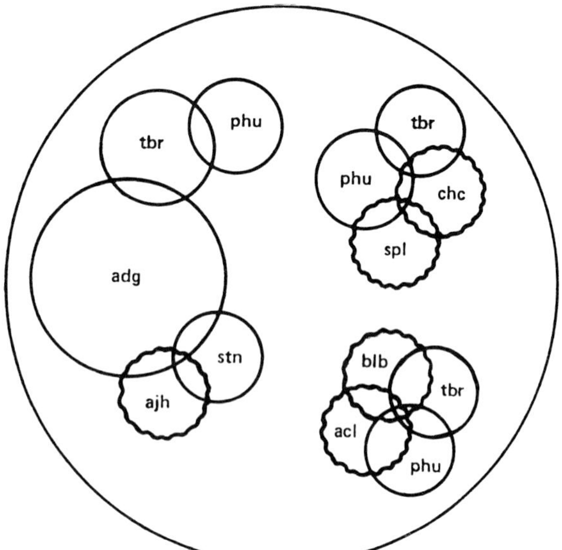

Fig. 2. *Solanum* gene pool in CIP's breeding populations [After Mendoza and Sawyer (7)].

(Fig. 1). After a few cycles of selection, the gene pool of various species was brought together (Fig. 2). From this combination of genetic resources, CIP's breeding goals became realized. A cycle of recurrent selection including advanced and backup populations involves various steps (Fig. 3).

Advanced Populations

Gene pools of selected clones from CIP's breeding program, selected clones and wild species from CIP's world potato collection, and varieties and advanced lines from other breeding programs comprise advanced populations. These populations have been produced by four to six cycles of recurrent selection combined with screening for resistances and adaptation to environmental stresses. The selected clones in each cycle are entered in the group of advanced breeding stocks. This system preserves a wide genetic base, and the increase in gene frequencies, as well as the recombination of desirable attributes, has been significant. This fact can be readily seen by the high frequency of selectable clones when the more advanced materials are evaluated. In the earlier stages of recurrent selection, the frequency of selectable clones was much lower because of undesirable characters introduced from the primitive and wild germ plasm. These undesirable characters have not been totally eliminated because of the potato's tetraploidy and highly heterozygous genetic structure. Their frequency, however, has decreased significantly in the most advanced cycles.

Single and bulk cross hybrids are produced in the advanced crossing block (Fig. 3). The evaluation of single F_1 hybrids either in the field for agronomic performance and progeny uniformity or in the greenhouses for stress disease or pest resistances provides information on the specific combining ability (SCA) of the progenitors involved in the crosses. Also, it permits the selection of the best progenies for production of populations for export. The evaluation of the F_1 bulk-cross hybrids provides information on the general combining ability (GCA) of the clones utilized as females.

Only female clones with high GCA effects can produce superior progenies when pollinated with a mixture of pollen of several male clones. Clones with high GCA effects for traits of yield, heat tolerance, late blight resistance, frost resistance and PLRV resistance have been identified. For use of true potato seed (TPS) in commercial potato production, it is also very important that parental clones with a high GCA for yield and tuber uniformity be identified.

Back-up Populations

The world potato collection of primitive cultivars and wild species is used to develop back-up populations. The genetic variability for several traits is immense in this material, but

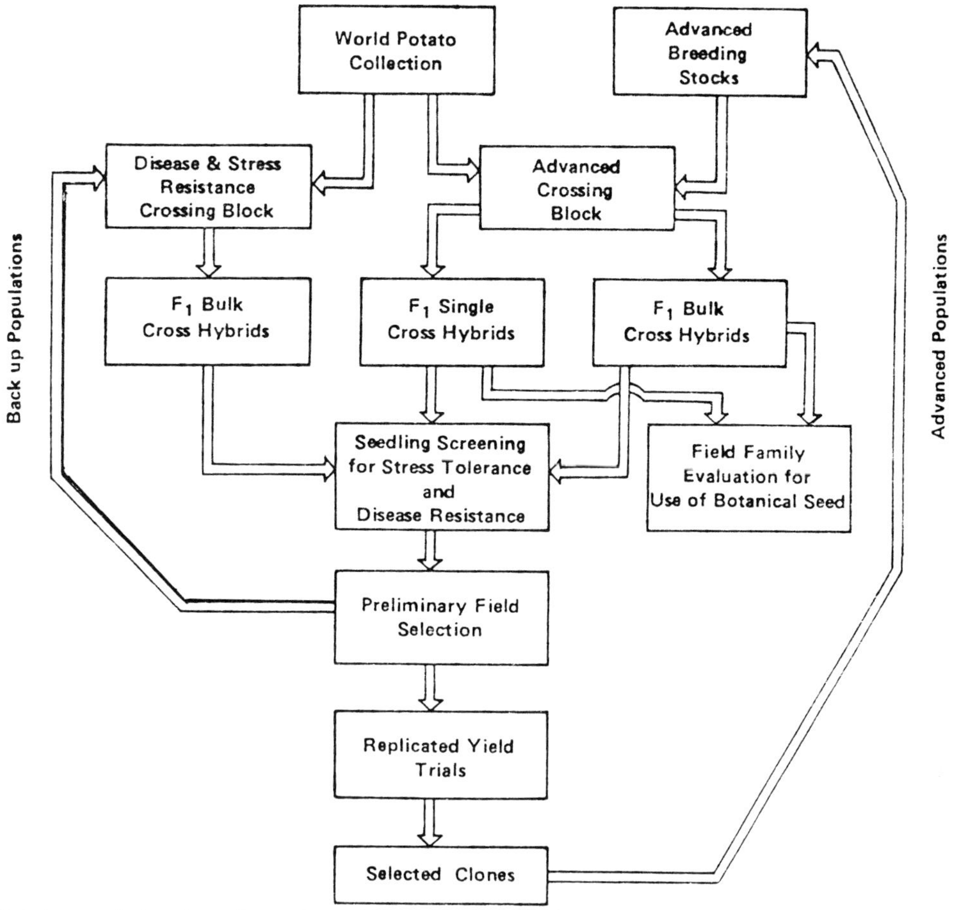

Fig. 3. Potato population breeding scheme [After Mendoza (4)].

the genetic distance from the most advanced populations is also considerable. Therefore, the objectives of population development by recurrent selection are 1) an increase in gene frequencies, 2) recombination of the genes controlling pest, disease, and stress tolerances, and 3) improvement, to some extent, of agronomic performance. This approach enables genetic distance between the primitive and wild germ plasm and the most advanced populations to be reduced. Whenever a new trait is needed, but not present in the advanced populations, it can be selected from the back-up populations, and the progress achieved in the other traits, which are already present in the most advanced materials, will not be harmed.

Environmental Orientation

The climates of the countries with which CIP is cooperating are so varied that from the beginning an effort was made to orient the breeding program environmentally. The potato crop is subjected to strong genotype-environment interactions that limit the possibilities of a wide range of adaptation. At the start of CIP's breeding program, three clearly defined populations were developed:

1. Highland tropic population - main traits were yield, resistance to late blight, cyst nematode, wart, and PVY; and tolerance to frost;

2. Lowland tropic population - main traits were yield, earliness, heat tolerance, and resistance to late blight, bacterial wilt, PVY, and root-knot;

3. Subtropical environment population (obtained and distributed from the research contracts) - main traits were yield and resistance to late blight, PVY, and PVX.

CIP's breeding program later developed more specific subpopulations with specific combinations of resistances and other attributes: earliness and leafroll virus resistance, late blight and early-blight resistance, frost resistance, and the ability to tuberize under long-day conditions.

Genetic materials are distributed according to the environmental characteristics of the

Table 1. Cultivated germ plasm tested for resistance to *M. incognita* [After Mendoza and Jatala (5)].

Germ plasm and No. of clones	Resistance %	Resistance Level
Primitive germ plasm (5865)	1.8	low
Other (750)	1.9	low
Cultivated diploids (1020)	2.7	low

Table 2. Wild *Solanum* germ plasm tested for resistance to *M. incognita* [After Mendoza and Jatala (5)].

Germ plasm and No. tested	Resistance %	Resistance Level
Species (65)	40.8	low to very high
Accessions (482)	18.6	low to very high
Genotypes (8060)	3.5	low to very high

Table 3. Reaction of several interspecific crosses between diploid material to *Meloidogyne incognita* [After Mendoza and Jatala (6)].

Combinations	No. of families	Percentages of individuals with each root-galling index[a] 1	2	3	4	5	No. of genotypes
Res × Res	20	51.32	10.77	35.12	2.36	0.00	1058
Res × Sus	12	27.86	10.15	51.17	10.15	0.00	768
Sus × Res	21	4.50	10.00	66.35	19.05	0.07	1370
Sus × Sus	9	.12	.24	44.07	51.28	3.05	819

[a]1 = No root galling and/or nematode reproduction.
5 = Very severe root galling and/or a very high nematode reproduction.

recipient countries and the yield constraints that may be present.

SEARCHING FOR RESISTANCE

The first step taken in searching for resistance to root-knot nematode was determination of the genetic variability for resistance present in the existing world collection of primitive *Solanum* cultivars as well as other cultivated diploids and tetraploid cultivars and breeding lines. Results of such screening are given in Table 1. The frequency of resistant clones and the level of resistance present in these populations were low. Therefore, it was necessary to test the wild tuber-bearing species from the genus *Solanum* (Table 2). From this group, three diploid species (*S. sparsipilum*, *S. chacoense*, and *S. microdontum*) were particularly interesting for their level of resistance and ability to cross with cultivated diploids.

GENETICS OF RESISTANCE

In order for the pattern of resistance transmission to be explored, the three diploid species were used to form hybrid combinations with the genotypes of the *Phureja-Stenotonum* population which expressed a certain level of resistance to *M. incognita*. Sixty-two hybrid families representing 4015 genotypes were evaluated for resistance (Table 3). The results suggest that two or more pairs of genes that interact with each other could be controlling the resistance to *Meloidogyne*. However, it appears that there is a certain maternal effect that calls for caution in the interpretation of results. Crosses of resistant × susceptible show that 27.8% of the individuals are highly resistant, while the reciprocal crosses gave only 4.5% of immune progenies. Results also clearly show the remarkable progress made in the selection for resistance using wild germ plasm as a source.

After the preliminary results, genetic experiments were conducted so the pattern of inheritance of resistance to *M. incognita* in *S. sparsipilum* could be determined. Four susceptible *S. phureja* diploid clones, two resistant *S. sparsipilum* clones and a resistant hybrid clone (*S. sparsipilum* × *S. phureja*) were crossed following a diallel mating design with reciprocals (Table 4). Forty-two hybrid families were obtained.

Evaluation for resistance was carried out under screenhouse conditions in two seasons with

Table 4. Parental clones used for studying resistance to root-knot nematode.

Clone	Species	Reaction
377106.1	*S. phureja*	Susceptible
377106.5	*S. phureja*	Susceptible
377149.1	*S. phureja*	Susceptible
66P18C-4	*S. phureja*	Susceptible
760147.2	*S. sparsipilum*	Resistant
760147.7	*S. sparsipilum*	Resistant
378929.2	(760147.7 × Phy bk LB) × (66P15.A × 760147)	Resistant

Table 5. Reciprocal differences in percent resistance to *Meloidogyne incognita*.

Crosses		% Resistance[a]	% Difference[b]
377106.1 × 760147.2	S×R	8.33	16.67
760147.2 × 377106.1	R×S	25.00	
377106.1 × 760147.7	S×R	12.50	15.8
760147.7 × 377106.1	R×S	28.3	
377106.1 × 378929.21	S×R	15.8	11.7
378929.21 × 377106.1	R×S	27.5	
377106.5 × 760147.2	S×R	15.83	20.87
760147.2 × 377106.5	R×S	36.7	
377106.5 × 760147.7	S×R	10.87	19.96
760147.7 × 377106.5	R×S	30.83	
377106.5 × 378929.21	S×R	12.5	12.7
378929.21 × 377106.5	R×S	25.2	
377149.1 × 760147.2	S×R	15.0	18.3
760147.2 × 377149.1	R×S	33.3	
377149.1 × 760147.7	S×R	17.5	14.1
760147.7 × 377149.1	R×S	31.6	
377149.1 × 378929.21	S×R	15.52	15.08
378929.21 × 377149.1	R×S	30.6	
66P18C.4 × 760147.2	S×R	19.17	16.66
760147.2 × 66P18C.4	R×S	35.83	
66P18C.4 × 760147.7	S×R	16.7	5.8
760147.7 × 66P18C.4	R×S	22.5	
66P18C.4 × 378929.21	S×R	11.3	13.7
378929.21 × 66P18C.4	R×S	25.0	

[a] Resistance is based on the root galling index of 1 and 2 on a scale of 1-5.
[b] Significant at 0.01 probability level.

mean temperatures of 24.5°C and 25.8°C, respectively. The experiments were conducted in a randomized complete block design replicated three times and having 40 seedlings per experimental unit. The resistance evaluation was according to the method described by Jatala (3).

Results indicate that variability due to GCA and reciprocal effects was highly significant. However, the GCA effects were much larger than reciprocal effects. A narrow-sense heritability estimate of $h^2 = .78$ was obtained. Reciprocal effects were highly significant indicating either a maternal effect of *S. sparsipilum* or an interaction between cytoplasmic elements with nuclear genes (Table 5).

Upon analysis of frequency of resistant and susceptible genotypes in the crosses of susceptible x resistant clones, several genetic models were studied, and the one appearing to fit the experimental data is described as follows. If three major complementary genes such as **A, B,**

Table 6. χ^2 test for the crosses of susceptible × resistant phenotypes assuming that the corresponding genotypes are aabbcc and AaBbCc, respectively.

Cross		Observed frequency		Expected frequency		χ^2_c[a]	χ^2_t, df=1[b]
		Res.	Sus.	Res.	Sus.		
$Y_{1.5}$	S×R	14	100	14.25	99.75	0.01	3.84
$Y_{1.6}$	S×R	12	104	14.5	101.5	0.49	3.84
$Y_{1.7}$	S×R	14	102	14.5	101.5	0.20	3.84
$Y_{2.5}$	S×R	16	88	13	91	0.79	3.84
$Y_{2.6}$	S×R	17	95	14	98	0.73	3.84
$Y_{2.7}$	S×R	12	95	13.375	93.625	0.16	3.84
$Y_{3.5}$	S×R	19	101	15	105	1.22	3.84
$Y_{3.6}$	S×R	220	100	15	105	1.81	3.84
$Y_{3.7}$	S×R	12	90	12.75	89.25	0.05	3.84
$Y_{4.5}$	S×R	17	103	15	105	0.30	3.84
$Y_{4.6}$	S×R	221	99	15	105	2.74	3.84
$Y_{4.7}$	S×R	15	105	15	105	0.0	3.84

[a] χ^2_c = calculated χ^2.
[b] χ^2_t = tabulated χ^2 for df = 1 and 0.01 probability level.

Table 7. Evolution of the frequency of resistance to *Meloidogyne incognita* through four cycles of phenotypic recurrent selection [After Mendoza and Jatala (5)].

Year	Selection cycle	No. tested	% Resistance
Before 1976	Existing genetic resources (both cultivated and wild)	15,695	3.5
1977	1	2688	4.5
1978	2 and 3	6482	27.2
1979	3 and 4	11,085	35.5

and **C** are assumed, then resistance would be determined by the presence of one of each of the dominant genes: i.e., a genotype such as **A-B-C-** would be resistant and **aabbcc** would be susceptible. The analysis of the data from each individual cross indicates that all three resistant clones had a genotype of **AaBbCc,** and the susceptibles were homozygous recessive. The expected segregation ratio for the mating of susceptible × resistant would be ⅛ resistant: ⅞ susceptible. In all the crosses, the calculated X^2 values were smaller than the tabulated X^2 at 0.01 probability level indicating that the experimental data agreed with the genetic hypothesis (Table 6). The reciprocal crosses, i.e., resistant × susceptible, were not analyzed under this hypothesis due to the cytoplasmic effects caused by *S. sparsipilum.*

BREEDING FOR RESISTANCE

The three wild species of *S. sparsipilum, S. chacoense,* and *S. microdontum* and the few resistant cultivated diploids were intercrossed for the start of a recurrent phenotypic selection scheme which after four cycles produced a significant increase in the percentage of resistant genotypes within the population (Table 7). Since these materials were only tested for resistance to race 3 of *M. incognita,* a Peruvian population, it was necessary that the plants' genetic variability in resistance to other important species and races of *Meloidogyne* be determined. Selected families from the 4th cycle of recurrent selection, in which 70% or more of the genotypes were resistant to the Peruvian population of *M. incognita,* were evaluated by the International *Meloidogyne* Project Research Center at Raleigh, North Carolina, for resistance to four races of *M. incognita,* two races of *M. arenaria,* and one race each of *M. javanica* and *M. hapla.*

The majority of the potato seedlings had resistance to one or several populations since only 15% of the seedlings tested had gall production by all *Meloidogyne* populations. Of the plants, 69% had no gall production in response

Table 8. High resistance (no galls per plant) to several species and races of *Meloidogyne* in selected CIP breeding lines [After Hartman et al. (1)].

Families tested	*M. incognita* races				*M. arenaria* races		*M. javanica*	*M. hapla*	Combined resistance[a]
	1	2	3	4	1	2			
378857	61	83	86	89	83	42	41	30	0
378875	56	64	70	78	74	56	65	11	2
378908	63	62	57	60	76	34	80	41	3
378911	54	57	77	87	89	22	57	20	2
378916	31	44	36	58	49	36	56	32	0
378924	60	52	71	76	57	48	81	40	5
378930	45	60	47	55	45	62	71	18	2
378950	52	50	33	56	40	49	60	29	0

[a] Percent resistant to all species and races.
After Hartman et al. (1).

Table 9. Reaction of field selected clones from 4X - 2X intercrosses and 4X - 2X × tropically adapted clones to *Meloidogyne incognita*.

Source	No. clones	Root galling index[a]				
		1	2	3	4	5
4X - 2X	345	1%	3%	68.6%	26.2%	1.1%
		4%				
4X - 2X × tropically	197	9.3%	6%	54.2%	29%	1.5%
		15.3%				

[a] 1 = No root galling and/or nematode reproduction to 5 = very severe root galling and/or a very high nematode reproduction.

to 5 or more root-knot populations, and only 14 plants had a high level of resistance (no galls) to all populations (Table 8).

Since the frequency of resistance in the 4th cycle of recurrent selection was very high, efforts were concentrated to transfer the resistance genes from this population to cultivated tetraploid forms. To do so, a series of 4X - 2X crosses were made with tropically adapted tetraploid clones (female parent) from CIP's breeding program and bulked pollen (male parent) of a selected number of the resistant clones mentioned above. Since the frequency of unreduced pollen was very low, the amount of seed obtained from these crosses was also low. Only 121 genotypes were obtained, and resultant seedlings were transplanted to the field for multiplication. From these, only 76 clones were harvested and then tested for resistance. A total of 10 resistant clones were selected with 3 of these having a high degree of resistance (i.e., no root galling or nematode reproduction). On the basis of these resistant tetraploid clones, two populations were created. First, the resistant clones were intercrossed, and second, the resistant clones were backcrossed to tropically adapted tetraploid clones so agronomic characteristics could be improved. The resulting genotypes were first field tested and selected for yield, earliness and other agronomically desirable attributes (Table 9). The selected clones were tested for resistance to *M. incognita* under greenhouse conditions.

Theoretically one would expect a higher frequency of resistant clones in the 4X - 2X intercrosses than in the 4X - 2X clones backcrossed to tropically adapted but *Meloidogyne*-susceptible clones. Since both populations mentioned in Table 9 were first field tested and selected for agronomic characteristics, it is conceivable that a large number of *M. incognita*-resistant clones were eliminated because of their poor agronomic characters. This was probably due to the prevailing contribution of poor genes from the wild genetic background of the resistant clones.

When all the results presented above are considered, it appears that genetic solutions to the problems created by the root-knot nematode are a reality. In the very near future, we expect to have the resistance identified in these studies transferred to commercial potato cultivars to be used in the lowland tropics or wherever *Meloidogyne* species are of economic importance.

Literature Cited

1. Hartman, K., J. N. Sasser, and P. Jatala. 1982. Evaluation of potato lines for resistance to major species and races of root-knot nematodes (*Meloidogyne* species), pp. 94-95 *In* Proc. Int. Congr. on Research for the Potato in the Year 2000. Int. Potato Center, Lima, Peru. 199 pp.
2. Jatala, P. 1978. Studies on the interrelationships of the plant parasitic nematodes and other organisms on potatoes, pp. 191-193 *In* Report of the Second Nematode Planning Conference. Int. Potato Center, Lima, Peru.
3. Jatala, P. 1982. An efficient method for screening large numbers of potato seedlings for resistance to *Meloidogyne* species, pp. 105-106 *In* Proc. Int. Congr. on Research for the Potato in the Year 2000. Int. Potato Center, Lima, Peru. 199 pp.
4. Mendoza, H. A. 1980. Philosophy paper on Organization and Program Development of the Breeding and Genetics Department, presented at the Planning Conference on Utilization of the Genetic Resources of the Potato. Int. Potato Center, Lima, Peru.
5. Mendoza, H. A. and P. Jatala. 1982. Breeding for resistance to root-knot nematodes, pp. 130 *In* Proc. Int. Congr. on Research for the Potato in the Year 2000. Int. Potato Center, Lima, Peru. 199 pp.
6. Mendoza, H. A. and P. Jatala. 1982. Breeding of potato for resistance to root-knot nematode, *Meloidogyne incognita*, pp. 143-153 *In* Proc. Third IMP Res. & Plann. Conf. on Root-knot Nematodes, *Meloidogyne* spp., Region II, March 22-26, 1982. Int. Potato Center, Lima, Peru. 233 pp.
7. Mendoza, H. A. and R. L. Sawyer. 1984. The breeding program at the International Potato Center. Plant Breeding Prog. Rev. In press.

THE NATURE OF RESISTANCE TO *MELOIDOGYNE INCOGNITA* (Kofoid & White, 1919) Chitwood, 1949

M. Canto-Sáenz

Universidad Nacional Agraria-La Molina
Lima, Perú

EVALUATION OF RESISTANCE AND TERMINOLOGY FOR PLANT RESPONSE

For a complete evaluation of plant response to nematodes, two parameters should be measured: nematode reproduction and damage caused by the nematode (39,51). Host efficiency is usually not correlated with damage (33,39,44,51), and a low level of parasite damage may be due either to a high level of resistance in the host or to a low level of parasitic ability in the parasite (59).

The host and the parasite are a single system, and the terminology about the system should be based on the system rather than on one or the other of its components (59). The terms non-host, non-efficient host (Pf/Pi<1) and efficient host (Pf/Pi>1) were proposed to describe the plant when only nematode reproduction is measured (39). When both parameters are measured, the terms used to describe the plant response are: immune (non-host, no damage), resistant (non-efficient host that suffers no damage), tolerant (efficient host that suffers no damage), and susceptible (efficient or non-efficient host that suffers damage). The term intolerant was proposed so a non-efficient host that suffers damage could be distinguished from an efficient host that suffers damage (14,20). The distinction between susceptible and intolerant is important, but intolerant, a negative term, cannot be used to differentiate one possibility in a combination of four, where each possibility has two components (De Morgan's law of logics). In other words, resistant or susceptible plants are also intolerant, and tolerant or susceptible plants are also non-resistant.

However, the terms resistant and susceptible are often used indiscriminately to describe plant response in relation to nematode reproduction or to describe damage caused by the nematode. The use of the same term to define two distinct phenomena is confusing and should be avoided. For example, the statement that the "resistance mechanism in oats apparently damages the plant as much as the susceptible response" (14) is a contradition that results when the term resistant is used only in relation to nematode reproduction. Some definitions of plant response in which the terms are used indiscriminately are: non-host (plants with pre-infectional resistance); immune (a plant able to prevent infection with no disease expression); resistant (plant that cannot prevent entrance by the parasite but is able to prevent, restrict or retard its development, or a plant in which penetration by juveniles does not occur), and tolerant (a plant that survives and gives satisfactory yield at a level of infection that causes economic loss on other varieties of the same species (4,24,70).

Host efficiency is measured by gall index, number of egg masses/g of root; females/g of root, eggs/g of root or index of reproduction (number of eggs developing in a resistant cultivar as a percentage of those developing on a susceptible cultivar) (23,24,71).

At high nematode densities, sometimes there is no gall formation and the egg masses may appear later. If the evaluation is based only on gall or egg mass index before 50 days, the results may be misinterpreted. Gall index is a useful criterion for general screening, but negative results should be retested (25,32,61).

Damage caused by the nematode is determined by relating pre-plant nematode densities (initial population) to growth and yield of annual crops (4). The minimal density that causes a measurable reduction in plant growth or yield is regarded as the damage threshold density (3), and when the cost of production and the value of a given crop are considered, the term economic threshold density (minimal nematode density

Table 1. Terms to describe plant response to nematodes.

	Nematode damage to the plant	
Host efficiency for nematode reproduction	**Statistically significant**	**Not statistically significant**
Efficient	Susceptible	Tolerant
Non-efficient	Hypersusceptible	Resistant

that causes economic loss) is used. Threshold density varies with nematode species, race, plant variety and environment (4). Infection of non-efficient or efficient hosts by low densities of *Meloidogyne* spp. may enhance growth and yield (44,50), have no effect (44), or cause severe damage (4).

Some damage threshold densities of *M. incognita* are as follows: 3000-81,000 eggs/plant for soybean, depending on temperature (48); 0.005-0.02 eggs/g soil for tomato (28); 0.005-0.03 eggs/g for sweet potato (28); 0.04-2 eggs/g soil for tomato (5); 5-10 eggs/g soil for potato (9).

Potato response confirms that host efficiency is not related to nematode damage and when only nematode reproduction is measured, the terms efficient and non-efficient hosts are proper since they do not require explanation or qualification. However, when both host efficiency and nematode damage are measured, only one term should be used to indicate the plant-nematode interaction. The most important feature of this interaction is that the behaviour of the host and the behaviour of the parasite are both treated as one system (59). I recommend the terms be redefined as follows: resistant (non-efficient host that suffers no statistically significant damage); tolerant (efficient host that suffers no statistically significant damage); susceptible (efficient host that suffers statistically significant damage). These terms would be important in plant breeding and agronomy because "a resistant variety must compare favorably with other available means of control in terms of yield in infested land and its effect on nematode population" (14). The term hypersusceptible (instead of intolerant) is proposed to define a non-efficient host that suffers statistically significant damage. In this interaction the plant is so susceptible that even though the nematode does not reproduce it causes significant damage. With the use of the term hypersusceptible, Dropkin and Nelson's table (20) would be modified (Table 1).

Cook (14) uses the terms tolerant and intolerant in relation to yield loss (host sensitivity) but mentions reproduction. The term sensitivity is widely used to define the tendency to express strong symptoms (1), but symptomatology is not always related to yield, and the term hypersensitivity describes a quick and drastic incompatible interaction. It is measured at the cellular or tissue level and may be related to susceptibility or resistance. Therefore, the statistical significance of nematode damage, in terms of plant yield and/or growth reduction, should be designated since necrosis of some cells is obviously harmful to those cells, but may not be measurable in terms of the whole plant.

When the plant response is examined at the cellular or tissue level, where neither nematode reproduction nor damage is measured, the terms compatible and incompatible response are useful.

THE NATURE OF INCOMPATIBLE RESPONSE TO *M. INCOGNITA*

Incompatible responses of plants to root-knot nematodes vary qualitatively and quantitatively even between cultivars of the same crops. Incompatible responses have been reported to act in the rhizosphere through toxic exudates (65) or by an effect on hatching (79), attraction (60) or penetration of the roots by the nematode. In some incompatible plants, the nematode either does not penetrate (62), or it penetrates for a short time (55), in low numbers (21) or in numbers equal to those penetrating compatible plants (10,21,57,78). Incompatible response affects individual nematodes differently. Some juveniles feed externally, penetrate, come out of the roots, or may migrate and feed inside. Some juveniles remain alive until seven days after inoculation (10).

After nematode penetration, several degrees of incompatibility have been reported (13,63). The most common response is hypersensitivity of cells damaged by migrating juveniles (22) or cells around the sedentary nematode (10,20,42,73,77). This type of response appears as early as 24 hr after inoculation, and the plant cells, as well as the undeveloped nematode, die (8,17,41,57,67). However, the degree of necrosis and time of appearance may differ even in plants of the same crop.

Some plants show no galling response (21,25), or galls may be small, inconspicuous (12) or few in number (45). Gall formation, however, does not indicate successful nematode development (26,32). Some galls contain only fragments of nematodes, whereas others have no detectable traces of juveniles and only necrotic areas are observed (10,46). The nematode disintegrates and disappears.

Incompatible plants are not always hypersensitive (19,30,46) since, in some cases, juveniles lie closely oppressed and parallel to the stele. Giant cells, if formed, may be abnormal and not fully developed (46). In such cells, the nematode will develop poorly or slowly with no eggs (46) or few eggs produced (13,20,45). The extent of giant cell development is related to the degree of nematode development (22,60). Senescent giant cells in incompatible plants may collapse leaving large cavities in the roots, and when few developed females occur, the giant cells are indistinguishable from those produced in compatible plants (46).

Identification of different types of incompatible responses in plants facilitates the selection for better types of resistance. The good growth and yield of a potato (E2) under high nematode density is probably related to its incompatible response which takes longer to appear and involves less extensive necrosis. This relation suggests that clones with similar types of response could be selected (10).

Biochemical Mechanisms of Incompatibility

Reports of biochemical mechanisms of incompatibility are numerous. Root diffusates repel (curcubitacin) or are toxic (margosa) to the nematode (72). In general, it appears that nematode secretion lacks some compounds; the effects of nematode secretion are counteracted, neutralized or blocked by chemicals in the plant (6,31); or nematode secretion blocks the synthesis of proteins and growth hormones in the plants (64). In the hypersensitive reactions, phenols are mainly involved (35,47,66), but other factors may be important (64). The death of the nematode after appearance of necrosis suggests that the plant responds to the presence of feeding of the nematode inside the roots by production or accumulation of toxic substances such as nicotine, proline, glyceollin, or phytoalexins such as hemigossypol, methoxyhemigossypol, gossypol and dimethoxygossypol. These substances may affect nematode behavior or cause the death and disintegration of the nematode (72). The incompatible response is faster in some incompatible plants than in others which suggests quantitative instead of qualitative differences since the final response is the same (10). Another interesting hypothesis is that hydroxyproline concentration increases and cells develop cyanide-insensitive respiration in incompatible plants when they are invaded by the nematode (80).

Correlation Between Parts of the Plant

It is important to determine if there is a correlation between the response of the different parts of the plant. Some potato tubers, for example, are not infected when artificially inoculated (41). There are several reports with other crops that incompatibility to root-knot nematodes is inherent to a particular part of the plant (either top or roots) and is not translocated (45,53,56). There is, however, a report to the contrary (29). Temperature requirements for symptom manifestation may also be different for the different parts of the plant (10).

FACTORS AFFECTING PLANT RESPONSE

Temperature

Temperature is one of the most important environmental factors that affect plant response to *M. incognita*. It affects nematode survival (7,17,52), distribution (49), embryogenesis and hatching (2,69,75), migration and penetration (15,38,54), development (16,48,52,69,74), and symptom expression in the plant (11,15,40).

Temperature requirements vary among *M. incognita* populations (thermotypes) (15,58,68), and with each host-parasite combination (11,15,52). Non-efficient hosts of *M. incognita* become progressively efficient as temperature rises (11,18,19,24,34,38). However, some plants and cultivars of the same plant species do not respond similarly (11,19,37). The equilibrium density for nematode reproduction may also be different among plants of the same crop (11). This response suggests the possibility of selecting for stable resistance at high temperatures.

The fact that plants become efficient hosts at high temperatures is probably due to three factors: 1) high temperatures are optimum conditions for nematode activity; 2) stress caused by high temperature makes the plants more vulnerable to nematode attack; and 3) chemicals responsible for cell necrosis may not be produced or may be neutralized or counteracted at high temperature.

In plants under environmental stress (except nutritional stress), nematode reproduction is higher (27,76). The optimum temperature for hatching was reported to be 27°C (7,69); embryogenesis is faster and migration and penetration by the nematode is favored by high temperatures (2,54,69,75). The nematode life cycle is also completed faster at high temperature; therefore, more generations are produced (16,52,69). Moreover, at high temperatures, fewer males develop (16).

Differential plant responses to nematodes at high temperatures are probably due to quantitative differences in the enzymatic reactions occurring in the plant-nematode relationship. The progressive increase in host efficiency with temperature increase supports this hypothesis.

Plant Age at Inoculation Time

Nematode effect on the growth of susceptible plants is influenced by plant age at inoculation. Older plants have more tissue already differentiated which the nematode usually does not penetrate (13). Thus, more root growth remains undamaged (36). Plant age also influences host efficiency for *M. incognita* and the resulting nematode damage in resistant plants of several crops (7.36,43,45), but not in all crops (11,16,27).

High nematode Pf in older plants is probably a function of this greater availability of roots and less individual competition (45). The scarce root systems of younger plants at transplanting time cause a concentration of juveniles around the root tips. When this situation occurs, root tip growth can stop, resulting in a dramatic reduction in size of root system (10). The number of eggs/g of root is usually higher in young plants. The response of young plants to root-knot nematodes, if not qualitatively, is sometimes quantitatively different from that of older plants. This difference should be taken into consideration in experiments with small seedlings where it is assumed that seedling response is the same as that of older plants (24,43). This assumption may be in relation to nematode reproduction or in relation to nematode damage to the plants.

Origin of Plants

Nematode reproduction in some susceptible plants is greatly affected by origin of plant. In potato, for example, the nematode Pf is higher in plants originating from cuttings and tubers, while number of eggs/g of root is higher in seedlings (11).

It seems likely that a combined effect of the nematode and high temperature would be of sufficient magnitude to kill the young plants or seedlings of non-efficient or efficient hosts. The planting of such crops in heavily infested land should be avoided.

Literature Cited

1. Ainsworth, G. C. 1971. Ainsworth and Bisby's dictionary of the fungi. Commonw. Agric. Bur., Butler & Tanner Ltd., London. 663 pp.
2. Barker, K. R., C. J. Nusbaum, and L. A. Nelson. 1969. Seasonal population dynamics of selected plant-parasitic nematodes as measured by three extraction procedures. J. Nematol. 1:232-239.
3. Barker, K. R. and C. J. Nusbaum. 1971. Diagnostic and advisory programs, pp. 281-301 *In* Plant parasitic nematodes, Vol. 1, ed. B. M. Zuckerman, W. F. Mai, and R. A. Rhode. Academic Press, New York. 345 pp.
4. Barker, K. R. and T. H. A. Olthof. 1976. Relationship between nematode population densities and crop responses. Annu. Rev. Phytopathol. 14:327-353.
5. Barker, K. R., P. B. Shoemaker, and L. A. Nelson. 1976. Relationships of initial populations densities of *Meloidogyne incognita* and *Meloidogyne hapla* to yield of tomato. J. Nematol. 8:232.
6. Barrons, K. C. 1939. Studies on the nature of root-knot resistance. J. Agric. Res. 58:263-271.
7. Bergeson, G. B. 1959. The influence of temperature on the survival of some species of the genus *Meloidogyne*, in the absence of a host. Nematologica 4:344-354.
8. Bird, A. F. 1962. The inducement of giant cells by *Meloidogyne javanica*. Nematologica 8:1-10.
9. Canto-Sáenz, M. and B. B. Brodie. 1984. Host efficiency of potatoes to *Meloidogyne* spp. and nematode damage threshold density on this crop. J. Nematol 16:In press.
10. Canto-Sáenz M. and B. B. Brodie. 1984. The nature of potato incompatible response to *Meloidogyne incognita* (Kofoid & White 1919) Chitwood 1949 and correlation be-

tween root and tuber response. J. Nematol. 16:In press.

11. Canto-Sáenz M. and B. B. Brodie. 1984. Factors affecting host efficiency of potatoes to *Meloidogyne incognita* (Kofoid & White 1919) Chitwood 1949. J. Nematol. 16:In press.
12. Christie, J. R. 1936. The development of root-knot nematode galls. Phytopathology 26:1-22.
13. Christie, J. R. 1949. Host parasite relationship of the root-knot nematode *Meloidogyne* spp. III. The nature of resistance in plants to root-knot. Proc. Helminthol. Soc. Wash. 16:104-108.
14. Cook, R. 1974. Nature and inheritance of nematode resistance in cereals. J. Nematol. 6:165-174.
15. Dao, F. D. 1970. Climatic influence on the distribution pattern of plant parasitic and soil inhabiting nematodes. Meded. Landbouhogesch. Wageningen 70-72:64-80.
16. Davide, R. G. and A. C. Triantaphyllou. 1967. Influence of the environment on development and sex differentiation of root-knot nematodes. I. Effect of infection density, age of the host plant and soil temperature. Nematologica 13:102-110.
17. Dean, J. L. and F. B. Struble. 1953. Resistance and susceptibility to root-knot nematodes in tomato and sweet potato. Phytopathology 43:290 (Abstr.).
18. Dropkin, V. H. 1963. Effect of temperature on growth of root-knot nematodes in soybeans and tobacco. Phytopathology 53:663-666.
19. Dropkin, V. H. 1969. The necrotic reaction of tomatoes and other hosts resistant to *Meloidogyne* reversed by temperature. Phytopathology 59:1632-1637.
20. Dropkin, V. H. and P. E. Nelson. 1960. The histopathology of root-knot nematode infections in soybeans. Phytopathology 50:442-447.
21. Dropkin, V. H. and R. E. Webb. 1967. Resistance of axenic tomato seedlings to *Meloidogyne incognita acrita* and to *Meloidogyne hapla*. Phytopathology 57:584-587.
22. Endo, B. Y. 1965. Histological responses of resistant and susceptible soybean varieties and backcross progeny to entry and development of *Heterodera glycines*. Phytopathology 55:375-381.
23. Fassuliotis, G. 1967. Species of *Cucumis* resistant to the root-knot nematode *Meloidogyne incognita acrita*. Plant Dis. Rep. 51:720-723.
24. Fassuliotis, G. 1979. Plant breeding for root-knot nematode resistance, pp. 425-453 *In* Root-knot nematodes (*Meloidogyne* spp.) systematics, biology and control, ed. F. Lamberti and C. E. Taylor. 477 pp.
25. Fassuliotis, G., J. R. Deakin and J. C. Hoffman. 1970. Root-knot nematode resistance in snap beans: Breeding and nature of resistance. J. Am. Soc. Hortic. Sci. 95:640-645.
26. Fassuliotis, G. and P. D. Dukes. 1972. Disease reactions of *Solanum melongena* and *S. sisymbrifolium* to *Meloidogyne incognita* and *Verticillium albo-atrum*. J. Nematol. 4:222.
27. Fawole, B. and W. F. Mai. 1979. Influence of plant age, light intensity, nematode inoculum level and their interactions on tomato growth and reproduction of *Meloidogyne hapla*. J. Nematol. 11:199-201.
28. Ferris. H. 1978. Development of nematode damage functions and economic thresholds using *Meloidogyne incognita* on tomatoes and sweet potatoes. J. Nematol. 10:286-287 (Abstr.).
29. Forster, A. R. 1956. Development of *Heterodera rostochiensis* and *Meloidogyne incognita* in cross-grafted solanaceous plants with different susceptibilities. Nematologica 1:283-289.
30. Giamalba, M. J., W. J. Martin and T. P. Hernández. 1963. Sweet potato varietal reaction to species and races of root-knot nematodes. Phytopathology 53:1187-1189.
31. Giebel, J. 1974. Biochemical mechanisms of plant resistance to nematodes: a review. J. Nematol. 6:175-184.
32. Golden, A. M. and T. Shafer. 1958. Unusual responses of *Hesperis matronalis* L. to root-knot nematodes (*Meloidogyne* spp.). Plant Dis. Rep. 42:1163-1166.
33. Hijink, M. J. and M. Oostenbrink. 1968. Vruchtwisseling ter bestrijdjing van plantenziekten en plagen. Versl. Meded. Plziektenk Dienst. Wageningen, Sep. ser. 368. 7 pp.
34. Holtzmann, O. V. 1965. Effect of soil temperature on resistance of tomato to root-knot nematode (*Meloidogyne incognita*). Phytopathology 55:990-992.
35. Hung, Ch. L. and R. A. Rohde. 1973. Phenol accumulation related to resistance in tomato to infection by root-knot and lesion nematodes. J. Nematol. 5:253-258.
36. Jaffe, B. A. and W. F. Mai. 1979. Growth

reduction of apple seedlings by *Pratylenchus penetrans* as influenced by seedling age at inoculation. J. Nematol. 11:161-165.
37. Jatala, P. and L. S. Morrison. 1967. Effect of temperature on the development of *Meloidogyne incognita* in sweet potato. Phytopathology 57:816 (Abstr.).
38. Jatala, P. and C. C. Russell. 1972. Nature of sweet potato resistance to *Meloidogyne incognita* and the effects of temperature on parasitism. J. Nematol. 4:1-7.
39. Jones, F. G. W. 1956. Soil populations of beet eelworm (*Heterodera schachtii*, Schm.) in relation to cropping. II. Microplot and field plot results. Ann. Appl. Biol. 44:25-56.
40. Joubert, T. G. and C. E. Rappard. 1971. Roodeplaat Albesto: eelworm-resistant tomato cultivar. Farm. S. Afric. 46(10):3, 11, 14.
41. Khana, M. L. and K. K. Nirula. 1964. Breeding potatoes for resistance to root-knot nematode. Current Sci. 33:314.
42. Kühn, H. 1958. Über die Abwehrnekrose eines Kartoffelbastardes gegen den Kartoffelnematoden (*Heterodera rostochiensis* Wr. in *Solanum tuberosum* sub-sp. *andigena* [Juz. et. Buk.] Hwk. x *Solanum tuberosum* L.). Z. Pflanzenkr. Pflanzenschutz 65:465-472.
43. Loos, C. A., 1953. *Meloidogyne brevicauda* n. sp. a cause of root-knot of mature tea in Ceylon. Proc. Helminthol. Soc. Wash. 20:83-91.
44. Madamba, C. P., J. N. Sasser, and L. A. Nelson. 1965. Some characteristics of the effects of *Meloidogyne* spp. on unsuitable host crops. N. C. Agric. Exp. Stn. Tech. Bull. 169:1-34.
45. McClure, M. A., K. C. Ellis and E. L. Nigh. 1974. Resistance of cotton to the root-knot nematode *Meloidogyne incognita*. J. Nematol. 6:17-20.
46. McClure, M. A., K. C. Ellis, and E. L. Nigh. 1974. Post infection development and histopathology of *Meloidogyne incognita* in resistant cotton. J. Nematol. 6:21-26.
47. Milne, D. L., D. N. Bashoft and P. W. Buchan. 1965. The nature of resistance of *Nicotiana repanda* to the root-knot nematode *Meloidogyne javanica*. S. Afr. J. Agric. Sci. 8:557-567.
48. Nardacci. J. F. and K. R. Barker. 1979. The influence of temperature on *Meloidogyne incognita* on soybean. J. Nematol. 11:62-70.
49. Nirula, K. K. and B. T. Raj. 1971. Vertical distribution of *Meloidogyne incognita* in potato fields. Indian Phytopathol. 24:375-378.
50. Olthof, T. H. A. and J. W. Potter. 1972. Relationship between population densities of *Meloidogyne hapla* and crop losses in summer maturing vegetables in Ontario. Phytopathology 62:981-986.
51. Oosterbrink, M. 1956. Over de invloed van verschillende gewassen op de vermeerdering van en de schade door *Pratylenchus pratensis* en *Pratylenchus penetrans* (Vermes, Nematoda). Tijdschr. Plantenziekten 62:189-203.
52. Peacock, F. C. 1957. Studies on root-knot nematodes of the genus *Meloidogyne* in the Gold Coast. Part I. Comparative development on susceptible and resistant host species. Nematologica 2:76-84.
53. Peacock, F. C. 1959. The development of a technique for studying the host parasite relationship of the root-knot nematode *Meloidogyne incognita* under controlled conditions. Nematologica 4:43-55.
54. Prot, J. C. and S. D. Van Gundy. 1979. Influence of soil type and temperature on the migration of *M. incognita* juveniles toward tomato roots, p. 26 *In* XI ONTA Meeting, Abstracts. Charleston, S. Carolina.
55. Reynolds, H. W., W. W. Carter and J. H. O'Bannon. 1970. Symptomless resistance of alfalfa to *Meloidogyne incognita acrita*. J. Nematol. 2:131-134.
56. Riggs, R. D. and N. N. Winstead. 1958. Attempts to transfer root-knot resistance in tomato by grafting. Phytopathology 48:344 (Abstr.).
57. Riggs, R. D. and N. N. Winstead. 1959. Studies on resistance in tomato to root-knot nematodes and on the occurrence of pathogenic biotypes. Phytopathology 49:716-724.
58. Ritter, M. 1973. Life cycles and development of *Meloidogyne* species. Eur. Mediterr. Plant Prot. Organ (OEPP/EPPO) Bull. 9:53-59.
59. Robinson, R. A. 1980. New concepts in breeding for disease resistance. Annu. Rev. Phytopathol. 18:189-210.
60. Rohde, R. A. 1972. Expression of resistance in plants to nematodes. Annu. Rev. Phytopathol. 10:233-252.
61. Santo, G. C., J. H. O'Bannon, A. M. Finley, and A. M. Golden. 1980. Occurrence and host range of a new root-knot nematode (*Meloidogyne chitwoodi*) in the Pacific Northwest. Plant Dis. 64:951-952.
62. Sasser, J. N. 1954. Identification and host-

parasite relationship of certain root-knot nematodes (*Meloidogyne* spp.). Univ. Md. Agric. Exp. Stn. Bull. A-77. 31 pp.
63. Sasser, J. N. and A. L. Taylor. 1952. Studies on the entry of larvae of root-knot nematodes into roots of susceptible and resistant plants. Phytopathology 42:474.
64. Sawhney, R. and J. M. Webster. 1979. The influence of some metabolic inhibitors on the response of susceptible/resistant cultivars of tomato to *Meloidogyne incognita*. Nematologica 25:86-93.
65. Scheffer, F., R. Kickuth and J. H. Visser. 1962. Die Wurtzelauscheidungen von *Eragrostis curvula* (Schrad.) Nees un ihr. Einfluss and Wurtzel-Knoten Nematoden Z. für Pflanzenernährung Düngung Bodenkunde 98:114-120.
66. Singh, B. and B. Choudhury. 1973. The chemical characteristics of tomato cultivars resistant to root-knot nematodes (*Meloidogyne* spp.). Nematologica 19:443-448.
67. Smart, G. C. and B. B. Brodie. 1977. Reaction to Katahdin and breeding line M-905-1-1 potato to *Meloidogyne incognita*. Nematropica 7:7 (Abstr.).
68. Thomason, I. J. and B. Lear. 1961. Rate of production of *Meloidogyne* spp. as influenced by soil temperature. Phytopathology 51:520-524.
69. Tyler, J. 1933. Development of the root-knot nematode as affected by temperature. Hilgardia 7:392-415.
70. Tyler, J. 1941. Plants reported resistant or tolerant to root-knot nematode infestation. Misc. Publ. U.S. Dep. Agric. 406. 91 pp.
71. Veech, J. A. 1978. An apparent relationship between methoxy substituted terpenoid aldehydes and the resistance of cotton to *Meloidogyne incognita*. Nematologica 24:81-87.
72. Veech, J. A. 1981. Plant resistance to nematodes, pp. 377-403 *In* Plant parasitic nematodes, Vol. III, ed. B. M. Zuckerman and R. A. Rohde. Academic Press, New York. 508 pp.
73. Veech, J. A. and B. Y. Endo. 1970. Comparative morphology and enzyme histochemistry in root-knot resistant and susceptible soybeans. Phytopathology 60:896-902.
74. Vrain, T. C., K. R. Barker and G. I. Holtzmann. 1978. Influence of low temperature on rate of development of *Meloidogyne incognita* and *M. hapla* larvae. J. Nematol. 10:166-171.
75. Vrain, T. C. and K. R. Barker. 1978. Influence of low temperature on development of *Meloidogyne incognita* and *M. hapla* eggs in egg masses. J. Nematol. 10:311-313.
76. Wallace, H. R. 1969. The influence of nematode numbers and of soil particle size, nutrients and temperature on the reproduction of *Meloidogyne javanica*. Nematologica 15:55-64.
77. Webster, J. M. and R. E. Pawson. 1972. An interpretation of the ultrastructural response of tomato roots susceptible and resistant to *Meloidogyne incognita* (Kofoid & White 1919) Chitwood, 1949. Eur. Mediterr. Plant Prot. Organ. (OEPP/EPPO) Bull. 6:33-39.
78. Williams, T. D. 1956. The resistance of potatoes to root eelworm. Nematologica 1:88-93.
79. Yadav, B. S. 1970. Tests for the nematocidal properties of some used plants on *Meloidogyne incognita*. Proc. India Sci. Congr. Assoc. 57:551.
80. Zacheo, G., G. Lamberti, R. Anigoni-Liso and O. Onigoni. 1977. Mitochondrial protein hydroxyproline content of susceptible and resistant tomatoes infected by *Meloidogyne incognita*. Nematologica 23:471-476.

CHAPTER 20

THE ROLE OF THE NEMATOLOGIST IN THE DEVELOPMENT OF RESISTANT CULTIVARS

George Fassuliotis

U.S. Vegetable Laboratory
Agricultural Research Service; USDA
Charleston, South Carolina, USA

The extent of damage to crop plants by root-knot nematodes (*Meloidogyne* spp.) has been well documented (29,38). Minimizing that loss through the use of nematicides has been effective in the past, but expensive. The cancellations of DBCP (1,2-dibromo-3-chloropropane) and EDB (ethylene dibromide), because of their potential hazards to human health, have reduced the available number of chemicals to the less effective nonfumigant and systemic nematicides. The lack of effective and inexpensive nematicides has made breeding for root-knot nematode resistance an attractive alternative. The potential return for the dollars invested in breeding for resistance justifies incorporation of root-knot-nematode resistance into the objectives of every plant-breeding program. Unlike chemical methods, nematode management with resistant cultivars requires no special equipment or extra capital investment by growers. Marketable yields from homozygous resistant cultivars are significantly greater than those from homozygous susceptible cultivars grown in nematode-infested soils (22,51). Target nematode species are managed as well as or better than with nematicides (8,51).

There is international interest in the breeding work of the U.S. Vegetable Laboratory because root-knot nematode resistance is included in all of the breeding programs in which resistant genes are available. Four of the five plant breeders at this station strive for root-knot nematode resistance in their breeding programs. Crops include tomato (*Lycopersicon esculentum*, Mill.), pepper (*Capsicum annuum* L.), bean (*Phaseolus vulgaris* L.), sweet potato (*Ipomoea batatas* (L.) Lam.), sweet corn (*Zea mays* L.), and cowpea (*Vigna unguiculata* (L.) Walp.). Cantaloupe (*Cucumis melo* L.) and okra (*Abelmoschus esculentus* (L.) Moench.) would be included if sexually compatible, resistant germ plasm were available to cross with these species.

The number of root-knot-nematode-resistant cultivars is increasing. Contributions to the lists of germ plasm and cultivars are being made by nematologists, pathologists, agronomists, and geneticists in state universities, USDA laboratories and commercial seed companies.

Some of the most significant discoveries with regard to root-knot-nematode resistance were made between 1900 and 1948 when all root-knot nematodes were regarded as one species. The era of genetic manipulation for resistance to root-knot nematodes began with the discovery that cowpeas exhibited genetic variability for susceptibility (48) and that resistance could be transferred genetically to a susceptible cultivar (33). During the following four decades, a considerable amount of work centered around screening plants for resistance to the root-knot nematode. Resistant germ plasm was identified in about a dozen major crops (16), some of which provided the foundation for many of the current resistant cultivars. In crops where resistance was developed through breeding, the resistant germ plasm was usually cross-compatible with the crop species. Resistance was readily transferred by a series of crosses, backcrosses and selections. However in tomato, root-knot-nematode susceptibility seemed to be fixed within the cultivated species, and genes for resistance could be transferred only through interspecific hybridization with *L. peruvianum* (L.) Mill., an incompatible wild species, by means of embryo culture techniques (43).

Prior to 1949, the root-knot nematodes were considered to be a single species, *Heterodera marioni*. Subsequent taxonomic revisions, however, established the existence of four major species: *Meloidogyne incognita*, *M. javanica*, *M. arenaria*, and *M. hapla* (5,45). The recognition of different *Meloidogyne* species provided workers with a sounder scientific and effective approach to plant breeding. An explanation was available for much of the discrepancy involving root-knot nematode reactions observed by various workers. Re-evaluation of resistance to the various species showed that some crop plants were resistant to two or more species whereas others carried resistance to only one species (16). For example, tomato with the *Mi* gene was found to carry resistance to *M. incognita*, *M. javanica* and *M. arenaria*. The common bean was resistant only to *M. incognita*; sweet potato was resistant to *M. incognita* and *M. javanica*; and pepper was found to have resistance to *M. incognita*, *M. arenaria* and *M. javanica* but not *M. hapla*. In a recently published check list of crop plants, over 450 cultivars in 13 botanical families were reported to carry resistance to at least one *Meloidogyne* species (40).

SPECIFIC RESPONSIBILITIES OF THE NEMATOLOGIST

In the chain of events that transpires in a breeding program for root-knot nematode resistance, there are a number of tasks that I believe should be the responsibility of a nematologist. As the project progresses, and an interrelationship develops between the nematologist and the plant breeder, each may extend his or her involvement into the other's sphere of expertise, depending on personal interests and objectives.

Survey and Assess the Magnitude of the Nematode Problem

Before initiation of a plant breeding program in which root-knot-nematode resistance is incorporated into a specific crop, the prevalence, distribution and importance of the nematodes should be assessed. Putting resources into breeding for resistance against a pathogen that is of minor importance makes little sense. After the nematode problem has been defined, an accurate identification of the nematode must be made. Species determination is made through microscopic examination of morphological characters with special attention paid to the morphology of the perineal pattern, stylet and head (10). Further identification is made by means of the North Carolina Differential Host Test (44). This test is useful for confirmation of the identification of the most widely distributed root-knot nematode species and race, e.g. *M. incognita* (4 races), *M. arenaria* (2 races), *M. javanica*, and *M. hapla*. An outline of the Differential Host Test has been published by Taylor and Sasser (44).

Establish Pure Cultures

Availability of a nematode population containing a single species is essential to the progress of a plant-breeding program. Natural field populations often consist of a mixture of species, and greenhouse populations often become contaminated by other nematode species. A population of a single nematode species can be prepared as follows. Individual egg masses are placed in water in a watch glass, and a slide is prepared of the perineal pattern of the corresponding females for microscopic examination. Some nematologists start a pure population from a single egg mass while others use several egg masses. The latter method is better because it enables retention of the natural variability that may be available within a field population. The egg masses are transferred to a 10-cm pot of sterile soil containing a susceptible tomato seedling. After 45-50 days, the root system is examined for galling and nematode reproduction. When the egg masses are a tan color, the infected roots are subdivided among other pots and replanted with susceptible tomato seedlings so the population can be increased. Once enough inoculum is available for the breeding project, it should be retested on the Host Plant Differentials, and perineal patterns of at least ten specimens made for confirmation of the initial identification to species and race. If the culture is subsequently maintained in the greenhouse on a suitable host plant, a continuous source of inoculum for screening plant material will be available (16).

Prepare Inoculum

Most workers currently involved in screening plants for resistance use egg inoculum. Collecting eggs for inoculum from infected roots is a relatively easy process. Perhaps one of the most significant contributions to nematode methodology was made by Hussey and Barker (27) when they developed the technique for extracting and collecting root-knot-nematode eggs. This technique is extremely useful in plant breeding. Infected roots are shaken vigorously in 200 ml of 20% commercial bleaching solution (approximately 1% sodium hypochlorite) in a one-liter container for 4 min. The bleaching solution is passed through a combination of 200- and 500-mesh sieves. The eggs collected on the 500-mesh sieve are rinsed thoroughly with tap water. Rinsing the remaining roots two more times with water removes additional eggs. The suspension is transferred to a beaker, and the egg concentration is calculated. Root-knot-nematode eggs extracted by this method are surface sterilized and relatively free of other organisms. The suspension is diluted with water so that when agitated, it contains about 1000 eggs/ml. Any pipetting apparatus that can be calibrated to deliver preset volumes of inoculum is satisfactory. In our work, we use the Pistolpet[R] (Manostat, 519 Eighth Ave., New York, NY) for tests involving large numbers of plants to be screened. This apparatus dispenses egg inoculum with rapidity and accuracy.

Infested soil containing infective second-stage juveniles or infected roots with egg masses can be used as inoculum, but they present certain disadvantages (16). Inoculum levels cannot be easily quantified, and fungal and bacterial disease organisms in the soil biosphere or rhizosphere may interact with the nematode to produce undesirable complexes (34).

Develop Screening Techniques

Various laboratory, field and greenhouse screening techniques have been reviewed (16). Screening techniques must be devised that optimize the amount of inoculum for differentiation of plant response among genotypes. Even the most resistant plant can be stressed if it is overwhelmed by a large number of infective nematode juveniles. It is important that the inoculum level be limited so that population pressures from extremely high dosages do not injure the root system and mask potentially useful genetic material. The screening procedures should be simple, not laborious, and allow virtually no plants to escape contact with the infective juveniles.

Screening should be done under conditions that are conducive for good plant growth. At our laboratory, all screening of plant material for resistance to *Meloidogyne* spp. is done in the greenhouse on benches or in Todd Planter Flats, which we refer to as speedling trays (Speedling Manufacturing Inc., P.O. Box 238, Sun City, Florida 33586). Greenhouse benches are filled with a steam-sterilized growing medium consisting of vermiculite and peat moss). Holes, 2-cm deep, are made at 4- to 5-cm intervals in rows 10 cm part. Before the seeds or seedlings are planted, each hole is inoculated with approximately 2000 eggs. Soil temperature is maintained between 22 and 26°C. A susceptible plant with a known root-knot nematode reaction is included with the plantings. Rutgers tomato is a good plant for indicating when plants should be dug and rated for resistance. When advanced breeding lines containing root-knot-nematode-resistant genes are being evaluated, the parental source of resistance, and that of susceptibility, should be included.

Speedling trays are useful for screening large plant populations in a minimum of space and soil (15). Each flat consists of 128, 3.8-cm square, inverted-pyramid-shaped root cells. Each cell is filled with a fertilized soil mixture, inoculated with 2000 eggs and then planted with either a seed or seedling. The trays are set on rails allowing roots that grow through the hole at the bottom of each cell to be air pruned. Because of the restrictive root growth within the cell, one can easily rate the root system after removal. We have found these trays suitable for screening corn, tomato, bean, and melon. R. S. Hussey and others (personal communication) have found this system extremely useful for screening soybean for cyst nematode resistance. When plants are screened for resistance, consideration must be given to the growth habit of the plant. The most desirable plant type is one that can be uprooted, evaluated and replanted so that seed or rhizomes may eventually be collected from a resistant plant (e.g. tomato, sweet potato). Plant types that do not usually survive transplanting (corn, bean, cowpea, soybean) must either be evaluated after seed have developed, or must be evaluated by special techniques that do not kill the roots (16,50). Advanced breeding lines are tested in replicated field trials. Before release, promising lines that show low galling and reproduction are retested in as many different geographic regions as possible so their adaptability and resistance to many *Meloidogyne* populations can be determined under different environmental conditions. Yield potential of resistant lines is compared with that of similar lines grown in the same soil mix, but treated with a nematicide (22,51).

Establish Rating Scheme for Resistance

In simplest terms, resistance can be defined as the character, or characters, of a plant that inhibit nematode reproduction (44). The various methods of rating plants for resistance to root-knot nematodes have been reviewed (16). Plants vary greatly in their response to root-knot nematodes. Upon infection, most genotypes show a galling response within 24 hours and reproduction is evident within 20 days. However, within some plant species, response to infection may vary greatly. Cultigens of beans, for example, may show extreme galling response to *M. incognita* or none at all. However, in both cases, reproduction will be similar. Other plant species may gall following penetration by juveniles, but nematode reproduction is significantly reduced (18,23,24). Most workers use galling and reproductive responses in evaluating plants for resistance. An index scale of 0-5 is most commonly used to indicate the severity of galling and degree of reproduction of the nematode: 0 =no galling or reproduction, 1 = trace of galling or reproduction (1% or less of the susceptible check plant), 2 = light galling or reproduction (2-10% of a susceptible plant), 3 = moderate galling or reproduction (11-25% of a susceptible plant), 4 = moderately heavy galling or reproduction (26-50% of a susceptible plant), and 5 = heavy galling or reproduction (51-100% of a susceptible plant). Plants falling in categories 0, 1, 2, and 3 are considered immune, highly resistant, very resistant and moderately resistant, respectively. Often susceptible plants adjacent to the edges of the screening bench do not show expected susceptible reactions because of differences in moisture and temperature stresses on the plants within the bench. As a result, such plants become classified as resistant. Fery and Dukes (21) used the outermost two rows around each bench as buffers to overcome this problem. When plants are reevaluated for resistance, it is extremely important that they be inoculated with the same population of root-knot nematode. Populations of the same species can vary in aggressiveness if they are from different sources. Viglierchio (46) found great variability in aggressiveness to tomato among California populations of *M. incognita*.

Identify Sources of Resistance

The most limiting factor in the expansion of food crops with root-knot-nematode resistance is the lack of genetic material among some plant species. It is obvious that a breeding program for root-knot-nematode resistance requires resistant germ plasm from which traits can be transferred to horticultural plants. Transfer of resistance is greatly simplified if resistant germ plasm can be found in adapted cultivars or in advanced breeding lines containing otherwise good horticultural qualities. Over the past 25 years, many tomato cultivars and breeding lines have been developed from which root-knot-nematode-resistant genotypes can be selected for use in a tomato breeding program (41). Development of a useful cultivar would be extremely prolonged if the program was started with the wild tomato, *L. peruvianum*, as the source of resistance (25).

Among other vegetable crops in which wide gene pools exist for genetic resistance to root-knot nematodes are cowpea, bean and sweet potato. When resistance does not exist in established cultivars, the breeder's working collection should be examined. If this search is fruitless, attention may then be turned to collections of primitive cultivars. Last of all, wild species that are closely or distantly related are evaluated if no other resistant germ plasm can be found. Several wild species have been identified with root-knot-nematode resistance (11,18), but their use in crop improvement cannot be realized until they can be genetically integrated into conventional breeding programs. Another disadvantage in the use of primitive or wild species is that many undesirable characters are expressed that can have an effect on quality, yield and agronomic suitability. In the development of the *M. incognita*-resistant snap bean 'NemaSnap,' repeated backcrossings to the horticultural types and selections were made for elimination of unacceptable characters present in the primitive cultivar (52,53). Introduction of nematode resistance into the tomato from *L. peruvianum* required a combination of embryo culture (43) and 15 years of work by plant breeders before a root-knot-nematode-resistant cultivar was released (36).

Cucurbitaceae constitutes a large botanical family consisting of the various types of melon, squash, pumpkin, etc. (49). No sources of resistance to root-knot nematodes are known in these cultivated species. Plant breeders and nematologists have been especially interested in utilizing resistance found in the African horned cucumber, *Cucumis metuliferus* Naud., for possible transfer to cantaloupe (13,14).

The ability to cross cantaloupe genotypes with related resistant wild species would have significant impact toward improvement of this crop since the genetic base is very narrow. Interspecific hybridization would introduce desirable genetic factors such as nematode and disease resistances into cultivated species (35). Such crosses are difficult to make. Cross pollination efforts over the past several years indicate that hybridization of *C. metuliferus* with *C. melo* is highly probable. *C. melo* pollen penetrated ovules of the wild species, and embryos developed in a small percentage of the ovules (13). The embryos, however, did not develop to maturity. Embryos have been rescued from the ovary and placed in culture media but none have been converted into viable transplants (14). The primary obstacle to further progress with embryo rescue techniques appears to be the lack of a defined medium on which the hybrid embryo can be cultured. The nutritional requirements of the immature embryos are much more complex than those of the mature embryo. Detailed studies are required if culture media and conditions for the conversion of immature embryos to viable transplants are to be defined. Our current work is primarily to define the nutritional requirements of immature *C. metuliferus* embryos for development into whole plants. It is likely that embryo culture can have as great an impact on cantaloupe improvement as it has had on tomato improvement (37).

FACTORS INFUENCING RESISTANCE

Several factors may alter the expression of resistance. Genetic resistance to *Meloidogyne* spp. is sensitive to soil temperatures above 28°C. Tomato, bean, and sweet potato lose resistance in elevated soil temperatures (6,20,26,28). High soil temperatures appear to be the main reason that root-knot nematode resistance is not effective in Florida, USA (47), and in many tropical countries. There is some evidence that reproduction of *M. incognita* at elevated soil temperatures may be race dependent. Recent results by Araujo et al. (1) indicate that race 4 reproduces better on resistant tomato genotypes at 32.5°C than does race 1.

In split-root experiments with resistant 'NC 95' tobacco, Eisenback (9) found that plant resistance to *M. incognita* can be modified by the presence of another root-knot nematode species on the same root system. This observation may explain one of the causes of resistance breakdown which was previously attributed to development of aggressive biotypes.

Dropkin et al. (7) found that application of exogenous kinetin to tomato seedlings altered their expression for resistance. There is some indication that tissue culture techniques which use phytohormones in the media for plant regeneration from various tissues may reverse the resistance of plants to root-knot nematodes (17).

FUTURE RESEARCH NEEDS AND OUTLOOK

How will we manage or suppress the nematode problems in the future if environmentally safe and inexpensive nematicides are not available? Clearly, at the present time intensive efforts must be devoted to development of genetic solutions to the problem. If resistant plants are included in cropping systems, target nematode populations can be suppressed (22,51). Also, resistant or tolerant cultivars make low dosage rates of less potent, nonfumigant nematicides a more practical and viable management option. These compounds applied in combination with resistant lines may make the difference between success and failure of a crop. Much research is needed to expand the list of crops and cultivars with resistance to root-knot nematodes. In crops in which resistance genes are readily available (e.g. sweet potato, pepper, cowpea, bean), conventional breeding methods can be used to produce new cultivars readily.

The recognition of races within *M. incognita* and *M. arenaria* opens new areas for research. Many, if not all of the cultivars already reported to have some resistance value will have to be reevaluated. Previously reported resistance may be race specific. Recently, McCleod et al. (31) reported the occurrence of several plant introductions (PI) of okra with resistance to *M. incognita* race 3. These same PIs, however, were susceptible to race 1 (G. Fassuliotis, unpublished); thus, resistance to *M. incognita* may be race specific. The nematologist and plant breeder are challenged to develop resistance to two or more of the important *Meloidogyne* species in those crops that carry resistance factors to none or only one species. The International *Meloidogyne* Project (IMP) has been instrumental in providing a worldwide view of root-knot nematodes and their distribution. From the data that has been collected, a crop having resistance to *M. incognita* and *M. javanica* would be resistant to 82% of the major *Meloidogyne* populations around the world. Tomato carrying resistance to *M. incognita, M. javanica* and *M. arenaria* would be resistant to 90% of the root-knot populations (39). The common bean which has resistance to *M. incognita* would have greater economic value if a cultivar could be developed with resistance to *M. javanica,* as well. Resistant genes to this nematode species have yet to be found.

The effect of temperature stress on the expression and reversal of resistance will have to be addressed further. Germ plasm with tolerance to soil temperatures above 28°C is needed so that resistant plants developed in the temperate climates will be useful in more tropical climates. If *L. peruvianum* lines are evaluated at soil temperatures above 28°C, germ plasm that retains its resistance under high temperature stress may possibly be identified.

Advances in plant biotechnology, often

referred to as "genetic engineering," have emerged which plant breeders hope can be used in the transfer of genetic factors into breeding lines (2). The degree to which this new technology will be used for plant breeding and, in particular, for transfer of resistance genes for root-knot nematodes is questionable at this time. The new biotechnology will have to be explored further for its application to nematode resistance.

Fassuliotis and Bhatt (17) proposed the use of somatic hybridization for transfer of resistance factor(s) from the wild species *Solanum sisymbriifolium* to the cultivated eggplant (*S. melongena* L.) since the species were sexually incompatible. Essential to the application of this technology is the isolation of protoplasts and the subsequent regeneration of whole plants from them (4). Both the eggplant and *S. sisymbriifolium* were regenerated from callus cultures and from protoplasts (3, 12, 19). Recently, S. Gleddie, Ottawa Research Station, regenerated hybrid plants by fusing their protoplasts. These plants have not yet been tested for root-knot-nematode resistance (personal communication). Although somatic hybridization may have scientific value in that genes may be transferred between incompatible species, it has not yet led to marketable plants. The practical use of this technique is still to be realized. There are four other reports (42) of somatic hybrids having been formed from the fusion of protoplasts of sexually incompatible plant species. One of these was a hybrid produced by protoplast fusion of potato and tomato; however, neither the fruit nor tubers were edible (32). Whether protoplast fusion can be utilized to transfer root-knot-nematode resistance to horticultural crops is yet to be determined. As the technology advances and as protoplast fusion and regeneration become standard routine laboratory techniques, the transfer of resistance from distantly related species seems more and more feasible: for example, transfer of root-knot nematode resistance to okra from Egyptian cotton, *Gossypium barbadense* L., (41) or development of resistant cucumbers and squash through the African horned cucumber.

Recent advances in recombinant DNA (gene splicing) technology may prove useful in plant protection (2). One day, movement of single genes or blocks of genes known to express resistance to root-knot and other nematodes may be possible by the recombination of this genetic information into plants at the molecular level. Crop plants with greater nematode resistance may also be developed through genetic engineering, such as the transfer of genetic pathways required to synthesize metabolites that provide resistance.

Evidence abounds that genetic variability is often generated with cell culture techniques (30). Although resistance to root-knot nematodes has been observed to be reversed after tissue culture, it is possible that some plants derived from tissue culture methods may yield variants with increased resistance and should be evaluated.

The new biotechnology is advancing rapidly. While many of the procedures may not be applicable to nematology today, nematologists and plant breeders may one day be able to utilize this technology to develop root-knot-nematode-resistant plant lines that have been impossible for scientists to produce with conventional breeding procedures. As the new biotechnology is evolving there is still much work to be done in plant breeding by conventional methods with the germ plasm on hand. Nematological involvement in a plant-breeding program for root-knot nematode resistance can be both challenging and rewarding. A great deal of gratification results from participation in the development of a cultivar that thrives in fields heavily infested with root-knot nematodes.

Literature Cited

1. Araujo, M. T., D. W. Dickson, J. J. Augustine, and M. J. Bassett. 1983. Reproduction of two races of *Meloidogyne incognita* in tomato plants grown at high temperature. J. Nematol. 15:640-641.
2. Barton, K. A. and W. J. Brill. 1983. Prospects in plant genetic engineering. Science 219:671-675.
3. Bhatt, D. P. and G. Fassuliotis. 1981. Plant regeneration from mesophyll protoplasts of eggplant. Z. Pflanzenphysiol. 104:81-89.
4. Carlson, P. S. and T. B. Rice. 1977. Genetic engineering and crop improvement. HortScience 12:135-137.
5. Chitwood, B. G. 1949. Root-knot nematodes. Part I. A revision of the genus *Meloidogyne* Goeldi, 1887. Proc. Helminthol. Soc. Wash. 16:90-104.

6. Dropkin, V. H. 1969. The necrotic reaction of tomatoes and other hosts resistant to *Meloidogyne*: Reversal by temperature. Phytopathology 59:1632-1637.
7. Dropkin, V. H., J. B. Helgeson and C. D. Upper. 1969. The hypersensitivity reaction of tomatoes resistant to *Meloidogyne incognita*: Reversal by cytokinins. J. Nematol. 1:55-61.
8. Dukes, P. D., R. L. Fery and M. G. Hamilton. 1979. Comparison of plant resistance and a nematicide for control of southern root-knot nematodes in southern peas, *Vigna unguiculata*. Phytopathology 69:526-527. (Abstr.).
9. Eisenback, J. D. 1983. Loss of resistance in tobacco cultivar 'NC 95' by infection of *Meloidogyne arenaria* or *M. hapla*. J. Nematol. 15:478. (Abstr.).
10. Eisenback, J. D., H. Hirschmann, J. N. Sasser, and A. C. Triantaphyllou. 1981. A guide to the four most common species of root-knot nematodes (*Meloidogyne* species) with a pictorial key. Coop. Publ. Deps. Plant Pathol. and Genet., North Carolina State Univ., and U.S. Agency Int. Dev., Raleigh, N.C. 48 pp.
11. Fassuliotis, G. 1970. Resistance of *Cucumis* spp. to the root-knot nematode, *Meloidogyne incognita acrita*. J. Nematol. 2:174-178.
12. Fassuliotis, G. 1975. Regeneration of whole plants from isolated stem parenchyma cells of *Solanum sisymbriifolium*. J. Am. Soc. Hortic. Sci. 100:636-638.
13. Fassuliotis, G. 1977. Self-fertilization of *Cucumis metuliferus* Naud. and its cross-incompatibility with *C. melo* L. J. Am. Soc. Hortic. Sci. 102:336-339.
14. Fassuliotis, G. 1977. Embryo culture of *Cucumis metuliferus* and the interspecific hybrid with *C. melo*, p. 861 *In* 4th Annu. Colloquium, Plant Cell and Tissue Culture, Sept. 6-9, 1977. Ohio State Univ., Columbus.
15. Fassuliotis, G. 1978. A system for screening large plant populations for root-knot nematode resistance. Nematropica 8:10. (Abstr.).
16. Fassuliotis, G. 1979. Plant breeding for root-knot nematode resistance, pp. 425-453 *In* Root-knot nematodes (*Meloidogyne* species) systematics, biology and control, ed. F. Lamberti and C. E. Taylor. Academic Press, New York. 477 pp.
17. Fassuliotis, G. and D. P. Bhatt. 1982. Potential of tissue culture for breeding root-knot nematode resistance into vegetables. J. Nematol. 14:10-14.
18. Fassuliotis, G. and P. D. Dukes. 1972. Disease reactions of *Solanum melongena* and *S. sisymbriifolium* to *Meloidogyne incognita* and *Verticillium albo-atrum*. J. Nematol. 4:222. (Abstr.).
19. Fassuliotis, G., B. V. Nelson, and D. P. Bhatt. 1981. Organogenesis in tissue culture of *Solanum melongena* cv. Florida Market. Plant Sci. Lett. 22:119-125.
20. Fassuliotis, G., J. R. Deakin and J. C. Hoffman. 1970. Root-knot nematode resistance in snap beans: breeding and nature of resistance. J. Am. Soc. Hortic. Sci. 95:640-645.
21. Fery, R. L. and P. D. Dukes. 1980. Inheritance of root-knot resistance in the cowpea (*Vigna unguiculata* (L.) Walp.). J. Am. Soc. Hortic. Sci.105:671-674.
22. Fery, R. L. and P. D. Dukes. 1984. Feasibility of breeding bell peppers (*Capsicum annum*) for resistance to the southern root-knot nematode, *Meloidogyne incognita*. HortScience 19:545 (Abstr.).
23. Fox, J. A. and L. I. Miller. 1973. Comparison of gall and egg-mass indices of two races of *Meloidogyne incognita* on differential hosts. Phytopathology 63:801. (Abstr.).
24. Golden, A. M. and T. Shafer. 1958. Unusual response of *Hesperis matronalis* L. to root-knot nematodes (*Meloidogyne* spp.). Plant Dis. Rep. 42:1163-1166.
25. Hawkes, J. G. 1977. The importance of wild germplasm in plant breeding. Euphytica 26:615-621.
26. Holtzmann, O. V. 1965. Effect of soil temperature on resistance of tomato to root-knot nematode (*Meloidogyne incognita*). Phytopathology 55:990-992.
27. Hussey, R. S. and K. R. Barker. 1973. A comparison of methods of collecting inocula of *Meloidogyne* spp. including a new technique. Plant Dis. Rep. 57:1025-1028.
28. Jatala, P. and C. C. Russell. 1972. Nature of sweet potato resistance to *Meloidogyne incognita* and the effect of temperature on parasitism. J. Nematol. 4:1-7.
29. Lamberti, F. and C. E. Taylor, eds. 1970. Root-knot nematodes (*Meloidogyne* species) systematics, biology and control. Academic Press, New York. 477 pp.
30. Larkin, P. J. and W. R. Scowcroft. 1981. Somaclonal variation—a novel source of variability from cell cultures. Theor. Appl. Genet. 60:197-214.

31. McLeod, J. M., W.Witcher, and W. M. Epps. 1983. Resistance of okra plant introductions to root-knot nematode and fusarium wilt. HortScience 18:249-250.

32. Melchers, G., M. D. Sacristan, and A. A. Holder. 1978. Somatic hybrid plants of potato and tomato regenerated from fused protoplasts. Carlsberg Res. Commun. 43:203-218.

33. Orton, W. A. 1913. The development of disease resistant varieties of plants. C. R. & Rapp. IV Conf. Int. Genet., Paris 1911:2455-2465.

34. Powell, N. T. 1971. Interaction of plant parasitic nematodes with other disease-causing agents, pp. 119-136 *In* Plant parasitic nematodes, Vol. II, ed. B. M. Zuckerman, W. F. Mai, and R. A. Rohde. Academic Press, New York. 347 pp.

35. Provvidenti, R. and D. Gonsalves. 1982. Resistance to papaya ringspot virus in *Cucumis metuliferus* and its relationship to resistance to watermelon mosaic virus 1. J. Hered. 73:239-240.

36. Rick, C. M. 1967. Exploiting species hybrids for vegetable improvement. Proc. XVII Int. Hortic. Congr. 17 (part 3): 217-229.

37. Rick, C. M. 1982. The potential of exotic germplasm for tomato, pp. 1-28 *In* Plant improvement and somatic cell genetics, ed. I. K. Vasil, W. R. Scowcroft and K. J. Frey. Academic Press, New York. 300 pp.

38. Sasser, J. N. 1980. Root-knot nematodes: A global menace to crop production. Plant Dis. 64:36-41.

39. Sasser, J. N. and C. C. Carter. 1982. Root-knot nematodes (*Meloidogyne* spp.): Identification, morphological and physiological variation, host range, ecology, and control, pp. 21-32 *In* Nematology in the southern region of the United States, ed. R. D. Riggs. South. Coop. Ser. Bull. 276. 206 pp.

40. Sasser, J. N. and M. F. Kirby. 1979. Crop cultivars resistant to root-knot nematodes, *Meloidogyne* spp., with information on seed sources. Coop. Publ. Dep. Plant Pathol., North Carolina State Univ., and U.S. Agency Int. Dev., Raleigh, N.C. 24 pp.

41. Shepherd, R. L. 1974. Breeding root-knot resistant *Gossypium hirsutum* L. using a resistant wild *G. barbadense* L. Crop Sci. 14:687-691.

42. Sink, K. C. 1984. Protoplast fusion for plant improvement. Proc. Am. Soc. Hortic. Sci. 19:33-37.

43. Smith, P. G. 1944. Embryo culture of a tomato species hybrid. Proc. Am. Soc. Hortic. Sci. 44:413-416.

44. Taylor, A. L. and J. N. Sasser. 1978. Biology, identification and control of root-knot nematodes (*Meloidogyne* species). Coop. Publ. Dep. Plant Pathol., North Carolina State Univ., and U.S. Agency Int. Dev., Raleigh, N.C. 111 pp.

45. Triantaphyllou, A. C. and J. N. Sasser. 1960. Variation in perineal patterns and host specificity of *Meloidogyne incognita*. Phytopathology 50:724-735.

46. Viglierchio, D. R. 1978. Resistant host responses to ten California populations of *Meloidogyne incognita*. J. Nematol. 10:224-227.

47. Walter, J. M. 1967. Hereditary resistance to disease in tomato. Annu. Rev. Phytopathol. 5:131-162.

48. Webber, H. J. and W. A. Orton. 1902. Some diseases of cowpea. II. A cowpea resistant to root-knot (*Heterodera radicicola*). U.S. Dep. Agric. Bur. Plant Ind. Bull. 17:23-28.

49. Whitaker, T. W. and G. W. Davis. 1962. The cucurbits: botany, cultivation and utilization. Leonard Hill Books Ltd., London. 250 pp.

50. Wyatt, J. E. and G. Fassuliotis. 1979. Methods for screening snap beans for resistance to root-knot nematodes. HortScience 14:27-28.

51. Wyatt, J. E., G. Fassuliotis and A. W. Johnson. 1980. Efficacy of resistance to root-knot nematode in snap beans. J. Am.Soc. Hortic. Sci. 105:923-926.

52. Wyatt, J. E., G. Fassuliotis, A. W. Johnson, J. C. Hoffman and J. R. Deakin. 1980. B4175 root-knot nematode resistant snap bean breeding line. HortScience 15:530.

53. Wyatt, J. E., G. Fassuliotis, J. C. Hoffman and J. R. Deakin. 1983. 'NemaSnap' snap bean. HortScience 18:776.

PRELIMINARY AND ADVANCED EVALUATION OF NEMATICIDES

D. P. Schmitt

Department of Plant Pathology
North Carolina State University
Raleigh, North Carolina, USA

The first effective nematicides (D-D and ethylene dibromide) were soil fumigants which were discovered in the early 1940's. Most of the recently developed nematicides are nonfumigants. The introduction of effective nematicides in the 1940's facilitated demonstration of the crop-damage potential of plant-parasitic nematodes. Surprisingly, there are few fumigant nematicides and only a small number of nonfumigants currently registered.

Nematicides are used primarily to reduce nematode population densities below damaging levels. The fumigants are rated to be approximately 50-90% effective in reducing populations. Nonfumigants may not kill nematodes at recommended usage rates, but cause behavioral changes that limit their parasitic activity.

Rotation with a nonhost or use of a resistant cultivar is generally more effective in nematode control than are nematicides. However, crop rotation may not be economical or feasible, and resistant cultivars often are unavailable. Thus, nematicides frequently are an attractive alternative for controlling pathogenic nematodes in order to grow a high yielding, good quality crop.

Currently available nematicides generally are not broad-spectrum biocides. They have differential effects on species of nematodes, and their activity is affected by physical and chemical properties of the soil, other pesticides, and the environment. Factors such as rainfall and soil texture have a major impact on the efficacy of these chemicals.

Developing a marketable nematicide is a long-term and expensive process. Many chemical compounds are screened for nematicidal properties, but only a few have sufficient activity to justify further testing. Those chemicals that are promising candidates as nematicides usually must be studied for toxicological and environmental impacts. These requirements may slow the registration process.

The development process of a nematicide will be discussed in three phases: 1) exploratory screening, 2) preliminary testing, and 3) advanced testing (Table 1). Some general comments are offered about statistical design, nematode assay, data collection and analysis, and reporting. These areas will be covered in depth in Chapters 1 and 13 of Volume II.

EXPLORATORY SCREENING OF CHEMICALS FOR NEMATICIDAL PROPERTIES

Essentially all initial screening of chemicals for activity against various types of pests is done by industry. Chemicals synthesized by chemists are evaluated for a wide range of biological effects. The target organisms in these evaluations may include weeds, insects, fungi, bacteria, nematodes, rodents, and perhaps some others. Only a small percentage of these chemicals ever show significant activity against nematodes.

The exploratory screening of candidate compounds typically is limited to one or rarely a few species of plant-parasitic nematodes. *Meloidogyne incognita* is the most frequently used test organism and the one most recommended (9). This economically important pest is an excellent test animal because it induces easily recognizable symptoms (root galls), is easy to rear, and produces large quantities of eggs. Good candidate chemicals, however, may be overlooked in tests using one or two nematode species because of the specific activity of some chemicals (20, 27).

It is impractical to use a number of nematode species as test organisms in this early stage of nematicide assessment. Nevertheless, it may be

Table 1. Major aspects for consideration in evaluating a chemical for nematicidal activity.

Level of investigation	Methods
Discovery:	*Laboratory screening of many chemicals.*
Preliminary:	*Greenhouse & Growth Chamber Soil Treatment:* 1. Infest soil, using sufficient treatments to include 3-4 nematode species. 2. Treat soil with candidate nematicides. 3. Assay soil and root for nematodes.
	Greenhouse & Growth Chamber Foliar Treatments: 1. Establish plants in infested soil. 2. Spray foliage not allowing any soil contact by the nematicide. 3. Perform nematode assay of soil and/or foliage.
	Field Evaluation for Fumigants: 1. Place infested soil and/or infected root tissue in cheesecloth bags in trenches in the field at 7.5-cm intervals for 2 m. 2. Fumigate at mid-point. 3. Remove bags from field and bioassay soil in the greenhouse.
	Field Evaluation for Nonvolatile Chemicals: 1. Plant susceptible seedlings in a 1-m^2 area of soil infested with an endoparasite. 2. Treat soil. 3. Carefully dig roots, wash and stain.
Advanced Testing:	*Evaluation of Formulated Material* *Field Crop Evaluation:* 1. Select site. 2. Characterize the site physically, chemically, and biologically. 3. Assay each plot prior to treatment, midseason, and harvest. 4. Test chemicals at several rates, methods of application and incorporation. 5. Determine crop response. 6. Evaluate and search for nontarget effect.
	Greenhouse Soil Nematode Evaluation: Follow the preliminary greenhouse-and growth-chamber treatment for pot culture; field-crop-evaluation treatment (advanced testing) for residual soil.
	Greenhouse Foliar Nematode Evaluation: 1. Establish plants and inoculate foliage. 2. Spray foliage or treat soil. 3. Assay foliage.

beneficial to test several species and determine which ones are good indicator organisms of nematicidal properties. Much experience has been gained in the last four decades in the evaluation of nematicides and candidate compounds for nematicidal activity. This experience should facilitate more effective evaluation of test compounds. Knowledge about key characteristics of a chemical, such as solubility in water, adsorption properties, volatility, toxicity, and systemic activity should facilitate the evaluation process.

PRELIMINARY TESTING OF CANDIDATE COMPOUNDS

Preliminary testing usually is done in greenhouses and growth chambers, and rarely under field conditions. It is practical to have broad, general guidelines because of different experimental conditions and markets. Suggested specific guidelines for comparative purposes involve inoculation of tomato with *Meloidogyne incognita*, soil treatment with experimental chemicals, and then assessment of root penetration and galling after a predetermined period (9).

Three to four economically important species with different metabolic, sensory, and cuticular characteristics will indicate the range of nematicidal activity of any given chemical. The chemical's efficacy is based on how well it permeates physical barriers such as the cuticle, intestinal wall, internal membranes and egg membranes (30) and how well it affects the sensory system or metabolism. Some nematodes are able to metabolize a pesticide without any adverse effect, whereas others are affected and may die. Four nematode taxa are suggested as test animals: *Caenorhabditis elegans* (20), *Meloidogyne* spp. (9), *Heterodera* spp. (27), and *Aphelenchus avenae* (16).

The classical method of rearing nematodes for nematicide evaluations has been in greenhouse pot cultures. The species suggested herein can be cultured gnotobiotically, but this method is not practical for most plant-parasitic nematode species. The major consideration in preparing inoculum is to assure that the nematodes are healthy and vigorous. Eggs or infective vermiform stages can be used as inoculum (see Vol. II, Chapter 2). Population densities that cause damage should be used.

Inoculum placement into the growing media can be achieved several ways. It is preferable to distribute the inoculum uniformly without injuring the nematodes. Eggs can be dispersed by limited, gentle mechanical mixing. The vermiform stages are much more sensitive to mixing and may need to be layered as uniformly as possible and covered with 2-3 cm of soil.

Evaluation of chemicals for nematicidal activity should include the technical and formulated products. Soil systems are used most commonly in the evaluation of nematicides, but soil-free solutions can also be utilized (13). Test nematodes can be placed directly into a range of concentrations of nematicide solutions. The nematodes can be observed directly for mortality and/or differential behavior. Soil-free test systems with growing plants also have merit (22). Seeds of a susceptible plant can be placed in or on absorbent growth media (paper pouches or towels) that allow germination and root growth. As roots emerge, an aliquant of *Meloidogyne* juveniles can be placed on the developing roots, and the paper growth medium treated with nematicides. Since galls develop quickly, chemical efficacy can be evaluated within 4 to 8 days.

The placement of the test chemical into soil is dependent upon its volatility, solubility and/or adsorptive properties. If the activity is unknown, it may be necessary to evaluate the materials as a volatile and nonvolatile nematicide. Volatile chemicals generally diffuse upward and laterally in the soil system. Such materials can be applied with a syringe or placed by aliquants at some predetermined distance into the infested soil.

Nonvolatile chemicals should be incorporated into the soil mechanically or with irrigation. In addition, systemic candidates ought to be evaluated in a foliar test for basipetal movement (8, 29). The foliage of the plant is sprayed carefully to ensure that no material reaches the soil. The plant can be tilted on its side, sprayed, allowed to dry, and then set upright in the greenhouse or growth chamber, or the soil can be covered with a barrier that would prevent any contact between the soil and runoff from the leaves.

Testing for systemic activity by soil application is more difficult. Additional research is needed on methods by which acropetal systemic activity may be determined, but feeding bioassays offer possibilities. For a foliar feeding bioassay, foliar-feeding nematodes are placed on the leaves and the soil treated with test chemicals. The number of foliar-feeding nematodes should be determined 3-4 wks later. With a split-root technique, soil in one-half of the root system is infested with nematodes, and the other half is treated with the candidate nematicide. Significant translocation within the plant will suppress or kill the nematodes.

Field evaluation is usually limited to advanced testing of candidate compounds. A satisfactory method for fumigants was developed by Christie (5). The method involves digging a trench 15 cm deep and 2 meters long. Infested soil is placed in cheesecloth bags at 7.5-cm intervals in the trench and covered with soil. The fumigant is injected 15 cm deep at the midpoint of each trench. The bags can be removed at a predetermined interval and a bioassay performed (see Vol. II, Chapter 2). The advantage of this method is the relative ease with which the efficacy of many compounds in a small area and under field conditions can be compared. In addition, considerable data can be generated in a short period of time.

A simple, rapid field test is not readily available for nonvolatile materials because nematodes generally are not killed at the concentrations used in the field. Because these nematicides primarily interfere with orientation during the infective and reproductive processes, a possible test might involve the planting of susceptible seedlings in small areas of treated soil infested with an endoparasitic nematode such as *Meloidogyne* or *Heterodera*. At some specified time (5-10 days after planting), plants are dug gently from the area that was treated and the roots stained (see Vol. II, Chapter 3) so the amount of infection can be assessed. In the case of *Meloidogyne* spp., the number of galls could be counted or indexed.

Preliminary testing usually involves a relatively large number of chemicals. Thus, it is important that the moisture and temperature of the soil medium (characterized for nutrient status and texture) be as near optimal as possible if proper growth of the host and activity of the nematode are to be ensured.

ADVANCED TESTING

Once a chemical is known to have potential as a nematicide, it should be tested in replicated field trials and on crops of anticipated usage. This usage pattern is largely dictated by market potential. Consequently, crops with high value and large acreage are normally chosen. Close attention must be given to experimental design, replication and type of analysis in order to increase the probabilities of successful experiments (18, 19).

Characterization of the Formulated Material

An effective nematicide is of little value if it cannot be delivered to the target. Therefore, handling features, flow characteristics, chemical changes occurring in the delivery system, effects on equipment (e.g. corrosion), water solubility, or other aspects of the material that inhibit its delivery to the soil or to the target should be noted so that formulation chemists can improve it.

Field Crops

Field testing typically requires considerable time and space. There are usually more problems associated with field testing than with growth chambers and greenhouses. Special precautions will ensure that valid data are attained in field tests (14).

The most critical aspect in testing materials in the field is site selection (3, 14). A site should be infested with numbers above the damage threshold of the desired test species. Each plot should be sampled so the initial population density can be determined because there is considerable variation in the vertical and horizontal distribution (2, 32). Soil is not uniform in any field, and major soil type differences within the experimental site should be avoided. The soil also should have physical and chemical characteristics that will support good crop growth. Weeds, pathogens and insects can be confounding factors and should be controlled. Characterization of the site for soil physical and chemical properties, nematodes and other pathogens, weeds, and insects is an important aspect of nematicide research. The crop should be a suitable, sensitive host but not hypersensitive.

Documentation of edaphic and environmental factors is valuable because chemical behavior may vary in different soils. The variable behavior of pesticides in different soils necessitates that methods of application, placement, and rates be tested. The ultimate, effective use of the product is dependent upon obtaining reliable data on all of these aspects.

Placement and Incorporation

Placement depends somewhat on the type of crop and nematicide (volatile or nonvolatile) being tested. Generally, volatile materials will become a gas that is lighter than air and will diffuse upward in the soil profile. However, it is possible that the gas may be somewhat heavier than air, thus placement near the surface is necessary. Suggested placement depths are 7.5, 15, and 30 cm. Volatile materials should be tried in the row and 10 to 15 cm on one or both sides of the row. The need for evaluating nematicides in several soil types cannot be overemphasized. For example, DBCP placed shallow or deep has effectively controlled *Hoplolaimus columbus* in a Marlboro soil (17). This pesticide controlled *Meloidogyne incognita* better when placed 20 cm deep than 40 cm deep in a Fuquay soil (17). Control of *Heterodera glycines* was consistent if EDB was placed at least 25 cm deep, whereas shallower placement resulted in erratic results (Schmitt, unpublished).

The method of nematicide application is dependent on the volatility and formulation of the product (23). It is essential to test methods of incorporation, the depth of placement, and the various rates on several types of soils and several crops (11, 28).

The placement of nonvolatile materials is dependent upon their solubility and adsorption properties. These chemicals should be tested through placement in the furrow and on the soil surface in bands ranging from 6 to 30 cm wide on soils tilled level or prepared with high beds. The banded treatments should be incorporated to several depths up to 15 cm, and compared to an unincorporated treatment. Aldicarb increased yields when it was incorporated in a clay loam soil infested with *Heterodera goettingiana*, regardless of the depth of incorporation, but did not give any benefit when it was applied in the furrow (31). However, incorporation of aldicarb to a 15-cm depth in a sandy loam gave the greatest yield increases. Implements vary in their ability to incorporate or distribute materials in soils, thus affecting the nematicide's efficacy. For example, control of *Globodera pallida* and *G. rostochiensis* was better when aldicarb was incorporated with rotavators than with a rotary harrow (31). Readily soluble nematicides also can be incorporated by irrigation (12).

Chemical root dips of nonphytotoxic nematicides may be tested on plant species propagated vegetatively and infected with endoparasites. Several rates and times of exposure should be evaluated. The principal element of chemical root dips is the absorption of sufficient concentrations to prevent or minimize infection (7, 21). Treated root stock are planted along with appropriate control plants in sterile soil for determination of treatment effectiveness. This test involves an assay or bioassay after 4-8 wks for assessment of host infection and nematode reproduction. Noninfected plants should also be treated and planted into infested soil so it can be determined whether a sufficient concentration of the nematicide(s) was absorbed to prevent or retard infection.

Any newly developed systemic nematicides are worthy of being evaluated for foliar application (8). An effective foliar-applied nematicide results in greater ease of application and possibly lower rates.

Rates

Much confusion in delivery of information to growers can be avoided if a standard is adopted for the reporting of data (10, 14). The most common recommendation is weight or volume of active and formulated material per thousand feet of row, or per 100 or 1000 square feet. Treatments may also be expressed in grams or milliliters of active and formulated product per 100 meters for row and band treatments and in grams and milliliters of active and formulated product per square meter for broadcast treatments.

Advanced nematicide trials provide the most useful information if rates ranging from zero to excessively high levels are used. At least three rates and preferably five or more spaced equally or in a gradient are recommended. This type of design will aid analysis and interpretation of data because response models can be used (18). Such interpretations of data are more meaningful than the separation of means or a statement that one treatment is significantly different from another (18).

Nematode mortality and disease parameters should be determined (see Chapter 1). The proportion of a soil-living nematode population killed by a treatment with a chemical depends on the concentration and time of exposure (26). The concentration × time parameter must be measured. This factor is not reported very frequently (26), probably because it is not easily estimated, especially with nonvolatile nematicides.

Greenhouse Crops

Greenhouse management has advanced to the stage where nematodes are rarely a limiting factor in the soil. The soil is generally treated with methyl bromide (6) or is steam sterilized; also sterile, artificial media often are used (D. Strider, personal communication). Conversely, foliar-feeding nematodes often pose a problem to greenhouse crops.

When nematicide application is tested in the greenhouse, refer to the section on preliminary testing of crops grown in pots, or the section on field crops for those grown in residual soil but covered by a greenhouse. If nematicidal activity of chemicals against foliar nematodes is to be determined, plants should be established in growing media before the foliage is inoculated with the pathogen. Several concentrations of the nematicides may be used either as foliar spray or soil treatments (9, 29). Foliar-applied nematicides must not come in contact with the soil.

Environmental Effects on Nematicidal Activity

Many edaphic and environmental factors influence the behavior of nematicides, the most important being soil type and moisture (23, 32). Soil temperature, pH, organic matter, nutrient status, and soil physical characteristics, especially porosity and texture are among the numerous factors that may also influence efficacy of nematicides. Interpretation of nematicide data is facilitated when these factors and the soil profile are characterized. With such information, inferences can be made about the adsorption properties of the clay and organic matter of the soil in relation to the chemical used. A certain range of air space and moisture is important for the activity of the fumigants and indirectly important for the nonfumigant (nonvolatile) nematicides. Moisture is important for activation of nonfumigants, and it affects the dispersal of fumigants. For example, adsorption of a fumigant is reduced by an increase in temperature and a decrease in soil moisture (15). The concentration-time product is inversely related to soil porosity (4). The pH may also affect nematicidal activity (1, 24).

Nematode Population Dynamics

Any effective nematicide must necessarily affect the population dynamics of pathogenic species in a nematode community. It may alter the structure of the entire community. Thus, it is essential that adequate data be collected so that major changes in the population throughout a growing season can be characterized. Important times to sample are prior to treatment, soon after treatment, at midseason, and at harvest (See Vol. II, Chapter 1).

Crop Response

Yield and quality of the crop are the ultimate measurements in a nematicide trial. In addition, other important variables to be measured during the growing season at various phenological stages of plant growth are plant vigor, disease, root galling and necrosis. These parameters can be used to make inferences about the activity of nematicides over the season. Dry matter accumulation from planting until harvest is useful for interpreting biological and nematicidal interactions that affect a nematicide's performance.

Ecological Impact

Increasing concern for the environment warrants assessment of the nontarget ecological impact of pesticides. Comprehensive studies of this type require a major commitment and are usually carried out by environmental scientists. However, certain nematicide-induced adverse effects can be readily detected. The researcher should look for impacts such as death or illness of wildlife, shifts in the microflora and microfauna, death or stress of desirable vegetation, and any other aspects. Related information or data may be useful in our attempts to protect the ecosystem in which we are trying to grow crops as well as those distally removed.

Pesticide Interactions

Many classes of pesticides are applied to a given crop during the growing season. Potential interactions, whether antagonistic or synergistic, should be determined during the advanced stages of nematicide evaluation. Researchers should use chemicals (herbicides, insecticides, fertilizers, etc.) that a farmer is likely to use in growing the crop. Major problems like increased or decreased toxicity to the crop are readily detected (25). The subtle effects are usually overlooked except in experiments designed specifically for pest and pesticide interactions.

EXPERIMENTAL DESIGN

The experimental design for nematicide testing should be simple and balanced (18, 19). Satisfactory designs for nematicide testing include the randomized-complete-block design with treatments factorially arranged, the Latin square for a small number of treatments, and the lattice design for experiments with a large number of treatments. A split-plot design is also practical in many cases. Replication is partially dependent on the design and expected variation, but sufficient replications should be used to assure an estimate of the experimental error and provide a fair level of precision for an estimate of the mean and sensitivity of test of significance. Each experiment has its own requirement for replications and cannot be dictated. Generally, the greater number of replications there are, the better the precision will be, but there is a point of diminishing returns (18). Typically, 6-8 replicates are required for reliable results in field tests, whereas 4-6 replicates may be satisfactory for greenhouse and growth chamber experiments (See Vol. II, Chapter 13).

DATA ANALYSES

The analysis of data in well-designed experiments is relatively straightforward. There are several advantages to statistical analysis of a nematicide experiment: 1) it aids in the separation of treatment and random effects; 2) it facilitates the condensing of data for presentation; and 3) it provides a standard error that gives a measure of the precision of the experiment(s). An investigator should not set a significance level of probability for the determination of the effectiveness of a material. It is better to use response models or develop an estimate of the magnitude of mean differences (18). Refer to Vol. II, Chapter 13, for types of analyses (19).

REPORTING RESULTS

Effective communication of results is achieved by a well-written, concise report. The materials and methods should include all the important details: e.g., site description (soil texture, soil analysis, etc.), test organisms, test procedures, types of data and statistical procedures. The results should be a summarization of the interpretated data. Large tables of unsummarized data with little or no interpretation are a poor means of communication of research findings. Tables and graphs should be clear and simple but provide sufficient information to be self explanatory.

Literature Cited

1. Adams, P. B., H. L. Warren, and C. J. Tate. 1971. Potassium azide: fungicidal activity as affected by pH. Plant Dis. Rep. 55:177-181.
2. Barker, K. R. (Chairman). 1978. Determining nematode population responses to control agents, pp. 114-125 *In* Methods for evaluating plant fungicides, nematicides, and bactericides, ed. E. I. Zehr. Am. Phytopathol. Soc., Minneapolis, Minn.
3. Bird, G. W. (Chairman). 1978. Site selection procedures for field evaluations of nematode control agents, pp. 108-113 *In* Methods for evaluating plant fungicides, nematicides, and bactericides, ed. E. I. Zehr. Am. Phytopathol. Soc., Minneapolis, Minn.
4. Call, F. and N. G. M. Hague. 1962. The relationship between the concentration of ethylene dibromide and nematicidal effects in soil fumigation. Nematologica 7:186-192.
5. Christie, J. R. 1945. Some preliminary tests to determine the efficacy of certain substances when used as soil fumigants to control the root knot nematode, *Heterodera marioni* (Cornu) Goodey. Proc. Helminthol. Soc. Wash. 12:14-19.
6. Coolen, W. A., G. J. Hendricks, and C. J. Deherde. 1972. Nematode control in glasshouse rose culture by means of methyl bromide and dibromochloropropane. Rijk. Voor Nematol. en Entomol. Marelbke. 22 pp.
7. Dale, P. S. and G. J. Vandermespel. 1972. Control of root knot nematodes on Chinese gooseberry (Kiwi fruit), *Actinidia chinensis*, by chemical bare root dip. Plant Dis. Rep. 56:850-851.
8. DiSanzo, C. P. 1982. Effect of foliar applications of carbofuran and a related compound on plant parasitic nematodes under microplot and field conditions. J. Nematol. 14:208-212.
9. DiSanzo, C. P. (Chairman). 1978. Guidelines for evaluating nematicides in greenhouses and growth chambers for the control of root knot nematodes, pp. 101-103 *In* Methods for evaluating plant fungicides, nematicides and bactericides, ed. E. I. Zehr. Amer. Phytopathol. Soc., Minneapolis, Minn.
10. Good, J. M., J. N. Sasser, and L. I. Miller. 1963. A suggested guide for reporting experiments on nematocidal chemicals. Plant Dis. Rep. 47:159-163.
11. Johnson, A. W. 1978. Test materials and environmental conditions for field evaluation of nematode control agents, pp. 106-108 *In* Methods for evaluating plant fungicides, nematicides, and bactericides, ed. E. I. Zehr. Am. Phytopathol. Soc., Minneapolis, Minn.
12. Johnson, A. W., J. R. Young, and B. J. Mullinex. 1981. Applying nematicides through an overhead sprinkler irrigation system for control of nematodes. J. Nematol. 13:154-159.
13. Johnson, D. E. and B. Lear. 1962. A method for assaying nematocidal activity for experimental chemicals. Plant Dis. Rep. 10:742-743.
14. Lear, B. and D. E. Johnson. 1972. Suggested procedures for evaluation of experimental nematicides in field applications. Fungicide and Nematicide Tests (Am. Phytopathol. Soc.) 27:1-2.
15. Lewis, G. D. and W. F. Mai. 1963. Temperature and moisture in relation to the toxicity of 1,3-dichloropropene to *Ditylenchus dipsaci* (Kuhn) Filipjev in organic soils. Plant Dis. Rep. 47:1097-1101.
16. Marks, C. F., I. J. Thomason, and C. E. Castro. 1968. Dynamics of the permeation of nematodes by water, nematicides and other substances. Exp. Parasitol. 22:321-337.
17. Minton, N. A., M. B. Parker, O. L. Brooks, and C. E. Perry. 1979. Effects of nematicide placement on nematode populations and soybean yield. J. Nematol. 11:150-155.
18. Nelson, L. A. 1978. Use of statistics in planning, data analysis, and interpretation of fungicide and nematicide tests, pp. 2-14 *In* Methods for evaluating plant fungicides, nematicides, and bactericides, ed. E. I. Zehr. Am. Phytopathol. Soc., Minneapolis, Minn. 141 pp.
19. Nelson, L. A. 1985. Design of greenhouse and field experiments for nematode investigations, pp. 177-186 *In* An Advanced

treatise on *Meloidogyne*, Vol. II: Methodology, ed. K. R. Barker, C. C. Carter, and J. N. Sasser. Coop. Publ. Dep. Plant Pathol., North Carolina State Univ., and U.S. Agency Int. Dev., Raleigh, N. C.
20. Ohba, K. and N. Ishibashi. 1984. A nematode, *Caenorhabditis elegans*, as test organism for nematicide evaluation. J. Pestic. Sci. (Nihon Noyaku Gakkaishi) 9:91-96.
21. Prassad, K. S. K. and Y. S. Rao. 1976. Chemotherapy of the root knot nematode (*Meloidogyne graminicola*) in rice. II. Effects of root dip treatment with nematicides on invasion and development of the nematode. J. Plant Dis. & Prot. 83:730-735.
22. Prieser, F. A., J. R. Babu, and A. A. Haidri. 1981. A soil free system for assaying nematicidal activity of chemicals. J. Nematol. 13:535-537.
23. Sasser, J. N. 1976. Nematode diseases of plants and their control including the important role of nematicides. pp. 187-253 *In* Pesticides in the environment, Vol. II, ed. R. White-Stevens. Marcel-Dekker, Inc., New York. 629 pp.
24. Schmitt, D. P. 1979. The relationship of soil pH to nematicidal efficacy in soybeans. Phytopathology 69:1-A9 (Abstr.).
25. Schmitt, D. P., F. T. Corbin, and L. A. Nelson. 1983. Population dynamics of *Heterodera glycines* and soybean response in soils treated with selected nematicides and herbicides. J. Nematol. 15:432-437.
26. Seinhorst, J. W. 1973. Dosage of nematicidal fumigants and mortality of nematodes. Neth. J. Plant Pathol. 79:180-188
27. Steele, A. E. 1978. Tests for nematicidal efficacy using larvae of *Heterodera schachtii*, pp. 105-106 *In* Methods for evaluating plant fungicides, nematicides and bactericides, ed. E. I. Zehr. Am. Phytopathol. Soc., Minneapolis, Minn.
28. Taylor, A. L. 1939. Efficient spacing of soil fumigants for field applications. Proc. Helminthol. Soc. Wash. 6:62-66.
29. Thirugnanam, M. 1978. Evaluation of nematicides for systemic eradication of root knot nematodes, pp. 103-104 *In* Methods for evaluating plant fungicides, nematicides and bactericides, ed. E. I. Zehr. Am. Phytopathol. Soc., Minneapolis, Minn.
30. Van Gundy, S. D. and M. V. McKenry. 1977. Action of nematicides, pp. 263-283 *In* Plant disease: An advanced treatise, Vol. I., ed. J. G. Horsfall and E. B. Cowling. Academic Press, New York. 488 pp.
31. Whitehead, A. G., R. H. Bromilow, K. A. Lord, S. R. Moss, and J. Smith. 1975. Incorporating granular nematicides in soil, pp. 133-144 *In* Proc. British Insecticide and Fungicide Conference (Aith, Brighton), Nov. 17-20, 1975. Vol. I. Brit. Insect. & Fungicide Counc., London. 372 pp.
32. Whitehead, A. G. and A. L. Winfield. 1978. Chemical control, pp. 283-301 *In* Plant nematology, ed. J. F. Southey. Minist. Agric., Fish. and Food Tech. Bull. 7. Her Majesty's Stationery Office, London. 440 pp.

THE ROLE OF NEMATICIDES IN NEMATODE MANAGEMENT

A. W. Johnson[1]

Supervisory Research Plant Pathologist
USDA-ARS
Coastal Plain Experiment Station
Tifton, Georgia, USA

Several events have had a profound influence on opinion in the United States regarding the importance of plant-parasitic nematodes. One important event that educated the farmers on damage caused by nematodes was the introduction of the soil fumigants D-D (1,3-dichloropropene and 1,2-dichloropropane and related hydrocarbons) (1), EDB (ethylene dibromide), and DBCP (1,2-dibromo-3-chloropropane) (3), making available for the first time nematicides that could be used effectively and practically on a field scale. The rapidity and extent to which these and other fumigants came into use is one of the most interesting and surprising responses in the history of pesticides.

Nematicides are one of the most important and reliable means of controlling a wide variety of nematodes. Commercial use of nematicides began about 1945, and suitable application methods have been devised for several highly effective nematicides. Their use increased slowly for the first several years, but has increased rapidly during the last 15 years. The principal use of nematicides is for control of nematode populations in the soil before planting crops. Nematicides are used to protect a number of highly susceptible crops from nematode infection. These chemicals are used on approximately 1.7 million acres, which is less than one percent of the cultivated agricultural land in the United States.

TYPES OF NEMATICIDES

There are two types of nematicides for control of nematodes on crops—soil fumigants and nonfumigants. Nematicides that are available are listed in Table 1, but many of these are registered for use on only a few crops.

Soil Fumigants

The search for chemicals effective for controlling nematodes has been going on for many years, but was not pursued vigorously until after the introduction of D-D soil fumigant in 1943. The success of D-D and ethylene dibromide, which came into use in 1951, stimulated widespread interest in soil fumigation. The first nematicides developed were liquids which are injected beneath the soil surface. These nematicides volatilize to produce vapors that kill nematodes. Vapors from soil fumigants diffuse through the soil, are dissolved in soil water, and enter nematode bodies through the cuticle.

Various authors have postulated the properties that make a chemical an ideal soil fumigant. According to Hollis (7), an ideal nematicide will confer protection to crop plants for prolonged periods (4 months to 3 years); possess a high proportion of active ingredients and effectiveness against plant-parasitic nematodes at dosage levels of 10 to 60 pounds per acre on an overall application basis; move independently of the water phase in the soil at vapor pressures in the range of 0.5 to 5 mm Hg at 30°C; be soluble in the range of 0.09 to 0.41 percent for the halogenated hydrocarbons; possess requisite characteristics of non-phytotoxicity and low mammalian toxicity and be free from the residual properties which might render food crops unfit for consumption; confer prolonged protection to crop plants by a complete kill of nematodes in all morphological stages; be chemically stable or possess a relatively long chemical half-life (60 to 90 days) in arable soils of tropical and subtropical regions; and kill nematodes by the action of its

[1]All research on intensive cropping systems mentioned in this paper was conducted cooperatively with Dr. C. C. Dowler, Dr. N. C. Glaze, Dr. D. R. Sumner, and Dr. R. B. Chalfant at the Coastal Plain Experiment Station, Tifton, Georgia.

Table 1. Nematicides available on world markets.

Common name	Registered trade name and manufacturer	Chemical name	Formulation and classification
Aldicarb	(TEMIK) Union Carbide Corp.	2-methyl-2-(methylthio) propionaldehyde *O*-(methyl-carbamoyl) oxime	Granular nematicide/ insecticide
Carbofuran	(FURADAN) Niagra Chem. Div., FMC Corp.	2,3-dihydro-2,2-dimethyl-7-benzofuranyl methyl-carbamate	Granular and flowable nematicide/insecticide
Chloropicrin	Great Lakes Chem. Corp.	trichloronitromethane	Liquid fumigant nematicide/ insecticide
1,3-D	(TELONE II) Dow Chem. Co.	1,3-dichloropropene and related chlorinated C_3 hydrocarbons	Liquid fumigant nematicide
DD Mixture[a]	(DD) Shell Dev. Co. (VIDDEN-D) Dow Chem. Co.	1,3-dichloropropene and 1,2-dichloropropane and related chlorinated hydrocarbons	Liquid fumigant nematicide
EDB[b]	(DOWFUME W-85) Dow Chem. Co. (SOILBROM 90) Great Lakes Chem. Corp.	ethylene dibromide	Liquid fumigant nematicide
Ethoprop	(MOCAP) Rhone-Poulenc Chem. Co.	*O*-ethyl *S,S*-dipropyl phosphorodithioate	Granular or emulsifiable liquid nematicide/insecticide
Fenamiphos	(NEMACUR) Chemagro Agri. Div., MOBAY Chem. Co.	ethyl 4-(methylthio)-*m*-tolyl isopropylphosphoramidate	Granular or emulsifiable liquid nematicide
Fensulfothion	(DASANIT) Chemagro Agri. Div., MOBAY Chem. Co.	*O,O*-diethyl *O*[*p*-(methylsul-finyl)phenyl]phosphoro-thioate	Granular nematicide
Methyl Bromide	(DOWFUME MC-2) Dow Chem. Co. (BROM-O-GAS) Great Lakes Chem. Corp.	bromomethane	Broad spectrum soil fumigant
Oxamyl	(VYDATE) E. I. du Pont de Nemours and Co.	methyl-*N'*,*N'*-dimethyl-*N*-[(methylcarbamoyl)oxy]-1-thiooxamimidate	Emulsifiable liquid nematicide/insecticide

[a] Not available after March 1984.
[b] EDB suspended as a soil fumigant September 30, 1983.

chemical breakdown products. It is becoming increasingly obvious that there is no such thing as an ideal nematicide or even an ideal soil fumigant. A chemical with properties that would make it ideal for one purpose or use under one set of conditions might make it quite unsuitable for another purpose or use under another set of conditions.

Nonfumigants

Some recently developed nematicides are water soluble and are called "nonfumigants." This relatively new group of chemicals represents an entirely new concept in chemical control of nematodes. These chemicals dc not fumigate the soil. They are distributed into the soil by incorporation and by percolation of water so that they come in contact with the nematodes. These chemicals generally are not toxic to the plants on which they are recommended and can be used safely at planting or as a post-plant application.

Systemics

The newest types of nematicides are systemic. They can be taken up by the roots from the soil, or by foliage after being sprayed with the material, then translocated to roots where they kill nematodes that feed on the roots or prevent nematodes from feeding on the roots. Presumably, they are taken into the nematodes during feeding, but they might also enter nematode bodies through the cuticle in contact with soil and plant tissue.

APPLYING NEMATICIDES

Nematicides can be applied as overall, row, strip, spot, side-dress, and irrigation treatments to crops.

Overall Treatment

Overall treatment of soil consists of applying nematicides evenly to the entire field. In an overall treatment, soil fumigants are generally injected into furrows or slits spaced 8 to 12 inches apart. Chisels work best when the shanks are staggered on the tool bar so that adjacent chisels are several inches apart from front to back. This staggering of shanks helps keep the chisels from becoming clogged with crop residue. A bed shaper following the chisel tool bar will seal the soil and prevent rapid volatilization of fumigants from the soil. Areas treated with methyl bromide + chloropicrin or other chemicals with high vapor pressure should be covered with polyethylene and the edges sealed immediately after application. Fumigant nematicides may also be applied overall with a moldboard plow. Chemical delivery tubes are attached to discharge the nematicide into the bottom of an open furrow which is then covered immediately with soil from the following plow.

Granular or liquid nonfumigants may also be applied overall. They are spread or sprayed uniformly to the soil surface and incorporated into the upper four inches of soil. A disk harrow may also be used to incorporate nematicides into the soil. Thorough mixing of the nonvolatile nematicides with the soil is desirable.

New nematicide application technology is being developed for nemagation (application of nematicides through irrigation systems). The nematicide is injected by pump into the water as it passes through the irrigation system. Several advantages of nemagation include: 1) uniformity of application; 2) prescription application; 3) timeliness of application; 4) easy incorporation and activation; 5) reduction of soil compaction; 6) reduction of operator hazards; 7) reduced nematicide requirements; 8) less risk for environmental contamination; 9) effectiveness; and 10) economic benefits.

Row Application

Row application of nematicides may be used in crop production. Pressurized or gravity-flow equipment can be used to apply liquid nematicides. In pre-plant row application, the chemical should be placed six to eight inches deeper than the original soil surface, then covered with a listed bed of soil formed by two discs mounted behind the fumigation chisels. These beds tend to seal volatile fumigants in the soil.

After nematicide application, the beds should remain undisturbed for 7 to 21 days, depending on the type of fumigants used, dosage, weather and soil conditions. The bed is leveled at planting with a blade, built-up sweep, board or opening plows and shovels which move the untreated top portion of the bed into the furrow. During seeding or transplanting, it is important that only treated soil be used to close the furrow.

Granular nematicides may be applied behind an opening shovel, in furrow, in a band across the seed furrow at planting, or in a band and incorporated into the top 4-inch soil layer.

Some fumigants can be applied under the row at planting time. When such chemicals are used, they should be placed at a minimum depth of six inches and at least two inches beneath and to the side of the seed. This procedure minimizes the possibility of injury to germinating seed and young seedlings as well as interference with seed placement.

Strip or Bed Treatment

Strip or bed treatment is a modified overall treatment. It is limited to wide row crops, such as orchards, vineyards, groves and some nursery planting. Most of the equipment, methods and chemicals used for overall treatment can be used for strip or bed treatment. The treated strip or bed is centered on the future planting site and is usually 4 to 10 feet wide. The treated area varies from one-third to one-half the land area, depending on width of the strips or beds and the number of strips or beds per acre.

The advantage of strip or bed application is the saving in time and chemical required per acre. The chief disadvantage is the possibility of introducing nematodes soon after planting by the movement of soil from untreated areas.

Spot Treatment

Spot treatment can be used where nematode damage is confined to a well-defined spot or spots in the field. If surrounding areas are relatively free of plant-parasitic nematodes, these spots can be treated individually. The moldboard-plow applicator can be used for spot treatment by simply turning on the fumigant applicator when the plow passes through the infested area. Also, nematicides suitable for application during planting can be released as the planter moves through the heavily infested areas. Spots can also be treated with an overall chisel applicator after plowing but before planting. Overall application is usually best for spot treatment.

Side-dressing Treatment of Established Plants

Side-dressing treatment of established plants can be used where nematodes become a problem

Table 2. Effect of film mulch and nematicides on root-gall indices of crops.

Treatment	Nematicide rate (kg a.i./ha)	Root-gall indices[a] Slicing cucumber	Summer squash	Okra	Pole bean
Unmulched					
Control	—	4.58 a	5.00 a	4.01 a	3.41 a
DD-MENCS	376	2.35 b	2.50 b	2.20 c	1.00 c
Film mulch					
Control	—	4.35 a	5.00 a	3.63 a	2.82 b
DD-MENCS	376	1.00 c	1.00 c	1.30 d	1.00 c
MBR-CP	376	N.I.	N.I.	1.01 d	1.00 c
Ethoprop	8.96	N.I.	N.I.	2.85 bc	1.33 c

[a] Means for the same crop followed by unlike letters are significantly different ($P = 0.05$) according to Duncan's multiple range test; NI: Treatment not included in test.

soon after an orchard; vineyard or nursery is established. This type of treatment will often restore vigor and increase yields, but it is a poor substitute for preplant treatment. Annual crops usually do not benefit from post-planting treatment.

Nematicides are used to protect principally tobacco (*Nicotiana tabacum* L.), cotton (*Gossypium hirsutum* L.), soybean (*Glycine max* [L.] Merr.), peanut (*Arachis hypogaea* L.), corn (*Zea mays* L.), sugar beet (*Beta vulgaris* L.), potato (*Solanum tuberosum* L.), pineapple (*Ananas cosmosus* [L.] Merr.) and a number of fruits and vegetables—all highly susceptible to nematode attack (6).

For the purpose of this paper, I will use selected examples of crop management practices and discuss the role of nematicides in nematode management. I have restricted most of my presentation to work conducted in the Southeastern Coastal Plain of the United States.

Film Mulch and Drip Irrigation

A series of studies were conducted at the Coastal Plain Experiment Station, Tifton, Georgia, on the influence of film mulch, drip irrigation and nematicides on nematode populations and yield of high value crops. The soil was naturally infested with *Meloidogyne incognita* (Kofoid & White) Chitwood. Soil was prepared, nematicides applied, drip-irrigation tubing laid, and all plots covered and sealed with film mulch. Beds were left undisturbed for 1 to 3 weeks. Holes were then cut in the film mulch to accommodate row spacing and seeding rates for each crop and to allow aeration. Seeds of various crops were hand planted through each hole 1 week later.

Soil was assayed for parasitic nematodes before treatment, at planting, and at various intervals after planting. Twenty cores (2.5 x 15cm) of soil were collected from the root zone of plants in the center 6 m of each plot, composited and mixed thoroughly. A 150-cm^3 aliquant was processed by a centrifugal-flotation method (9). Plants were indexed for galls caused by root-knot nematodes during the growing season and after the final harvest. Individual plants were rated on a 1-5 scale (1 = no galls, 2 = 1-25%, 3 = 26-50%, 4 = 51-75%, and 5 = 76-100% roots galled).

Numbers of *M. incognita* juveniles in the soil were positively correlated (correlation coefficient r = 0.75 to 0.92) with root-gall indices; therefore, only root-gall indices are reported in this paper. Root-gall indices were lower in unmulched plots treated with DD-MENCS (20% methyl isothiocyanate + 80% chlorinated C_3 hydrocarbons) and in film-mulched plots treated with DD-MENCS, MBR-CP (98% methyl bromide + 2% chloropicrin), and ethoprop (*O*-ethyl *S*, *S*-dipropyl phosphorodithioate) than in untreated plots (Table 2).

Yield increases from DD-MENCS-treated plots ranged from 128% for pole bean (*Phaseolus vulgaris* L.) to 1175% for cucumber (*Cucumis sativus* L.) (Table 3). Yields of all crops from film-mulched plots were always greater than those from unmulched plots. Yield increases from crops in film-mulched plots treated with soil fumigants ranged from 8% (pole bean—MBR-CP) to 581% (sweet corn—DD-MENCS). Ethoprop applied to film-mulched plots increased yield of okra (*Abelmoschus esculentus* L.) 38% over the control, but decreased yield of pole bean 18% over the control. Detailed results of these studies have been reported (2, 8, 16, 19).

Table 3. **Effect of film mulch and nematicides on marketable yield of crops.**

Treatment	Rate (kg a.i./ha)	Percent yield increase over controls[a] Slicing cucumber	Summer squash	Sweet corn	Okra	Okra	Pole bean
Unmulched							
DD-MENCS	376	1175	180	220	507	173	128
Film mulch							
DD-MENCS	376	208	53	581	454	235	17
MBR-CP	336	N.I.	N.I.	N.I.	N.I.	208	8
Ethoprop	8.96	N.I.	N.I.	N.I.	N.I	38	—18

[a] NI: Treatment not included in test.

Table 4. **Total marketable yield of cucumber as influenced by film mulch and soil chemical treatments.**

Treatment	Rate (ka.i./ha)	Marketable yield (quintals/ha)[a] Crop 1	Crop 2	Crop 3	Total	Marketable yield as % of total yield
Unmulched						
Bensulide (B), control	6.70	58 d	0 d	120 c	178 d	58 c
DD-MENCS + B	376.0 + 6.70	332 b	6 d	273 b	611 b	71 cd
Film mulch						
Bensulide (B), control	6.70	187 c	89 ab	180 c	456 c	75 bc
DD-MENCS	376.0	646 a	77 bc	371 a	1094 a	79 ab
Sodium azide (SA)	33.6	263 a	51 c	174 c	488 bc	71 cd
Chloroneb + B	26.9 + 6.70	250 bc	62 bc	138 c	450 c	71 cd
SAA + ethoprop	33.6 + 8.96	330 b	55 c	128 c	513 bc	68 d
Ethoprop	8.96	238 bc	52 c	160 c	450 c	70 cd
MBR-CP	336.0	703 a	118 a	387 a	1208 a	81 a

[a] Data are means for four replicate plots. Within each column, values followed by the same letter are not significantly different ($P = < 0.05$) according to Duncan's multiple-range test. Absence of letters indicates that values in the column are not significantly different.

Use of trickle irrigation for high-value vegetable crops has increased greatly in the southeastern USA in the past five years. This practice, especially in conjunction with soil fumigation and plastic-film mulching has resulted in remarkable yield increases for many crops (2, 8, 16, 19, 37). Because of the high initial costs of this production system, there is considerable interest in multiple cropping (growing two or three crops in succession) and in reducing the amount of fumigant.

Pepper (*Capsicum frutescens* L.) and summer squash [*Cucurbita pepo* var. *melopepo* (L.) Alef.] were grown in immediate succession in undisturbed beds using trickle irrigation and DD-MENCS (21). The highest combined yield in metric tons/ha for both crops was obtained with film mulch + DD-MENCS (117.6) followed by film mulch (112.3), DD-MENCS (93.4) and control (69.0). Yield from the second crop was negatively correlated (r = -0.87) with the degree of plant infection with *M. incognita*. Multiple cropping of pepper and squash with trickle irrigation has an excellent potential provided nematodes and other soil-borne pathogens can be adequately controlled.

More recently three crops of cucumber were grown in succession in beds by use of trickle irrigation, plastic film mulch, and soil chemical treatments over a 17-month period, including a fallow winter season (18). Total yield of the three crops was highest (1208 quintals/ha) in film-mulched plots treated with MBR-CP, and the next highest in film-mulched plots treated with DD-MENCS (1094 quintals/ha); total yield was only 456 quintals/ha in film-mulched control (untreated) plots (Table 4). Plant growth and yields were greatest when populations of root-knot nematodes were suppressed to very low levels (Table 5). Data from this study indicate that it is possible to control root-knot nematodes and increase yields of three con-

Table 5. Growth and root-gall of cucumber as influenced by film mulch and soil chemical treatment.

Treatment	Nematicide rate (kg a.i./ha)	Growth index[a]	Root-gall index[b] Crop 1 125 days after planting	Crop 2 100 days after planting	Crop 3 80 days after planting
Unmulched					
Bensulide (B), control	6.70	1.75 d	3.90 ab	4.00 ab	5.00 a
DD-MENCS + B	376.0 + 6.70	3.50 bc	1.00 c	1.10 c	2.53 b
Film mulch					
Bensulide (B), control	6.70	2.75 c	4.86 a	3.40 b	5.00 a
DD-MENCS	376.0	4.25 ab	1.03 c	1.00 c	1.48 c
Sodium azide (SA)	33.6	4.50 a	4.81 a	4.53 a	4.98 a
Chloroneb + B	26.9 + 6.70	2.75 c	4.38 a	3.83 ab	5.00 a
SA + ethoprop	33.6 + 8.96	4.50 a	3.26 b	4.33 a	5.00 a
Ethoprop	8.96	3.50 bc	3.27 b	3.83 ab	4.85 a
MBR-CP	336.0	5.00 a	1.00 c	1.13 c	1.00 c

[a] 1-5 scale: 1 = plants small, chlorotic, nonvigorous; and 5 = plants large, dark green, vigorous.

[b] Index on a scale of 1-5, in which 1 = no galls and 5 = 76-100% roots galled. Data are means for four replicate plots. Within each column, values followed by a letter in common are not significantly different ($P = < 0.05$) according to Duncan's multiple-range test.

secutive crops of cucumber with one application of a soil fumigant such as DD-MENCS or MBR-CP before film mulching. The study was discontinued after cultivation of three crops, but since populations of nematodes were still reduced at the last harvest, a fourth crop could probably have been grown successfully in fumigated soil. Intensive production in this manner allows fixed costs to be defrayed over several crops, thus increasing the magnitude of return per dollar invested.

Since multiple cropping has been successful with film mulching, trickle irrigation, and soil fumigation (18, 21), there has been considerable interest in the concurrent reduction of soil fumigant rate under that crop production system. A study was conducted to evaluate DD-MENCS at various rates in a cucumber-squash-cucumber cropping system with film mulch and trickle irrigation (17). The results of that study indicate that growth and yield of all crops were greatest when nematodes and soil-borne fungi were controlled. *M. incognita* control, in plots treated with DD-MENCS at 215 kg/ha was greater ($P = 0.05$), based on root-gall indices, than in untreated plots 54 days after the third crop (cucumber) was planted. The recommended rate of DD-MENCS (376 kg/ha) for control of nematodes and fungi under conventional production practices can be reduced by 43% and still maintain adequate nematode control and yield increase of vegetable crops under the film-mulch and trickle irrigation production system.

A series of studies to evaluate drip irrigation as a vehicle for applying nematicides was conducted in Florida by Overman from 1974 to present. Current information indicates a drip irrigation system can be an efficient, convenient means of distributing nematicides into a crop area (30). Emulsifiable or water-miscible materials are most readily distributed through drip irrigaion. Broad spectrum soil fumigants and fumigant nematicides have been applied to the soil with minimal amounts of water (23, 24, 25, 26, 27, 28, 29, 31). Efficacy of several chemicals has compared favorably with standard machine application procedures by injection, incorporation, or drench into the soil. In Florida, metham (sodium methyl dithiocarbamate), oxamyl [methyl *N'*, *N'*-dimethyl-*N*-{(methylcarbamoyl) oxy}-1-thiooxamimidate], and ethylene dibromide in drip irrigation water have increased yield of *Chrysanthemum* over water alone. Methyl bromide and emulsifiable formulations of ethylene dibromide alone or combined with chloropicrin have improved yield of tomato over nontreated controls when the chemicals were injected at labeled rates into drip irrigation.

ANNUAL AND PERENNIAL CROPS

The benefits resulting from use of nematicides are clearly documented in an FAO analysis of 779 experimental field tests in which nematodes were controlled (38). There is interest in the use of broadspectrum soil fumigants to control soil-borne pests (primarily nematodes and fungi), and recent research indicates spring application of certain fumigants improves marketable yields and reduces plant pathogens below damaging levels (22). Growers have been hesitant to use these chemicals because of high cost, phytotoxicity with early season application, and reinfestation of the treated soil. An evaluation study was conducted on the fall application of methyl bromide-chloropicrin (67-33%) gas mixture at 392 kg/ha, methyl bromide-chloropicrin (67-31%) gel at 246 kg/ha, DD-MENCS at 187 and 327 liters/ha, metham at 748 liters/ha, fenamiphos (ethyl 4-(methylthio)-*m*-tolyl isopropylphosphoramidate) at 9 kg/ha, and ethylene dibromide (85%) at 56 liters/ha in vegetable transplant production. Methyl bromide-chloropicrin gas mixture and the two rates of DD-MENCS increased marketable transplant yields of pepper, tomato (*Lycopersicon esculentum* Mill.) and cabbage (*Brassica oleracea* var. *capitata* L.) by 1235%, 118%, and 29%, respectively, over that of the nontreated plots. These treatments also increased the fresh weights of plants and reduced populations of root-knot nematodes. Metham and methyl bromide chloropicrin gel were generally less effective than methyl bromide-chloropicrin gas or DD-MENCS. Neither fenamiphos nor ethylene dibromide reduced root-galls on tomato below detectable levels or improved the marketable yield of the three crops. Fall fumigation eliminated phytotoxicity.

More recent reports of the benefits of nematicides in crop production have been published in *Plant Disease Reporter, Plant Disease, Journal of Nematology, Phytopathology,* and *Fungicide and Nematicide Tests.* With this volume of literature available, no further consideration will be given the annual application of nematicides for nematode control in crop production.

INTENSIVE CROPPING SYSTEMS

Millet (*Pennisetum americanum* [L.] Leek), milo (*Sorghum vulgare* Pers.), soybean (*Glycine max* [L.] Merr. 'Bragg'), crotalaria (*Crotalaria spectabilis* Roth.), and pigeon pea (*Cajanus cajan* [L.] Millsp. 'Norman') were used in conjunction with clean fallow and a nematicide (fensulfothion, *O,O*-diethyl *O*-(*p*-(methylsulfinyl)phenyl) phosphorothioate) for managing root-knot nematode populations in the production of tomato transplants (14). Clean fallow was the most effective treatment in suppressing nematode numbers. After two years in tomato, root-knot nematode populations increased to damaging levels, and fallow was no longer effective for complete control (i.e. no evidence of nematode infection) even in conjunction with fensulfothion. The rank of crops and cultural practice from the least effective to the most effective for control of *M. incognita* was milo (*Sorghum* sp.), soybean, crotalaria, millet, pigeon pea, and fallow. Although root-knot nematode control was nearly 99% in fallowed plots treated with fensulfothion annually for four years, this degree of control did not meet the standards of Georgia certification regulations which state that there be no evidence of nematode infection. In other cropping systems, the integrated use of fensulfothion and cultural practices such as clean fallow may be beneficial where control of root-knot nematodes need not be complete.

In a four-year study (1971-1974), various combinations of eight crop plants—turnip (*Brassica rapa* L.), corn, snap bean (*Phaseolus vulgaris* L.), peanut, soybean, cabbage, cucumber, and southern pea (*Vigna sinensis* [Torner] Savi)—were planted in six annual intensive cropping systems in soils naturally infested with root-knot nematodes (ca. 90% *M. incognita* and 10% *M. hapla* Chitwood) and other nematodes (15). The cropping systems were: Turnip-corn-snap bean (T-C-SNB), turnip-peanut-snap bean (T-P-SNB), turnip-corn-turnip (T-C-T), turnip-peanut-turnip (T-P-T), snap bean-soybean-cabbage (SNB-SB-CB), and turnip-cucumber-southern pea-turnip (T-CU-SP-T). A nematicide (ethoprop 10G) was broadcast at 9 kg a.i./ha and incorporated immediately into the top 15 cm of soil with a tractor-mounted rototiller once each spring just before planting corn, peanut, snap bean, and cucumber.

Root-gall indices of crops in the T-C-SNB, T-P-SNB, T-C-T, T-P-T, and SNB-SB-CB systems were not significantly affected by ethoprop. Root-gall indices of cucumber in the T-CU-SP-T system were lower each year in ethoprop-treated plots than in untreated plots (Table 6). Root-gall indices of southern pea following cucumber in plots treated with ethoprop were lower than those in untreated plots in 1971 and 1972, but not in 1973 and 1974.

Application of ethoprop before planting corn, peanut, and snap bean did not increase yields, but yields of cucumber were consistently increased 85% (range 50-118%) (Table 7). More details of this study are presented in the chapter "Specific crop rotation effects combined with cultural practices and nematicides."

The effects of annual applications of methyl bromide-chloropicrin (MBR-CP), DD-MENCS, and nonvolatile nematicides (ethoprop and fenamiphos) were studied in an integrated

Table 6. Effect of ethoprop on root-gall indices for a turnip-cucumber-southern pea-turnip cropping system, 1971-74.[a]

Nematicide treatment[b]	Root-gall Indices[c] Turnip	Cucumber	Southern pea	Turnip	Turnip	Cucumber	Southern pea	Turnip
	1971				1972			
No	1.02	2.01 a	2.41 a	1.09	1.00	2.92 a	2.30 a	2.32
Yes	1.02	1.15 b	1.57 b	1.05	1.00	1.39 b	1.41 b	1.80
	1973				1974			
No	1.35 a	3.02 a	2.50	1.11	1.00	2.28 a	2.51	1.12
Yes	1.09 b	1.76 b	2.19	1.21	1.00	1.33 b	2.37	1.09

[a] Root-gall index on 1-5 scale: 1 = no galling, 2 = 1-25%, 3 = 26-50%, 4 = 51-75%, and 5 = 76-100% roots galled.
[b] Ethoprop 10G was applied (8.96 kg a.i./ha) and incorporated into the soil 15 cm deep with a tractor-powered rototiller just before planting cucumber each year. Ethoprop was not applied before planting other crops.
[c] Within each column, values followed by different letters are significant ($P = <0.05$) according to Duncan's multiple-range test. Absence of letters indicates that values in the column are not significantly different.

Table 7. Effects of ethoprop on yield of crops in a turnip-cucumber-southern pea-turnip cropping system, 1971-74.

Nematicide treatment[a]	Yield (kg/ha)[b] Turnip	Cucumber	Southern pea	Turnip	Turnip	Cucumber	Southern pea	Turnip
	1971				1972			
No	16007	6024 b	4170	17168	12358	3519 b	1120	15795
Yes	16458	9035 a	3892	18291	13147	7679 a	1410	22121
	1973				1974			
No	8483	5024 b	1707	9331	9074	4502 b	842	7531
Yes	7028	10123 a	1901	7875	8723	8390 a	814	9147

[a] Ethoprop 10G was applied (8.96 kg a.i./ha) and incorporated into the soil 15 cm deep with a tractor-powered rototiller just before planting cucumber. Ethoprop was not applied before planting other crops.
[b] Within each column, values followed by different letters are significant ($P = < 0.05$) according to Duncan's multiple-range test. Absence of letters indicates that values in the column are not significantly different.

pest management program under irrigation with three intensive multi-cropping systems. The cropping systems were: turnip, peanut, cucumber, turnip, cucumber, soybean (T-P-CU-T-CU-SB); sweet corn, soybean, wheat (*Triticum aestivum* L.), soybean, spinach (*Spinacia oleracea* L.) (SC-SB-W-SB-S); and turnip, field corn, and southern pea (T-FC-SP). The first two systems were two-year cycles and the last, a one-year cycle.

Numbers of root-knot nematode juveniles were lowest in the SC-SB-W-SB-S cropping system, intermediate in the T-P-CU-T-CU-SB system, and highest in the T-FC-SP system (Figs. 1,2,3). Generally, numbers of root-knot nematodes in plots treated with MBR-CP and DD-MENCS were near or below detectable levels on the first two crops, but increased on a third susceptible crop. Fenamiphos was more effective than ethoprop in suppressing numbers of root-knot nematodes. Average yield increases across all crops in all cropping systems from plots treated with MBR-CP, DD-MENCS, and ethoprop/fenamiphos were 47%, 14%, and 1%, respectively (Table 8).

Economic data presented in Table 9 indicate that treatments with MBR-CP and DD-MENCS were not economically feasible. The ethoprop treatment, though not feasible on the T-P-CU-T-CU-SB system, was successful on the SC-SB-W-SB-S and T-FC-SP cropping systems. More details of this study are presented in Chapter 25.

NEMATICIDE APPLICATION THROUGH SPRINKLER IRRIGATION

In the southeastern Coastal Plain where nematodes cause severe damage to many crops,

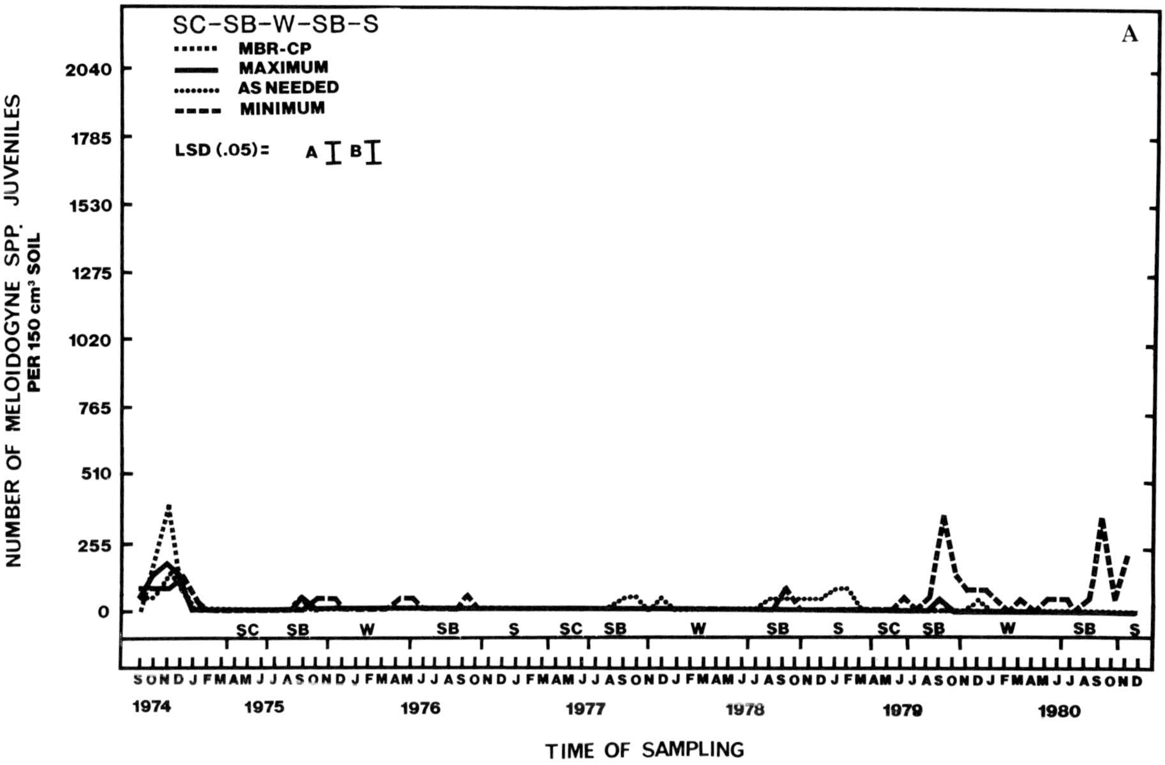

Fig. 1. Effect of four pest management levels on population densities of *Meloidogyne* spp. in three intensive cropping systems: A) SC-SB-W-SB-S = sweetcorn-soybean-wheat-soybean-spinach; B) T-P-CU-T-CU-SB = turnip-peanut-cucumber-(fallow)-turnip-cucumber-soybean-(fallow); C) T-FC-SP = turnip-field corn-southern pea-(fallow). A = L.S.D. 0.05 for different pest management levels on the same sampling date; B = L.S.D. 0.05 for the same pest management level on different sampling dates. MBR-CP = 98% methyl bromide + 2% chloropicrin, 392 kg/ha injected 25 cm deep into the soil with chisels 20 cm apart, applied in November or December each year. Maximum = DD-MENCS, 376.3 kg/ha applied in November or December each year plus ethoprop, 8.96 kg a.i./ha applied by spreading granules on the soil surface and incorporating 15 cm deep with a tractor-driven rototiller before planting each crop in 1975 and 1976; after 1976, the application of ethoprop was discontinued. As Needed = ethoprop was applied as described above before planting each crop from 1975-1978 when numbers of root-knot juveniles in the soil exceeded 25/150 cm^3 soil on the preceding crop and fenamiphos (8.96 kg a.i./ha) was used in a similar manner in 1979 and 1980. Minimum = control, nematicides were not used.

(B and C on following page)

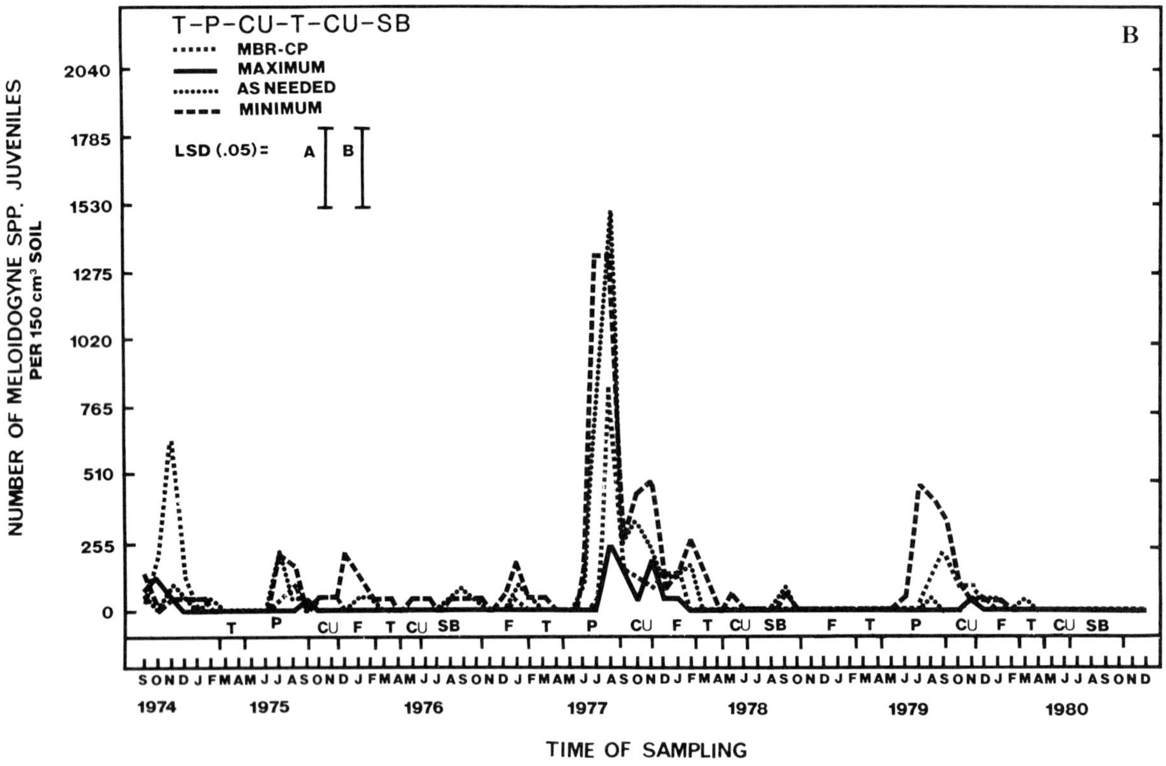
T-P-CU-T-CU-SB
B
MBR-CP
MAXIMUM
AS NEEDED
MINIMUM
LSD (.05)=
A
B
NUMBER OF MELOIDOGYNE SPP. JUVENILES
PER 150 cm³ SOIL
2040
1785
1530
1275
1020
765
510
255
0
T P CU F T CU SB F T P CU F T CU SB F T P CU F T CU SB
S O N D J F M A M J J A S O N D J F M A M J J A S O N D J F M A M J J A S O N D J F M A M J J A S O N D J F M A M J J A S O N D J F M A M J J A S O N D
1974
1975
1976
1977
1978
1979
1980
TIME OF SAMPLING

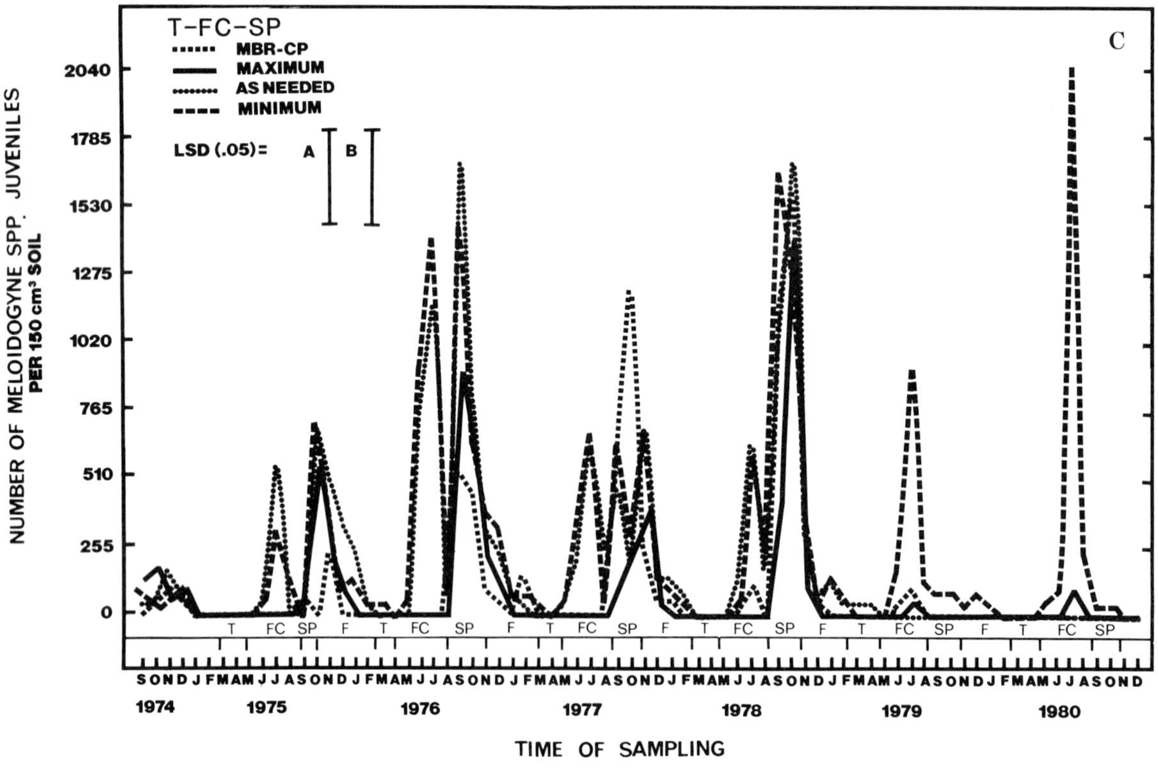
T-FC-SP
C
MBR-CP
MAXIMUM
AS NEEDED
MINIMUM
LSD (.05)=
A
B
NUMBER OF MELOIDOGYNE SPP. JUVENILES
PER 150 cm³ SOIL
2040
1785
1530
1275
1020
765
510
255
0
T FC SP F T FC SP F T FC SP F T FC SP F T FC SP F T FC SP
S O N D J F M A M J J A S O N D J F M A M J J A S O N D J F M A M J J A S O N D J F M A M J J A S O N D J F M A M J J A S O N D J F M A M J J A S O N D
1974
1975
1976
1977
1978
1979
1980
TIME OF SAMPLING

Table 8. Effect of three intensive cropping systems and nematicides on yield of crops averaged across each system, 1975-1980.

Pest management level[a]	CS1[b] Turnip (T/A)	Peanut (lb/A)	Cucumber (lb/A)	Turnip (T/A)	Cucumber (lb/A)	Soybean (bu/A)	% increase over control
MBR-CP	8.5	5263	11904	7.2	29162	49.6	87
DD-MENCS	6.2	5365	5982	4.3	13413	44.4	21
Ethoprop/ Fenamiphos	6.6	3449	5939	5.0	8257	39.4	1
Control	5.0	3142	7236	5.8	7733	43.9	—
	CS2 Sweet corn (T/A)	**Soybean (bu/A)**	**Wheat (bu/A)**	**Soybean (bu/A)**	**Spinach (T/A)**	**% increase over control**	
MBR-CP	9.3	49.6	24	50.5	5.4	15	
DD-MENCS	9.3	36.4	23	46.4	5.2	4	
Ethoprop/ Fenamiphos	8.6	39.0	28	40.1	4.4	1	
Control	7.8	36.0	28	41.0	4.9	—	
	CS3 Turnip (T/A)	**Cabbage (bu/A)**	**Southern pea (lb/A shelled) 'Pinkeye Purple Hull'**	**Southern pea (lb/A shelled) 'Worthmore'**	**% increase over control**		
MBR-CP	8.3	200	1341	2188	29		
DD-MENCS	5.6	200	1336	2113	16		
Ethoprop/ Fenamiphos	5.0	187	963	2092	1		
Control	5.9	183	883	1938	—		
		total % yield increase due to PML					
PML		MBR-CP	47				
		DD-MENCS	14				
		Ethoprop/ Fenamiphos	1				

[a] Pest management level: MBR-CP = 98% methyl bromide + 2% chloropicrin, applied at 392 kg/ha annually in November or December; DD-MENCS = 20% methyl isothiocyanate + 80% chlorinated C_3 hydrocarbons, applied at 376.3 kg/ha annually; Ethoprop/Fenamiphos = ethoprop applied at 8.96 kg a.i./ha before planting each crop from 1975-1978 and fenamiphos applied at 8.96 kg a.i./ha before planting each crop in 1979 and 1980; and Control = no nematicide.

[b] CS1 = Cropping system 1, two-year cycle; CS2 = Cropping system 2, two-year cycle; and CS3 = Cropping system 3, one-year cycle.

almost all commercial vegetables and many agronomic crops are grown under overhead-sprinkler irrigation with a planned pest management program. Application of chemicals through sprinkler irrigation water is a promising method of pest management. The application of nematicides through sprinkler irrigation is still in its infancy. Only limited information is available on the response of crops treated at planting and efficacy of nematicides applied through sprinkler irrigation.

Tomato

Johnson (10) reported that fenamiphos and carbofuran increased growth and yield of direct-seeded tomatoes grown for transplants and decreased root-galling induced by root-knot nematodes. These results were obtained when the chemicals were applied at 11.2 kg a.i./ha through a sprinkler irrigation system with risers equipped with an eductor, needle valve, and siphon hose. Nematicides were premixed with 3.8 liters of water. The flow of nematicides from the container through the siphon hose into the riser was controlled with the needle valve to allow the nematicide-water mixture to be siphoned from the container as the sprinkler head (0.80-cm opening diam.) completed ten revolutions (approximately 1.24 cm water/ha). Immediately after nematicide application, all plots were irrigated with about 1.2 cm

Table 9. Mean returns and costs ($/acre) or crops and pest control inputs by pest management levels in three intensive cropping systems.[a]

Cropping system[b]	Pest management level	Returns: Gross	Returns: Net	Pesticide costs [c]
T-P-CU-T-CU-SB	MBR-CP [d]	2440.45	— 614.83	1514.86
(1975-76)	DD-MENCS	1480.10	— 838.97	931.96
	Ethoprop/Fenamiphos	1291.31	— 101.04	260.14
	Control	1636.47	333.02	150.78
SC-SB-W-SB-S	MBR-CP	1694.53	—1026.70	1499.65
(1975-76)	DD-MENCS	1750.12	— 284.84	767.50
	Ethoprop/Fenamiphos	1580.26	353.93	99.16
	Control	1491.00	382.56	48.03
T-FC-SP	MBR-CP	1014.18	— 352.92	823.16
(1975-79)	DD-MENCS	864.37	— 192.29	490.14
	Ethoprop/Fenamiphos	773.35	136.43	47.30
	Control	779.61	269.51	29.78

[a] All computations are based on representative annual input and output prices.
[b] T = turnip, P = peanut, CU = cucumber, SB = soybean, SC = sweet corn, W = wheat, S = spinach, FC = field corn, and SP = southern pea.
[c] Includes costs of scouting and pesticide application.
[d] Pest management level: MBR-CP = 98% methyl bromide + 2% chloropicrin, applied at 392 kg/ha annually in November or December; DD-MENCS = 20% methyl isothiocyanate + 80% chlorinated C_3 hydrocarbons, applied at 376.3 kg/ha annually; Ethoprop/Fenamiphos = ethoprop applied at 8.96 kg a.i./ha before planting each crop from 1975-1978 and fenamiphos applied at 8.96 kg a.i./ha before planting each crop in 1979 and 1980; and Control = no nematicide.

Table 10. Growth, yield, and root-gall indices of tomato transplants as influenced by application of nematicides through a sprinkler irrigation system.

Nematicide treatment[a]	Growth index[b]	Yield (1000's/ha)[c]: Marketable	Cull	Total	Root-gall index[d]
Control	3.75 b	1100 c	162 b	1262 b	2.14 a
Fenamiphos	5.00 a	1647 a	306 b	1953 a	1.18 b
Ethoprop	2.00 c	365 d	772 a	1137 b	1.18 b
Carbofuran	5.00 a	1459 b	216 b	1675 a	1.42 b

[a] Nematicides applied at 11.2 kg a.i./ha.
[b] 1-5 scale: 1 = poor and 5 = excellent growth.
[c] Small letters indicate Duncan's multiple range groupings of treatments that do not differ significantly at the 5% level of probability.
[d] 1-5 scale: 1 = no galls, 2 = 1-25%, 3 = 26-50%, 4 = 51-75%, and 5 = 76-100% of roots galled.

water/ha. Approximately 0.3 cm of water was applied daily for 15 days to improve seedling emergence. Ethoprop, at the same dosage applied in a similar manner, was phytotoxic, but reduced the number of root galls when compared with untreated controls (Table 10). None of the nematicides reduced root-knot nematodes *M. incognita* below detectable levels, which is necessary in tomato transplant production. The results of this application concept were encouraging and warranted further study.

Cucumber

The principal use of nematicides is for control of nematode populations in soil before planting annual crops (39). Preliminary evaluations of fenamiphos (6.7 kg a.i./ha) applied to cucumber at the blossom stage (one month after planting) through a center pivot irrigation system with 8 mm of water was not phytotoxic. When applied this way, it appeared to inhibit plant damage from root-knot nematodes, even though yields were reduced from nematode infection prior to the application (Johnson, unpublished data).

Snap Bean

Smittle and Johnson (36) reported that fenamiphos applied at 6.7 kg a.i./ha through a solid set or a center pivot irrigation system with

Table 11. Effect of tillage methods and a nematicide on root-gall index, pod yield, and crop value of 'Eagle' snap bean in the spring of 1980[a].

Treatment	Root-gall index[b]	Pod yield (MT/ha)	Pod yield (g/plant)	Crop value ($/ha)
Tillage				
Moldboard plow	2.1 a	12.8 a	110 a	1789 a
Subsoil-plant	2.0 a	13.0 a	97 a	1837 a
Nematicide				
Control	3.1 a	10.5 b	82 b	1481 b
Fenamiphos	1.0 b	15.2 a	124 a	2145 a

[a] Means within columns followed by the same letter are not significantly different ($P = 0.05$) according to Duncan's multiple range test. No letters indicates no significant difference.
[b] 1-5 scale: 1 = no galls, 2 = 1-25%, 3 = 26-50%, 4 = 51-75%, and 5 = 76-100% of roots galled.

Table 12. Effect of tillage methods and a nematicide on root-gall index, pod yield, and crop value of 'Eagle' snap bean in the fall of 1980.[a]

Treatment	Number *M. incognita* per 150 cm³ soil[b]	Root-gall index 11 wk after planting[c]	Pod yield MT/ha	Pod yield g/plant	Crop value ($/ha)
Tillage					
Disk	124 a	3.0 a	1.4 a	37 a	200 a
Subsoil-bed	75 a	3.1 a	3.2 a	48 a	426 a
Nematicide					
Control	134 a	4.5 a	1.2 b	36 b	157 b
Fenamiphos	65 b	1.7 b	3.5 a	49 a	468 a

[a] Means within columns followed by the same letter are not significantly different according to Duncan's multiple-range test, ($P =$ 0.05). No letters indicate no significant difference.
[b] Samples collected 16 July before incorporation of corn residue.
[c] 1-5 scale: 1 = no galls, 2 = 1-25%, 3 = 26-50%, 4 = 51-75%, and 5 = 76-100% of roots galled.

2.8 cm of water/ha effectively controlled *M. incognita* and significantly increased snap bean yields regardless of growing season, tillage method, or cover crop system. The value ($/ha) of the spring crop treated with fenamiphos was $664 greater than the value of the untreated crop (Table 11). Although the value of the fall crop was lower than the value of the spring crop, the application of fenamiphos increased crop value by $311 over the untreated control (Table 12).

Squash, Corn, and Southern Pea

Johnson et al. (20) compared two methods of application of ethoprop and fenamiphos and one method of application of carbofuran at 6.7 kg a.i./ha on summer squash, corn, and southern pea for control of *M. incognita*: (i) granules of ethoprop and fenamiphos were applied over the soil surface and incorporated 15 cm deep with a tractor-drawn rototiller (designated System Ro); and (ii) emulsified concentrate formulations of ethoprop, fenamiphos, and carbofuran (2,3-dihydro-2,2-dimethyl-7-benzofuranyl methyl carbamate) were injected through irrigation system with 123,000 liters water/ha (designated System Ir). Root-gall indices of all crops recorded 30 days after planting were lower in treated plots than in untreated plots (Table 13). Root-gall indices of squash and corn were lower in plots treated with fenamiphos than in plots treated with ethoprop. Root-gall indices of squash were lower in Ro plots than in Ir plots. However, there were no significant differences in nematode population densities and root-gall indices in plots treated with carbofuran (Ir) vs. fenamiphos (Ir) or in the fenamiphos vs. ethoprop and Ir vs. Ro interactions (Table 13).

Yield of squash was 196% greater from treated plots than from untreated plots (Table 14). Yield of southern pea was not affected by treatments compared with untreated controls, but yield from plots treated with fenamiphos

Table 13. Root-gall indices on squash, corn, and southern pea as influenced by nematicides and method of application 30 days after planting.

Treatment and method of application[a]	Root-gall index[b] Squash		Corn		Southern pea	
Control (CK)	3.95		2.18		1.33	
Fenamiphos (P)/Ir	1.70		1.03		1.05	
Fenamiphos (P)/Ro	1.10		1.05		1.03	
Ethoprop (E)/Ir	2.75		1.48		1.05	
Ethoprop (E)/Ro	1.78		1.15		1.00	
Carbofuran (C)/Ir	2.00		1.10		1.05	
Comparisons:						
Ck vs other treatments	3.95	1.87	2.18	1.16	1.33	1.04
C(Ir) vs. P(Ir)	2.00	1.70	1.10	1.03	1.05	1.05
P(Ir + Ro) vs E(Ir + Ro)	1.30	2.27	1.04	1.32	1.04	1.03
Ir(P + E) vs Ro(P + E)	2.23	1.44	1.25	1.10	1.05	1.01
Interactions:						
P vs E and Ir vs Ro	1.74	1.93	1.09	1.26	1.03	1.04

[a] Nematicides applied at 6.7 kg a.i./ha. Ir = injected via sprinkler irrigation system; Ro = granules incorporated 15-cm deep with tractor-powered rototiller.
[b] 1-5 scale: 1 = no galls, 2 = 1-25%, 3 = 26-50%, 4 = 51-75%, and 5 = 76-100% of roots galled. Means underlined by continous line are not significantly different.

Table 14. Yield of squash, southern pea and corn as affected by nematicides and method of application.

Treatment and method of application[a]	Yield (kg/ha)[b] Squash		Southern pea		Corn (grain)	
Control (CK)	2659		1317		2372	
Fenamiphos (P)/Ir	8054		1433		3306	
Fenamiphos (P)/Ro	9204		1189		3959	
Ethoprop (E)/Ir	7884		1067		3405	
Ethoprop (E)/Ro	6403		1382		3603	
Carbofuran (C)/Ir	7810		901		3965	
Comparisons:						
Ck vs other treatments	2659	7871	1317	1194	2372	3648
C(Ir) vs. P(Ir)	7180	8054	901	1433	3965	3306
P(Ir + Ro) vs E(Ir + Ro)	8629	7144	1311	1225	3633	3504
Ir(P + E) vs Ro(P + E)	7969	7804	1250	1286	3356	3781
Interactions:						
P vs E and Ir vs Ro	7229	8544	1408	1128	3455	3682

[a] Nematicides applied at 6.7 kg a.i./ha. Ir = injected via sprinkler irrigation system; Ro = granules incorporated 15-cm deep with tractor-powered rototiller.
[b] Means underlined by continous line are not significantly different.

Table 15. Effect of fenamiphos (6.7 kg a.i./ha) on root-gall indices and yield of squash, 1981.[a]

Treatment[b]	Water (cm)[c]	Root-gall index[d] May 25		July 13		Total mkt. yield kg/ha	
		AP[e]	PP[e]	AP	PP	AP	PP
Control	0.64	4.90 a	4.80 a	5.00 a	5.00 a	1616 c	874 b
PI	0.25	1.33 b	3.45 a	2.95 c	4.90 a	7938 a	5062 a
PI	0.64	1.43 b	4.20 a	3.48 b	4.75 a	7044 ab	4655 a
PI	1.27	1.50 b	4.50 a	2.85 c	4.80 a	5624 bc	5590 a
PI	1.91	1.33 b	4.63 a	3.10 bc	4.85 a	6871 ab	6088 a
PG	0.64	1.33 b	4.28 a	3.30 bc	5.00 a	9574 a	5001 a
	Mean	1.97 z	4.31 y	3.45 z	4.88 y	6445 y	4545 z

[a] Data in columns followed by the same letter and data underscored by the same line in rows are not significantly (P = 0.05) different according to the Waller-Duncan K-ratio t-test.
[b] PI = fenamiphos applied through an irrigation simulator;
PG = fenamiphos granules spread on surface of soil and incorporated 15 cm deep with a tractor-powered rototiller AP, and granules spread on surface of the soil and not incorporated PP.
[c] Applied through an irrigation simulator.
[d] 1-5 scale: 1 = no galls, 2 = 1-25%, 3 = 26-50%, 4 = 51-75%, and 5 = 76-100% of roots galled.
[e] AP = Immediately after planting and PP = 2 weeks after planting.

(Ir) was 59% greater than yield from plots treated with carbofuran (Ir). Yield of corn (grain) was 54% greater from treated plots than from untreated plots. Yields of all crops were unaffected by method of nematicide application.

Fenamiphos at 6.7 kg a.i./ha was applied through an irrigation simulator with 0.25, 0.64, 1.27, and 1.91 cm water (11, 12). Root-gall indices of plants from plots treated with fenamiphos immediately after planting (AP) were lower than those from untreated plots, but were not affected by different amounts of water in most plots (Table 15). Root-gall indices of plants from plots treated 2 weeks after planting (PP) were not different from controls on either sampling date. Across all treatments applied, root-gall index means from AP were lower than those from PP plots on both sampling dates.

Yields of squash from AP plots treated with fenamiphos in irrigation water were 325% (range 248%-391%) greater than yield from non-treated plots (Table 15). Yield from plants in plots treated with nematicide granules was 492% greater than yield from plants in non-treated plots, but was not different from most nematicide/water treatments. Yield from plants in PP plots treated with fenamiphos in irrigation water was 512% (range 433%-597%) greater than yield from plants in nontreated plots. The yield mean across all AP treatments was 42% greater than the mean across all PP treatments.

A similar test was established on corn, 'Pioneer X 304C,' following squash, except the PP treatments were applied one week after planting. Corn was planted on July 28, and the AP treatments were applied the next day. Root-gall indices of corn were lower (P = 0.05) in all treated plots than in controls following both AP and PP treatments on both sampling dates (Table 16). The mean root-gall indices across all treatments were lower from AP plots than those from PP plots.

Yields from plants in plots treated with fenamiphos AP and PP were 160% (range 128%-181%) and 147% (range 116%-181%) greater than yields from nontreated plots, respectively. The average yield from plants in plots treated with fenamiphos AP was 15% greater than that of plants from plots treated with the nematicide one week after planting.

Potato

The Columbia (*M. chitwoodi*) and northern (*M. hapla*) root-knot nematodes are a severe problem in potato production in the Pacific northwest because they reduce the quantity and quality of the potato tuber (34). Approximately 90% of the potato acreage in Washington is grown under overhead sprinkler irrigation systems with 60-70% of this acreage receiving some type of nematicide treatment to control root-knot nematodes. The nematicides most commonly used have been D-D and Telone II (1,3-dichloropropene) soil fumigants. However,

Table 16. Effect of fenamiphos (6.7 kg a.i./ha) on root-gall indices and yield of field corn, 1981.[a]

Treatment[b]	Water (cm)[b]	Root-gall index[c] Sept. 10 AP[d]	Sept. 10 PP[d]	Dec. 16 AP	Dec. 16 PP	Yield grain (kg/ha) AP	Yield grain (kg/ha) PP
Control	0.64	2.00 a	2.38 a	2.15 a	2.65 a	1182 c	1081 c
PI	0.25	1.05 b	1.70 b	1.13 b	1.18 b	3323 a	3033 a
PI	0.64	1.00 b	1.48 b	1.00 b	1.13 b	3006 a	2520 b
PI	1.27	1.00 b	1.43 b	1.13 b	1.60 b	2696 b	2333 b
PI	1.91	1.00 b	1.55 b	1.03 b	1.28 b	3221 a	2790 ab
	Mean	1.21 z	1.71 y	1.29 z	1.57 y	2686 z	2351 y

[a] Data in columns followed by the same letter and data underscored by the same line in rows are not significantly (P = 0.05) different according to the Waller-Duncan K-ratio t-test.
[b] PI (fenamiphos) and water both applied through an irrigation simulator.
[c] 1-5 scale: 1 = no galls, 2 = 1-25%, 3 = 26-50%, 4 = 51-75%, and 5 = 76-100% of roots galled.
[d] AP = 1 day after planting and PP = 1 week after planting.

Table 17. Control of *Meloidogyne hapla* on alfalfa with metham applied through a center pivot system.

Metham (liters/ha)	*M. hapla* juveniles/250 cc soil[a] Pre-treatment Oct. 78	Post-treatment Mar. 79	Post-treatment Apr. 79	Alfalfa roots[a] Apr. 79
Nontreated	440	42	280	95
374	390	0	0	0
468	410	0	0	0
701	420	0	0	0

[a] Values represent means of five samples, except for nontreated means, which consisted of 15 samples.

the use of metham has become more popular in recent years for controlling nematodes and other pests and diseases on potato since Gerstl et al. (4) demonstrated that metham was most effective in controlling *Verticillium dahliae* on potato when applied to soil in irrigation water.

Metham was applied October, 1978, for control of *M. hapla* on a 3-year-old alfalfa field by injection into the mainline of a center pivot irrigation system (33). 'Russet Burbank' potatoes were planted in April, 1979, following alfalfa (*Medicago sativa* L.). The results showed that metham effectively controlled *M. hapla* in soil and alfalfa roots (Table 17). Nematodes were not found in treated plots as compared with the nontreated controls. Tubers examined for nematode infection at late harvest (October 24) showed that *M. hapla* females were not detected in the 468 and 701 liter/ha treatments, whereas 19 and 69% of the tubers were infected in the 374 liter/ha treatment and control plots, respectively (Table 18).

Grapes

Several species of nematodes are of economic importance in grape (*Vitis californica* Benth.) production in Washington. Yield increases of 13% to 44% have been achieved the second year after application of aldicarb (2-methyl-2-(methylthio) propionaldehyde *O*-(methylcarbamoyl) oxime), carbofuran, and fenamiphos (35). Approximately 40% of the grapes in Washington are grown under overhead sprinkler irrigation. In 1981, oxamyl was applied via solidset overhead sprinkler system and evaluated for control of ring (*Criconemella xenoplax* Raski) and root-knot *M. hapla* nematodes. Results of this study indicate no significant (P= 0.05) differences in numbers of nematodes and yield among treatments (Table 19). These treatments were repeated in 1982, and yields from plants in plots treated with oxamyl (1.12 and 2.24 kg a.i./ha) were greater than those from untreated plots. Significant yield increases have been obtained on wine grapes in California with multiple applications of oxamyl by sprinkler irrigation the first year (35). Data from nonvolatile nematicide plots on Concord grapes in Washington indicate that yield increases do not occur until at least the second year after application.

Table 18. Effect of metham on *Meloidogyne hapla* infection of 'Russet Burbank' potato tubers, 1979.

Metham (liters/ha)	Percent tubers infected with *M. hapla*[a] October 24
Nontreated	69
374	19
468	0
701	0

[a] Values represent means of five samples, except for nontreated means, which consisted of 15 samples.

Table 19. Yields of Concord grapes as affected by oxamyl applied through solid-set overhead sprinklers, 1981-1983.

Treatments	Rate kg a.i./ha	Yield (MT/ha)[a] 1981	1982	1983
Nontreated	0	22.0	41.1	37.5
Oxamyl	0.56	20.8	45.9	37.7
Nontreated	0	19.7	40.6 b	39.2
Oxamyl	1.12	20.8	46.7 a	41.1
Nontreated	0	20.4	41.3 b	35.5
Oxamyl	2.24	26.0	52.3 a	39.4

[a] Within each column, values for nontreated vs oxamyl at different rates followed by different letters are significant ($P = <0.05$) according to Duncan's multiple-range test. Absence of letters indicates that values in the column are not significantly different.

DETRIMENTAL EFFECTS OF NEMATICIDES

Limited information exists on the long-term ecological effects of frequent or repeated use of nematicides on agricultural soils. A review by Good (5) includes certain detrimental effects of halogenated soil fumigants.

Several years ago DBCP was found to affect adversely the fertility of workers in a factory manufacturing and packaging DBCP (40). Subsequently, DBCP was found in a well on that pesticide manufacturer's property and also in both agricultural and domestic wells in the San Joaquin Valley in California where that chemical had been used extensively for control of nematodes on grape, peach, and citrus. As a result of these and other reports, the Environmental Protection Agency (EPA) suspended registration of DBCP in 1977.

More recently, the carbamate oxime insecticide/nematicide, aldicarb, was found in ground water on Long Island in New York, and in Wisconsin where it was used for Colorado potato beetle and/or golden nematode (*Globodera rostochiensis* Wollenweber) control. Several states have suspended use of aldicarb for control of nematodes on certain crops. In addition, EDB has been found in ground water and has been suspended for use as a soil fumigant since October 1983.

Rohde et al. (32) reported that following application of ethoprop at 8.96 kg a.i./ha, concentrations of the nematicide in the 0-15-cm soil layer were near 6 μg/g of soil immediately after application, but were < 1 μg/g of soil 5 days later on corn and southern pea and 30 days later on turnip. Johnson et al. (13) reported that following application of fenamiphos at 4.5, 6.7, 9.0, and 18.0 kg a.i./ha via injection into a sprinkler irrigation system, concentration of the nematicide was < 1 μg/g of soil in the top 0-15 cm soil layer.

RESEARCH NEEDS ON NEMATICIDES

Much additional information is needed on nematicides in order to improve their effectiveness, reduce the amount of chemical applied, and minimize potential environmental hazards in soil, ground water, and crop residues. Improvements can be made in application equipment, in the timing of application in relation to dosage needed, and time of planting crops especially in relation to soil and weather conditions that affect nematode populations. The status of sequential or simultaneous application of nematicides with other pesticides is currently being studied so the interactions among pesticides, on phytotoxicity to crops, and on soil and plant residues can be determined. Additional information is needed on factors governing the movement and persistence of nematicides that can be applied as therapeutants to established crops, especially perennials, after nematode damage becomes apparent. Development of such chemicals will reduce the amount of nematicides applied to soil.

The greatest need in the area of nematicidal research is the development of improved methods of forecasting potential nematode damage so that use of nematicides will be recommended only when nematode populations are predicted to reach economic threshold levels.

For the foreseeable future nematicides will play a major role in nematode population management for the production of a number of food and fiber crops and for use in regulatory and quarantine programs. Increased capitalization and specialization of farms dictate an ever-increasing shift toward mono-cropping and/or intensive cropping systems that invariably leads to increased crop losses by nematodes. Therefore, use of nematicides is a necessary adjunct to continuous cropping of land, or to a greatly narrowed crop rotation base (5). In such a system, more highly effective nematicides must be developed that will control nematodes with lower amounts of active ingredient per unit of land.

Increased reliance on nematicides will be necessary for the production of many crops in the United States, unless integrated control systems are devised. Knowledge of the separate components of an integrated control system is sufficiently advanced that the establishment of integrated control systems for plant-parasitic nematodes can begin. Managing nematode populations by integrated systems may employ all or parts of the following components: production of nematode-free seed and propagating materials; destruction of nematode-infected plant materials immediately after harvest; use of clean fallow and flooding where appropriate; use of resistant and susceptible crop cultivars in rotations; use of nonhost and trap crops; and additions of biodegradable organic matter to soil to encourage biological control where practical. High level control can be obtained for many species of nematodes using the above nonchemical means; then nematicides need only be used to supplement the above control practices. These are intriguing challenges to modern scientific agriculture.

Literature Cited

1. Carter, W. 1943. A promising new soil amendment and disinfectant. Science 97:383-384.
2. Chalfant, R. B., C. A. Jaworski, A. W. Johnson, and D. R. Sumner. 1977. Reflective film mulches, millet barriers, and pesticides: Effects on watermelon mosaic virus, insects, nematodes, soil-borne fungi, and yield of summer squash. J. Am. Soc. Hortic. Sci. 102:11-15.
3. Christie, J. R. 1959. Plant nematodes, their bionomics and control. Fla. Agric. Exp. Stn., Gainesville, Fla. 256 pp.
4. Gerstl, Z., U. Mingelgrin, J. Krikun, and B. Yaron. 1977. Behavior and effectiveness of Vapam applied to soil in irrigation water, pp. 42-50 *In* Proc. Israel-France Symp. on Behavior of Pestic. Soil, 1975, ed. M. Horowitz. Agric. Res. Organ. Spec. Publ. (Bet-Dagan) 82. 98 pp.
5. Good, J. M. 1972. Bionomics and integrated control of plant parasitic nematodes. J. Environ. Qual. 1:382-386.
6. Good, J. M. 1972. Management of plant parasitic nematode populations, pp. 109-127 *In* Tall Timbers Conf. on Ecol. Anim. Control by Habitat Management No. 4. Tallahassee, Fla., February 24-25.
7. Hollis, John P. 1958. Specifications for ideal nematocides. Plant Dis. Rep. 42:291-307.
8. Jaworski, C. A., A. W. Johnson, R. B. Chalfant, and D. R. Sumner. 1974. A system approach for production of high value vegetables on southeastern Coastal Plain soils. Ga. Agric. Res. 16(2):12-15.
9. Jenkins, W. R. 1964. A rapid centrifugal-flotation technique for separating nematodes from the soil. Plant Dis. Rep. 43:692.
10. Johnson, A. W. 1978. Effect of nematicides applied through overhead irrigation on control of root-knot nematodes on tomato transplants. Plant Dis. Rep. 62:48-51.
11. Johnson, A. W. 1981. Application of nematicides through an overhead sprinkler irrigation system for control of nematodes on crops, pp. 89-95 *In* Proc. of the First Nat. Symp. on Chemigation, Tifton, Ga., 1981.
12. Johnson, A. W. 1982. Application of nematicides through an overhead sprinkler irrigation system for control of nematodes on crops, pp. 69-73 *In* Proc. of the Second Nat. Symp. on Chemigation, Tifton, Ga., 1982.
13. Johnson, A. W. 1982. Soil distribution of phenamiphos applied by overhead sprinkler irrigation to control *Meloidogyne incognita* on vegetables. Plant Dis. 66:489-491.
14. Johnson, A. W., and G. M. Campbell. 1980. Managing nematode population densities on tomato transplants using crop rotation and a nematicide. J. Nematol. 12:6-19.
15. Johnson, A. W., C. C. Dowler, N. C. Glaze, and D. R. Sumner. 1983. Effects of intensive cropping systems and pesticides on nematodes and crop yields. Agric. Res. Results, USDA-ARS. ARR-S-14/May. 36 pp.
16. Johnson, A. W., C. A. Jaworski, D. R. Sumner, and R. B. Chalfant. 1979. Effects of film mulch, trickle irrigation, and soil pesticides on nematodes and yield of polebean. Plant Dis. Rep. 63:360-364.
17. Johnson, A. W., D. R. Sumner, and C. A.

Jaworski. 1979. Effect of film mulch, trickle irrigation, and DD-MENCS on nematodes, fungi, and vegetable yields in a multicrop production system. Phytopathology 69:1172-1175.
18. Johnson, A. W., D. R. Sumner, and C. A. Jaworski. 1979. Effects of management practices on nematode and fungus populations and cucumber yield. J. Nematol. 11:84-93.
19. Johnson, A. W., D. R. Sumner, C. A. Jaworski, and R. B. Chalfant. 1977. Effects of management practices on nematode and fungi populations and okra yield. J. Nematol. 9:136-142.
20. Johnson, A. W., J. R. Young, and B. G. Mullinix. 1981. Applying nematicides through an overhead sprinkler irrigation system for control of nematodes. J. Nematol. 13:154-159.
21. Kays, S. J., A. W. Johnson, and C. A. Jaworski. 1976. Multiple cropping with trickle irrigation. HortScience 11:135-136.
22. McCarter, S. M., C. A. Jaworski, and A. W. Johnson. 1978. Effect of continuous plant culture and soil fumigation on soilborne plant pathogens and on growth of tomato transplants. Phytopathology 68:1475-1481.
23. Overman, A. J. 1974. Nematicides in linear drip irrigation for full-bed mulch. Soil Crop Sci. Soc. Fla. Proc. 34:197-200.
24. Overman. A. J. 1975. Efficacy of nematicides applied through micropore tape. Nematropica 5:1. (Abstr.).
25. Overman, A. J. 1976. Efficacy of soil fumigants applied via a drip irrigation system. Proc. Fla. State Hortic. Soc. 89:143-145.
26. Overman, A. J. 1977. Crop response to nematicides and drip irrigation on sandy soil, pp. 172-179 *In* Proc. 7th Int. Agric. Plastics Congr.
27. Overman. A. J. 1978. Crop response to SMDC applied through a drip irrigation system, pp. 128-136 *In* Proc. 14th Nat. Agric. Plastics Congr.
28. Overman, A. J. 1980. Management of nematodes in drip-irrigated row crop, pp. 47-48 *In* Eur. Soc. Nematol XVth Int. Nematol. Symp. (Abstr.).
29. Overman, A. J. 1982. Soil fumigation via drip irrigation under full-bed mulch culture for row crops. Soil Crop Sci. Soc. Fla. Proc. 41:153-155.
30. Overman, A. J. 1984. Drip irrigation as a vehicle for nematicides. Nematropica 14:92-96. (Abstr.).
31. Overman, A. J. and J. F. Price. 1983. Application of pesticides via drip irrigation to control nematodes and foliar arthropods. Soil Crop Sci. Soc. Fla. Proc. 42:92-96.
32. Rohde, W. A., A. W. Johnson, C. C. Dowler, and N. C. Glaze. 1980. Influence of climate and cropping patterns on the efficacy of ethoprop, methyl bromide, and DD-MENCS for control of root-knot nematodes. J. Nematol. 12:33-39.
33. Santo, G. S. 1982. Nemagation on Irish potato and grapes, pp. 64-67 *In* Proc. of the Second Nat. Symp. on Chemigation, Tifton, GA, 1982.
34. Santo, G. S. and J. H. O'Bannon. 1981. Effect of soil temperature on the pathogenicity and reproduction of *Meloidogyne chitwoodi* and *M. hapla* on Russet Burbank potato. J. Nematol. 13:483-486.
35. Santo, G. S. and R. P. Ponti. 1981. Systemic nematicides increase yields of Concord grapes. Proc. Wash. Grape Soc., 1980, 10:56-57.
36. Smittle, D. A., and A. W. Johnson. 1982. Effects of management practices on *Meloidogyne incognita* and snap bean yield. J. Nematol. 14:63-68.
37. Sumner, D. R., A. W. Johnson, C. A. Jaworski, and R. B. Chalfant. 1978. Influence of film mulches and soil pesticides on root diseases and populations of soilborne fungi in vegetables. Plant and Soil 49:267-283.
38. Taylor, A. L. 1967. Principles of measurement of crop losses: nematodes, pp. 225-237 *In* Proc. FAO Symp. on Crop Losses, Rome, Italy, October 2-6.
39. Taylor, A. L. and J. N. Sasser. 1978. Biology, identification and control of root-knot nematodes (*Meloidogyne* species). Coop. Publ. Dep. Plant Pathol., North Carolina State Univ., and U.S. Agency Int. Dev. Raleigh, N. C. 111 pp.
40. Thomason, I. J. 1982. Conference directions and objectives, pp. 5-7 *In* Proc. of a Conf. - Pesticides in Soil and Groundwater, Davis, Calif. 1982.

MODE OF ACTION OF NEMATICIDES[1]

Harvey W. Spurr, Jr.[2]
Oxford Tobacco Research Laboratory
Oxford, North Carolina, USA

Nematicides are taken for granted: that is, they are accepted and used, often without enthusiasm, and not credited for the large measure of nematode control they provide. Nematologists are primarily biologists; thus, little research is devoted to studies of nematicides and their mode of action. This presentation is a brief introduction to nematicides and their mode of action. Perhaps this knowledge will stimulate an interest in the major method of nematode control. Also, it is important that we know something about the products on which U.S. growers spend more than $100 million per year. This awareness could assist the future discovery and development of new nematicides and other chemicals that control nematodes.

Nematicides by definition are toxic to nematodes. Death results from contact with a lethal dose or concentration of nematicide. The application of nematistatic chemicals may decrease crop losses also. In this presentation, nematicidal mode of action refers to the lethal effect resulting from an interaction between a chemical applied for control and a specific site within a nematode. Several reviews are available (2,3,4,10,15,19,26,28). A recent text, *Introduction to Biochemical Toxicology* edited by Hodgson and Guthrie (7), has an excellent discussion of insecticide toxicology and mode of action that is pertinent to nematicidal mode of action (12,13).

[1] Cooperative investigations of Agricultural Research Service, U. S. Department of Agriculture, and North Carolina State University, Department of Plant Pathology, Raleigh. Use of trade names does not imply its endorsement by the USDA or the North Carolina Agricultural Research Service of the products named or similar ones not mentioned.

[2] Research Plant Pathologist, Southern Region, ARS, USDA, Oxford Tobacco Research Laboratory, Oxford, NC 27565 also, Professor of Plant Pathology, North Carolina State University, Raleigh 27695-7616.

FUMIGANT AND NONFUMIGANT NEMATICIDES

Nematicides for the control of plant-parasitic nematodes are usually applied to soil. A few nematicides are applied to plants as foliar sprays. Most plant-parasitic nematodes are soil inhabitants during a part of their life. Since some nematicides are systemic, soil applications can serve as a means for control of certain foliar nematodes also. Nematicides can be classified according to chemical, physical or biological properties; however, a two category classification system has developed over time with reference to nematicide movement in soil following application. Thus, nematicides are usually labeled as fumigants or nonfumigants (Table 1). Developing a knowledge of this classification is useful in understanding nematicidal mode of action.

Widespread use of fumigants started about 1950. Fumigant nematicides are usually formulated as liquids which vaporize and move through air spaces in soil after vaporizing. They tend to move deep in soil, have a broad spectrum of biocidal activity (may kill fungi and seeds), and are applied at high rates. Since they are also phytotoxic, except for 1,2-dibromo-3-choropropane (DBCP), a waiting period of two-three weeks is required between application and planting. Fumigants are organic halides or organic sulfurs and may be referred to as alkyl halides.

Nonfumigant and systemic nematicides were introduced commercially in 1970. These nematicides are often formulated on granules but some are available as liquids for spraying on soil or foliage. Compared to fumigants, nonfumigants move by percolation in soil water, do not move deeply in soil, tend to have a narrower spectrum of biocidal activity, are nonphytotoxic at suggested use rates, and are active at lower dosages. Many commonly used nonfumi-

Table 1. Common nematicides and probable modes of action.

Nematicide	Mode of action
Fumigants	
Water: (required to activate toxicity in soil)	
MIT (methyl isothiocyanate) metham (Vapam) dazomet (Mylone)	MIT may react with amino acids, oxidases and nucleophilic sites on proteins; metham and dazomet form MIT
Air: (required for movement in soil)	
1,3-D (1,3-dichloropropene) EDB (ethylene dibromide) DBCP (Nemagon) MB (methyl bromide) chloropicrin	alkyl halides may react with nucleophilic sites on proteins and oxidize iron porphyrins and hemeproteins
Combinations or multipurpose fumigants:	
Vorlex (15% MIT + 85% D-D) Terr-O-Cide(s) (EDB + chloropicrin) Telone C17 (dichloropropene + chloropicrin)	see above
Nonfumigants	
Phosphates:	
ethoprop (Mocap) fensulfothion (Dasanit) fenamiphos (Nemacur)	irreversible binding of acetylcholinesterase, esterase inhibition and various pharmacologic actions
Carbamates:	
carbofuran (Furadan) aldicarb (Temik) oxamyl (Vydate)	reversible binding of acetylcholinesterase, esterase inhibition and various pharmacologic actions

gants are organic phosphates or carbamates, and many are systemic in plants (Table 1).

In 1984, three fumigants [methyl bromide (MB); 1,3-dichloropropene (1,3-D); and methyl isothiocyanate (MIT)] and two nonfumigants (aldicarb and ethoprop) will account for about 90% of nematicides sold commercially. Once in the soil, nematicides are subjected to chemical, physical and biological processes simultaneously. These processes mitigate activity or, stated another way, influence efficacy. Nematicidal efficacy is also influenced by factors such as length of exposure, stage of nematode development, location of nematode to the chemical toxicant, temperature, moisture, and soil composition.

In order to kill or inactivate, a nematicide must come in contact with a nematode. Direct penetration of the cuticle or ingestion during feeding are two methods of contact. Available evidence indicates that fumigants such as ethylene dibromide (EDB) (14) and nonfumigants such as aldicarb (18,19) both are able to penetrate the nematode cuticle directly and reach an equilibrium at similar rates or within about 30 minutes. Direct penetration of the cuticle by nematicides appears to be the main route to vital sites.

NEMATICIDAL MODE OF ACTION

Mode of action will be covered according to the outline in Table 1. It is important to realize the difficulties encountered in doing research to prove that a specific biochemical interaction between a nematicide and a specific site or structure within a nematode is the direct cause of death or inactivation. In most instances, nematicides interact at numerous sites or may have a multiplicity of effects. The examples presented are based on results derived from the research of many investigators.

Fumigants

Metham and dazomet decompose in water to form MIT (Fig. 1) which directly penetrates the cuticle and reacts with amino acids, oxidases and nucleophilic sites on proteins (11). The exact mechanisms have not been rigorously demon-

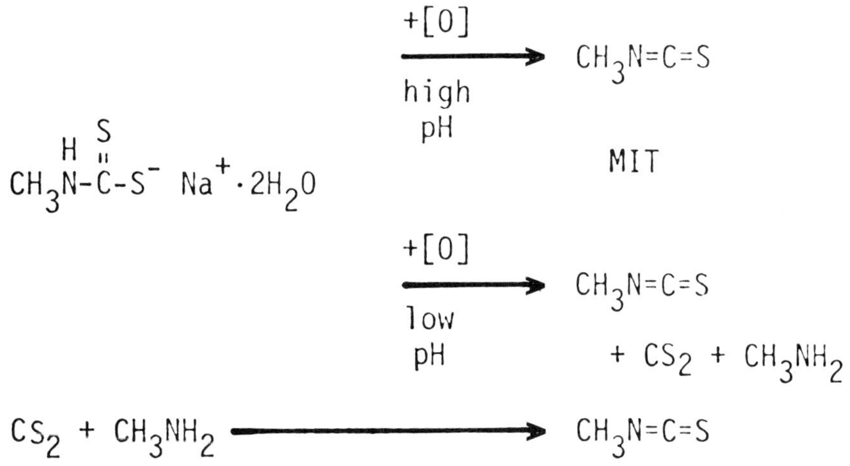

Fig. 1. Reaction of metham in water to form methyl isothiocyanate, methylamine and carbon disulfide.

strated. The breakdown and spread of MIT from metham and dazomet in warm, well-aerated soil is rapid.

Alkyl halides (MB; DBCP; 1,3-D) also directly penetrate the cuticle. Moje (16) observed a correlation between toxicity and relative activity of alkyl halides in biomolecular nucleophilic substitution reactions (NH_2, SH, OH) on protein surfaces. These reactions are considered relatively slow, and more recently, Wade and Castro (27) suggested that the rapid action of these toxicants could be explained by the oxidation of iron porphyrins and hemeproteins. Iron porphyrins are components of hemeprotein molecules and are found in terminal oxidases or mitochondrial oxidases. The oxidation of heme centers in the respiratory system is very rapid. Many enzymes or metabolic sites are inhibited simultaneously thus stopping several vital processes and leading to a rapid death. It was suggested in a review by Evans (4) that the toxicity of 1,3-D and related compounds may relate to their solvolysis in water.

Nonfumigants

Phosphate (e.g. ethoprop) and carbamate (e.g. aldicarb) nematicides directly penetrate nematodes and inhibit acetylcholinesterase (AChE), cholinesterase and other esterase enzymes. Hydrolysis of acetylcholine by acetylcholinesterase is a vital part of neurotransmission at cholinergic synapses in the nervous system (13). Cholinesterase and other esterases also function in various metabolic systems, the importance of which is less well understood. These nematicides are developed as a "spin-off" from the development of insecticides. The inhibition of AChE in insects has been extensively documented (10). Organophosphates irreversibly phosphorylate active sites on AChE. Carbamoylation of AChE tends to be more reversible.

The existence of AChE in plant-parasitic nematodes was first demonstrated by Rohde (21) using a site-specific acetylthiocholine substrate reaction that was inactivated by known cholinesterase inhibitors. The staining sites observed included the nerve ring and associated ganglia, amphids, phasmids and papillae on the lips and body surface. Nerve rings were stained more consistently by an improved staining technique following an ultrasonic pretreatment (29). A pretreatment with acetone for 12 hours prior to staining with the thiocholine direct-coloring method resulted in coloring throughout the body of *Panagrellus redivivus* with the anterior nerve ring most intensely stained (Alan Ayers, unpublished). *Criconemella ornata* did not respond to the acetone pretreatment in the same way, possibly due to its thicker cuticle.

Narcosis and Behavioral Effects

Narcotic and behavioral changes in plant-parasitic nematodes have been described for both fumigant and nonfumigant nematicides. These are the symptomatic changes observed and may be referred to also as pharmacologic actions of nematicides. An example is the effect of nematicides on the rate of nematode movement. Hyperactivity may be an initial reaction to contact with an alkyl halide and may be followed by decreased activity and paralysis. Similar changes in movement were observed in relation to contact with carbamates. Responses vary with chemical structure. Nelmes (17) studied the effect of aldicarb on behavior of *Globodera rostochiensis*. Aldicarb rapidly in-

Table 2. Toxicity of aldicarb, aldicarb metabolites, and related structures to plants, insects, mites, and nematodes.

						Toxicity Rating[b]						
Compound no.	Structure	Metabolites[a]				Insects[c]				Mite		Nema
		S	P	A	I	Ad	Aw	MB	Fy	A	E	
1	$CH_3SC(CH_3)_2CHNOC(O)NHCH_3$	x	x	x	x	5	5	3	5	5	5	5
2	$CH_3SOC(CH_3)_2CHNOC(O)NHCH_3$	x	x	x	x	5	5	5	5	5	1	4
3	$CH_3SO_2C(CH_3)_2CHNOC(O)NHCH_3$	x	x	x	x	1	1	5	5	5	3	4
4	$CH_3SO(CH_3)_2CHNOC(O)NH_2$					1	1	1	5	1	1	3
5	$CH_3SC(CH_3)_2CHNOC(O)NHCH_2OH$					3	1	1	5	5	1	1
6	$CH_3SO_2C(CH_3)_2CHNOC(O)NHCH_2OH$					1	1	1	3	1	1	1
7	$CH_3SC(CH_3)_2CHNOH$			x		1	1	1	1	1	1	1
8	$CH_3SOC(CH_3)_2CHNOH$					1	1	1	1	1	1	1
9	$CH_3SO_2C(CH_3)_2CHNOH$					1	1	1	1	1	1	1
10	$CH_3SC(CH_3)_2CN$					1	1	1	1	1	1	1
11	$CH_3SOC(CH_3)_2CN$	x	x	x	x	1	1	1	3	1	1	1
12	$CH_3SO_2C(CH_3)_2CN$		x	x	x	1	1	1	3	1	1	1
13	$CH_3SC(CH_3)_2CONH_2$					1	1	1	1	1	1	1
14	$CH_3SOC(CH_3)_2CH_2OH$		x			1	1	1	1	1	1	1
15	$CH_3SC(CH_3)_2COOH$		x			1	1	1	1	1	1	1
16	$CH_3SOC(CH_3)_2COOH$		x			1	1	1	1	1	1	1
17	$CH_3SO_2C(CH_3)_2COOH$		x			1	1	1	1	1	1	1

[a] Compounds which have been identified as metabolites of aldicarb in S=soil, P=plants, A=animals, I=insects.
[b] Toxicity ratings from 1 = no control to 5 = complete control.
[c] Abbreviations and rates (a.i.), insect, mite and nematode laboratory tests:

Ad	Bean Aphid, *Aphis fabae* Scopoli	100 ppm
Aw	Southern Armyworm, *Spodoptera (Prodenia) eridania* Cramer	1000 ppm
MB	Mexican Bean Beetle, *Epilachna varivestis* Mulsant	100 ppm
Fy	Housefly, *Musca domestica* L.	1000 ppm
A	Two-spotted Spider Mite, Adult	1000 ppm
E	Two-spotted Spider Mite, Egg *Tetranychus urticae* Koch	
Nema	Root-knot Nematode, *Meloidogyne incognita* (Kofoid and White)	84 kg/ha

hibited body activity and stimulated abnormal stylet movement following contact. A 24-hr post-nematicide water treatment allowed effective recovery of body vigor and cessation of abnormal stylet movement of the nematodes treated with lower concentrations. At higher concentrations, the nematodes remained paralyzed, the toxic effect being apparently irreversible. Nelmes suggested the inhibition of body activity may result from AChE inhibition that was irreversible at higher concentrations due to continued binding of the enzyme. Various other effects observed include the prevention of or decrease in egg hatching, decreased motility, and delays in the molting processes (8,9). Many of these behavior changes may relate to inhibition of various esterase enzymes. A detailed explanation of cholinesterase inhibition is described by Main (12,13).

NEMATICIDE DETOXIFICATION

As previously stated, immediately following application, physical, chemical and biological interactions occur between a nematicide and the constituents in its environment. These interactions include metabolism by microorganisms and other living systems. The net result is a breakdown or degradation of the nematicide molecule and its detoxification. The breakdown products or metabolites are integrated with time into the environment. Often the environment can potentiate nematicidal activity. For example, the nematicidal activity of vapam and dazomet depends upon reactions with water to initiate the formation of toxic MIT and ensuing reactions. The nonfumigant nematicide aldicarb is metabolized to its sulfoxide, which is equally potent as a nematicide, and the nitrile sulfoxide which is non-nematicidal. Over time the aldicarb sulfoxide is hydrolyzed to the oxime sulfoxide and nitrile sulfoxide which are non-toxic (Fig. 2, Table 2) (1,25).

Both insecticidal and nematicidal activity depend upon the carbamate structure being intact.

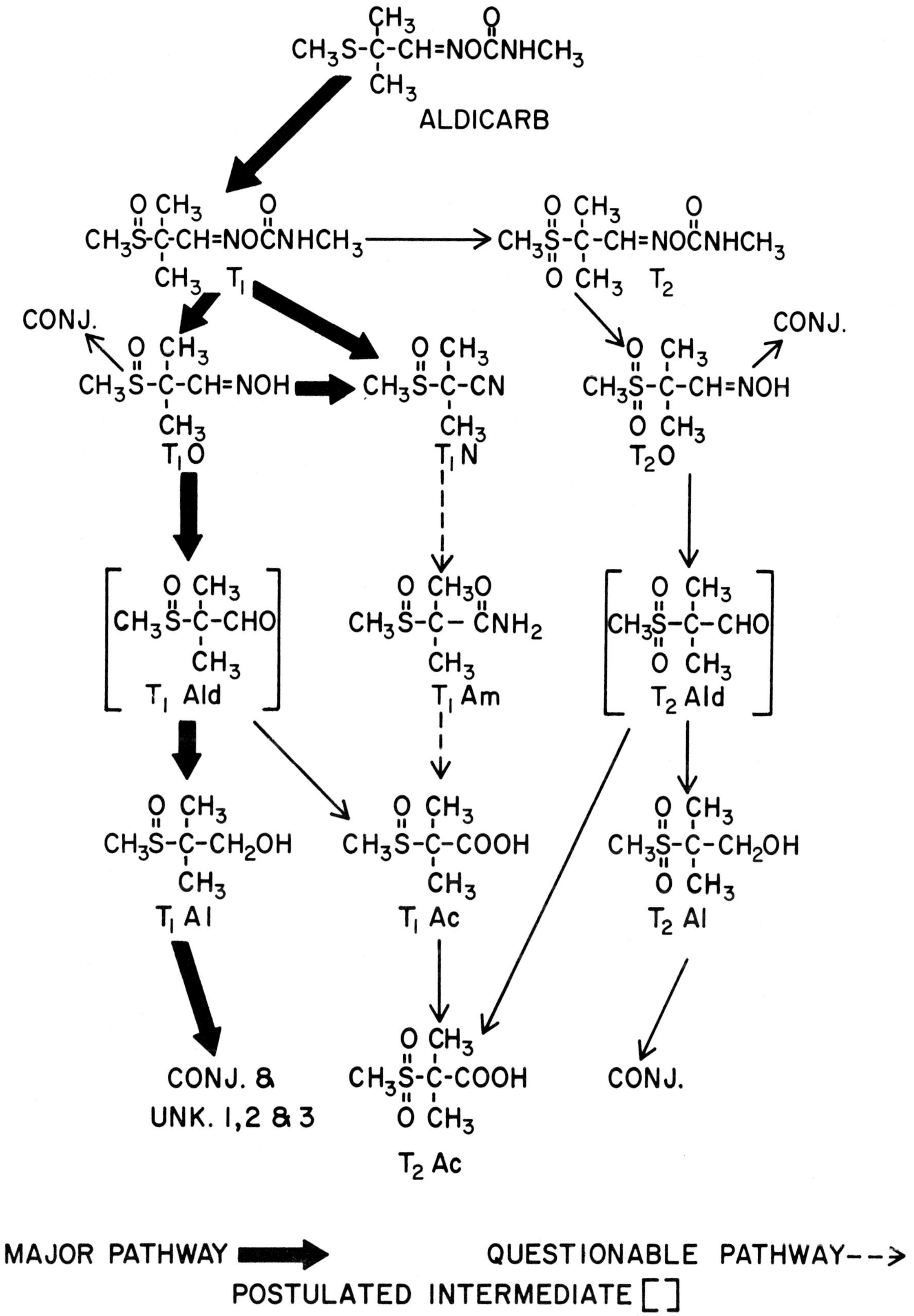

Fig. 2. A metabolic pathway of aldicarb in cotton plants.

Table 3. Inhibition ($I_{50}M$) of three esterases separated by ammonium sulfate precipitation (20, 50, 70% saturation) from *Panagrellus redivivus*.

Inhibitor	Enzyme fraction and substrate used: 20% Acetylcholine[a]	50% Acetylcholine[a]	70% o-Methoxyphenyl acetate[b]
Aldicarb 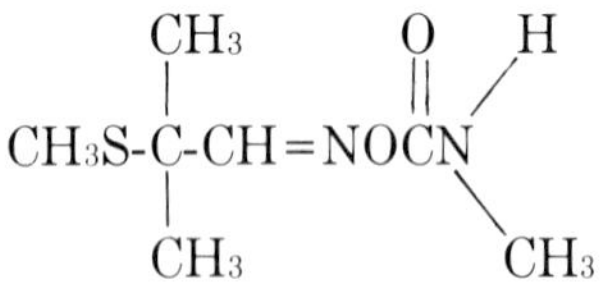	6.8×10^{-6}	1.2×10^{-5}	1.0×10^{-4}
Eserine 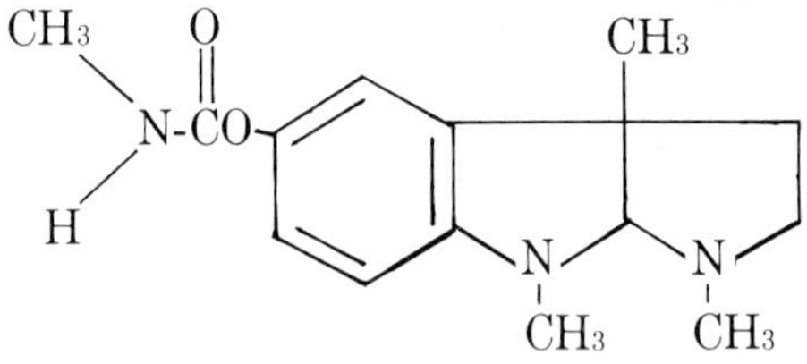	2.9×10^{-8}	3.9×10^{-7}	2.6×10^{-5}
Bidrin 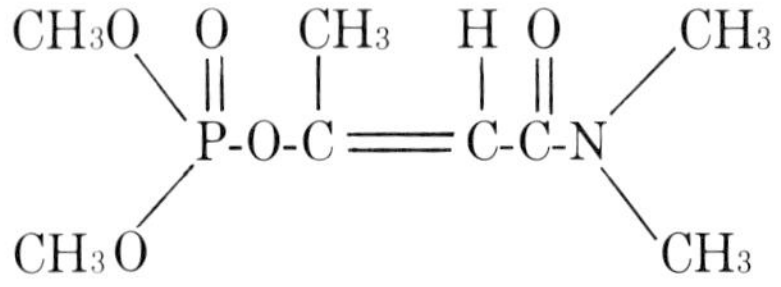	9.0×10^{-7}	4.1×10^{-5}	5.2×10^{-5}

[a] 0.01 M acetylcholine used as substrate
[b] 0.001 M o-Methoxyphenyl acetate used as substrate

Hydrolysis of carbamates and phosphates removes the portion of the molecule that occupies the esteratic site of AChE thereby precluding inhibition and thus toxicity (10,13). Figure 2 illustrates the metabolism of aldicarb in the cotton plant, which is fairly similar to metabolism in soil or other biological systems. All of the carbamates, compounds with O_2CNHCH_3, are active insecticides-nematicides, and the other compounds are metabolites and are essentially inactive as shown in Table 2.

Spurr and Chancey (23,24) demonstrated the effectiveness of oxime carbamate nematicides to inhibit esterase enzymes using the free-living nematode *Panagrellus redivivus* as a model. Three enzymes capable of hydrolyzing ester bonds were separated by ammonium sulfate fractionation from cell-free preparations and differed in substrate specificity and molecular weight. Two of the enzymes were cholinesterases (acetylcholinesterase or cholinesterase), and the other did not hydrolyze choline substrates but did hydrolyze aliphatic and aromatic esters. These enzymes were effectively inhibited by phosphate and carbamate insecticides (Table 3). Aldicarb inhibited acetylcholine hydrolysis at $I_{50} = 6.5 \times 10^{-6}M$, a concentration 1000 times less than that required to inhibit this nematode in a water contact test ($ED_{50} = 2.6 \times 10^{-3}M$). This finding demonstrated the effectiveness of these structures for the inhibition of esterase activity and their correlated toxicity to nematodes.

The importance of the mode of action studies, such as those on cholinesterase inhibition and associated behavioral changes, is often overlooked or not understood. These studies provide information on how the chemical structure relates to toxicity, enzyme inhibition, and behavior. Also, these studies provide information on how chemical structures can be modified to improve penetration of the cuticle, overall inhibition, and finally the efficacy of nematicides. Most studies of nematicide metabolism relate to studies of mode of action (1). This process in turn leads to following nematicidally active metabolites and their degradation or detoxification and integration into the environment after nematode control is achieved (25). Mechanisms other than enzyme inhibitions should also be considered and have not been studied in relation to nematicidal mode of action. A review of the fundamental principles of toxicant-receptor interactions may provide the insight for future explorations in this area (12).

SUMMARY AND CONCLUSIONS

Modern nematicides must fulfill many requirements including the control of nematodes to be acceptable for widespread application. The knowledge generated by studies of chemistry and biology together can lead to many improvements in this important technology. The giant increase in nematode control, following the widespread introduction of fumigants in 1950 and organic phosphates and carbamates in 1970 can be matched by the development of sophisticated management techniques employing chemical tools.

Even as this report is being written, excitement is being generated by the results of a new class of insecticide-nematicide, the avermectins. They have demonstrated a broad spectrum of antihelmintic activity. Promising results for the control of root-knot nematodes have been reported (6,22). Avermectins, produced by the soil microbe *Streptomyces avermitilis*, block the transmission of electrical potentials between interneurons and excitatory motoneurons in the ventral nerve cord. They inhibit transmission between inhibitory motoneurons and muscle by opening the chloride channels on the membranes but have little effect on excitatory neuromuscular junctions. The block of interneuron to excitatory motoneuron transmission can be reversed by picrotoxin, a known gamma-aminobutyric acid (GABA) antagonist. Avermectins cause release of GABA from brain synaptosomes in rats (20). Interference with the chloride gate was observed by Fritz (5).

Other potential control chemicals for the future also differ in mode of action from chemicals now in use. These include antifeedants, pheromones and other hormones. Their potential as insecticides of the future is presented in a review by Hummel (10). The important role of chemicals for the control of nematodes will continue in the future.

Literature Cited

1. Bartley, W. J., N. R. Andrawes, E. L. Chancey, W. P. Bagley, and H. W. Spurr. 1970. The metabolism of Temik aldicarb pesticide [2-methyl-2-(methylthio)-propionaldehyde O-(methyl-carbamoyl) oxime] in the cotton plant. J. Agric. Food Chem. 18:446-453.
2. Bunt, J. A. 1975. Effect and mode of action of some systemic nematicides. Meded. Landbouwhogesch. Wageningen 75-10:1-127.
3. Castro, C. E. and I. J. Thomason. 1971. Mode of action of nematocides, pp. 289-296 *In* Plant parasitic nematodes, Vol. 2, ed. B. M. Zuckerman, W. F. Mai, and R. A. Rohde. Academic Press, New York. 347 pp.
4. Evans, A. A. F. 1973. Mode of action of nematicides. Ann. Appl. Biol. 75:469-473.
5. Fritz, L. C., C. C. Wang, and G. Gorio. 1979. Avermectin B_1a irreversibly blocks postsynaptic potentials at a lobster neuromuscular junction by reducing muscle membrane resistance. Proc. Nat. Acad. Sci. U.S.A. 76:2062-2066.
6. Garabedian. S. and S. D. Van Gundy. 1983. Use of avermectin for the control of *Meloidogyne incognita* on tomatoes. J. Nematol. 15:503-510.
7. Hodgson, E. and F. E. Guthrie, eds. 1980. Introduction to biochemical toxicology. Elsevier, New York. 437 pp.
8. Hough, A. and I. J. Thomason. 1975. Effects of aldicarb on the behavior of *Heterodera schachtii* and *Meloidogyne javanica*. J. Nematol. 7:221-229.
9. Huang, S. P. and S. D. Van Gundy. 1978. Effects of aldicarb and its sulfoxide and sulfone on the biology of *Tylenchulus semipenetrans*. J. Nematol. 10:100-106.
10. Hummel, H. E. 1983. Insecticides and their design. J. Nematol. 15:615-639.
11. Kaufmann, D. D. 1967. Degradation of carbamate herbicides in soil. J. Agric. Food Chem. 15:577-582.
12. Main, A. R. 1980. Toxicant-receptor interactions: fundamental principles, pp. 180-192 *In* Introduction to biochemical toxicology, ed. E. Hodgson and F. E. Guthrie. Elsevier, New York. 437 pp.
13. Main. A. R. 1980. Cholinesterase inhibitors, pp. 192-223 *In* Introduction to biochemical

toxicology, ed. E. Hodgson and F. E. Guthrie. Elsevier, New York. 437 pp.
14. Marks, C. F., I. J. Thomason, and C. E. Castro. 1968. Dynamics of permeation of nematodes by water, nematicides, and other substances. Exp. Parasitol. 22:321-327.
15. McKenry, W. V. 1981. The nature, mode of action, and biological activity of nematicides, pp. 59-73 *In* CRC Handbook of pest management in agriculture, Vol. 3., ed. D. Pimentel. Boca Raton, Fla. 656 pp.
16. Moje, W. 1960. The chemistry of nematicidal activity of organic halides. Adv. Pest Control Res. 3:183-187.
17. Nelmes, A. J. 1970. Behavioral responses of *Heterodera rostochiensis* larvae to aldicarb and its sulfoxide and sulfone. J. Nematol. 2:223-227.
18. Nelmes, A. J. 1971. The permeability of nematodes to aldicarb, a nematicidal oxime carbamate. Meded. Fac. Landbouwwet. Rijksuniv. Gent. 36:904-914.
19. Nelmes, A. J., D. L. Trudgill, and D. C. M. Corbett. 1973. Chemotherapy in the study of plant parasitic nematodes. Symp. Br. Soc. Parasitol. 11:95-112.
20. Pong, S. S. and C. C. Wang. 1980. The specificity of high affinity binding of Avermectin B_1a to mammalian brain. Neuropharmacology 19:311-317.
21. Rohde, R. A. 1960. Acetylcholinesterase in plant-parasitic nematodes and an anticholinesterase from asparagus. Proc. Helminthol. Soc. Wash. 27:121-123.
22. Sasser, J. N., T. L. Kirkpatrick, and R. A. Dybas. 1982. Efficacy of avermectins for root-knot control in tobacco. Plant Dis. 66:961-965.
23. Spurr, H. W., Jr. and E. L. Chancey. 1967. Phosphate and carbamate inhibition of three nematode esterases. Phytopathology 57:832 (Abstr.).
24. Spurr, H. W., Jr. and E. L. Chancey. 1985. Aldicarb metabolism in relation to esterase inhibition and nematicidal efficacy. J. Nematol. In press.
25. Spurr, H. W., Jr. and A. A. Sousa. 1974. Potential interactions of aldicarb and its metabolites on nontarget organisms in the environment. J. Environ. Qual. 3:130-133.
26. Van Gundy, S. D. and M. V. McKenry. 1977. Action of nematicides, pp. 263-283 *In* Plant disease: an advanced treatise, Vol. 1, ed. J. G. Horsfall and E. B. Cowling. Academic Press, New York. 465 pp.
27. Wade, R. S. and C. E. Castro. 1973. The oxidation of heme proteins by alkyl halides. J. Am. Chem. Soc. 95:226-231.
28. Wright, D. J. 1981. Nematicides: mode of action and new approaches to chemical control, pp. 421-449 *In* Plant parasitic nematodes, Vol. 3, ed. B. M. Zuckerman and R. A. Rohde. Academic Press, New York. 508 pp.
29. Wright, D. J. and F. A. Awan. 1976. Acetylcholinesterase activity in the region of the nematode nerve ring: improved histochemical specificity using ultrasonic pretreatment. Nematologica 22:326-331.

CROPPING SYSTEMS RESEARCH AND ROOT-KNOT NEMATODE CONTROL

S. A. Raymundo

International Potato Center
Lima, Peru

Root knot incited by *Meloidogyne* spp. is a major problem particularly under tropical conditions. Its severity increases when susceptible crops are planted together or in succession in the same ground. The use of nematicides to control root knot is well known. In many instances, however, cost of nematicides is beyond the means of the average crop grower. Consequently, other methods of nematode control, including appropriate cropping systems should be identified to give the crop grower a range of options for nematode management.

In this paper, cropping system is defined as the sequence of growing crops along with the required technologies for their production. Sequence can be of two kinds: spatial and temporal.

SPATIAL CROP SEQUENCE AND ROOT-KNOT NEMATODE CONTROL

Spatial sequence refers to the arrangement of different crop species in a piece of land. Such terms as mixed cropping, row intercropping and strip cropping connote such arrangment. Temporal sequence refers to distribution of crops over a period of time. Such terms as crop rotation, relay cropping, discontinuous planting and fallow rotation give connotation of this distribution.

Studies on crop diversity through mixed cropping and insect pest damage relationships are available (2,20,24). Similar work relating to nematodes are wanting. As is true with insect pests, a number of investigators have found reduced damage when root-knot-nematode susceptible hosts are mixed with other crops (4,7,8,13,19,26). Possible explanations for such reductions have also been offered. Intercropping with marigold (*Tagetes* sp.) adversely affected various nematodes, including *Meloidogyne* sp., that would have severely damaged susceptible hosts in mixtures. It was suggested that marigold acted as an antagonistic crop in that its roots released chemicals that repelled or killed nematodes (26).

Hackney and Dickerson (8) recorded that the association of tomato (*Lycopersicon* sp.) with castor bean (*Ricinus communis*) resulted in considerable reduction in the number of galls in tomato. When tomato was associated with marigold or *Chrysanthemum*, however, it was apparent that tomato was the more hospitable crop as shown by the consistently larger galls compared with the galls on marigold and *Chrysanthemum*. The nematode-suppressing properties of marigold could have been due to nematicidal compounds like terthienyl (2, 2′, 5′, 2′ thienyl) which have been claimed to be present on such plants (7). In interplanting studies by Choudhury (4), the number of root knots and mature females in the susceptible brinjal crop (*Solanum melongena*) was significantly fewer in plots planted with both brinjal and *Tagetes* than in plots where brinjal was grown alone.

In crop association studies, it has been found that there is considerably less severe galling by *M. incognita* on potato (*Solanum tuberosum*) roots intercropped with onion (*Allium* sp.) and corn (*Zea mays*) compared with the galling found on potato alone (19). Potatoes intercropped with corn have been found to be extremely vigorous with well-developed root systems. A possible physiological basis for this improved plant growth was offered by Midmore (13).

The highly favorable environment for potato growth and development, which resulted from intercropping with corn, may have been responsible for the reduced root damage on potato. With optimum growth and a well-developed root system, chances for the nematodes to infect and establish themselves successfully were likely to have been remarkably reduced.

Composites of crops in the mixture determine extent of root-knot nematode damage. In the Ivory Coast, cassava is grown simultaneously along with other plants. In this way nematode attacks are not serious except when cassava (*Manihot esculenta*) is grown mixed with suitable hosts such as okra (*Abelmoschus esculentus*) or eggplant (11). In the latter instance, a continuum of susceptible hosts are simultaneously present in the same ground thus resulting in the rapid spread of infection.

TEMPORAL CROP SEQUENCE AND ROOT-KNOT NEMATODE CONTROL IN DEVELOPING COUNTRIES

Crop rotation is the best known method of temporal sequence management to reduce damage by nematodes. The principle behind crop rotation is that nematode multiplication is restricted and damage correspondingly reduced when continuous cultivation of root-knot susceptible crops is avoided.

Crop sequences which give good control of root-knot nematodes have been reported by various investigators. Smit and Bos (23) reported that wheat (*Triticum* sp.) and onion reduced high root-knot population levels when used as rotation crops. Merny (12) mentioned that one year of tomato and two years of rice (*Oryza sativa*) had given effective control of *Meloidogyne* populations and facilitated commercial tomato production in Ivory Coast. Ogunfowora (17) stated that inclusion of rice, maize, groundnut (*Arachis hypogaea*) and *Stylozanthes* in rotation sequences controlled root-knot nematodes and increased yields of the main crop, tomato.

In crop rotation trials in Senegal, it was found that a crop of peanut (groundnut) or strawberry (*Fragaria* sp.) grown prior to tomato in soil heavily infested with root-knot nematodes greatly increased yields of tomato as compared to those of a previous crop of tomato (16). In the wet highlands of Sri Lanka, rotation of root-knot susceptible crops like tomato with leek (*Allium porrum*) reduces both *M. incognita* and *M. javanica* (22). Other crops are claimed to be effective in rotations in Sri Lanka, such as onion, groundnut and the legume, sun hemp (*Crotolaria juncea*). Groundnut or corn in rotation with sweet potato decreased numbers of *M. incognita* to low levels in second rotation (3). Meanwhile, cabbage (*Brassica oleracea capitata*) as a monocrop and groundnut/corn and sweet potato/cabbage as multicrops suppressed numbers of *M. incognita* to low levels in the third rotation. Cabbage in rotation with corn, however, did not decrease the number of nematodes in the third rotation.

In Indonesia, Widjaja and Windrich (27) have shown that marigold (*Tagetes patula*), *Glycine javanica*, *Crotolaria usaramoensis*, corn and sweet potato (*Ipomoea batatas*) are useful plants for reducing *Meloidogyne* spp. density in the soil. For cereal-based cropping systems, the following crop sequences for root-knot nematode control are recommended in the Philippines: rice-mung bean (*Phaseolus aureus*)-corn-cabbage-rice, rice-tobacco (*Nicotiana tobacum*)-rice, rice-cotton-rice and rice-watermelon (*Citrullus vulgaris*)-rice (6). Myint (15) wrote that kohlrabi (*Brassica oleracea caulorapa*)-tomato-bitter gourd (*Momordica charantia*)-sweet potato-tomato rotation dramatically reduced root-knot damage.

On tobacco, Muro (14) reported that such crops as peanut, corn, wheat and *Sorghum* sp. when planted prior to tobacco significantly reduced the incidence of *M. incognita*. A similar observation was recorded by Cortado and Davide (5). In Chile, cultural rotation using grain crops such as wheat and corn have helped in reducing root-knot nematode populations in the soil (21). In Morocco, turnip and carrots have been reported to inhibit nematode development (1).

From the foregoing studies, one can see that the use of crop rotation to control root-knot nematodes occurs worldwide. Most of the works cited involve planting of root-knot susceptible crops in alternate seasons with more resistant species. A rotation which seems to be quite common is that of solanaceous crops with cereals.

CONSIDERATIONS TO IMPROVE EFFICIENCY OF CROP ROTATION SYSTEM

Good root-knot nematode control is one of several criteria for identification of useful crop rotation systems. For the efficiency of the system to be increased, several factors should be understood thoroughly. Some of these include: nematode interaction with other soil organisms (fungi, bacteria, viruses, other nematodes, etc.), the ability of such interaction to influence root-knot nematode development, and its direct or indirect damage on the rotation crops. It is not the purpose of this paper to describe such interactions which have already been covered in a number of publications (10,18,25).

In addition to the effects of other soil organisms on root-knot nematodes, various other factors are equally important in the development of efficient crop rotation systems. The factors that I consider to be of major importance are discussed in the following paragraphs.

Usefulness of the Crop

The crops in a rotation scheme should be of economic value to the user of the system. Marigold suppresses nematode multiplication, but it does not have economic value in the Philippines. Consequently, any rotation system with marigold as one of its components would not serve much use under conditions in the Philippines. On the other hand, marigold would be a welcome component if the system were used in Peru because anthocyanin is extracted from this crop and used as an ingredient for locally processed poultry feeds.

A crop rotation utilizing different crop species could be a useful tool for root-knot control for some growers. To others, however, rotations of varieties (of the same crop species) with different genes for resistance (gene rotation) could better serve their purpose.

The gene rotation approach could be particularly useful in places where monoculture is indispensable. In Nigeria, monocropping of tomato has aggravated damage by *Meloidogyne* spp. so much that yields have declined remarkably (23). However, due to the high price of tomatoes, the growers were reluctant to plant other crops because even the low tomato yields gave them good economic returns.

The concept of plant disease control through gene rotation relies heavily on the precise identification of nematode race, pathotype or strain (9). Much work has been done on pathogenic variation among *Meloidogyne* species. It is known that variation exists within species. However, for the gene rotation approach to be used efficiently, much research is needed to clarify the nature and origin of intra-*Meloidogyne* species variation.

Adaptability of Crops to Local Environment

A rotation crop not only should help reduce or stabilize nematode populations but also should yield satisfactorily. Local conditions including short day length, poor soil type and extremes of temperature are all detrimental for normal growth and development and hence adversely affect the yields of previously selected rotation crops. It is important, therefore, that these crops should be grown in the field not only so their reaction to the nematode can be determined but also their total adaptability to the local enviroment.

Effective Control of Other Pests

Various above-ground pests that can harm selected rotation crops are ever present. If meaningful use of these crops is expected, effective control for such pests should be at hand. Planting rice to reduce *Meloidogyne* populations is meaningless if the rice plants are blighted by the blast fungus. Similarly, onion may inhibit root-knot nematodes, but the crop is of no use if it is killed by thrips. Likewise, poor ear formation due to attack by maize streak virus could negate the usefulness of corn in a root-knot control rotation scheme.

Several below-ground pests, completely distinct and unrelated to root-knot infection, can be highly devastating, and an effective method of control may or may not be available. Root-knot-resistant potato can be rendered unfit for use if the tubers are severely damaged by tuber moth. Likewise, garlic (*Allium sativum*) could reduce root-knot damage but would be disadvantageous as a rotation crop if *Sclerotium* bulb rot were left uncontrolled. Root-knot-resistant tomato cultivars would be doubly useful if they were resistant to the devastating foliar pathogen *Cladosporium fulvum.*

Applicability of Rotation Scheme to Local Conditions

Rotation schemes incorporating a fallow period have been found effective against root-knot nematodes. This approach would be feasible were adequate land available. With the diminishing acreage of cultivable land, however, pressure to put every available area under cultivation continues. For this reason, the need to identify suitable cultivars to be used for either an inter-species or intra-species crop rotation scheme becomes even more acute.

Certain crop rotation systems incorporate grass or weed fallow. The system would be attractive in localities where livestock form a part of the farming system. Otherwise, adoption of the rotation scheme is questionable.

Availability of Resources to Meet Normal Growth Requirements of Rotation Crops

Crops like rice need a lot of water; paddy rice requires more than upland rice. Such requirements can be met in one of two ways: use of artificial irrigation or dependence upon natural precipitation. Either of the two should appear certain before a decision is made in favor of including rice or other water-loving crops in a rotation cycle. Certain crops also have specific nutritional requirements. One should ensure that sources of such nutrients are guaranteed before these crops are used in a rotation scheme.

From the above, it is clear that the total physical and biological environment, as well as the accompanying socio-economic forces, are of utmost importance in designing meaningful crop rotation systems. Disciplines must be integrated in order for such systems to be designed. What is good for the nematologist may not be good for the agronomist or the economist. If the system evolves with a multidisciplinary approach, the likelihood of its being adopted by the intended users is almost certain.

Literature Cited

1. Agadr, S., M. Ammati, and A. Fougeroux. 1979. Research on root-knot nematodes in Morocco, pp. 113-126 *In* Proc. Second IMP Res. & Plann. Conf. on Root-knot Nematodes, *Meloidogyne* spp., Region VII, Nov. 26-30, 1979. Athens, Greece. 146 pp.
2. Buranday, R. P. and R. S. Raros. 1975. Effects of cabbage-tomato inter-cropping on the incidence and oviposition of the diamond-back-moth, *Plutella xylostella*. Philipp. Entomol. 2:369-374.
3. Charas, C. 1981. Progress report on the root-knot nematode studies in Thailand, pp. 97-105 *In* Proc. Third IMP Res. & Plann. Conf. on Root-knot Nematodes, *Meloidogyne* spp., Region VI, July 20-24, 1981. Jakarta, Indonesia. 241 pp.
4. Choudhury, B. C. 1981. Root-knot nematode problem on various crop plants in Bangladesh, pp. 142-145 *In* Proc. Third IMP Res. & Plann. Conf. on Root-knot Nematodes, *Meloidogyne* spp., Region VI, July 20-24, 1981. Jakarta, Indonesia. 241 pp.
5. Cortado, R. V. and R. G. Davide. 1968. Survey, identification and pathogenicity test of nematodes associated with tobacco in the Cagayan Valley and Ilocos Region. Philipp. Agric. 51:779-801.
6. Davide, R. G. and A. Zorilla. 1983. National Crop Protection Center (NCPC) Leaflet Series No. 2. Los Baños, Laguna, Philippines
7. Gommers, F. J., J. Bakker, and Liesbeth Smits. 1980. Effects of singlet oxygen generated by the nematicidal compund α-terthienyl from *Tagetes* on the nematode *Aphelenchus avenue*. Nematologica 6:369-375.
8. Hackney, R. W. and O. J. Dickerson. 1975. Marigold, castor bean, and chrysanthemum as controls of *Meloidogyne incognita* and *Pratylenchus alleni*. J. Nematol. 7:84-90.
9. International Rice Research Institute. 1982. Evaluation of the gene rotation concept for rice blast control. Int. Rice Res. Inst. (IRRI), Los Baños, Laguna, Philippines. 130 pp.
10. Jatala, P., E. R. French, and L. Gutarra. 1975. Interrelationships of *Meloidogyne incognita acrita* and *Pseudomonas solanacearum* on potatoes. J. Nematol. 7:325 (Abstr.).
11. Luc, M. 1968. Nematological problem in the former French African tropical territories and Madagascar, pp. 93-112 *In* Tropical nematology, ed. G. C. Smart, Jr. and V. G. Perry. Univ. Florida Press, Gainesville. 153 pp.
12. Merny, G. 1976. Root-knot problems in Ivory Coast, pp. 60-65 *In* Proc. IMP Res. & Plan. Conf. on Root-knot Nematodes, *Meloidogyne* spp., June 7-11, 1976. IITA, Ibadan, Nigeria. 83 pp.
13. Midmore, D. 1983. Shading and its influence in intercropping systems with a potato base. Training Course on Agronomy for the Potato in the Hot Tropics. International Potato Center, Lima, Peru. 8 pp.
14. Muro, A. D. 1975. Root-knot nematode (*Meloidogyne incognita*) control by two- and three-year crop rotation. Ann. Ist. Sper. Tab. (Rome) 2:89-91.
15. Myint, Y. Y. 1981. Country report on root-knot nematode in Burma, pp. 163-170 *In* Proc. Third IMP Res. & Plann. Conf. on Root-knot Nematodes, *Meloidogyne* spp., Region VI, July 20-24, 1981. Jakarta, Indonesia. 241 pp.
16. Netscher, C. and D. P. Taylor. 1976. *Meloidogyne* research at O.R.S.T.O.M., Dakar, pp. 66-79 *In* Proc. IMP Res. & Plann. Conf. on Root-knot Nematodes, *Meloidogyne* spp., June 7-11, 1976. IITA, Ibadan, Nigeria. 83 pp.
17. Ogunfowora, A. O. 1978. Further studies on control of root-knot nematodes, *Meloidogyne* spp., pp.31-37 *In* Proc. Second IMP Res. Plann. Conf. on Root-knot Nematodes, *Meloidogyne* spp., Region IV,

Abidjan, Ivory Coast. IITA, Ibadan, Nigeria. 93 pp.

18. Pitcher, R. S. 1963. Interrelationships of nematodes and other pathogens of plants. Helminthol. Abstr. 34:1-17.
19. Raymundo, S. A. 1983. Major pests of potatoes in hot humid climates and their control. Training Course on Agronomy for Potato in the Hot Tropics. International Potato Center, Lima, Peru. 7 pp.
20. Raymundo, S. A. and J. Alacazar. 1983. Control of potato tubermoth (*Phthorimaea operculella*) through crop association. Proc. Sixth Symp. Int. Soc. Trop. Root Crops, February 20-25. Lima, Peru. 180 pp.
21. Rodriguez, H. C. 1983. Informe de Chile, pp. 12-15 *In* Proc. Third IMP Res. & Plann. Conf. on Root-knot Nematodes, *Meloidogyne* spp., Region II, March 22-26, 1982. CIP, Lima, Peru. 233 pp.
22. Sivapalan, P. 1981. Report from Sri Lanka, pp. 9-19 *In* Proc. Third IMP Res. & Plann. Conf. on Root-knot Nematodes, *Meloidogyne* spp., Region VI, July 20-24, 1981. Jakarta, Indonesia. 241 pp.
23. Smit, J. J. and W. S. Bos. 1976. Root-knot in northern Nigeria, pp. 41-43 *In* Proc. IMP Res. & Plann. Conf. on Root-knot Nematodes, *Meloidogyne* spp., June 7-11, 1976. IITA, Ibadan, Nigeria. 83 pp.
24. Suryatna, E. S. and R. R. Harwood. 1976. Nutrient uptake of two traditional intercrop combination and insect and disease incidence in three intercrop combinations. IRRI Saturday Seminar, February 28. 11 pp.
25. Taylor, C. E. 1979. *Meloidogyne* interrelationships with microorganisms, pp. 375-398 *In* Root-knot nematodes (*Meloidogyne* spp.) systematics, biology and control, ed. F. Lamberti and C. E. Taylor. Academic Press, London. 477 pp.
26. Visser, T. and M. K. Vithilingham. 1959. The effects of marigold and some other crops on the *Pratylenchus, Meloidogyne* population in the soil. Tea Q. 30:30-38.
27. Widjaja, W. dan Rustman & Windrich. 1974. Studies on nematological problems in horticulture, pp. 393-402 *In* Agricultural Cooperation Project Indonesia The Netherlands. Research Reports Section II. Technical Contribution Bogar.

SPECIFIC CROP ROTATION EFFECTS COMBINED WITH CULTURAL PRACTICES AND NEMATICIDES

A. W. Johnson[1]

Supervisory Research Plant Pathologist
USDA-ARS
Coastal Plain Experiment Station
Tifton, Georgia, USA

Crop rotation is one of the oldest and most important methods for managing nematodes that feed on roots of annual crop plants (5). Strategies utilizing crop rotations for managing nematodes are based on the resistance, susceptibility, or tolerance of crops to the predominant species of pathogenic nematodes in a specific area. In an optimum crop rotation system, the preceding crop prevents damage to the following crop by the target nematode species and suppresses the nematode population without increasing other species of nematodes that may be parasitic to the future crop. Other species of nematodes that might cause economic damage under a favorable crop rotation system also must be considered. For example, corn (*Zea mays*) will suppress populations of the northern root-knot nematode (*Meloidogyne hapla*), whereas population densities of some other root-knot, stubby-root, lesion, sting, lance, and ring nematodes may increase when corn is planted frequently. Severe crop damage may result from certain nematodes when susceptible crops are monocultured too frequently or when a rotation system with nonhost or poor-host crops is too short. Although investigations of crop rotation systems continue, our knowledge of their effects on nematode population dynamics is advanced enough for making management recommendations. Numerous cropping systems for managing nematodes have been reviewed (5, 6, 20, 23-25, 29).

Integrated pest management (IPM) programs use a variety of tactics to manage one or more pests. The objective of coordinating different tactics to manage plant pests increases the efficient production of food and fiber while reducing the overall unfavorable effects of the various practices. IPM systems vary with pests, crops, locations, and producers; recommendations may vary for a given crop at a given location because of changes in pests, management techniques, weather conditions, and economic circumstances. Thus, IPM is a complex, dynamic group of practices that defies precise description (4). Integrated pest management is rapidly becoming accepted as a crop production system that integrates methodologies across crop production disciplines in a fashion that is compatible with the crop production system. Therefore, IPM could also mean "integrated production management." Although much has been written and spoken about integrated control and pest management (1), little information is currently available on integrated management of nematodes by means of specific crop rotations, cultural practices, nematicides, and other parameters (2, 4, 5, 9-11, 13, 14, 18, 26, 28, 29).

Intensive cropping is feasible in the southern United States because of the long growing season; thus many farmers produce two or three crops each year on the same land area (intensive or multiple-cropping) to maximize utilization of the land, resources, and profits. However, little is known about the effects of intensive cropping on nematode populations.

1. A STUDY USING EIGHT CROPS PLANTED IN SIX ANNUAL CROPPING SEQUENCES

Materials and Methods

In a 4-year study (1971-1974), various com-

[1]All research on intensive cropping systems mentioned in this paper was conducted cooperatively with Dr. C. C. Dowler, Dr. N. C. Glaze, Dr. D. R. Sumner, and Dr. R. B. Chalfant at the Coastal Plain Experiment Station, Tifton, Georgia.

binations of eight crop plants were grown in six annual intensive cropping systems on plots of Tifton, Carnegie, and Dothan loamy sands (about 76% sand, 16% silt, and 9% clay, pH 6.0-6.7, OM< 1%). The soils were naturally infested with mixed populations of *Meloidogyne incognita* (Kofoid & White) Chitwood (90%) and *M. hapla* Chitwood (10%); *Pratylenchus zeae* Graham (85%) and *P. brachyurus* (Godfrey) Filip. & Sch.-Stek. (15%); *Paratrichodorus minor* (Colbran) Siddiqi, *Macroposthonia ornata* (Raski) De Grisse & Loof, and *Helicotylenchus dihystera* (Cobb) Sher. Only data related to *Meloidogyne* spp. are presented in this manuscript. Data on other nematodes have been published (14).

The land had been in cultivation more than 30 years, primarily with corn, tobacco, soybean, and small grain. A split-split-plot randomized complete-block design of 5 replications was used. Whole plots included the following six annual cropping systems: turnip-corn-snap bean (T-C-SNB); turnip-peanut-snap bean (T-P-SNB); turnip-corn-turnip (T-C-T); turnip-peanut-turnip (T-P-T); snap bean-soybean-cabbage (SNB-SB-CB); and turnip-cucumber-southern pea-turnip (T-CU-SP-T). The crops, cultivars, and planting dates are listed in Table 1. Crops in each system were grown in the same plots from February 1971 through January 1975. Subplots included nematicide vs. no nematicide, and sub-subplots included herbicide vs. cultivation.

The nematicide *O*-ethyl *S,S*-dipropyl phosphorodithioate (ethoprop 10G) was applied overall (9 kg a.i./ha) during the spring just prior to planting corn, peanut, snap bean, or cucumber. It was incorporated immediately into the top 15 cm of soil with a tractor-mounted rototiller. The herbicides applied to all crops are listed in Table 2. Each pesticide treatment and cropping system was maintained on the same sub-plot for the duration of the experiment.

The soil was disk-harrowed and turned 25-30 cm deep with a moldboard plow after each crop was harvested. The following crop was planted 2 to 7 days after incorporating residue from the preceding crop. Plots were clean-fallowed after the fall crops of snap bean (90 days) and turnip (60 days) were harvested.

Soil samples (20 cores, 2.1 by 20 cm) containing both rhizospheric and nonrhizospheric soil were collected monthly from within the rows of plants beginning February 1971 and extending through January 1975 (except during December 1971, January and December 1972, and January 1973) for information on nematode population dynamics. The soil from each plot was mixed thoroughly, and a 150-cm^3 aliquant was processed by a centrifugal-flotation method (7) to separate nematodes from the soil.

At harvest, 20 plants were removed from each plot, except those planted to corn and peanut, and rated on a scale of 1 to 5 for percentage of roots galled by root-knot nematodes, as follows: 1 = no galling, 2 = 1-25, 3 = 26-50, 4 = 51-75, and 5 = 76-100%. Turnip, corn, soybean, peanut, snap bean, and southern pea were harvested mechanically, and cucumber and cabbage were hand-harvested. Yields were recorded in kilograms per hectare.

The experimental variables used in this study were limited to ethoprop vs. no nematicide, herbicides vs. cultivation, and cropping systems. The effects of herbicides on nematode populations and crop response are not included in this paper.

Results and Discussion—1971-1974

Population densities of root-knot nematodes were influenced by different cropping systems and crops within each system (Figs. 1, 2). The greatest numbers occurred in the SNB-SB-CB and T-CU-SP-T systems. Numbers of root-knot nematodes in the SNB-SB-CB system were greatest on soybean in October of each year. Numbers of these nematodes on snap bean were low to moderate (< 130/150 cm^3 soil). Application of ethoprop immediately before snap bean was planted did not suppress root-knot nematodes or prevent increase of nematode populations on succeeding crops. Cabbage supported low numbers (< 130/150 cm^3 soil) of root-knot nematodes.

Numbers of these nematodes in the T-CU-SP-T sequence were greatest in July and August of each year on southern pea following cucumber and lowest following the two consecutive crops of turnip. Application of ethoprop immediately before cucumber was planted caused a significant reduction in numbers of root-knot nematodes in plots planted to southern pea in 1971 and 1972, but not in 1973 and 1974. Population densities of root-knot nematodes in the T-C-SNB and the T-C-T systems did not exceed 140/150 cm^3 soil during 1971 and 1972, but they increased rapidly to near 700/150 cm^3 soil on corn in August in 1973 and 1974 in both cropping systems and to 850/150 cm^3 soil on snap bean following corn in 1973. Population densities of these nematodes were near or below detectable levels in the T-P-SNB and T-P-T systems on most sampling dates.

Root-galling of crops in the T-C-SNB, T-P-

Table 1. Cultivars and planting dates for crops in intensive cropping systems, 1971-74.

Crop	Cultivar	1971	1972	1973	1974
Turnip[a]					
Spring		Feb. 18	Feb. 24	Feb. 23	Feb. 26
Fall		Sept. 9	Sept. 15	Sept. 18	Sept. 19
Snap bean					
Spring	'GV50'	Mar. 18	Mar. 21	Mar. 21[b]	Mar. 20[c]
Fall	'GV50'	Sept. 8	Sept. 14	Sept. 14	Sept. 18
Corn	'Funks G4761'	Apr. 10	Apr. 18	Apr. 10	Apr. 10
Peanut	'Tifspan'	Apr. 10	Apr. 10	Apr. 10	Apr. 10
Soybean	'Davis'	June 1	May 9	June 20	June 20
Cucumber	'Chipper'	Apr. 12	Apr. 17	Apr. 18[d]	Apr. 24
Cabbage	'Market Prize'	Oct. 25	Nov. 3	Oct. 25	Oct. 25
Southern pea	'Pinkeye Purple Hull'	June 30	June 14	June 20	June 20

[a] The following cultivars were used in spring and fall crops: 'Purple Top' and 'White Globe', 1971-72; 'Shogoin', 1973-74.
[b] Replanted Apr. 18
[c] Replanted Apr. 23
[d] Replanted May 1.

Table 2. Common and chemical names, rate and method of application for herbicides used on various crops in intensive cropping systems—1971-74.

Crop	Common name	Chemical	Rate (kg a.i./ha)	Method of application
Corn	butylate	*S*-ethyl diisobutylthio-carbamate	3.36	inject at planting
	atrazine + phytoblend oil	2-chloro-4-(ethylamino)-6-(isopropylamino)-*s*-triazine	1.68	over top of 10-cm tall corn
	ametryne	2-(ethylamino)-4-(isopropyl-amino)-6-(methylthio)-*s*-triazine	1.12	as directed postemergence spray in nitrogen solution 6 weeks after planting
Peanut	benefin	*N*-butyl-*N*-ethyl-α,α,α-trifluoro-2,6-dinitro-*p*-toluidine	1.4	preplant incorporated
	vernolate	*S*-propyl dipropylthio-carbamate	2.24	inject at planting
	dinoseb	2-*sec*-butyl-4,6-dinitrophenol	0.56	over top of crop as five repeated sprays
Soybean	vernolate	*S*-propyl dipropylthio-carbamate	2.24	inject at planting
	chloroxuron	3-[*p*-(*p*-chlorophenoxy) phenyl]-1,1-dimethylurea	1.12	directed spray to 8-cm tall soybean
	dinoseb	2-*sec*-butyl-4,6-dinitrophenol	0.56	directed spray to 15-20-cm tall soybean
	linuron	3-(3,4-dichlorophenyl)-1-methoxy-1-methylurea	1.12	directed spray to 38-cm tall soybean
Turnip	DCPA	dimethyl tetrachloroterephthalate	8.96	preemergence
Southern pea & snap bean	trifluralin	α,α,α-trifluoro-2,6-dinitro-*N*,*N*-dipropyl-*p*-toluidine	0.56	preplant incorporated
	dinoseb	2-*sec*-butyl-4,6-dinitrophenol	1.68	at ground cracking
Cabbage	DCPA	dimethyl tetrachloroterephthalate	8.96	at transplanting
Cucumber	nitralin	4-(methylsulfonyl) 2,6-dinitro-*N*,*N*-dipropylaniline	0.84	premergence + irrigation

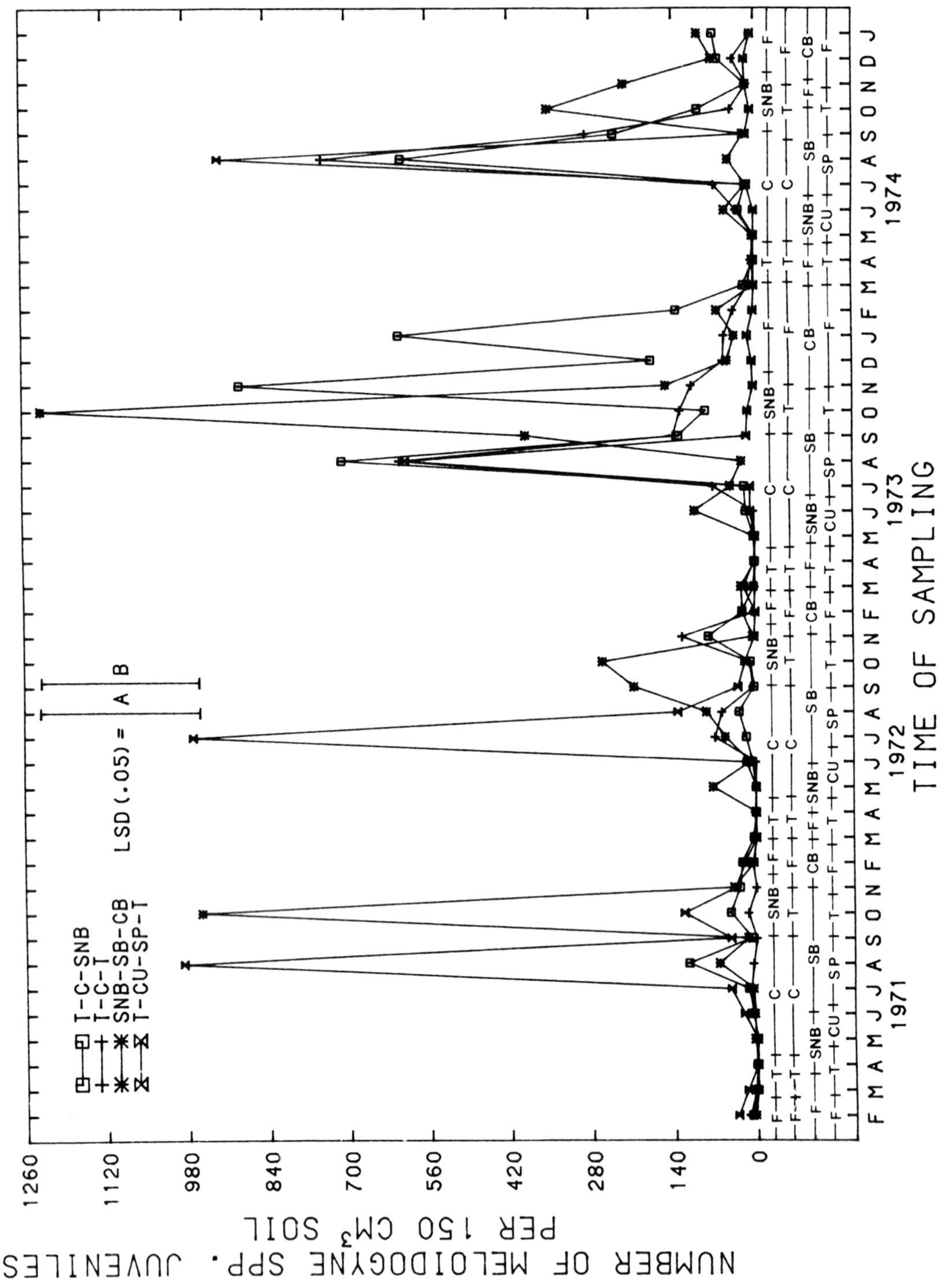

Fig. 1. Effect of four intensive cropping systems on population densities of *Meloidogyne* spp. as determined at monthly intervals for 4 years. A = L.S.D. 0.05 for different cropping systems on the same sampling date. B = L.S.D. 0.05 for the same cropping system on different sampling dates. T, C, SNB, SB, CB, CU, SP, and F = turnip, corn, snap bean, soybean, cabbage, cucumber, southern pea, and fallow, respectively.

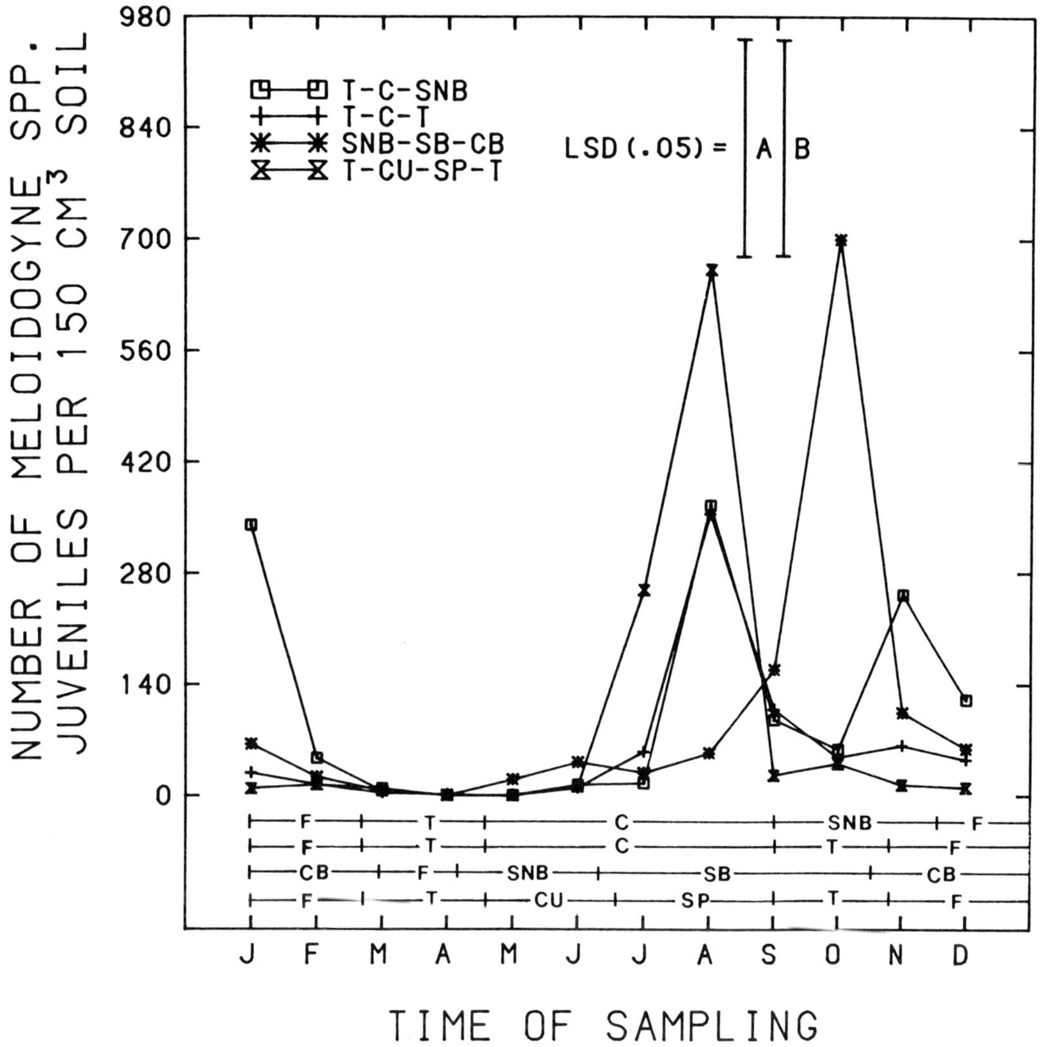

Fig. 2. **Effect of four intensive cropping systems on population densities of *Meloidogyne* spp. as determined by plotting means of five replicates for each month over 4 years. A = L.S.D. 0.05 for different cropping systems on the same sampling date. B = L.S.D. 0.05 for the same cropping system on different sampling dates. T, C, SNB, SB, CB, CU, SP, and F = turnip, corn, snap bean, soybean, cabbage, cucumber, southern pea, and fallow, respectively.**

SNB, T-C-T, T-P-T, and SNB-SB-CB systems was not affected significantly by ethoprop. Root-gall indices of cucumber in the T-CU-SP-T system were lower each year in ethoprop-treated plots than in untreated plots (Table 3). Root-gall indices of southern pea following cucumber in plots treated with ethoprop were lower than those in untreated plots in 1971 and 1972, but not in 1973 and 1974.

Data from sub-plots (nematicide vs. no nematicide) indicate that an annual application of ethoprop before corn, peanut, and snap bean were planted did not increase yields of these crops or succeeding crops (Table 4). Yields of snap bean in the SNB-SB-CB system were lower in ethoprop-treated plots than in untreated plots in 1972, but yields were not affected by the nematicide during the other three years. Yield of cucumber in the T-CU-SP-T system was higher each year in ethoprop-treated plots than in untreated plots, but yields of succeeding crops were not affected by the nematicide (Table 5).

Application of ethoprop before corn, peanut,

Table 3. Effect of ethoprop on root-gall indices for a turnip-cucumber-southern pea-turnip cropping system, 1971-74.

Nematicide treatment[b]	Root-gall indices[a] Turnip	Cucumber	Southern pea	Turnip	Turnip	Cucumber	Southern pea	Turnip
	1971				1972			
No	1.02	2.01 a	2.41 a	1.09	1.00	2.92 a	2.30 a	2.32
Yes	1.02	1.15 b	1.57 b	1.05	1.00	1.39 b	1.41 b	1.80
	1973				1974			
No	1.35 a	3.02 a	2.50	1.11	1.00	2.28 a	2.51	1.12
Yes	1.09 b	1.76 b	2.19	1.21	1.00	1.33 b	2.37	1.09

[a] Root-gall index on 1-5 scale: 1 = no galling, 2 = 1-25%, 3 = 26-50%, 4 = 51-75%, and 5 = 76-100% roots galled. Means followed by unlike letters are significantly different ($P = 0.05$) according to Duncan's multiple-range test. Absence of letters indicates nonsignificance.

[b] Ethoprop 10G was applied (8.96 kg a.i./ha) and incorporated into the soil 15 cm deep with a tractor-powered rototiller just before planting cucumber each year. Ethoprop was not applied before planting other crops.

and snap bean were planted did not suppress nematode popualtions consistently during the season or increase yields. These results indicate that the nematode populations were below the threshold level or that ethoprop was not lethal to the nematodes. Rohde et al. (22) reported that numbers of *M. incognita* in the turnip-field corn-southern pea cropping system were not suppressed on field corn and southern pea when treated with ethoprop (8.96 kg a.i./ha).

Numbers of *M. incognita* juveniles in plots following fall and spring turnip were near or below detectable levels when cucumber was planted; however, application of ethoprop consistently reduced root-galling in cucumber and increased yields 85% (range 50-118%). We conclude that a single application of ethoprop (8.96 kg a.i./ha) before planting a tolerant crop (corn) or a nonhost crop (peanut) to *M. incognita* is not beneficial for control of nematodes and does not improve yields when numbers of *M. incognita* juveniles in the soil are less than 25/150 cm^3 soil at time of planting. However, with similar nematode inoculum levels, a single application of the nematicide before planting a susceptible crop (cucumber) to *M. incognita* will provide protection from root-knot nematodes and increase yields.

The lack of residual nematode control by ethoprop in succeeding crops was expected. We previously reported that ethoprop is not a persistent chemical in Coastal Plain soils and that the half-life ranges from 3 to 15 days under field conditions (22).

Table 6 lists the suitability ratings, based on the nematode population data in Figures 1 and 2, of each intensive cropping system for use in suppressing populations of *M. incognita*. The T-P-SNB and T-P-T systems were rated good because numbers of *M. incognita* juveniles in the soil were less than 140/150 cm^3 soil on all sampling dates. The T-C-SNB and T-C-T systems were rated good/poor because numbers of *M. incognita* increased with time in both cropping systems. Numbers of *M. incognita* juveniles in the T-C-SNB and T-C-T systems were less than 140/150 cm^3 soil on all sampling dates through June 1973 but were above 420/150 cm^3 soil in 1973 and 1974. Both SNB-SB-CB and T-CU-SP-T systems were rated poor because numbers of *M. incognita* juveniles exceeded 420/150 cm^3 soil on soybean and southern pea, respectively. These results demonstrate the value of integrating nonhost crops (peanut) and poor hosts (turnip) with susceptible crops (cucumber, soybean) to minimize damage caused by root-knot nematodes. Different populations of *M. incognita* and cultivars of plants may respond differently than reported in this manuscript.

2. A STUDY USING THREE MULTI-CROPPING SYSTEMS AND FOUR NEMATICIDE TREATMENTS UNDER IRRIGATION

Materials and Methods

Nematode populations were studied in an integrated pest management program under irrigation with three intensive multi-cropping systems and four nematicide treatments. Plots were established in September 1974 on Tifton sandy loam (75% sand, 10% silt, 15% clay) naturally infested with root-knot nematodes, *Meloidogyne* spp. (ca. 90% *M. incognita* (Kofoid & White) Chitwood and 10% *M. hapla*

Table 4. Effects of ethoprop on yields (kg/ha) of crops, 1971-74.

	1971 yield			1972 yield[c]		
Nematicide[a]	**Turnip**	**Corn**	**Snap bean**	**Turnip**	**Corn**	**Snap bean**
No	18804	8788	2474	14943	7374	3135
Yes	18316	8382	2401	14879	8128	3058
	Turnip	**Peanut**	**Snap bean**	**Turnip**	**Peanut**	**Snap bean**
No	17953	1932	2468	14360	824	3223
Yes	17112	2172	2358	13215	1011	3372
	Turnip	**Corn**	**Turnip**	**Turnips**	**Corn**	**Turnip**
No	17529	8946	22409	14543	8039	23107
Yes	18179	9068	20829	15388	8793	24214
	Turnip	**Peanut**	**Turnip**	**Turnip**	**Peanut**	**Turnip**
No	17331	1770	22673	13618	1111	24703
Yes	17712	2087	22787	13093	1221	23478
	Snap bean	**Soybean**	**Cabbage[b]**	**Snap bean**	**Soybean**	**Cabbage**
No	5658	3163	——	3852 a	1044	2663 a
Yes	4954	3102	——	2469 b	790	2318 b

	1973 yield			1974 yield		
Nematicide[a]	**Turnip**	**Corn**	**Snap bean**	**Turnip**	**Corn**	**Snap bean[d]**
No	10511	6723	1129	14061	6475	——
Yes	9392	6970	1044	14799	7210	——
	Turnip	**Peanut**	**Snap bean**	**Turnip**	**Peanut**	**Snap bean**
No	7192	659	1504	14719	1082	——
Yes	9248	738	1443	13987	946	——
	Turnip	**Corn**	**Turnip**	**Turnip**	**Corn**	**Turnip**
No	8931	6747	11786	12902	6545	9986
Yes	9123	7174	9492	12709	6946	11380
	Turnip	**Peanut**	**Turnip**	**Turnip**	**Peanut**	**Turnip**
No	7107	1138	11356	9083	1302	9120
Yes	6460	1346	10855	9986	1290	7006
	Snap bean	**Soybean**	**Cabbage**	**Snap bean**	**Soybean**	**Cabbage**
No	3093	1306	——	1436	1391	31424
Yes	3215	1202	——	1119	1278	30374

[a] Ethoprop 10G was applied (8.96 kg a.i./ha) and incorporated into the soil 15 cm deep with a tractor-powered rototiller just before planting corn, peanut, and the spring crop of snap bean each year. Ethoprop was not applied before planting other crops.
[b] Crop destroyed by freeze in 1971 and 1973.
[c] Means followed by unlike letters are significantly different ($P = 0.05$) according to Duncan's multiple range test. Absence of letters indicates nonsignificance.
[d] No snap bean yield data available for 1974.

Chitwood); lesion nematodes, *Pratylenchus* spp. (ca.65% *P. scribneri* Steiner, 25% *P. brachyurus* (Godfrey) Filip. & Sch.-Stek., and 10% *P. zeae* Graham); stubby-root nematodes, *Paratrichodorus minor* (Colbran) Siddiqi; ring nematodes, *Macroposthonia ornata* (Raski) De Grisse & Loof; and spiral nematodes, *Helicotylenchus dihystera* (Cobb) Sher. Soil pH was maintained between 6.0 and 6.7 as measured in a saturated paste. Each plot contained three beds of 1.8 x 7.6 m. The experimental design was a split-plot in strips with treatments replicated six times. Treatments are referred to as 1) MBR-CP, 2) maximum, 3) optimum/as needed, and 4) minimum, respectively.

Table 5. Effects of ethoprop on yields (kg/ha) of crops, 1971-74.

Nematicide treatment[a]	Turnip	Cucumber	Southern pea	Turnip	Turnip	Cucumber	Southern pea	Turnip
	1971 yield[b]				1972 yield			
No	16007	6024 b	4170	17168	12358	3519 b	1120	15795
Yes	16458	9035 a	3892	18291	13147	7679 a	1410	22121
	1973 yield				1974 yield			
No	8483	5024 b	1707	9331	9074	4502 b	842	7531
Yes	7028	10123 a	1901	7875	8723	8390 a	814	9147

[a] Ethoprop 10G was applied (8.96 kg a.i./ha) and incorporated into the soil 15 cm deep with a tractor-powered rototiller just before planting cucumber. Ethoprop was not applied before planting other crops.
[b] Means followed by unlike letters are significantly different ($P = 0.05$) according to Duncan's multiple-range test. Absence of letters indicates nonsignificance.

Table 6. Suitability ratings of cropping systems for use in managing *Meloidogyne incognita* populations, 1971-74.

Cropping system[a]	Ratings based on numbers of nematodes/150 cm^3 soil[b]
T-C-SNB	Good[c], Poor[d]
T-P-SNB	Good
T-C-T	Good[c], Poor[d]
T-P-T	Good
SNB-SB-CB	Poor
T-CU-SP-T	Poor

[a] T, C, SNB, P, SB, CB, CU and SP = turnip, corn, snap bean, peanut, soybean, cabbage, cucumber, and southern pea, respectively.
[b] Good = < 140; Fair = 140-420; Poor = > 420.
[c] Suitability rating for 1971-72.
[d] Suitability rating for 1973-74.

1) MBR-CP Between 14 November and 8 February each year, soil was treated with 98% methyl bromide + 2% chloropicrin (MBR-CP), 392 kg/ha injected 25 cm deep with chisels 20 cm apart. The soil surface was shaped with a bedding attachment, and covered with black polyethylene (152 μm thick) for 48 hours. No other soil pesticides were applied, and weeds were removed by hand in each crop.
2) MAXIMUM (DD-MENCS) Soil was fumigated between 14 November and 17 December each year with 20% methyl isothiocyanate + 80% chlorinated C_3 hydrocarbons (DD-MENCS), 376.3 kg/ha, by the same methods described for MBR-CP, except plots were not covered with plastic. Ethoprop (8.96 kg a.i./ha) was applied by spreading granules on the soil surface and incorporating them 15 cm deep with a tractor-driven rototiller before planting each crop in 1975 and 1976. A maximum control of weeds and insects was maintained by the application of herbicides and insecticides on each crop.
3) AS NEEDED (ETHOPROP/FENAMIPHOS) Ethoprop was applied before planting each crop from 1975-1978 when numbers of root-knot nematode juveniles in the plots exceeded 25/150 cm^3 on the preceding crop, and fenamiphos (8.96 kg a.i./ha) was used in a similar manner in 1979 and 1980. Herbicides and cultivation and insecticides were used as needed for weed and insect control.
4) MINIMUM (NO NEMATICIDES) Pesticide usage was limited. Nematicides were not used. One herbicide and cultivation were used on each crop to control weeds, and insecticides were applied as determined by scouting reports.

The three cropping systems evaluated were: 1) Turnip-Peanut-Cucumber-Turnip-Cucumber-Soybean (T-P-CU-T-CU-SB); 2) Sweet corn-Soybean-Wheat-Soybean-Spinach (SC-SB-W-SB-S); and 3) Turnip-Field corn-Southern pea (T-FC-SP). The first two systems were two-year cycles, and the last cropping system was a one-year cycle. The crops, cultivars, and planting dates are listed in Table 7.

The soil was assayed monthly for plant-parasitic nematodes during the six years. Twenty cores of soil, 2.5 x 15 cm deep, were collected from each subplot, composited, and thoroughly mixed. A 150-cm^3 sample was processed by the centrifugal-flotation method (7). Plant roots of most crops were examined for galls induced by root-knot nematodes (*M. incognita* or *M. hapla*) one month after planting and after harvest each year. Twenty plants selected at random were uprooted and rated for galls on a 1-5 scale as described earlier.

Results and Discussion

Numbers of root-knot nematode juveniles were lowest in the SC-SB-W-SB-S cropping

Table 7. Cultivars and planting dates for crops in intensive cropping systems, 1975-80.

Crop	Cultivar	Planting date					
		1975	1976	1977	1978	1979	1980
Turnip	'Shogoin'	26 Feb.	13 Feb.	22 Feb.	15 Feb.	15 Feb.	20 Feb.
Peanut	'Florunner'	23 Apr.	——	15 Apr.	——	10 Apr.	——
Cucumber[a]		16 Sept.	5 Apr.	2 Sept.	17 Apr.	7 Sept.	25 Apr.
Soybean	'Coker' 136	17 June	4 June	14 June	9 June	18 June	17 June
Sweet corn	'Merit'	21 Mar.	——	28 Mar.	——	2 Apr.	——
Wheat	'Oasis'	12 Nov.	——	2 Nov.	——	29 Oct.	——
Spinach	'Grandstand'	——	29 Oct.	——	17 Oct.	——	15 Oct.
Corn[b]		23 Apr.	2 Apr.	15 Apr.	12 Apr.	11 Apr.	25 Apr.
Southern pea[c]		21 Aug.	10 Aug.	10 Aug.	9 Aug.	8 Aug.	14 Aug.

[a] 'Chipper' (1975-78) and 'Lipso' (1979-80)
[b] 'Pioneer 3369A' (1975-78) and 'Funk's 4507' (1979-80).
[c] 'GA 226' (1975, 1976), 'Pinkeye Purple Hull' (1977, 1978), and 'Worthmore' (1979-80).

system, intermediate in the T-P-CU-T-CU-SB system, and highest in the T-FC-SP system (Fig. 3A-C). Root-knot nematodes increased more on soybean than on other crops in the SC-SB-W-SB-S system.

Numbers of *M. hapla* juveniles increased on peanut in the T-P-CU-T-CU-SB system (Fig. 4). Generally, numbers of nematodes were highest on peanut in August or September and declined to near or below detectable levels on cucumber following turnip.

Numbers of *Meloidogyne incognita* increased on corn in August and on Pinkeye Purple Hull southern pea in October or November in the T-FC-SP system (Fig. 5). Numbers of these nematodes were greater on Pinkeye Purple Hull than on Worthmore southern pea (Fig. 6). Turnip, wheat, sweet corn, and spinach were poor hosts for *M. incognita* and *M. hapla*.

Numbers of *M. incognita* and *M. hapla* in plots treated with MBR-CP and DD-MENCS were near or below detectable levels on turnip and corn, but increased rapidly on southern pea following corn. Numbers of nematodes in plots treated with ethoprop on an as needed basis were not different ($P = 0.05$) from those in untreated plots on most sampling dates; however, numbers of nematodes in plots treated with fenamiphos were similar to those in plots treated with MBR-CP and DD-MENCS (Fig. 7).

Root-gall indices of turnip, soybean, and Worthmore southern pea were low in all plots (Table 8). Root-gall indices of peanut, cucumber, spinach, and Pinkeye Purple Hull southern pea were lower in plots treated with MBR-CP and DD-MENCS than in other plots.

Average yields across all crops in all cropping systems from plots treated with MBR-CP, DD-MENCS, and ethoprop/fenamiphos were 47%, 14%, and 1% greater, respectively, than yields of crops from untreated plots. Average yields across all crops within the various cropping systems from plots treated with MBR-CP, DD-MENCS, and ethoprop/fenamiphos, respectively, were greater than those from untreated plots as follows: T-P-CU-T-CU-SB, 87%, 21%, and 1%; SC-SB-W-SB-S, 15%, 4%, and 1%; and T-FC-SP, 29%, 16%, and 1% (Table 9).

Economic data including gross and net returns and pesticide costs are presented in Table 10. Based on net returns, MBR-CP and DD-MENCS were not economically feasible. Ethoprop/fenamiphos was not feasible on the T-P-CU-T-CU-SB system but was on the SC-SB-W-SB-S and T-FC-SP cropping systems. The greatest net return in each of the three cropping systems resulted when no nematicides were applied. The net returns per year varied among and within cropping systems. The least net return for the three cropping systems resulted from the T-C-SB system, and the greatest net return resulted from the T-FC-SP cropping system (Table 11).

In summary, the low numbers of *M. incognita* in the SC-SB-W-SB-S cropping system were expected. Sweet corn supports low (0 to 50/150 cm^3 soil) numbers of *M. incognita*, but in other tests, yield (weight, number of ears, and weight per ear) has not been inversely correlated with numbers of *M. incognita* juveniles in the soil (8). 'Coker 136' soybean is resistant to *M. incognita* (19). Spinach (27) and wheat (21) are poor hosts for *M. incognita* when soil temperatures are less than 18°C.

The increase of *M. hapla* on peanut in the T-P-CU-T-CU-SB sequence was unexpected because only *M. incognita* had been identified from these plots previously. Yields of the second crop of peanut were reduced 51% when *M. hapla* was not controlled.

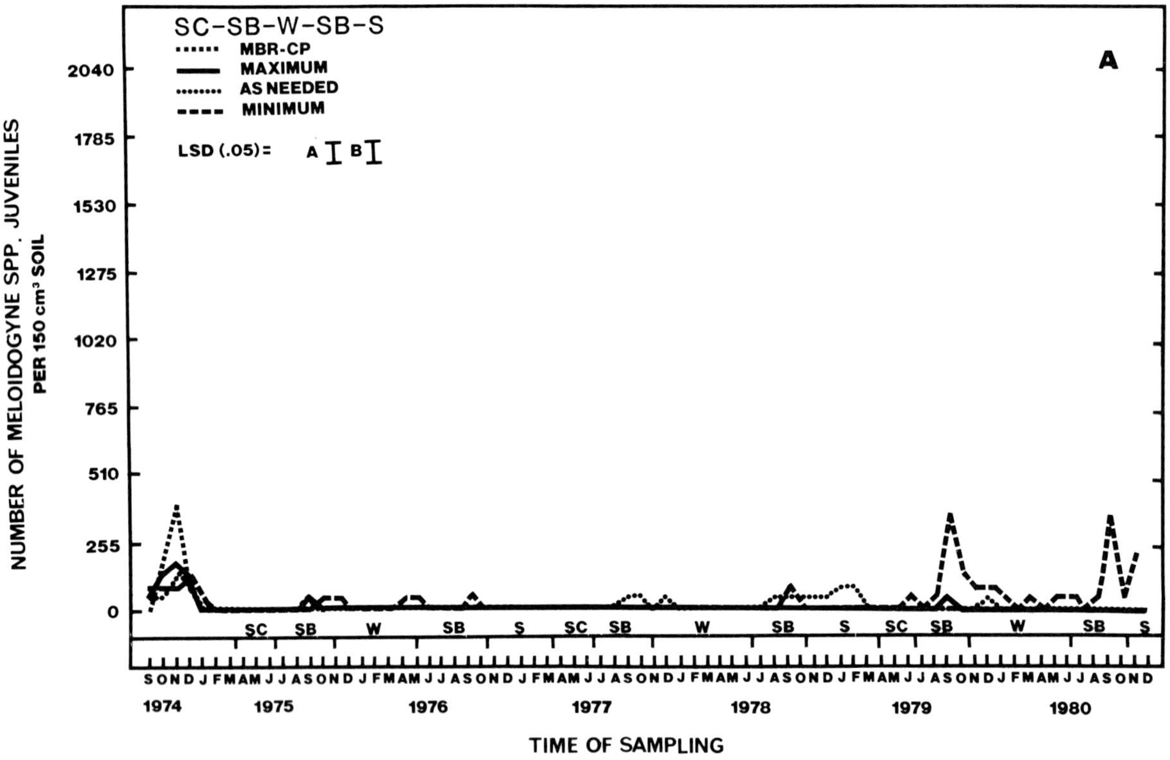

Fig. 3 A-C. Effect of four pest management levels on population densities of *Meloidogyne* spp. in three intensive cropping systems: 3A) SC-SB-W-SB-S = sweet corn-soybean-wheat-soybean-spinach; 3B) T-P-CU-T-CU-SB = turnip-peanut-cucumber-turnip-cucumber-soybean; 3C) T-FC-SP-F = turnip-field corn-southern pea-fallow; A = L.S.D. 0.05 for different pest management levels on the same sampling date; B = L.S.D. 0.05 for the same pest management level on different sampling dates; MBR-CP = 98% methyl bromide + 2% chloropicrin, 392 kg/ha injected 25 cm deep into the soil with chisels 20 cm apart, applied in November or December each year; Maximum = DD-MENCS, 376.3 kg/ha applied in November or December each year plus ethoprop, 8.96 kg a.i./ha applied by spreading granules on the soil surface and incorporating 15 cm deep with a tractor-driven rototiller before planting each crop in 1975 and 1976. After 1976 the application of ethoprop was discontinued; As Needed = ethoprop was applied as described above before planting each crop from 1975-1978 when numbers of root-knot nematode juveniles in the soil exceeded 25/150 cm^3 soil on the preceding crop and fenamiphos (8.96 a.i./ha) was used in a similar manner in 1979 and 1980; and Minimum = control, nematicides were not used.

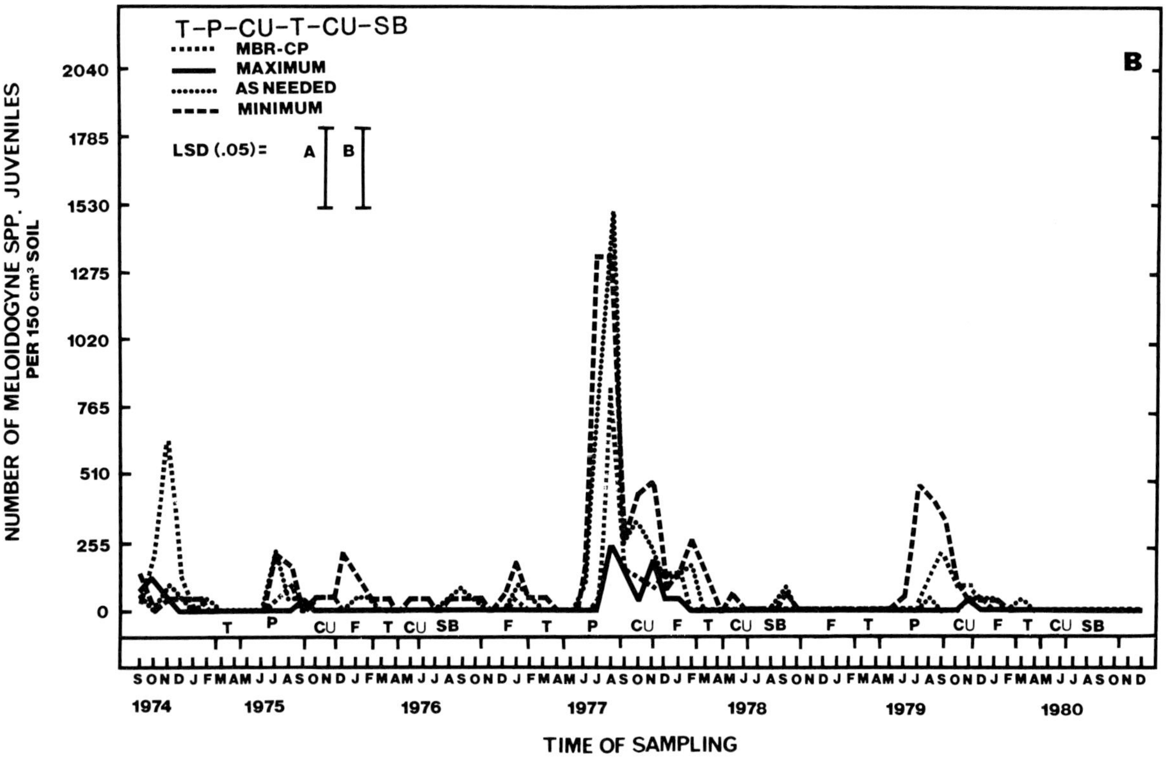
T-P-CU-T-CU-SB
MBR-CP
MAXIMUM
AS NEEDED
MINIMUM
LSD (.05)= A B
B
NUMBER OF MELOIDOGYNE SPP. JUVENILES PER 150 cm³ SOIL
2040
1785
1530
1275
1020
765
510
255
0
T P CU F T CU SB F T P CU F T CU SB F T P CU F T CU SB
S O N D J F M A M J J A S O N D J F M A M J J A S O N D J F M A M J J A S O N D J F M A M J J A S O N D J F M A M J J A S O N D J F M A M J J A S O N D
1974 1975 1976 1977 1978 1979 1980
TIME OF SAMPLING

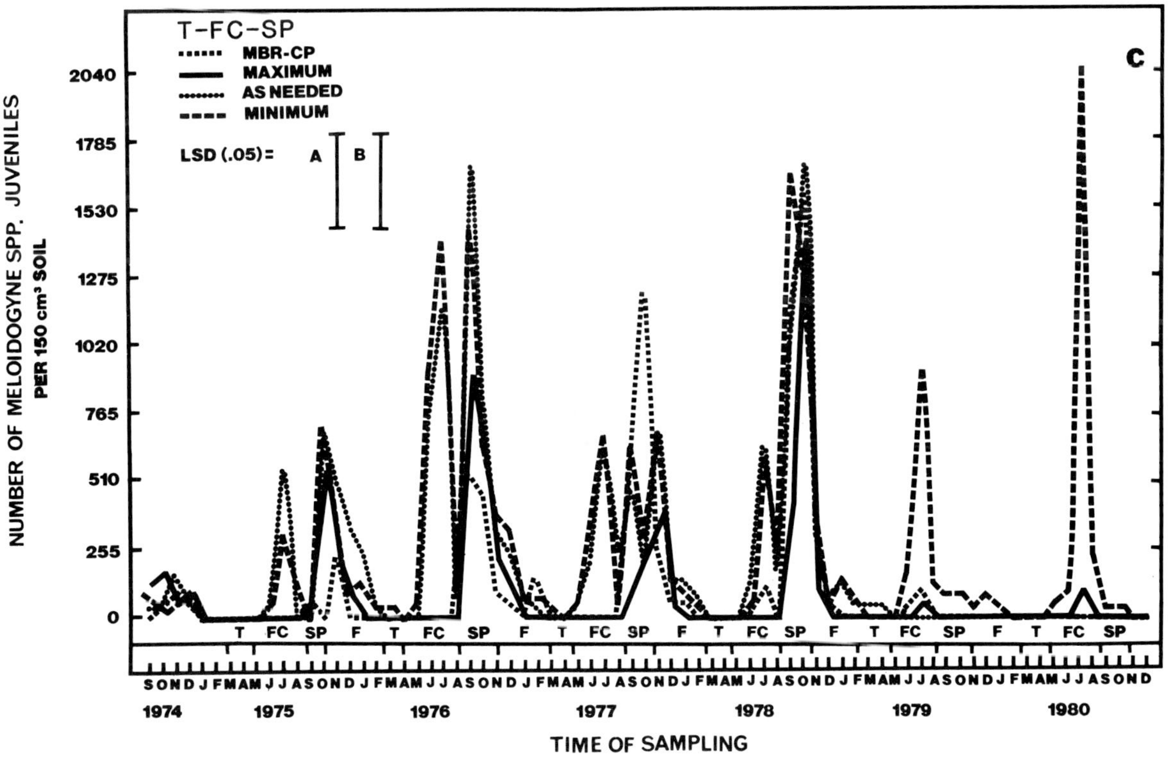
T-FC-SP
MBR-CP
MAXIMUM
AS NEEDED
MINIMUM
LSD (.05)= A B
C
NUMBER OF MELOIDOGYNE SPP. JUVENILES PER 150 cm³ SOIL
2040
1785
1530
1275
1020
765
510
255
0
T FC SP F T FC SP F T FC SP F T FC SP F T FC SP F T FC SP
S O N D J F M A M J J A S O N D J F M A M J J A S O N D J F M A M J J A S O N D J F M A M J J A S O N D J F M A M J J A S O N D J F M A M J J A S O N D
1974 1975 1976 1977 1978 1979 1980
TIME OF SAMPLING

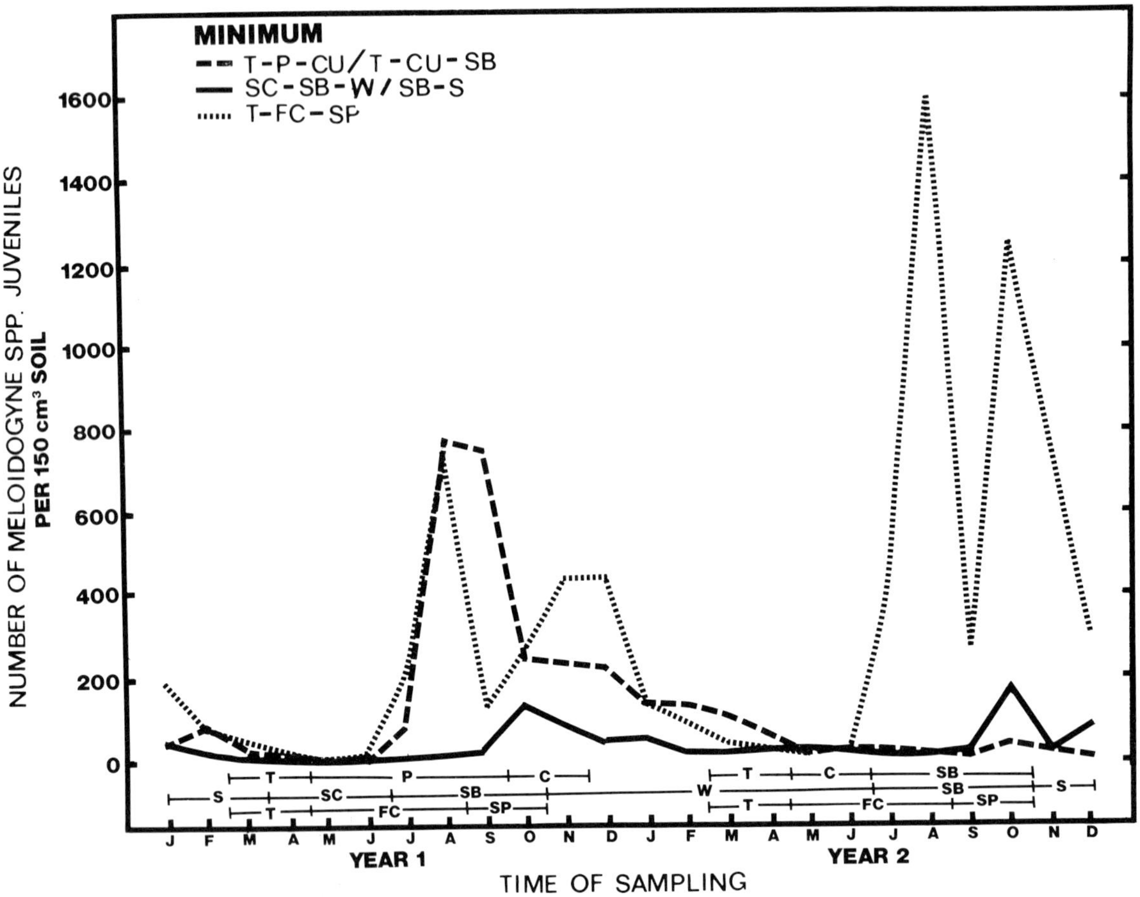

Fig. 4. Effect of three intensive cropping systems on population densities of *Meloidogyne* spp.: T-P-CU/T-CU-SB = turnip-peanut-cucumber/turnip-cucumber-soybean; SC-SB-W/SB-S = sweet corn-soybean-wheat/soybean-spinach; and T-FC-SP = turnip-field corn-southern pea. Curves for *Meloidogyne* spp. were obtained by plotting means of six replicates for each month over six years.

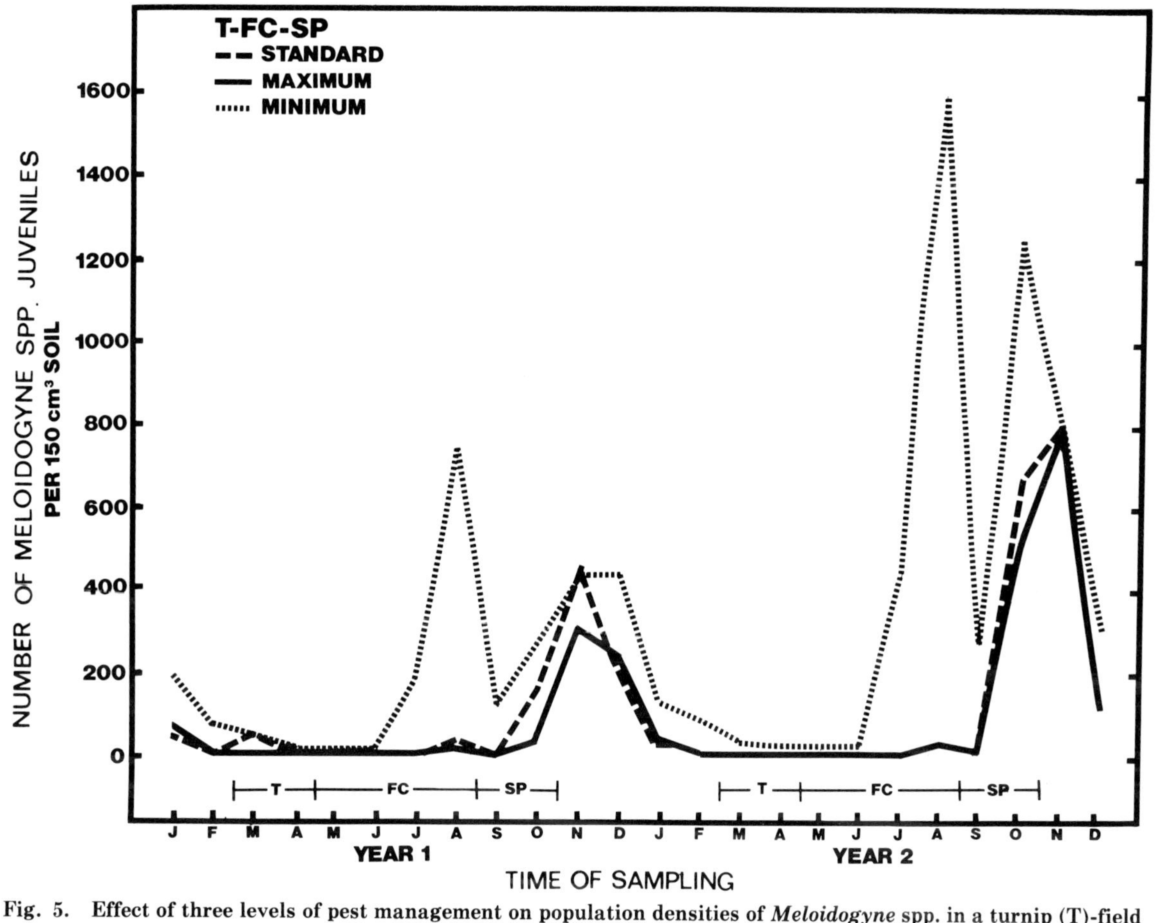

Fig. 5. Effect of three levels of pest management on population densities of *Meloidogyne* spp. in a turnip (T)-field corn (FC)-southern pea (SP) cropping system: Standard = 98% methyl bromide + 2% chloropicrin, 392 kg/ha injected 25 cm deep into the soil with chisels 20 cm apart applied in November or December each year; Maximum = DD-MENCS, 276.3 kg/ha applied in November or December each year plus ethoprop, 8.96 kg a.i./ha applied by spreading granules on the soil surface and incorporating 15 cm deep with a tractor-driven rototiller before planting each crop in 1975 and 1976. After 1976, the application of ethoprop was discontinued; Minimum = control, nematicides were not used. Curves for *Meloidogyne* spp. were obtained by plotting means of six replicates for each month over six years.

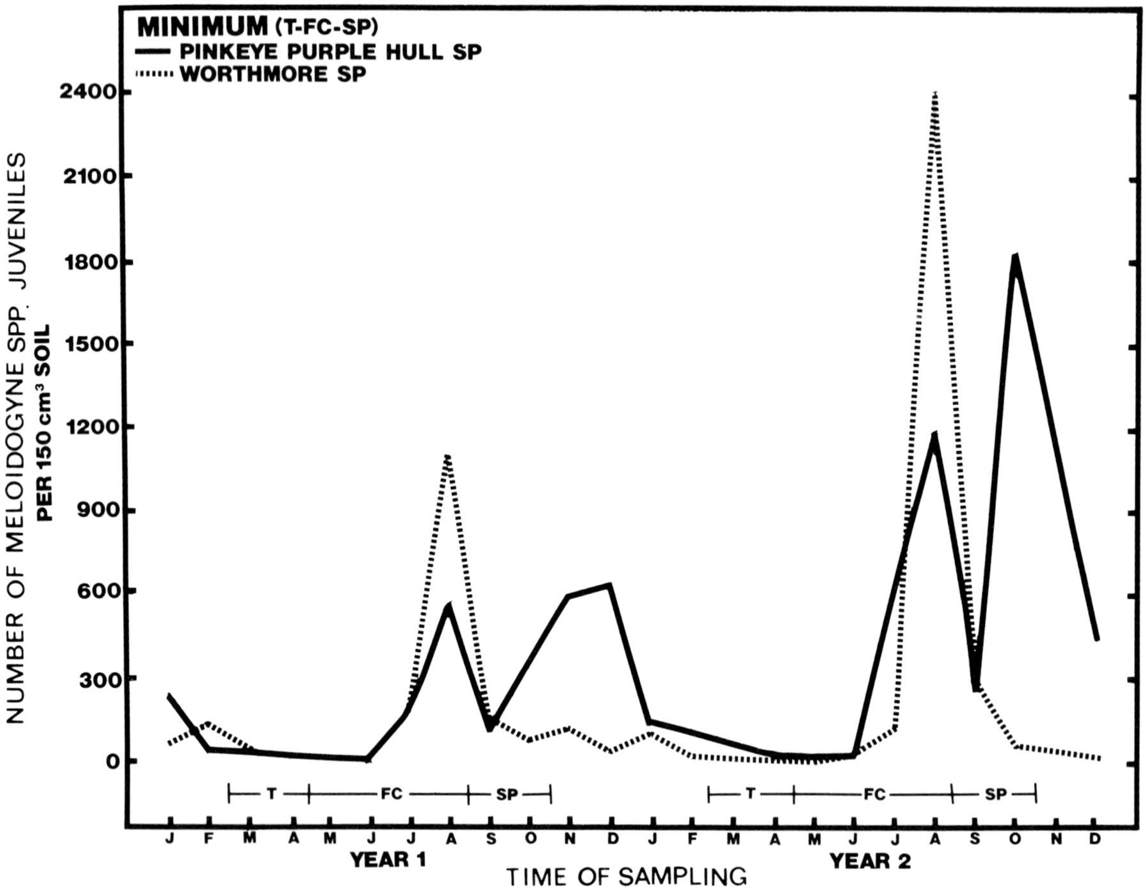

Fig. 6. Effect of two cultivars of southern pea on population densities of *Meloidogyne* spp. in a turnip (T)-field corn (FC)-southern pea (SP) intensive cropping system: 'Pinkeye Purple Hull' (susceptible) and 'Worthmore' (resistant). Curves for *Meloidogyne* spp. were obtained by plotting means of six replicates for each month over four years for 'Pinkeye Purple Hull' and two years for 'Worthmore.'

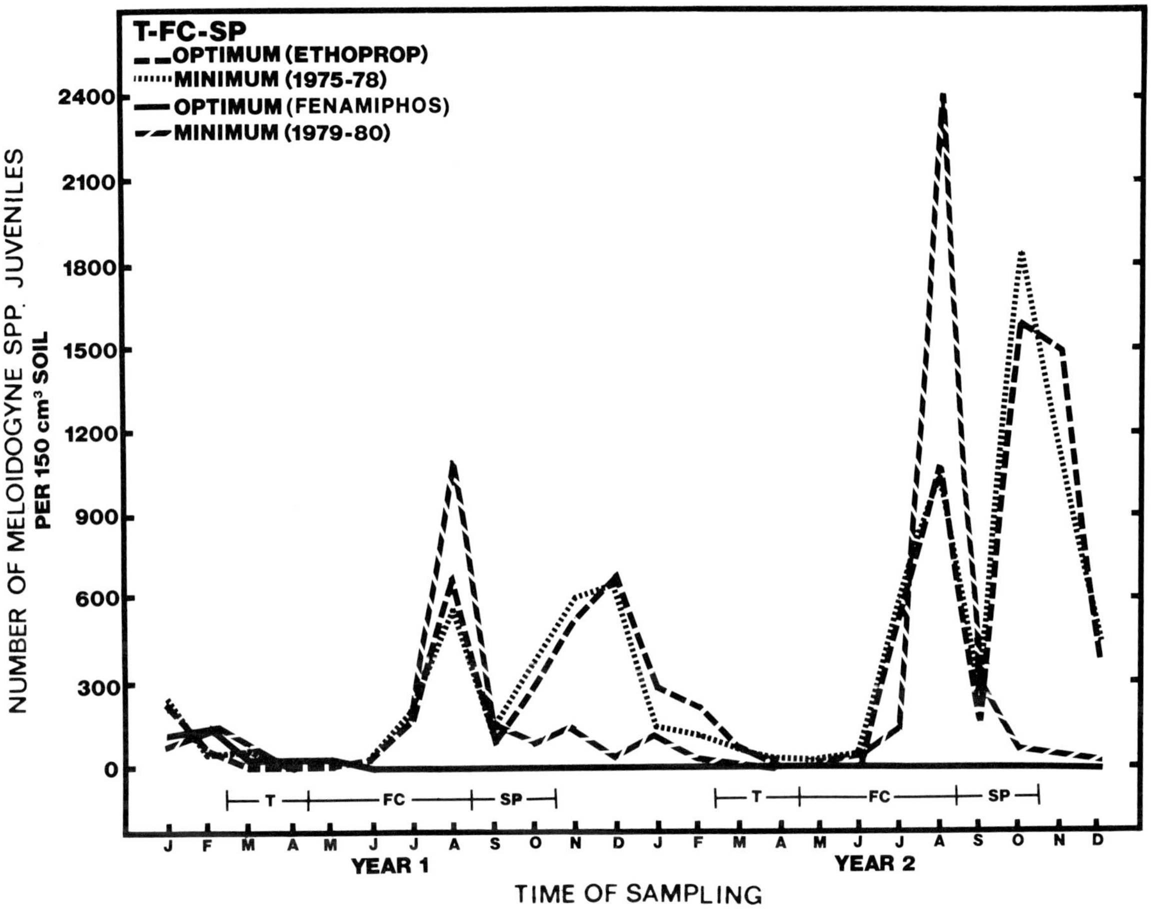

Fig. 7. Effect of ethoprop (8.96 kg a.i./ha) and fenamiphos (8.96 kg a.i./ha) on population densities of *Meloidogyne* spp. in a turnip (T)-field corn (FC)-southern pea (SP) intensive cropping system. Curves for *Meloidogyne* spp. were obtained by plotting means of six replicates for each month over four years for ethoprop vs. minimum (1975-1978) and two years for fenamiphos vs. minimum (1979-1980). Minimum = control, nematicides were not used.

Table 8. Effect of intensive cropping and nematicides on root-gall indices of crops, 1975-1980.

Pest management level[a]	Root-gall indices[b] within each cropping system[c]					
	Turnip	Peanut	Cucumber	Turnip	Cucumber	Soybean
MBR-CP	1.00	1.05 b	1.96	1.00	1.00 b	1.00
Maximum	1.00	1.10 b	1.46	1.00	1.00 b	1.00
As Needed	1.10	1.30 a	1.64	1.00	2.45 a	1.05
Minimum	1.20	1.40 a	1.76	1.10	2.50 a	1.05
	Sweet corn	Soybean	Wheat	Soybean	Spinach	
MBR-CP	——	1.00	——	1.00	1.00 b	
Maximum	——	1.00	——	1.00	1.02 b	
As Needed	——	1.07	——	1.05	1.60 a	
Minimum	——	1.17	——	1.05	1.52 a	
			Southern pea			
	Turnip	Field corn	'Pinkeye Purple Hull'	'Worthmore'		
MBR-CP	1.00 b	1.00 b	1.38 b	1.00		
Maximum	1.00 b	1.00 b	1.52 b	1.00		
As Needed	190 a	1.30 a	2.08 a	1.00		
Minimum	2.22 a	1.56 a	2.10 a	1.00		

[a] MBR-CP = 98% methyl bromide + 2% chloropicrin applied at 392 kg/ha annually in November or December each year; Maximum = DD-MENCS (20% methyl isothiocyanate + 80% chlorinated C_3 hydrocarbons) applied at 376.3 kg/ha annually; As Needed = ethoprop applied at 8.96 kg a.i./ha before planting each crop from 1975-1978 and fenamiphos applied at 8.9 kg a.i./ha before planting each crop in 1979 and 1980; and Minimum = no nematicides.

[b] 1-5 Scale: 1 = no galls, 2 = 1-25%, 3 = 26-50%, 4 = 51-75%, and 5 = 76-100% galled. Means for the same crop followed by unlike letters are significantly (P = 0.05) different according to Duncan's multiple-range test. Absence of letters indicates nonsignificance.

[c] Turnip-Peanut-Cucumber-Turnip-Cucumber-Soybean, two-year cycle; Sweet corn-Soybean-Wheat-Soybean-Spinach, two-year cycle; and Turnip-Field corn-Southern pea, one-year cycle.

The rapid increase of *M. incognita* on Pinkeye Purple Hull southern pea and field corn was expected since both crops are good hosts for this nematode (3, 15). The generally poor performance of ethoprop in managing root-knot nematodes in these intensive cropping systems was unexpected because the nematicide has been reported to control root-knot nematodes on many crops (12, 16, 17) in monocultural (one crop per year) systems. Poor performance of ethoprop has been reported for other intensive cropping systems (14).

The resistance to *M. incognita* in Worthmore southern pea was beneficial. Worthmore is more suitable for fresh market than for processing and may prove valuable in home gardens where damage from root-knot nematodes is usually severe.

The excellent control of root-knot nematodes by fenamiphos was expected. Since this chemical suppresses nematode populations and does not require a waiting period before planting, it has potential use in intensive crop production systems.

Our data indicate root-knot nematode populations can be manipulated by intensive cropping systems and by crops within each system. Furthermore, they confirm and extend previously reported results (14, 15). In addition to determining the suitability of certain crop plants as hosts to *M. incognita* and *M. hapla*, we have established the importance of growing agronomic and vegetable crops in succession to produce three crops each year or five crops in two years from the same land area.

These integrated pest management studies utilized cultural practices such as destruction of crop residues immediately after harvest, clean fallow from time of harvest until the succeeding crops were planted, crop rotation, nonhost crops, trap crops, resistant cultivars, and nematicides to reduce nematode survival and thus reduce inoculum levels for the subsequent crops. These crops were grown under "optimum

Table 9. Effect of intensive cropping systems and nematicides on yield of crops averaged across each system, 1975-1980.

Pest management level[a]	Yield within each cropping system[b] Turnip (T/A)	Peanut (lb/A)	Cucumber (lb/A)	Turnip (T/A)	Cucumber (lb/A)	Soybean (bu/A)	% increase over control
MBR-CP	8.5	5263	11904	7.2	29162	49.6	87
Maximum	6.2	5365	5982	4.3	13413	44.4	21
As Needed	6.6	3449	5939	5.0	8257	39.4	1
Minimum	5.0	3142	7236	5.8	7733	43.9	—

	Sweet corn (T/A)	Soybean (bu/A)	Wheat (bu/A)	Soybean (bu/A)	Spinach (T/A)	% increase over control
MBR-CP	9.3	49.6	24	50.5	5.4	15
Maximum	9.3	36.4	23	46.4	5.2	4
As Needed	8.6	39.0	28	40.1	4.4	1
Minimum	7.8	36.0	28	41.0	4.9	—

	Turnip (T/A)	Field corn (bu/A)	Southern pea (lb/A shelled) 'Pinkeye Purple Hull'	Southern pea (lb/A shelled) 'Worthmore'	% increase over control
MBR-CP	8.3	200	1341	2188	29
Maximum	5.6	200	1336	2113	16
As Needed	5.0	187	963	2092	1
Minimum	5.9	183	883	1938	—

[a] MBR-CP = 98% methyl bromide + 2% chloropicrin applied at 392 kg/ha annually in November or December each year; Maximum = DD-MENCS (20% methyl isothiocyanate + 80% chlorinated C_3 hydrocarbons) applied at 376.3 kg/ha annually; As Needed = ethoprop applied at 8.96 kg a.i./ha before planting each crop from 1975-1978 and fenamiphos applied at 8.96 kg a.i./ha before planting each crop in 1979 and 1980; and Minimum = no nematicides.

[b] Turnip-Peanut-Cucumber-Turnip-Cucumber-Soybean, two-year cycle; Sweet corn-Soybean-Wheat-Soybean-Spinach, two-year cycle, and Turnip-Field corn-Southern pea, one-year cycle.

production practices" that involved the most current technology available on seeding rates, fertilization and irrigation. These conditions are conducive for maximum plant and root development which subsequently supports large numbers of root-knot nematodes on susceptible plants, but which could also partially mask the damage caused by root-knot nematodes.

In addition, our data demonstrate that maximum pest (nematode, soil-borne insect, fungus, and weed) control with MBR-CP and DD-MENCS is not economically feasible with the dosages that were used on these intensive cropping systems. The minor differences in yield between the As Needed (ethoprop/fenamiphos) and Minimum (no nematicide) treatments indicate that the application of ethoprop in intensive cropping systems is not economically beneficial. These results also demonstrate that cultural practices such as destruction of crop residues immediately after harvest, clean fallow, crop rotation, nonhost crops, trap crops, resistant cultivars, and optimal fertilization and irrigation prevented maximum crop damage by root-knot nematodes.

Our data can be used as a guide in selecting crops and cultivars for intensive cropping systems in root-knot-nematode-infested soil in the southeastern Coastal Plain of the United States. Additional studies are currently being conducted at the Coastal Plain Experiment Station, Tifton, GA, on the effect of tillage methods, crop rotations, nonhost crops, resistant cultivars, winter cover crops, trap crops, and chemicals on managing nematode populations in irrigated intensive cropping systems.

Table 10. Mean returns and costs ($/acre) of crops and pest control inputs by pest management levels in three intensive cropping systems.[a]

Cropping system[b]	Chemical treatment[c]	Returns		Pesticide costs[d]
		Gross	Net	
T-P-CU-T-CU-SB	MBR-CP	2440.45	—614.83	1514.86
(1975-76)	Maximum	1480.10	—838.97	931.96
	As Needed	1291.31	—101.04	260.14
	Minimum	1636.47	333.02	150.78
SC-SB-W-SB-S	MBR-CP	1694.53	—1026.70	1499.65
(1975-76)	Maximum	1750.12	—284.84	767.50
	As Needed	1580.26	353.93	99.16
	Minimum	1491.00	382.56	48.03
T-FC-SP	MBR-CP	1014.18	—352.92	823.16
(1975-79)	Maximum	864.37	—192.29	490.14
	As Needed	773.35	136.43	47.30
	Minimum	779.61	269.51	29.78

[a] All computations are based on representative annual input and output prices.
[b] T = turnip, P = peanut, CU = cucumber, SB = soybean, SC = sweet corn, W = wheat, S = spinach, FC = field corn, and SP = southern pea.
[c] MBR-CP = 98% methyl bromide + 2% chloropicrin, applied at 392 kg/ha annually in November or December; Maximum = DD-MENCS (20% methyl isothiocyanate + 80% chlorinated C_3 hydrocarbons) applied at 376.3 kg/ha annually; As Needed = ethoprop applied at 8.96 kg a.i./ha before planting each crop from 1975-1978 and fenamiphos applied at 8.96 kg a.i./ha before planting each crop in 1979 and 1980; and Minimum = no nematicide.
[d] Includes costs of scouting and application.

Table 11. Mean returns and costs ($/acre) of crops with minimum pest control inputs in three intensive cropping systems by year.[a]

Cropping system[b]	Year	Returns		Pesticide costs[c]
		Gross	Net	
T-P-CU	1975	958.26	216.68	119.85
T-CU-SB	1976	678.21	116.34	30.93
SC-SB	1975	825.20	157.97	27.65
W-SB-S	1976	665.80	244.59	20.38
T-FC-SP	1975-79	779.61	269.51	29.78

[a] All computations are based on representative annual input and output prices.
[b] T, P, CU, SB, SC, W, S, FC, and SP = turnip, peanut, cucumber, soybean, sweet corn, wheat, spinach, field corn, and southern pea, respectively.
[c] Includes costs of scouting and pesticide application.

Literature Cited

1. Apple, J. L. and R. F. Smith. 1976. Integrated pest management. Plenum Press, New York. 200 pp.
2. Bertrand, D. E., A. W. Johnson, C. C. Dowler, N. C. Glaze, R. B. Chalfant, D. R. Sumner, S. C. Phatak, and J. E. Epperson. 1981. Nematode population dynamics in an integrated pest management program with intensive cropping sequences. Phytopathology 71:860 (Abstr.).
3. Chalfant, R. B. and A. W. Johnson. 1972. Field evaluation of pesticides applied to the soil for control of insects and nematodes affecting southern peas in Georgia. J. Econ. Entomol. 65:1711-1713.
4. Council for Agricultural Science and Technology. 1982. Integrated pest management. Counc. Agric. Sci. and Technol. Rep. 93. 105 pp.
5. Good, J. M. 1968. Relation of plant parasitic

nematodes to soil management practices, pp. 113-38 *In* Tropical nematology, ed. G. C. Smart, Jr. and V. G. Perry. Univ. Fla. Press, Gainesville. 152 pp.

6. Good, J. M. 1972. Management of plant parasitic nematode populations, pp. 109-127 *In* Proc. Tall Timbers Conf. on Ecol. Anim. Control by Habitat Management, No. 4. Tallahassee, Fla., February 24-25.
7. Jenkins, W. R. 1964. A rapid centrifugal-flotation technique for separating nematodes from soil. Plant Dis. Rep. 48:692.
8. Johnson, A. W. 1975. Resistance of sweet corn cultivars to plant-parasitic nematodes. Plant Dis. Rep. 59:373-376.
9. Johnson, A. W. 1983. Managing nematode populations in crop production, pp. 193-203 *In* Nematology in the southern region of the United States, ed. R. D. Riggs. South. Coop. Ser. Bull. 276. 206 pp.
10. Johnson, A. W. and G. M. Campbell. 1977. Influence of cropping systems and a nematicide on root-knot nematodes in tomato transplant production. J. Nematol. 9:273-274. (Abstr.).
11. Johnson, A. W. and G. M. Campbell. 1980. Managing nematode population densities on tomato transplants using crop rotation and a nematicide. J. Nematol. 12:6-19.
12. Johnson, A. W., P. D. Dukes, and R. A. Flowers. 1973. Organic pesticides for control of root-knot nematodes on tobacco. Tob. Sci. 17:153-154.
13. Johnson, A. W., C. C. Dowler, J. E. Epperson, D. R. Sumner, N. C. Glaze, R. B. Chalfant, and S. C. Phatak. 1981. Effects of management practices on nematodes and yield in an intensive cropping system. J. Nematol. 13:445 (Abstr.).
14. Johnson, A. W., C. C. Dowler, N. C. Glaze, and D. R. Sumner. 1983. Effects of intensive cropping systems and pesticides on nematodes and crop yields. Agric. Res. Results, USDA-ARS, ARR-S-14/May. 36 pp.
15. Johnson, A. W., C. C. Dowler, and E. W. Hauser. 1974. Seasonal population dynamics of selected plant-parasitic nematodes in four monocultured crops. J. Nematol. 6:187-190.
16. Johnson, A. W., and S. A. Harmon. 1974. Cantaloupe yield and grade increased by chemical control of *Meloidogyne incognita*. Plant Dis. Rep. 58:746-749.
17. Johnson, A. W., and S. A. Harmon. 1974. Lima bean yield increased by chemical control of *Meloidogyne incognita*. Plant Dis. Rep. 58:749-753.
18. Johnson, A. W., D. R. Sumner, C. C. Dowler, and N. C. Glaze. 1976. Influence of three cropping systems and four levels of pest management on populations of root-knot and lesion nematodes. J. Nematol. 8:290-291. (Abstr.).
19. Minton, N. A., M. B. Parker, and R. A. Flowers. 1975. Response of soybean cultivars to *Meloidogyne incognita* and to the combined effects of *M. arenaria* and *Sclerotium rolfsii*. Plant Dis. Rep. 59:920-923.
20. Nusbaum, C. J. and H. Ferris. 1973. The role of cropping systems in nematode population management. Annu. Rev. Phytopathol. 11:424-40.
21. Roberts, P. A., S. D. Van Gundy, and H. E. McKinney. 1981. Effects of soil temperature and planting date of wheat on *Meloidogyne incognita* reproduction, soil populations, and grain yield. J. Nematol. 13:338-345.
22. Rohde, W. A., A. W. Johnson, C. C. Dowler, and N. C. Glaze. 1980. Influence of climate and cropping patterns on the efficacy of ethoprop, methyl bromide, and DD-MENCS for control of root-knot nematodes. J. Nematol. 12:33-39.
23. Ross, J. P. 1962. Crop rotation effects on the soybean cyst nematode population and soybean yields. Phytopathology 52:815-818.
24. Sasser, J. N. 1951. Population dynamics of nematode parasites of tobacco in certain crop rotations. Phytopathology 41:31.
25. Slack, D. A., R. D. Riggs, and M. L. Hamblen. 1981. Nematode control in soybeans: Rotation and population dynamics of soybean cyst and other nematodes. Arkansas Agric. Exp. Stn. Rep. Ser. 263:1-36.
26. Smittle, D. A. and A. W. Johnson. 1982. Effects of management practices on *Meloidogyne incognita* and snap bean yield. J. Nematol. 14:63-68.
27. Sumner, D. R., S. J. Kays, and A. W. Johnson. 1976. Etiology and control of root diseases of spinach. Phytopathology 66:1267-1273.
28. Sumner, D. R., A. W. Johnson, N. C. Glaze, and C. C. Dowler. 1975. Disease, nematode, and weed control in intensive cropping systems. Ga. Agric. Res. 16(4):1-16.
29. Todd, F. A. 1971. System control: A prescription for flue-cured tobacco disease. N. C. Agric. Ext. Serv. Ext. Circ. 530:1-14.

BIOLOGICAL CONTROL OF NEMATODES

P. Jatala

Head, Department of Nematology & Entomology
International Potato Center
Lima, Perú

A strong movement to determine the potential of biological control agents in nematode management has occurred over the past several years. The impetus behind this movement has been largely due to the recent advances in the use of toxic pesticides and an awareness of their danger, the time required for development of resistant cultivars, and the economic pressure on land use which limits the use of rotation and other cultural methods. Since many of the most commonly used nematicides are expensive and being taken out of the market because of their harmful effect on humans as well as their persistence in the soil or contamination of the water table, efforts are concentrating on integration of biological control agents into overall nematode management strategies.

Although several fungi, bacteria and sporozoans are known to reduce nematode populations under laboratory and greenhouse conditions, results of field trials have been inconclusive and, for most cases, disappointing. However, the nematologists at the International Potato Center (CIP) have discovered a fungus *Paecilomyces lilacinus* (2,3,4,5) that consistently and efficiently controls the population of root-knot nematodes *Meloidogyne* spp., a cosmopolitan pest of many important food and fiber crops. This discovery has provided CIP with a comparative advantage in testing the potential of this organism as an alternative method to chemical control of nematodes.

THE RESEARCH APPROACH

During the past several decades, the major research emphasis on the utilization of biological entities for control of plant-parasitic nematodes has been focused principally on the application of nematode-trapping fungi. The action of these organisms is passive and accidental, and they kill only a limited number of nematodes. The nematodes must pass through the trapping rings or be captured by the sticky hyphal network. Nematodes are then killed by the intrusion of globose structures, and the body contents is consumed by the assimilative hyphae arising from these globose bodies (1).

The interaction between nematode-trapping fungi and nematodes is complex. Activities of these fungi may be influenced by soil, pH, moisture, and temperature (6,7,10). Therefore, their establishment and multiplication in the agricultural soil become limiting due to their lack of adaptability and competitiveness as well as their slow multiplication and dissemination rate. As a result, their field performance has been inconsistent.

Egg parasites, on the other hand, are more dramatic in reducing the nematode population. Nematode eggs of the group Heteroderidae and those deposited in a gelatinous matrix are more vulnerable to attack by these organisms than are those of migratory parasites. Once in contact with the cysts or egg masses, the fungus grows rapidly and eventually parasitizes all the eggs that are in the early embryonic developmental stages. The discovery of fungi parasitic on the eggs of plant-parasitic nematodes has been very recent. Several organisms have been reported to parasitize eggs of plant-parasitic nematodes (9,11,12,13). Although the majority of these organisms are associated with the eggs and cysts of important plant-parasitic nematodes, their efficiency in and adaptability to different soil and environmental conditions vary. Some have limited reproductive potential (8,12) while others require enriched laboratory media for sporulation or have limited saprophytic growth (8).

Discovery of *P. lilacinus* as an effective parasite of *Meloidogyne* eggs and its further confirmation as an organism having many attributes of a successful biological control agent led to further greenhouse and field experimen-

Table 1. International *Meloidogyne* project regions and/or countries that received cultures of *Paecilomyces lilacinus*.

Central America Caribbean	North & South Americas	Asia & Pacific	Africa	Europe
Costa Rica	Argentina	Bangladesh	Canary Islands	Cyprus
Cuba	Bolivia	India	Ivory Coast	England
Guadeloupe	Brazil	Indonesia	Egypt	France
Guatemala	Canada	Jordan	Malawi	Germany
Jamaica	Chile	Korea	Morocco	Portugal
Mexico	Colombia	Malaysia	Nigeria	
Panama	Ecuador	Nepal	Tanzania	
Puerto Rico	Peru	Pakistan	Zimbabwe	
Trinidad	Uruguay	Philippines		
	U.S.A.	S. Arabia		
	Venezuela	Tahitii		
		Thailand		
		Viet Nam		

tations. After several successful greenhouse experiments, field application of *P. lilacinus* was initially carried out in Peru to control *M. incognita* on potato (*Solanum tuberosum*) and later was extended to test its efficiency in controlling nematodes in other countries.

Methodology and the approach in field application of this organism had to be evolved in such a way that the maximum benefit could be obtained with limited complexity in methodology adaptable to areas with minimum resources. Field applications were carried out in farmers' fields with the farmer assuming responsibilities in the normal care of the crop and in field operations. Field performance was evaluated by the CIP Nematology staff in collaboration with the members of the CIP Region 1. Results of several field evaluations were consistent in indicating the great potential for use of biological entities in controlling important plant-parasitic nematodes.

Experiments were carried out in different countries so that the efficiency and adaptability of *P. lilacinus* in controlling *Meloidogyne* species under different climatological and soil environmental conditions could be determined. Similarly, a simple methodology, requiring minimal equipment had to be developed for the culture and multiplication of this organism. In addition to a simple multiplication method, a simple procedure for inoculation and evaluation was necessary to provide uniformity in the type of data collected as well as to simplify the interpretation of data.

The nematology program at CIP collaborates with the International *Meloidogyne* Project (IMP), organized and directed by Dr. J. N. Sasser of North Carolina State University. Since the IMP has over 100 collaborators in over 70 different countries, we decided to utilize the IMP as a vehicle for distribution of *P. lilacinus* to the nematologists collaborating with that program. Cultures of *P. lilacinus* were sent to nematologists in 46 countries around the world for testing (Table 1).

It is hoped that data collected by the investigators in different countries will be pooled and evaluated so the effectiveness of *P. lilacinus* in managing *Meloidogyne*, *Globodera*, *Tylenchulus*, and other nematode species affecting important crops under diverse climatological and soil conditions can be determined.

RESULTS

P. lilacinus infected eggs of *M. incognita* and destroyed the embryos within 5 days in laboratory tests. Percent infection of the eggs was directly correlated with the length of time that they were exposed to the fungus. In greenhouse tests, a reduction of up to 70% root galls and egg masses was observed in the plants from fungus-treated tubers. Since up to 60% of the eggs within egg masses were infected with the fungus, the size of subsequent nematode generations was reduced.

In a field test conducted in 1979, plants grown in plots inoculated with the fungus had a significantly lower root galling index than did those grown in the nematicide-treated plots (Table 2). Similarly, 86% of the egg masses collected from the roots of fungus-treated plots were infected with *P. lilacinus*, and over 54% of the eggs within these egg masses were destroyed. Similar on-farm trials were carried out in seven different farmer's fields in the

Table 2. Effect of different treatments on the reaction of potato cv. 'Cuzco' to *Meloidogyne incognita*.

Treatment	Average root galling index[a]	% egg mass infection	% eggs destroyed
Paecilomyces lilacinus	2.93	86.0	54.7
Temik 10% G (25 kg/ha)	3.36	0	0
Nemacur 5% G (50 kg/ha)	3.44	0	0
Furadan 5% G (50 kg/ha)	3.57	0	0
Organic matter (10 ton/ha)	3.62	0	0
Control	3.60	0	0
LSD 0.01	0.19		
0.05	0.14		

[a] Based on 120 observations. 1 = no root galling to 5 = very severe root galling.

Table 3. Effect of field application of *Paecilomyces lilacinus* and Temik on *Meloidogyne incognita*: Effect on yield of potato.

Treatment	Kg total yield/ 400 m^2 plot	Kg infested tuber yield/400 m^2 plot	Kg non-infested tuber yield/400m^2 plot
Control	814.61	691.92	122.69
P. lilacinus	721.63	419.04	302.59
Temik 10% G (25kg/ha)	656.88	509.76	147.12
LSD 0.05	104.62	27.60	85.10
0.01	139.32	36.72	113.28

coastal region of Peru. Consistent results indicating the potential of *P. lilacinus* as an alternative to nematicides were obtained. The fungus significantly reduced the tuber and root galling due to *Meloidogyne* infection.

Yield of marketable potatoes (non-infected tubers) was significantly higher in the fungus-treated plots than in those of the control and the nematicide-treated plots. Representative data on the effect of *P. lilacinus* on the yield of marketable potatoes was taken (Table 3).

Results of field experiments conducted in three consecutive years indicated consistent reduction of *M. incognita* populations as expressed by tuber galling. Additionally, the fungus spread rapidly within the experimental plots, infesting the adjacent control plots. Although this spread confounds the result, it is a desirable characteristic of a successful and competitive biological control agent. It is noteworthy to mention that in the field where this fungus was originally isolated, after 3 years of consecutive cropping of *M. incognita*-susceptible hosts, no nematode could be found infecting the root system. Artificial inoculation of this field by *M. incognita* was not successful in establishing the nematode after this period. The continued presence of fungus in the field was confirmed by its isolation from the soil.

In an experiment at La Molina, Temik (10 percent G at 25 kg/ha) and Temik plus fungus combined with rotation reduced nematodes and increased tuber quality in fungus and fungus-plus-Temik plots. Yield of corn (*Zea mays*), a rotation crop, in fungus- and fungus-plus-Temik-treated plots was double that in control plots. Yield in fungus-treated plots was higher than in the plots treated only with Temik.

The effect of *P. lilacinus* on the multiplication rate of *Globodera pallida* under laboratory conditions shows a correlation of infection rate with that of exposure period (Figure 1). There was a stimulation of hatching for up to 25 days after which the reduction in hatching rate was correlated with egg infection.

In 1982, on-farm trials to test the effect of *P. lilacinus* in the control of nematodes such as *M. incognita*, *Radopholus similis*, *Globodera rostochiensis*, and *G. pallida* in potato and other crops were conducted in different countries. In the province of Herrera, Panama, the control of *M. incognita* on tomato by *P. lilacinus* was tested on a large-scale basis. The fungus-treated plots had higher yield than did the control plots and those treated with carbofuran (Table 4). Experiments on control of *G. rostochiensis* in Cerro Punta, Panama, resulted in yields that were similar to those of the previous experi-

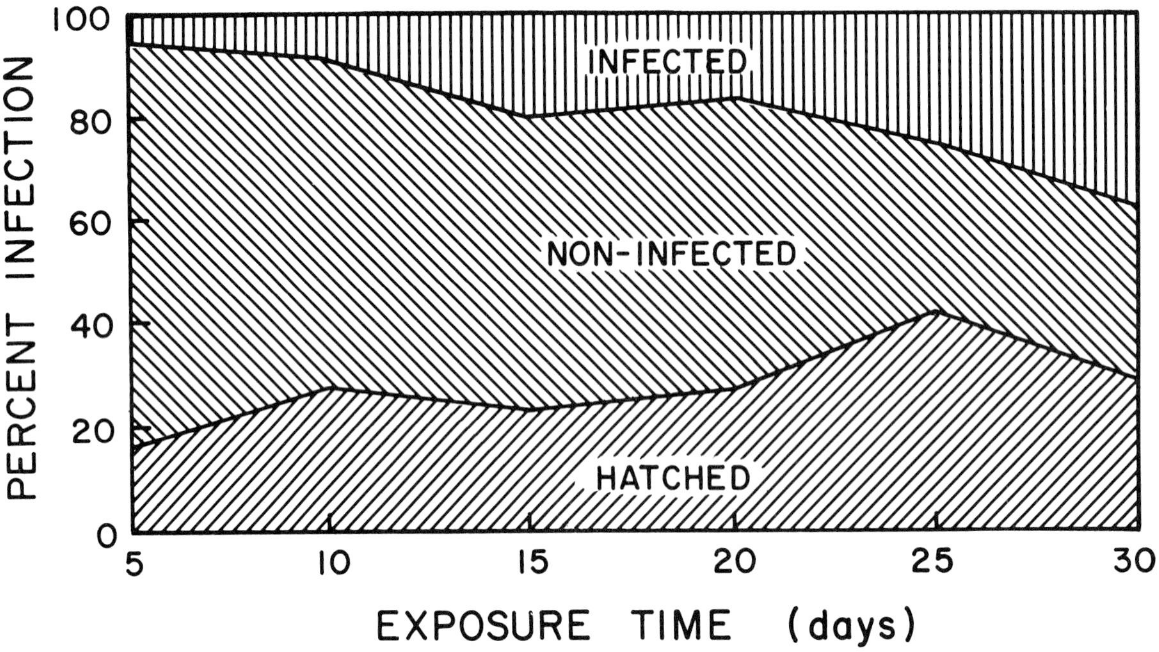

Fig. 1. Infection of *Globodera pallida* eggs by *Paecilomyces lilacinus*: Effect on hatching.

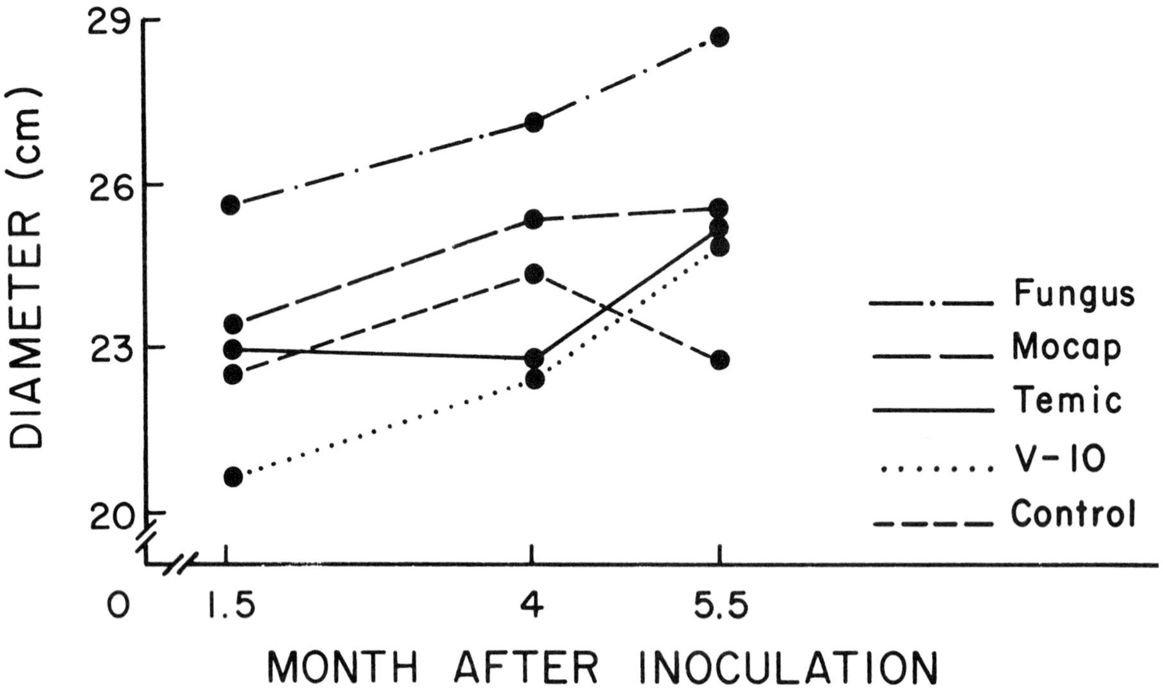

Fig. 2. Effect of *Paecilomyces lilacinus* on *Tylenchulus semipenetrans*: Effect on the orange fruit diameter.

Table 4. Effect of *Paecilomyces lilacinus* in controlling *Meloidogyne incognita* on tomatoes: Effects on yield. Panama—1982.[a]

	Kg yield/plant			
Treatments	1st Harvest	2nd Harvest	3rd Harvest	Total yield
P. lilacinus	0.95	0.65	0.53	2.13
Carbofuran (2.5 kg a.i./ha)	0.85	0.55	0.45	1.85
Control	0.85	0.43	0.24	1.52
L.S.D. 0.05	N.S.	0.16	0.16	0.38
L.S.D. 0.01	N.S.	0.23	0.22	0.54

[a] A CIP-IDIAP (Panama) collaborative study

Table 5. Biological control of *Globodera rostochiensis* on potato cv. 'Alpha' by *Paecilomyces lilacinus*. Cerro Punta, Panama—1982.[a]

	Ton/ha yield		
Treatments	Commercial tubers	Total yield	No. cysts/ 50 cc Soil
P. lilacinus	20.58	26.64	58.88
Carbofuran (3.0 kg a.i./ha)	15.72	20.52	70.35
Control	19.68	24.90	67.33
L.S.D. 0.05	N.S.	N.S.	5.73
L.S.D. 0.01	N.S.	N.S.	7.63

[a] A CIP-IDIAP (Panama) collaborative study

ment. Similarly, there were significantly lower numbers of cysts in the fungus-treated plots than in the nematicide-treated or control plots (Table 5).

In greenhouse tests, the effects of various inoculum concentrations used on *G. pallida* cysts in two different inoculation periods were determined. Results showed that a significantly higher percentage of *G. pallida* eggs hatched in the control (non-inoculated) cysts as compared to those inoculated with the fungus. Approximately 50 percent of the apparently healthy *G. pallida* eggs were either infected or damaged by *P. lilacinus* during these two experimental periods.

Efficacy of the fungus is now being tried on other crops. In 1983, tests were conducted in a citrus grove infested with *Tylenchulus semipenetrans* in Huaral, Peru. Fruit diameter of the orange cv. 'Valencia' was significantly affected by applications of three nematicides and *P. lilacinus*. The fruit diameter from fungus-treated trees was greater than that from trees treated with Temik, Vydate and Mocap (Fig. 2). Similarly, the number of nematodes in roots of and soil around citrus trees inoculated with *P. lilacinus* was significantly lower than that in the non-treated or the nematicide-treated trees. Reduction of nematode populations was correlated with increased healthy root growth and increased fruit diameter.

LIKELY FUTURE IMPACT

The reduction of tuber and root galling in fungus-treated plots of farmers' fields has been the first evidence of a successful biological control agent for nematodes. Farmers who have participated in the study and have seen these results are now eager to use biological control against nematodes in other crops such as tomato, bean and cotton. A similar impact has been created in the United States as a consequence of the results obtained by the International *Meloidogyne* Project in management of root-knot nematodes infecting various crops. Researchers are now involved in studies to improve certain aspects related to the mass production and field application of the fungus. However, the most important achievement in the use of *P. lilacinus* for controlling nematodes so far has been in the Philippines. Since effectiveness of the fungus for root-knot nematode control was within the range of the efficacy of a number of nematicides (91%), further studies were conducted. Results on substrate study showed that the fungus grows abundantly in sterilized mashed potato, chopped water lily plants (Nymphaeaceae), ipil ipil (*Leucaena glauca*) leaves and corn grits. Farmers are being advised to introduce the fungus with these readily available substrates into the soil for nematode control.

In similar studies with the potato cyst nematode (*Globodera rostochiensis*), the efficacy of *P. lilacinus* has been comparable to that of the most common nematicides (48-55%). Although the yield was lower than that of chemically treated plants, there was still an increase of about 2-4 tons/ha over the controls which is remarkable considering that there is practically no additional cost in producing the fungus.

Since successful results have been obtained from tests of *P. lilacinus* for control of *Meloidogyne* species in different countries (Argentina, Ecuador, Panama, Peru, Philippines, USA, etc.), the potential of this organism for large-scale application is quite evident. It is noteworthy to mention that the impact of the biological control of nematodes with *P. lilacinus* will not be restricted to potato. In fact, we expect extensive use of *P. lilacinus* for control of root-knot nematodes in tropical regions where this nematode is one of the most important pests of not only potato but also several major crops as well. This technology solves an important nematode problem of many major crops, and will become an immensely useful and advantageous alternative to nematicides in many developing country programs.

Improved potential and prospects for such effective biological control have given the field of biological nematode control a new dimension, which can be summarized in the opinion given by Dr. R. Davide from the Philippines. He has stated that with the discovery of the effectiveness of *P. lilacinus* against nematodes, our biological control studies seem to be paying dividends. He believes that this fungus is by far the most promising and practicable biological control agent for management of root-knot and cyst nematodes.

Literature Cited

1. Drechsler, C. 1937. Some Hyphomycetes that prey on free-living terricolous nematodes. Mycologia 29:447-552.
2. Franco, J., P. Jatala, and M. Bocangel. 1981. Efficiency of *Paecilomyces lilacinus* as a biocontrol agent of *Globodera pallida*. J. Nematol. 13:438-439.
3. Jatala, P., R. Kaltenbach, and M. Bocangel. 1979. Biological control of *Meloidogyne incognita acrita* and *Globodera pallida* on potatoes. J. Nematol. 11:303.
4. Jatala, P., R. Kaltenbach, M. Bocangel, A. J. Devaux, and R. Campos. 1980. Field application of *Paecilomyces lilacinus* for controlling *Meloidogyne incognita* on potatoes. J. Nematol. 12:226-227 (Abstr.).
5. Jatala, P., R. Salas, R. Kaltenbach, and M. Bocangel. 1981. Multiple application and long-term effect of *Paecilomyces lilacinus* in controlling *Meloidogyne incognita* under field conditions. J. Nematol. 13:445.
6. Johnson, L. F. 1962. Effect of the addition of organic amendments to soil on root-knot of tomatoes. II. Relation of soil temperature, moisture and pH. Phytopathology 52:410-413.
7. Mankau, R. 1968. Reduction of root-knot disease with organic amendments under semifield conditions. Plant Dis. Rep. 52:315-319.
8. Mankau, R. 1981. Microbial control of nematodes, pp. 475-494 *In* Plant-parasitic nematodes, Vol. III, ed. B. M. Zuckerman and R. A. Rhode. Academic Press, New York. 508 pp.
9. Nigh, E. A. 1979. *Acremonium strictum* and *Fusarium oxysporum*, two fungal egg parasites of *Heterodera schachtii*, the sugar-beet cyst nematode. Ph.D. thesis. University of California, Riverside, California. 79 pp.
10. Stirling, G. R. 1977. Effect of temperature on parasitism of *Meloidogyne incognita* eggs by *Dactylella oviparasitica*. Nematologica 25:104-110.
11. Stirling, G. R. and R. Mankau. 1978. *Dactylella oviparasitica* a new fungal parasite of *Meloidogyne* eggs. Mycologia 70:774-783.
12. Tribe, H. T. 1977. Pathology of cyst nematodes. Biol. Rev. Camb. Philos. Soc. 52:477-508.
13. Tribe, H. T. 1979. Extent of disease in populations of *Heterodera* with special reference of *H. schachtii*. Ann. Appl. Biol. 92:61-72.

THE PRACTICAL APPLICATION OF A NEMATODE ADVISORY SERVICE

J. L. Imbriani

Agronomic Division
North Carolina Department of Agriculture
Raleigh, North Carolina, USA

This chapter is primarly concerned with the operation of the Nematode Advisory Service (NAS) in North Carolina. Although the subject of this treatise is the root-knot nematode, advisory services usually concern themselves with all phytoparasitic nematodes (a few notable exceptions deal only with cyst nematodes), and this discussion will not be limited to *Meloidogyne spp.* Several of the components of nematology that affect the function of an advisory service (such as sampling, assay methods and the relationship of nematode populations to crop growth) are discussed in detail elsewhere in the treatise and will not be elaborated upon here.

Advisory Service assays for nematodes fall into two groups—predictive and diagnostic. Diagnostic assays include the identification of plant disease problems involving nematodes and the suggestion of corrective measures for future crops in the case of annuals or appropriate therapy for perinnials. Predictive assays are used to estimate the likelihood of nematode damage to a future crop and to recommend action to avoid or minimize loss. The sampling strategies, assay methodologies, interpretation of assay results, and recommendations differ with the two groups. These differences will be discussed in the appropriate sections of this chapter.

HISTORY AND CURRENT STATUS

The prospect of developing an advisory service in North Carolina was envisaged in 1953 (Nusbaum, personal communication). The practicality of such a program was predicated on the assumptions that (1) reliable sampling and assay techniques could be developed, (2) assay data would be of practical use to the grower, and (3) facilities and staff could be provided to handle the anticipated work load. A basic research program supporting the development of such a program was started in 1957 when W. M. Powell and C. J. Nusbaum began a comprehensive study on sampling methods, storage of samples, and laboratory procedures. Impetus was given to the development of the program in 1958 by the discovery that root-knot nematodes were profoundly affecting the results of research station tobacco plot experiments. In its infancy (1960 to 1963), the advisory program functioned largely to aid researchers in the elimination of nematodes as interfering factors in agronomic studies on several crops (4).

A 5-year plan for the supporting research and development of a NAS for the state was proposed by Nusbaum in 1964, and in July 1965, a bill appropriating funds for this development was approved by the General Assembly. Included in this bill were salaries for a Nematologist and 3 additional personnel. K. R. Barker accepted the postition of Nematologist and continued the investigations started earlier by Nusbaum and others. As the result of the successful development of the pilot program, the Nematode Advisory Service was created, administration transferred to the North Carolina Department of Agriculture and new facilities occupied in April 1974.

The Nematode Advisory Section is one of three comprising the Agronomic Division of the N. C. Department of Agriculture. The other two are soil testing and plant tissue analysis. The number of samples assayed per year has increased from about 4500 during the first year of operation to current levels of about 30,000. Personnel consist of a Nematologist, Laboratory Supervisor, four technicians and three laboratory helpers, two of which are part-time. Samples for assay are accepted from all N. C.

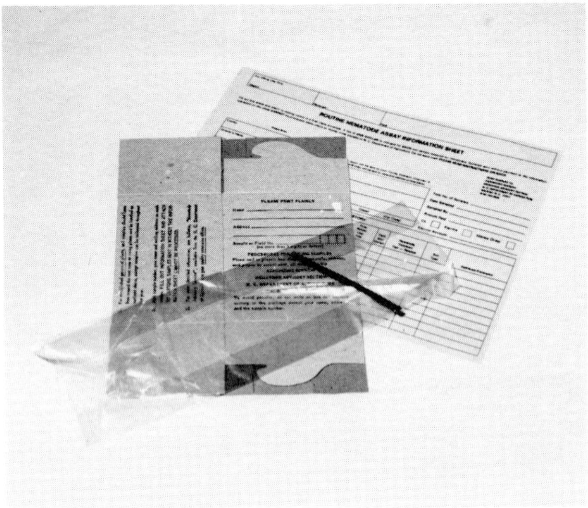

Fig. 1. Nematode sampling kit provided by the North Carolina Nematode Advisory Service.

residents, and there is a one dollar fee per sample. Some data on crops and acreage sampled have been given elsewhere (1).

ASSAY PROCEDURES

Sampling Procedures

Users of the advisory service are provided with most of the materials required for sample submission (Fig. 1). Plastic bags, twist ties, shipment cartons, and assay information sheets (Appendices 1 and 2) are made available without charge through a number of sources. Sampling procedures are detailed on the boxes and information sheets. The recommended methods for sampling differ for diagnostic and predictive assays.

PROBLEM DIAGNOSIS It is recommended that samples consist of 20 one-inch-diameter cores taken with a soil probe from the root zone of plants in the margin of the affected area (Appendix 2). Instructions emphasize the need for inclusion of root material, and it is suggested that comparative samples from unaffected areas be submitted also.

PREDICTIVE ASSAYS Recommended sampling time is late summer and early fall. This time is selected because the nematode populations are highest with an increased probability of detection of moderate to low nematode populations. This sampling period is also more practical for high volume laboratories. Adequate lead time is needed to complete the large numbers of assays for growers to have the results with sufficient time to make management decisions for the coming season.

For sample collection, use of a soil sampling probe and the pattern in Appendix 1 is recommended. Each sample should be a composite of at least twenty cores representing an area no greater than 5 acres. For larger acreage it is recommended that the fields be divided into 4- or 5-acre blocks and that at least 50% of these blocks be sampled.

HANDLING OF SAMPLES Diagnostic and predicitive assay samples are handled identically. They are placed in plastic bags, sealed with a twist tie, and then shipped to the laboratory. Instructions for the handling of samples are given on the assay information sheets, and emphasis is placed on the importance of keeping the sample from overheating or freezing.

Laboratory Procedures

RECEIVING OF SAMPLES Upon receipt the samples are checked to be certain that the number of samples received is the same as total submitted and that the field identification codes listed by the grower on the information sheet match those on the sample box. If there are any discrepancies, the samples are held and growers notified by form letter of the nature of the problem. A prenumbered tag is attached to each sample along with labels indicating the need for any special extraction procedures. Then samples are stored in incubators at 10° C until processed.

EXTRACTION OF NEMATODES All samples are extracted using a semiautomatic elutriator (2) (Fig. 2). A 500-ml portion of the sample is placed in a plastic beaker along with the attached labels. If less than 500 ml was received, the actual volume processed is noted on the tag, and counts are later corrected to a 500-ml volume. Screens are then selected according to sample requirements indicated on the tag. Roots and organic debris are collected on a 24-mesh screen placed just beneath the spout, and the nematodes are collected on a 400-mesh screen. Sugar flotation is then used to separate the nematodes from fine soil particles. Material from the 400-mesh screen is washed into a 150-ml beaker and then allowed to settle for 60 minutes. About ¾ of the supernatant is discarded, and then the remaining soil and nematodes are washed into 50-ml centrifuge tubes and spun at 420 *g* for 5 minutes. The resulting supernatant is discarded and the pellet resuspended in 1.33 M (454 g/liter) sucrose before being centrifuged 1 minute at 420 *g*. The sucrose solution

Fig. 2. Semi-automatic elutriator.

Fig. 3. Stereoscopic microscope fitted with mechanical stage to allow use of rectangular plastic dish for counting nematodes.

is sieved through a 635-mesh screen. The nematodes are quickly washed from the screen into a beaker and then counted.

Diagnostic samples are assayed similarly. Roots from the 24-mesh screen are examined for symptoms typical of nematode damage, weighed, cut into pieces about 2 cm long, and then placed in beakers in water and aerated with a stream of air for 48 hrs. The nematodes are collected on a 635-mesh screen, baermannized on tissue paper (if a clean sample is necessary), and then counted. Cysts are collected on a 60-mesh screen placed below the 24-mesh screen in the elutriator. The cysts are sugar floated and counted using a procedure similar to that described for soil except the sucrose solution is 2.66 M (907 g/liter).

COUNTING The use of rectangular dishes and mechanical stages greatly facilitates counting (Fig. 3). The dishes are tops and bottoms of plastic boxes 7.30 × 3.02 × 2.54 cm (2 7/8 × 1 3/16 x 1 in) (#575 Stockline Service, Creative Packaging Co., 4411 Hollins Road, Roanoke, Va. 24012, USA) that have 50% of the area scribed into four lines for counting. Counts are recorded on a tiered tally meter (2 tiers with 6 positions

Table 1. Nematode management recommendations in North Carolina.[a]

Alphabetical recommendation	Hazard level	Corrective action
A. Production of the crop to be grown should not be affected by the kinds and numbers of nematodes detected in this assay.	Level 1 None to minor (<10% risk)	Rotation when warranted
B. The population of nematodes found may cause crop damage, and chemical soil treatment (CST) may be profitable, especially should growing conditions be unfavorable.	Level 2 Slight to moderate (10 - 50% risk)	Rotation Resistant cultivar Nonfumigant CST Fumigant CST Multipurpose CST
C. The nematode population found indicates that chemical soil treatment (CST) would be profitable.	Level 3 Severe (>50% risk)	Rotation Resistant cultivar Fumigant CST Multipurpose CST
D. Nematode-resistant variety is recommended.	Used at all Levels	Resistant cultivar
E. Crop rotation with non-host crops is suggested.	Used at all Levels	Rotation

[a] Partially from Barker & Imbriani (1).

each; Forestry Suppliers, Inc, Box 8397, Jackson, Mississippi 39204, USA), and then listed on a data recording sheet attached to the grower-supplied information form. All data are corrected to represent a 500-ml volume. Counts from root fractions are expressed as the number per 3 g of root.

IDENTIFICATION OF SPECIES AND RACES In a large-volume lab it is not feasible to key all phytoparasites to species; therefore, such identifications are done only when necessary. Many plant-parasitic nematodes are both widespread and common in North Carolina so generic identification is sufficient. When possible, *Pratylenchus* is identified through the dissecting microscope to species on the basis of presence of males, size of adult female, position of vulva, and tail shape. Slides are prepared occasionally for study under the compound microscope. Populations of lance nematodes (lacking males) suspected of being *Hoplolaimus columbus* are identified to species through the compound microscope. Cropping history (including varieties) is used routinely for the presumptive identification of species of root-knot nematodes. Actual species identification is based on male-head characteristics (3). This method appears to be reliable for the populations common in North Carolina but must be used cautiously. For example, males recovered from any particular plant only indicate the species which infected that particular plant and may not indicate all of the species found in the field. This point is important of course since mixtures are common. No attempt is made to characterize races of root-knot nematodes.

Soybean cyst nematode resistance screening tests are conducted on request. Actual race identifications are not attempted, but the efficacy of available resistant varieties against populations of the nematode is determined. The tests currently consist of bioassays using susceptible control plants and varieties with resistance to races 1, 3 or 4 of the nematode.

Recommendations

PREDICTIVE ASSAYS A series of threshold guides have been prepared for the important nematodes species on most crop plants grown in the state. The 1984 guide for peanuts is given in Appendix 3 as an example. These guides, based on results of field and microplot tests by crop specialists, are revised as new data become available and are completely reviewed yearly. The major nematode problems, threshold values, and corresponding recommendations are listed. Recommendations are made by comparing assay results and threshold numbers with a consideration of soil type and cropping history. The recommendations are general and roughly correspond to three hazard levels (Table 1). These hazard levels relate to the probability (expressed as a percent risk) of significant yield reduction because of nematode damage. Recommendations for the use of resistant varieties and crop rotation are made when warranted for all three hazard levels.

DIAGNOSTIC ASSAYS Recommendations are based on root symptoms and presence of relatively high populations of a nematode species of known pathogenicity for the particular host plant. In the case of endoparasites, the nematodes found in the root assay are especially important. When separate samples representing areas with good or poor growth are submitted, nematode numbers are compared. These comparisons must be made cautiously, however, because if the root system is severely damaged, fewer nematodes may be recovered from areas of poor growth than from areas of normal growth. The same alphabetical recommendations are used for diagnostic and predictive assays, but remarks with additional information are typed on the upper right portion of the assay report (Fig. 4).

DATA STORAGE AND REPORT PREPARATION

Upon completion of the assay and recommendations, the data (names, addresses, assay results and recommendations) are keyed into a computer, edited and arranged into intermediate records. These intermediate records are used to print the assay reports and mailing labels. The master file contains all the data for samples completed during the year, may be updated at any time, and is retired annually into a history file. Data summaries, reports, or statistical analyses can be obtained from either the current or history master files by the use of SAS or Easytrieve statements.

THE NEMATODE ASSAY REPORT AND SUPPORTING DOCUMENTS

The assay report (Fig. 4) lists field numbers, cropping history, assay results and recommendations. In the case of diagnostic assays, the results of the root extraction are listed immediately below the corresponding soil assay. Potential or current nematode problems are identified using asterisks: a single asterisk for hazard level 2 (slight to moderate) and two asterisks for hazard level 3 (severe). In the recommendation column, the alphabetical recommendations and, when appropriate, nema note numbers are listed. The nema notes are a series of documents containing information beyond the general recommendations. They may list alternative crops, resistant varieties, nematicides (and rates of application) or other pertinent information. The nema note for tobacco is given as an example in Appendix 4.

PROBLEMS ENCOUNTERED AND PROSPECTS FOR NEMATODE ADVISORY SERVICES

The status of and prospects for advisory services in the United States have been recently reviewed and some of the difficulties discussed (1, 5). Also, the problems associated with obtaining representative samples, selecting appropriate assay procedures, and relating assay results to crop response are discussed in other chapers of this treatise. However, some of the more common problems encountered by North Carolina's Nematode Advisory Service merit discussion here.

Sample quality is a weak link. Although guidelines are provided for proper sampling and handling of the sample, it is not possible to control the sample quality. Considerable educational effort is made to assure that samples are properly taken and handled. During processing, notations are made for those samples that appear to be of poor quality (lacking roots(diagnostic) or too small, dry or wet (diagnostic and predictive)). A nema note detailing the problem is used to inform the grower about the nature of the sample, and resampling is suggested. The grower is also urged to interpret the assay results from such samples cautiously.

Sample quality is also estimated on the basis of the assay. When samples are taken from fields in which good host crops for nematodes have been grown (such as corn or soybean), resampling is suggested if no phytoparasitic nematodes are initially found. It appears, however, that the majority of the samples assayed in North Carolina are of good quality when judged by the above criteria. However, in many cases, the acreage represented by a sample exceeds that which is recommended (4 to 5 acres). Development of more efficient sampling

Report # 010000

NEMATODE ASSAY REPORT

AGRONOMIC DIVISION, N. C. DEPARTMENT OF AGRICULTURE
BLUE RIDGE ROAD CENTER, RALEIGH, N. C. 27611
PHONE: (919) 733-2656

To:
SASSER JOSEPH N.
628 GROVE AVENUE
RALEIGH, NC. 27606

Copies sent to:

Farm Location (County): WAKE

Lab No.	Field No.	CROP HISTORY			Root-Knot (Meloidogyne)	Lesion (Pratylenchus)
		Crop Before Last Year	Crop Last Year	Crop to be Grown		
11122	001	GARDEN	GARDEN	GARDEN	8950 **	20
11123	002	SOYBEAN	CORN	PEANUT	250	120 *
11124	003	PEANUT	PEANUT	PEANUT	1950 **	400 **
11125	004	SOYBEAN	SOYBEAN	SOYBEAN	200	60
11126	005	CORN	SOYBEAN	SOYBEAN		50
P 11127	BAD	COTTON	SOYBEAN	COTTON	20	80 **
T 11127	BAD	ROOT EXTRACTION FOR ABOVE FIELD				450 **
P 11128	OK1	COTTON	SOYBEAN	COTTON	10	20
T 11128	OK1	ROOT EXTRACTION FOR ABOVE FIELD				

Recommendations

A — Production of the crop to grown should not be affected by the kinds and numbers of nematodes detected in this assay.

B — The population of nematodes found may cause crop damage and chemical soil treatment may be profitable, especially should growing conditions be unfavorable.

The recommendations based upon assays of the soil samples submitted a
assistance in selecting nematode control treatments, contact your local Coun

Fig. 4. Computer printed nematode assay report.

Nematologist Remarks:

Relative Abundance of Nematodes Recovered Per Pint of Soil

Stunt (Tylenchorhynchus)	Spiral (Helicotylenchus)	Stubby-root (Trichodorus)	Dagger (Xiphinema)	Ring (Criconemoides)	Cyst (Heterodera)	Sting (Belonolaimus)	Lance (Hoplolaimus)	Others – I	Others – II	Recommendations (See Explanation Below)	Nema Note (No. Listed Below)
100		40		30						C	
20	60	80			10					B	
80	180	300 **		40		450 **				C	7-3
650	340	280			20 **					D+E	2-1 2-4
540	700				350 **					C/E	2-3
100				40						C	
10	3										
250				90						A	
4	6										

C – The nematode population found indicates that chemical soil treatment would be profitable.

D – Nematode–resistant variety is recommended.

E – Crop rotation with non–host crops is suggested.

ffered only as a guide in helping you plan your nematode control program. For
xtension Office or the Agronomic Division Regional Agronomists.

methods will help encourage growers to sample fields more adequately.

Occasionaly incomplete assay information sheets are a major problem. Since presumptive identification of root-knot species is based on cropping history (and varieties grown), incomplete history data (or, worse yet, incorrect data) make recommendations tenuous. In instances where use of resistant varieties is the major management strategy but where populations of nematodes are commonly found against which this resistance is not effective (for example soybean cyst nematode), then accurate knowledge of varietal history is critical. Several nema notes have been developed to address the problem of resistance efficacy.

In a laboratory, such as North Carolina's, that assays large numbers of samples, all procedures must be efficient. Assay methodologies that produce the highest quality assay in minimum time are utilized. Procedures cannot be used that considerably lengthen the time required to assay a sample without significantly increasing its quality. Also, some procedures (such as greenhouse bioassay for root-knot-nematode-species identification) are prohibitively expensive and time consuming and require too much space to be used routinely.

A more precise understanding of the relationship of preplant nematode populations to plant response and the factors that alter this relationship is required. The concepts of damage and economic thresholds are valuable, but additional data and refinement are needed to increase their practical utility. Models are useful tools for the understanding of nematode population dynamics and ecology but currently lack the precision required for advisory work.

There is a need for feedback from users of advisory services. It is only through such information that the required "fine tuning" can be accomplished. The N. C. Department of Agriculture currently employs Regional Agronomists who through their association with growers provide extremely valuable information used in program adjustment.

The prospects for advisory services are good, but there is a need for addressing the problems discussed (and others) in order to realize fully the potential of nematode advisory programs.

Literature Cited

1. Barker, K. R. and J. L. Imbriani. 1984. Nematode advisory programs—status and prospects. Plant Dis. 68:735-741.
2. Byrd, D. W., Jr., K. R. Barker, H. Ferris, C. J. Nusbaum, W. E. Griffin, R. H. Small, and C. A. Stone. 1976. Two semi-automatic elutriators for extracting nematodes and certain fungi from soil. J. Nematol. 8:206-212.
3. Eisenback, J. D. and H. Hirschmann. 1981. Identification of *Meloidogyne* species on the basis of head shape and stylet morphology of the male. J. Nematol. 13:513-521.
4. Nusbaum, C. J. 1963. Nematode assays—a program to aid agricultural workers. Down to Earth 19:2.
5. Rickard, D. A. and K. R. Barker. 1982. Nematode assays and advisory services, pp. 8-20 *In* Nematology in the southeastern region of the United States, ed. R. D. Riggs. South. Coop. Ser. Bull. 276. 206 pp.

Appendix 1. Front and back of information sheet used for predictive soil assays.

For Office Use Only Report	Received	Paid

ROUTINE NEMATODE ASSAY INFORMATION SHEET

Fill out this sheet and attach to mailing carton in a first class envelope. A fee of **ONE DOLLAR** is charged for **EACH** soil sample assayed for nematodes. Samples sent without payment or the information requested on this sheet **CANNOT** be processed. Make check or money order payable to N. C. Department of Agriculture. Do not send cash. **PLEASE READ INSTRUCTIONS ON BACK.**

SEND SAMPLES TO:
AGRONOMIC DIVISION
NEMATODE ADVISORY SECTION
N. C. DEPARTMENT OF AGRICULTURE
RALEIGH, N. C. 27611
(919) 733-2655

(Please Print)

County Phone Number	A copy of your report will be sent to your County Extension Chairman. If desired one additional copy will be sent to name and address printed below.
Grower's Name	
(Last) (First)	(Name)
(Address)	(Address)
(City) (Zip Code)	(City) (State) (Zip Code)

Total No. of Samples ______

Date Sampled ______

Sampled By ______

Amount Paid ______

Ck ☐ Escrow ☐ Money Order ☐

Escrow Account ______

FIELD HISTORY

Lab Number (Leave blank)	Your Field Number	Crop To Be Grown Crop — Variety	Crop Last Year Crop — Variety	Crop Before Last Year	Field Size (Acres)	Nematicide Treatment Year — Material	Soil Type	Additional Comments

PROCEDURES FOR TAKING SOIL SAMPLES FOR NEMATODE ASSAYS

Great care should be exercised in taking soil samples for nematode assays. *Results are no better than samples submitted.* The procedures listed below should be carefully followed in obtaining and handling soil samples for nematode assays.

WHEN TO SAMPLE

For *annual crops* (tobacco, tomato, peanut, soybean, etc.), samples collected in the FALL provide more reliable information in predicting nematode development and crop response than spring-collected samples. Therefore, growers are encouraged to collect soil samples in late summer or early fall. For established *perennial plants* (ornamentals, turfgrasses, peach, etc.), soil samples can be collected throughout the year.

TAKING THE SAMPLES

Take samples only when the soil is in good working condition, neither too wet, dry, nor frozen.

For annual crops:

1. In each field to be sampled, take the sample from an area with common crop history and in which a single crop will be planted.
2. If the soil in the area to be sampled is fairly uniform and is 4 acres or less in size, one sample will suffice. If the field is larger than 4 but less than 8 acres, divide the field into two blocks of approximately equal size and take one sample from each block. When fields are larger than 8 acres select 4-acre blocks representing at least 50% of the field and take a sample from each block
3. If the soil in the area to be sampled is variable, such as having heavy clay soil in one portion and a sandy soil in the remainder, take one sample from each soil type.
4. Take samples from plowed layer of soil (6 to **8 inches) with a soil sampling tube (1-inch** bore). Take at least 20 cores of soil in a systematic pattern for each sample area (see diagram). Collect the cores in a bucket, then **mix thoroughly and transfer enough soil to** fill a one-quart plastic bag. The remaining soil can be submitted with the appropriate information sheet to the Agronomic Division Soil Test Laboratory for soil fertility analysis.

For perennial crops:

1. Send in separate soil samples for each crop or plant species.
2. Soil should come from the root zone of declining plants. *Do not sample directly around dead plants.* Turfgrass samples should come from the edge of damaged areas and to a depth of 4 inches.
3. Soil samples should be collected and mixed as for annual crops.

HANDLING SAMPLES

1. Place each sample in plastic bag, *seal bag tightly* to keep soil moist, and put in a quart soil-sample box. Write the field number in the space provided on each box. These numbers are the only means by which your samples can be identified in the laboratory, and they must correspond to the column 'Your Field Number' on the Nematode Assay Information Sheet.
2. *Samples must be protected from both overheating and freezing. Do not place samples in direct sunlight, the trunk of a car or a freezer.*
3. Record information requested on field history on the Nematode Assay Information Sheet. The information on cropping history, including the variety grown, is essential for accurate prediction of nematode hazard levels.
4. Send the samples to the Assay Laboratory immediately.

Diagram to Follow in Collecting Field Samples

CAUTION: Since nematodes are not uniformly distributed, *entire field must be sampled.*

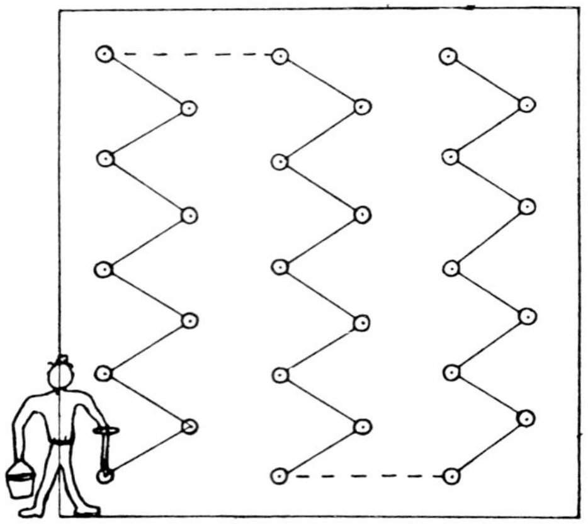

Appendix 2. Front and back of information sheet used for diagnostic assays.

Form AD5

(Please Print)

County	Phone Number
Grower's Name (Last)	(First)
(Address)	
(City)	(Zip Code)

NEMATODE-PROBLEM DIAGNOSIS INFORMATION SHEET

Accurate problem diagnosis is possible only when *all* questions are answered.

Fill out this sheet and attach to mailing carton in a first class envelope. A fee of ONE DOLLAR is charged for EACH soil sample assayed for nematodes. Samples sent in without payment or the information requested on this sheet CANNOT be processed. Make check or money order payable to N. C. Department of Agriculture, we request that cash not be submitted by mail.

Mail samples to:

Agronomic Division
Nematode Advisory Section
N. C. Department of Agriculture
Raleigh, N. C. 27611
(919) 733-2656

Total number of samples sent ________

Dollars enclosed ________

Sampled by ________ Date ________

In addition to the Nematode Assay Report you will receive, copies are sent to your County Extension Chairman and an Extension Specialist in the Department of Plant Pathology at N. C. State University. If you desire others to receive a copy, enclose their names and addresses.

Make sure the numbers in the column 'Your Field Number' correspond to the numbers on the soil-sample boxes. These numbers are the only means by which your samples can be accurately identified in the laboratory. The other information requested on this sheet is needed to aid in interpreting assay data.

Instructions for taking SOIL and ROOT SAMPLES are printed on the reverse side. For additional information on the Nematode Advisory Service, see bulletin, 'A Nematode Diagnostic and Advisory Service for North Carolina' available from the N. C. Department of Agriculture or your County Extension Office.

Lab No. (Leave Blank)	Your Field No.	Current Crop (Variety)	Date Planted	Crop Last Year (Variety)	Nematicide Applied (Material - Year)	Soil Type	Plant Symptoms			Pattern of Affected Plants within Field	
							Stunted	Yellow Color	Dead	Entire Field	Occurring in Patches
							☐	☐	☐	☐	☐
							☐	☐	☐	☐	☐
							☐	☐	☐	☐	☐
							☐	☐	☐	☐	☐

Additional Comments:

Have samples been sent to:
☐ Soil Test Laboratory
☐ Plant Analysis Laboratory
☐ Plant Disease and Insect Clinic (North Carolina State University)

HAVE YOU ENCLOSED A SAMPLE OF ROOTS FROM AFFECTED PLANTS WITH THE SOIL?

PROCEDURE FOR TAKING PROBLEM FIELD SAMPLES FOR NEMATODE ASSAYS

This Nematode-Problem Diagnostic Information Sheet is for use with nematode assay samples collected from problem fields where the existing crop is exhibiting symptoms of poor growth. More information is requested than for the Routine Nematode Assay Information Sheet; the additional information requested on this form is required for accurate diagnosis of suspected nematode problems.

Great care should be exercised in taking root and soil samples for nematode assays. *Results are no better than sample submitted.* The procedures listed below should be carefully followed in obtaining and handling all samples for nematode assays.

COLLECTING ROOT AND SOIL SAMPLES

Soil samples for problem diagnosis should be collected from around the margin of the affected area, where plants are exhibiting moderate to severe symptoms. Samples should not be collected from the most severely affected area if these plants are dying or dead. If the field has more than one affected area for each sample collect samples from several such areas. For each sample collect at least twenty (20) soil cores from the top 8 to 12 inches of soil (the root zone). Fill a one quart plastic bag approximately three-quarters full with soil that has been throughly mixed.

In addition to a soil sample, for problem diagnosis it is important to have a plant root sample. Collect root samples from plants exhibiting moderate to severe symptoms but *not* from dead plants. In collecting the roots the plant should be carefully removed from the soil with a shovel or spade and not pulled from the ground. After carefully shaking off the adhereing soil, some of the smallest fibrous roots should be collected. The roots should be collected from a minimum of five (5) plants taken from the same areas of the field where soil samples were collected. Place a small handful of roots in the plastic bag with the corresponding soil sample, covering the roots with a small amount of soil. Care must be taken when collecting root samples to prevent the roots from drying out; be sure to seal the plastic bag tightly.

After collecting the root and soil sample, excess soil can be submitted in the one-pint Soil Test box with the appropriate information sheet to the Agronomic Division's Soil Testing Laboratory for soil fertility analysis.

HANDLING SAMPLES

1. Place each sample in plastic bag, *seal bag tightly* to keep soil moist, and put in a quart soil-sample box. Write the appropriate field number in the space provided on each box. These numbers are the only means by which your samples can be accurately identified in the laboratory, and they should correspond to the numbers in the column 'Your Field Number' on the Nematode Assay Information Sheet. Send the samples to the Assay Laboratory *promptly.*

2. *Keep samples out of the direct sunlight to avoid overheating.* Samples may also be damaged by heat if they are kept in the trunk of a car.

3. Record *all* information requested on field history, crop variety, symptoms, and pattern of affected areas on the Nematode Problem Diagnosis Information Sheet. This information is absolutely necessary for accurate diagnosis of *your* problem. If samples have also been sent to the Soil Testing Laboratory, Plant Tissue Analysis Laboratory, or to the Plant Disease and Insect Clinic (North Carolina State University) for diagnosis, please so indicate on the information sheet. With this information we can compare the results of all tests and make a more accurate diagnosis of the problem.

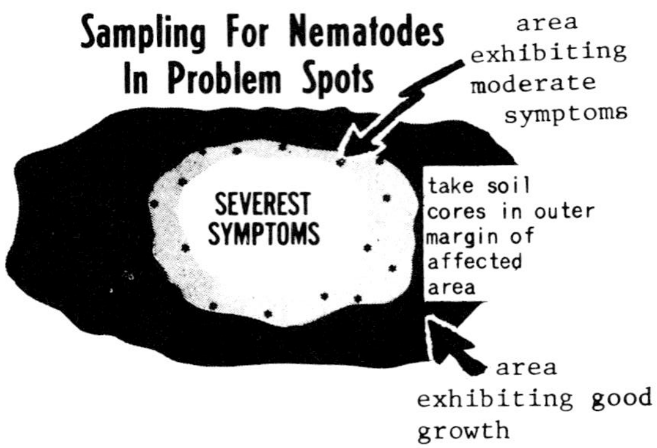

Collect representative root and soil samples from areas where plants are exhibiting moderate symptoms.

Appendix 3. 1983 nematode threshold guide for peanuts.

NEMATODE THRESHOLD DENSITIES (PER PINT OF SOIL) AND RECOMMENDATIONS

Crop: Peanuts Important Nematodes: 1. Root-knot 2. Sting 3. Lesion 4. Ring

Crop History	Crop To Be Grown	Root Knot	Lesion	Stunt	Spiral	Stubby-Root	Dagger	Ring	Cyst	Sting	Lance	Others	Recommendations
Legumes/ Tobacco	Peanuts	* 500–990											B
		** 1000+											C
Cotton/ Corn		No Problem											
			* 10- 250	P. brachyurus									B
			** 260+										C
			* 500+	P. zeae									B 008
						* 250+							B 008
							* 300+						B 008
								* 60–250					B
								** 260+					C
										** 10+			C 7-3

Recommendations:

A - The kinds and numbers of nematodes present are believed inadequate to affect production of the crop to be grown.
B - The population of nematodes found may be hazardous and soil treatment may be profitable, especially under unfavorable growing conditions.
C - The nematode population found indicates that soil treatment would be profitable on the crop to be grown.
D - Nematode-resistant variety is recommended.
E - Crop rotation is suggested.

Remarks:

1. *M. hapla* most common root-knot species on peanuts. *M. arenaria* in N.C. is usually race 2 which does not attack peanuts.
2. Use nema note 7-3 when sting nemas are found.
3. Low thresholds for lesion are based on pod rot complex with *P. brachyurus* being most serious.

Appendix 4. Nema Note 003 utilized for tobacco.

N.C.D.A.
NEMA
NOTES

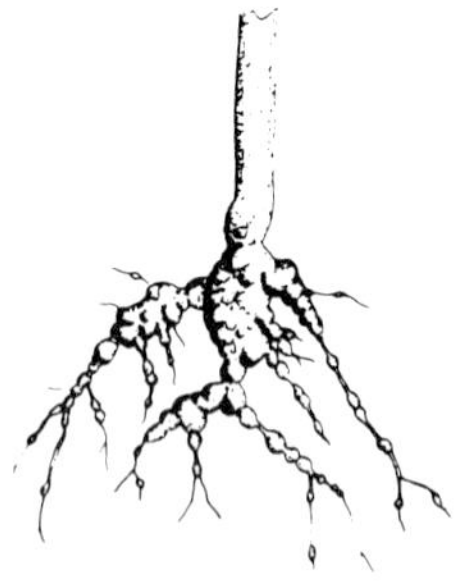

AGRONOMIC DIVISION
BLUE RIDGE ROAD CENTER
RALEIGH, NORTH CAROLINA 27611
Phone: (919) 733-2655

NEMATODE ADVISORY SECTION
Note No. 003

Root-Knot Nematode Management Recommendations For Tobacco

Root-knot nematodes are a serious problem on tobacco. To manage root-knot damage to your tobacco crop you should consider implementing the appropriate control practice from those listed below.

	Root-knot Population Fall Sample	Root-knot Population Spring Sample	Hazard Level	Recommendations
3-1	0	0	None	No root-knot nematodes were detected and a nematicide is not recommended, but a chemical soil treatment may be required for control of soil-borne diseases such as black shank or granville wilt.
3-2	10-200	10-20	Very Low	Root-knot nematodes will not cause economic damage at this level. A root-knot resistant variety should be considered to prevent the numbers from increasing unless root-knot damage has been observed on a root-knot resistant variety. If so, such varieties will be of limited value. A chemical soil treatment may be required, however, for control of other soil-borne diseases.
3-3	201-1000	21-100	Low	Use of resistant variety or a nematicide treatment with a rating of at least 4 is recommended unless root-knot damage has been observed on resistant tobacco in this field in the past. If resistant varieties have been affected a nematicide with a rating of 1 or 2 is suggested.
3-4	1001-3000	101-300	Moderate	Use of resistant variety and a nematacide treatment with a rating of at least 3 is recommended unless root-knot damage has been observed on resistant tobacco in this field in the past. If resistant varieties have been affected a nematicide with a rating of 1 or 2 is suggested.
3-5	3000+	300+	High	Use of a resistant variety and a nematicide treatment with a rating of 1 or 2 is recommended. If damage by root-knot nematodes on resistant varieties has been seen in this field in the past resistance will not be effective and use of a nematicide with a rating of 1 or 2 is recommended.

AGRONOMIC DIVISION
BLUE RIDGE ROAD CENTER
RALEIGH, NORTH CAROLINA 27611
Phone: (919) 733-2655

N.C.D.A. NEMA NOTES

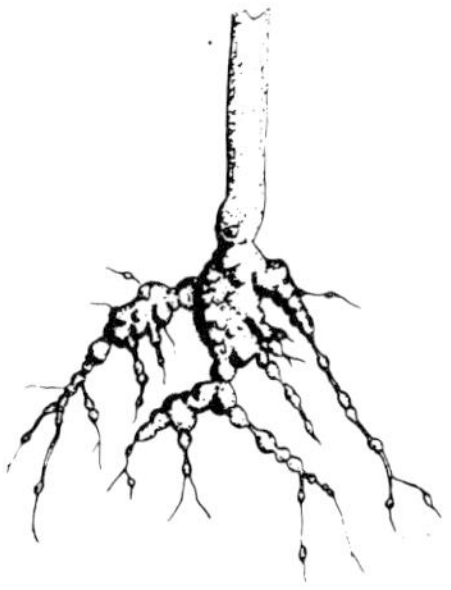

NEMATODE ADVISORY SECTION
Note No. 003

Root-Knot Nematode Management Recommendations For Tobacco

3-6 The variety of the previous tobacco crop was not listed on the information sheet. If a resistant variety was used it was not effective and use of a nematicide with a rating of 1 or 2 is recommended.

3-7 Currently available root-knot resistant tobacco varieties will not be effective in controlling this population of nematodes. A nematicide treatment with a rating of 1 or 2 is recommended.

3-8 Tobacco Beds — Use of a soil fumigant is recommended for tobacco beds regardless of the nematode population to control weeds and soil-borne diseases.

A listing of tobacco varieties having resistance to root-knot nematodes can be found in the latest edition of Tobacco Information Bulletin published by the North Carolina Agricultural Extension Service. The Tobacco Information also contains the rating of the currently recommended nematicide treatments. This publication should be consulted when planning your tobacco nematode and disease control program. Copies of Tobacco Information may be obtained from your County Agricultural Extension Office.

Assistance in interpreting the nematode control recommendations can also be obtained from the Regional Agronomists of the North Carolina Department of Agriculture. See the reverse side of this page for their names and addresses.

SOIL TESTING PLANT ANALYSIS NEMATODE ADVISORY

Telephone: 058 27 63133, extn. 2259

NEMATOLOGY DEPARTMENT

ROTHAMSTED EXPERIMENTAL STATION

Harpenden, Hertfordshire AL5 2JQ

England

With the Compliments

of

Dr.A.R.Stone

Part Six

Nematology in Developing Countries

CHAPTER PAGE

REPORT ON THE STATUS OF *MELOIDOGYNE* RESEARCH IN MEXICO, CENTRAL AMERICA AND THE CARIBBEAN COUNTRIES

Carlos Sosa-Moss

Centro de Fitopatología
Colegio de Postgraduados
Chapingo, México

Region I of the International *Meloidogyne* Project includes Mexico, the countries of Central America, the Caribbean Islands and also Surinam (Fig. 1). Although Region I is located between 4°N and 33°N and should have tropical and subtropical climates, it really varies from glacially cold to temperate or tropical with low to high humidity. This variation is due to the volcanic range of mountains, crossing the entire length of this region, from a continuation of the South American Andes to the two Sierras that run parallel to the two coasts of Mexico. As a consequence, topography is subject to abrupt changes in elevation; there are plain valleys and steep mountain slopes.

The diversity of soils and climatological conditions permits a great variety of crops to be cultivated in Region I countries. Thus, we can talk about temperate or tropical agriculture, irrigated or nonirrigated areas, and modern mechanized or traditional subsistence agriculture where, in many cases, a few rows constitute the entire farm.

This type of agriculture is so complex that for the analysis of cropping practices, sociological aspects must be taken into account when solutions to the growers' problems are considered. Different solutions are required for each zone, and application of technology cannot be general for all the countries in the region or, in many cases, even for the areas within any single country.

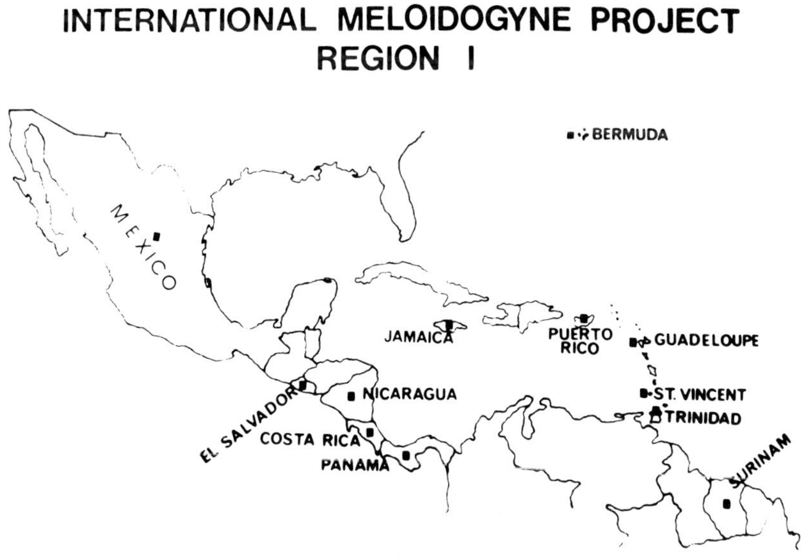

Fig. 1. International *Meloidogyne* Project Region I.

Plant pathological problems are also complex and difficult to work out; the ones caused by nematodes are not the exception. Some countries geographically included in Region I of the International *Meloidogyne* Project have not been represented, and in others, there are few activities due to their political instability. For this reason there is a lack of information from some countries, and from some others, it is very sporadic.

CURRENT STATUS OF *MELOIDOGYNE* RESEARCH IN REGION I

Species Identified

Nematodes of the genus *Meloidogyne* are widely distributed in Mexico, Central America and the Caribbean Islands. They affect many cultivated and wild plants, including herbaceous annual crops and woody perennial shrubs. Because root galling is obvious and spectacular, the presence of *Meloidogyne* has been known for many years and has been assumed to be responsible for yield reduction (1, 2, 8, 9, 20, 21, 22, 24, 26, 27, 34, 36, 37, 39, 44, 46, 48, 52, 58, 59, 63, 73, 74, 75, 79, 84, 87, 91, 96, 101, 102, 103, 127, 128, 129, 130, 138, 139, 144, 153, 161, 163, 164, 169, 170, 180, 181, 183, 203, 204, 206, 207, 208, 209, 212, 215, 220, 221, 225, 231, 232).

MELOIDOGYNE INCOGNITA To date, the majority of *Meloidogyne* populations identified in Region I have been *M. incognita*. This species has a very wide host range and distribution. It has been reported on many annual or perennial crops, among which are amaranthus (*Amaranthus* sp.), apple (*Malus* sp.), avocado (*Persea americana*), banana (*Musa* sp), bean (*Phaseolus* sp.), beet (*Beta vulgaris*), bitter gourd (*Momordica charantia*), black pepper (*Piper nigrum*), buckwheat (*Fagopyrum esculentum*), cabbage (*Brassica oleracea* var. *capitata*), cardamom (*Elettaria cardamomum*), carrot (*Daucus carota* var. *sativa*), cashew (*Anacardium* sp.), cassava (*Manihot esculenta*), celery (*Apium graveolens* var. *dulce*), citrus (*Citrus* sp.), coffee (*Coffea arabica*), corn (*Zea mays*), cucumber (*Cucumis sativus*), chayote (*Sechium edule*), eggplant (*Solanum melongena*), ginger (*Zingiber officinale*), grape (*Vitis vinifera*), kenaf (*Hibiscus cannabinus*), lettuce (*Lactuca sativa*), luffa (*Luffa* sp.), melon (*Citrullus* sp.), naranjilla (*Solanum quitoense*), okra (*Abelmoschus esculentus*), onion (*Allium* sp.), papaya (*Carica papaya*), parsley (*Petroselinum crispum*), peach (*Prunus persica*), pepper (*Capsicum* sp.), pigeon pea (*Cajanus cajan*), plantain (*Musa* sp.), portulaca (*Portulaca* sp.), potato (*Solanum tuberosum*), pumpkin (*Cucurbita* sp.), sorrel (*Rumex* sp.?), soybean (*Glycine max*), sugar cane (*Saccharum officinarum*), sweet potato (*Ipomoea batatas*), tobacco (*Nicotiana* sp.), tomato (*Lycopersicon esculentum*), watermelon (*Citrullus vulgaris*), winged bean (*Psophocarpus tetragonolobus*), yam (*Dioscorea* sp.) and yardlong bean (*Vigna sesquipedalis*). Also, it has been observed in many ornamentals.

Meloidogyne incognita has been recorded in Mexico (14, 20, 21, 34, 36, 52, 86, 90, 113, 122, 132, 142, 156, 168, 183, 184, 186, 187, 188, 194, 195, 196, 197, 211, 213, 217, 218, 219, 230, 232), El Salvador (1, 86, 211), Nicaragua (24), Costa Rica (10, 51, 70, 71, 86, 106, 107, 109, 146, 211), Panama (42, 43, 86, 111, 205, 208, 209, 210, 211, 214), Surinam (86, 165, 166, 211), Trinidad and Tobago (86, 169, 170, 173, 174, 177, 178, 211), St. Vincent (179), Guadeloupe and Martinique (11, 86, 92, 211), Jamaica (44, 72, 80, 86, 87, 211), Puerto Rico (3, 4, 17, 18, 19, 86, 157, 158, 211) and Bermuda (23, 211).

Races of this species have been identified using the North Carolina Differential Host Test. The known distribution of *M. incognita* races in Region I is as follows: in Mexico, races 1 and 3 (62, 68, 86, 89, 100, 104, 184, 211); in El Salvador, races 1, 2 and 3 (86, 211); in Costa Rica, only race 1 (86, 211); in Panama, races 1 and 2 (86, 211); in Surinam, only race 1 (86, 211); in Trinidad and Tobago, races 1 and 2 (86, 211); in Guadeloupe, only race 1 (86, 211); in Jamaica, races 1, 2 and 3 (86, 211) and in Bermuda, only race 1 (211). Puerto Rico is the only country in Region I where all four known races of *M. incognita* have been recorded (6, 18, 19, 86, 211).

MELOIDOGYNE JAVANICA This species is present in all the countries of Region I: Mexico (65, 156, 184, 211), El Salvador (1), Nicaragua (24), Costa Rica (10, 86, 160, 211), Panama (43, 209, 210, 220), Surinam (166), Trinidad (86, 176, 211), Guadeloupe (86, 176, 211), Jamaica (44, 73, 75, 211), and Puerto Rico (86, 17, 19). However, it has been reported affecting fewer crops and showing a more localized distribution than *M. incognita*. The susceptible hosts (annual or perennial) are banana, black peper, carnation (*Dianthus* sp.), carrot, cocoa (*Theobroma cacao*), coffee, gerbera (*Gerbera* sp.), grape, luna bean (?), papaya, pigeon pea, sugar cane, tobacco, tomato and wild potato (*Solanum* sp.).

MELOIDOGYNE ARENARIA This root-knot species has been recorded in the following countries of Region I: Mexico (97, 115, 156, 162, 184, 188, 195,

213, 218), Belize (211), Surinam (86, 166, 211), Guadeloupe (86, 211), Jamaica (44, 73, 75), Puerto Rico (17, 18, 19, 211) and Bermuda (211). In all of these countries, except in Jamaica, some populations of *M. arenaria* have been identified as race 2, by means of the North Carolina Differential Host Test. It has been observed galling roots of beet, black pepper, cabbage, carnation, coleus, cucumber, dasheen (*Colocasia esculenta*), eggplant, heliconia (*Heliconia* sp.), mint (*Mentha* sp.), okra, papaya, peanut (*Arachis hypogaea*), pepper, pineapple (*Ananas comosus*), potato, pumpkin, sugar cane, tobacco and tomato.

MELOIDOGYNE HAPLA This species seems to be the least distributed of the four most important *Meloidogyne* species. Its populations are localized in temperate areas of high elevation in the tropical or subtropical countries of Region I. It has been identified as a parasite of a very restricted group of cultivated plants, such as beet, cabbage, cardamom, carnation, carrot, celery, cucumber, daisy (*Chrysanthemum* sp.?), lettuce, peach, potato, strawberry (*Fragaria* sp.), sugar beet and tomato. A few populations of *M. hapla* have been recorded in Mexico (35, 36, 188, 195, 211, 217, 218, 219), El Salvador (86), Costa Rica (10, 51, 86, 107, 110, 211) and possibly Jamaica (44, 73, 75).

OTHER *MELOIDOGYNE* SPECIES In addition to the four most common species of *Meloidogyne*, several others have been identified. *Meloidogyne exigua* has been reported in coffee and several other important crops from Mexico, Nicaragua, Costa Rica, Panama, Trinidad and Puerto Rico (10, 24, 43, 46, 51, 58, 59, 129, 159, 176). In Mexico, *M. kikuyensis* was found on pineapple, and recently it has been observed causing abundant galls on avocado and papaya (183; C. Sosa-Moss personal communication). Another species newly identified in Mexico, based only on male morphology, is *M. chitwoodi*, but the host is unknown (J. D. Eisenback, personal communication; 184). *Meloidogyne oryzae* has been detected galling roots of rice (*Oryza sativa*) and plantain in Surinam, and it has been observed that both populations reproduce well on tomato cv. 'Large Chico,' but galls are inconspicuous (165, 166). Another possible new species that also causes severe damage in rice has been recorded in Costa Rica and Panama (10, 47, 108, 210).

Mixed populations of *Meloidogyne* spp. are frequently observed in Region I countries; there are several reports about concomitant species of root-knot nematodes, especially *M. incognita* with *M. javanica* or *M. arenaria*. Also, weeds are efficient hosts for *Meloidogyne* sp. (44, 74, 79, 87, 166). Even in Mexico, well-developed females of *M. incognita* have been observed in the stems of *Orobanche ramosa*, a parasitic plant of tobacco roots (183).

Damage Evaluation

In many countries of Region I, attempts have been made to measure damage and losses due to *Meloidogyne* spp. In Mexico, the damage caused by *M. incognita* has been evaluated in bean, tomato, potato, coffee and corn. Yields were reduced 30% to 100%, depending on the crop and initial population used as inoculum (38, 100, 168, 183, 186, 188, 190, 196, 218). In another study, two varieties of tomato were infected with *M. incognita*, *M. arenaria* and *M. hapla*, and the damage was measured (217, 219).

In Costa Rica, the effect of variable initial populations of *M. incognita* on yield of lettuce was determined. Yield reductions have also been estimated in cucumber, cardamom, tobacco and other crops. Damage to rice by a possible new species of *Meloidogyne* has also been demonstrated (10, 40, 47, 64, 107),

In Panama, losses caused by different levels of *M. incognita* populations on several varieties of rice were evaluated. It was found that in some varieties grain weight was highly reduced. Similar studies carried out in tobacco have produced similar results (210).

In Surinam, damage to rice by *M. oryzae* has been evaluated. Nematode infection has been found to reduce the yield of paddy rice (166).

In Jamaica, crop losses caused by species of *Meloidogyne* alone or in conjunction with other nematodes were evaluated. *M. incognita* was found to reduce yield of carrot, cucumber, lettuce, okra, pothos (*Scindapsus* sp.), pumpkin, squash (*Cucurbita* sp.), tobacco and tomato. This reduction varied from 20 to almost 100% (73, 76, 77).

In Puerto Rico, the host-parasite relationships between several populations of *Meloidogyne* spp. and some cultivated plants were investigated. Progressive inoculum levels of *M. incognita*, *M. arenaria*, *M. javanica* and *M. exigua* were found to reduce growth and production of coffee, pepper, pineapple, sugar cane and tomato, as well as root crops and ornamentals (5, 19, 125, 216, 224).

It is important to mention that in *Meloidogyne* damage evaluation studies conducted in Region I countries, there has not been a systematic attempt to standardize initial inoculum levels, which no doubt accounts for a great deal of the variation in measuring yield losses.

Histopathology

Few studies in Region I focus on describing damage at a histological level and understanding the nature of *Meloidogyne* spp. parasitism.

In Mexico, a *Meloidogyne incognita* population that does not cause galling in maize (*Zea mays*) roots was found to form giant cells and induce hyperplasia around the vascular tissue in the maize cv. 'Criollo Chalqueno.' The affected cells did not form a real gall; instead, the symptom was root swelling (131). Also the histological disturbances caused by single and mixed populations of *M. incognita* race 1 and *Nacobbus aberrans* on roots of tomato cv. 'Ace VF 55' were studied (104). In another histopathological study, bean cv. 'Black Valentine,' which is susceptible to *M. incognita* race 1, was compared with the resistant cv. 'Nemasnap.' In the roots of the resistant cultivar, giant cells developed but hyperplasia was absent (89). Finally, alterations, caused by *M. arenaria*, in tissues of immature peanut pods were investigated (115).

In Trinidad, a histopathological study indicated differences in sensitivity among tomato varieties (176).

In Puerto Rico, a great number of giant cells that had been induced by *Meloidogyne incognita* were found in the xylem, parenchyma and cortex of the roots of coffee cv. 'Borbon.' In some cases, it was observed that the xylem vessels were completely occluded by the giant cells. Also, the histopathology of *M. incognita* in association with *Fusarium oxysporum* f. sp. *coffeae* has been studied in detail (10, 126).

Interaction With Other Pathogens

The combined effects of *Meloidogyne* spp. and other pathogens have also been investigated in several countries of Region I.

In Mexico, interactions among *M. incognita* race 3, *Fusarium oxysporum* f. sp. *lycopersici* and *Phytophthora capsici* have been studied on tomato cv. 'Ace VF 55.' Catastrophic damage was observed when the three organisms were present. Also, interactions of *M. incognita* with *Fusarium solani* f. sp. *phaseoli* and/or *Rhizoctonia solani* on snap bean cv. 'Black Valentine' have been investigated, and similar results have been reported (66, 69, 95, 98, 99, 100, 188, 191, 226).

In Costa Rica, pathogenic interactions were found between *Meloidogyne* spp. and *Fusarium oxysporum* f. sp. *pisi* in peas (*Pisum* sp.) (135).

Similarly, in Panama, interactions were detected between *M. incognita* and *Pseudomonas solanacearum* in tomato, and it was observed that when both pathogens were together, the nematode activity accentuated the severity of bacterial wilt (30, 210).

Also in Jamaica and Guadeloupe, *Meloidogyne* spp., especially *M. incognita*, have been found associated pathogenically with *Pseudomonas solanacearum* in tomato and pepper. Moreover, in Guadeloupe, multiple interactions among the root-knot nematode, the abovementioned bacteria and *Fusarium* species were found (11,73).

In Puerto Rico, the association between *M. incognita* and *Fusarium solani* in pepper was demonstrated to be synergistic only when both organisms were inoculated simultaneously to the plants. In coffee trees, the interaction of *M. incognita* and *Fusarium oxysporum* f. sp. *coffeae* was investigated. It was found that chlorosis, wilting and stunting were greater when both pathogens were present. Also on coffee, *M. exigua* and *Fusarium* sp. were very often synergistically associated.

In another study, *M. incognita* and *Fusarium oxysporum* f. sp. *dioscoreae* did not show any pathogenic association on yam (17, 18, 19, 124, 126, 223; J. Toro and N. Acosta, personal communication).

Meloidogyne species mixed with other nematode genera and even mixtures of two or more root-knot species are very common in Region I countries. However, only in Trinidad was research conducted on the association between *Meloidogyne incognita* and *Rotylenchulus reniformis* in soybean. Increased damage and yield reduction was detected when mixed populations of both nematodes were inoculated to the plant roots (175).

Ecological Studies

Some studies on the ecology of *Meloidogyne* species have been undertaken in Region I of the IMP.

In Mexico, it was reported that *Meloidogyne* sp. developed better in soil with a pH of 7. Also, it was observed that applications of Metalaxyl to control *Phytophthora cinnamomi* on avocado increased the frequency of *Meloidogyne* sp. galls on the roots of the treated trees. In another study, some soil types were found to suppress *Meloidogyne* sp. infection (15, 55, 56, 57, 61, 154, 155).

In Costa Rica, the effect of soil type on the development and pathogenicity of *M. incognita* in lettuce was measured. Also, the spatial distributions of *M. incognita* in vegetable crops, and *Meloidogyne* spp. in burley tobacco, rice fields and in orchards of avocado, cashew, citrus and papaya were studied. An erratic non-uniform distribution was found (10, 64, 106, 107, 108).

In Surinam, the damage of *Meloidogyne oryzae* in flooded and dry pots was investigated. Reduction in grain yield and plant weight were detected in both types of pots. The effect of soil type on the reproduction of *M. incognita* was also studied, and higher production of egg masses was found in sandy soil with pH 6.0 than in a shallow soil with pH 7.0 (166).

In Trinidad it was observed that the eggs of *Meloidogyne* sp. survive in the soil for several days. Also, seasonal variations in field populations of *M. incognita* have been recorded and the effect of soil type and crop variety on the nematode population have been measured (171, 176, 178).

In Puerto Rico, the impact of temperature and soil type on *Meloidogyne incognita* associated with bean has been studied. Results indicate that nematode reproduction is higher at 25°C in the soil type locally referred to as Fraternidad (227; J. Toro and N. Acosta personal communication).

Control Measures

The variety of methods to control *Meloidogyne* spp. that have been tested in Region I countries can be separated into the following categories: chemical control, genetic resistance, crop rotation, biological control, and others.

CHEMICAL CONTROL In Mexico, Central America and the Caribbean countries, chemical control with nematicides has been used in an attempt to reduce losses caused by *Meloidogyne* species.

In Mexico, Costa Rica, Panama, Surinam, Trinidad, St. Vincent, Jamaica and Puerto Rico, nonfumigants, such as aldicarb, carbofuran, ethoprop, fensulfothion, isazofos, oxamyl, fenamiphos and sulfocarb, have been found effective in the control of root-knot nematodes. These products were generally used in granular form (10 or 15G). Also, in small areas, fumigants such as DD, 1-3-D, MIT and methyl bromide, have been injected into the soil.

All of the above-mentioned nematicides have been employed against several *Meloidogyne* species, especially *M. incognita*, in a variety of crops: including, banana, bean, cabbage, carrot, cardamom, cauliflower (*Brassica oleracea* var. *botrytis*), celery, coffee, cucumber, grape, lettuce, melon, okra, pea, peanut, pepper, pigeon pea, pineapple, plantain, potato, rice, sugar cane, sweet potato, tobacco, tomato and watermelon.

The nematicide rates that were used varied largely (from 0.5 to 11.2 kg a.i. per ha) depending on the soil type, crop, and method of application (16, 17, 25, 32, 33, 41, 45, 46, 49, 50, 51, 54, 67, 76, 77, 78, 82, 107, 112, 114, 118, 119, 120, 132, 133, 134, 139, 143, 145, 147, 148, 149, 150, 151, 152, 158, 166, 167, 172, 173, 174, 176, 179, 210, 222, 229).

In Region I, a different way of chemical control has been investigated. This method consists of treating seeds with a low concentration of nematicide solutions such as carbofuran, fenamiphos and oxamyl. Microdoses of active ingredient protect the plants during the first weeks after germination and reduce yield losses due to *Meloidogyne* spp.

In Mexico such chemicals have been applied to seeds of bean and tomato, and protection against *M. incognita* has been observed (116, 117, 140, 141, 184, 187, 192).

In Surinam, seed treatment of bean, bitter gourd, cucumber, rice and watermelon has given good protection against *M. oryzae* and *M. incognita* (166).

Another method of nematicide application for control of *Meloidogyne* sp. has been tested in Region I, but to a lesser extent. This method is the drench application of fenamiphos, used in Puerto Rico to reduce damage of *Meloidogyne* in coleus (3).

GENETIC RESISTANCE Resistant varieties may be the best choice to reduce *Meloidogyne* spp. populations in Region I, when the cost of nematicides is considered. Up to date, many different crop varieties have been screened for resistance.

In Mexico, bean, potato and tomato were tested for resistance to *M. incognita*. In tomato, the French cv. 'Piernita' showed resistance, as did the lines derived from the crosses between this variety and other commercial ones. Also, differences in response of tomato varieties to *M. incognita*, *M. arenaria* and *M. hapla* were measured. In potato, of 20 varieties screened, 3 behaved as resistant ('Lopez,' 'Patrones,' and 'Leona') and 2 as tolerant ('Procura' and 'Marijke'). In other investigations, some of the bean varieties included for selection showed resistance to *M. incognita* race 1, and in the northwest part of Mexico, four rootstocks of grape were evaluated for resistance to *M. javanica* (12, 13, 14, 62, 65, 121, 136, 137, 182, 185, 186, 188, 217, 219, 228).

In El Salvador, wild varieties of *Lycopersicon pimpinellifolium* have shown resistance to

Meloidogyne spp., and a resistant variety of tomato has been selected (53, 85).

In Costa Rica, ten varieties of bean and ten of maize were screened for resistance to *M. incognita*. Differences in host-parasite interaction were found among varieties of both crops. In coffee, the effect of resistant root-stocks on *M. exigua* was evaluated, but the grafted trees did not yield more than the normal susceptible ones. Also, rice varieties showed differnces in susceptibility to *Meloidogyne* sp. (10, 47, 71, 105, 107).

In Panama, varieties of bean, soybean, tomato and vigna were evaluated for resistance to *M. incognita*. Some resistant varieties were observed in the three groups of these plants. Two varieties of vigna, generally considered to be resistant, turned out to be susceptible in Panama. Also, commercial varieties of tomato, resistant to *M. incognita* and tolerant to *Pseudomonas solanacearum*, were derived from crosses between *Lycopersicon pimpinellifolium* and *L. esculentum*. Seven lines or varieties of rice were inoculated with a possible new species of *Meloidogyne*, and differences in susceptibility or resistance were observed (28, 30, 93, 94, 208, 209, 210).

In Surinam, several varieties of cabbage, bean, red pepper and tomato were infected with *Meloidogyne* spp. Most of the cultivars included were good hosts for the root-knot nematodes, but some showed various degrees of resistance, especially tomato (165, 166).

In Guadeloupe, genetic resistance has received special attention; bean varieties resistant to *M. incognita* have been developed, using 'Manoa Wonder' as progenitor. Tomato and pepper varieties resistant to *M. javanica*, *M. incognita* and *M. arenaria* have also been developed; some of these varieties expressed resistance to *Pseudomonas solanacearum*, *Fusarium* sp. and *Phytophthora infestans* (11, 91).

In Jamaica, a list of plant crops were tested for resistance to *Meloidogyne* sp. Some varieties of amaranthus, sorrel, sweet corn, sweet pepper, sweet potato and vigna showed some tolerance or resistance (73, 75).

In Puerto Rico, bean, pepper, pigeon pea, soybean, tomato and vigna were evaluated for resistance to *M. incognita* and *M. javanica*. Resistance was observed only in some varieties of these crops (5, 7, 17, 19; J. Toro and N. Acosta, personal communication).

CROP ROTATION Many of the countries in Region I have mentioned in official publications that crop rotation might greatly improve crop yield in soils infested with *Meloidogyne* spp. In Panama, Trinidad, Guadeloupe and St. Vincent, there has been a systematic attempt to implement programs of crop rotation or crop combinations to reduce *Meloidogyne* populations (11, 176, 179, 208).

However, it is in Jamaica where more complete research on crop rotation has been accomplished. Cropping systems for each region of the country were conducted, including not only the predominant *Meloidogyne* species, but also the predominant races. These programs are very complete and include from six to nine different crops in each rotation system. Amaranthus, cassava, grasses, onion and sweet potato have demonstrated usefulness because populations of some *Meloidogyne* species were reduced after these crops were planted (73, 75, 80, 81).

BIOLOGICAL CONTROL This method of root-knot nematode control has only recently received attention in Region I; in consequence, there are few results available. The fungal parasite of *Meloidogyne* species' eggs, *Paecilomyces lilacinus*, is being tested in some countries, especially in Mexico and Panama, without definite results yet. Another biological agent, *Dactylaria leptospora*, has been observed trapping *Meloidogyne* juveniles in Mexico, but formal practical research has not been undertaken (29, 31, 68, 123, 189).

OTHER CONTROL MEASURES Some unusual methods of control for *Meloidogyne* species have been sporadically investigated in Region I.

In Mexico, the effect of some insecticides used as nematicides against *Meloidogyne* spp. in tomato, by drench application to infested soils, has been studied (193). Also in Mexico, molasses in water solution has been added to artificially infested soil under greenhouse conditions in an attempt to control *M. incognita* (197).

Other research has been conducted in Mexico on the effect of nitrogen fertilizers, such as ammonium and urea, on the behavior of *M. incognita* juveniles "in vitro." In the greenhouse, the effect of these nitrogen compounds on juvenile penetration of tomato roots has been investigated (198, 199, 200, 201, 202).

The efficacy of some plants with nematicidal properties has been sparsely investigated in Mexico (60, 88, 230, 232).

The influence of soil amendment with organic manure on the control of *Meloidogyne* sp. has been studied in Mexico and Guadeloupe (11, 61,).

In Trinidad, other cultural methods such as liming following plastic mulch and solarization resulted in a reduction of root-knot nematode populations, and marked yield response was observed in cabbage, cauliflower, cucumber, tomato and watermelon (172, 176).

Finally, in Region I countries, only one attempt of integrated control has been reported. It was done in Costa Rica, where selected practices for control of *Meloidogyne incognita* in celery were combined successfully (83).

RESEARCH PROJECTS ON *MELOIDOGYNE* UNDERWAY IN REGION I

Research on *Meloidogyne* in progress in Mexico, Central America and the Caribbean countries could be considered as a continuation of the investigations presented above.

Projects in which diverse aspects of *Meloidogyne* species are under study can be summarized in the following manner:

Survey and identification of *Meloidogyne* species and its races;

Crop loss evaluation with respect to *Meloidogyne* spp. populations and to their parasitic interaction with fungi, bacteria and other concomitant species of nematodes;

Chemical control by means of nematicides in granular application, and by seed treatment with systemic nematicides soluble or miscible in water;

Selection and breeding for resistance;

Crop rotation systems and miscellaneous cultural methods;

Evaluation of *Paecilomyces lilacinus* for biological control and search for other local agents useful for that purpose.

With respect to the survey and identification of *Meloidogyne* species and its races, almost all the countries included in Region I of the IMP have permanent programs in place. The scope of each of these programs depends on the country, the economic importance of *Meloidogyne* spp. to its local or national agriculture, and the facilities and qualified personnel available.

The emphasis within current projects to assess crop loss due to variable populations of *Meloidogyne* spp. and interactions with other plant pathogens is being placed on economically important crops. *M. incognita* receives the most attention because it is the most widely distributed species and affects the majority of cultivated plants. Nevertheless, in some countries *M. javanica*, *M. arenaria*, *M. hapla* or other *Meloidogyne* species are investigated with greater emphasis, depending on local environmental conditions and the predominant economic crops: for instance, *M. oryzae* in Surinam.

Chemical control is now being studied in practically all the countries of Region I using the same nematicide compounds. Up to date, only in Mexico and Surinam are research projects on seed treatment with chemicals underway.

Selection for resistance is receiving great emphasis in Mexico, Costa Rica, Panama, Puerto Rico and especially Guadeloupe. French geneticists are actively working on breeding for resistance to *M. incognita* populations present in Guadeloupe. Within two years, they will release varieties of eggplant, green bean, and tomato resistant to root-knot nematodes and other plant pathogens. These varieties could be profitable to the rest of Region I countries.

With regard to cultural practices for control of *Meloidogyne* populations, crop rotation is being intensively evaluated in Mexico, Costa Rica, Puerto Rico, Guadeloupe, and especially in Jamaica, and more information about its great potential and benefit for underdeveloped countries is being generated.

Research on biological control has been enthusiastically initiated in the last years. However, the egg-parasitic fungus *Paecilomyces lilacinus* is the only organism that so far receives attention. At present, only Mexico, Costa Rica, Panama and Surinam are involved in projects with this biological agent.

RESEARCH NEEDS AND PRIORITIES WITH RESPECT TO *MELOIDOGYNE* RESEARCH IN REGION I

Much has been said about the objectives and goals of the International *Meloidogyne* Project. Many of these objectives have largely been met in Region I countries, and the benefits can be evaluated from different points of view. The systematic and well-oriented study of one of the most harmful pests in the world will have a very important impact on the economies of underdeveloped countries in terms of food production in the coming years. The benefit is economical not only because the IMP represents a new philosophy that needs to be applied to the solution of common agricultural problems over the world, but also because it can be used as a model to study other plant pathogens.

Countries in Region I need to continue research on *Meloidogyne* species, but some priorities should be established based on the knowledge and experience we now have. Survey and identification of species and races of *Meloidogyne*, crop loss evaluation, and studies of interactions with other plant pathogens should be permanent activities in each country because *Meloidogyne* species are widely distributed and are a very dynamic group of pathogens. The great variation of environmental conditions and agricultural practices are responsible for the modalities of *Meloidogyne* problems in each area or country of the entire Region I. This type of research will provide useful ecological and economic information that is necessary if decisions about convenient and effective control measures are to be made.

Due to the economic status of the countries included in Region I of the IMP, research directed toward control of *Meloidogyne* species should have the following priorities: crop rotation, resistance or tolerance using local and international germ plasm, biological control and chemical control.

Chemical control cannot be considered as a priority control method because it is extremely expensive in Mexico, Central America and the Caribbean countries. Also, the chemicals used as nematicides very often are not readily available for continued use, and sometimes farmers tend to misuse them with the result that accidents occur. In addition, chemical distributors often benefit more themselves by the testing of these chemicals than do the farmers.

However, chemical nematicides have their place in control programs in countries of Region I and research about them should be carried out. New methods of application such as seed treatment, drench application and others developed to reduce cost and contamination could be useful to complement the other more convenient control measures mentioned above.

Acknowledgements:

I would like to express my gratitude to the following investigators for the information they have so kindly provided me. They are: Marywbska Calderon from Nicaragua, Adrian Figueroa and Luis Salazar from Costa Rica, Julio Lara from Panama, Rinie Segeren from Surinam, Alain Kermarrec from Guadeloupe, Sharon Eason-Heath from Jamaica and Jorge Toro from Puerto Rico.

Also I am indebted to my comrades: Drs. Sebastian Romero and Peter Hahn, Ings. Victor Manuel Ruiz and Mario Alberto Hernandez for their invaluable help in the preparation of this manuscript. Finally, I would like to thank Misses Sahara Flores, Lidia Morales and Angeles Morales for their excellent typing.

Literature Cited

1. Abrego, L. 1976. Fitonematología en El Salvador, pp. 22-23 *In* Memorias de la Conferencia Regional de Planeamiento del Proyecto Internactional Meloidogyne, Region I, 26-30 abril 1976. Facultad de Agronomía, Univ. Panamá, Panamá.
2. Aceves, J. J. y D. Téliz O. 1982. Enfermedades del duraznero en Tetela del Volcán, Morelos, pp. 184-185 *In* Avances en la Investigación en el Colegio de Postgraduados. Chapingo, México.
3. Acosta, N. 1976. Control of *Meloidogyne incognita* on *Coleus blumei* by drench application of fensulfothion. Nematropica 6:23.
4. Acosta, N. and A. Ayala. 1975. Pathogenicity of *Pratylenchus coffeae, Scutellonema bradys, Meloidogyne incognita* and *Rotylenchulus reniformis* on *Dioscorea rotundata.* J. Nematol. 7:1-6.
5. Acosta, N. and J. A. Negron. 1984 . Susceptibility of various tomato lines to a population of *Meloidogyne incognita.* J. Agric. Univ. P. R. In press.
6. Acosta, N. and J. A. Negron. 1982. Susceptibility of six soybean cultivars to *Meloidogyne incognita* race 4. Nematropica 12:181-187.
7. Acosta, N. and J. A. Negron. 1982. Susceptibility of various tomato lines to a population of *Meloidogyne incognita.* Nematropica 12:173-180.
8. Aguilar, B. G., V. G. Almaguer, y S. Palacios A. 1978. Observaciones preliminares sobre el ataque de nemátodos noduladores de raíces de los géneros *Meloidogyne* (Goeldi) y *Nacobbus* (Thorne y Allen) en trigo sarraceno *Fagopyrum escullentum* L., en el área de Chapingo, México, pp. 19-20 *In* Resumenes VIII Congr. Nac. Fitopathol. Oaxtepec, Mor. México.
9. Alcocer, G. L. 1959. Morfologia y biología de un nemátodo aislado de las raíces nodulares de plantas de jitomate *Lycopersicon esculentum* Mill., de las principales regiones agrícolas de México. Tesis profesional. Escuela Nacional de Ciencias Biológicas, Inst. Politéc. Nac., México.
10. Alvarado, M., C. Sancho, J. Castro, L. Salazar, y R. López. 1982. Informe sobre el trabajo realizado para el Proyecto Internacional Meloidogyne en la Universidad de Costa Rica en el período de 1979-1981, pp. 26-34 *In* Proc. Third IMP Res. Plann. Conf. on Root-Knot Nematodes, *Meloidogyne* spp., Region I, Jan. 11-15, 1982. Inst. In-

vest. Agropecuarias de Panamá, Panamá. 180 pp.

11. Anais, G. 1982. Research on *Meloidogyne* in Guadeloupe, pp. 57-60 *In* Proc. Third Res. Plann. Conf. on Root-knot Nematodes, *Meloidogyne* spp., Region I, Jan. 11-15, 1982. Inst. Invest. Agropecuarias de Panamá, Panamá. 180 pp.
12. Armenta, C. F. X. y C. Sosa-Moss. 1975. Resistencia genética de vegetales al ataque de nemátodos, p. 79 *In* Avances en la Enseñanza y la Investigación en el Colegio de Postgraduados. Chapingo, México.
13. Armenta, C. F. X. y C. Sosa-Moss. 1976. Resistencia genética de papa al ataque de nemátodos, p. 94 *In* Avances en la Enseñanza y la Investigación en el Colegio de Postgraduados. Chapingo, México.
14. Armenta, C. F. X. y C. Sosa-Moss. 1978. Resistencia al nemátodo nodulador de las raíces *Meloidogyne incognita* (Kofoid and White), de veinte variedades de papa (*Solanum tuberosum* L.). Tesis profesional, Dep. Parasitol. U.A.CH., Chapingo, México.
15. Avila, A. L. N., R. Montes B., y E. N. Becerra. L. 1980. Efectos del pH sobre los nemátodos asociados al cultivo del plátano, usando al frijol (*Phaseolus vulgaris* L.) como planta indicadora, p. 16 *In* Resumenes IX Congr. Nac. Fitopatol. Uruapan, Mich. México.
16. Ayala, A. 1972. Control químico de los nemátodos del tomate pimiento y quimbombó, en suelos arenosos. Resumenes V Reunión Anual OTAN. México.
17. Ayala, A. 1976. State of knowledge of *Meloidogyne* in Puerto Rico, pp. 93-95 *In* Proc. Reg. Plann. Conf. of the Int. Meloidogyne Project, Region I, Apr. 26-30, 1976, Facultad de Agronomia, Facultad de Agronomia, Univ. Panamá, Panamá.
18. Ayala, A. 1981. Annual report of the International Meloidogyne Program in Puerto Rico, 1977-1978, pp. 219-229 *In* Proc. Second Reg. Plann. Conf. of the Int. Meloidogyne Project, Region I, Sept. 4-8, 1978, San José, Costa Rica. Facultad de Agronomia, Univ. Panamá, Panamá. 246 pp.
19. Ayala, A., N. Acosta., J. Negron, and J. Toro. 1982. Meloidogyne research progress in Puerto Rico, pp. 75-83 *In* Proc. Third IMP Res. Plann. Conf. on Root-Knot Nematodes, *Meloidogyne* spp., Region I, Jan. 11-15, 1982, Inst. Invest. Agropecuarias de Panamá, Panamá. 180 pp.
20. Becerra, L. E. N. y C. Sosa-Moss. 1977. El nemátodo enquistado del maíz *Punctodera chalcoensis* (Stone, Sosa-Moss and Mulvey) y su relación con otros microorganismos del suelo, pp. 133-134 *In* Avances en la Enseñanza y la Investigación en el Colegio de Postgraduados. Chapingo, México.
21. Becerra, L. E. N. y C. Sosa-Moss. 1977. Se reporta *Meloidogyne incognita* (Kofoid and White, 1919) Chitwood 1949, atacando *Zea mays* (L) en México, sin causar agallas radiculares. Resumenes IX Reunión de Nematólogos de los Trópicos Americanos. Lima, Perú.
22. Bee-Rodríguez, D. and A. Ayala. 1977. Nematodes associated with sorghum in Puerto Rico. Nematropica 7:16.
23. Burpee, L. L. 1981. Studies on root-knot nematodes in Bermuda, p. 233 *In* Proc. Second Reg. Plann. Conf. of the Int. Meloidogyne Project, Region I, Sept. 4-8, 1978, San José, Costa Rica. Facultad de Agronomía, Univ. Panamá, Panamá. 246 pp.
24. Calderón V., M. 1982. Informe sobre la situación nematológica en Nicaragua, pp. 66-68 *In* Proc. Third IMP Res. Plann. Conf. on Root-knot Nematodes, *Meloidogyne* spp., Region I, 11-15 enero 1982. Inst. Invest. Agropecuarias de Panamá, Panamá. 180 pp.
25. Calvo, B. y R. López. 1980. Combate químico de *Meloidogyne incognita* en dos cultivares de tabaco Burley. Agron. Costarric. 4:175-182.
26. Campos, V.A. 1965. Nemátodos del género *Meloidogyne*, en la papa. Agric. Téc. Méx. 2(5):205-206.
27. Campos, V. A. 1965. Nemátodos del género *Meloidogyne*, en la papa. El Campo 30:34-38.
28. Candanedo, L. E. 1981. Desarrollo de lineas de tomate industrial con resistencia a *Pseudomonas solanacearum* y evaluación de su resistencia a cinco poblaciones de *Meloidogyne incognita*, pp. 52-60 *In* Memorias de la Segunda Conferencia Regional de Planeamiento del Proyecto Internacional Meloidogyne, Region I, 4-8 sept. 1978, San José, Costa Rica. Facultad de Agronomia, Univ. Panamá, Panamá. 246 pp.
29. Candanedo, L. E., J. Lara, P. Jatala, y F. González. 1982. Evaluación preliminar del comportamiento de *Paecilomyces lilacinus*

como controlador biológico del nemátodo nodulador, *Meloidogyne incognita*, en tomate industrial. Nematrópica 12:154.

30. Candanedo, L. E., R. Lasso G. y J. M. Osorio. 1978. Evaluación preliminar de la resistencia o tolerancia a *Pseudomonas solanecearum* y cinco poblaciones de nemátodos del género *Meloidogyne*, en lineas de tomate industrial. Cienc. Agropecuaria, Panamá, 1:99-104.
31. Candanedo, L. E., R. Rodríguez, J. Lara, G. de León, F. González, and P. Jatala. 1982. Biological control of *Meloidogyne* spp. and *Globodera* spp. by *Paecilomyces lilacinus* in Panama, pp. 147-149 *In* Proc. Third IMP Res. Plann. Conf. on Root-Knot Nematodes *Meloidogyne* spp., Region I, Jan. 11-15, 1982. Inst. Invest. Agropecuarias de Panamá, Panamá. 180 pp.
32. Cárdenas, D. F. 1971. Efecto del 1-3 dicloropropeno en el control del nemátodo *Meloidogyne* spp. en el cultivo de la papa *Solanum tuberosum*. Tesis M. C., Inst. Tec. Estud. Superiores de Monterrey, Monterrey, N. L. México.
33. Carrillo, A. M. A. y R. López. 1979. Respuesta del tabaco estufado a la aplicacion de nematicidas. Nematrópica 9:129-134.
34. Castillo, F. D. 1980. Análisis de los métodos taxonómicos usados en la identificación de las especies del género *Meloidogyne* (Goeldi, 1887) Chitwood, 1949, p. 13 *In* Resumenes del IX Congr. Nac. Fitopatol., Uruapan, Mich. México.
35. Castillo, F. D. y C. Sosa-Moss. 1977. Estudio taxonómico de algunas poblaciones de *Meloidogyne* que atacan cultivos de importancia económica en México, p. 221 *In* Avances en la Enseñanza y la Investigación en el Colegio de Postgraduados. Chapingo, México.
36. Castillo, F. D. y C. Sosa-Moss. 1978. Estudio taxonómico de algunas poblaciones de *Meloidogyne* que atacan cultivos de importancia económica en México, p. 204 *In* Avances en la Ensenanza y la Investigación en el Colegio de Postgraduados. Chapingo, México.
37. Castillo, P. G. 1977. *Meloidogyne* spp. nemátodo que parasita diversos cultivos en la Comarca Lagunera. Centro de Investigaciones Agricolas del Noroeste. Inst. Nac. Invest. Agric., Semin. Téc. 4(13):10.
38. Castillo, P. G. 1977. Estudio y evaluación de daños causados por el nemátodo nodulador de las raices en semilleros de café (*Coffea arabica* L.) en condiciones de invernadero. Tesis de licenciatura, Facultad de Biología, Univ. Veracruzana, Jalapa, Ver. México.
39. Castillo, P. G. 1980. Algunos aspectos sobre nemátodos asociados al cultivo del cafeto, p. 12 *In* Resumenes IX Congr. Nac. Fitopatol. Uruapan, Mich. México.
40. Castro Z., J. A. y R. López Ch. 1981. Respuesta de dos cultivares de lechuga (*Lactuca sativa* L.) a densidades crecientes de inóculo de *Meloidogyne incognita* (Kofoid y White) Chitwood. Agron. Costarric. 5:65-73.
41. Cauich, B. M., R. Montes B. y E. N. Becerra L. 1980. Efecto de la aplicación de productos químicos de acción general y selectiva, sobre las poblaciones de nemátodos fitoparásitos y hongos fitopatógenos y el desarrollo y producción del cultivo de frijol (*Phaseolus vulgaris* L.) en la Chontalpa, Tabasco, p. 25 *In* Resumenes IX Congr. Nac. Fitopatol., Uruapan, Mich. México.
42. Cralley, E. M. 1954. Las enfermedades de las plantas en Panamá. Ministerio de Agricultura, Comercio e Industrias. Servicio Interamericano de Cooperación Agricola en Panamá. (Mimeografiado).
43. Cuellar, R. 1960. Phytonematology in Panama and Central America. Soil Crop Sci. Soc. Fla. Proc. 19:430-435.
44. Dixon, W. B. and R. Latta. 1965. Nematological investigations 1958-1961. Jam. Minist. of Agric. and Lands, Bull. No. 59.
45. Eason-Heath, S. A. E., D. G. Hutton, and P. L. Coates-Beckford. 1982. Yield response of Irish potato (*Solanum tuberosum*) to control of *Meloidogyne* sp., pp. 123-126 *In* Proc. Third IMP Res. Plann. Conf. on Root-Knot Nematodes, *Meloidogyne* spp., Region I, Jan. 11-15, 1982. Inst. Invest. Agropecuarias de Panamá, Panamá. 180 pp.
46. Figueroa, M. A. 1976. El nematodo *Meloidogyne* sp. en Costa Rica, pp. 24-30 *In* Memorias de la Conferencia Regional de Planeamiento del Proyecto Internacional Meloidogyne, Region I, 26-30 abril 1976. Facultad de Agronomía, Univ. Panamá, Panamá.
47. Figueroa, M. A. 1981. Informe de labores realizadas para el Proyecto Internacional Meloidogyne por el Ministerio de Agricultura y Ganadería de Costa Rica, pp. 23-26 *In* Memorias de la Segunda Conferencia Regional de Planeamiento del Proyecto Internacional Meloidogyne,

Region I, 4-8 sept. 1978, San José, Costa Rica. Facultad de Agronomía, Univ. Panamá, Panamá. 246 pp.

48. Figueroa, M. A. 1982. Reconocimiento de géneros de nemátodos parásitos del Cafeto, pp. 21-22 *In* Proc. Third IMP Res. Plann. Conf. on Root-Knot Nematodes, *Meloidogyne* spp., Region I, 11-15 enero 1982. Inst. Invest. Agropecuarias de Panamá, Panamá. 180 pp.
49. Figueroa, M. A. 1982. Efectos del carbofuran y oxamyl en el café Caturra, p. 23 *In* Proc. Third IMP Res. Plann. Conf. on Root-Knot Nematodes, *Meloidogyne* spp., Region I, Jan. 11-15, 1982. Inst. Invest. Agropecuarias de Panamá, Panamá. 180 pp.
50. Figueroa, M. A. 1982. Evaluaciones del nematicida Aldicarb en parcelas demostrativas de café Caturra, pp. 24-25 *In* Proc. Third IMP Res. Plann. Conf. on Root-Knot Nematodes, *Meloidogyne* spp., Region I, 11-15 enero 1982. Inst. Invest. Agropecuarias de Panamá, Panamá. 180 pp.
51. Figueroa, M. A. y F. Perlaza. 1982. Investigación sobre *Meloidogyne* en Costa Rica, pp. 12-15 *In* Proc. Third IMP Res. Plann. Conf. on Root-Knot Nematodes, *Meloidogyne* spp., Region I, 11-15 enero 1982. Inst. Invest. Agropecuarias de Panamá, Panamá. 180 pp.
52. Figueroa, T. M. G. y J. S. Camacho G. 1982. Exploración nematólogica preliminar de nemátodos asociados al cultivo de la ocra (*Hibiscus esculentus* L.), en el Valle de Iguala Gro. Tesis profesional, Facultad de Cienc., Univ. Nac. Autonoma de Mexico, México.
53. Fischnaler, D. F. A. 1972. Una nueva variedad de tomate resistente a *Meloidogyne* sp. obtenida en El Salvador. Resumenes V Reunión Anual OTAN. México.
54. Fischnaler, D. F. A. 1972. Determinación de la mejor frecuencia de aplicación del nematicida experimental D-1410 para el control de *Meloidogyne* spp. en tomate (*Lycopersicon esculentum* Mill.). Nematrópica 2:17.
55. Franco, F. E. y R. García E. 1983. Comparación de la incidencia de *Meloidogyne* sp. de aguacatero entre cuatro Agro-Ecosistemas distintos. Nematrópica 13:111.
56. Franco, F. E. y R. García E. 1983. Efecto sobre la incidencia de *Meloidogyne* sp. de tratamientos contra *Phytophthora cinnamomi* en Aguacatero. Nematrópica 13:111-112.
57. Franco, F. E., R. García E., N. Marbán M. y D. Téliz O. 1983. Dinámica de población de *Meloidogyne* sp. bajo distintos manejos del agroecosistema de aguacatero (*Persea americana* Mill), y patogenicidad de *Pythium* spp. a este frutal. Tesis de maestría, Colegio de Postgraduados, Chapingo, México.
58. Gandara, G. 1906. La anguilula del cafeto. Secr. Agric. y Fomento, Comisión de Parasitol. Agríc. Circ. No. 51:1-15.
59. Gandara, G. 1920. Parásitos animales del naranjo, pp. 15-17 *En* Enfermedades y plagas del naranjo. Secr. Agric. y Fomento. Dirección de Agric. Bol. No. 111.
60. García, E. R. 1978. *Chenopodium ambrosioides* L., planta con uso potencial en el combate de nemátodos fitoparasitos: Información preliminar, pp. 76-77 *In* Resumenes VIII Congr. Nac. Fitopatol. Soc. Mex. Fitopatol., Oaxtepec, Mor. México.
61. García, E. R. 1982. Efecto de la aplicación de estiércol, metalaxyl o ambos en el daño inducido por *Meloidogyne* sp. a raíces de aguacatero, p. 173 *In* Avances en la Investigación en el Colegio de Postgraduados. Chapingo, México.
62. García-Gómez, F. A. y C. Sosa-Moss. 1984. Comportamiento de treinta y seis variedades y lineas de frijol, al ataque de *Meloidogyne incognita*, raza 1. Tesis profesional, Dep. Parasitol., U.A.CH., Chapingo, México.
63. González, F. L. 1979. Nemátodos fitoparasitos asociados con la caña de azúcar en varias zonas de Costa Rica. Nematrópica 9:32-35.
64. González, F. L. y R. López. 1980. Efecto de densidades de inóculo y características del suelo, sobre la patogenicidad de *Meloidogyne incognita* en lechuga. Agron. Costarric. 4:155-163.
65. Guerrero, R. C. y M. Jiménez L. 1982. Evaluación de patrones de vid al ataque de nemátodos en un suelo infestado, en la Costa de Hermosillo. Ciclo 1980/81, p. 98 *In* Resumenes X Congr. Fitopatol., Soc. Mex. Fitopatol., Culiacán, Sin. México.
66. Hernandez, A. H. 1977. Interacción de *Meloidogyne incognita* (Kofoid and White) Chitwood, *Fusarium solani* (Mart) Sacc. f. sp. *phaseoli* (Burk) Snyder and Hansen y *Rhizoctonia solani* Kuhn. en frijol ejotero, p. 91 *In* Memorias VII Congr. Nac. Fitopatol., México.

67. Hernández, H. C., F. A. García-Gómez y C. Sosa-Moss. 1984. Identificación de *Meloidogyne* sp. causante de la "jicamilla del mélon," en Izúcar de Matamoros, Pueb. y evaluación de cinco nematicidas, para su control. Tesis profesional. Dep. Parasitol., U.A.CH., Chapingo México.
68. Hernández Macias, B. y C. Sosa-Mss. 1984. Efecto de tres tipos de suelo en la actividad parasítica de *Paecilomyces lilacinus* (Thom.) Samson, como posible agente de control biológico de *Meloidogyne incognita* (Kofoid & White) Chitwood, en cultivo de frijol (*Phaseolus vulgaris*) var. Black Valentine. Tesis profesional. Escuela Nacional de Estudios Profesionales, Biologia, Iztacala, U.N.A.M. México.
69. Hernández, R. S. y C. Sosa-Moss. 1984. Evaluación del daño causado a frijol ejotero variedad Black Vallantine, por el nemátodo agallador *Meloidogyne incognita* raza 1 y por el hongo *Fusarium solani* f. sp. *phaseoli*, separadamente y en interacción. Tesis profesional, Escuela Superior de Agricultura, Univ. Autonoma de Nayarit, México.
70. Hidalgo, L. y R. López. 1980. Caracterización morfométrica de cuatro poblaciones de *Meloidogyne incognita* de Costa Rica. Turrialba 30:129-135.
71. Hidalgo, L. y R. López. 1980. Susceptibilidad de diez cultivares de maíz (*Zea mays* L.) a cuatro poblaciones de *Meloidogyne incognita*. Turrialba 30:316-322.
72. Hutton, D. G. 1974. The current nematode situation on vegetables in Jamaica. Vegetable Seminar. Jam. School of Agric., Twickenham Park, Jamaica.
73. Hutton, D. G. 1976. Country report on the status of root-knot nematode problems and research, pp. 67-79 *In* Proc. Reg. Plann. Conf. of the Int. Meloidogyne Project, Region I, Apr. 26-30, 1976. Facultad de Agronomia, Univ. Panamá, Panamá.
74. Hutton, D. G. 1979. The plant nematodes found associated with various crop and weed plants in Jamaica, 1970-1977. Jam. Minist. Agric. Bull. 67 (New Series):23-28.
75. Hutton, D. G. 1981. Report on work carried out in Jamaica for the International Meloidogyne Project, from May 1976 to July 1978, pp. 190-212 *In* Proc. Second Reg. Plann. Conf. of the Int. Meloidogyne Project, Region I, Sept. 4-8, 1978, San José, Costa Rica. Facultad de Agronomía, Univ. Panamá, Panamá. 246 pp.
76. Hutton, D. G. 1982. Effect of parasitic nematodes on establishment, growth and production of tobacco, pp. 114-117 *In* Proc. Third IMP Res. Plann. Conf. on Root-Knot Nematodes, *Meloidogyne* spp., Region I, Jan. 11-15, 1982. Inst. Invest. Agropecuarias de Panamá, Panamá. 180 pp.
77. Hutton, D. G. 1982. Effect of *Meloidogyne incognita* on yield of cucumber, pp. 120-122 *In* Proc. Third IMP Res. Plann. Conf. on Root-Knot Nematodes, *Meloidogyne* spp., Region I, Jan. 11-15, 1982. Inst. Invest. Agropecuarias de Panamá, Panamá. 180 pp.
78. Hutton, D. G. and D. C. Chung. 1973. Effects of post-planting applications of the nematicide DBCP to plantain. Nematropica 3:46-50.
79. Hutton, D. G., P. L. Coates-Beckford and S. A. E. Eason-Heath. 1982. Parasitic nematodes associated with various plant species in Jamaica, 1949-1981, pp. 92-108 *In* Proc. Third IMP Res. Plann. Conf. on Root-Knot Nematodes, *Meloidogyne* spp., Region I, Jan. 11-15, 1982. Inst. Invest. Agropecuarias de Panamá, Panamá. 180 pp.
80. Hutton D. G., P. L. Coates-Beckford and S. A. E. Eason-Heath. 1982. Three-year crop rotation for control of *Meloidogyne incognita* on the south-central plains of Jamaica, pp. 136-146 *In* Proc. Third IMP Res. Plann. Conf. on Root-Knot Nematodes *Meloidogyne* spp., Region I, Jan. 11-15, 1982. Inst. Invest. Agropecuarias de Panamá. Panamá. 180 pp.
81. Hutton, D. G., P. L. Coates-Beckford, and S. A. E. Eason-Heath. 1983. Management of *Meloidogyne incognita* populations by crop rotation in a small-scale field trial and nematode pathogenic effects on selected cultivars. Nematropica 13:153-163.
82. Hutton, D. G., S. A. E. Eason-Heath, and P. L. Coates-Beckdord. 1982. Control of *Meloidogyne incognita* affecting coffee, pp. 109-113 *In* Proc. Third IMP Res. Plann. Conf. on Root-Knot Nematodes, *Meloidogyne* spp., Region I, Jan. 11-15, 1982. Inst. Invest. Agropecuarias de Panamá, Panamá. 180 pp.
83. Incer, A. A. y R. Lopez. 1979. Evaluación de prácticas selectas para el combate integrado de *Meloidogyne incognita* en apio. Nematrópica 9:140-146.
84. Innes, R. F, P. D. Manser, and G. F. Clarke. 1958. Yield decline in sugar cane on

part of Worthy Park. Jam. Assoc. Sugar Tech. J. 21:1-8.

85. Interiano, M. J. D. y J. D. Quintanilla A. 1972. Respuesta de especies mejoradas de tomate silvestre (*Lycopersicon pimpinellifollium*) al parasitismo del nematodo nodulador (*Meloidogyne* spp.). Nematrópica 2:19.
86. Int. Meloidogyne Project. 1981. Summary of *Meloidogyne* species and races identified from Region I, pp. 243-245 *In* Proc. Second Reg. Plann. Conf. of the Int. Meloidogyne Project, Region I, Sept. 4-8, 1978, San José, Costa Rica. Facultad de Agronomía, Univ. Panamá, Panamá. 246 pp.
87. Jam. Dep. Agric. 1951. Investigations 1949-50. Jam. Dep. Agric. Bull. 47:35-37.
88. Jarquín, B. H. 1983. Efecto de extractos vegetales sobre poblaciones de nemátodos en vieros (sic) de café. Nematrópica 13:116.
89. Jarquin-Nieto, A., C. Sosa-Moss, y E. Cárdenas S. 1984. Reacción histopatológica de dos variedades de frijol, una susceptible (Black Valentine) y otra resistente (Nemasnap), al ataque de *Meloidogyne incognita* raza 1 y *Nacobbus aberrans*, separadamente y en interacción. Tesis profesional, Dep. Parasitol., U.A.CH., Chapingo, México.
90. Juárez, C., C. Sosa-Moss y E. Rodríguez Ch. 1975. Evaluación del daño de algunas especies de nemátodos a cultivos agricolas, p. 81 *In* Avances en la Enseñanza y la Investigación en el Colegio de Postgraduados. Chapingo, México.
91. Kermarrec, A. 1976. Investigaciones en *Meloidogyne* spp. en Guadeloupe, p. 46 *In* Memorias de la Conferencia Regional de Planeamiento del Proyecto Internacional Meloidogyne, Region I, 26-30 abril 1976. Facultad de Agronomía, Univ. Panamá, Panamá.
92. Kermarrec, A. 1981. Report on the status of *Meloidogyne* in Guadeloupe, pp. 234-242 *In* Proc. Second Reg. Plann. Conf. of the Int. Meloidogyne Project, Region I, Sept. 4-8, 1978, San Jose, Costa Rica. Facultad de Agronomía, Univ. Panamá, Panamá. 246 pp.
93. Lara, J. 1982. Detección de resistencia a *Meloidogyne incognita* raza 1 en variedades y lineas de tomate, frijol, caupi y soya. Nematrópica 12:159.
94. Lasso, G. R. y L. Aruz S. 1982. Desarrollo y evaluación de lineas de tomate con tolerancia a *Pseudomonas solanacearum* y resistencia a *Meloidogyne incognita*, Proc. Third IMP Res. Plann. Conf. on Root-Knot Nematodes, *Meloidogyne* spp., Region I, 11-15 enero 1982. Inst. Invest. Agropecuarias de Panamá, Panamá. 180 pp.
95. López, A. G., H. Hernández A. y C. Sosa-Moss. 1976. Relación entre algunas especies de nemátodos y hongos fitoparásitos, p. 95 *In* Avances en la Enseñanza y la Investigación en el Colegio de Postgraduados. Chapingo, México.
96. López, F. M. C., I. Cid del Prado V., y C. Sosa-Moss 1977. Estudio morfofisiológico del nemátodo nodulador *Meloidogyne* sp., en el cultivo del clavel, pp. 135-136 *In* Avances en la Enseñanza y la Investigación en el Colegio de Postgraduados. Chapingo, México.
97. López, F. M. C. y C. Sosa-Moss. 1978. Identificación morfofisiológica del nemátodo nodulador *Meloidogyne* sp. que ataca al clavel, p. 202, 218 *In* Avances en la Enseñanza y la Investigación en el Colegio de Postgraduados. Chapingo, México.
98. López, H. S. R. y C. Sosa-Moss. 1977. Estudio sobre la interacción de hongos fitapatógenos del suelo con *Meloidogyne incognita*, en el cultivo de tomate en el Estado de Morelos, pp. 132-133 *In* Avances en la Enseñanza y la Investigación en el Colegio de Postgraduados. Chapingo, México.
99. López, H. S. R. y C. Sosa-Moss. 1978. Interacción de hongos fitopatógenos del suelo con *Meloidogyne incognita*, en el cultivo de tomate, en el Estado de Morelos, p. 215 *In* Avances en la Enseñanza y la Investigación en el Colegio de Postgraduados. Chapingo, México.
100. López, H. S. R. y C. Sosa-Moss. 1980. Estudio de la interacción nemátodo nodulador y hongos fitopatógenos del suelo, en el cultivo de tomate del Estado de Morelos. Tesis de maestría, Colegio de Postgraduados, Chapingo, México.
101. López, M. G. y D. Téliz. O. 1982. Identificación y dinámica de fitonemátodos asociados al aguacate en agroecosistemas de Atlixco y Tochimilco, Pue, p. 204 *In* Avances en la Investigación en el Colegio de Postgraduados. Chapingo, México.
102. López, M. G. y D. Téliz O. 1983. Nemátodos fitoparásitos asociados al cultivo del aguacate en agroecosistemas de Atlixco y Tochimilco, Pue. Nematrópica 13:116-117.
103. López, M. G. y D. Téliz O. 1983. Nemátodos fitoparásitos asociados al cultivo del

aguacate en agroecosistemas de Atlixco y Tochimilco, Pue. Tesis profesional, Dep. Parasitol., U.A.CH., Chapingo, México.

104. López Portillo, A., C. Sosa-Moss y E. Cárdenas S. 1984. Estudio histopatológico de raíces de jitomate atacadas por *Nacobbus aberrans* y *Meloidogyne incognita* raza 1, separadamente y en interacción. Tesis profesional, Dep. Parasitol., U.A.CH., Chapingo, México.
105. López, R. 1980. Susceptibilidad comparativa de diez cultivares de frijol común (*Phaseolus vulgaris* L.) al ataque de *Meloidogyne incognita*. Agron. Costarric. 4:69-73.
106. López, R. 1981. Observaciones sobre la distribución espacial de *Meloidogyne incognita* después de la cosecha en dos plantios de tabaco Burley. Turrialba 31:11-14.
107. López, R. 1981. Informe sobre las investigaciones realizadas en la Universidad de Costa Rica como parte del "Proyecto Internacional Meloidogyne," pp. 16-22 *In* Memorias de la Segunda Conferencia Regional de Planeamiento del Proyecto Internacional Meloidogyne, Region I, 4-8 sept. 1978, San José, Costa Rica. Facultad de Agronomía, Univ. Panamá, Panamá. 246 pp.
108. López Ch., R. 1981. Distribución espacial de nemátodos del arroz después de la cosecha en el sureste de Costa Rica. Agron. Costarric. 5:49-53.
109. López Ch., R. y J. Azofeifa Ch. 1981. Reconocimiento de nemátodos fitoparásitos asociados con hortalizas en Costa Rica. Agron. Costarric. 5:29-35.
110. López, R., y L. A. Salazar F. 1978. Morfometria y algunos hospedantes de *Meloidogyne hapla* en la Cordillera Volcánica Central de Costa Rica. Agron. Costarric. 2:29-38.
111. Lucas, G. B. 1958. Tobacco diseases in Panama. Plant Dis. Rep. 42:1301.
112. Magallon, A. J. M. 1974. Evaluación de control del nemátodo *Meloidogyne hapla* con Vydate (DPX-1410) en el cultivo de papa *Solanum tuberosum* en la región de Navidad, Nuevo León. Tesis profesional, Univ. Agraria Antonio Narro, Saltillo, Coah. México.
113. Manzo, B. H. y A. A. Pedroza M. 1982. Necrosis radicular del plátano *Musa paradisiaca* por *Meloidogyne incognita* en el Valle de Apatzingan, p. 110 *In* Resumenes X Congr. Nac. Fitopatol., Soc. Mex. Fitopatol. Culiacán, Sin. México.
114. Marban, M. N. y V. Zamudio. 1982. Control químico de *Nacobbus aberrans* y *Meloidogyne incognita* en jitomate (ACE/55 VS) de Tecamachalco, Puebla, p. 205 *In* Avances en la Investigación en el Colegio de Postgraduados. Chapingo, México.
115. Mariscal, M. E., C. Sosa-Moss, y E. Cárdenas S. 1984. Alteraciones histopatólogicas causadas a la vaina del cacahuate por *Meloidogyne arenaria*. Tesis profesional, Dep. Parasitol., U.A.CH., Chapingo. México.
116. Marroquín, A. L. M. y E. Zavaleta M. 1980. Tratamiento a semilla mediante nematicidas sistémicos para el control del nemátodo agallador. Tesis profesional, Escuela Nacional de Ciencias Biológicas, Inst. Politéc. Nac., México.
117. Marroquín, A. L. M., E. Zavaleta M., y C. Sosa-Moss. 1978. Uso de nematicidas sistémicos en tratamientos a semillas para el control de la enfermedad llamada "Jicamilla" (*Meloidogyne* spp.), pp. 199-200 *In* Avances en la Enseñanza y la Investigación en el Colegio de Postgraduados. Chapingo, México.
118. Martínez, A. J. 1970. Distribución, identificación y efectividad de tres nematicidas para el control de la jicamilla (*Meloidogyne* spp.) del melón en el Valle de Apatzingan, Mich. Tesis profesional, Facultad de Agrobiologia, Univ. Michoacana de San Nicolás de Hidalgo, Uruapan, Mich. México.
119. Martínez, Z. G. y C. Sosa-Moss. 1972. Prueba comparativa de cuatro nematicidas para el control de *Meloidogyne incognita* en frijol ejotero. Tesis profesional, Dep. Parasitol., Escuela Nac. Agric., Chapingo, México.
120. Mattey, F. J. y R. López. 1978. Evaluación de nematicidas y de métodos de aplicación en el combate de nemátodos fitoparásitos y en la producción y calidad de la lechuga. Turrialba 28:15-18.
121. Montes, B. R. 1982. Comportamiento de 10 genotipos de frijol al ataque de *Meloidogyne* sp., p. 111 *In* Resumenes X Congr. Nac. Fitopatol., Soc. Mex. Fitopatol. Culiacán, Sin. México.
122. Mundo, O. M. 1978. Nemátodos asociados con el cultivo de plátano en Tabasco, Méx., p. 40 *In* Resumenes VIII Congr. Nac. Fitopatol., Soc. Mex. Fitopatol. Oaxtepec, Mor. México.
123. Mundo, O. M. 1978b. Estudio sobre el comportamiento como predator de nemátodos

de *Dactylaria leptospora*, pp. 77-79 *In* Resumenes VIII Congr. Nac. Fitopatol., Soc. Mex. Fitopatol. Oaxtepec, Mor. México.

124. Negron, J. A. 1980. Interaction of *Meloidogyne incognita* and *Fusarium oxysporum* f. sp. *coffeae* in *Coffea arabica* L. Ms. thesis. Univ. P. R.
125. Negron, J. A. and N. Acosta. 1984. Host-parasite relationships of *Meloidogyne incognita* and *Coffea arabica*. J. Agric. Univ. P. R.:In press.
126. Negron, J. A. and N. Acosta. 1981. Histopapotathology (sic) of *Meloidogyne incognita* and *Fusarium oxysporum* f. sp. *coffeae* infecting coffee in Puerto Rico. Nematropica 11:83-84.
127. Nicar. Minist. Agric. Ganad. 1975. Informe anual del laboratorio de fitopatología. Nicar. Minist. Agric. Ganad. Managua, Nicaragua.
128. Nicar. Minist. Desarrollo Agropecuario y Reforma Agraria. 1978. Informe anual del laboratorio de fitopatología. Nicar. Minist. Desarrollo Agropecuario y Reforma Agraria. Managua, Nicaragua.
129. Nicar. Minist. Desarrollo Agropecuario y Reforma Agraria. 1979. Misión técnica alemana. Resultados obtenidos en Fitopatología desde el año 1953 hasta noviembre 1979, en la República de Nicaragua. Nicar. Minist. Desarrollo Agropecuario y Reforma Agraria. Managua, Nicaragua.
130. Nicar. Minist. Desarrollo Agropecuario y Reforma Agraria. 1980. Informe anual del laboratorio de fitopatología. Nicar. Minist. Desarrollo Agropecuario y Reforma Agraria. Managua, Nicaragua.
131. Nieto, A. L. G. y C. Sosa-Moss. 1979. Determinación del ciclo de vida del nemátodo nodulador de raíces (*Meloidogyne* spp.), en maiz (*Zea mays* L.) e histopatología de las células hospedantes. Tesis profesional, Facultad de Biología, U.N.A.M. México.
132. Ojeda, M. R. y D. Cerecer S. 1982. Ensayos con Nemacur en vid, p. 78 *In* Resumenes X Congr. Nac. Fitopatol., Soc. Mex. Fitopatol. Culiacán, Sin. México.
133. Padilla, C. A. y R. López. 1978. Efecto de la aplicación de nematicidas granulados en el combate de *Meloidogyne* spp. en arveja. Nematrópica 8:20.
134. Padilla, C. A. y R. López. 1979. Evaluación de nematicidas granulados para el combate de *Meloidogyne* spp. en arveja (*Pisum sativum* L.). Agron. Costarric. 3:89-95.
135. Padilla, C. A., R. López y E. Vargas. 1980. Interacción entre *Meloidogyne* spp. y *Fusarium oxysporum* f. sp. *pisi* en arveja. Agron. Costarric. 4:55-60.
136. Palacios, A. A. y C. Sosa-Moss. 1971. Resistencia genética de algunas variedades de tomate (*Lycopersicum* {*sic*} *esculentum*) al ataque de *Meloidogyne* spp. Agrociencia 9:119-125.
137. Palacios, A. A. y C. Sosa-Moss. 1971. Resistencia genética de algunas variedades de tomate (*Lycopersicon esculentum*) al ataque de *Meloidogyne* spp. Nematrópica 1:4.
138. Palacios, A. S. y D. Téliz. 1967. La jicamilla del jitomate en el Estado de Morelos, p. 8 *In* Resumenes V Congr. Nac. Soc. Mex. Fitopatol. México, D. F. México.
139. Palacios, A. S. y D. Téliz. 1970. Distribución, identificación y control quimico del nemátodo nodulador de raices (*Meloidogyne* spp.) causante de la jicamilla del jitomate en el Estado de Morelos. Tesis profesional, Dep. Parasitol., Escuela Nac. Agric. Chapingo, México.
140. Parisi, C. R. A., C. J. Torres G. y C. Sosa-Moss. 1971. Incorporación de un nematicida sistémico a la planta de frijol, por immersión de semillas, p. 103 In Resumenes IV Reunión Anual. Organ. Nemátol. Trop. Am. San Salvador, El Salvador.
141. Parisi, C. R. A., C. J. Torres G. y C. Sosa-Moss. 1971. Incorporación de un nematicida sistémico a la planta de frijol (*Phaseolus vulgaris* L.) por immersión de semillas. Nematrópica 2:22-23.
142. Pedroza, M. A. A. 1980. Detección de especies de *Meloidogyne* spp. en las áreas meloneras del Estado de Michoacan, p. 9 *In* Resumenes IX Congr. Nac. Fitopatol. Uruapan, Mich. México.
143. Pérez, A. A. 1970. Evaluación de 6 insecticidas-nematicidas sistémicos para el control del nemátodo *Meloidogyne* spp. y pulgones en la papa en la regíon de Navidad, Nuevo León. Tesis profesional, Escuela Superior de Agric. "Antonio Narro," Saltillo, Coah. México.
144. Pérez, M. M. y R. Rodríguez M. 1973. Exploración nematológica en el cultivo de alfalfa (*Medicago sativa* L.) en el Estado de México. Nematrópica 3:51.
145. Perlaza, F. 1982. Evaluación de cuatro nematicidas para el combate de *Meloidogyne incognita* en tomate, pp. 16-17 *In* Proc. Third IMP Res. Plann. Conf. on Root-Knot Nematodes, *Meloidogyne* spp., Region I, 11-15 enero 1982. Inst. Invest.

Agropecuarias de Panamá, Panamá. 180 pp.
146. Perlaza, F. 1982. Morfometría y respuesta de hospedantes diferenciales de una población de *Meloidogyne* sp., pp. 18-20 *In* Proc. Third Res. Plann. Conf. on Root-Knot Nematodes, *Meloidogyne* spp., Region I, 11-15 enero 1982. Inst. Invest. Agropecuarias de Panamá, Panamá. 180 pp.
147. Perlaza, F., R. López, y E. Vargas. 1978. Efecto de la aplicación combinada de nematicidas y fungicidas en el combate de *Meloidogyne incognita*, *M. hapla* y *Alternaria* sp. en lechuga. Fitopatología 13(2):90-96.
148. Perlaza, F., R. López y E. Vargas. 1979. Combate químico de *Meloidogyne* spp. y *Alternaria* sp. en zanahoria (*Daucus carota* L.). Turrialba 29:263-267.
149. Ramírez, A. 1971. Algunos aspectos sobre el control químico de los nemátodos en tomate. Tesis profesional, Facultad de Agronomía, Univ. Costa Rica.
150. Rivera, R. y R. López. 1978. Combate químico de *Meloidogyne incognita* en dos cultivares de apio. Nematrópica 8:21.
151. Rivera B., R. y R. López Ch. 1982. Evaluación de nematicidas para el combate de *Meloidogyne incognita* en dos cultivares de *Apium graveolens*. Turrialba 32:67-73.
152. Robalino, G., J. Roman y M. Cordero. 1983. Efecto del nematicida-insecticida Oxamyl aplicado al suelo y a las axilas de las hojas del bananero. Nematrópica 13:135-143.
153. Rocha, P. M. A. y M. Mundo O. 1978. Nemátodos fitoparásitos asociados al cultivo del frijol (*Phaseolus vulgaris* L.) en la Chontalpa, Tabasco, pp. 93-94 *In* Resumenes VIII Congr. Nac. Fitopatol. Soc. Mex. Fitopatol. Oaxtepec, Mor. México.
154. Rodríguez, G. P. y R. García E. 1981. Supresividad a *Meloidogyne* sp. en suelos de agroecosistemas tradicionales, p. 233 *In* Avances en la Enseñanza y la Investigación en el Colegion de Postgraduados. Chapingo, México.
155. Rodríguez, G. P. y R. García E. 1983. Supresividad a *Meloidogyne* sp. en suelos de agroecosistemas tradicionales. Nematrópica 13:120.
156. Rodríguez, B. R., J. A. Quiñonez F. y C. Sosa-Moss. 1984. Identificación de especies de *Meloidogyne* en tabaco y otros cultivos de importancia agrícola, en la zona de influencia del Río Santiago, en el Edo. de Nayarit. Tesis profesional, Escuela Superior de Agricultura, Univ. Autónoma de Nayarit. México.
157. Román, J., X. Rivas y J. Rodríguez. 1974. Control de los nemátodos del plátano por rotación con yerba Pangola. Nematrópica 4:4-5.
158. Román, J., X. Rivas y J. Rodríguez. 1974. Control químico de los nemátodos del plátano. Nematrópica 4:5.
159. Sales, L. A. y E. Echandi. 1961. Parasitic nematodes in coffee plantations of Costa Rica. Coffee 3(8):6-9.
160. Salazar, F. L. A. 1980. Variaciones morfométricas y respuesta de nueve hospedantes diferenciales a tres poblaciones de *Meloidogyne javanica* de Costa Rica. Turrialba 39:344-351.
161. Salazar, F. L. A. y R. López. 1980. Caracterización morfométrica y ámbito de hospedantes diferenciales de diez poblaciones de *Meloidogyne* spp. de Costa Rica. Agron. Costarric. 4:21-31.
162. Sánchez, R. J. L. e I. Icaza C. 1982. Nemátodos fitoparásitos en Hierbabuena (*Mentha piperita*) en la region de Nativitas, Tlax., p. 39 *In* Resumenes X Congr. Nac. Fitopatol., Soc. Mex. Fitopatol. Culiacán, Sin. México.
163. Santacruz, U. H. y A. Pedroza M. 1980. Identificación de géneros de nemátodos y evaluación de sus poblaciones en el Estado de Michoacán. Resumenes IX Congr. Nac. Fitopatol. Soc. Mex. Fitopatol. Uruapan, Mich. México.
164. Santacruz, U. H. y A. Pedroza M. 1983. Estudio preliminar de la nematofauna en el Estado de Michoacan. Nematrópica 13:120-121.
165. Segeren van den Oever, R. 1981. Root-knot nematode problems in Suriname, pp. 230-231 *In* Proc. Second Reg. Plann. Conf. of the Int. Meloidogyne Project, Region I, Sept. 4-8, 1978, San Jose, Costa Rica. Facultad de Agronomía, Univ. Panamá, Panamá. 246 pp.
166. Segeren van den Oever, R. 1982. Research on *Meloidogyne* in Suriname, pp. 69-74 *In* Proc. Third Res. Plann. Conf. on Root-knot Nematodes, *Meloidogyne* spp., Region I, Jan. 11-15, 1982. Inst. Invest. Agropecuiarias de Panamá, Panamá. 180 pp.
167. Segreda Van Deer Laat, A. 1960. Ensayos para combatir la hernia del tomate causada por nemátodos del género *Meloidogyne*, con fumigantes del suelo.

Tesis profesional, Facultad de Agronomía, Univ. Costa Rica.
168. Serrano, E. A. 1966. Poblaciones del nemátodo *Meloidogyne incognita* var. *acrita* Chitwood y su influencia en el contenido de nitrógeno y fósforo en el frijol. Tesis profesional, Escuela de Agric. y Ganad., Univ. Sonora, México.
169. Singh, N. D. 1973. A note on plant parasitic nematodes associated with sugarcane in Trinidad. Nematropica 3:54-55.
170. Singh, N. D. 1973. Preliminary report of plant parasitic nematodes associated with important crops in Trinidad. Nematropica 3:56-61.
171. Singh, N. D. 1974. Preliminary investigations on the parasitic nematodes associated with tobacco in Trinidad. Nematropica 4:11-12.
172. Singh, N. D. 1975. Effects of chemicals and plastic mulch on nematode populations and yield of tomato. Nematropica 5:6-10.
173. Singh, N. D. 1975. Effects of nematicides on the nematode populations and yield of tobacco. Nematropica 5:13-17.
174. Singh, N. D. 1975. Influence of oxamyl application on *Meloidogyne incognita* and *Rotylenchulus reniformis* penetration into roots of tomato, lettuce and pigeonpeas. Nematropica 5:29.
175. Singh, N. D. 1976. Interaction of *Meloidogyne incognita* and *Rotylenchulus reniformis* on soybean. Nematropica 6:76-81.
176. Singh, N. D. 1976. A summary of root-knot nematode investigations in the Caribbean, p. 87 *In* Proc. Reg. Plann. Conf. of the Int. Meloidogyne Project, Region I, Apr. 26-30, 1976. Facultad de Agronomía. Univ. Panamá, Panamá.
177. Singh, N. D. 1979. A nematological investigation of carrots in St. Vincent. Caribbean Agric. Res. Dev. Inst., U.M.I. Campus, St. Augustine, Trinidad.
178. Singh, N. D. 1981. Report on the work carried out for the International Meloidogyne Project in Trinidad, pp. 213-218 *In* Proc. Second Reg. Plann. Conf. of the Int. Meloidogyne Project, Region I, Sept. 4-8, 1978, San José, Costa Rica. Facultad de Agronomía, Univ. Panamá, Panamá. 246 pp.
179. Singh, N. D. 1982. Root-knot nematode (*Meloidogyne incognita*) investigation on carrot in St. Vincent, W.I., pp. 84-91 *In* Proc. Third IMP Res. Plann. Conf. on Root-knot Nematodes, *Meloidogyne* spp. Inst. Invest. Agropecuarias de Panamá, Panamá . 180 pp.
180. Smith, F. E. V. 1936. Jam. Dep. Sci. & Agric. Annu. Rep. 1936:57-58.
181. Smith, F. E. V. 1937. Jam. Dep. Sci. & Agric. Annu. Rep. 1937:51-52.
182. Sosa-Moss, C. 1973. Resistencia génetica de vegetales al ataque de nemátodos, p. 159 *In* Avances en la Enseñanza y la Investigación en el Colegio de Posgraduados. Chapingo, México.
183. Sosa-Moss, C. 1976. Situación actual del conocimiento sobre los nemátodos del género *Meloidogyne* en México, pp. 32-35 *In* Memorias de la Conferencia Regional de Planeamiento del Proyecto International Meloidogyne, Region I, 26-30 abril 1976. Facultad de Agronomía, Univ. Panamá, Panamá.
184. Sosa-Moss, C. 1982. Reporte de investigaciones sobre *Meloidogyne* spp. en México, pp. 61-65 *In* Proc. Third IMP Res. Plann. Conf. on Root-knot Nematodes *Meloidogyne* spp., Region I, 11-15 enero 1982. Inst. Invest. Agropecuarias de Panamá, Panamá. 180 pp.
185. Sosa-Moss, C. y F. S. Armenta C. 1978. Resistencia génetica de papa al ataque de nemátodos, p.213 *In* Avances en la Enseñanza y la Investigación en el Colegio de Postgraduados. Chapingo, México.
186. Sosa-Moss, C. y M. J. Bautista. 1972. Respuesta de 3 variedades de *Lycopersicon esculentum* a 7 niveles de *Meloidogyne incognita*. Resumenes V Reunión Anual OTAN. México.
187. Sosa-Moss, C. y J. S. Camacho G. 1973. Protección del frijol ejotero a *Meloidogyne incognita* por tratamiento químico de la semilla. Nematrópica 3:12-13.
188. Sosa-Moss, C. y D. Castillo F. 1981. Avances de investigación sobre el género *Meloidogyne* en México, pp. 6-15 *In* Memorias de la Segunda Conferencia Regional de Planeamiento del Proyecto Internacional Meloidogyne, Region I, 4-8 sept. 1978, San José, Costa Rica. Facultad de Agronomía, Univ. Panamá, Panamá. 246 pp.
189. Sosa-Moss, C., A. Jarquin N. y A. López Portillo. 1982. Control biológico de nemátodos fitoparásitos, p. 203 *In* Avances en la Investigación en el Colegio de Postgraduados. Chapingo, México.
190. Sosa-Moss, C. y C. Juárez. 1976. Evaluación del daño de *Meloidogyne incognita* al cultivo de papa, pp. 95-96 *In* Avances en la

Enseñanza y la Investigación en el Colegio de Postgraduados. Chapingo, México.
191. Sosa-Moss, C., G. López A., y H. Hernández A. 1973. Relación entre algunas especies de nemátodos y hongos fitopatógenos, p. 163 *In* Avances en la Enseñanza y la Investigación en el Colegio de Postgraduados. Chapingo, México.
192. Sosa-Moss, C. y M. Osuna A. 1982. Immersión de semillas en soluciones de nematicidas sistémicos, para reducir el daño por nemátodos fitoparásitos, p. 201 *In* Avances en la Investigación en el Colegio de Postgraduados. Chapingo, México.
193. Sosa-Moss, C. and R. Rodriguez. 1970. Effect of some insecticides used as nematocides against *Meloidogyne* spp. in tomato, p. 183 *In* Abstr. 10th Int. Nematol. Symp., Eur. Soc. Nemtol., Pescara, Italy.
194. Sosa-Moss, C. y E. Rodriguez. 1973. Evaluación del daño de algunas especies de nemátodos a cultivos agricolas, pp. 191-192 *In* Avances en la Enseñanza y la Investigación en el Colegio de Postgraduados. Chapingo, México.
195. Sosa-Moss, C. y J. N. Sasser. 1978. Proyecto Internacional de *Meloidogyne*, pp. 219-220 *In* Avances en la Enseñanza y la Investigación en el Colegio de Postgraduados. Chapingo, México.
196. Sosa-Moss, C. y J. M. Torres. 1973. Respuesta del frijol ejotero (*Phaseolus vulgaris* L. var. Black Valentine) a siete niveles de población de *Meloidogyne incognita* (Kofoid y White 1919) Chitwood 1949 (Nematoda: Heteroderidae). Nematrópica 3:17-18.
197. Sosa-Moss, C. y H. Weihs A. 1973. Uso de la melaza de caña de azucar para combatir *Meloidogyne incognita* (Kofoid y White) Chitwood, (Nematoda: Heteroderidae). Nematrópica 3:18-19.
198. Talavera, C. y N. Marban. 1980. Efecto de diferentes fuentes y dosis de nitrogeno sobre el daño causado por *Meloidogyne incognita* a plantas de jitomate, bajo condiciones de invernadero, pp. 233-234 *In* Avances en la Enseñanza y la Investigación en el Colegio de Postgraduados. Chapingo, México.
199. Talavera, C., N. Marban y C. Sosa-Moss. 1980. Efecto de diferentes fuentes de nitrógeno a diferentes dosis, sobre el daño causado por *Meloidogyne incognita* a plantas de jitomate, bajo condiciones de invernadero, p. 16 *In* Resumenes IX Congr. Nac. Fitopatol., Soc. Mex. Fitopatol. Uruapan, Mich. México.
200. Talavera, C., N. Marban y C. Sosa-Moss. 1980. Efecto de diferentes fuentes de nitrogeno a diferentes dosis, sobre el daño causado por *Meloidogyne incognita* a plantas de jitomate. Tesis de maestría, Colegio de Postgraduados, Chapingo, México.
201. Talavera, C., N. Marban, y C. Sosa-Moss. 1982. Efecto directo del sulfato de amonio, nitrato de amonio y urea, sobre los juveniles de *Meloidogyne incognita* "in vitro" y sobre el daño a jitomate en invernadero, p. 97 *In* Resumenes X Congr. Nac. Fitopatol., Soc. Mex. Fitopatol. Culiacán, Sin. México.
202. Talavera, C., N. Marban y C. Sosa-Moss. 1983. Efectos de diferentes fuentes y dosis de nitrógeno sobre el daño causado por *Meloidogyne incognita* a plantas de jitomate. Nematrópica 13:121-122.
203. Tarjan, A. C. 1967. Some plant nematode genera associated with citrus and other crops in Costa Rica and Panama. Turrialba 17:280-283.
204. Tarte, R. 1965. Nemátodos parásitos de plantas en Panamá. Minist. Agric. Comercio e Ind., República de Panamá, Invest. Agropecuarias No. 4. 16 pp.
205. Tarte, R. 1966. El nemátodo *Meloidogyne incognita acrita* Chitwood 1949, parásito de las raíces de okra (*Hibiscus esculentus* L.) en Panamá. Minist. de Agric., Comercio e Ind., República de Panamá, Invest. Agropecuarias No. 10. 4 pp.
206. Tarte, R. 1970. Reconocimiento de nemátodos asociados con diversos cultivos en Panamá. Turrialba 20:401-406.
207. Tarte, R. 1971. Estudio sobre la distribución y poblaciones de nemátodos en fincas de hortalizas intensamente cultivads. Turrialba 21:34-37.
208. Tarte, R. 1976. Estudios sobre *Meloidogyne* spp. en Panamá, pp. 41-44 *In* Memorias de la Conferencia Regional de Planeamiento del Proyecto Internacional Meloidogyne, Region I, 26-30 abril 1976. Facultad de Agronomía, Univ. Panamá, Panamá.
209. Tarte, R. 1976. El nemátodo nodulador *Meloidogyne* sp.; determinación de especies existentes en Panamá, daños ocasionados al cultivo del arroz y búsqueda de resistencia en tomate industrial (Proyecto Internacional de Meloidogyne), pp. 420-446 *In* Invest. Agropecuarias. Facultad de Agronomía, Univ. Panamá, Panamá.
210. Tarte, R. 1981. Informe sobre el progreso de la investigación para el Proyecto Internacional Meloidogyne en Panamá 1976-

1978, pp. 27-51 *In* Memorias de la Segunda Conferencia Regional de Planeamiento del Proyecto Internacional Meloidogyne, Region I, Sept. 4-8, 1976, San José, Costa Rica. Facultad de Agronomía, Univ. Panamá, Panamá. 246 pp.

211. Taylor, A. L., J. N. Sasser and L. A. Nelson. 1982. Relationship of climate and soil characteristics to geographical distribution of *Meloidogyne* species in agricultural soils. Coop. Publ. Dep. Plant Pathol., North Carolina State Univ., and U.S. Agency Int. Dev. Raleigh, N.C. 65 pp.

212. Tellez, O. 1912. Las anguilulas de las raíces del cafeto. La Hacienda 7:119-121, 147-148.

213. Toledo, L. R., C. Sosa-Moss, y M. G. Figueroa. 1983. Identificación y dinámica de población de las especies de *Meloidogyne* que atacan la ocra en Iguala y Santa Teresa, Gro. Tesis profesional, Colegio Superior Agropecuario del Edo. de Guerrero, México.

214. Toler, R. W., R. Cuellar, y J. B. Ferrer. 1959. Preliminary survey of plant diseases in the Republic of Panama, 1955-1958. Plant Dis. Rep. 43:1201-1203.

215. Toro, J. 1978. Relaciones hospedero-parásito y la variabilidad de catorce poblaciones del nemátodo nodulador *Meloidogyne* spp. en Puerto Rico. Tesis de maestría, Univ. P. R., Puerto Rico.

216. Toro, J. 1984. Host-parasite relationships and the variability of fourteen populations of the root-knot nematodes, *Meloidogyne* spp. in Puerto Rico. J. Agric., Univ. P. R.:In press.

217. Torres, P. F. J. y C. Sosa-Moss. 1978. Estudio del comportamiento de dos variedades de tomate (*Lycopersicon esculentum*) frente al ataque de tres especies de *Meloidogyne*, pp. 220-221 *In* Avances en la Enseñanza y la Investigación en el Colegio de Postgraduados. Chapingo, México.

218. Torres, P. F. J. y C. Sosa-Moss. 1978. Comportamiento de dos variedades de tomate (*Lycopersicon esculentum*) al ataque de *Meloidogyne incognita* (Nematoda: Meloidogynidae), p. 201 *In* Avances en la Enseñanza y la Investigación en el Colegio de Postgraduados. Chapingo, México.

219. Torres, P. F. J. y C. Sosa-Moss. 1979. Efecto de tres especies de *Meloidogyne* (Nematoda: Meloidogynidae), sobre dos variedades de tomate (*Lycopersicon esculentum* M.). Tesis profesional, Facultad de Cience., Univ. Nac. Autónoma de México, México.

220. Valiela, F. M. V. 1964. Desenvolvimiento de la fitopatología en los diferentes países latinoamericanos, pp. 5, 22, 34, 44, 47, 55 *In* Memorias de la VI Reunión Latinoamericana de Fitotecnia. Lima, Perú.

221. Valle, G. P. 1977. *Meloidogyne* spp., nemátodo que parasita diversos cultivos en la Comarca Lagunera. Centro de Invest. Agric. del Noroeste, Inst. Nac. Invest. Agricolas, Sem. Tec. 4(13):10.

222. Valle, L. S. and A. Ayala. 1976. Control of plantain nematodes with contact nematicides. Nematropica 6:55-59.

223. Varela, F. 1982. Interacciones de *Meloidogyne incognita* y *Fusarium solani* en pimiento (*Capsicum annum* L.). Tesis de maestría, Univ. P. R., Puerto Rico.

224. Varela, F., A. Ayala y J. Toro. 1983. Relación huesped-parásito de *Meloidogyne incognita* en los cultivares de pimiento, Blanco del País y Cubannelle, en Puerto Rico. Nematrópica 13:122.

225. Vázquez, I. y C. Sosa-Moss. 1982. Identificación y dinámica de población de nemátodos fitoparásitos en especies del género *Pinus* de la Región Central de México, p. 200 *In* Avances en la Investigación en el Colegio de Postgraduados. Chapingo, México.

226. Vejar, L. J. M. 1966. Relación *Meloidogyne* sp. *Rhizoctonia solani* y su influencia en el desarrollo de frijol, bajo condiciones de invernadero. Tesis profesional, Escuela de Agric. y Ganad., Univ. Sonora, México.

227. Vicente, N. 1982. Efecto de algunos factores ecológicos en la variabilidad y relación hospedero-parástito de *Meloidogyne incognita*. Tesis de maestría, Univ. P.R., Puerto Rico.

228. Villanueva, N. E. y A. Palacios A. 1971. Comportamiento de los cruzamientos de 6 variedades comerciales de tomate por una fuente de resistencia a nemátodos (*Meloidogyne* spp.). Tesis, Escuela Nac. Agric., Chapingo, México.

229. Zamudio, G. V. 1983. Control de nemátodos en el cultivo del jitomate (*Lycopersicon esculentum* Mill) en el Valle de Valsequillo, Puebla. México. Nematrópica 13:123.

230. Zavaleta, M. E. 1978. Busqueda de plantas con propiedades nematicidas, p. 198 *In* Avances en la Enseñanza y la Investigación en el Colegio de Postgraduados. Chapingo, México.

231. Zavaleta, M. E. y R. Montes. 1978. Exploración nematológica en los campos cultivados con caña de azúcar (*Saccharum of-*

ficinarum L.) en el municipio de Huimanguillo, Tab. y pruebas de patogenicidad con diferentes niveles de población de nemátodos fitoparásitos en cana de azucar, p. 23 *In* Resumenes VIII Congr. Nac. Fitopatol., Soc. Mex. Fitopatol. Oaxtepec. Mor. México.

232. Zavaleta, M. E. y C. Sosa-Moss. 1978. Control integrado para la "Jicamilla" (*Meloidogyne* spp.) del jitomate, pepino y frijol ejotero, pp. 198-199 *In* Avances en la Enseñanza y la Investigación en el Colegio de Postgraduados. Chapingo, Mexico.

SUMMARY REPORT ON THE CURRENT STATUS, PROGRESS AND NEEDS FOR *MELOIDOGYNE* RESEARCH IN SOUTH AMERICA

Enrique Cabanillas
Universidad Nacional Pedro Ruiz Gallo
Lambayeque, Peru

The South America region includes Argentina, Bolivia, Chile, Colombia, Ecuador, Paraguay, Peru, Uruguay and Venezuela. The research on root-knot nematodes in several of these countries has advanced, but further research on the development of effective strategies is necessary in order to control *Meloidogyne* species.

Meloidogyne species constitute a major and complex group of plant-pathogenic nematodes affecting crop production throughout this region. Regardless of the few people working on this important problem and their difficulties in doing research in this region, cooperators of the International *Meloidogyne* Project (IMP) in each of these countries are making significant and valuable contributions. Because some of this information is summarized in the IMP Proceedings, only a general report on the relevant problems and achievements is presented here.

REPORTS BY COUNTRY

Argentina

In recent years, nematology has been receiving increased attention in this country. Research emphasis has been on identification of *Meloidogyne* species and races; varietal resistance in tobacco, tomato and pepper; and the use of chemical control on tobacco, potato and tomato. At present, work is in progress on the evaluation of bean for resistance and estimation of population thresholds on vegetable crops. Table 1 shows the different species of *Meloidogyne* and the affected economic crops in this country.

As a result of the *Meloidogyne* research, other problems were found such as mixed populations of *Meloidogyne* species and *Nacobbus* which can lead to a wrong diagnosis due to similar root symptoms. Furthermore, varieties of tomato resistant to *M. incognita* became susceptible to *Nacobbus aberrans*, and tobacco and tomato resistance to *M. incognita* was broken by mixed populations of *M. incognita*, *M. javanica*, *M. arenaria*, and *Nacobbus* spp. As extension work, field trials on crop rotation are being included to demonstrate the benefit derived from control practices.

Bolivia

Research on root-knot nematodes has recently become established in Bolivia, and some progress has been made in identifying *Meloidogyne* species and in ascertaining their distribution on different crops in different ecological zones of this country. The *Meloidogyne* species that have been identified include *M. hapla*, *M. incognita* and *M. incognita acrita*. These species are mainly distributed in the eastern and southeastern parts of Bolivia, particularly in the Departments of Santa Cruz and Tarija. These locations are tropical and subtropical zones with average temperatures ranging from 20 to 33° C and altitudes between 450 and 1000 m above sea level (a.s.l.). *Meloidogyne* spp. were found throughout these zones, but mainly in sandy soils and on crops such as potato (*Solanum tuberosum*), cotton, sugar cane, lettuce, banana, tobacco, tomato, grape (*Vitis vinifera*), etc. Also, *Meloidogyne* spp. were found at elevations of up to 3400 m a.s.l. in potato (*Solanum andigenum*) in the Department of Potosí. Also, in the Department of Betanzos, mixed populations of *M. incognita* and *Nacobbus aberrans* were found on potato crops. *M. acronea* has been found associated with a leguminosae plant at an altitude of 2000 m a.s.l.

Table 1. Different species of *Meloidogyne* on some important crops and weeds (*Echinochloa* sp.) in Argentina.

Meloidogyne	Crops[a]								Weed
species	1	2	3	4	5	6	7	8	9
M. incognita	x	x	x	x	x	x	x	—	—
M. javanica	x	x	x	—	x	x	x	x	—
M. arenaria	x	—	x	—	x	x	—	—	—
M. hapla	—	—	—	—	—	x	—	—	—
M. decalineata	—	—	—	—	x	—	—	—	—
M. (or *Hypsoperine*) *ottersoni*	—	—	—	—	—	—	—	—	x

[a] 1 = tomato, 2 = soybean, 3 = tobacco, 4 = banana, 5 = bean, 6 = potato, 7 = squash, 8 = peach, 9 = *Echinochloa* sp.

Chile

Here, special attention has been given to the identification of *Meloidogyne* species, host range, use of systemic nematicides and varietal resistance in fruit crops such as peaches and grapes. Currently different rates of Temik are being tested on grapes, and grape rootstocks are being evaluated for resistance or tolerance to root-knot nematodes. The species most prevalent in Chile are *M. incognita, M. hapla, M. javanica,* and *M. naasi.* The most pathogenic is *M. incognita* which is commonly found associated with *Agrobacterium, Pythium,* and *Rhizoctonia* on fruit trees (almond, peach, grape) and annual crops such as tomato, bean, and carnation. Use of wheat and corn in crop rotations has helped reduce root-knot populations in the soil, but populations of *Pratylenchus* sp. have increased in large proportions. Use of Furadan 10%G and Temik 10% G has led to effective control of root-knot nematodes on fruit trees, nurseries, orchards, and vineyards. Recently, research on biological control of *Meloidogyne* spp. and *Globodera* spp. with the fungus *Paecilomyces lilacinus* has been tested in Chile.

Colombia

In this country, the following research has been emphasized: *Meloidogyne* identification, chemical control and crop resistance, especially tomato (*Lycopersicon esculentum*), lulo (*Solanum quitoense*), potato (*Solanum tuberosum* —sub-spp: *tuberosum*, and *andigenum*), bean (*Phaseolus vulgaris*), and carnation (*Dianthus caryophillus* L.). At present, work is being done on the incorporation of the Mi-gene to native varieties of tomato such as 'Licato.' The common *Meloidogyne* species found in Colombia are *incognita, javanica,* and *arenaria,* the most prevalent being *M. incognita.* Races 1, 2 and 4 of *M. incognita* have been found in the Department of Antioquia in Colombia. Future plans involve a continuation of research on the development of root-knot resistant cultivars for economic crops.

Ecuador

Research emphasis has been on identification of *Meloidogyne* species and their distribution in the different geographical and ecological zones of this country. A survey carried out by the IMP local cooperator in 1976-77 revealed *M. incognita* race 1 to be the predominant species in Ecuador and *M. javanica, M. arenaria* and *M. hapla* to be restricted to certain crops and zones. Mixed populations of *M. incognita* and *M. arenaria* are common. Economic crops such as alfalfa, banana, naranjilla (*Solanum quitoense*), tomato-tree (*Cyphomandra betacea*), citrus, potato, tomato, and carrot are affected by *Meloidogyne* spp.

At present, the research continues on identification of *Meloidogyne* spp. and their distribution in the different zones of Ecuador. Furthermore, control studies focus on crop rotation, plant resistance and chemical and agronomic measures.

In the future, the geographical distribution of *Meloidogyne* spp. in the coastal areas will be studied. Also, evaluation will continue for root-knot resistant varieties of the main crops such as banana, coffee, cacao, soybean and cotton. In addition, work will continue on biological, chemical and crop rotation methods to control root-knot nematodes in the coast, the mountains and western parts of Ecuador.

Paraguay

Research on the identification of *Meloidogyne* species has been emphasized in this country. *Meloidogyne javanica, M. exigua, M. incognita,* and *M. arenaria* have all been found. Root-knot nematodes damage crops such as tomato, tobacco, pepper, carrot, lettuce, banana, sugar cane, cotton, etc. *M. javanica* has been found infecting the wild potato (*Solanum chacoense*). Mixed populations of *M. javanica* and *M. incognita* have also been found.

Meloidogyne spp. have been controlled here primarily by application of soil fumigants to seedbeds of tomato, tobacco and pepper. Crop rotation of strawberries (*Fragaria* sp.) with susceptible vegetable crops has been very successful in soils infested with *M. javanica* and *M. incognita*.

At present, Paraguay's only plant pathology laboratory is located at the Faculty of Agricultural Engineering in Asunción. No separate plant nematology course is offered, so the subject is covered in the Phytopathology course.

Peru

The emphasis on *Meloidogyne* in Peru involves breeding and screening for resistance in potato at the International Potato Center, in tomato at the National Agrarian University in Lima, and in bean at the National Bean Program in Lambayeque. Root-knot nematodes have been found attacking several important crops such as potato, bean, cotton, banana, tobacco, coffee, sugar cane, passion fruit, papaw tree, and different vegetables (tomato, lettuce, carrot, pepper, etc.). Regardless of the lack of excellent information on crop losses due to nematodes in some areas of this country, the average yield is below optimum when compared with the world averages for most economic crops. The greater part of the loss is borne by small farmers, particularly those in warm areas with sandy soils. Present information indicates that *M. incognita* is the most common *Meloidogyne* species in Peru. Research now in progress includes the role of cropping systems in nematode control on potato, use of the fungus *Paecilomyces lilacinus* and nematicides, and cross-trials of crop rotation including *Tagetes patula*. Of the bean cultivars and lines screened for resistance to *M. incognita* in Lambayeque, five were found to be resistant and were recommended for inclusion in bean improvement programs. Recently, we got involved with nematological problems affecting the passion fruit or *maracuya (Passiflora edulis* var. *flavicarpa)* in the localities of Motupe and Olmos on the northern coast of Peru. As of now there are 1300 hectares planted with this promising crop, some in areas where banana plantations have been removed due to Panama Disease interacting with nematodes. In 1982, a complete plantation of passion fruit had to be burned because of phytosanitary problems including nematodes such as *Rotylenchulus*, *Paratylenchus* and *Meloidogyne*. Research on chemical control with nematicides and the fungus *Paecilomyces lilacinus* is being conducted in these areas. In Trujillo, research on screening bean cultivars and lines for resistance to *M. incognita* is being conducted. Also, recommendations are being made to the farmers of this area to rotate vegetable crops, for instance leek (resistant) with celery (susceptible), to control root-knot nematodes.

Venezuela

In the past, international nematology courses that have helped train people in plant nematology have been offered in Venezuela. Research emphasis has been on identification of *Meloidogyne* species and their control. The *Meloidogyne* species that have been identified are *hapla*, *incognita*, and *javanica*.

Actually, the root-knot nematodes cause severe damage in certain crops such as tomato, tobacco, potato, banana and cotton. *Meloidogyne* spp. have been associated with rice, bean, coffee, cacao, sugar cane, citrus, papaw, pineapple, cotton, pepper, lettuce, onion, carrot, and other crops. The damage to these crops by *Meloidogyne* has been recognized but not assessed economically. Research on methods to control *Meloidogyne* spp. has emphasized chemicals, crop rotation, trap crops and screening for resistance. The results indicate that the use of chemical products is promising in some crops, but the high cost is a limiting factor. Results from screening for resistance against *M. incognita* show that there are some promising resistant bean varieties like 'IBRN 13' ('Mexico 309') and 'IBRN 14' ('Turrialba 1 (S-19N)'). Also, there has been research on determination of the threshold levels of *M. incognita* on bean (*Phaseolus vulgaris* L.).

Uruguay

Research on root-knot nematodes has started recently in this country, and some progress has been made toward determining the effect of *Meloidogyne* species on certain crops and their control. Research emphasis has been on recognition of the crops affected by root-knot nematodes, screening of tobacco cultivars for resistance and testing of nematicides as a means of control. Results indicate that *Meloidogyne* spp. affect several crops such as tobacco, tomato, potato, pepper, beet, soybean, carrot, peach, bean, sweet potato, etc. Among these crops, tobacco and tomato show severe infection by *Meloidogyne* spp. Both *M. incognita* and *M. javanica* have been identified on tobacco. From 1971 to 1975, several chemical products were tested for efficacy to control root-knot nematodes in tobacco.

In 1971-72, a test on susceptibility of tobacco cultivars to *Meloidogyne* spp. was carried out. Results based on a scale rated from 0 to 10 showed that there were some promising resistant tobacco cultivars such as 'C 254,' 'Speight,' 'C 258,' and 'NC 95.' Results from a survey indicated that the gramineous weed *Setaria geniculata* was heavily attacked by *Meloidogyne* spp., and larvae of a microhymenopterous species were found associated with the egg masses of the root-knot nematodes.

CONCLUSIONS AND RECOMMENDATIONS

In the South American region, advances have been made in root-knot nematode research by means of the methods and supplies that the IMP program has provided to this region. Our countries can still benefit from the accumulated experience and findings of nematologists in developed nations. The establishment of a realistic and functional plant nematology program to provide diagnostic and advisory services with well-trained nematologists is necessary. In most of these countries, the available resources, both human and financial, are limited. Therefore, it is advisable to establish priorities in the research program with a more effective organization which will allow us to gain people's confidence to accomplish our responsibilities and achieve the immediate goals. Also, it is necessary to develop research along with teaching and extension programs in plant nematology. By now, the IMP cooperators are aware of the magnitude of the problem that plant-parasitic nematodes can cause to economic crops, but farmers, crop specialists and other people are often unaware that nematodes exist, so it is necessary to create this awareness in our own locality. Furthermore, it is a priority to initiate and/or support the establishment of introductory, college-level courses in nematology. Finally, with respect to research in this region, there should be close cooperation, yet each country must extablish its own priorities and develop the best combination of available management strategies to prevent the continuous buildup of serious pests and diseases, including root-knot nematodes. Management strategies should be integrated in order to improve actual production and yields of the economic crops to meet the increased demands of a growing population.

Acknowledgment

The author wishes to thank Dr. J. N. Sasser, Professor and Principal Investigator of the IMP and the following who have reviewed this report and have made valuable contributions: Dra. Julia Amaya de Guerra, Dr. Manuel Canto, Ing. Rafael Navarro, Mr. Mauricio Jimenez, Mr. Hector Gonzales, Mr. Pedro Gallo, Ms. Graciela M. de Sisler, Mr. Miguel Costilla, and Ing. Omar Guerrero.

SUMMARY REPORT ON THE CURRENT STATUS, PROGRESS AND NEEDS FOR *MELOIDOGYNE* RESEARCH IN BRAZIL (REGION III)

Silamar Ferraz

Department of Plant Pathology
Federal University of Vicosa
Vicosa—MG—Brazil

Brazil covers about 3,287,200 square miles, nearly half the area of all South America and about equal to that of Europe. The fifth largest country in the world, Brazil is surpassed in size only by Russia, China, Canada and the United States. The country extends from about 5° north latitude to 34° south latitude being enclosed mostly between the Equator and the Tropic of Capricorn.

There are three major geographical regions:

—The Amazon Basin or Amazonia, in northern and western Brazil, almost entirely covered by tropical rain forest. The chief products are rubber, Brazil and babassu nuts, cabinet wood, fibers and plants used in medicine and industry.

—Northeastern Brazil, with a large dry area, in the central part, and a fertile coastal plain. Farmers grow sugar cane, cocoa, cotton, bean, cassava, corn, tobacco, banana, rice, coffee and others.

—The central and southern uplands, where most of Brazil's agriculture and industry are centered. Coffee, soybean, orange, sugar cane, banana, rice, cotton, corn, bean and many others are grown in large areas.

Although about nine-tenths of Brazil is in the tropical zone, the climate is highly varied, being modified by winds, altitude and distance from the ocean. All of these variations in soil type, precipitation and climate contribute to the presence of many specific nematode problems throughout the country.

HISTORY OF MELOIDOGYNE RESEARCH IN BRAZIL

Since the discovery and description of *Meloidogyne exigua* in 1887 by Goeldi, eleven other species have been found in Brazil: *M. incognita, M. javanica, M. arenaria, M. hapla, M. coffeicola, M. bauruensis, M. inornata, M. acrita, M. thamesi, M. lordelloi,* and *M. elegans.* The most widespread and important ones are, by far, *M. incognita* and *M. javanica. M. exigua* is also very common in the coffee-growing areas.

The nematological research was initiated in the 1950's, and in 1974, the Brazilian Society of Nematology was established. The Society promotes annual meetings and has been a key factor in the development of nematology in Brazil. The membership is in the vicinity of 150, one third of the members being active researchers. More than 70% of all nematological papers published in Brazil are in some way related to *Meloidogyne,* which gives an indication of the great importance of this genus in our agriculture.

In the last few years, more emphasis has been given to the following research lines: general surveys, host-parasite relationships, biology and histopathology, interactions with other plant pathogens and control. Control comprises most of the nematological research, and different strategies have been studied. Besides the screening of nematicides, other products have been tried. Among them are "manipueira" (a by-product of the cassava flour industry), residues of sugar industry, cakes and others. A major activity has also been the screening for resistance. In soybean, for instance, hundreds of cultivars and lines have been tested each year against *M. javanica* and *M. incognita.* A source of resistance has also been sought for *M. incognita* on coffee. Even though *M. exigua* is the most disseminated species on coffee, *M. incognita* is the most devastating one. From 1976 to 1977, more than 3 million coffee transplants were destroyed by government inspectors due to the presence of this nematode. In the field, there is no way to maintain a profitable crop in infested land, the situation being worse on sandy soil. If a resistant variety is found, it could be used in a breeding program and/or for grafting purposes.

Dry beans (*Phaseolus vulgaris*) constitute a very important staple crop and a basic diet for Brazilian people. Yields are very low due to diseases and pests, among them *M. javanica* and *M. incognita*. Being a typical low income crop, the best way to solve the nematode problem would be the use of resistant varieties. Researchers have been screening the available material, but so far without success. An interesting fact is that some inoculated bean varieties show only a few small galls but a high number of eggs is always present.

Attention to nematode problems on sugar cane has increased since the government decision to use this crop as an alternative source of energy. Nematicide application on infested land has been shown to be profitable. The crop expansion program has incorporated large areas of poor soil, enhancing the loss caused by nematodes. Trials on varietal resistance and rotation with poor host crops to control *Meloidogyne* and other important nematodes are under way.

Meloidogyne species are also limiting factors for other crops such as cotton, tobacco, potato, tomato, fig, melon, carrot, okra and many others.

RESEARCH NEEDS FOR BRAZIL

Extensive Surveys

By this method, the entire country could get a better knowledge of the prevalence and geographic distribution of the different species. The information obtained in such surveys will permit the orientation of research toward the most important problems in the whole area or in specific regions.

Determination of Meloidogyne Races

This procedure is essential to any program of screening for resistance and also will yield information as to which races are the most important ones infecting any particular crop in the area.

Search for Resistant Lines and Cultivars

Several Brazilian researchers are already engaged in this activity. Foreign material has also been introduced and tested. Hopefully, the results obtained will be used directly by the farmers, or will serve as a basic step in breeding programs. It is necessary, however, to improve the screening methods available in order to permit handling of a large quantity of plant material and, at the same time, to prevent the loss of good germ plasm.

Search for Poor Hosts or Antagonistic Plants

So far, good field results have been obtained with *Stizolobium aterrimum (Mucuna aterrima)* and *Crotalaria spectabilis*, but many other promising species are available for testing. Crop rotation now seems to be one of the best alternative means of nematode control in Brazil.

Biological Control

Researchers are starting to evaluate the effectiveness of *Paecilomyces lilacinus* under our conditions. Another interesting possibility would be *Bacillus penetrans*. This organism is often found on different species of plant-parasitic nematodes from many areas in Brazil.

Development of Integrated Crop Protection Systems

This is an important objective of the International *Meloidogyne* Project and is highly needed in any county.

Other research areas that deserve attention in the near future are population dynamics, economics of chemical control, seed treatment, and utilization of mycorrhizae to diminish nematode effects.

THE INTERNATIONAL *MELOIDOGYNE* PROJECT IN REGION IV: CURRENT STATUS, PROGRESS AND FUTURE OUTLOOK

Olufunke A. Egunjobi

Department of Agricultural Biology
University of Ibadan
Ibadan, Nigeria

The International *Meloidogne* Project (IMP) objectives include,among other things, a commitment to alleviation of *Meloidogyne* damage on crops in the developing nations. Region IV, representing West Africa, forms a substantial part of this developing world, and perhaps is one of the most problematic areas in terms of actual *Meloidogyne* damage to crops and in deficiencies in available means to combat this damage.

Today, there are 14 countries in West Africa. These are Senegambia, Guinea Bissau, Guinea, Mali, Sierra Leone, Liberia, Ivory Coast, Upper Volta, Ghana, Togo, Benin Republic, Cameroun, Niger Republic and Nigeria. Of these, only six countries carry out research in nematology, and these six are actively involved in the International *Meloidogyne* Project. These are Benin Republic, Senegambia, Liberia, Ivory Coast, Ghana and Nigeria. Institutes actually involved are thirteen. These are Office de la Recherche Scientifique et Technique Outre-Mer (O.R.S.T.O.M.) in Dakar, Senegambia, and in Abidjan, Ivory Coast; United Nations Development Project (UNDP) in Monrovia, Liberia; University of Science and Technology in Kumasi, Ghana; Crop Research Institute (CRI) also in Kumasi, Ghana; Institute of Agricultural Research in Zaria, Nigeria; the University of Ife in Ile-Ife, Nigeria; the University of Nigeria, Nsukka, Nigeria; the National Root Crops Research Institute (NRCRI) in Umuahia, Nigeria; and the National Cereals Research Institute (NCRI), International Institute of Tropical Agriculture (IITA), National Institute of Horticultural Research (NIHORT), and University of Ibadan in Ibadan, Nigeria (Fig. 1).

Geographically, Region IV covers a range of ecological strata varying from the swamp-forest coastal region through rainforest guinea savana and from Sudan savana to semi-arid parts of Niger Republic and Senegambia. All kinds of cropping systems are practiced there, although mixed and shifting cultivation systems are traditional to the area.

Physical factors in the region are suitable for rapid population buildup of *Meloidogyne* spp.; temperature and moisture are adequate. In fact, it is believed that 10-12 generations of these nematodes may be occurring each year, based on a life cycle period of 21-28 days. Hence, Region IV appears to form a "haven of peace" for *Meloidogyne* spp.

Major staple food and fiber crops include rice, cowpea (protopea), yam, cassava, cocoa, corn, tobacco, kenaf, jute, plantain, groundnut (peanut), oil palm and coconut palm.

With reference to IMP, three meetings have been held in Region IV: in 1976, 1978 and 1981. Progress reports of the component areas, countries and/or institutes have been published in proceedings emanating from these conferences. Therefore, in this summary report an attempt will be made to highlight the current status of IMP in this region, summarize our achievements, indicate the current direction of our research efforts and identify our areas of research needs.

CURRENT STATUS OF IMP IN REGION IV

Following the birth of IMP in 1975, there emerged in the countries of Region IV a reawakening of interest in the root-knot nematodes as a group of serious pests of crops and in nematological research in general. The International *Meloidogyne* Project thus began to serve another unintended purpose, that of a

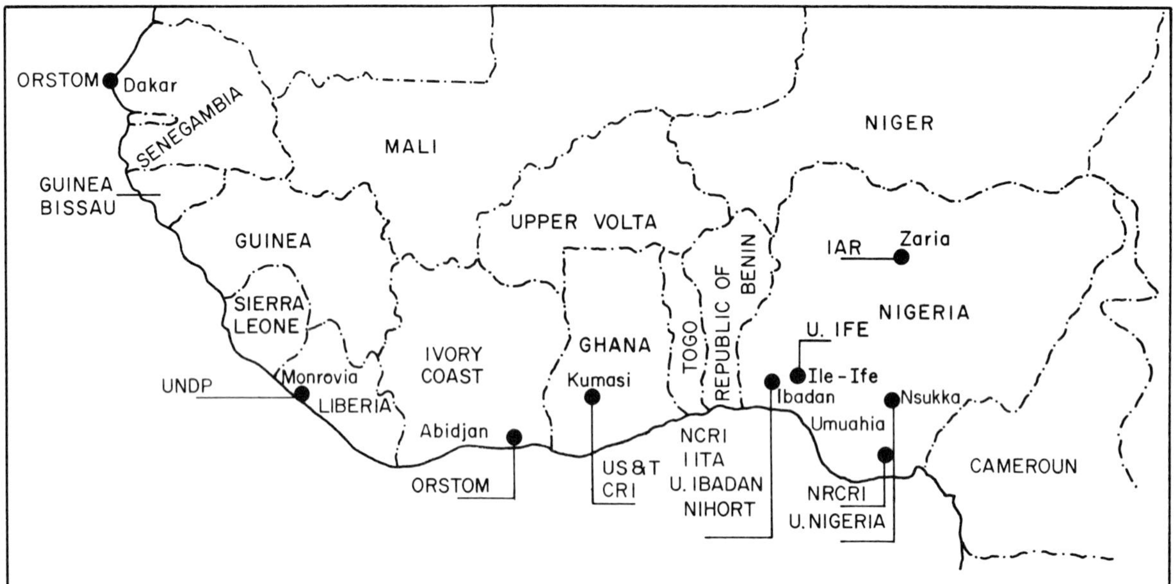

Fig I. INTERNATIONAL MELOIDOGYNE PROJECT – REGION IV
● Indicates Participating Institutions and Research Stations

propaganda agent for nematology. Should IMP survive another few years, I believe that the people and governments of this region would start to look towards her for solution to their *Meloidogyne* problems.

I consider it pertinent to mention at this point that this degree of success has been made possible by the presence of a few dedicated nematologists that are scattered across the area. In countries where nematologists are nonexistent, nothing is known about IMP. In fact, nematode problems do not exist in such countries as it is obvious that nematode problems can be identified only by eyes that are trained to detect them.

THE MAIN ACHIEVEMENTS OF REGION IV

Early research work that was pertinent to IMP can be summarized as follows:

1. root-knot nematode identification and distribution,
2. host range studies,
3. the biology of *Meloidogyne* spp.,
4. economic importance and crop loss assessments, and
5. Control measures, including
   (a) chemical control,
   (b) screening for resistance,
   (c) farming systems/crop rotation,
   (d) breeding for resistance.

Nematode Identification and Distribution

No discrepancies exist in the literature as to the presence of three species of the genus *Meloidogyne* Goeldi in this region. The known species are *M. incognita, M. javanica* and *M. arenaria* (7,9,11,12,19,22,24, 26-30,32,38,44). Of these, the most widespread is *M. incognita* (24,39).

Evidence also occurs in the literature to support the existence of at least one other species. Fortuner (19) referred to a "rice *Meloidogyne*." Diomande (personal communciation) specified the presence of an unidentified species found on peanut in western part of the Ivory Coast. Others simply referred to a *Meloidogyne* sp. (38) or spp.(30).

These four species appear to be unrestricted in their distribution in all the countries investigated. Although there is reason to believe that *M. incognita* and *M. arenaria* may be most important in the rainfed areas along the coast, *M. javanica* predominates in the savanna areas (8,12,15,). Analysis of 79 nematode samples from this region at North Carolina State University revealed *M. incognita* occurring in 65% of the samples.

Geographic distribution studies have been conducted extensively in Nigeria (8, 9, 12, 25, 40, 41) and to a lesser extent in Ghana (19, 24,), the Ivory Coast, and Senegambia. A more recent distribution study by Fortuner in 1978 (19) was restricted to upland rice fields. In this study, an

Table 1. *Meloidogyne* spp. identified in Region IV.

Countries	*M. incognita*	*M. javanica*	*M. arenaria*	*M.* sp.
	***Meloidogyne* spp. present**			
Nigeria	+	+	+	+
Ghana	+	+	+	?
Ivory Coast	+	+	+	+
Liberia	+	+	?	?
Senegambia	+	+	+	?

unidentified species was found to be most concentrated along the coastal area, occurring wherever rice was growing. Olowe in 1978 (42) made a similar survey of the cowpea-growing areas of Nigeria. These surveys agreed on the fact that *Meloidogyne* spp. are most commonly encountered in the southern parts of these countries, as well as in irrigated agricultural lands of the northern savanna areas.

Host Range Studies

Another area of *Meloidogyne* research that has been fairly well considered in Region IV is that of host range studies. In 1962, Wilson (45) recorded 108 host plants of *Meloidogyne* spp. in Nigeria. In 1967, Caveness (10) found 116 plants hosting *Meloidogyne* spp. in Nigeria. In Ghana, Addoh (1) listed 126 plants as good hosts of the root-knot nematode. Over the years, the number of plants known to host *Meloidogyne* spp. in Region IV, including weeds, food and fiber crops, has increased to over 150 (12, 16, 21, 22, 26, 28, 33, 35, 38, 44). Today, almost all food and fiber crops grown in this area have been known to host one or more species of this genus.

The Biology of Meloidogyne Spp.

Studies relating to the biology of these species that have been carried out in this region can be conveniently described under two headings: (a) life cycle studies, and (b) other biological studies.

LIFE CYCLE STUDIES The life cycle of the root-knot nematodes has been studied mainly on cowpea and tomato plants by various authors. It is now well established that the complete life cycle from second-stage juveniles to the egg-laying stage occurs within 21-23 days, depending on the host plant and the prevailing temperature conditions. In the recent past (1982), Nwauzor (31) found that *M. incognita* takes 17, 20, 22 and 31 days to develop from second-stage juveniles to adult males, adult females, egg-producing females and a new generation of second-stage juveniles, respectively.

Earlier, Caveness (9) had studied the life cycle of *M. incognita*, *M. javanica*, and *M. arenaria* on cowpea. *M. incognita* completed the life cycle in 21 days, and the latter two species took 23 days. In Ghana, the complete life cycle of *M. incognita* on tomato and cowpea took 28 days (24).

OTHER BIOLOGICAL STUDIES Literature is scanty on other aspects of the biology of these species in Region IV. Ogbuji and Okonkwo (36) in 1972 studied some aspects of the effect of pH on *M. incognita* developing on tomato. Results from this study serve as a pointer to the need for similar studies on a regional basis, as results from one geographical location may be inapplicable in another.

That *Meloidogyne* juveniles can migrate 50 cm vertically in 3 days was demonstrated at ORSTOM in Dakar, Senegambia (26). Netscher & Taylor (28) confirmed that root exudates enhance their migration in the soil. They also found that *M. arenaria* juveniles enter the roots of resistant peanut and cause hypersensitivity which in turn results in the death of the nematodes. Treatments with nitrogen and phosphorus minimize root galling and increase grain yield of cowpea (42).

Economic Importance and Crop Loss Assessments

It is yet impossible to be precise concerning annual losses caused by *Meloidogyne* spp. in this Region. However, some estimates of individual crop losses have been made by various authors. Bridge (9) put general crop loss at 40% in irrigated areas of northern Nigeria. In a few cases, total crop losses in the field have been reported: for example, with okra cv. 'Early Dwarf' (37) and tomato (30). Yield losses of 20-59% have been reported for cowpea (8,10, 38,39,43).

Yams when galled have a market value reduction of 40% (31). If stored in barns for 16 weeks, 25% of the edible part of the yam was lost to *M. incognita* and *M. javanica* (31).

In 1962, Wilson (45) estimated a possible loss of up to 75% in the yield of tomato in a field near Samaru, northern Nigeria, resulting from *Meloidogyne* spp. infestation.

In Benin Republic, general yield losses in tomato, associated with the nematodes were 55% (47). Ogunfowora (38) also reported losses ranging from 10-89% depending on the tomato cultivar and level of root-knot nematode infestation of the field soil. In upland rice, loss is put at 75% in the Ivory Coast (17). In parts of Ghana, potato production was said to be reduced by about 33% (24).

Control Measures

Research into the management of root-knot nematodes has been concentrated on four major areas: chemical control, screening for resistance, farming systems/crop rotation, and breeding for resistance.

CHEMICAL CONTROL Suffice it to say that most of the standard chemical nematicides released have been tried by one author or another in this region. These chemicals include metam-sodium (Vapam), oxamyl (Vydate), carbofuran (Furadan), phorate (Thimet), Miral, and Nemacur (2, 5, 39; Afolami, personal communication). Most of these, e.g. oxamyl, carbofuran and phorate, have been associated with increased yield of vegetables like lima bean, okra, soybean (2) and of cash crops like rice (5).

However, the resultant increases do not always pay their way. For this reason, substitutes to the chemical have been sought in lipid antioxidants like PB (piperonyl butoxide), and AA (ascorbic acid) (18). This research has opened up a new field, and results so far give some promise.

Thinking along the same line, Amosu (4) tried root exudates of *Eupatorium odoratum*, *Ricinus communis*, *Lycopersicon esculentum* and *Vigna unguiculata*. These extracts inhibited larval hatch in *M. incognita* when tested *in vitro*. The practicability of field application is yet untested. Soil amendments as a means of cultural control have also been investigated (9), but have not yet been found to be practicable.

However, it is apparent that, although some chemical control measures are effective, they are uneconomical for the majority of our subsistence farmers (8,9,26,38). Small farm holdings and limited financial resources pose special constraints in this regard. Therefore, more attention is now being paid to screening for resistance among crops and to a reevaluation of our traditional cropping systems. The latter approach resulted from general observations of lighter nematode infections on crops grown in traditional cropping systems than on those grown in modern monocultural systems (34,42). It is envisaged that a combination of these two methods may go a long way in solving the *Meloidogyne* problems of the smallscale farmers in Region IV.

SCREENING FOR RESISTANCE Screening for resistance among local crops is currently gaining ground in many areas. In the Ivory Coast, upland rice and cassava are being screened (17). Both improved and unimproved cultivars of *Oryza sativa* are found to be generally more susceptible than *O. glaberrima* cultivars. Improved *O. sativa* cultivars are more tolerant than the local ones. In Nigeria, Babatola (5) identified four rice cultivars that are resistant to *Meloidogyne incognita*.

So far, the two most widely grown clones of cassava in the Ivory Coast appear to be relatively free from *Meloidogyne* infestation (Diomande, personal communication). On the contrary, Caveness (13) found no resistance in any of the 56 cassava lines exposed to *M. incognita* race 2 and *M. javanica* in the greenhouse.

A number of resistant cowpea lines have also been identified. In Nigeria, Idowu (personal communication) reported 31 and 21 highly resistant and resistant lines, respectively. Zannou (46) found lines that were highly resistant to *M. incognita* race 2 and two lines that were similarly resistant to *M. javanica*. Several other lines possessed various degrees of resistance. Olowe (43) found only 1 cowpea line out of 73 tested to possess high resistance to *M. incognita* races 1, 3 and 4. The same line was found to be mildly resistant to *M. arenaria*. He found that cultivar 'Acc 7002' combined tolerance/resistance to root-knot nematode, virus and fungal blotch diseases. Ogunfowora (40) reported slight resistance in 2 of 31 lines of cowpea exposed to *M. incognita* in the greenhouse. Amosu (3) classified 36 cultivars and lines as highly resistant to *M. incognita*.

In sweet potato (*Ipomea batatas*), 55 of 414 lines showed good resistance to *M. incognita* race 2 and to *M. javanica* (15). All 21 lines of cocoyam (*Xanthosoma sagittifolium*) examined were highly resistant to *M. incognita* race 2 and to *M. javanica* (14, 15).

In yam (*Dioscorea* spp), Atu (personal communication) found three fairly resistant cultivars. Nwauzor (31) recorded one cultivar of *D. cayenesis* and one of *D. dumetorum* as immune to *M. incognita*. Caveness (14,15) confirmed a cultivar of *D. dumetorum* as highly resistant.

Ten of 287 lines of lima bean possessed moderate resistance to *M. incognita* race 2 (13). Soybean cvs. 'Hardee' and 'Hampton' are highly resistant and resistant, respectively, to *M. incognita* (39). Resistance has also been found among some leafy vegetables (39) and in hybrids of cocoa, *Theobroma cacao* (Afolami, personal communication).

One line of tomato (*Lycopersicon esculentum*) was found to possess some resistance to a mixture of *M. incognita* and *M. javanica* in Ghana (23). In Nigeria, cvs. 'Nematex,' 'Atkinson,' 'Rossol,' 'Enterprizer,' 'VFN 8,' 'Ife 1' (improved line 76) and 'Ronita' were shown to possess some degree of resistance or tolerance to root-knot nematode *Meloidogyne* spp. (3,39).

To date, peanut (*Arachis hypogaea*) remains the only commercial crop that has been unanimously declared highly resistant to most *Meloidogyne* spp. in this Region (6,9,20,22,39).

FARMING SYSTEMS / CROP ROTATION RESEARCH Some cropping sequences have been investigated for their effects on *Meloidogyne* population dynamics and yield of susceptible crops in Nigeria. Zero-tillage farming practices have also been investigated. Cover crops of *Pueraria* and *Glycine* were found to reduce populations of *Meloidogyne* spp. effectively (14,15). Inter-cropping of maize with leucaena also reduced populations, but maize/cassava mixtures encouraged population build-up. It has also been shown that growing non-host or resistant crops such as maize, rice, groundnut (peanut) or *Stylozanthes gracilis* prior to growing susceptible tomato cultivars (Ife 1) often greatly reduces nematode field infection with economic increases in yield of tomato (40). Continuous bushfallow gives some control of root-knot nematodes and good yield of tomato. Continuous stylo has been shown to give the best nematode control and also good yield of tomato. Continuous tomato cultivation, however, gives very poor control of nematodes and results in poor yield of tomato(40).

BREEDING FOR RESISTANCE Apart from the release at IITA of the cowpea 'Vita-3' by Singh (13), very little progress has been recorded in this area of research. Currently, a breeding program is underway in the Departent of Agricultural Biology of the University of Ibadan. A cultivar of cowpea 'AGRIB-VI' is in the pipeline. Ladeinde (personal communication) is optimistic that a cultivar resistant to *M. incognita* will emerge from these efforts.

Also breeders at the National Cereals Research Institute in Nigeria are embarking on a nematode resistance breeding program using cv. '64298,' which is known to possess high resistance to races 1, 3 and 4 of *M. incognita* and *M. arenaria* but which is not widely accepted, and the elite cv. 'Acc 73001,' which is acceptable but susceptible to all local species of *Meloidogyne* (43).

DIRECTION OF CURRENT RESEARCH EFFORTS

Without disregarding basic research, there appears to be a shift in emphasis from the basic to a more applied type of research. In recent years, most of the emphasis has been on locating resistance in economic crops. It is believed that resistant varieties will contribute significantly to solving our *Meloidogyne* problems.

For some time to come, West Africa will continue to lean heavily on the subsistence farmers for most of its agricultural needs. These farmers still believe in and indeed practice mixed cropping/mixed farming systems. Therefore, cropping systems research as it affects *Meloidogyne* infection capabilities and crop yield responses is beginning to feature more prominently than ever before in our research efforts. Cultural practices like tilling and no-tilling, cover cropping with different cover crops, and mulching are being investigated as possible means of reducing *Meloidogyne* infestation on economic crops. It is hoped that results from these investigations may form a basis for an integrated control strategy for *Meloidogyne* spp. in the region.

AREAS OF RESEARCH NEEDS

Substantial progress has been made in *Meloidogyne* research in this region, especially since IMP came into existence. Judging from the increase in activities evidenced by the increase in number and improvement in the quality of published data from the region over the past eight years, IMP appears to have provided a necessary catalyst.

However, much more research still needs to be done. Efforts must be intensified in the areas of farming systems research and breeding for resistance but especially in the areas of ecology and biology, for it is only by properly understanding such aspects of these pests that we can hope to find a lasting solution to the *Meloidogyne* problem.

It must also be reiterated that information is scarce from most countries in this region because of lack of trained personnel. There is an urgent need to train more nematologists in the region. In those countries where nematologists are present, financial limitations mitigate against their productivity.

At this point, the formation of an action committee or a re-activation of the committees organized at the 1981 IMP Region IV conference may be considered. Such a committee or committees will be mandated to coordinate research results with a view to making them more meaningful. It/they will identify areas of research priorities. For instance, such a committee/committees could study closely all available data on resistant germ plasm in each crop with a view to finding a basis for breeding locally acceptable materials for the region. These steps could lead to a cooperative and coordinated research effort in the region. In essence, IMP must live on so that its aims may be realized.

Literature Cited

1. Addoh, P. G. 1970. The root-knot nematode problem in Ghana: host and non-host plants of *Meloidogyne* species. Ghana J. Agric. Sci. 3: 3-12.
2. Adesiyan, S. O. 1981. Comparison of methods and rates of application of some short residual toxicants for the control of *Meloidogyne incognita* attacking some vegetable crops, pp. 157-160 *In* Proc. Third IMP Res. Plann. Conf. on Root-knot Nematodes, *Meloidogyne* spp., Regions IV & V, Nov. 16-20, 1981. Int. Inst. Trop. Agric., Ibadan, Nigeria. 275 pp.
3. Amosu, J. O. 1976. Report of *Meloidogyne* research at University of Ife, Ile-Ife, pp. 38-40 *In* Proc. First IMP Res. Plann. Conf. on Root-knot Nematodes, *Meloidogyne* spp., Region IV, June 7-11, 1976. Int. Inst. Trop. Agric., Ibadan, Nigeria. 83 pp.
4. Amosu, J. O. 1981. Control of root-knot nematodes by cultural practices, pp. 259-264 *In* Proc. Third IMP Res. Plann. Conf. on Root-knot Nematodes, *Meloidogyne* spp., Regions IV & V, Nov. 16-20, 1981. Int. Inst. Trop. Agric., Ibadan, Nigeria. 275 pp.
5. Babatola, J. O. 1981. Root-knot nematode problems on rice in Nigeria, pp. 135-139 *In* Proc. Third IMP Res. Plann. Conf. on Root-knot Nematodes, *Meloidogyne* spp., Regions IV & V, Nov. 16-20, 1981. Int. Inst. Trop Agric., Ibadan, Nigeria. 275 pp.
6. Bos, W. S. 1978. Root-knot nematodes in the Nigerian savanna zones with special reference to root-knot problems of irrigated crops in the Kano River Project at Kadawa, pp. 38-46 *In* Proc. Second IMP Res. Plann. Conf. on Root-knot Nematodes, *Meloidogyne* spp., Region IV, Feb. 20-24, 1978, Abidjan, Ivory Coast. Int. Inst. Trop. Agric., Ibadan, Nigeria. 93 pp.
7. Bridge, J. 1972. Nematode problems with yams *Dioscorea* spp. in Nigeria. Pest Artic. News Summ. (PANS) 18(1):89-91.
8. Bridge, J. 1972. Plant parasitic nematodes of irrigated crops in the northern states of Nigeria. Inst. Agric. Res. (Nigeria), Samaru Misc. Paper 42.
9. Caveness, F. E. 1967. End of tour progress report on the nematology project. Minist. of Agric. and Natur. Resources, Western Region, Nigeria. 135 pp.
10. Caveness, F. E. 1967. Shadehouse host ranges of some Nigerian nematodes. Plant Dis. Rep. 51:33-37.
11. Caveness, F. E. 1976. The root-knot nematodes, *Meloidogyne* spp. in Nigeria, pp. 24-37 In Proc. First IMP Res. Plann. Conf. on Root-knot Nematodes, *Meloidogyne* spp., Region IV, June 7-11, 1976. Int. Inst. Trop. Agric., Ibadan, Nigeria. 83 pp.
12. Caveness, F. E. 1978. Nematode attack on cowpeas in Nigeria, p. 302 *In* Proc. of the First Int. Inst. Trop. Agric. (IITA), Grain Legume Improvement Workshop, Oct. 29-Nov. 2, 1977. Ibadan, Nigeria.
13. Caveness, F. E. 1978. Root-knot nematodes on selected food crops in Nigeria, pp. 74-78 *In* Proc. Second IMP Res. Plann. Conf. on Root-knot Nematodes, *Meloidogyne* spp., Region IV, Feb. 20-24, 1978, Abidjan, Ivory Coast. Int. Inst. Trop. Agric., Ibadan, Nigeria. 93 pp.
14. Caveness, F. E. 1980. Root and tuber improvement program. Nematology Subprogram Report, Int. Inst. Trop. Agric., Ibadan, Nigeria.
15. Caveness, F. E. 1982. Root and tuber improvement program. Nematology Subprogram Report, Int. Inst. Trop. Agric., Ibadan, Nigeria.
16. Caveness, F. E. and G. F. Wilson. 1975. The effects of root-knot nematode on growth and development of *Celosia argentea* L., pp. 71-73 *In* Proc. ISHS 4th Afr. Symp. on Hortic. Crops, Kumasi, Ghana.
17. Diomande, M. 1981. Root-knot nematodes on upland rice (*Oryza sativa* and *O. glaberrima*) and cassava (*Manihot esculenta*) in Ivory Coast, pp. 37-45 *In* Proc. Third IMP Res. Plann. Conf. on Root-knot Nematodes, *Meloidogyne* spp., Regions IV & V, Nov. 16-20, 1981, Int. Inst. Trop. Agric., Ibadan, Nigeria. 275 pp.
18. Fawole, B. 1981. Lipid anti-oxidant control of root-knot nematodes, pp. 168-171 *In* Proc. Third IMP Res. Plann. Conf. on Root-knot

Nematodes, *Meloidogyne* spp., Regions IV & V, Nov. 16-20, 1981. Int. Inst. Trop. Agric., Ibadan, Nigeria. 275 pp.

19. Fortuner, R. 1978. *Meloidogyne* spp. on upland rice in Ivory Coast, pp. 8-10 *In* Proc. Second IMP Res. Plann. Conf. on Root-knot Nematodes, *Meloidogyne* spp., Region IV, Feb. 20-24, 1978, Abidjan, Ivory Coast. Int. Inst. Trop. Agric., Ibadan, Nigeria. 93 pp.
20. Hemeng, B. M. S. 1978. The International *Meloidogyne* Project, pp. 21-23 *In* Proc. Second IMP Res. Plann. Conf. on Root-knot Nematodes, *Meloidogyne* spp., Region IV, Feb. 20-24, 1978, Abidjan, Ivory Coast. Int. Inst. Trop. Agric., Ibadan, Nigeria. 93 pp.
21. Hemeng, B. M. S. 1981. The root-knot nematode problems in Ghana: past, present and future, pp. 57-69 *In* Proc. Third IMP Res. Plann. Conf. on Root-knot Nematodes, *Meloidogyne* spp., Regions IV & V, Nov. 16-20, 1981. Int. Inst. Trop. Agric., Ibadan, Nigeria. 275 pp.
22. Hemeng, O. B. 1978. Progress report on the International *Meloidogyne* Project, pp. 24-26 *In* Proc. Second IMP Res. Plann. Conf. on Root-knot Nematodes, *Meloidogyne* spp., Region IV, Feb. 20-24, 1978, Abidjan, Ivory Coast. Int. Inst. Trop. Agric., Ibadan, Nigeria. 93 pp.
23. Hemeng, O. B. 1981. International *Meloidogyne* Project Report for 1981, pp. 46-56 *In* Proc. Third Res. Plann. Conf. on Root-knot Nematodes, *Meloidogyne* spp., Regions IV & V, Nov. 20-24, 1981. Int. Inst. Trop. Agric., Ibadan, Nigeria. 275 pp.
24. Hemeng, O. B. and B. M. S. Hemeng. 1976. Root-knot nematodes in Ghana, pp. 48-52 *In* Proc. First IMP Res. Plann. Conf. on Root-knot Nematodes, *Meloidogyne* spp., Region IV, June 7-11, 1976. Int. Inst. Trop. Agric., Ibadan, Nigeria. 83 pp.
25. Idowu, A. A. 1981. The distribution of root-knot nematodes (*Meloidogyne* spp.) in relation to elevation and soil type in vegetable growing areas of upper northern Nigeria, pp. 128-134 *In* Proc. Third IMP Res. Plann. Conf. on Root-knot Nematodes, *Meloidogyne* spp., Regions IV & V, Nov. 16-20, 1981. Int. Inst. Trop. Agric., Ibadan, Nigeria. 275 pp.
26. Merney, G. 1976. Root-knot problem in Ivory Coast, pp. 60-65 *In* Proc. First IMP Res. Plann. Conf. on Root-knot Nematodes, *Meloidogyne* spp., Region IV, June 7-11, 1976. Int. Inst. Trop. Agr, Ibadan, Nigeria. 83 pp.
27. Netscher, C. 1978. Morphological and physiological variability of species of *Meloidogyne* in West Africa and implications for their control, pp. 11-15 *In* Proc. Second IMP Res. Plann. Conf. on Root-knot Nematodes, *Meloidogyne* spp., Region IV, Feb. 20-24, 1978, Abidjan, Ivory Coast. Int. Inst. Trop. Agric., Ibadan, Nigeria. 93 pp.
28. Netscher, C. and D. P. Taylor. 1976. *Meloidogyne* research at O.R.S.T.O.M., Dakar, pp. 66-71 *In* Proc. First IMP Res. Plann. Conf. on Root-knot Nematodes. *Meloidogyne* spp., Region IV, June 7-11, 1976. Int. Inst. Trop. Agric., Ibadan, Nigeria. 83 pp.
29. Nieuwenhuyzen, A. A. van. 1976. Report on the root-knot nematode situation in Liberia, pp. 56-59 *In* Proc. First IMP Res. Plann. Conf. on Root-knot Nematodes, *Meloidogyne* spp., Region IV, June 7-11, 1976. Int. Inst. Trop. Agric., Ibadan, Nigeria. 83 pp.
30. Nwauzor, E. C. 1979. Life cycle of the root-knot nematode, *Meloidogyne javanica* (Treub, 1885) Chitwood, 1949 in tomato (*Lycopersicon esculentum* L. cv Ife 1) and response of some tomato cultivars to population density of *M. javanica*. Ms. thesis, Dep. of Agric. Biol., Univ. Ibadan.
31. Nwauzor, E. C. 1982. Identification, biology and control of root-knot nematodes (*Meloidogyne* spp.) on edible yams (*Dioscorea* spp) in eastern Nigeria. Ph.D. thesis, Dep. of Agric. Biol., Univ. Ibadan.
32. Odihirin, R. A. 1976. *Meloidogyne* research at the University of Ibadan, pp. 20-23 *In* Proc. First IMP Res. Plann. Conf. on Root-knot Nematodes, *Meloidogyne* spp., Region IV, June 7-11, 1976, Int. Inst. Trop. Agric., Ibadan, Nigeria. 83 pp.
33. Odihirin, R. A. and T. O. Adesida. 1975. Locations and situations in which plant parasitic nematodes survive the dry season in Nigeria. II. The role of weeds in carrying root-knot nematodes over the dry season in southern Nigeria. Nigerian Soc. Plant Prot. Occas. Publ. No. 1.
34. Ogbuji, R. O. 1976. The status of root-knot nematodes in the eastern states of Nigeria, pp. 44-47 *In* Proc. First IMP Res. Plann. Conf. on Root-knot Nematodes, *Meloidogyne* spp., Region IV, June 7-11, 1976. Int. Inst. Trop. Agric., Ibadan, Nigeria. 83 pp.
35. Ogbuji, R. O. 1978. Some weed hosts of a population of *Meloidogyne incognita* in Nigeria, pp. 27-30 *In* Proc. Second IMP Res. Plann. Conf. on Root-knot Nematodes,

Meloidogyne spp., Region IV, Feb. 20-24, 1978, Abidjan, Ivory Coast. Int. Inst. Trop. Agric., Ibadan, Nigeria. 93 pp.

36. Ogbuji, R. O. 1981. Death in Early Dwarf okra caused by the root-knot nematode, *Meloidogyne incognita*, pp. 174-181 *In* Proc. Third IMP Res. Plann. Conf. on Root-knot Nematodes, *Meloidogyne* spp., Regions IV & V, Nov. 16-20, 1981. Int. Inst. Trop. Agric., Ibadan, Nigeria. 275 pp.
37. Ogbuji, R. O. and J. C. Okonkwo. 1977. Influence of soil pH on reproduction of *Meloidogyne incognita*. Niger. J. Plant Prot. 3:136-138.
38. Ogunfowora, A. O. 1976. Research on *Meloidogyne* at the Institute of Agricultural Research and Training, University of Ife, Moor Plantation, Ibadan, pp. 9-14 *In* Proc. First IMP Res. Plann. Conf. on Root-knot Nematodes, *Meloidogyne* spp., Region IV, June 7-11, 1976. Int. Inst. Trop. Agric., Ibadan, Nigeria. 83 pp.
39. Ogunfowora, A. O. 1978. Further studies on control of root-knot nematodes, pp. 31-37 *In* Proc. Second IMP Res. Plann. Conf. on Root-knot Nematodes, *Meloidogyne* spp., Region IV, Feb. 20-24, 1978, Abidjan, Ivory Coast. Int. Inst. Trop. Agric., Ibadan, Nigeria. 93 pp.
40. Ogunfowora, A. O. 1981. Root-knot nematodes on cowpea and some selected vegetable crops, pp. 72-84 *In* Proc. Third IMP Res. Plann. Conf. on Root-knot Nematodes. *Meloidogyne* spp., Regions IV & V, Nov. 16-20, 1981. Int. Inst. Trop. Agric., Ibadan Nigeria. 275 pp.
41. Olowe, T. O. 1976. Research work on root-knot nematodes at the National Cereals Research Institute, pp. 15-19 *In* Proc. First IMP Res. Plann. Conf. on Root-knot Nematodes, *Meloidogyne* spp., Region IV, June 7-11, 1976. Int. Inst. Trop. Agric., Ibadan, Nigeria. 83 pp.
42. Olowe, T. O. 1978. International *Meloidogyne* Project progress report 197 pp. 58-69 *In* Proc. Second IMP Res. Plan Conf. on Root-knot Nematode *Meloidogyne* spp., Region IV, Feb. 20-2 1978, Abidjan, Ivory Coast. Int. Inst. Tro Agric., Ibadan, Nigeria. 93 pp.
43. Olowe, T. O. 1981. Importance of root-kno nematodes on cowpea *Vigna unguiculat* (L.) Walp. in Nigeria, pp. 85-109 *In* Pro Third IMP Res. Plann. Conf. on Root-kno Nematodes, *Meloidogyne* spp., Regions IV V, Nov. 16-20, 1981. Int. Inst. Trop. Agric. Ibadan, Nigeria. 275 pp.
44. Taylor, D. P. and C. Netscher. 1978 *Meloidogyne* research at O.R.S.T.O.M. Dakar, pp. 16-20 *In* Proc. Second IMP Res Plann. Conf. on Root-knot Nematodes *Meloidogyne* spp., Region IV, Feb. 20-24 1978, Abidjan, Ivory Coast. Int. Inst. Trop Agric., Ibadan, Nigeria. 93 pp.
45. Wilson, W. R. 1962. Root-knot nematode (*Meloidogyne* spp.) in northern Nigeria, pp 16-17 *In* Proc. First InterAfrican Nematol Conf., Kikuyu, 24-28 October, 1960, OCTA Publ. No. 86.
46. Zannou, T. A. 1981. Resistance to root-knot nematodes, *Meloidogyne javanica* (Treub, 1885) Chitwood, 1949 and *M. incognita* (Kofoid & White, 1919) Chitwood, 1949 in cowpea, *Vigna unguiculata* (L.) Walp. Thesis for the degree of Ing. Agron. of the Univ. Nat. de Benin, Fac. des Sci. Agron., Abomey Calavi—R P Benin and Univ. Ibadan, Fac. Agric. and Forest., Ibadan, Nigeria, in co-operation with the Int. Inst. Trop. Agric.
47. Zannou, T. A. and E. H. Dodego. 1981. *Meloidogyne* research programme in Benin Republic, pp. 70-71 *In* Proc. Third IMP Res. Plann. Conf. on Root-knot Nematodes, *Meloidogyne* spp., Regions IV & V, Nov. 16-20, 1981. Int. Inst. Trop. Agric., Ibadan, Nigeria. 275 pp.

MELOIDOGYNE SPP. RESEARCH IN REGION V OF THE INTERNATIONAL *MELOIDOGYNE* PROJECT (IMP)

V. W. Saka

Crop Production Department
University of Malawi, Bunda College of Agriculture
Lilongwe, Malawi

In order to appreciate the major research emphasis, current status and problems of *Meloidogyne* spp. research in Region V, I think it is very important to trace briefly the influence of the past political history of the Region on plant nematology in general and *Meloidogyne* spp. in particular. Historically, countries currently grouped in Region V by IMP were, at one time or another, under Britain. Until 1977, the three countries Kenya, Uganda and Tanzania formed what was known as the East African Community. These countries funded and worked within the East African Agriculture and Forestry Research Organisation (E.A.A.F.R.O.) based at Muguga in Kenya. It served as a central agricultural research facility where problems of common importance were investigated. Obviously, plant-parasitic nematodes had to be studied on an interterritorial basis. Similarly, up to 1963, in what used to be known as British Central Africa, Nyasaland (Malawi) and the two Rhodesia (Zambia and Zimbabwe) were federated. Although there was no research organization similar to the one in East Africa, the center for nematology work was in Salisbury, now Harare. Perhaps what is of interest is that both centers emphasized surveys, identification and control of *Meloidogyne* spp. In case of Botswana, very little is known about nematology work.

RESEARCH EMPHASIS

Surveys

Much of our present knowledge of the distribution and the economic importance of *Meloidogyne* spp. in Region V is based on the surveys and identifications made by De Grisse (7), Whitehead (26,27,28) and Martin (9,10,11). In Kenya, De Grisse (7) described a root-knot nematode *Meloidogyne kikuyensis* n. sp. found attacking Kikuyu grass (*Pennisetum clandestinum* Hochst) and reported that it closely resembled *M. africana* Whitehead. However, he did not report the pathogenicity of this species, and the hosts and the life cycle of *M. kikuyensis* are still not well known.

Subsequent surveys of *Meloidogyne* spp. in cultivated soils of many parts of East Africa revealed that *M. javanica* and *M. incognita* were the most common species on food crops (26,28). *M. hapla* commonly attacked pyrethrum at higher altitudes and arabica coffee was attacked by several other species such as *M. megadora*, *M. decalineata*, *M. kikuyensis* and *M. africana*. The frequencies of *M. javanica* and *M. incognita* were not correlated with soil texture. However, there seemed to be some relationship between altitude and moisture with the distribution of these two species in that *M. javanica* was the dominant species at high altitudes with wetter soils whereas *M. incognita* was more successful in hotter soils, and both species were equally abundant at intermediate altitudes with moderate rainfall. Whitehead (27) also reported that *M. javanica* and *M. incognita* very rarely reproduce on coffee in tropical Africa but that *M. megadora*, *M. africana* and *M. decalineata* do; he considered these species to be indigenous to tropical Africa. However, the pathogenicity and economic importance of these species were unknown.

In Central Africa, the economic importance of root-knot nematodes was first reported by Jack (8) and Smee (23) early this century. However, it was Martin (9,10) in Zimbabwe who identified four root-knot species, *M. javanica*, *M. incognita acrita*, *M. hapla* and *M. arenaria*. More recent work has been carried out by Way (25) on the populations identified by Martin. Results of differential host tests and perineal patterns has revealed the occurence of *M. javanica*, *M. incognita* races 1 and 3 and *M. hapla* but apparently leaves the existence of *M. arenaria* much in doubt.

Martin (10) found *M. javanica* and *M. incognita* to be widely spread on many hosts. A survey of root-knot-infected tobacco in Zimbabwe and Zambia revealed that 99% of the soil samples were infested with *M. javanica* and 1% with *M. incognita acrita*. He also observed that *M. incognita acrita* infestations tended to be located in low-lying spots and subject to a certain amount of water-logging and speculated that the species was presumably introduced into the country whereas *M. javanica* may be indigenous.

In a separate survey carried out in the southern region of Malawi (Nyasaland), Martin (11) found *Meloidogyne* spp. to be widely distributed in the tea nurseries. In some areas, occasional plants were infected, and in others many plants were heavily infected with galls, mainly on tap roots. He also noted that only small populations of females were viable on the roots and egg-mass production was generally poor. *M. javanica* was found in 70% of the infected plants, the rest being what he identified as *M. arenaria* and, in one case, *M. incognita acrita*. Only three tea plants were lightly galled by *M. javanica*. These results apparently supported what had been reported in Sri Lanka by Sivapalan (22) who observed that susceptibility of tea to root-knot nematodes decreased with age and that this resistance was markedly observed between 8 and 15 months of growth. He found that juveniles of some root-knot nematodes were able to penetrate young feeder roots of mature tea bushes but failed to induce any galls. The development from second to third stage was inhibited apparently by an unknown mechanism.

Subsequent surveys in Malawi by various workers (2,3,4,12,18) have all reported that *M. javanica* and *M. incognita* are widely distributed and cause a lot of economic damage on many crops in the country. Bridge and Page (3) reported the incidence of *M. javanica* in burley and flue-cured tobacco estates to be as high as 96 to 100%. Infestations of this extent have been observed in other crops such as potato, okra, banana, pigeon pea and cowpea (Saka, unpublished). In a separate survey, Bridge et al. (2) reported an isolated case of *M. acronea* in cotton in the Shire Valley in the southern region of Malawi. The nematode has been observed to reduce growth and cause low production of flowers in cotton. Pathogenicity tests by Bridge et al. (2) have shown that maize (*Zea mays*), pigeon pea (*Cajan cajanus*), sorghum (*Sorghum bicolor*), groundnut (*Arachis hypogaea*), leucaena, guar bean (*Cyamopsis psoraloides*), and tomato (*Lycopersicon esculentum*) are hosts whereas bulrush millet (formerly *Pennisetum typhoides*, now *P. americanum*) and finger millet (*Eleusine coracana*) are nonhosts. Further surveys for *M. acronea* in different ecological zones have shown that it is not widely distributed in the country (Saka, unpublished).

Disease Interaction

Relatively very little emphasis has been put on interaction of *Meloidogyne* spp. with other disease-causing agents in Region V. Perry (16), after a survey in cotton-growing areas of Uganda and Tanganyika, reported the interaction of *M. incognita* with *Fusarium oxysporum vasinfectum* on cotton. Inoculation experiments confirmed a rather synergistic relationship between the two organisms, and these results were supported when soil fumigation reduced root-knot intensity from 64% to 18%, and a corresponding decrease in wilt incidence from 14% to 2% occurred. Shepherd (21) has reported the interaction of *Rhizoctonia solani* with *M. javanica* in tobacco in Zimbabwe where he observed that galling increased root invasion by the fungus; however, there was no evidence of a further reduction in yield due to the fungal attack. Bridge and Page (3) have reported the interaction of *Fusarium oxysporum* v. *nicotianae* with *M. javanica* in tobacco estates that they surveyed in Malawi as a relationship in which nematodes predispose the plants to invasion by the fungus.

CONTROL

The objective of the major emphasis on the survey and identification of *Meloidogyne* spp. was to provide a basis for effective control measures against these nematodes. So far, two methods, namely cultural and chemical, are widely used in various countries.

Cultural Control

In Zimbabwe, Shepherd (21) has reported early planting of tobacco on ridges after early ploughing to be the major cultural control method against root-knot nematodes. Under Zimbabwean conditions, there is reduction of root-knot nematode populations in the top layer of soil particularly in sandy soils. The principle behind this method is based on Daulton &

Nusbaum (6) who reported that Zimbabwean *M. javanica* eggs were killed at 36°C after 12 hours or longer. Shepherd (21) has reported that temperatures up to such levels in the upper 50 mm of soil in tobacco fields are quite often recorded in Zimbabwe.

Crop Rotation

Rotation is probably the most inexpensive control measure against root-knot nematodes, but because of the nematode's wide host range, design of an effective rotation system is often rather difficult. After pot experiments with various crops, Whitehead (27) suggested the use of some resistant varieties of maize, sweet potato, groundnut, cotton, sorghum, velvet bean and soybean in order to control *M. javanica* and *M. incognita* in Kenya. Whether further investigations were done on these crops is unknown.

Tobacco being the biggest cash crop foreign exchange earner in Malawi and Zimbabwe, research on rotation invariably has been aimed at the devising of an effective system to protect this crop from nematodes. Consequently many studies have involved the identification of suitable crops for inclusion in pasture rotation. In Zimbabwe, Daulton (5) extensively investigated the inclusion of pasture grasses in tobacco-growing areas and reported a yield increase as well as a reduction of root-knot damage with a three-year rotation of weeping lovegrass (*Eragrostis curvula* cv. 'Ermelo'), Rhodes grass (*Chloris gayana cv. 'Katambora'*), and panic grass (*Panicum maximum* cv. 'Sabi'). Similar results were later supported by Shepherd (21) who observed that *Pratylenchus zeae*, *P. brachyurus* and *Scutellonema brachyurum* were capable of reproducing on these grasses and speculated that the reduction of *M. javanica* by these pasture grasses was due to starvation rather than nematicidal chemicals.

In Malawi, Norse (15) tested six widely grown maize varieties in a three-year maize-groundnut-tobacco rotation and reported root-knot nematode reduction with the two cultivars 'Bingo' and 'Muthenga' and claimed that they were heterozygously resistant to *M. javanica*. It is not understood how he arrived at this conclusion. However, quite recently Bridge & Page (3) reported in a survey of tobacco-growing areas in the central region of Malawi that the Rhodes grass-maize-groundnut-tobacco rotation was inefficient because many solanaceous and other nematode weed hosts were allowed to grow during the grass cycle of the rotation. These hosts sustained some nematode populations which later infected tobacco. Therefore, Bridge & Page (3) recommended a longer rotation, preferably 5-6 years, and regular weeding.

Chemical Control

The use of chemicals to control root-knot nematodes, particularly in tobacco, is not at all new in the Region. Recently Shepherd (21), in a review of nematode pests of tobacco in southern Africa, reported that attempts with chemical control started in the 1920's, but it was almost 30 years ago that D-D (dichloropropane-dichloropropene), MB (methyl bromide), and EDB (ethylene dibromide) were recommended for the seedbeds and the fields. Currently, the nematicides most widely used in tobacco in Zimbabwe and Malawi are methyl bromide and ethylene dibromide in seedbeds and fields, respectively.

Satisfactory results have been reported by many tobacco farmers in Malawi with ethylene dibromide although Bridge & Page (3), after visiting some tobacco estates where EDB was often used, reported that there was only a 40% nematode reduction. They attributed the poor performance mainly to faulty injection guns as well as poor application techniques. They suggested further investigations into time and depth of application of the chemical in tobacco.

In Zimbabwe, Shepherd (20,21) reported that metam-sodium and dazomet have been extensively tested but that the results have been unsatisfactory. Use of the chemicals was not to be recommended in tobacco. Similarly in Malawi, nematicidal trials have been carried out with the newer organophosphates and carbamates in tobacco, banana, potato and tomato fields. The results are also unsatisfactory for a recommendation on any of these crops (Saka, unpublished).

CURRENT RESEARCH PROJECT

In general, much of the research work currently underway in the Region is concerned with identification of *Meloidogyne* spp., testing of different crop cultivars, screening of various nematicides and evaluation of different cropping systems.

Uganda

Bafokuzara (1) reported that research work would concentrate on the effect of cropping patterns and organic amendments against *Meloidogyne* spp. identified through the use of perineal patterns and differential host tests. He also planned to test different crops as well as

Table 1. Agro-climatic zones of Uganda

Zone	District	Agricultural system
I	Busoga/Bukedi	Banana, millet and cotton system, with outliers of the main coffee-banana system.
II	Bugisu/Sebei	Montane systems: Arabica coffee, banana (wheat and maize in Sebei).
III	Teso	Teso systems: finger millet, cotton and cattle keeping (mixed agriculture).
IV	Karamoja	Pastoral system—cattle keeping.
V	Lango/Acholi	Northern systems: finger millet, cotton, tobacco (some mixed agriculture also).
VI	West Nile/Madi	West Nile systems: basic agriculture like Zone V but with predominance of cassava as staple food.
VII	Bunyoro/Toro	Arabica and Robusta coffee and banana system, Montane systems: heterogenous agriculture but basically banana, coffee, tea.
VIII	Ankole	Montane systems in the west: Pastoral to the east: Arabic and Robusta coffee, tea, banana, cattle.
IX	Kigezi	Montane systems but with larger annual crop acreage than other montane systems. Sorghum is major staple. Arabica coffee, tea.
X	Lake Victoria Crescent	Main Robusta coffee and banana system: Robusta coffee, banana, tea, cocoa, sugar.
XI	Northern Buganda	Western extension of the banana-millet-cotton system, but now largely taken up by big ranching projects.

Source: Planning Cell, Ministry of Agriculture.

breed tomato and tobacco varieties against root-knot nematodes.

Recently, Bafokuzara (personal communication) has reported that survey studies to investigate the occurrence, distribution and intensity of different *Meloidogyne* spp. in the 11 different agro-climatic zones of Uganda (Table 1) are underway. The main objective of these studies is to relate nematode species prevalence in a given area to ecological factors such as cultivation practices, vegetation, major soil types and the weather. The results should ultimately facilitate design of control programs.

Bafokuzara is also carrying out host range studies with *Meloidogyne incognita* race 1 which is commonly found in the central region of Uganda. The nematode is being tested against major food and cash (economic) crops of the country which have apparently been identified and arranged into families. Presently Leguminosae plants (1-3 weeks old) are being inoculated with 400 second-stage juveniles of *M. incognita* race 1, and each cultivar is replicated ten times. The plants are analyzed for galling, egg mass production or other root abnormalities and rated as resistant, intermediate or susceptible eight weeks after inoculation. The results of these studies, of course, will be reported later as these and other plants belonging to other families are re-evaluated in control programs in the greenhouse and in field tests, with a view to making recommendations to farmers (Bafokuzara, personal communication).

Tanzania

Swai (personal communication) has reported that *M. hapla* is widely distributed in pyrethrum fields and nurseries in the southern highlands of Tanzania. Losses of up to 50% have been observed under severe infestations. *M. hapla* also infects local weeds such as *Bidens pilosa*, *Conyza* sp., *Sonchus oleraceus*, *Dichrocephala integrifolia*, *Gnaphalium luteo album*, *Grassocephalum rubens*, *Pycnostachys* sp., *Ageratum conyzoides* and *Sonchus bipontinii*.

In order to identify some resistant common food crops of pyrethrum-growing areas, Swai has tested three maize cultivars ('H632,' 'H6302' and 'H614'), two wheat cultivars ('Mamba' and 'Tai'), and two potato cultivars under greenhouse conditions. Maize and wheat cultivars have been reported to be immune whereas potato cultivars showed what Swai calls high-level resistance. In a separate experiment in the greenhouse, three nematicides-fenamiphos (Nemacur), carbofuran (Furadan), and aldicarb (Temik)—have been screened against *M. hapla*. At the rate of 10kg/ha, Temik 10G and Nemacur 5G prevented root galling. Furadan 5G inhibited galling when the chemical was used at the rate of 44kg/ha. Further experiments are being conducted with these chemicals under field conditions.

Zimbabwe

Shepherd & Way (personal communication) have reported that work to find a bioassay method suitable to decide the economic threshold (24) still continues. Apparently, the results obtained so far are unsatisfactory, although final data have yet to be analyzed.

In separate experiments, three cassava cultivars, five Rhodes grass (*Chloris gayana*) cultivars, three legumes (Siratro, fine stem stylo and silverleaf *Desmodium*), two *Brassica* species (*B. napus* cvs. 'Regent' and 'Altex' and *B. campestris* cv. 'Candle'), and ten maize cultivars were tested against *M. javanica* and *M. incognita* races 1 and 3. Results revealed that all three cassava cultivars, all Rhodes grass cultivars and all maize cultivars were more susceptible to *M. incognita* races 1 and 3 than to *M. javanica*. Silverleaf *Desmodium* was a host to *M. javanica*. All three nematode populations produced very low ratings on Siratro and fine stem stylo. Silverleaf *Desmodium* was not infected heavily by either race of *M. incognita*. It was also observed that tomatoes grown in soil mixed with the chopped roots of legumes also had low infection ratings after six weeks, presumably because the organic amendments suppressed the nematodes.

Results with maize cultivars showed that only a small number of females with eggs were present in the roots compared with many hundreds in a similar sample of susceptible tobacco. As a result of this experiment, a field trial is underway to study whether an increased period (1,2,3 and 4 years) of maize in rotation with tobacco can sufficiently reduce *M. javanica* to the point that no fumigation is necessary in normal years. It is hoped that if an increased period under maize is sufficiently effective, then food production could be increased on tobacco farms. In cooperation with the plant breeders, the testing of resistant tobacco lines is being carried out in the greenhouse. With agronomic trials of promising lines, resistant flue-cured tobacco possibly can be released in the near future.

Shepherd (personal communication) has also observed that if the rate of EDB per m^2 is effective, then the depth of the chemical application is not important, and he also found that spike application of water-miscible EDB in the field for oriental tobacco was more effective than use of a watering can; however, both methods are good for the growth of the crop. In an experiment designed to study the effect of EDB/aldicarb combinations (0, 1.5 and 3.0 ml/station EDB), Shepherd observed an increase of nematode control in heavily infested soil at all levels of treatments although phytotoxicity has occurred in some cases and resulted in loss of plants at the highest rate (4kg ai/ha aldicarb).

In Zimbabwe, future work will concentrate on the following areas of research: (a) comparing methods for recovery of nematodes from soil, (b) studying the movement of *M. javanica* juveniles in soil to elucidate how fumigated areas are re-infested, (c) screening possible tolerant or resistant plants for use in rotation with susceptible plants, and (d) testing new nematicides as well as investigating improved chemical control methods.

Botswana

Busang (personal communication) has reported that *Meloidogyne* spp. have been a problem mainly on vegetable crops where loss of yield and market value has been observed by both commercial and small farmers. However, due to the absence of nematologists, or even plant pathologists, the problem was for some time neglected. Quite recently two species of root-knot nematodes, *M. javanica* and *M. incognita*, were found on pigweed (*Amaranthus thunbergii*), prostrate globe (*Amaranthus gomphrena celoides*), flannel weed (*Sida cordifolia*), *Pavonia* spp. and *Kyphorcarpa angustifolia*.

Several projects are currently underway. The first involves the identification of root-knot nematodes, although *M. incognita* (Kofoid and White) race 2 has already been identified from beet (*Beta vulgaris*) and spinach (*Spinacia oleracea*) by IMP at Raleigh, North Carolina. Host range and pathogenicity studies constitute the second group of projects. Several vegetable crops, flowers, and food crops, such as bean, sorghum and millet, are being tested in the

Table 2. Incidence of *F. udum, M. javanica* and *F. udum/M. javanica* in pigeon pea (500 samples).

Pathogen	Number of plants infected
F. udum Butler, only	275
M. javanica, only	125
F. udum Butler/*M. javanica*	100
Total	500

greenhouse, and so far indications of variabilities in susceptibility and resistance have actually been observed among the different crops. Finally, the effects of soil temperature on the pathogenicity of *M. incognita* in sunflower is being studied.

Kenya

Ngundo (13,14) reported some damage on beans caused by *M. incognita* and screened 23 local and imported bean cultivars (*Phaseolus vulgaris*) against *M. incognita* and *M. javanica*. Of the 23 cultivars tested, seven were found resistant and none were immune. Interestingly, one cultivar PI 165426 that was reported to be highly resistant to *M. incognita* was susceptible to a mixed population of *M. incognita* and *M. javanica*. Gichure (personal communication) reported that projects currently underway are concerned with (a) a survey of *M. africana* Whitehead in coffee, (b) screening of different nematicides in tomato and potato, and (c) study of disease complexes of *M. javanica* with *Fusarium* wilt in pyrethrum and of *M. javanica* with bacterial wilts in various potato cultivars.

Malawi

Work on screening ICRISAT pigeon pea and local lines against *Fusarium udum* Butler and *M. javanica* still continues after observations that there is an interaction between these organisms (Table 2) (17). In a pot experiment in the greenhouse, severe wilting was observed in plants inoculated with the two organisms. Currently three screening plots have been established at different ecological locations with the objective of identifying resistant germ plasm for the small-holder farmers who are the victims of these two organisms.

In addition to pigeon pea, other crops such as bean, cowpea and groundnut (peanut) are being screened against root-knot nematodes in the greenhouse as well as the field. The resistant cultivars from these tests eventually will be evaluated in the many but varied cropping systems that exist in Malawi.

In cooperation with the University of Florida Departments of Agronomy, Entomology and Nematology, a study has been initiated to study the degree to which *Meloidogyne* spp. limit biological nitrogen fixation in forage legumes in the tropics. Other activities include a comparison of EDB with aldicarb, oxamyl, carbofuran, and fenamiphos in such crops as potato, banana, tomato and tobacco as well as an investigation of the effects of different cropping systems on the population dynamics of *Meloidogyne* spp. and other plant-parasitic nematodes.

FUTURE NEEDS

Sasser and Krishnappa (19) quite rightly pointed out that "the sound and orderly growth of any science depends on the wisdom, dedication and direction given during its infancy." From the foregoing summary, it is obvious that nematology in Region V had a good infancy although the departure of expatriate nematologists has left a vacuum that has taken long to be filled in most countries. However, it is pleasing to note that currently every country has a trained nematologist involved in research because of the present demand for information on how to control nematodes by the smallholder farmers.

The victims of *Meloidogyne* spp. damage in Region V are the small growers who simply cannot afford to pay the high prices for chemicals and have too little land for rotation. Therefore, it is very important that the future research should emphasize the screening of local and foreign cultivars as well as the evaluation of the various cropping systems currently practiced by small growers.

Admittedly since the countries of Region V are developing countries, our resources for research are limited, and funds are badly needed for conducting research of this nature. There is also a need to train more nematologists either locally or abroad. Both of these activities will need an association or participation with one such as that which currently exists between Region V and the International *Meloidogyne* Project.

Finally, on behalf of my fellow cooperators from Region V, I would like to thank Professor J. N. Sasser sincerely for his insight into *Meloidogyne* spp. problems in developing countries. The creation of the International *Meloidogyne* Project has given the cooperators in the Region a chance to interact among themselves as well as with other workers in the Project and the opportunity to keep abreast with recent research as well as publish our results. I would also like to express my gratitude to the U.S.A. Government for funding this important project, and I sincerely hope this assistance will continue in the future.

Literature Cited

1. Bafokuzara, N. D. 1982. Report on the research work carried out in Uganda as part of the International *Meloidogyne* Project, pp. 8-12 *In* Proc. of the Third IMP Res. & Plann. Conf. on Root-knot Nematodes, *Meloidogyne* spp., Regions IV & V, Nov. 16-20, 1981. Int. Inst. Trop. Agric. (IITA), Nigeria. 275 pp.
2. Bridge, J. G., E. Jones, and L. J. Page. 1976. *Meloidogyne acronea* associated with reduced growth of cotton in Malawi. Plant Dis. Rep. 60: 5-7.
3. Bridge, J. G. and L. J. Page. 1977. Assessment of the importance and control of plant parasitic nematodes in Malawi. ODM Report of Scientific Liaison Officer.
4. Corbett, D. C. M. 1967. Nematodes as plant parasities in Malawi. PANS (Pest. Art. and News Summ.) 13:151-162.
5. Daulton, R. A. C. 1962. Survey of land use and plant parasitic nematode control practices in the southwestern flue-cured tobacco areas of Rhodesia. Rhod. J. Agric. Res. 59:216-217.
6. Daulton, R. A. C. and C. J. Nusbaum. 1961. The effect of soil temperature on the survival of the root-knot nematodes *Meloidogyne javanica* and *M. hapla*. Nematologica 6:280-294.
7. De Grisse, A. 1960. *Meloidogyne kikuyensis* n. sp. a parasite of Kikuyu grass *Pennisetum clandestinum* in Kenya. Nematologica 5:303-308.
8. Jack, R. W. 1927. Root gallworm or root-knot eelworm (*Heterodera radicicola*). Rhod. J. Agric. Res. 24:957-964.
9. Martin, G. C. 1955. Plant and soil nematodes of the Federation of Rhodesia and Nyasaland. Preliminary investigations. Nematodes catalogued under hosts or associated plants. Rhod. J. Agric. Res. 52:346-361.
10. Martin, G. C. 1957. Four kinds of root-knot nematodes. Rhod. J. Agric. Res. 54:324-326.
11. Martin, G. C. 1960. Plant parasitic nematodes associated with tea production in Southern Province of Nyasaland, p. 11 *In* Proc. 1st Inter-African Plant Nematol. Conf. I.A.P. S.C. Publ. No. 86.
12. Mughogho, K. K. and E. B. Choo. 1969. List of plant-parasitic nematodes in Malawi. Minist. Agric. Tech. Bull. No. 1. Zomba, Malawi.
13. Ngundo, B. W. 1976. Reports by regional investigation on the state of knowledge concerning root-knot nematodes in East Africa, p. 11 *In* Proc. of the Res. & Plann. Conf. on Root-knot Nematodes, *Meloidogyne* spp. Jan. 12-16, 1976. Raleigh, NC. 106 pp.
14. Ngundo, B. W. 1977. Screening of bean cultivars for resistance to *Meloidogyne* spp. in Kenya. Plant Dis. Rep. 61:991-993.
15. Norse, D. 1972. Nematode populations in maize-groundnut-tobacco rotation and the resistance of maize varieties to *Meloidogyne javanica*. Trop. Agric. 49:355-360.
16. Perry, D. 1960. The interaction of *Fusarium oxysporum* f. *vasinfectum* and *Meloidogyne* spp. on cotton. Report by D. C. M. Corbett to Nyasaland Govt. of the 1st Inter-African Plant Nematol. Conf. held at Muguga, Kenya. 32 pp.
17. Saka, V. W. 1982. International *Meloidogyne* Project Report in Malawi, pp. 31-36 *In* Proc. of the Third IMP Res. & Plann. Conf. on Root-knot Nematodes, *Meloidogyne* spp., Regions IV & V, Nov. 16-20, 1981. Int. Inst. Trop. Agr. (IITA), Ibadan, Nigeria. 275 pp.
18. Saka, V. W. and M. A. Siddiqi. 1979. Plant parasitic nematodes associated with plants in Malawi. Plant Dis. Rep. 63:945-948.
19. Sasser, J. N. and K. Krishnappa. 1980. The development of nematology on a world basis. J. Nematol. 12:154-158.
20. Shepherd, J. A. 1982. Report to the Third Regional Conference on root-knot nematode research held at the International Institute of Tropical Agriculture, Ibadan, Nigeria, pp. 21-27 *In* Proc. of the Third IMP Res. & Plann. Conf. on Root-knot nematodes, *Meloidogyne* spp., Regions IV & V, Nov. 16-20, 1981. Int. Inst. Trop. Agric. (IITA), Ibadan, Nigeria. 275 pp.
21. Shepherd, J. A. 1982. Nematode pests of tobacco, pp. 58-68 *In* Nematology in Southern Africa, ed. D. P. Keetch and J. Heyns. Repub. S. Afr. Dep. Agric. Fish. Sci. Bull. No. 400. 170 pp.
22. Sivapalan, P. 1972. Nematode pests of tea,

pp. 285-311 *In* Economic nematology, ed. J. M. Webster. Academic Press, N. Y. 563 pp.

23. Smee, C. 1928. Root gallworm in tobacco seedbeds in Nyasaland (*Heterodera radicicola,* Greef). Bull. Nyasaland Dep. Agric. Entomol. Ser. No. 3.
24. Swai, I. S. 1982. Root-knot nematodes, *Meloidogyne* species, in Tanzania, pp. 28-30 *In* Proc. of the Third IMP Res. & Plann. Conf. on Root-knot Nematodes, *Meloidogyne* spp., Regions IV & V, Nov. 16-20, 1981. Int. Inst. Trop. Agric. (IITA), Ibadan, Nigeria. 275 pp.
25. Way, J. I. 1982. *Meloidogyne* species in Zimbabwe—Susceptible crops other than tobacco, taxonomy and resistant crops, pp. 13-20 *In* Proc. of the Third IMP Res. & Plann. Conf. on Root-knot Nematodes *Meloidogyne* spp., Regions IV & V, Nov. 16-20, 1981. Int. Inst. Trop. Agric. (IITA), Ibadan, Nigeria. 275 pp.
26. Whitehead, A. G. 1968. Taxonomy of *Meloidogyne* Goeldi 1887 (Nematodea: Heteroderidae) with descriptions of four new species. Trans. Zool. Soc. Lond. 31:263-401.
27. Whitehead, A. G. 1969. The distribution of root-knot nematodes (*Meloidogyne* spp) in tropical Africa. Nematologica 15:315-333.
28. Whitehead, A. G. and L. Kariuki. 1961. Root-knot nematode surveys of cultivated areas in East Africa. East Afr. Agric. For. J. 26:87-91.

SUMMARY REPORT ON THE CURRENT STATUS, PROGRESS AND NEEDS FOR *MELOIDOGYNE* RESEARCH IN REGION VI

R. G. Davide
University of the Philippines at Los Baños
College, Laguna, Philippines

It is indeed a great pleasure for me to present this summary report of Region VI which comprises Bangladesh, Burma, Fiji Islands, Indonesia, Japan, Malaysia, Nepal, Pakistan, Philippines, Republic of China (Taiwan), South Korea, Sri Lanka and Thailand. When I received the letter of Dr. J. N. Sasser on January 11, 1983, requesting me to do this job, I thought it would be a difficult assignment not only because I did not have much time to do it but also because I felt that I wouldn't be able to get the necessary information from the country cooperators who are very far from the Philippines. Neverthcless, I immediately established contact, through airmail letters, with my colleagues in the region asking for their support and assistance in making this report truly a product of our joint efforts. To my surprise, I received instantaneous and enthusiastic responses from them, including my colleagues in the Philippines Dr. R. B. Valdez and Dr. T. T. Reyes. Only then, did I realize how successfully IMP is bridging the information gap among the nematologists of the region, bringing them much closer, and making them more helpful to each other. I think this is just one of the tangible accomplishments of IMP in the region. We have developed among ourselves a strong bond of friendship and fellowship.

However, before paying tribute to the achievements of IMP in the region, I feel that it is fitting and proper to give due recognition to the man who dreamed, conceived and gave birth to the idea and ideals of IMP. This man, as we all know, is Dr. J. N. Sasser who is imbued with a missionary spirit to spread the gospel of nematology to the Third World so that the small and poor farmers can have more food by reducing the damage caused by root-knot nematodes on their food crops. As his present disciples of nematology, we are gathered here together for two-and-one-half weeks with a mission to know more of the latest approaches and technologies on root-knot nematode studies so that we can be more effective in teaching our farmers on how to control these nematodes in order to increase their food supply. These pioneering efforts of Dr. Sasser are deeply appreciated by all of us.

IMP is indeed a unique scientific cooperative venture which is in itself an experiment of world-wide proportions testing the hypothesis that nematologists from the developed and the developing countries can cross political, social, and cultural boundaries and harmoniously work together in solving a global pest problem of crops, namely crop losses due to root-knot nematodes. As far as I know, this project is the first of its kind in the world today, and the impact of this program has now been felt in many of the developing countries. As far as Region VI is concerned, IMP generally has had a strong influence on the development of nematology. It has stimulated the growth of the science and served as a catalyst in creating awareness among various farming sectors and in agricultural policy-making bodies of governments concerning the seriousness of the threat of root-knot nematodes to crop production. The project also challenges the nematologists in the region to assert their role strongly in the national food production campaign. Above all, it encourages us, the nematologists, to be more productive in research and to see to it that our research products reach and benefit the end-users—our farmers. Of course, there are other ways in which the IMP has benefited the region that cannot be quantified or expressed in appropriate words. We cannot fully describe our feelings on how lucky we are to have gotten involved in this project.

Having read the individual reports submitted, I am pleased to inform this body, that we in Region VI have already substantially advanced our research work on various aspects of the IMP objectives. These advances are in the areas of root-knot nematode survey, species and race identification, host range, varietal screening and breeding for resistance, and control through cultural, biological, and chemical methods.

SURVEY, IDENTIFICATION AND HOST RANGE

As far as the survey and identification of *Meloidogyne* species are concerned, many countries in the region, particularly Japan, South Korea, Philippines, Thailand, Sri Lanka, Indonesia and Pakistan, have already done considerable studies, and they know quite extensively the distribution of *M. incognita, M. javanica, M. arenaria* and *M. hapla*. A list of crops attacked by these species is already available in these countries. However, some member countries, particularly those that joined the group later, such as Nepal, Burma, Fiji Islands and Bangladesh, are still doing survey and host-range studies.

SCREENING AND BREEDING FOR RESISTANT VARIETIES

In breeding crops for resistance to root-knot nematodes, Japan, Thailand, Taiwan, Philippines and South Korea have already made substantial progress. A number of resistant cultivars have been identified. For instance, Dr. Toshio Shiga and Dr. Tomokis Takomata in Japan have screened 617 sweet-potato clones and found 67 to have resistance to *M. incognita*. Likewise, Dr. Sontirat of Thailand reported that Cuban kenaf cvs. 'W.A. Gahana' and 'ER-75-C-100' and Hawaiian cultivars of tomato like 'Healani' and 'Daruma' have resistance to *M. incognita*. In Korea, Dr. Choi found that tobacco cv. 'Osibeubtcho-E' and line 6601-8 have resistance to *M. incognita*. A number of tomato lines in Taiwan such as CL106-5-1-0, L97, L274, L4109, L4126, and L313 were reported by Dr. J. E. Yuen to be highly resistant to all races of *M. incognita, M. javanica* and *M. hapla*, and this was confirmed by Dr. J. N. Sasser here in North Carolina.

In the Philippines, Dr. T. T. Reyes reported that 19 pepper cultivars, two cabbage cvs. 'Leo #80' and 'K-K,' and radish cv. 'Ramgo' were resistant to *M. incognita*. Dr. Reyes also reported that such tobacco cultivars as 'Wisconsin Seedling,' 'W-8C,' 'Ramcol 70-2' and 'H-225' have resistance to *M. javanica*, while 'W-8C,' 'Ramcol 70-2' and 'Consum' also have resistance to *M. incognita*. On the other hand, Dr. R. B. Valdez found a number of tomato lines, namely 'Hawaii,' 'Kalohi,' 'Kowalo,' 'Atkinson,' 'Healani,' 'BWN-18,' 'Romulus VFN' and 'RV7,' to be resistant to both *M. incognita* and *M. javanica*. In addition, he also observed that some breeding lines of sponge gourd, *e.g.* lines 81-6, 81-8, 81-17, 81-19 and 81-20, are resistant to *M. incognita*.

Likewise, in Indonesia, Mr. Widjaja Hadisoeganda has screened 88 cultivars of white potato and found some with slight resistance to *M. incognita* race 1; these include 'Nahdo,' 'Nordstern,' 'Cosima,' 'Nicola,' 'Colidan,' 'Exton,' 'Thung 151,' 'Rapan 104,' '6-52-9,' and 'Rapan 181.'

Other countries in the region are also currently doing varietal screening for resistance, and they will report their findings in the near future. I know that in Bangladesh, work on varietal screening of rice cultivars against the rice root-knot nematode has already been initiated. Dr. Choudhury is also currently screening tomato and eggplant varieties against this nematode. Likewise in Pakistan, according to Dr. M. Saeed and Dr. R. Ahmad, some screening work is being done on solanaceous crops. On the other hand, nematologists in Nepal, Burma, Sri Lanka, Malaysia and Fiji Islands are still in the initial planning stages of the study. With IMP's catalyzing influence, I am confident that we will soon have more cultivars resistant to root-knot nematodes in the region.

CROPPING SYSTEMS AND OTHER CULTURAL PRACTICES FOR ROOT-KNOT NEMATODE CONTROL

Since our third planning conference in Jakarta in July, 1981, there has been considerable progress in our research studies on cropping systems and other cultural methods for root-knot nematode control. This is particularly true in the Philippines, Thailand, Japan, Sri Lanka, Taiwan, Indonesia and Bangladesh. In the Philippines, as a result of our studies and observations on various existing cropping systems, we have already recommended and encouraged our farmers to continue or apply the following cropping systems to control root-knot nematodes:

1) rice-tobacco-rice,
2) corn-corn-tomato,
3) rice-cotton-rice,
4) rice-watermelon-rice,
5) tomato-asparagus-radish,
6) rice-corn-tomato,

7) rice-onion-cabbage, and/or
8) garlic-onion-tomato.

These cropping sytems are being studied further so their effects on the build-up of other species of nematodes and insects can be determined. So far, no serious problems have been encountered by the farmers.

In Thailand, Dr. Sontirat observed that two consecutive years of growing a non-host crop are more effective than planting a susceptible crop followed by a non-host crop. Likewise in Sri Lanka, Dr. Nalini Gnanapragasam reported that cropping patterns with peanut, *Crotalaria juncea*, onion and gingelly (*Sesamum indicum*) could greatly reduce populations of *M. incognita* and *M. javanica*. In Fiji Islands, Mr. M. Heinlein and Mr. Vilsoni showed that a rotation cropping of cassava-dalo-ginger was effective against the nematode, while in Bangladesh, Dr. Choudhury observed that three consecutive seasons of peanut or other resistant crops reduced the root-knot nematode population below the economic threshold level so that a subsequent susceptible crop, e.g. brinjal (eggplant) could be grown successfully. Other countries in the region are now giving more emphasis to cropping systems research, and more information should be forthcoming from them later.

BIOLOGICAL CONTROL

Studies on the biological control of root-knot nematodes have not yet been given full attention in the region. However, the recent discovery of the fungus *Paecilomyces lilacinus* by Dr. P. Jatala of Peru as a promising biological control agent against *Meloidogyne* spp. has sparked our interest in testing the effectiveness of the fungus not only against *Meloidogyne* spp. but also on other species like the cyst nematodes on potato and *Radopholus similis* on banana. We have recently demonstrated in the Philippines that *P. lilacinus* can significantly control root-knot nematodes on tomato and other crops and result in yield increases comparable to the effects of some nematicides. Likewise, it has been found effective against *Globodera rostochiensis* on potato and has resulted in a yield increase of 1 to 4 tons per hectare. In addition to the Peruvian isolate, we were able to obtain four Philippine isolates whose effectiveness against root-knot and the potato cyst nematodes is comparable to that of the Peruvian isolate. These isolates can easily be cultured in the Philippines on rice hull, coir dust, corn grit, mashed potato, chopped leaves of water hyacinth (*Eichhornia crassipes*) and leaves of ipil-ipil (*Leucaena leucocephala*). Any of these substrates, particularly the rice hull, are easily available in the rural areas and can be utilized by our farmers at no extra cost. The use of *Paecilomyces lilacinus* is now recommended in the Philippines to control both root-knot and potato cyst nematodes. Our search for more local isolates of *P. lilacinus* is in progress, and we hope to find more efficient isolates for nematode control.

In Pakistan, Dr. M. Saeed reported that the *P. lilacinus* from Dr. Jatala and some *Bacillus* spp., when applied as seed treatment on cucurbits, legumes and other crops, proved promising in laboratory tests. I understand that Dr. Jatala is sending *P. lilacinus* cultures to other IMP regions, and we hope that others too will get good results from such experiments. We hope that the IMP can make a general recommendation regarding the use of *P. lilacinus* against root-knot nematodes. I think that the technology of its application can easily be transferred to the farmers' fields at minimum cost.

CHEMICAL CONTROL

On chemical control studies, the region in general has not done much. This is probably due to a number of reasons. First, considerable studies have already been done by the developed countries that transport and sell nematicides to the Third World markets. What we developing countries can do is to determine the economics and adaptability to our conditions. A second reason for the lack of research in this area is the relatively high cost of the imported nematicides and their restrictions and hazardous effects to humans. Thirdly, research emphasis is given primarily to the development of resistant varieties, cropping systems and biological control. This approach is in line with IMP objectives to explore first the non-costly and less hazardous control methods which could benefit our poor farmers.

However, we are not totally discouraging the use of nematicides. In fact, they should be made available to growers who can afford to use them. In the Philippines, we are continuously evaluating nematicides for use in large plantations of banana, sugar cane, pineapple and other crops. Nematicides such as oxamyl, fenamiphos, aldicarb, ethoprop and carbofuran are locally available. However, only carbofuran has so far been registered for general use while the others are registered for institutional or plantation-company use only.

Other countries in the region have reported results of experiments showing significant effects of nematicides against root-knot nematodes. In Japan, Dr. H. Inagaki showed that *M. incognita* on sweet potato was effectively controlled by soil treatment with D-D, EDB and chloropicrin. In Indonesia, Mr. Hadisoeganda reported that oxamyl and carbofuran were most effective against *Meloidogyne* spp. on tomato. Likewise, in Bangladesh, carbofuran was effective against root-knot nematodes on tomato, while in South Korea, Dr. Choi demonstrated that chemical control of *M. hapla* on peanut with Telone C-17 had significantly increased crop yield. In Malaysia, fenamiphos application on black pepper infected with root-knot nematodes resulted in 55% yield increase. In the Fiji Islands and Nepal, research studies on the chemical control of root-knot nematodes are in progress. Fiji Islands has a research project on chemical dipping of ginger rhizomes and soil treatment in the field, while in Nepal Dr. Neupane is conducting chemical control of root-knot nematodes in eggplant and okra.

Going through the country reports of Region VI, I can generally conclude that considerable achievements have already been made in the attainment of the goals of IMP. Technologies can now be generated from these research findings in cropping systems, resistant varieties, and biological and chemical control. We hope to package these technologies and work out strategies for their effective dissemination to the farmers' fields.

RESEARCH RECOMMENDATIONS

Since we are still far from accomplishing the objectives of IMP, each member-country in the region has eagerly indicated its research needs and has summarized them as follows:

1. More survey studies should be conducted so the distribution and extent of damage of crops caused by different species and races of *Meloidogyne* can be determined;
2. Screening and breeding for crop resistance against *Meloidogyne* spp. should be intensified by the identification of breeders and nematologists and the encouragement to work as a team;
3. More studies should be done on cropping systems by taking into consideration the existing cropping patterns that farmers are already using and by modifying these patterns to make them more effective against root-knot nematodes;
4. Biological control studies through the use of *P. lilacinus* and other microorganisms should be thoroughly explored;
5. Chemical control studies may be pursued further if there are new chemicals available for testing; and
6. Studies should be done on how research findings can be delivered effectively to the farmers' fields.

Finally, I wish to conclude this report by expressing my sincere thanks and appreciation to the delegates of the member countries in Region VI who enthusiastically responded to my call for assistance in making this report. To Dr. Sasser and his staff who worked so hard to bring us together in this workshop, we wish to thank you very much and we assure you of our full support and cooperation in achieving the objectives and goals of IMP.

THE STATUS OF ROOT-KNOT NEMATODES IN THE MIDDLE EAST, REGION VII OF THE INTERNATIONAL *MELOIDOGYNE* PROJECT

I. K. A. Ibrahim

Department of Plant Pathology
College of Agriculture, Alexandria University
Alexandria, Egypt

The Middle East or Mediterranean Region, which is Region VII of the International *Meloidogyne* Project (IMP), extends from the Persian Gulf to the Atlantic Ocean and from the Arab Peninsula and North Africa to South Europe. This region includes about 18 countries: namely, Algeria, Cyprus, Egypt, Greece, Iran, Iraq, Italy, Jordan, Libya, Morocco, Portugal, Saudi Arabia, Spain, Sudan, Syria, Tunisia, Turkey and Yemen which are distributed in the continents Africa, Asia and Europe.

Region VII of the IMP comprises many geographical areas (mediterranean, coastal plains, Sahara desert, oases, mountains, river valleys, etc.) which differ in climate and other characteristics. Most countries of this region have a mediterranean climate with moist, mild winters and dry, hot summers. The temperature rarely drops below 30° F in winter, and summer temperatures, although high, are moderated to some extent by cool winds from the north. The leading agricultural crops are wheat, cotton, barley, corn, rice, peanut, potato, bean, pea, tomato, olive, orange, grape and date. Most of the known vegetable, fruit and ornamental crops are grown also.

Previous studies revealed that the root-knot nematodes (*Meloidogyne* spp.) are of widespread occurrence and great economic importance in many localities of Region VII. These nematodes are causing considerable losses of many agricultural crops, especially in light sandy soil. About eight species of root-knot nematodes have been found in countries of this region. In order of occurrence and importance, these species are *M. javanica, M. incognita, M. arenaria, M. hapla, M. naasi, M. artiellia, M. exigua* and *M. africana.* Generally, *M. javanica* and *M. incognita* are the most widespread species, whereas *M. arenaria* and *M. hapla* are of limited occurrence. On the other hand, the other species are rather rare. In Italy and Maltese islands, *M. naasi* occurs on durum wheat, barley and sweet vetch (Lamberti, 1978, 1979). In Greece, *M. artiellia* and *M. exigua* have been found on wheat and peach roots, respectively (Koliopanos, 1978). Decker et al. (1980) reported that *M. africana* occurs in Sudan.

ROOT-KNOT NEMATODES IN COUNTRIES OF REGION VII

Cyprus

In 1978, Philis indicated that root-knot nematodes are well established in the island of Cyprus and are favored by the warm climate of the mediterranean region (17). These nematodes attack many plants and cause serious yield losses. Tomato, one of the main vegetable crops, is devastated by *Meloidogyne* spp. An extensive survey study revealed that there are two species of root-knot nematodes, *M. javanica* and *M. incognita.* Rapid nematode development occurs from April to October, and nematode activities almost cease from December to February. Also, Philis showed that tomato cvs. 'VFN-8,' '70T82,' and '72T51' were resistant to *M. javanica* and produced good crops in infested soil (17). He reported yield losses of several vegetable crops caused by root-knot nematodes and indicated that losses can easily reach 35-50%, compared to non-infested fields (18).

Egypt

In Egypt, nematological studies have shown that root-knot nematodes are of widespread occurrence and are becoming a real threat to agriculture especially in localities with light sandy soil (7-11). Three species, namely, *M.*

javanica, M. incognita and *M. arenaria* were found in the surveyed agrarian areas. The first two species are widely distributed while *M. arenaria* is of limited occurrence (10). Results of the differential host test showed the existence of two races (Races 1 & 3) in the *M. incognita* group and only one race (Race 1) of *M. arenaria.* Host range investigations revealed that about 115 plant species were recognized as hosts of the root-knot nematodes. The host plants black mustard, casuarina, common balm, date palm, guamia, London rocket, mint, pudding-pipe tree, saw thistle, sedge, sugar beet, sugar cane, Washington palm, wheat and willow may be considered new hosts in Egypt for *Meloidogyne* spp. The species *M. incognita* was isolated from 103 host plants while *M. javanica* and *M. arenaria* were found on 65 and 20 host plants, respectively.

The reactions of certain plant cultivars to *M. javanica, M. incognita* (Race 1 and Race 3) and *M. arenaria* Race 1 were determined in the greenhouse. Results of these studies indicated that tomato cvs. 'Amex-VFN,' 'Monita,' 'Patriot,' 'VFN-Bush' and 'VFN-8;' the rice cvs. 'IR 28,' 'IR 459,' and 'Philip. 24;' the citrus rootstocks sour orange (*Citrus aurantium*), lime (*C. aurantifolia*), Cleopatra mandarine (*C. reticulata*), grapefruit (*C. paradisi*), rough lemon (*C. jambhiri*), trifoliate orange (*C. trifoliate*), and troyer citrange (*Poncirus trifoliate* X *C. sinensis*) were resistant to all nematode populations. The timber trees *Casuarina glauca, Cupressus sempervirens* and *Eucalyptus camaldiolensis* were resistant to the tested nematodes, except *C. glauca* was susceptible to *M. incognita* Race 1.

Oteifa and Elgindi tested the resistance of 79 cultivars of 21 economic crops to certain biotypes of *M. incognita* and *M. javanica* and showed that peanut cvs. 'Giza 4,' 'No. 219' and 'No. 296' and rice cvs. 'Giza 159,' 'Giza 171,' 'Giza 172,' 'Sakha 1' and 'Sakha 2' were resistant or immune to all the tested nematodes (16). The wheat cvs. 'Giza 155,' 'Giza 156,' 'Shenab' and 'Strok-7' and barley cvs. 'Bonus,' 'Giza 119' and 'Giza 121' were relatively resistant to most of the tested nematodes.

Greece

Koliopanos showed that the root-knot nematodes were first described in Greece in 1935, and their galls were observed on roots of tomato, okra, potato, tobacco, lettuce, almond, apricot, dahlia and eggplant (12). Until 1979, more than 85 plants had been reported as hosts for root-knot nematodes and most of these plants are of economic importance (13). The root-knot nematode species *M. incognita, M. javanica, M. arenaria, M. hapla, M. artiellia, M. thamesi* and *M. exigua* have been found in Greece. However, the most prevalent species are *M. incognita, M. javanica* and *M. arenaria,* which have been detected on many host plants. The other species have been reported on few hosts: *M. artiellia* on wheat, *M. hapla* on cyclamen and *M. exigua* on peach (12, 13).

Iran

In Iran, four *Meloidogyne* species (*M. javanica, M. incognita, M. arenaria* and *M. hapla*) are present on a wide range of crop plants and weeds. These nematodes have been isolated from many economic crops growing in different regions of the country. About 63 plants have been found infected with the root-knot nematodes in Iran (1). Abivardi et al. indicated that the introduction of root-knot-nematode-infected tomato seedlings to the new areas may be an important factor for the nematode spread (2). They noticed that the different climatic conditions in the surveyed counties of Fars Province did not seem to affect the presence or absence of *M. incognita* and *M. javanica,* while *M. hapla* was only observed in cold regions.

Iraq

In Iraq, root-knot nematodes are considered serious pests on many crops, especially vegetables, tobacco, sugar beet and date palm trees. The three species *M. javanica, M. incognita* and *M. arenaria* have been reported in Iraq. *M. javanica* is the most widely distributed and causes severe damage to tomato, eggplant, tobacco and date palm trees. Studies on the use of several nematicides to control root-knot nematodes showed that ethylene dibromide and D-D mixture were very effective. DBCP, Nemacur and Furadan ranked next in effectiveness against these nematodes on tomato and tobacco crops (23, 24).

Italy

In this country, five species of root-knot nematodes (*M. incognita, M. javanica, M. arenaria, M. hapla* and *M. naasi*) have been found attacking cultivated plants (6, 14). The most common species is *M. incognita* which is distributed through all the country and causes remarkable crop losses on some fruit trees (peach and olive) and various vegetable, industrial and flower crops. Among the most damaged crops are tomato, eggplant, cantaloupe and watermelon on which yield reductions of up to 50% have been observed. Other plants frequently attacked and severely affected are carnation, cucumber, celery, artichoke, carrot, garden beet, anemone and white clover. *Meloidogyne javanica* is more widespread in the southern regions of Italy where it attacks tomato, eggplant and carnation. The two species *M. arenaria* and *M. hapla* are rather rare. *M. arenaria* has been found on peach, tobacco and carnation, while *M. hapla* infects mainly winter crops such as carrot and potato. *M. naasi* was present in 17% of declining durum wheat fields sampled in southern Italy (14).

Jordan

Root-knot nematodes have been known to occur in the Jordan Valley since the mid 1950's. Vegetable crops and banana trees grown under irrigation exhibited root galling. Almost all of the irrigated lands in the Jordan Valley and the elevated highlands are infested with root-knot nematodes, and the problem is one of national dimensions (3, 4). Two species of *Meloidogyne* occur in Jordan: namely, *M. javanica* and *M. incognita*. Results of the differential host test indicate the presence of one race (Race 1) of *M. incognita*. On vegetable crops, *M. javanica* predominates in the southern Jordan Valley while *M. incognita* predominates the irrigated highlands. It was observed that *M. javanica* has an optimum temperature that is 4-5° C higher than that of *M. incognita*. Results of screening tests under field conditions showed that tomato cvs. 'Mont Favet 75-26 Inra,' 'Mistral,' 'Engeria VFN,' 'Piersol VFN,' 'VFN-8,' 'Calmart' and 'Nemasol' were resistant to highly resistant to *M. javanica*. Tomato cvs. 'Maramand Trianon VFN' and 'Marette' were susceptible to *M. javanica* but exhibited resistance to *M. incognita* (4).

Portugal

In Portugal, root-knot nematodes were first observed in 1882. The results obtained since the establishment of the Middle East Region VII of the IMP show the presence of the root-knot nematode species *M. incognita*, *M. javanica*, *M. arenaria* and *M. hapla* on many host plants (19,20). More than 60 plant species have been recorded as hosts for *Meloidogyne* spp. Important host plants include grape, tomato, potato, pepper, tobacco, peanut, sugar beet, melon, carrot, strawberry, fig, celery, olive, peach, eggplant, broad bean, rose and alfalfa.

Sudan

Root-knot nematodes are considered one of the most important parasites of many agricultural crops in Sudan (26). These nematodes have been found to be widely distributed in river silt along the Nile bank and also in light alluvial soils of western and southern sectors of Sudan, but not under the Gezira heavy clay soil. In these localities, root-knot nematodes can infect many crop and weed plants, but those most severely affected are vegetables and tobacco. Species of root-knot nematodes identified in Sudan are *M. javanica*, *M. incognita*, *M. arenaria*, *M. africana* and *M. megadora* (5, 26). The most economically important species are *M. javanica* and *M. incognita*. Yassin reported that *M. javanica* can inflict heavy damage especially on tomato and tobacco in the northern sector, and *M. incognita* is more prevalent in the south on tomato and eggplant (26). Up to 70-100% of tobacco and tomato plants in pilot farms around Zalingei in western Sudan might be completely devastated by *M. javanica*. In southern Sudan, *M. incognita* is considered a limiting factor in the production of vegetable crops to supply a canning factory in Wau District. The tomato cultivars 'VFN-8,' 'VFN-Bush,' 'Motaba,' 'Rossol,' 'Piersol,' 'Ronita,' and 'Marsol' exhibit resistance to *M. javanica* in Wad Medani.

Turkey

The root-knot nematode disease is one of the most important problems to many agricultural crops in Turkey. The root-knot nematode species: *M. incognita*, *M. javanica*, *M. arenaria*, *M. hapla* and *M. thamesi* have been found on many cultivated plants and weeds. Crops such as tomato, cucumber, bean, squash, tobacco, eggplant, pepper, sugar beet, peach, grape, fig and plum were found to be susceptible to *Meloidogyne* spp. (27, 28).

Yemen

Severe damage by root-knot nematodes has been observed in Yemen on tomato, potato, eggplant, pepper, squash and banana. Results of a survey study indicated that root-knot nematodes were present in about 30% of 110 fields sampled. Two species, *M. incognita* and *M. javanica*, were identified. *M. incognita*, the most common, was found in 60% of the tested populations, followed by M. javanica. Tomato seed beds were often severely infested with both *M. incognita* and *M. javanica*. It was evident that certain adverse factors, in addition to severe *M. incognita* attack, were responsible for heavy losses in banana production (21, 22).

IDENTIFICATION OF ROOT-KNOT NEMATODE ISOLATES

A total of 172 isolates (populations) of root-knot nematodes were collected from galled roots

Table 1. Species and races of root-knot nematodes found in 172 isolates collected from 18 countries of the Middle East and Mediterranean region as determined by the host differential test.

	No. of isolates[a]							
Country	MJ	MI R 1	MI R 2	MI R 3	MA R 1	MA R 2	MH	Total
Algeria	1	1	—	—	—	—	—	2
Cyprus	19	—	—	—	—	—	—	19
Egypt	26	9	—	1	2	—	—	38
Greece	2	3	—	—	—	—	—	5
Iran	11	2	—	—	—	3	—	16
Iraq	3	—	—	—	—	—	—	3
Italy	1	1	—	—	—	—	2	4
Jordan	6	—	—	—	—	—	—	6
Libya	6	1	—	—	—	—	1	8
Malta	1	1	—	—	—	—	—	2
Morocco	12	1	—	—	1	—	1	15
Portugal	2	—	1	1	—	2	1	7
Saudi Arabia	3	2	1	—	—	—	—	6
Spain (Canary Island)	8	6	—	1	—	—	—	15
Sudan	5	4	1	—	—	—	—	10
Syria	1	1	—	—	—	—	—	2
Turkey	4	2	—	—	—	1	—	7
Yemen	4	2	—	1	—	—	—	7
Total	115	36	3	4	3	6	5	172

a MJ: *M. javanica*, MI: *M. incognita*, MA: *M. arenaria*, MH: *M. hapla*, R: race.

of infected host plants grown in 18 countries of the Middle East region. The nematode isolates, in the form of egg masses, were sent to the IMP at North Carolina State University for identification to species and race by morphological characters and the differential host test (25). Data from these studies are presented in Table 1. The results showed that 115 isolates (about 66.8% of the total samples) belong to *M. javanica*, 43 to *M. incognita* (36 race 1, 3 race 2 and 4 race 3), 9 to *M. arenaria* and 5 to *M. hapla*. It is clear that *M. javanica* is prevalent and has been found in samples from all countries of Region VII. *M. incognita* race 1, which occurred in samples from most participating countries, is the next most common. In contrast, the other root-knot nematode species and races were of limited occurrence. The colder climate species *M. hapla* was detected only in samples from Italy, Libya, Morocco and Portugal.

CONCLUSION

The present studies indicate that in Region VII of the IMP *M. javanica* and *M. incognita* race 1 are dominant whereas the other species and races of root-knot nematodes are rare or limited in occurrence.

It is evident that *Meloidogyne* spp. are by far the most important nematode pathogens on many agricultural crops in most countries of the Middle East and Mediterranean Region. Serious crop losses attributed to root-knot nematodes have been reported from many countries of this region. The main factors which contribute to the economic importance of these nematodes are: the great reductions in crop yield which occur when susceptible plant cultivars are grown in heavily infested soil, their wide host range, their persistence in a viable state in soil for long periods of time, their high reproductive capacity (rate) in the presence of the growing host plant, and the absence of generally acceptable chemical control measures. Although all recognized nematicides give satisfactory control of root-knot nematodes, chemical control is a technical and economic problem in most developing countries. On the other hand, the use of crop rotation, resistant varieties and biological methods of control normally offer the cheapest and best means of root-knot nematode control.

Present changes occurring in agricultural production in developing countries, such as the use of intensive cropping systems with one or more crops in the same field, necessitate a more thorough examination of certain aspects of the root-knot nematode problem, i.e., economic thresholds, population dynamics and resistance of plant cultivars. There is an urgent need for more research on the biology, ecology and control of the root-knot nematodes under the natural field conditions of developing countries, in spite of the fact that the root-knot nematode disease has been studied for several years in some of these countries.

Literature Cited

1. Abivardi, C. 1978. Status of root-knot nematodes, *Meloidogyne* spp. in Iran, pp. 26-35 *In* Proc. First IMP Res. Plann. Conf. on Root-knot Nematodes, *Meloidogyne* spp., Region VII, Jan. 29-Feb. 2, 1978. Cairo, Egypt. 85 pp.
2. Abivardi, C., M. Shahcheraghi, and M. Sharafeh. 1979. Recent studies on distribution and control of the root-knot nematodes (*Meloidogyne* spp) in southern Iran, pp. 40-51 *In* Second IMP Res. Plann. Conf. on Root-knot Nematodes, *Meloidogyne* spp., Region VII, Nov. 26-30, 1979. Athens, Greece. 146 pp.
3. Abu-Gharbieh, W. I. 1978. The root-knot nematode, *Meloidogyne*, in Jordan, pp. 16-20 *In* Proc. First IMP Res. Plann. Conf. on Root-knot Nematodes, *Meloidogyne* spp., Region VII, Jan. 29-Feb. 2, 1978. Cairo, Egypt. 85 pp.
4. Abu-Gharbieh, W. I. 1983. The root-knot nematodes, *Meloidogyne* spp., in Jordan, pp. 39-58 *In* Third IMP Res. Plann. Conf. on Root-knot Nematodes, *Meloidogyne* spp., Region VII, Sept. 13-17, 1982. Coimbra, Portugal. 196 pp.
5. Decker, H., A. M. Yassin and E. M. El-Amin. 1980. Plant nematology in Sudan, a review article. Z. Angew. Zool. 67:1-20.
6. Di Vito, M. 1979. Status of research on the biology and control of the root-knot nematodes in Italy, pp. 135-138 *In* Proc. Second IMP Res. Plann. Conf. on Root-knot Nematodes, *Meloidogyne* spp., Region VII, Nov. 26-30, 1979. Athens, Greece. 146 pp.
7. Elgindi, D. M. and F. F. Mousa. 1979. Distribution and identity of *Meloidogyne species and their biotypes in Egypt, pp. 58-61 In* Proc. Second IMP Res. Plann. Conf. on Root-knot Nematodes, *Meloidogyne* spp., Region VII, Nov. 26-30, 1979. Athens, Greece. 146 pp.
8. Ibrahim, I. K. A. 1978. The current status of root-knot nematodes in northern Egypt, pp. 36-37 *In* Proc. First Res. Plann. Conf. on Root-knot Nematodes, *Meloidogyne* spp., Region VII, Jan. 29-Feb. 2,1978. Cairo, Egypt. 85 pp.
9. Ibrahim, I. K. A. 1979. Studies on root-knot nematodes in northern Egypt, pp. 52-57 *In* Proc. Second IMP Res. Plann. Conf. of Root-knot Nematodes, *Meloidogyne* spp., Region VII, Nov. 26-30, 1979. Athens, Greece. 146 pp.
10. Ibrahim, I. K. A. 1983. Species and races of root-knot nematodes and their relationships to economic host plants in northern Egypt, pp. 66-84 *In* Proc. Third IMP Res. Plann. Conf. on Root-knot Nematodes, *Meloidogyne* spp., Region VII, Sept. 13-17, 1982. Coimbra, Portugal. 196 pp.
11. Ibrahim, I. K. A., M. A. Rezk, and M. A. El-Saedy. 1976. Occurrence and distribution of plant parasitic nematodes in northern Egypt. Alexandria J. Agric. Res. 24:93-101.
12. Koliopanos, C. N. 1978. The problem of the root-knot nematodes in Greece, pp. 20-24 *In* Proc. First IMP Res. Plann. Conf. on Root-knot Nematodes, *Meloidogyne* spp., Region VII, Jan. 29-Feb. 2, 1978. Cairo, Egypt. 85 pp.
13. Koliopanos, C. N. 1979. Contribution to the study of the root-knot nematode (*Meloidogyne* spp.) in Greece, pp. 35-39 *In* Proc. Second IMP Res. Plann. Conf. on Root-knot Nematodes, *Meloidogyne* spp., Region VII, Nov. 26-30, 1979. Athens, Greece. 146 pp.
14. Lamberti, F. 1978. Root-knot nematodes in Italy, pp. 74-77 *In* Proc. First IMP Res. Plann. Conf. on Root-knot Nematodes, *Meloidogyne* spp., Region VII, Jan. 29-Feb. 2, 1978. Cairo, Egypt. 85 pp.
15. Lamberti, F. 1979. Economic importance of *Meloidogyne* spp. in subtropical and mediterranean climates, pp. 341-357 *In* Root-knot nematodes (*Meloidogyne* species), systematics, biology and control, ed. F. Lamberti and C. F. Taylor. Academic Press, London. 477 pp.
16. Oteifa, B. A. and D. M. Elgindi. 1983. Relative susceptibility of certain commer-

cially important cultivars to existing biotypes of *M. incognita*, and *M. javanica*, in Nile-Delta, Egypt, pp. 157-169 *In* Proc. Third IMP Res. Plann. Conf. on Root-knot Nematodes, *Meloidogyne* spp., Region VII, Sept. 13-17, 1982. Coimbra, Portugal. 196 pp.

17. Philis, J. 1978. Report on root-knot nematodes in Cyprus, pp. 43-46 *In* Proc. First IMP Res. Plann. Conf. on Root-knot Nematodes, *Meloidogyne* spp., Region VII, Jan. 29-Feb. 2, 1978. Cairo, Egypt. 85 pp.
18. Philis, J. 1979. IMP: Technical research report—Region VII, pp. 67-74 *In* Proc. Second IMP Res. Plann. Conf. on Root-knot Nematodes, *Meloidogyne* spp., Region VII, Nov. 26-30 1979. Athens, Greece. 146 pp.
19. Santos, M. S. N. de A. 1978. Report on the state of knowledge concerning root-knot nematodes in Portugal, pp. 10-15 *In* Proc. First IMP Res. Plann. Conf. on Root-knot Nematodes, *Meloidogyne* spp., Region VII, Jan. 29-Feb. 2, 1978. Cairo, Egypt. 85 pp.
20. Santos, M. S. N. de A. and I. M. de O. Abrantes. 1979. Root-knot nematodes in Portugal, pp. 17-23 *In* Proc. Second IMP Res. Plann. Conf. on Root-knot Nematodes, *Meloidogyne* spp., Region VII, Nov. 26-30, 1979. Athens, Greece. 146 pp.
21. Sikora, R. A. 1978. Status of root-knot nematodes, *Meloidogyne* spp., in the Yemen Arabic Republic, pp. 50-57 *In* Proc. First IMP Res. Plann. Conf. on Root-knot Nematodes, *Meloidogyne* spp., Region VII, Jan. 29-Feb. 2, 1978. Cairo, Egypt. 85 pp.
22. Sikora, R. A. 1979. Observations on *Meloidogyne* with emphasis on disease complexes and the effect of host plant on morphometrics, pp. 93-104 *In* Proc. Second IMP Res. Plann. Conf. on Root-knot Nematodes, *Meloidogyne* spp., Region VII, Nov. 26-30, 1979. Athens, Greece. 146 pp.
23. Stephan, Z. A. 1978. Report on the status of root-knot nematodes, *Meloidogyne* spp., in Iraq, pp. 42-43 *In* Proc. First IMP Res. Plann. Conf. on Root-knot Nematodes, *Meloidogyne* spp., Region VII, Jan. 29-Feb. 2, 1978. Cairo, Egypt. 85 pp.
24. Stephan, Z. A. 1979. The survey program of IMP project in Iraq, pp. 64-66 *In* Proc. Second IMP Res. Plann. Conf. on Root-knot Nematodes, *Meloidogyne* spp., Region VII, Nov. 26-30, 1979. Athens, Greece. 146 pp.
25. Taylor, A. L. and J. N. Sasser. 1978. Biology, identification and control of root-knot nematodes (*Meloidogyne* species). Coop. Publ. Dep. Plant Pathol., North Carolina State Univ., and the U. S. Agency Int. Dev., Raleigh, N.C. 111 pp.
26. Yassin, A. M. 1978. Root-knot nematodes in the Sudan, pp. 46-50 *In* Proc. First IMP Res. Plann. Conf. on Root-knot Nematodes, *Meloidogyne* spp., Region VII, Jan. 29-Feb. 2, 1978. Cairo, Egypt. 85 pp.
27. Yuksel, H. S. 1978. Status of root-knot nematode problems and research in Turkey, pp. 38-42 *In* Proc. First IMP Res. Plann. Conf. on Root-knot Nematodes, *Meloidogyne* spp., Region VII, Jan. 29-Feb. 2, 1978. Cairo, Egypt. 85 pp.
28. Yuksel, H. S. 1979. Some studies on differential host tests for root-knot nematodes in Turkey, pp. 62-63 *In* Proc. Second IMP Res. Plann. Conf. on Root-knot Nematodes, *Meloidogyne* spp., Region VII, Nov. 26-30 1979. Athens, Greece. 146 pp.

NEMATOLOGY IN DEVELOPING COUNTRIES INDIA—IMP REGION VIII

K. Krishnappa
University of Agricultural Sciences
Bangalore, India

A BRIEF HISTORY OF NEMATOLOGY IN INDIA

The study of nematodes in India dates back as early as 1901 when Barber (13) reported the root-knot nematode on tea from the southern part of the country. Some of the important historical developments during the eight decades that followed are enumerated in Table 1. However, a few of the highlights are outlined as follows.

Training Programmes

An International Training Course was held in New Delhi in 1964 in collaboration with The International Agricultural Centre, Wageningen, The Netherlands. Beginning in 1967, the South-East Asia post-graduate nematology course was organized for a few years at New Delhi and Aligarh. In 1973, an all India short term course on "Recent advances in nematological techniques and control" was conducted at the University of Agricultural Sciences, Bangalore. The "Summer Institute in Phytonematology" was held at Allahabad in 1976.

Teaching

Separate divisions of nematology have been established at the Indian Agricultural Research Institute, New Delhi; Orissa University of Agriculture and Technology, Bhubaneshbar; Haryana Agricultural University, Hissar; and Tamilnadu Agricultural University, Coimbatore. A few others have initiated separation from plant pathology, entomology, zoology, botany or biology divisions. This movement has strengthened the post-graduate teaching and research programmes.

Professional Society

The Nematological Society of India was founded in 1969 and has been publishing the *Indian Journal of Nematology* since 1971. The present membership is around 200.

Publications

Four books have been published exclusively on plant nematology in India (23, 86, 113, 116), in addition to chapters on nematology in other books (45).

All India Coordinated Project

The Indian Council of Agricultural Research has established an All India Coordinated Research Project on Nematode Pests of Crops and Their Control at various centres and has been conducting All India Nematology Workshops since 1972.

Some of the economically important plant-parasitic nematodes of India include *Meloidogyne* spp. on vegetables, pulses, cereals etc.; *Radopholus similis* on banana and black pepper; *Pratylenchus coffeae* on coffee; *Heterodera avenae* on wheat and barley; *Globodera* spp. on potato; *Rotylenchulus reniformis* on many crop plants; *Tylenchulus semipenetrans* on citrus; *Anguina tritici* on wheat; and *Hirschmanniella oryzae* on rice. The available estimated crop losses due to some of these parasites are given in Table 2.

STUDY OF *MELOIDOGYNE* IN INDIA

Root knot caused by *Meloidogyne* spp. is one of the most economically important nematode diseases on crop plants in India. An attempt has been made thus far to assess the root-knot problem both qualitatively and quantitatively. This information has been generated from various sources, including personal communications and cooperation among nematologists. Most of the research is the outcome of as many as 21 agricultural universities, research institutes under Indian Council of Agricultural

Table 1. Historical nematological developments in India.

1901	Root-knot nematode on tea in South India
1913-1919	Ufra disease of rice in Bengal, *Ditylenchus angustus*
1917	Tundu disease of wheat, *Anguina tritici*
1922-1934	Root-knot on various crop plants in Madras State
1936	White tip of rice, *Aphelenchoides besseyi* in Central India
1959	Cereal cyst nematode, *Heterodera avenae*
1961	Citrus nematode, *Tylenchulus semipenetrans*
1961	Golden cyst nematode of potato, *Heterodera rostochiensis*
1964	First international nematology course at I.A.R.I., New Delhi
1966	Burrowing nematode on banana, *Radopholus similis*
1967	First Southeast Asia post-graduate nematology course at I.A.R.I., New Delhi
1969	The Nematological Society of India
1970	An Introduction to Plant Nematology (book)
1971	Indian Journal of Nematology, First All India Nematology Symposium, I.A.R.I., New Delhi
1971	Plant Parasitic and Soil Nematodes of India (book)
1972	First All India Nematology Workshop, New Delhi
1973	All India Short Term Course on Recent Advances in Nematological Techniques and Control, Bangalore
1976	Summer Institute in Phyto-nematology, Allahabad
1983	Plant Nematology (book)
1984	Plant Parasitic Nematodes in India (book)

Research, different centres under the All India Coordinated Research Project on nematode pests of crops and their control, ad hoc projects and various other institutions.

It may be that during the summarization of the vast data, some of the findings might not have found a place in this text. However, every effort was made to compile and include as much information as possible in the form of various tables.

Host Range and Distribution

So far ten species have been reported attacking a large number of host plants belonging to different genera (113) (Table 3). Out of the ten species reported only three, *M. incognita*, *M. javanica* and *M. graminicola*, are predominant, accounting for approximately 48.0, 15.0, and 5.5% of total *Meloidogyne* populations, respectively (Table 4). *M. incognita* and *M. javanica* have been found to attack mostly the vegetables (Tables 5, 6, 7, 8), whereas *M. graminicola* is specific to rice (Table 9) and restricted in its distribution to only a few states. *M. arenaria* occurs rarely and attacks only a few selected crops (22) (Table 10). However, the existence and identity of the remaining six reported species [*M. hapla* (55), *M. lucknowica* (110), *M. brevicauda* (77), *M. africana* (18), *M. exigua* (118), and *M. thamesi* (97)] at present is doubtful (Table 11). With the reported variations within the species and the current improved technology for identification, these species should be examined again since some of them were reported many years ago. Similarly, in 22% of the studies, the exact species was never clearly indicated.

Yield Losses

Although root-knot nematodes are serious pests, reports on the estimation of yield losses have been infrequent, vague, and imprecise (15, 31, 46, 85).

Host-Parasite Relationship

The majority of the work has concentrated on the establishment of the pathogenic inoculum level of *Meloidogyne* on various crops (128). Research of this nature has been done on vegetables (tomato, eggplant, okra, cucurbits), pulses (mung bean, cowpea), cereals (rice,

Table 2. Estimated crop losses due to various nematodes in India.

Nematode	Crop	Area or % area Infected	Loss in Rupees	Distribution
Heterodera avenae	Wheat	—	32 million	Rajasthan
	Barley		25 million	Punjab, Haryana Himachal Pradesh, Jammu and Kashmir
Anguina tritici	Wheat	1-80%	70 million	Northern India
Globodera rostochiensis	Potato	2800 hectares		Niligiri Hills
Pratylenchus coffeae	Coffee	1000 hectares	20 million	Southern India
Radopholus similis	Banana	—	–	Southern India
	Pepper			
	Coffee			
	Coconut			
	Arecanut			
Tylenchulus semipenetrans	Citrus	15%		Throughout India
Meloidogyne spp.	Tomato	39.7-46.2%	—	Throughout India
	Okra	28.08-90.90%		
	Brinjal	33.68-44.87%		
	French bean	43.48%		
	Cowpea	28.60%		
Rotylenchulus reniformis	Cotton	14.9%		Throughout India
	Maize	6.0%		
	Finger millet	4.8%		
	Cowpea	13.2%		
	Black gram	8.7%		

Table 3. Species of *Meloïdogyne* with their host genera so far reported from India.

Species	Host genera
M. incognita	232
M. javanica	144
M. graminicola	21
M. arenaria	15
M. hapla	9
M. africana	4
M. exigua	3
M. thamesi	3
M. lucknowica	2
M. brevicauda	1

maize) and fruits (grapes, passion fruit). Water absorption capability of tomato roots and photosynthetic characters in eggplant and bengalgram have been found to be adversely affected by the attack of *M. incognita* (1).

Seed Transmission

Root-knot juveniles in soil clods and plant debris mixed with true seeds may play a vital role in transmission (53).

Biology

Amino acids like α and β-alanine, serine, proline, tryptophan, etc., have been found to affect the hatching of juveniles considerably (36). Even some of the fungicides were effective in reducing the emergence of *M. incognita* juveniles (124). Ultrasonics and electrolysis have also been found to have lethal effects (77). Hydrogen-ion concentrations of 7.1 to 7.25 were favourable for the development of the nematode

Table 4. Studies made with different species of *Meloidogyne* on various crop plants in India.

	Crops								
Species	**General**	**Vegetables**	**Pulses**	**Rice**	**Commercial**	**Plantation & Medicinal**	**Cereals & millets**	**Fibre**	**Others**	**Total**
M. incognita	98	187	40	—	6	12	12	—	16	371
M. javanica	46	43	12	—	5	—	—	5	8	119
M. incognita & M. javanica	34	12	3	—	6	—	—	—	1	56
M. graminicola	—	—	—	43	—	—	—	—	—	43
M. arenaria	—	5	2	—	—	1	—	—	1	9
M. hapla	—	1	—	—	—	—	—	—	—	1
M. brevicauda	—	—	—	—	—	1	—	—	—	1
M. lucknowica	—	2	—	—	—	—	—	—	—	2
M. africana	1	—	—	—	—	—	—	—	—	1
M. exigua	—	—	—	—	—	1	—	—	—	1
M. thamesi	—	1	—	—	—	1	—	—	—	2
M. spp[a]	71	59	—	—	24	7	—	—	9	170
Total	250	310	57	43	41	23	12	5	35	776

[a] *M.* spp. includes the studies made on more than one species at a time and also where the exact species was not mentioned.

Table 5. Numbers of studies made on various aspects of *M. incognita* on different crop plants in India.

Aspects studied	Crops							
	General	Vegetables	Pulses	Plantation & medicinal	Cereals	Commercial	Others	Total
General	1	—	—	—	—	—	—	1
Methodology	1	—	—	—	—	—	—	1
Host range & distribution	49	—	—	—	—	—	—	49
Yield losses	—	3	—	—	—	—	—	3
Host-parasite relationship	—	10	3	1	2	1	6	23
Biology	5	3	—	—	—	—	—	8
Cytology	1	—	—	—	—	—	—	1
Ecology	3	6	—	—	—	—	—	9
Physiological races	3	—	—	—	—	—	—	3
Interaction	—	29	9	1	5	—	1	45
Histopathology	—	5	—	—	—	—	—	5
Histochemistry	—	4	1	—	—	—	—	5
Biochemistry	—	18	3	—	—	—	—	21
Control								
General	5	—	—	—	—	—	—	5
Physical	—	2	—	—	—	1	—	3
Cultural	—	2	1	—	—	—	—	3
Crop rotation	7	—	—	—	—	—	—	7
Biological	13	4	—	—	—	—	—	17
Organic amendments	3	13	—	—	2	—	—	18
Inorganic fertilizers	—	5	—	—	—	1	1	7
Chemical								
In vitro	3	—	—	—	—	—	—	3
Seed treatment	—	4	1	—	—	—	—	5
Plant and soil application	4	49	10	6	—	2	3	74
Host resistance	—	28	12	4	2	1	5	52
Integrated	—	2	—	—	1	—	—	3
Total	98	187	40	12	12	6	16	371

Table 6. Numbers of studies made on various aspects of *M. javanica* on different crop plants in India.

	Crops						
Aspects studied	**General**	**Vegetables**	**Pulses**	**Commercial**	**Fibre**	**Others**	**Total**
Host range and distribution	29	—	—	—	—	—	29
Yield losses	—	2	—	—	—	—	2
Host-parasite relationship	—	2	2	—	—	—	4
Biology	1	1	—	—	—	—	2
Interaction	—	3	3	4	—	—	10
Histopathology	1	3	—	—	1	1	6
Histochemistry	—	1	—	—	1	—	2
Control							
Cultural	2	—	—	—	—	—	2
Biological	—	2	—	—	—	5	7
Organic amendments	9	7	5	—	—	—	21
Inorganic fertilizers	—	2	—	—	—	—	2
Chemical							
In vitro	2	—	—	—	—	—	—
Seed treatment	—	—	1	—	—	—	1
Plant and soil application	1	11	—	—	2	—	14
Host resistance	1	9	1	1	1	2	15
Total	46	43	12	5	5	8	119

Table 7. Numbers of studies made on various aspects of *M. incognita* and *M. javanica* together on different crop plants in India.

Aspects studied	Crops					
	General	Vegetables	Commercial	Pulses	Others	Total
General	1	—	—	—	—	1
Host range and distribution	29	—	—	—	—	29
Biology	1	1	—	—	—	2
Interaction	—	—	—	—	1	1
Control						
Crop rotation	1	—	—	—	—	1
Biological	2	—	—	—	—	2
Chemical	—	5	4	1	—	10
Host resistance	—	6	2	2	—	10
Total	34	12	6	3	1	56

(9). Studies on the effect of seasons on the duration of the life cycle showed that life cycles are longest during winter and shortest in summer (48). Cobalt^{60} radiation has been shown to reduce the growth and development of *M. graminicola* in rice roots (72). Rao & Israel (81) reported that *M. graminicola* required about 26 to 51 days to complete the life cycle at different periods of the year and that penetration of infective juveniles occurred within five hours of inoculation.

Cytology

Three distinct populations of *M. incognita* with chromosome numbers 2n = 27, 2n = 36 and 2n = 44 (45) have been identified, and differences in chromosome numbers of populations have been attributed to polyploid origin (78). Krishnappa (44) compared the three physiological races cytologically and found that there was no correlation regardless of their geographical origin and the original host on which they were collected.

Ecology

Several ecological factors like soil aeration, temperature, manuring, vegetation, and pH have been shown to affect population and disease development (69). Neutral pH and soil moisture at field capacity retard disease to a great extent (35). Temperature also affects the vertical distribution of root-knot populations in the field. In potato fields, it has been observed that the nematodes congregate more in the upper layers of the soil during summer and deeper during winter months (63). Nematodes are more prevalent in sandy soils than in clay soils and are more evenly distributed in cultivated soils with more intensity of root galling as opposed to virgin soils (57). Rao & Israel (80) have reported the existence of a linear relationship between the activity of nematodes and percentage of sand content in soil. They also have found that application of nitrogen alone or in conjunction with phosphorus increases the degree of *M. graminicola* infestation in the standing crop (80). Soil temperatures of 18.0°C to 23.5°C have been found to be congenial for root-knot formation and temperatures from 18.5°C to 20.5°C congenial for eggmass production (79).

Physiological Races

With varied agroclimatic conditions, different cropping patterns and non-specific agricultural practices in India, there is a high possibility of variation within the species. However, only a few reports have indicated the possible occurrence of biotypes within *M. incognita* (75, 87, 88) and *M. graminicola* (82). Three physiological races of *M. incognita* have been identified based on differential host reaction (44). Race 1 has been found to occur in all the agroclimatic zones irrespective of the original host. Race 2 is present only in Karnataka zone and race 3 in Karnataka and Tamil Nadu (Tables 12 & 13). One of the possible reasons for the occurrence of all

Table 8. Numbers of studies made on various aspects of *Meloidogyne* spp.[a] on different crop plants in India.

Aspects studied	Crops: General	Vegetables	Commercial	Plantation and medicinal	Others	Total
General	9	5	1	3	1	19
Methodology	3	—	—	—	—	3
Host range and distribution	44	—	—	—	—	44
Yield losses	—	1	1	—	—	2
Host parasite relationship	1	5	—	—	1	7
Seed transmission	—	—	—	—	1	1
Biology	—	1	—	—	—	1
Ecology	2	—	2	—	—	4
Interaction	1	5	—	—	1	7
Histopathology	—	2	—	1	—	3
Histochemistry	—	—	—	—	1	1
Biochemistry	—	2	—	—	—	2
Control						
General	2	1	2	—	—	5
Physical	—	—	—	—	1	1
Cultural	1	2	3	—	—	6
Crop rotation	1	—	3	—	—	4
Biological	1	3	—	—	1	5
Organic amendments	—	7	1	—	2	10
Chemical	6	11	9	3	—	29
Host resistance	—	14	2	—	—	16
Total	71	59	24	7	9	170

[a]*M.* spp. includes the studies made on more than one species at a time and also where the exact species was not mentioned.

three of the physiological races in Karnataka region may be due to the testing of large numbers of populations from many localities and crop plants. Once large numbers of root-knot populations in other regions are tested, a truer picture may exist with regard to the effect of specific ecological factors on the occurrence and frequency of physiolocical races. Such insights may lead to improved management practices especially through crop rotation and screening for host resistance.

Interaction

Association of root-knot nematodes with other microorganisms is quite prevalent. It has been found that wherever there is such an association, the effect is synergistic, resulting in more damage to the host. Several fungi (*Rhizoctonia*, *Sclerotium*, *Fusarium*, *Phomopsis*, *Aspergillus*, *Cortecium*, *Curvularia*, *Pythium* etc.) are known for such associations (16, 40, 76). *R. bataticola* and *M. incognita*, when inoculated simultaneously in soil, reduce the germination of seeds in okra and eggplant (16). A similar observation has been made by Shukla & Swarup (102) concerning tomato. The presence of *Phomopsis vexans* has reduced the development of root-knot index on eggplant and also the nematode population (40).

Similarly a few viruses (maize, bottle-gourd,

Table 9. Numbers of studies made on various aspects of *M. graminicola* on rice in India.

Aspects studied	Number of Publications
General	1
Yield losses	2
Host range and distribution	2
Host parasite relationship	3
Biology	3
Ecology	5
Physiological races	1
Interaction	2
Control	
Physical	1
Cultural	1
Organic amendments	1
Chemical	
Seed treatment	2
Plant and soil application	8
Host resistance	7
Histopathology	3
Biochemistry	1
Total	43

eggplant and tobacco mosaic, as well as tomato leaf curl) and mycoplasma-like organisms (little leaf of eggplant) have been found to interact with *Meloidogyne* spp. (26, 50, 126). When mosaic virus has been associated with root-knot, the incubation period for mosaic symptoms has been shorter and the reproduction of the nematode has been greater than when either has been inoculated alone (41). The plant growth and virus multiplication has not been affected but the nematode population has been suppressed in tobacco when *M. incognita* was inoculated preceeded by the virus (26). In case of eggplant and bottle gourd, however, root-knot development has been enhanced by the presence of mosaic virus (51).

Pani & Das (64) have reported the association of *Meloidogyne* spp. with bacterial wilt of tomato.

Association of *M. javanica* with the stem borer (*Stomopteryx nertaria*) has been found to cause more damage than either separately on *Phaseolus aureus* (73).

Interestingly the cumulative effect of different plant-parasitic nematodes has been observed in the case of *Meloidogyne* spp. with *Tylenchorhynchus* sp., *Radopholus similis*, *Rotylenchus* sp., *Hoplolaimus* sp., and *Heterodera oryzicola*, and also within the species of *Meloidogyne* (37, 83, 100). Population build-up of *Hoplolaimus indicus* around tomato roots has been inhibited by *M. incognita*, perhaps due to competition between the two nematode species or possibly the physico-chemical changes brought out by root-knot infection (28). In the case of maize, the presence of lower numbers of *T. vulgaris* and *M. incognita*

Table 10. Studies made on various aspects of *M. arenaria* on different crop plants in India.

Aspect Studied	Crops					
	Chillies	Peanut	Brinjal	Betelvine	Bolsom	Total
Host range and distribution	—	—	1	1	—	2
Host parasite relationship	1	2	—	—	—	3
Control						
Chemical	1	—	—	—	—	1
Host resistance	2	—	—	—	—	2
Histopathology	—	—	—	—	1	1
Total	4	2	1	1	1	9

Table 11. Numbers of studies made on other minor species of *Meloidogyne* in India.

Aspects studied	**Species**						
	M. hapla	***M. lucknowica***	***M. brevicauda***	***M. africana***	***M. exigua***	***M. thamesi***	**Total**
Morphology and taxonomy	—	1 (sponge gourd)	—	—	—	—	1
Host range and distribution	1	—	—	1	1 (coffee)	2	5
Histopathology	—	1 (eggplant)	—	—	—	—	1
Control	—	—	1 (tea)	—	—	—	1
Total	1	2	1	1	1	2	8

Table 12. Occurrence of races of *Meloidogyne incognita* in certain regions of India.

Region and State	No. of populations	Identification by differential host reaction		
		Race 1	Race 2	Race 3
Northern zone (Haryana and Punjab)	13	13	—	—
Northeastern zone Assam and West Bengal)	7	7	—	—
Eastern coastal zone (Andhra Pradesh and Orissa)	13	13	—	—
Deep southern zone (Kerala and Tamilnadu)	10	9	—	1
Karnataka zone (Karnataka)	84	44	31	9
Total	127	86	31	10

Table 13. Frequency of occurrence of races of *M. incognita* on different crops in India (based on reaction of differential hosts).

Original host	Numbers of populations of *M. incognita*		
	Race 1	Race 2	Race 3
Brinjal (eggplant)	14	4	1
Tomato	13	3	2
Bhendi	10	1	1
Green chillies	5	3	—
Root crops	8	4	—
Other vegetables	5	4	1
Cole crops	5	2	—
Leafy vegetables	2	2	2
Spices and commercials	8	2	1
Pulses	5	2	1
Fruit crops	5	2	1
Fibre and miscellaneous	6	2	—
Total	86	31	10

had a synergistic effect on cyst production by *Heterodera zeae*, whereas gall formation by *M. incognita* has been adversely affected by presence of both the other nematode species. The final population of *T. vulgaris* has been found to be reduced by low levels of *H. zeae* and by both high (1000 juveniles) and low (500 juveniles) levels of *M. incognita* (33).

Nodule formation by legumes has also been affected by the interaction of *M. incognita* and *Rhizobium* sp. (106).

Histopathology

Root-knot infection in different crops has resulted in various histological changes like formation of multinucleate giant cells with thickened cell walls (112), poor development of xylem with disrupted flow of water, broken epidermis, cork and cortical cavities (108). Haustorical appendages have been observed in the giant cells of tomato, brinjal and chillies infected with *M. incognita* (14). In rice also, the giant cells have been seen in the vascular region accompanied by distorted xylem and a large number of syncytia and associated with egglaying females (89). The development of the first three races of *Meloidogyne* has been recently compared by histopathological studies (44).

Histochemistry

The chemistry of plant cells has been shown to be altered by root-knot infection. Increased carbohydrate content in cowpea infected with *M. incognita* (98), increased proteins and ascorbic acid levels in tomato (123), increased amino acids in sugar beet (84), and pollen sterility in eggplant (119) have been noticed. Free amino acids and indole compounds have also been seen in okra and jute infected with *M. javanica*. Infection with *Meloidogyne* spp. has caused localization of total lipids in papaya.

Biochemistry

Various compounds are responsible for imparting resistance to root-knot infection. The concentration of various phenols and orthodihydroxy phenols has been found to be highest in a resistant tomato variety (12). The IAA-oxidase system and peroxidase have been more active in the resistant tomato varieties (24). In tomato, the quantity of DNA and RNA has been found to be higher in infected plants than in healthy plants (52). Mote & Dasgupta (54) have reported the significance of phenylalanine ammonia lyase in the resistance response of tomato to *M. incognita*. Similarly the total proteins also increased in the *M. incognita*-infected carrot (8).

In the rice cv. 'TKM-6,' the multiplication of *M. graminicola* was inhibited indicating the presence of a biochemical defense mechanism against the nematode (94).

Control

Attempts to control this nematode began on economic crop plants as early as 1933 (10).

PHYSICAL METHODS Incubation of potato tubers at 45°C for 48 hr has been shown to kill about 98.9% of *M. incognita* without affecting tuber viability (62). Krishnamurty & Elias (42) showed that drying of *M. incognita*-infested soil reduces the intensity of primary galls in tobacco. Physiotherapy of rice plants against *M. graminicola* has been recommended by Prasad & Rao (70).

CULTURAL Various cultural practices like changes in dates of planting (25), water stress (56), rabbing tobacco seed bed with paddy husk (43), removal of the knotted root portion in tobacco (67), deep ploughing and fallowing have considerably reduced the nematode population.

CROP ROTATION Successful results have been achieved by rotation with non- and poor hosts in reducing the root-knot nematode population levels to a considerable extent (4, 68, 90, 99, 120).

BIOLOGICAL Use of weed plants (59), plant extracts (58, 61), antagonistic plants like *Tagetes* (3) and garlic (7), root exudates (2), fungal filtrates (17, 93) and bacteria (74) have given some encouraging results.

ORGANIC AMENDMENTS One of the important aspects of root-knot nematode research in India which has attracted several scientists is the use of organic amendments in nematode management. A considerable amount of work has been carried out at Pantnagar and Aligarh. Out of several amendments tested, oil cakes like neem, groundnut, etc. (39, 109) have given successful results. Others like sawdust (115), green leaves (32), decaffinated tea waste and water hyacinth (91), prawn exoskeleton (21), rice-hull ash (95) and several other indigenous materials have given encouraging results.

INORGANIC FERTILIZERS Effects of different levels of certain inorganic elements on root-knot incidence have been studied (65). Results from India indicate that the addition of nitrogen increases galling by *M. incognita* in tobacco plants (122). However, increased levels of potash have significantly reduced the number of galls by *M.*

javanica in tomato (27) and *M. incognita* in castor (30).

CHEMICAL/IN VITRO STUDIES Some chemicals like fatty acids (114), fungicides (103), antihelminth antibiotics (5), piperazine citrate (29) and homoeopathic drugs (101) have been tested for their effect on hatching and mortality in the laboratory.

CHEMICAL SEED TREATMENT Considerable reduction in root-knot incidence has been obtained by treatment of gram seeds with carbofuran and fenamiphos (34); okra (117) and sugar beet (125) with carbofuran and aldicarb sulfone; tomato with oil cakes (111); and rice with oxamyl, phorate and fensulfothion (71), and ceresan dry (92).

PLANT AND SOIL APPLICATION OF CHEMICALS Several chemicals including halogenated hydrocarbons (DBCP, DD, EDB, MBr), organophosphates (phorate, dichlofenthion, ethoprop, thionazin, parathion, chlorophonium, fenamiphos), dithiocarbamates (aldicarb, carbofuran, methomyl, metam-sodium), amino acids, antihelminth antibiotics and several others have been tested for their efficacy at different dosages and methods of application (66, 127). Aldicarb at 1 kg/ha and carbofuran at 2 kg/ha decreased the root-knot index and larval population and significantly increased grain yield of moth bean, *Vigna acontifolia* (127). Fenamiphos and aldicarb gave good control of *M. incognita* and higher yields in french bean (105). Similarly DBCP, DD and dichlofenthion were very effective in controlling *M. incognita* and increasing yield of green gram (107). Aldicarb at 0.4 g/m^2 was found to be highly effective in protecting tomato seedlings against root-knot nematodes (6). However, most of these chemicals are neither available nor practical in India, and only carbofuran and aldicarb seem to be widely used.

HOST RESISTANCE A number of attempts have been made to identify the sources of resistance among both the cultivated and weed hosts for several crop plants. However, a wide range of reactions was encountered under different populations and sets of conditions in cucurbits (38), tomato (19, 49 104), brinjal (60), cowpea (20), rice (Y.S. Rao, personal communication) and several other crops (86), some of which are indicated in Table 14. Although several of these cultivars have been found to be resistant to at least one *Meloidogyne* species, they have to be rigorously screened against several isolates of each species under field conditions before being recommended (44).

INTEGRATED CONTROL Attempts have been made to integrate methods of control for effective management of root-knot nematodes attacking vegetables (121), sorghum (96), and brinjal (47).

FUTURE NEEDS

More emphasis in the future needs to be placed on the following apsects:

1. Precise estimation of crop losses at different population levels under field conditions for correlation with crop health and determination of thresholds for important crops.
2. Collection and characterization of as many *Meloidogyne* populations as possible from different host plants in each state for identification of species and races and assessment of frequency and distribution.
3. Identification of sources of resistance among important food and fibre crops to *Meloidogyne* species/races common in various agroclimatic zones.
4. Better utilization of indigenous materials like oil cakes and other economical amendments.
5. Development of integrated nematode-management models for various crops.
6. Strengthening of graduate teaching and research programmes by establishment of separate departments in all the institutions.
7. Creation of increased awareness among farmers concerning nematode problems through extension services, literature, audiovisuals etc., in local languages, demonstrations in farmers' fields, and farmers' training programmes; strengthening of the various extension programmes in nematology under Lab-to-Land and Training-and-Visit systems.

Acknowledgements

The cooperation and help of several nematologists working in various institutions in India in providing their findings is greatly appreciated. Except through their encouragement, it would have not been possible to compile this vast information.

Table 14. Cultivars/lines which have shown various kinds of resistant reactions to different species of *Meloidogyne*.

Crop	Species of *Meloidogyne*	Resistant cultivar/line
Tomato	*M. incognita*	SL-120, Pelican, Nematex, NMR-1, VFN Bush, New Yorker, Healani, Kalohi, Small Fry, West Virginia, LA 121, EC 118270, EC 118272, EC 118276, EC 118277, Hybrid Karataka, Hawaii 7746, Hawaii 7747, Anahu.
	M. javanica	Healani, Kalohi, Anahu, Hawaii 7526, Atkinson, Nematex, Y-207, Y-240.
	M. incognita *M. javanica* *M. arenaria*	Kalohi, 7544, 7540, Healani, LP_1, LP_2, 66N1, 569N-10, Manalucie, Atkinson.
	Meloidogyne spp.	Nematex, VFN-8, 65N215-1, 65N255-1, SL-120.
Potato	*M. incognita*	HC 294, A-2708, Great Scot, VTn^2.
Brinjal	*M. incognita*	Vijaya, Black Beauty, T_3, S_1, S_4, T_2(UP), Banaras Giant.
	Meloidogyne spp.	S96-2, S419, Pol Baigan.
Cucurbits	*M. incognita*	Bikaner, Jaipuri.
Sugar beet	*M. incognita*	Maribo-N, Maribo-autopoly.
Capsicum	*M. incognita*	G_4,Mirch-1, Red Long, K_2, NP-46A, CA(P)63.
Cowpea	*M. incognita*	C-152, 821-1-B, IC-9641, 23/3/1, P-1327, A(338-R), IC-2902, EC-107163, 29-5, 761-1122, IC-2441, RS9.
	M. javanica	HFC-12, HFC-311, C-28, RS9.
Green pea	*M. incognita* *M. javanica*	T44, T_2.
Guar	*M. incognita* *M. javanica*	B-19-1-55, RGC-114.
Tobacco	*Meloidogyne* spp.	Speight G-28.
Rice	*M. graminicola*	Hamas, IR-5-47-2, Manoharsali, TKM-6, Garem, Basanti.

Literature Cited

1. Alam, M. M., N. Hasan, and S. K. Saxena. 1974. Effect of root-knot nematode, *Meloidogyne incognita* on water absorption capacity of tomato roots. Indian J. Nematol. 4:244-246.
2. Alam, M. M., M. R. Kirmani, and A. M. Khan. 1976. Studies on the role of root exudates for nematode control by the interculture of mustard and rocket-salad wheat and barley. Fert. Technol. 13:289-292.
3. Alam, M. M., S. K. Saxena, and A. M. Khan. 1977. Influence of interculture of marigold and margosa with some vegetable crops on plant growth and

nematode population. Acta Bot. Indica 5:33-39.
4. Alam, M. M., S. K. Saxena, and A. M. Khan. 1981. Soil population of plant parasitic nematodes under various cropping sequences. Nematol. Mediterr. 9:49-56.
5. Anonymous. 1972. Eighteenth Annual Report. Univ. Agric. Sci., Bangalore, India. 191 pp.
6. Anonymous. 1981. Biennial Report of the All India Co-ordinated Research Project on Nematode Pests of Crops and Their Control. G.A.U., Anand.
7. Arya, M. and B. Tiagi. 1978. Biological control of carrot root-knots incited by *Meloidogyne incognita* (Kofoid & White) Chitw. J. Indian Bot. Soc. 57:102. (Abstr.).
8. Arya, M. and B. Tiagi. 1982. Changes in total proteins in three carrot cultivars infested with *Meloidogyne incognita*. Indian J. Nematol. 12:398-399.
9. Atwal, A. S. and A. Mangar. 1971. Biology, host range and abundance of *Meloidogyne incognita* (Kofoid & White, 1919). J. Zool. Soc., India 23:113-118.
10. Ayyar, R. K. 1933. Some experiments on the control of the root-gall nematode *Heterodera radicicola* Greeff) in S. India. Madras Agric. J. 21:97-107.
11. Azmi, M. I., M. S. Jairajpuri, and M. Sajid. 1976. Studies on nematode behaviour, lethal effects of ultrasonics and temperature on the juveniles of *Meloidogyne incognita*. Indian J. Nematol. 6:183-184.
12. Bajaj, K. L. and R. Mahajan. 1977. Phenolic constituents of tomato resistant and susceptible to *Meloidogyne incognita* (Kofoid & White) Chitwood. Nematol. Mediterr. 5:329-333.
13. Barber, C. A. 1901. Tea eelworm diseases in South India. Bull. Gov. Madras, Dep. Land Rec. Agric. 45:227-234.
14. Bhatnagar, A., P. C. Trivedi, and B. Tiagi. 1977. Giant cells induced by nematode, *Meloidogyne incognita* on *Capsicum annuum* and *Abelmoschus esculentus*. Indian J. Mycol. & Plant Pathol. 7:95-96.
15. Biswas, H. and Y. S. Rao. 1970. Incidence of *Meloidogyne graminicola* and its influence on yield of rice. Oryza 8:101-102.
16. Chhabra, H. K. and J. K. Sharma. 1981. Combined effect of *Meloidogyne incognita* and *Rhizoctonia bataticola* on preemergence damping off of okra and brinjal. Sci. & Cult. 47:256-257.
17. Chhabra, H. K., I. Singh, and A. S. Sidhu. 1979. Effect of culture filtrate of *Rhizoctonia solani* on egg hatch and larval mortality of *Meloidogyne incognita*. Indian J. Mycol. & Plant Pathol. 9:111-112.
18. Chitwood, B. G. and M. C. Toung. 1960. *Meloidogyne* from Taiwan and New Delhi. Paper presented at the 52nd Annu. meeting of the Am. Phytopathol. Soc., Aug. 1960. Phytopathology 50:631-632. (Abstr.).
19. Dahiya, R. S. and D. S. Bhatti. 1980. Evaluation of varieties of some vegetable crops for resistance to root-knot nematodes. Haryana J. Hortic. Sci. 9:95-97.
20. Darekar, K. S. and B. D. Patel. 1981. Reaction of some cowpea varieties to root-knot nematode. Indian J. Nematol. 11:83-84.
21. Das, S. N. and S. N. Mahapatra. 1983. Effectiveness of prawn exoskeletons as soil amendments for controlling root-knot nematodes in tomato. Indian Sci. Congr. Assoc. Mysore.
22. Dhruj, I. and M. U. Vaishnav. 1981. Pathogenicity of root-knot nematode on groundnut. Indian J. Nematol. 11:217-218.
23. Edward, J. C. and S. L. Misra. 1970. An introduction to plant nematology. Central Book Depot, Allahabad, India. 80 pp.
24. Ganguly, S. and D. R. Dasgupta. 1981. Protein patterns in resistant and susceptible tomato varieties inoculated with the root-knot nematode, *Meloidogyne incognita*. Indian J. Nematol. 11:180-188.
25. Gaur, H. S., S. D. Misra, and V. C. Sood. 1979. Effect of date of sowing on the relation between population density of the root-knot nematode, *Meloidogyne incognita* and the growth of three varieties of chickpea, *Cicer arientinum*. Indian J. Nematol. 9:152-159.
26. Goswami, B. K. and V. V. Chenulu. 1974. Interaction of root-knot nematode *Meloidogyne incognita* and tobacco mosaic virus in tomato. Indian J. Nematol. 4:69-80.
27. Gupta, D. C. and M. C. Mukhopadhyaya. 1971. Effect of N, P and K on the root-knot nematode, *Meloidogyne javanica* (Treub) Chitwood. Sci. and Cult. 37:246-247.
28. Hasan, N. and M. M. Alam. 1975. Studies on the influence of *Meloidogyne incognita* on the population of *Hoplolaimus indicus* around tomato roots. Geobios (Jodhpur) 2:121-122.
29. Husain, S. I. and A. Masood. 1974. Studies on the effect of piperazine citrate on nematode mortality and larval hatching of

Meloidogyne incognita. Indian J. Nematol. 4:102-104.

30. Ismail, W. and S. K. Saxena. 1980. Potassium nutrition of castor in relation to infestation by *Meloidogyne incognita* and *Rotylenchulus reniformis*. Indian J. Nematol. 10:1-8.
31. Jain, R. K. and D. S. Bhatti. 1978. Effect of *Meloidogyne javanica* on the yield of brinjal and tomato. Nematol. Mediterr. 6:243-246.
32. Kali Ram and D. C. Gupta. 1982. Efficacy of plant leaves, nematicides and fertilizers alone and in combination against *Meloidogyne javanica* infecting chickpea (*Cicer arientinum* L.). Indian J. Nematol. 12:221-225.
33. Kaul, R. K. and C. L. Sethi. 1982. Interaction amongst *Heterodera zeae*, *Meloidogyne incognita* and *Tylenchorhynchus vulgaris* on *Zea mays*. Indian J. Nematol. 12:91-98.
34. Kaushik, H. D. and H. K. Bajaj. 1981. Control of root-knot nematode, *Meloidogyne javanica* infesting mung bean and gram by seed treatment. Haryana Agric. Univ. J. Res. XI:106-108.
35. Khaleel, S. and H. M. Shah. 1979. Effect of soil moisture and pH on root-knot disease of tobacco. Tob. Res. 5:129-134.
36. Khan, A. M. and S.. K. Saxena. 1968. Factors influencing larval hatching in the root-knot nematode, *Meloidogyne incognita* (Kofoid and White) Chitwood. Indian Phytopathol. 21:62-65.
37. Khan, A. M., S. K. Saxena, and S. H. Anjum. 1981. Self interaction of *Meloidogyne incognita* on tomato and *Tylenchorhynchus brassicae* on cauliflower. Acta Bot. Indica 9:60-63.
38. Khan, A. M., S. K. Saxena, and Z. A. Siddiqi. 1971. Reaction of some cultivated and wild species of *Lycopersicon* and cucurbits to *Meloidogyne incognita* (Kofoid and White) Chitwood. Indian Phytopathol. 24:197-199.
39. Khan, M. W., A. M. Khan, and S. K. Saxena. 1973. Influence of certain oilcake amendments on nematodes and fungi in tomato yield. Acta Bot. Indica 1:49-54.
40. Khan, R. M., S. K. Saxena, M. W. Khan, and A. M. Khan. 1980. Interaction of *Meloidogyne incognita* and *Phomopsis vexans*. Indian J. Nematol. 10:242-244.
41. Khurana, S. M. P., B. K. Goswami, and S. P. Raychoudhuri. 1970. Interaction of maize mosaic with root-knot nematode, *Meloidogyne incognita* (Kofoid and White) Chitwood in maize (*Zea mays* L.) Phytopathol. Z. 69:267-272.
42. Krishnamurty, G. V. G. and N. A. Elias. 1969. Effect of drying root-knot infested soil on survival of *Meloidogyne incognita*. Indian Phytopathol. 22:262-264.
43. Krishnamurty, G. V. G., K. Nagarajan, and Ramji Lal. 1979. Multiple advantages of rabbing with paddy husk on tobacco seedbeds. Tob. News 2(7&8):17-20.
44. Krishnappa, K. 1982. Studies on races of root-knot nematode, *Meloidogyne incognita* in India with particular reference to Karnataka. Ph.D. thesis. Univ. of Agric. Sci., Bangalore, India. 197 pp.
45. Krishnappa, K. and J. C. Edward. 1983. Nematodes, pp. 70-97 *In* Applied soil biology and ecology, ed. G. K. Veeresh and D. Rajagopal. Sharada Publ. Bangalore, India. 407 pp.
46. Krishnappa, K., K. G. H. Setty, and K. S. K. Prasad. 1981. Crop loss assessment in brinjal due to root-knot nematode *Meloidogyne incognita*. Indian J. Nematol. 11:95 (Abstr.).
47. Kuriyan, J. K. an M. S. Sheela. 1981. Integrated control of *Meloidogyne incognita* on brinjal. Indian J. Nematol. 11:129.
48. Lall, B. S., K. K. Don, and R. B. S. Birut. 1965. Studies on the bionomics of root-knot nematode, *Meloidogyne javanica* (Treub) Chitwood (Nematoda: Heteroderidae). Indian J. Entomol. 27:46-50.
49. Mahajan, R., S. Ahuja, J. S. Kanwar, and K. S. Nandpuri. 1979. Resistance in tomato to root-knot nematode, *Meloidogyne incognita*. Veg. Sci. 6:57-60.
50. Mahajan, R. and H. K. Chhabra. 1977. Combined control of root-knot nematode and leaf curl virus disease in tomato. Nematol. Mediterr. 5:141-143.
51. Mahmood, K., S. Qumar, A. Naqui, and M. M. Alam. 1974. The influence of bottle-gourd mosaic virus and brinjal mosaic virus on the development of root-knot caused by *Meloidogyne incognita* (Kofoid and White, 1949) Chitwood 1949. Indian Phytopathol. 27:610-611.
52. Masood, A. and S. K. Saxena. 1980. Nucleic acid changes in three tomato varieties infected with *Meloidogyne incognita*. Indian J. Nematol. 10:102-104.
53. Mathur, V. K., K. C. Sanwal, R. V. Singh, and A. Lal. 1981. Role of soil clods and plant debris mixed with true seed in dissemination of plant parasitic nematodes. Indian J. Nematol. 11:92-94.

54. Mote, U. N. and D. R. Dasgupta. 1979. Significance of phenylalanine ammonia lyase as resistance response in tomato to the root-knot nematode, *Meloidogyne incognita.* Indian J. Nematol. 9:66-68. (Abstr.).
55. Mukherjee, T. D. 1960. Two & a Bud 7:5-7.
56. Mukherjee, T. K. and B. Tiagi. 1978. Effect of water stress on root galls incited by *Meloidogyne incognita* in *Lagenaria leucantha.* Indian Phytopathol. 31:216-219.
57. Mukherjee, T. K. and Jagmohan. 1980. Soil factors on the incidence of root-knot nematodes. Indian J. Mycol. & Plant Pathol. 10:XXX.
58. Nandal, S. N. and D. S. Bhatti. 1982. Phytotherapy of root-knot nematode—I. Effect of some physico-chemical factors on the nematicidal activity of certain weeds/shrubs extracts against *Meloidogyne javanica.* Nematol. Mediterr. In press.
59. Nandal, S. N. and D. S. Bhatti. 1983. Preliminary screening of some weed shrubs for their nematicidal activity against *Meloidogyne javanica.* Indian J. Nematol. 13:123-127.
60. Nandwana, R. P., B. S. Yadav, M. K. Varma, and A. Lal. 1980. Susceptibility of certain brinjal varieties to the root-knot nematode, *Meloidogyne incognita.* Indian J. Mycol. & Plant Pathol. 10:181-182.
61. Nath, R., M. M. Khan, R. S. K. Wamshi, and R. P. Dwivedi. 1982. Effect of *Argemone mexicana* on *Meloidogyne javanica* in okra (*Abelmoschus esculentus*). Indian J. Nematol. 12:205-208.
62. Nirula, K. K. and K. K. Bassi. 1965. Thermotherapy for root-knot nematode (*Meloidogyne incognita*) in potato tubers. Indian Potato J. 7:9-11.
63. Nirula, K. K. and B. T. Raj. 1971. Vertical distribution of *Meloidogyne incognita* in potato fields. Indian Phytopathol. 24:375-378.
64. Pani, A. K. and S. N. Das. 1972. Studies on etiological complexes in plant diseases. I. Association of root-knot nematodes in bacterial wilt of tomato. J. Res. OUAT. 2:54-59.
65. Pant, V., S. Hakim, and S. K. Saxena. 1983. Effect of different levels of NPK on the growth of tomato cv. Marglobe and on the morphometrics of the root-knot nematode, *Meloidogyne incognita.* Indian J. Nematol. 13:110-112.
66. Patel, G. J., H. M. Shah, D. J. Patel, and G. B. Valand. 1981. Effective duration of some propene nematicides in control of root-knot disease in beedi tobacco nursery. Indian J. Nematol. 11:74-77.
67. Patel, G. J., H. M. Shah, D. J. Patel, and G. B. Valand. 1981. Effect of removal of knotted root portion on the establishment and yield of beedi tobacco. Indian J. Nematol. 11:126.
68. Patel, G. J., S. H. Patel, N. M. Patel, and D. J. Patel. 1979. Feasibility of successful cultivation of beedi tobacco after summer bajra. Tob. Res. 5:37-42.
69. Prasad, J. and K. K. Jha. 1969. Variation in nematode populations as affected by soil conditions of Bihar. Indian Phytopathol. 22:314-321.
70. Prasad, K. S. K. and Y. S. Rao. 1976. Physiotherapy of the rice plant against the root-knot nematode (*Meloidogyne graminicola*). Proc. Indian Nat. Acad. Sci. 42:295-298.
71. Prasad, K. S. K. and Y. S. Rao. 1977. Effects of soaking seeds in chemicals on the control of root-knot nematode, *Meloidogyne graminicola.* Int. Rice Res. Newsletter 2:19.
72. Prasad, K. S. K., Y. S. Rao, and K. Krishnappa. 1982. Effect of 60 Cobalt radiation on the growth and development of *Meloidogyne graminicola* in rice roots. Indian J. Nematol. 12:171-173.
73. Prasad, S. K., M. L. Chawla, S. Kumar, and H. P. Saxena. 1971. Root-knot nematode, *Meloidogyne javanica* (Treub, 1885) Chitwood, 1949 and stem borer, *Stomopteryx nertaria* Meyrick affecting green gram, *Phaseolus aureus.* Indian J. Entomol. 33:55-60.
74. Prasad, S. S. S. V., K. V. B. R. Tilak, and K. V. Gollakota. 1972. Role of *Bacillus thuringiensis* var. thuringiensis on the larval survivability and egg hatching of *Meloidogyne* spp., the causative agent of root-knot disease. J. Invertebr. Pathol. 20:377-378.
75. Raja, A. and J. S. Gill. 1982. Studies on physiological specialisation in some populations of the root-knot nematode *Meloidogyne incognita.* Indian J. Nematol. 12:345-351.
76. Ram Nath and R. P. Dwivedi. 1981. Effect of root-knot nematode on development of gram caused by *Fusarium oxysporum* f. *ciceri* and root-rot by *Rhizoctonia* sp. Indian J. Mycol. & Plant Pathol. 11:46-49.
77. Rao, G. N. 1970. Tea pests in southern India and their control. PANS (Pest Art. and News Summ.) 16:667-672.

78. Rao, S. V. and S. S. Hussaini. 1981. Cytology of root-knot nematode, *Meloidogyne incognita*. Tob. Res. 7:158-161.
79. Rao, Y. S. and P. Israel. 1972. Effect of temperature on the hatching of eggs of the root-knot nematode (*Meloidogyne graminicola*). Oryza 9:73-75.
80. Rao, Y. S. and P. Israel. 1972. Influence of soil type on the activity of the rice root-knot nematode, *Meloidogyne graminicola* Golden & Birchfield. Indian J. Agric. Sci. 42:744-747.
81. Rao. Y. S. and P. Israel. 1973. Life history and bionomics of *Meloidogyne graminicola*, the rice root-knot nematode. Indian Phytopathol. 26:333-336.
82. Rao, Y. S., R. N. Jena, and K. S. K. Prasad. 1977. Infectivity of two isolates of *Meloidogyne graminicola* in rice. Indian J. Nematol. 7:98-99.
83. Rao, Y. S. and J. S. Prasad. 1981. Interaction of *Heterodera oryzicola* and *Meloidogyne graminicola* in artificial inoculation in rice. Indian J. Nematol. 11:104.
84. Rashid, A.,T. N. A. Farooqi, S. R. Misra, and K. Singh. 1981. Pathogenicity of root-knot nematode *Meloidogyne incognita*, on sugarbeet and amino acid changes in the infected roots. Indian J. Entomol. 43:130-136.
85. Reddy, D. D. R. 1981. Analysis of crop losses in tomato due to *Meloidogyne incognita*. Indian J. Nematol. 11:95 (Abstr.).
86. Reddy, P. P. 1983. Plant nematology. Agricole Publ. Acad., India. 287 pp.
87. Reddy, P. P., K. G. H. Setty, and H. C. Govindu. 1972. Association of the root-knot nematode with the southern blight fungus in causing a possible disease complex in flue-cured tobacco in Mysore. Mysore J. Agric. Sci. 6:357-358.
88. Routray, D. N. and S. N. Das. 1982. Occurrence of two races in *Meloidogyne incognita* around Bhubaneswar (India). Indian J. Nematol. 12:407-409.
89. Roy, A. K. 1976. Pathological effects of *Meloidogyne graminicola* on rice and histopathological studies on rice and maize. Indian Phytopathol. 29:359-362.
90. Roy, A.K. 1978. Effectiveness of rotation with non-host or fallow on reducing infestation of *Meloidogyne graminicola*. Indian J. Nematol. 8:156-158.
91. Roy, A. K. 1979. Control of root-knot of jute with decaffeinated tea waste and water hyacinth compost. Indian Phytopathol. 32:365-368.
92. Sahu, S. C. and Y. S. Rao. 1982. Seed treatment as prophylaxis against root-knot nematode in rice. Int. Rice Res. Newsl. 1:17.
93. Sakhuja, P. K., I. Singh, and S. K. Sharma. 1979. Effect of some fungal filtrates on the hatching of *Meloidogyne incognita*. Indian Phytopathol. 31:376-377.
94. Sampath, S., Y. S. Rao, and J. K. Roy. 1970. The nature of pest resistance in an indica rice variety TKM-6. Curr. Sci. (Bangalore) 39:162-163.
95. Sen, K. and M. K. Dasgupta. 1981. Effect of rice hull ash on root-knot nematodes in tomato. Trop. Pest. Manage. 27:518-519.
96. Seshadri, A. R. 1979. Integrated approach to control of nematodes infesting corn and sorghum in the tropics and subtropics. Paper presented at IX Int. Congr. Plant Prot., Aug. 5-11, 1979, Washington, D. C., U.S.A.
97. Sethi, C. L., J. S. Gill, and G. Swarup. 1964. A note on the prevalence of *Meloidogyne* spp. in vegetable crops. Indian Phytopathol. 27:69-70.
98. Sharma, N. K. and C. L. Sethi. 1976. Influence of *Meloidogyne incognita* and *Heterodera cajani* on carbohydrate content of cowpea. Indian J. Nematol. 6:171-173.
99. Sharma, S. K., I. Singh, and P. K. Sakhuja. 1980. Influence of different cropping sequences on the population of root-knot nematode, *Meloidogyne incognita* and performance of the subsequent mung bean crop. Indian J. Nematol. 10:53-58.
100. Sheela, M. S. and T. S. Venkitesam. 1981. Interrelationships of infectivity between the burrowing and root-knot nematodes in black peper, *Piper nigrum*,. Indian J. Nemtol. 11:105 (Abstr.).
101. Sheela, T. 1982. Effect of homoeopathic drugs on the mortality of *Meloidogyne incognita*. Indian J. Nematol. 12:395-396.
102. Shukla, V. N. and G. Swarup. 1970. Interrelationships of the root-knot nematode, *Meloidogyne incognita* (Kofoid and White, 1919) Chitwood, 1949 and *Sclerotium rolfsii* Sacc., on the emergence of tomato seedlings. Bull. Indian Phytopathol. Soc. 6:52-54.
103. Shukla, V. N. and G. Swarup. 1974. Effect of some fungicides on hatch of *Meloidogyne incognita*. Indian J. Nematol. 4:234-235.
104. Singh, B. and B. Choudhury. 1974. Screen-

ing tomato cultiars for resistance to *Meloidogyne* species. PANS (Pest Art. and News Summ.) 20:319-322.

105. Singh, D. B. and P. P. Reddy. 1981. Note on the chemical control of root-knot nematodes infecting french bean. Indian J. Agri. Sci. 51:534-535.

106. Singh, I., V. P. S. Chahal, P. K. Sakhuja, and J. S. Chohah. 1977. Effect of different levels of *Meloidogyne incognita* on the presence or absence of *Rhizobium phaseoli* on *Phaseolus aureus*. Indian J. Nematol. 7:172.

107. Singh, I., S. K. Sharma, P. K. Sakhuja, and N. K. Sharma. 1981. Efficacy and persistence of some nematicides against root-knot nematode, *Meloidogyne incognita* on pulses. Indian J. Nematol. 11:67-69.

108. Singh, L. and R. Gill. 1975. Pathology of *Meloidogyne javanica* (Treub, 1885) Chitwood, 1949 in *Hibiscus cannabinus* Linn. Geobios 2:112-114.

109. Singh, R. S. and K. Sitaramaiah. 1973. Control of plant parasitic nematodes with organic amendments of soil. Final. Tech. Report. Effect of organic amendments, green manuring and inorganic fertilizers on root-knot of vegetable crops. Res. Bull. Exp. Stn. and College of Agric. G. B. Pant Univ. Agric. and Technol., Pantnagar 6. 289 pp.

110. Singh, S. P. 1969. A new plant parasitic nematode *Meloidogyne lucknowica* n. sp. from the root galls of *Luffa cylindrica* (Sponge gourd) in India. Zool. Anz. 182:259-270.

111. Singh, S. P., M. Ahmad, A. M. Khan, and S. K. Saxena. 1980. Effect of seed treatment with certain oil cakes and nematicides on the growth of tomato and on rhizosphere population of nematodes and fungi. Nematol. Mediterr. 8:193-198.

112. Singhal, A. P. and C. M. Govil. 1978. Root-knot studies in brinjal. Indian Bot. Soc. First Bot. Con. Dec. 28-30. Meerut, India. J. Indian Bot. Soc. 57 (Suppl.):15 (Abstr.).

113. Sitaramaiah, K. 1984. Plant parasitic nematodes of India. Today and Tomorrow's Print. and Publ., New Delhi. 292 pp.

114. Sitaramaiah, K. and R. S. Singh. 1977. Response of *Meloidogyne javanica* and other nematodes to fatty acids. Indian J. Nematol. 7:58-65.

115. Sitaramaiah, K., R. S. Singh, and R. G. Choudhury. 1976. Control of root-knot (*Meloidogyne javanica*) of vegetable crops, pp. 113-114 *In* Annu. Rep. Res. 1974-75, U. B. Pant. Univ. Agric. & Tech. India, Directorate of Exp. Stn.

116. Sitaramaiah, K., R. S. Singh, K. P. Singh, and R. A. Sikora. 1971. Plant parasitic and soil nematodes of India. Exp. Stn. Bull. No. 3. (U.P. Agric. Univ. Pantnagar and U.S.D.A.). 70 pp.

117. Sivakumar, C. V., S. Palanisamy, and T. G. Naganathan. 1976. Persistence of nematicidal activity in seed treatment of okra with carbofuran and aldicarb sulfone. Indian J. Nematol. 6:106-108.

118. Srinivasan, C. S. and G. I. D'Souza. 1965. A note on the occurrence of the Brazilian pyroid coffee nematode *Meloidogyne exigua* Goeldi in South India. Indian Coffee 29:13.

119. Srivastava, J. N. and S. V. S. Chauhan. 1981. Morphological and histochemical studies on pollen abortion in *Solanum melongena* plants infested by *Meloidogyne incognita*. Indian Phytopathol. 34:43-47.

120. Sundaresh, H. N. and K. G. H. Setty. 1977. Crop rotation as an effective and practical means of controlling root-knot nematode (*Meloidogyne incognita* Chitwood). Curr. Res. (Bangalore) 6:157-158.

121. Sundaresh, H. N., K. G. H. Setty, and H. C. Govindu. 1977. Integrated control of root-knot nematode (*Meloidogyne incognita* Chitwood). Mysore J. Agric. Sci. 11:540-543.

122. Tandon, H. L. S. 1973. The crop nutrition pest incidence complex in India. PANS (Pest Art. and News Summ.) 19:372-380.

123. Tiagi, A., B. V. Ratnam, and K. Tiagi. 1981. Histochemical studies of tomato root galls caused by *Meloidogyne incognita*. Indian Phytopthol. 34:96-99.

124. Trivedi, P. C., A. Bhatnagar and B. Tiagi. 1978. Effect of fungicides, light and temperature on larval emergence of *Meloidogyne incognita*. Symp. Plant Dis. Prob., Oct. 1-3. 69 pp.

125. Varaprasad, K. S. and V. K. Mathur. 1980. Efficacy of carbofuran and aldicarb sulfone seed treatment on plant growth and against *Meloidogyne incognita* on sugarbeet. Indian J. Nematol. 10:130-134.

126. Varma, M. K. and J. P. Agnihotri. 1978. Efficacy of an antibiotic and a nematicide in controlling little leaf and root-knot disease complex of brinjal. Indian J. Nematol. 8:144-145.

127. Varma, M. K., R. C. Sharma, and J. R. Mathur. 1981. Efficacy of granular

nematicides against *Meloidogyne incognita* in mothbean (*Vigna acontifolia*). Nematol. Mediterr. 9:109-110.

128. Yadav, B. S. and A. C. Varma. 1981. Effect of various inoculum levels of root-knot nematode *Meloidogyne incognita* on brinjal, *Solanum melongena* L. Indian J. Nematol. 11:97 (Abstr.).

TRANSFER OF AGRICULTURAL TECHNOLOGY

Cathy Cameron Carter
Department of Plant Pathology
North Carolina State University
Raleigh, North Carolina, USA

Technology transfer is a broad topic encompassing many theories of technical assistance. In some cases, technology transfer and technical assistance have been considered to be synonymous. In other cases, one has been viewed as only a single component of the other.

TECHNICAL ASSISTANCE VS. TECHNOLOGY TRANSFER

Technical assistance has been defined as the transfer of knowledge and skills from one group of people to another with a view to fostering development (18). When defined in this manner, technical assistance represents a one-way process. Definitions of technology transfer, although similar to those of technical assistance, generally connote two-way communication.

Basche et al. (3) and Goulet (19) define technology transfer as an exchange of technical knowledge among countries that promotes education, international understanding, and mutual respect. The word "exchange" implies two-way communication, an element that often is missing in definitions of technical assistance. Capriles (7) defines technology transfer as "a process in which a country is free to choose autonomously, from among different alternatives of scientific and technical knowledge, those best suited to its natural conditions, development objective, capacity for assimilation, and pattern of living." The concept of freedom of choice distinguishes this definition from that of technical assistance. Historically, two-way communication between donor and recipient and the choice of appropriate technology by the recipient have not been consistently recognized as necessary components of technical assistance.

Diffusion Theory

In the late 1940's and 1950's, technical assistance often operated on the principle that the technology available in developed countries could be taken and applied to less developed countries (29). This practice grew out of the diffusion model or "trickle down" theory, which presumed effective dissemination of technical knowledge (20). Since agricultural research stations were nonexistent in most developing countries, technical agricultural knowledge was drawn from experiment stations and extension services in developed countries.

This theory ignored several factors. First, agricultural techniques and materials used in temperate climates cannot be transferred directly to agriculture in the tropics (9). About 85% of the world's scientific agricultural research occurs in temperate-zone countries and, therefore, is not necessarily directly transferable to the tropics (15,36). In spite of this fact, much of the technical assistance of the 1950's fell into the "direct-transfer" category and consisted primarily of exports of seeds, animals, equipment, and temperate-zone cultural practices (13,20). Some of this technical assistance was useful, but it was often designed specifically for production techniques and consumption standards of western, industrialized countries (9,14,21,44). For example, in many cases, it was assumed erroneously that fertilizer, irrigation, or mechanized farm equipment would be available to the developing country farmer, or that food qualities favored by developed countries would be acceptable to developing country farmers and consumers.

Problems arising from the application of the diffusion model led to a reassessment of international technical assistance (29). Technology from developed countries had to be deployed with an appreciation of the problems of agriculture in the tropics. It became obvious that successful generation and transfer of appropriate technology were going to require open communication between scientists in developed and developing countries.

WHAT IS TRANSFERRED

Technology transfer can be examined in three phases: material transfer, design transfer, and capacity transfer (20). Many of the early efforts in technical assistance consisted entirely of material transfer: goods or commodities such as seeds, plants, animals, books, and machines. Design transfer, on the other hand, includes instruction in laboratory procedures, transfer of equipment design or blueprints, or other methods that explain the use or operation of existing technology. The third category, capacity transfer, is the ultimate form of technology transfer. It encompasses all the steps leading to the formation of a core of indigenous, professional people with the necessary educational background to develop their own new and appropriate technology.

In a sense, capacity transfer can be considered more important than material or design transfer because it leads to the development of indigenous agricultural research and/or training programs. Two common methods of capacity transfer include 1) the employing of scientific personnel to perform specialized jobs and/or to instruct local people in the performance of these jobs, and 2) the sending of students to universities in developed countries or to International Agriculture Research Centers for advanced education.

In theory, this second approach is a particularly appropriate way to transfer technology to developing countries in tropical climates because the students have an opportunity to select and adapt the technology they learn to meet the needs of agriculture in their own countries (37). If this approach is to be most effective, students must be familiar with the agricultural problems in their own countries before they go overseas to study; otherwise, they may find themselves inadequately equipped when they return home.

In 1979, the National Association for Foreign Student Affairs conducted a survey on the relevance of U.S. college graduate programs to students from developing countries (2). Over 87% of the agriculture professors questioned indicated that foreign graduate students would face a substantial number of difficulties in applying their newly acquired education in their professional careers at home due to a lack of adequate equipment and technical facilities, cultural differences and problems in translating theoretical knowledge into practical application. Over 60% of the professors indicated a willingness to change course requirements, thesis topics, and special assignments to make degree programs more relevant to foreign students. In spite of this willingness on the part of the professors, only 39% of them said that foreign students were likely to inquire about tailoring a thesis research problem to relate to a professional career at home. Unless more effort is taken to make agricultural graduate study programs relevant to foreign students, capacity transfer will not be achieved. A shortage of scientific personnel in most developing countries is the major factor limiting the establishment and maintenance of a research organization large enough to deal with existing agricultural problems.

The events leading to the establishment of the International Agriculture Research Center CIMMYT form a classic example of successful capacity transfer (6). In 1943, none of the Mexicans working in agricultural sciences had a Ph.D. degree and only a few had Master's degrees. In 1945, the Office of Special Studies in Mexico employed 7 scientists from the Rockefeller Foundation to start an agricultural research program, and 25 Mexicans were hired into internship positions. The staff increased to 20 scientists by the late 1950's, and in 1960, the Office of Special Studies was dissolved and merged into a new National Institute of Agricultural Research under Mexican direction. In 1963, the Institute became the International Center for Maize and Wheat Improvement. Over 700 Mexicans worked as interns at the Institute for one or more years in the period from 1943 to 1963; approximately 250 Mexicans received fellowships to universities in the U.S. or other developed countries; 156 Mexicans received Master's degrees in agricultural sciences; and 85 received Ph.D.'s in agricultural sciences (6). In the year CIMMYT was established, all but four of these trained personnel were employed in agricultural sciences in Mexico, so there was a good demand for these highly trained specialists.

WHO BENEFITS FROM TECHNOLOGY TRANSFER

In addition to identifying the type of technology to be transferred, pinpointing the anticipated recipient of such aid is also important. This approach is necessary to ensure that the benefits derived from new technology reach the intended persons. For example, the development of the new high-yielding varieties of wheat and rice that launched the Green Revolution were considered at the time to be examples of unbiased, directly transferable technology with the potential to benefit all farmers, including subsistence farmers and those with small landholdings. In fact, however, the high-yielding varieties were not designed to be introduced alone, but as one component of a package involving many complementary inputs: fertilizers, adequate water, and effective control of plant diseases, insects and weeds, not to mention the presence of large markets, extension services, credit institutions, and farm machinery. The necessity for inputs such as these made the use of the new varieties time consuming, expensive, and impractical for many farmers and completely inappropriate for subsistence farmers (44). As a result, it was often the wealthier farmers with more land and resources who would benefit the most from new technology (31).

In recent years, significant increases in agricultural production in developing countries have occurred largely in export crops rather than in crops or livestock grown primarily for the internal market (11). Such occurrences are understandable because export crops are necessary to solve balance-of-payment problems. However, if the aim of technology transfer is not merely to increase yields, but also to reduce the extent of hunger and improve standards of living within the developing countries, then the small farmer who is not generally an exporter is the person who needs to benefit from advances in agricultural technology (23,27). Unfortunately, many agricultural advances serve only to make the larger, mechanized farms more profitable and to make the subsistence farmer poorer.

HOW TECHNOLOGY IS TRANSFERRED

Transfer of agricultural technology is dependent on the existence of an effective institutional framework (4). In the United States, the land grant system serves this function by linking the university to farmers as well as to private and public institutions. In the past, technical assistance often included efforts to establish academic and professional organizations, similar to those which exist in the United States, within other societies where similar institutional infrastructure did not exist (6). Such efforts tended to burden the developing country with an overextended, and often ineffectual, academic bureaucracy (41). This senario occurred in India several decades ago. At that time, agricultural training programs were based on British systems. Many instructors that were trained in Britain received little or no education in dealing with the agricultural situation in India. As a result, instructors returning to India lectured on agricultural principles and theory but lacked the practical knowledge of essential farming skills.

Other countries faced different kinds of problems. One of these was the lack of an education program to carry effective new technology to the farmers and/or a credit program to enable the purchase of required inputs (12). These problems indicate that the existing institutional framework has an important function in technology transfer. Three aspects of that framework will be discussed here: the research program, the extension service, and the farmer.

Role of Research

Research organizations have two major purposes: 1) the generation of new ideas and technologies and 2) the training of emerging generations of scientists (10). Long-term (15-20 year) goals for research should be established before program development begins. Once the overall goals are envisioned, short-term plans can be developed to fit the framework (39).

The first step in an agricultural development program should be the initiation of a substantial, highly integrated research program that adapts basic research to solve farm problems (1). There is a great need for applied research because many problems of significance to developing countries have not been studied sufficiently elsewhere in the world. Such problems include management and fertility of tropical soils, ecology of major pests and diseases, including nematodes, and soil-plant-water relationships under subtropical and tropical conditions. Research should emphasize relevance to the existing agricultural situation so that small farmers can benefit from the proposed solutions (16). Development of new technology is not going to be well founded unless the physical and socio-economic conditions of the regions where it is likely to be used are given due consideration. Developed countries that are labor scarce often do not realize soon enough how inappropriate are their solutions for labor-intensive societies (44).

In the late 1960's, the Development Assistance Committee of the Organization for Economic Cooperation and Development (OECD) in Paris proposed an international system for agricultural research (28). Under this system, selected "world centres" would concentrate on broad basic and applied research in the fields of crop and animal production. The centres would develop new plant types; study the problems of disease control, soil fertility, and other specific problems; and train scientists who could then work in "regional, national, and/or provincial centres." Certainly, the Consultative Group of International Agriculture Research Centers falls well within this definition of "world centre." These institutions generate improved agricultural technology for increased food production as evidenced through their publications, recommendations, improved and diverse crop germ plasm, and trained personnel (38,42).

The "regional centres," as defined by the OECD, are to serve large areas defined by similar ecological characteristics. Smaller countries within a certain region might draw largely on the work of the regional centres, but larger countries would need to establish national research systems working in connection with the regional centres. Provincial centres naturally would be part of the national research system within some of the larger countries.

THE INTERNATIONAL *MELOIDOGYNE* PROJECT The organization of the International *Meloidogyne* Project (IMP) serves as a model of the guidelines proposed by the OECD. The IMP is a USAID-funded research program, in operation since 1975, which works toward the development of management strategies for root-knot nematodes. The project center in Raleigh, N.C., can be categorized as a world center, each of eight project regions (Central America, South America, Brazil, West Africa, East Africa, Asia, Middle East and India) as regional centers, and each of the regions as comprised of various national and provincial centers. Research planning conferences have been held periodically in most of these project regions, and they have given scientists opportunities to share research findings, suggest new approaches, and pinpoint important regional nematode problems.

The IMP has relied heavily on the cooperation of its worldwide network to generate new technology. These scientists have conducted extensive sample surveys in their countries for the presence of root-knot nematodes. Sample populations sent to the project center have been identified and maintained in greenhouse culture. These populations now comprise the world's most extensive collection of *Meloidogyne* species and are vital to all aspects of the IMP's research program.

In addition to conducting research and generating technology, the IMP emphasizes transfer of technology. Periodic regional conferences, the compilation and distribution of research findings, graduate training at the project center, and an intensive three-week international workshop have been sponsored to further this goal. The emphasis is on education because scientists need periodic refresher training to continue effective research. It has been estimated that within 4 to 7 years after graduation, most of the knowledge a scientist acquires during his advanced education becomes obsolete (1). Therefore, research personnel need to attend workshops and conferences, take sabbaticals, and stay abreast of the current literature to avoid becoming out of date. The emphasis of the IMP on education, accompanied by small research grants for the purchase of specialized equipment, books, and subscriptions to nematological journals, has served to increase the research capabilities of scientists in developing countries.

Role of Extension

In the United States, extension services disseminate the agricultural technology produced by the universities. In other countries, similar duties may be the responsibility of the government, the university, farmer's associations, banks, or statutory bodies. In the United States, the agricultural extension service evolved with the existing economic and political framework and adapted itself to the needs of the U.S. farmer. Since this system was so highly successful here, the adoption of this model was previously advocated in developing countries receiving U.S. aid. However, despite strong efforts to "transplant" extension services into other social, economic, agronomic, and political conditions, success has not been common. For extension services to be successful they not only must have strong links to research, but also must operate with an awareness of local conditions, cultural traditions, and an understanding of the ways of life and thinking of the farmers.

What then is meant by the role of extension in technology transfer? Though the entire institution of U.S. agricultural extension may not be universally adaptable, its functions are indispensable to the transfer of agricultural technology. These functions include introduction of practical applications of research results, instruction of growers in new skills, and encouragement in the acceptance of new ideas.

Extension cannot provide all the services needed by agriculturists, but it sometimes can stimulate or coordinate activities of public agencies, or even of public enterprises, to provide goods and services necessary to implement technological change. Perhaps, more importantly, extension can expand the individual farmer's capacity to serve himself. In addition to benefiting the farmer, extension-type services establish a link between research/teaching institutions and farmers that helps these institutions maintain their relevance to the real problems and concerns of the growers. Here again, it is the concerns of the small farmers that need to be heard in order for the problems of rural poverty and hunger to be addressed.

Experts in the field have recommended a ratio of one agricultural extension agent to every 500 families (40). However, some of the services established in the Andes Mountains have only one agent to every 10,000 farm families (35). In order to serve so many families, agents have to travel extensively and are unable to gain an understanding of local conditions or develop the necessary rapport with individuals and groups. In such situations, the tendency is for the agent to concentrate on the more easily accessible farmers with large holdings and adequate resources. When this course of action is followed, new technology may be adopted and used on relatively large acreages, but the benefits bypass subsistence farmers.

TRAINING AND VISIT SYSTEM OF AGRICULTURAL EXTENSION The Training and Visit (T&V) system of extension developed by Daniel Benor has had some success in overcoming the previously mentioned problem (5). This system is characterized by a management structure with a clear line of responsibility. It introduces relatively unsophisticated techniques, and concentrates initial efforts on a few specified major problems. Although ratios of one village extension worker to every 300 to 500 farm families have been viewed as optimum, adequate results have been obtained with ratios as low as 1 in 1200. The village extension worker (VEW) maintains his effectiveness under such conditions by getting certain farmers to volunteer to serve as contacts with other farmers. The VEW sets up a system, such as that used in West Bengal, whereby he visits each of eight groups of farmers for a full day once every two weeks. He always visits a given group on the same day of the week, so that group always knows when he should be with them. On each visit, he meets at least with the contact farmers. Interested farmers who are unable to attend a meeting then rely on the contact farmers for their information. In addition to his eight days with the farmers, the VEW has two days of in-service training every two weeks from agriculture extension officers and various subject matter specialists.

The T&V system has proven itself to be an effective method of extension. It has been introduced primarily in countries such as Turkey, India, Pakistan, Burma, Nepal, Sri Lanka, and Thailand, where there are large numbers of farmers working on small farms. Bangladesh is also beginning to adopt this approach, and Indonesia's extension service already incorporates many principles of the T&V approach.

To survive, however, the T&V system must establish contact and exchange with the agricultural research system already functioning in each particular country. In India, this linkage has been accomplished by the formation of headquarter and district-level technical committees composed of researchers and extension specialists who communicate with each other on their problems and progress. Establishing such a link may well be the most difficult step in setting up a successful T&V program, and methods of going about it will no doubt vary from country to country. Starting out small and making adjustments along the way can lead to the organizational framework best suited to the particular situation.

Role of the Farmer

The farmer is probably the most undervalued participant in the technology transfer process (8). Researchers and extension personnel have often been frustrated by the apparent unwillingness of farmers to use the new technology being made available to them. As a result, the farmer may be labeled as uncooperative and be denied credit for his years of agricultural experience. The farmer certainly has something to say about the kinds of improvements he views as necessary. He is not likely to accept without reservation advice offered by intruders who may not be aware of pertinent circumstances and who don't have to take the risks. Agricultural scientists and developing country farmers both want to increase the efficiency of agricultural production, but each group copes with a different set of problems and has a unique perception of the situation. The solution, therefore, is to bring farmers and scientists into meaningful communication (15,24-27).

THE INTERNATIONAL POTATO CENTER & THE FARMER Dr. Robert E. Rhoades, agricultural anthropologist for the International Potato Center, has suggested seven key questions to be asked of farmers throughout the processes of technology generation, adaptation, and transfer (33).

1) Is the problem to be solved important to them? This question should prevent the researcher from projecting his own values or preferences into the farmers' circumstances. Often, when this question is not asked, the researcher goes on to solve "problems" that are not important to farmers.

2) Do farmers understand the trials? Many technologies interesting to scientists may be alien

to farmers, so sufficient attention should be given to explaining new practices.

3) Do farmers have time, inputs, and labor required by the improved technology? Growing a crop "under the farmer's conditions" may entail much more than doing a demonstration trial in a farmer's field. Methods of transportation and availability of markets are also important components of "farmer's conditions."

4) Does the proposed technology make sense within the present farming system? All aspects of a new technology must be evaluated for its effects on the ecosystem and compared critically with the indigenous farming practices.

5) Is the mood favorable for investing in certain crops in a region? An understanding of the farmer's attitude toward investment or innovation in crop production is at stake here. In regions where there is the possibility of replacing a traditional crop by a commercially more attractive crop, this question is particularly appropriate.

6) Is the proposed change compatible with local preferences, beliefs, or community sanctions? Things to be considered here include taste and color preferences regarding food crops, local customs concerning planting and harvesting dates, or, in some cases, government regulations concerning field rotation, irrigation, or types of crops to be planted. Some discretion and tact are necessary when local customs are analyzed for the presence of a rational basis.

7) Do farmers believe the technology will hold up over the long term? Demonstration trials usually last only a season or two, but farmers will want to know whether the new technology will stand the test of time. Farmers may use low-yielding varieties consistently when high-yielding varieties are available simply because they do not know how the new varieties will hold up under drought, early frost, or other conditions which the farmer knows he can expect at regular intervals.

These seven common-sense questions help scientists to think like farmers and give them a good idea of the technology's potential acceptability. New technology that conflicts with socio-cultural elements should not be cast aside completely, but rather the most limiting factor to its success should be addressed. Increasingly, the development of varieties that perform well under less than ideal conditions, such as drought, low soil fertility, and chronic infestations of pests and diseases, is being researched. Adapting technology to a complex farming system may be easier than having farmers change their system to fit new technology (33).

FAILURE AND SUCCESS IN TECHNOLOGY TRANSFER

Several agriculturists have outlined reasons for failure in technology transfer: 1) introduction of techniques beyond the comprehension of the indigenous farmer; 2) insufficient or inappropriate basic research; 3) existing social barriers and ingrained traditional farming practices; 4) inactive bureaucracy in the client country; 5) failure of the host or client country to muster sufficient resources to complement the foreign experience; 6) lack of practical farm-level demonstrations on management of necessary farm operations; 7) ineffective farmer involvement in the transfer; and 8) inadequate training for village-level extension workers (8,17,27). The following examples of actual attempts to transfer agricultural technology may illustrate some of these points.

Dryland Farming Systems in North Africa and the Middle East

In the late 1960's, South Australian dryland farming practices were suggested for introduction to parts of North Africa and the Middle East (8). Up until this time, farming practices used in these areas were based on European and North American methods that were not designed for arid climates. Although these methods had not been very productive or successful in the Middle East, they were familiar. Politicians with the power to have work started on dryland farming projects were difficult to convince and did not give their approval until the mid-1970's. At that time, projects to introduce Australian dryland farming practices were begun in Libya, Algeria, and Tunisia. The major objective was to prove the technical feasibility of a rotation of alfalfa, a pasture plant, with cereal crops in a cereal-livestock program.

Researchers outlined the system in detail on paper and described everything from methods of pasture regeneration to soil conservation measures. The plan was to be so detailed that the farmers would be able to consult it for every decision that needed to be made. Unfortunately, many of the farmers concerned did not read, and no extension system was in operation to transfer the practice to the farmers.

In Tunisia, excellent alfalfa pastures on research farms produced greatly improved grazing and sufficient quantitites of nitrogen for a subsequent cereal crop. However, in the farmers' fields, cereal yields were actually lower than normal because the farmers had not been instructed in the operations of shallow cultivation and weed control. The farmers' reactions were so negative that they became unreceptive to further dryland farming proposals. Thus in Tunisia, the project was considered a failure. In Libya, however, farm managers were brought in from Australia to instruct the Libyan farmers in the new practices. As a result, successful introduction of the Australian dryland farming system did occur in Libya.

Potato Storage Experiments in Peru

In the late 1970's, an interdisciplinary team of social and biological scientists from the International Potato Center began experiments geared toward reduction of potato losses to disease in storage (34,39). They were looking for simple control methods that small-scale farmers in the Andes mountains could adapt to their current practices with as little inconvenience as possible. The biological scientists began research on storage principles used many years ago by European and American growers. The social scientists on the team went to the Mantaro Valley in the Central Peruvian mountains to talk to farmers.

Farmers were puzzled to hear about the research to reduce storage losses because, as far as they were concerned, there were no losses. All potatoes, even the small, shriveled, and spoiled ones, were used for some purpose, whether it was for animal feed, dehydration for long-term storage, or even immediate home consumption. In fact, the social scientists were informed that the shriveled, partly spoiled potatoes were preferred in some dishes because they had a "sweeter" taste.

The social scientists carried the news to the biological scientists that they were working on an imaginary problem. The post-harvest physiologists were somewhat reluctant to believe this, so another visit was made to the farmers. This time the farmers acknowledged that although there were no losses in terms of potatoes for animal or human consumption, there were losses in the category of seed potatoes. The potatoes were stored in darkness, usually in home cellars. Under these conditions, the potatoes produced long sprouts. Not only did the removal of these sprouts cost the farmer much time and labor, but also subsequent germination was reduced. The social, economic, and biological losses were found to be real, so the scientists redefined their research problem to concentrate on improvements in seed potato storage.

Diffused light storage was thought to be a satisfactory answer to the seed potato problem, and time was spent constructing model storage facilities using this principle. However, further conferences with the farmers ruled out the possibility of storage facilities separate from the house for reasons of security and convenience. Converting inner courtyards or verandas of houses for diffused light storage seemed to be a suitable compromise, so experiments were set up to test this possibility. Farmers showed interest in the new set-up but failed to adopt the method. The social scientists determined that it was the expense of the seed trays used by the researchers that the farmers had trouble accepting; therefore, the farmers were shown how to construct seed trays out of locally available materials. From this point on, the technology began to be adopted in principle though not always exactly as designed by the researchers.

Farmer-Back-To-Farmer

The potato storage experiment just described led the International Potato Center to formulate a model for technology transfer known as farmer-back-to-farmer (34). As the title indicates, farmers have a primary role in the process of regenerating, applying and transferring appropriate technology. Problems identified by the farmers receive first priority. An interdisciplinary team of researchers works closely with farmers to formulate a common definition of the problem. Once a potential solution is developed, it is tested and adapted under conditions as similar as possible to those of the farmers. Finally, the farmers evaluate the new technology. Based on their reactions and suggestions, the technology can be improved through the same farmer-back-to-farmer process.

The farmer-back-to-farmer model differs significantly from previously mentioned theories. The diffusion model or "trickle down" theory might be represented by the diagram in Fig. 1.

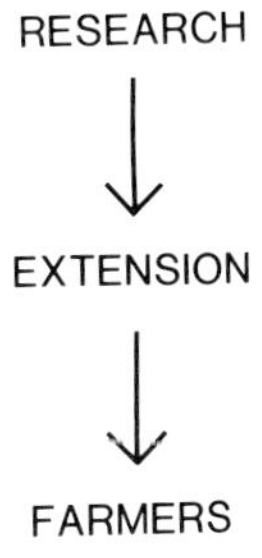

Fig. 1. Schematic representation of the diffusion model of technology transfer.

Fig. 2. Schematic representation of a dynamic model of technology transfer.

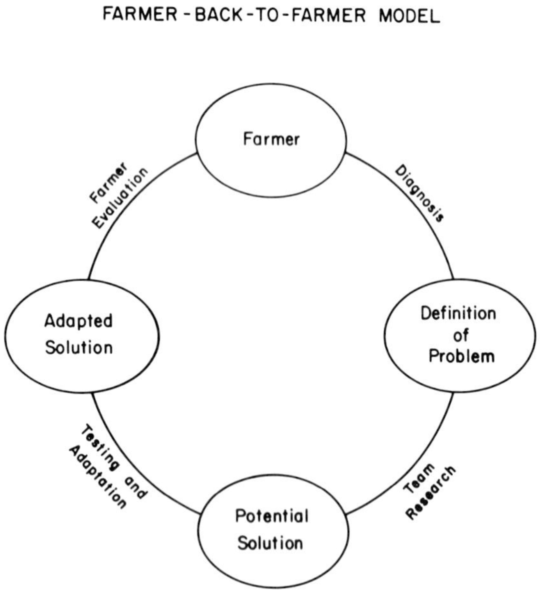

Fig. 3. Schematic representation of the farmer-back-to-farmer model of technology transfer [Adapted from Rhoades & Booth (34)].

Information developed by researchers is passed on to extension personnel and eventually trickles down to the farmer.

Once problems of the diffusion theory were recognized, models such as the one illustrated in Fig. 2 were adopted. Organizational structures like this one allowed two-way communications between research and extension and between extension and farmers. However, direct communication between researchers and farmers was still nonexistant, and the extension service acted as a buffer that effectively prevented such interaction. With the farmer-back-to-farmer model (Fig. 3), the farmer is given credit for his agricultural experience and assumes a responsible role in the technology transfer process. The researcher also benefits by gaining a better awareness of common agricultural problems (32).

ROLE OF TECHNOLOGY TRANSFER IN SOLVING WORLD FOOD PROBLEMS

In 1974, at the World Food Conference in Rome, it was stated that the following steps must be taken in order to fight world hunger:

1) Direct basic research toward the poor majority,

2) Accelerate the development and acceptance of appropriate technology,

3) Conduct comparative studies of the development experience, and

4) Design better international mechanisms for mutual cooperation for improving food production and nutrition (16,30).

Each of these directives stresses the importance of effective technology transfer to the goal of increased food production. If the population doubles in the next generation or so, effective technology transfer will be even more imperative because at least 90% of the extra food needed will be expected to come from increased yields per hectare (22). The usefulness of new technologies, however, will continue to be limited strongly by the land tenure system, social structure, food distribution systems, the distribution of power in the local area, and government agricultural development programs and policies (23,31,40,43).

In 1979, James P. Grant, president of the Overseas Development Council in Washington, D.C., stated that "if the world's food supply were to increase even by 50% over the next five years, this increase alone would not greatly affect the existing patterns of human deprivation. What is needed are changes in modes of production and patterns of distribution. World hunger in our times is less a problem of production than a consequence of the lack of purchasing power among the growing number of unemployed and underemployed in the developing countries" (16). Economists even point out that increases in agricultural production actually lead to price declines unless there is a corresponding increase in demand (6,11,22). In such cases, farmers may have no economic incentive for using new technologies to increase production.

In spite of the grim realities, there is much optimism that food production in developing countries can be doubled by the end of this century and thereby eliminate much hunger and malnutrition (16). To proceed toward this goal, future research must focus largely on the needs of the small-scale producer. In this way, new technological solutions will be more likely to be appropriate for subsistence farmers, not just the large land owners. This approach, combined with improved communication among researchers, extension personnel, and farmers, practically guarantees the development of appropriate agricultural technology. Although development of such technology alone cannot totally alleviate world hunger, it will undoubtedly be responsible for much progress toward successful achievement of that goal.

Literature Cited

1. Arnon, I. 1981. Modernization of agriculture in developing countries—resources, potentials and problems. John Wiley & Sons, New York. 565 pp.
2. Baron, M. 1979. The relevance of U.S. graduate programs to foreign students from developing countries. A report of a survey conducted by the National Association for Foreign Student Affairs (NAFSA) in Cooperation with the Council of Graduate Schools. NAFSA, Washington, D.C. 15 pp.
3. Basche, J. R., Jr. and M. G. Duerr. 1975. International transfer of technology: a worldwide survey of chief executives. The Conference Board, Inc., New York. 21 pp.
4. Behrman, J. N. 1979. International technology flows for development: suggestions for U.S. government and corporate initiatives, pp. 118-127 *In* Mobilizing technology for world development, ed. J. Ramesh and C. Weiss, Jr. Praeger Publ., New York. 234 pp.
5. Benor, D. and J. Q. Harrison. 1977. Agricultural extension—the training and visit system. World Bank, Washington, D.C. 55 pp.
6. Blase, M. G. (ed.). 1971. Institutions in agricultural development. Iowa State Univ. Press, Ames, Iowa. 247 pp.
7. Capriles, R. S. 1979. Technology transfer and the industrialists in Latin America, pp. 13-26 *In* Integrated technology transfer, ed. J. Richardson. Lomond Books, Mt. Airy, Maryland. 162 pp.
8. Chatterton, B. and L. Chatterton. 1982. Failures in technology transfer—are farmers overlooked? Food Policy 7:141-155.
9. Cochrane, W. W. 1969. The world food problem: a guardedly optimistic view. Thomas Y. Crowell Co., Inc., New York. 331 pp.
10. D'Ambrosio, U. 1979. Knowledge transfer and the universities: a policy dilemma, pp. 37-43 *In* Integrated technology transfer, ed. J. Richardson. Lomond Books, Mt. Airy, Maryland. 162 pp.
11. Dorner, P. 1971. Land tenure institutions, pp. 14-31 *In* Institutions in agricultural development. Iowa State University Press, Ames, Iowa. 247 pp.
12. Duncan, E. R. (ed.). 1977. Dimensions of world food problems. Iowa State Univ. Press, Ames, Iowa. 309 pp.
13. Eckaus, R. S. 1977. Appropriate technologies for developing countries. Nat. Acad. Sci., Washington, D.C. 140 pp.
14. Eklund, P. 1983. Technology development and adoption rates. Food Policy 8:141-153.
15. Gable, R. W. and J. F. Springer. 1976. Administering agricultural development in Asia—a comparative analysis of four national programs. Westview Press, Boulder, Colo. 398 pp.
16. Grant, J. P. 1979. An international challenge: science and technology for managing the world food problem, pp. 204-211 *In* Mobilizing technology for world development, ed. J. Ramesh and C. Weiss, Jr. Praeger Publ., New York. 234 pp.
17. Green, J. 1961. Success and failure in technical assistance. Human Organization 20:1-10.
18. Goulet, D. 1971. The cruel choice—a new concept in the theory of development. Atheneum, New York. 362 pp.
19. Goulet, D. 1977. The uncertain promise—value conflicts in technology transfer. IDOC/North America, New York. 320 pp.
20. Hayami, Y. and V. W. Ruttan. 1971. Agricultural development: an international perspective. John Hopkins Press, Baltimore/London. 367 pp.
21. Helleiner, G. K. 1979. International technology issues: southern needs and northern responses, pp. 84-97 *In* Mobilizing technology for world development, ed. J. Ramesh and C. weiss, Jr. Praeger Publ., New York. 234 pp.
22. Hungate, L. S. and R. W. Sherman. 1979. Food and economics. Avi Publ. Co., Inc., Westport, Conn. 244 pp.
23. International Food Policy Research Institute. 1977. Food needs of developing countries: projections of production and consumption to 1990. Int. Food Policy Res. Inst. (IFPRI) Res. Rep. No. 3. Washington, D.C. 157 pp.
24. International Potato Center. 1981. Farmer inputs turn direction of agricultural research. CIP Cir. Vol. IX, No. 3. 3 pp.
25. International Potato Center. 1981. Farmers help scientists change research objective. CIP Circ. Vol. IX, No. 4. 3 pp.
26. International Potato Center. 1981. The whys of technology acceptance by farmers. CIP Circ. Vol. IX, No. 5. 5 pp.
27. Jedlicka, J. 1981. Technology transfer to subsistence farmers—Management process

and behavioral technique. Interciencia 6:257-260.

28. Kristensen, T. 1968. The food problem of developing countries. Organ. Econ. Coop. Dev., Paris. 114 pp.
29. Nelson, R. R. 1974. Less developed countries—technology transfer and adaptation: the role of the indigenous science community. Econ. Dev. Cultural Change 23:61-77.
30. Phillips, J. 1967. The development of agriculture and forestry in the tropics: patterns, problems, and promise. Frederick A. Praeger, Publ., New York/Washington. 221 pp.
31. Poleman, T. T. 1981. A reappraisal of the extent of world hunger. Food Policy 6:236-252.
32. Rhoades, R. E. 1982. The art of the informal agricultural survey. Int. Potato Center, Soc. Sci. Dep. Training Document 1982-2. 40 pp.
33. Rhoades, R. E. 1982. Understanding small farmers: sociocultural perspectives on experimental farm trials. Int. Potato Center, Soc. Sci. Dep. Training Document 1982-3. 9 pp.
34. Rhoades, R. E. and R. H. Booth. 1982. Farmer-back-to-farmer: a model for generating acceptable agricultural technology. Agric. Admin. 11:127-137.
35. Rice, E. B. 1974. Extension in the Andes: an evaluation of official U.S. assistance to agricultural extension services in Central and South America. U.S. Agency Int. Dev., Washington, D.C. 552 pp.
36. Ross, D. N. 1977. Partners in agroeconomic development. The Conference Board, Inc., New York. 59 pp.
37. Spencer, D. L. 1970. Technology gap in perspective: strategy of international technology transfer. Spartan Books, New York/Washington. 172 pp.
38. Thurston, H. D. 1977. International crop development centers: a pathologist's perspective. Annu. Rev. Phytopathol. 15:223-47.
39. Werge, R. W. and M. Benavides. 1981. Investigation of farming in Peru by means of a multiple-visit survey. Int. Potato Center, Soc. Sci. Dep. Training Document 1980-5. 12 pp.
40. Whyte, W. F. 1975. Organizing for agricultural development: human aspects in the utilization of science and technology. Transaction Books, New Brunswick, New Jersey. 62 pp.
41. Willett, J. W. 1982. Research for agriculture in developing countries—the U.S. role. Food Policy 7:13-26.
42. Wortman, S. 1975. The world food situation: a new initiative. Working Papers of the Rockefeller Foundation. 35 pp.
43. Wortman, S. and R. W. Cummings, Jr. 1978. To feed this world—the challenge and the strategy. Johns Hopkins Univ. Press, Baltimore/London. 440 pp.
44. Yudelman, M., G. Butler, and R. Banerjii. 1971. Technological change in agriculture and employment in developing countries. Organ. Econ. Coop. Dev., Dev. Centre Studies, Employment Series No. 4, Paris. 204 pp.

Index

INDEX

Geographic Index

Host Plant Index

Subject & Taxonomic Index

An Advanced Treatise on *Meloidogyne*

Volume II: Methodology

edited by
K. R. Barker
C. C. Carter
and
J. N. Sasser

A Cooperative Publication of the
DEPARTMENT of PLANT PATHOLOGY
and the
UNITED STATES AGENCY for
INTERNATIONAL DEVELOPMENT

Printed by North Carolina State University Graphics.
1985

For information on the availability of
this publication, write to
Dr. J. N. Sasser
Department of Plant Pathology
North Carolina State University
Box 7616
Raleigh, N.C. 27695-7616
U.S.A.

Library of Congress Catalog Card Number
84-61978
ISBN 0-931901-02-2 (v.2)

North Carolina State University Graphics
Box 7226
Raleigh, North Carolina 27695-7226
U.S.A.

PREFACE

The world community has focused great attention over the past two decades upon the question of how we can feed our burgeoning human population. Even with the present famine in Africa, it is generally recognized that our world's natural resource endowments can support a population substantially larger than the present one. However for this increased support to occur, substantial areas of new land must be brought into crop production, and productivity must be increased by at least 3% annually on lands presently under cultivation. The proposed gain in productivity is a great challenge to crop-production specialists! If this rate of growth in productivity is to be realized, significant advances must be made in all components of the production systems of the developing world, including those of protecting crops against pests such as insects, pathogens, weeds and plant-parasitic nematodes.

Plant-parasitic nematodes are recognized in the United States and other developed countries as potentially serious constraints to crop productivity. This recognition did not occur generally, however, until after World War II when the study of plant nematodes began to receive attention concomitantly with the development of effective and economical management practices (including chemical control agents). The recognition of nematodes as serious deterrents to crop productivity did not occur simultaneously in the developing countries because there were very few trained nematologists to confirm the presence of damaging population levels and to initiate work toward development of appropriate management strategies and tactics.

The United States Agency for International Development (USAID) recognized the lack of attention to plant-parasitic nematodes in the developing nations in 1975 when it funded a project at North Carolina State University entitled "Research on Integrated Crop Protection Systems with Emphasis on the Root-Knot Nematodes (*Meloidogyne* spp.) Affecting Economic Food Crops in Developing Nations." The project was given a world-wide mandate and focused on the *Meloidogyne* species because they are: (a) widely distributed geographically, (b) adapted to a very wide host range, and (c) generally recognized as causing more economic damage to food crops than any other single group of plant-parasitic nematodes. Since problems caused by species of *Meloidogyne* are common to much of the developing world, it was reasoned that this concentration would attract great interest and commitment by scientists in those nations.

The activity, which came to be known as the International *Meloidogyne* Project (IMP), did draw the attention of scientists throughout the world as evidenced by the many collaborators primarily from more than 60 developing nations. A truly peer-type relationship was established between the collaborating country scientists and the IMP staff. Even though the IMP was conceptualized and funded before the Collaborative Research Support Programs (CRSP's) were established under Title XII, it embodies most of the CRSP characteristics. The collaborating scientists jointly planned the work that was to be done in the various regions of the world, and all of them worked on problems that were of common interest. In a few short years and with modest funding inputs through the IMP, the science of nematology was greatly advanced throughout the developing world. Scientists were recognized within their respective countries for their interest and expertise in nematology. Special in-service training was provided through workshops and conferences to provide a common focus and knowledge base for the projects. Work was initiated collaboratively within each participating country which demonstrated the need to manage plant-parasitic nematode populations below the damage threshold levels. Finally, various management strategies were evaluated as the basis for the identification of appropriate technology for each of the countries/regions.

In the process of conducting the project on a world-wide basis, scientists at North Carolina State University had the unique opportunity of assembling a vast amount of observational and research data on *Meloidogyne* spp., such as (a) geographic distribution patterns of species and races and the ecological factors that determine these patterns, (b) cytogenetic basis of species and their evolution, (c) morphology and taxonomy, (d) comparative biochemistry of species and races, and (e) sources of genetic resistance within food crops. The IMP provided the scientific base for the development of economical and ecologically sound management strategies for root-knot nematodes on a worldwide basis. Few groups of plant pathogens have been treated in such a comprehensive manner through a world network of collaborating scientists.

Perhaps, the most important product of the IMP is this publication *An Advanced Treatise on Meloidogyne* with Volume I devoted to *Biology and Control* and Volume II devoted to *Methodology*. These books will be of great value to crop-protection specialists, both researchers and extension specialists, throughout the world. They present state-of-the-art knowledge of this important group of plant pathogens, and a significant component of that knowledge is a product of the IMP. Both volumes represent the proceedings of the final conference conducted under the IMP at Raleigh, North Carolina in April, 1983. This Conference was attended by 95 scientists representing 39 countries.

North Carolina State University is proud of the achievements realized under the IMP. Appreciation is expressed to all of those who have collaborated on the project thereby making it an outstanding effort which will contribute much to enhanced food production throughout the world. We also express appreciation to those representatives of USAID who had the vision and determination to sustain the funding for the IMP during a 9-year period.

J. Lawrence Apple, Coordinator
International Programs
North Carolina State University

TABLE OF CONTENTS

Part One

Sampling and Extraction

CHAPTER 1

SAMPLING NEMATODE COMMUNITIES

K. R. Barker

Plant Pathology Department
North Carolina State University
Raleigh, North Carolina, USA

Sampling nematode communities has been the basis for the development of improved control tactics for more than 30 years. This approach came about as a result of the early research by N.A. Cobb (17) and others who developed reasonably reliable sampling and extraction procedures. One of the first quantitative studies in nematode sampling was conducted by Anscombe on the potato-cyst nematode (4). With the development of rapid, analytical computer programs, much effort has been directed toward improved sampling for various nematode taxa, especially *Meloidogyne* and *Heterodera* spp. These studies include the characterization of temporal and spatial distribution patterns of numerous species.

Various approaches have been employed to characterize the typically polyspecific nematode communities (1,8,24,28,50). Many investigations have documented that nematode populations and communities change over time, with the maximum numbers usually occurring near harvest for annual crops (7,8,11,12,14,23, 32,34). Only during the last 15 years, however, have the horizontal spatial patterns, reflecting the typically aggregated populations, been studied extensively (6,8,15,22,28,30,39,45). Most experiments have been based on the collection of composite samples with varying numbers of soil cores. Samples are collected with a sampling tube, and the populations of the taxa present are extracted and counted. A few studies of spatial patterns have been based on single core samples (28,49). Frequency distributions have been analyzed and fitted to various probability distributions, such as the negative binomial, Neyman Type A, and Poisson (1,4,15,28,49,57,58,63).

The mapping of nematode damage by aerial infrared photography also has potential as a means of determining within-field distributions (6,41,51,60,61). For example, this technique was used to assess sugarbeet damage caused by *Heterodera schachtii* with such a high degree of accuracy that the author suggested aerial surveys could be used as a basis for recommendations of fumigant nematicides (61). In contrast, detection of damage caused by *Pratylenchus penetrans* in flue-cured tobacco was less reliable by this method (41). Promising results have also been obtained with the seed-borne *Ditylenchus dipsaci* in alfalfa, where the size of infested foci were followed during initial years of infestation (6). This technique had sufficient accuracy that the non-random spatial pattern of this nematode could be described by a negative binomial in given fields (6).

Other rather unique methods for characterization of nematode infestations have been explored. The analyses of the relative ATP content in cysts, as indicated by measurement of bioluminesce with a photometer, was found to have promise for quantitative estimation of field populations of *Globodera pallida* and *G. rostochiensis* (5). Another interesting concept in nematode management is tare soil sampling for *Heterodera schachtii* (68). These soil samples, collected at the processing plant, have a high level of sensitivity by which relative infestation levels of this nematode can be detected. Bioassays, which are particularly sensitive when carefully handled soil and/or tissue samples are used, are very useful for *Meloidogyne* spp. (12,27,47,53,54). Thus, nematode-population assays to be used for research and/or management purposes can be approached in various ways.

This chapter is concerned with the recent progress in and potential of sampling for nematodes for various purposes. Specific areas discussed include the major objectives of sampling, consideration of the spatial and temporal patterns of nematodes in reference to sampling, importance of nematode biology and crop management, approaches to sample collection, care and management of samples, and the methods of characterizing nematode communities based on sample data. This discussion applies to soil as well as plant samples collected for nematode assay. The details of extraction and assay procedures are given in the following chapter.

OBJECTIVES OF SAMPLING

The major purposes of sampling for nematodes include population estimation for research, advisory or predictive programs, disease diagnoses and general surveys. The goals of sampling for nematodes vary with each type of experiment and other purposes of sampling. For integrated pest management programs, the ultimate goal of sampling is relating numbers and kinds of nematodes to crop performance as well as evaluation and selection of management tactics. In contrast, research programs may have as their primary goals the characterizing and understanding of nematode-population dynamics. The goal of sampling in quarantine programs is the detection of any nematodes present and prevention of spread.

The specific objectives of sampling, as well as any mathematical models to be used, should be determined before a sampling program is planned. This approach is particularly important since the objectives may dictate the sampling design and type and amount of data needed (8,18,66). In developing sampling procedures for a specific purpose, one should be familiar with the expected precision for the particular nematodes in question. Smaller sampling errors can be expected where nematodes occur randomly in the soil, leading to a Poisson frequency distribution (Table 1,2). In practice, nematodes occur in aggregated or contagious distributions which lead to much greater sampling errors. For example, sampling infestations of *Globodera rostochiensis* where the infestation level is 1 cyst/g is subject to a standard error of approximately 18%, as compared to the ex-

Table 1. Range of sampling errors for randomly distributed nematode populations[a].

Number of nematodes in sample (n)	Standard error $\sqrt{n}$	Precision (coefficient of variation) *SE % n*	95% Confidence limits $n \pm 2\sqrt{n}$
4	2	50	0-8
25	5	20	15-35
100	10	10	80-120
400	20	5	360-440

[a]Adapted from Jones and Kempton (34); SE = standard error.

Table 2. Chances of detecting nematodes in 500-g soil samples[a].

Average field population		Percentage chances	
Per acre	Per 500 g soil	For detection	For failure to detect
2×10^6	1	63	37
4×10^6	2	85	15
6×10^6	3	95	5
10×10^6	5	99	1

[a]After Southey (64), assuming sampling errors are near those of the Poisson distribution.

pected error of about 10% for a randomly distributed population (16,64).

The quarantine regulations in many countries require land to be certified free of specific nematodes, such as the potato-cyst nematode, on the basis of soil sampling (64,65). Unfortunately, a negative result from sampling only indicates that the population is unlikely to exceed a given density. Thus, when sampling is for quarantine purposes, the requirements for sampling precision become more critical at low population levels. Both the numbers and sizes of samples must be increased to ensure detection of low numbers of nematodes (64).

Simple detection by sampling still serves as the basis for determining geographic distributions of plant-parasitic nematodes. Although this type of sampling requires less rigor than that for quarantine purposes, the information on the geographic distribution of nematodes is still limited. Most of the available publications on distribution of these pests must be considered carefully since some data reflect the distribution of nematologists as well as the known distributions of nematodes.

Sampling for advisory purposes, diagnosis of disease problems, and crop-loss assessment is becoming increasingly important as pest-management programs become more prevalent (37,59). Even though initial numbers of nematodes can generally be related to the yield of annual crops, many factors, including errors in sampling, may affect this relationship (11,12, 24,52). This topic is discussed further in chapters 10 and 16. Since many nematodes are involved in disease complexes, sampling for diagnostic purposes requires that healthy as well as diseased plants be assessed for this purpose. Isolation of associated fungi and bacteria also may be required where it is suspected that these organisms are involved in disease etiology.

CONSIDERATION OF SPATIAL AND TEMPORAL PATTERNS

The spatial and temporal patterns of nematodes must be considered carefully in sampling for any purpose. The horizontal spatial patterns of plant-parasitic nematodes are typically patchy or contagious types (Fig. 1) (8,28, 42,45). This fact poses many difficulties in sampling as well as experimental design in reference to nematode research and control. The vertical spatial patterns of nematodes also vary greatly, depending on the crop, soil type, and the nematode species involved (13,14,23,

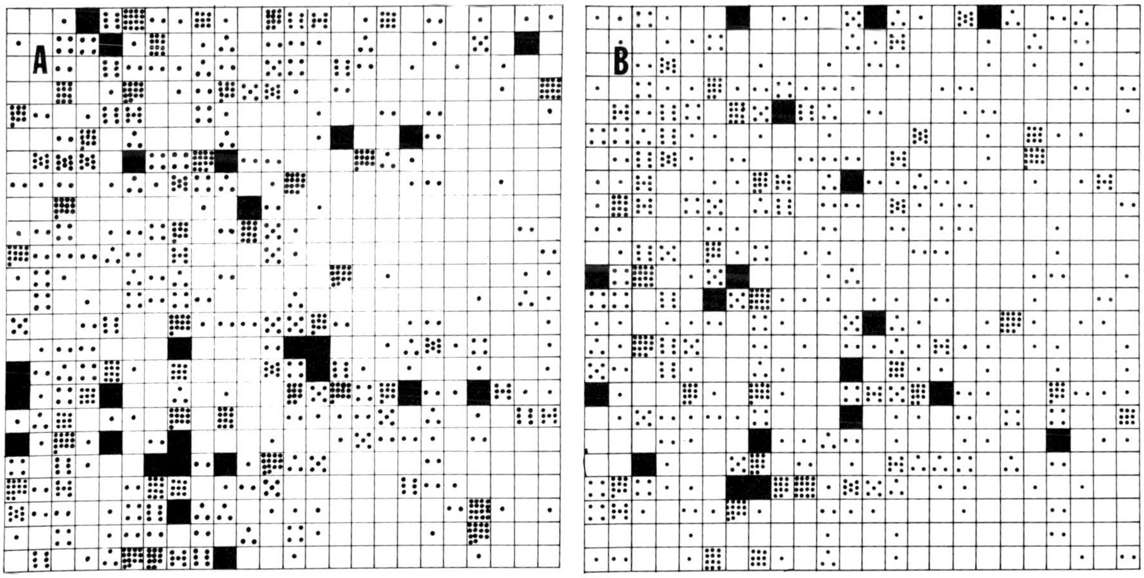

Fig. 1. Clustered horizontal spatial pattern of *Meloidogyne incognita* (spring samples) as determined in two samplings (A & B), those for "A" being collected on April 4, 1968, and those for "B" on April 25, 1968. Each dot in the 5 x 5-m quadrats represents 25 juveniles/500 cm^3 of soil; black quadrats had $\geq$ 400 juveniles for this soil volume; crop was a fescue cover crop following tobacco. Respective distribution parameters per sampling for A & B: σ^2 = 33.5, 22.0; $\bar{x}$ = 3.0, 2.3; k = 0.31(n.s.), 0.35; % zeros = 47.2, 46.4, and C. V. (%) 194, 206 (based on counts/20 cm^3 soil). [Distribution data from both sample sets could be fitted to Neyman Type A and Logarithmic distributions].

70). Equally important, the seasonal fluctuations or temporal patterns change with crop and nematode species (8,9,32).

The aggregation of nematodes is responsible for the typically patchy stunting associated with these pests. These aggregates vary over a field and within small areas of a field. Repeated sampling of 5 x 5-m quadrats, with 5-core composite samples, gives a similar general spatial pattern of *Meloidogyne incognita* in a field, but the assay results vary greatly for given quadrats (Fig. 1). This divergence apparently is due to the occurrence of small clusters of this nematode and to related sampling error. Spatial patterns of nematodes can be mapped in several ways. This mapping may be done by hand as in Figure 1 for *Meloidogyne incognita* or by various computer programs. Contour maps can be used to depict the densities of *Meloidogyne* spp. as in Figure 2, in a field that had been planted to tobacco (8,49). Note that in this case, the direction of the rows had a striking influence on the relative areas infested with this nematode. This relationship was true for *Meloidogyne* and *Criconemella* spp. in a recent study (49) (Fig. 2).

Thus sampling should be oriented across crop rows rather than with them. Three-dimensional graphs represent another type of computer mapping of nematode distributions (28, 29). These are especially useful in relating nematode numbers to edaphic factors such as soil texture (28).

Although horizontal nematode-population assay data have been fitted to a number of standard mathematical distributions, they are most often described by the negative binomial distribution (1,4,8,15,28,44,49,57,58,63). This distribution has a variance (s^2) to mean ($\bar{x}$) ratio that is greater than 1.0, indicating aggregation. The variance to mean ratio would equal 1.0 for randomly distributed populations. A regular distribution has an expected s^2 that is less than the $\bar{x}$. The negative binomial is especially suited for describing clustered populations. It is described by two parameters: the $\bar{x}$ and an aggregation index, the exponent k. The k parameter reflects the magnitude of clumping. The value of k also increases with the mean. Fractional k values indicate that the distribution is approaching the logarithmic series which occurs at $k = 0$. As k increases to infinity (for practical purposes, k values ≥ 8), the distribution becomes identical to the Poisson, indicating a near random distribution (20,66). A simple method for estimating k is:

$$k = \frac{\bar{x}^2}{s^2 - \bar{x}}$$

The accuracy of this estimation method is dependent on n, the sample size. In general, k should be estimated in this manner only when $n > 30$. Computer programs are available for testing the goodness-of-fit of the negative binomial distribution to sample data and for deriving estimates of k (25,28). Analyses of

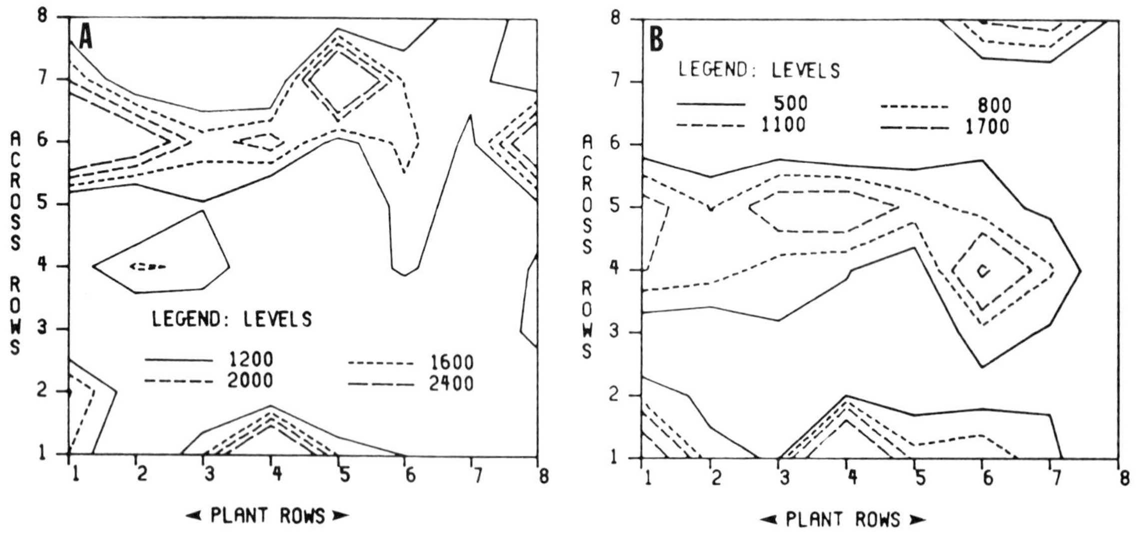

Fig. 2. Contour maps of relative levels of A) *Meloidogyne incognita* and B) *Criconemella* spp. population densities (nematode numbers/500 cm^3 soil) in a tobacco field [after Noe (49)].

sample data for *Meloidogyne* spp. have given a range of k values of 0.29 to 2.0, indicating a high degree of aggregation (28,49). Some data for this nematode cannot be fitted to the negative binomial distribution (47), but may be fitted to others (Fig. 1).

Taylor's power law (20,66,67) provides another index of aggregation for a given species:

$$s^2 = a(\bar{x})^b$$

where s^2 = samples variance; a = sampling factor; and b = index of aggregation. Values of $b > 1.0$ indicate clustering of the population, with aggregation increasing as b increases. This model has been used in the study of a few plant-parasitic nematodes (8,15,24). Various other indices of aggregation (1,8,18,20,49,50,66) also may be used to describe nematode populations.

Information on the vertical spatial patterns of plant-parasitic nematodes also is essential for reliable sampling. Typically, most plant-parasitic nematodes are closely associated with the root distribution. The greatest population densities of most species occur in the plow layer of the soil. This fact is true for most *Meloidogyne* spp., but some have been found at depths of 120 cm or greater (2,23,48). The vertical spatial patterns of *M. incognita* and other nematodes may change with time (32,48). Numbers of juveniles of this species may increase at depths of 30-45 cm in the spring while those at lesser depths decline (Fig. 3).

The timing of sampling for nematodes should be based on knowledge of the seasonal population dynamics of the target species. The fact that the life span of most *Meloidogyne* spp. is less than 6 months partially accounts for population declines after the harvest of annual plants (Fig. 4). In contrast, various life stages of *Globodera* and *Heterodera* spp. may survive for many years, giving more temporal stability to these nematode populations. Highly virulent species of *Meloidogyne*, such as *M. arenaria*, may cause enough damage to their host that the population declines and infestations are undetectable at planting time in the warmer or tropical regions. The management practices, including the use of nematicides and/or destruction of roots after harvest, also alter the spatial and temporal patterns of nematodes. The differential temporal changes of the predominant species in polyspecific communities often pose problems in sampling and should be considered in selecting sampling dates.

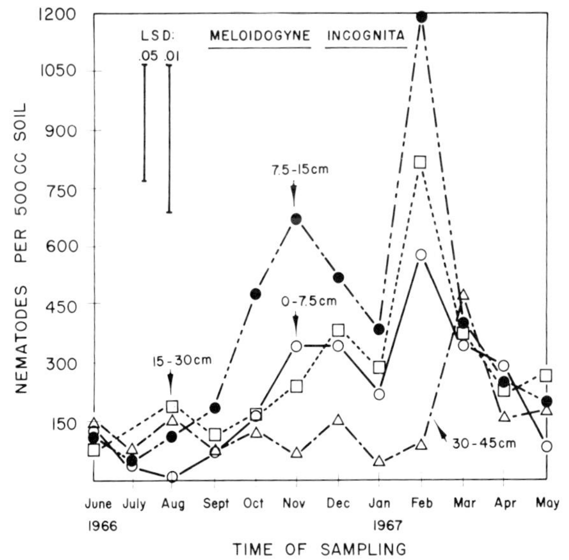

Fig. 3. Population changes of juveniles of *Meloidogyne incognita* at different soil depths during growing season (crop was flue-cured tobacco; sugar-flotation-sieving extraction (7); note increase in nematode numbers at 30-45 cm depth in spring).

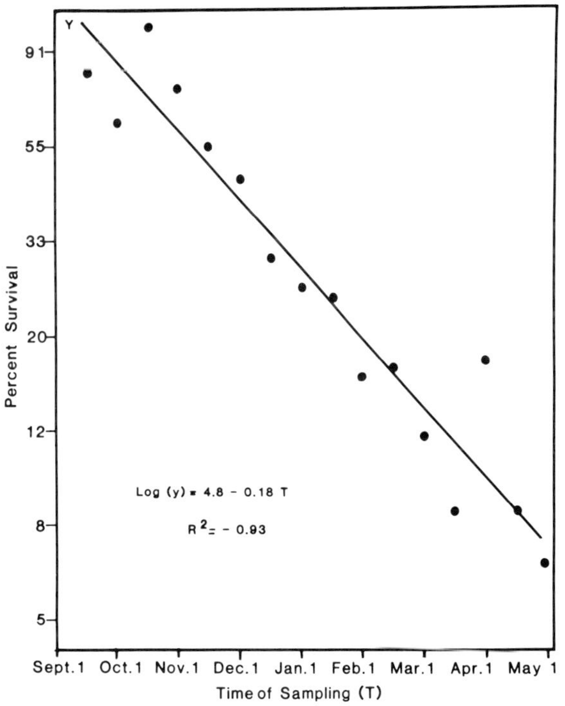

Fig. 4. Regression depicting winter population decline of *Meloidogyne incognita* after tobacco.

IMPORTANCE OF NEMATODE BIOLOGY AND CROP MANAGEMENT

One must be familiar with the biology and population dynamics of target species in developing sampling schedules and procedures. The parasitic habit, type and number of egg depositions by females is particularly important in considering sampling. Some species deposit eggs singly, some in masses, and some develop eggs that are retained in cysts. The life stages detectable may pose problems in sampling for nematodes. The eggs of most plant-parasitic nematodes are difficult to identify if they can be extracted. Similarly, juveniles typically can be identified only to the generic level.

The decision of whether to collect root as well as soil samples for nematode assay depends on the host-parasite relations of the nematodes in question. Polyspecific communities that contain both ectoparasitic and endoparasitic nematodes may require soil and root samples, particularly after mid-season for annual crops and for all samples of perennial crops. The specific timing of sample collection is dependent on the population fluctuations for each of the target species.

Crop phenology and management have important impacts on nematode populations which must be evaluated in sampling. These relationships, however, may be confounded by effects of initial population levels of nematodes. Normally, nematodes build up to maximum levels at midseason or near harvest on annual crops. Where high initial populations occur, however, the maximum populations may develop before midseason and then decline as a result of severe damage to the host. With perennial plants, populations fluctuate over time with greatest population levels occurring sometime after maximum plant growth.

Management practices often affect nematode populations and should be considered in collecting soil samples for nematode assays. Generally, row crops accentuate the patchiness or non-random distribution of nematodes. In contrast, broadcast seeding of crops such as alfalfa may minimize the contagious distribution of nematodes. Minimal or no-tillage cropping practices also allow nematodes to remain in greater numbers in the root zones of the old crop, resulting in a more patchy spatial pattern. In contrast, tillage of land after crop harvest disperses the nematodes from the old roots as well as increases death rates of the nematodes. Where possible, soil samples should be collected after plowing which results in the nematodes being more randomly distributed (34).

Edaphic factors greatly influence the local as well as regional distribution and population dynamics of nematodes. A few nematodes, such as *Belonolaimus longicaudatus* occur only in soils with a high sand content. The peak populations of nematodes such as *Meloidogyne incognita* may occur near the end of a tomato season in a fine-textured soil where the temperatures are relatively low, but at midseason in sandy soil in a warmer geographic location (11). These effects may be altered by management tactics such as the use of nematicides and/or resistant cultivars. The worldwide study of the relationship of edaphic factors to populations of *Meloidogyne* spp. by Sasser (62) should provide greater understanding on how these factors affect nematode distribution as well as crop response.

SAMPLE COLLECTION

Only limited developments have occured in the technology of soil-sampling devices. The classical, cylindrical, tube-type sampler and auger designed by soil scientists still serve as standard equipment for most nematologists. Improvements in the cylindrical soil-core sampling tube have been directed largely at reducing the effort required to probe the soil. Esser et al. (21) developed two types of cone-shaped tubes that facilitate more rapid and less strenuous sample collection than that allowed by the Hoffer or Oakfield sampling tubes. There are also various types of electric (40) and hydraulic-assisted (69) tubes that facilitate sampling in compacted soils and allow the collection of relatively undisturbed cores of soil to a depth of a meter or more. Sampling for roots with an auger was improved by the development of a screen-shaker unit for separating soil from roots (35).

Two rather unique sampling approaches have been utilized for cyst nematodes. One system involves a machine that operates on the vacuum principle (31). Another sampler is comprised of an all-terrain vehicle with buckets on a chain (19). The greatest need in sampling is for an automated tractor-mounted sampling device which could be used during land preparation. This would be particularly useful in the extensive IPM programs.

Timing of sample collection is dependent on the purpose of the nematode assay and the population dynamics of the target species. Survey samples for use in quarantines or in the characterization of nematode distributions should be made when nematode populations occur in maximum numbers. Evaluations of nematicides and resistant cultivars should be based on samples taken when the populations of non-treated, susceptible plants are near maximum, but before the effects of the management tactics have been depleted. The best time for sampling for advisory purposes in relation to crop performance is immediately prior to establishment of a crop. Some nematode populations, however, may be below detectable levels at that time of the year. Because of this fact, some nematode advisory programs utilize fall sampling and then extrapolate, based on expected mortality, from those population densities to those expected at the time the subsequent crop is established. This estimation can be done with simple regressions (Fig. 4) or more comprehensive models. Unfortunately, the slopes of such regressions vary with management practices (Fig. 5) and environment. Populations of *Heterodera glycines* have a 60-

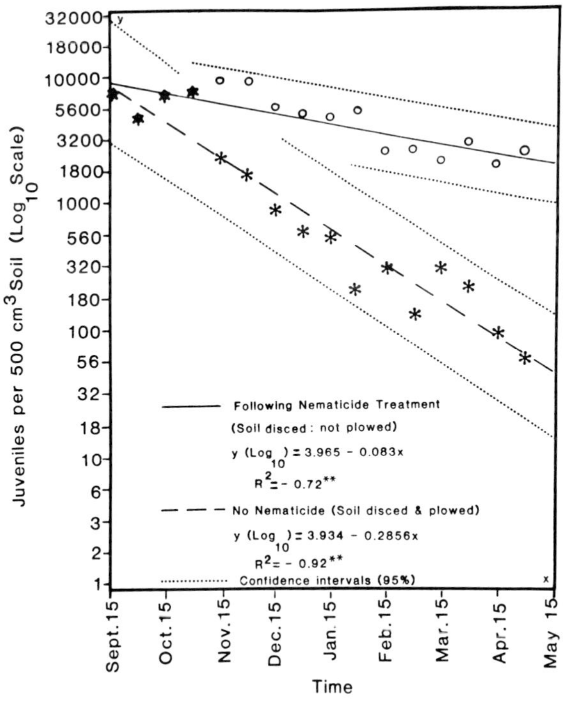

Fig. 5. Effects of management practices on survival of *Meloidogyne incognita* after a tobacco crop (nematodes extracted by centrifugal flotation).

70% mortality during some winters in North Carolina, but decline little during other years (D. P. Schmitt, personal communication).

Three to four different sample periods should be employed in the evaluation of nematicide efficacy and relative cultivar resistance. For nematicide treatments, a pre-treatment sample (P_i), within 1 week prior to treatment, an intermediate post-treatment (P_{ie}), a midseaon (P_m), and a final sample (P_f) should be used (7). Where nematicide treatments are applied in the fall to avoid delays in establishment of spring crops, the sample schedule should be modified accordingly. The collection of post-treatment samples in late fall and early spring are necessary if nematicide efficacy on perennials is to be determined. A population sample at harvest may be sufficient for evaluation of resistant cultivars, but more frequent assays are desirable even for this purpose. The optimum sampling times may be determined more precisely by calculation of degree hours or days (24), or cumulative indices of nematode activity (33) necessary for maximum population increase. Weekly or biweekly nematode and plant monitoring may be necessary in modeling experiments. This topic is discussed by Ferris in Chapter 16.

The relative size of sample area and numbers of cores collected per area for nematode assay differ greatly for different purposes and between assay and research programs. A sample area of 1-2 hectares or less is recommended for advisory purposes (47,59). Very large fields should be broken down on the basis of a grid, and representative areas with uniform soil and crop histories should be selected randomly to provide a basis for sampling. For regulatory purposes, the sample area should be much smaller, in the range of approximately 1/3 hectare, but this differs from country to country (64).

Sample size may be determined on the basis of relative population densities and the nematode spatial pattern (8,26,28,43,44,45,46). Nematode sample data for some fields do not fit any mathematical distribution (1,47). In these situations, empirical methods of computing relative errors for samples can be utilized (47).

The approach of measuring sample error may be expanded to include measuring that of crop-response (24), but this is yet to be accomplished for most nematode-crop combinations. A soil sample of 10 to 30 cores may give adequate information for estimates of a large population, but 100 or more cores may be necessary if low population levels are to be estimated or even detected. The collection of five composite samples with 20 cores each was found to give satisfactory or acceptable sample information for the nematodes infesting a 7-hectare alfalfa field (28,30).

Mathematical models have been developed by which the numbers of samples needed for given levels of precision can be estimated (36,66). Numbers of cores necessary for aggregated populations, which can be described by the negative binomial, may be calculated with the following equation:

$$n = (Z\alpha/2)^2 \frac{(1/\bar{x} + 1/k)}{E^2},$$

where n = number of samples; Z = upper $\alpha/2$ of the normal distribution; $\bar{x}$ = mean of the population; k = experimentally determined dispersion parameter of the negative binomial; and E = predetermined standard error as a decimal of the mean. This model, as described by Karindinos (36), is much more rigorous than those given by other authors (8,66). It does have the advantage of being more suitable for a single sample. This model will show that the recommended numbers of cores of 20-50 provides a population estimate within a range of 30-50% of the mean for most row crops.

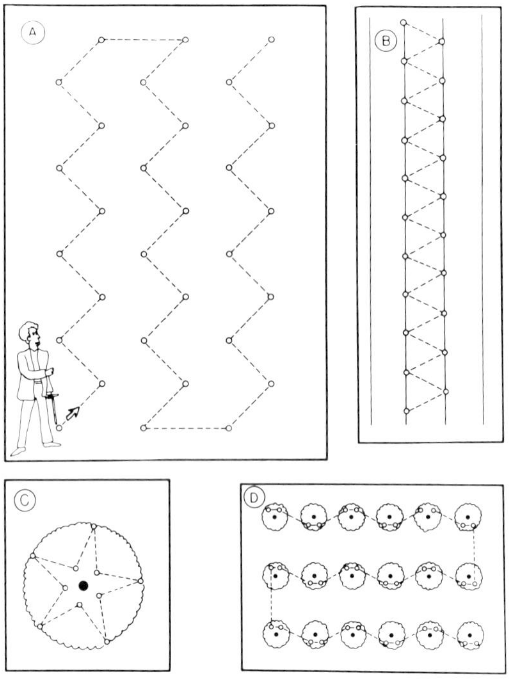

Fig. 6. Suggested diagrams for collecting soil samples for nematode assays. A) Recommended pattern of collecting a minimum of 20-30 cores for general assays in large plots or fields of 1-2 ha; B) Sampling pattern for collecting soil in two center rows of four-row experimental plots with performance data also being obtained from these two rows; C) Procedure for sampling single-plant plot; D) Patterns for sampling in feeder-root zone of established perennials (additional cores should be collected from each plant where small plots are used) [after Barker et al. (7)].

Sampling patterns that may be used for nematodes include random, stratified random, systematic, and two-stage sampling. The specific type of spatial pattern and relative density of the target nematode species should be considered in selection of the sampling procedure (Fig. 6). The nematode species being assayed also have an influence on the results of sampling experiments. Church et al. (16) suggested that a systematic approach to sampling of *Globodera rostochiensis* was preferable over random sampling. Samples were taken at the rate of 25-50 cores for fields up to 10 acres. The problems associated with the clustered distributions of nematodes as described herein have been documented by many investigators. In working with *Pratylenchus penetrans*, Proctor and Marks (57) concluded that 40-core samples gave more reliable results than 20-core samples of 0.01-ha plots. Recent studies by Seinhorst (63) also show that large samples may reduce some of the error in sampling for *Globodera rostochiensis* in very small plots.

Considerable attention has been given to sampling procedures for *Meloidogyne* spp. by a number of researchers (28-30,47,49,53,54). Although nematologists have obtained similar k values for *Meloidogyne* spp. on row crops, the estimated numbers of soil cores and samples needed for a given level of precision vary greatly (28, 43, 46; Barker et al., unpublished). This divergence is due to utilization of different models for calculation of these estimates. According to the Karindinos model (36), 30 to 60 cores are needed if a population estimate within 50% of the mean is to be obtained (Barker et al., unpublished). The less rigorous model, which deletes the $(Z\alpha/2)^2$ component of the model, gives core or sample estimates that are about 70% fewer than when this component is included. The precision attainable in solidly planted crops, such as alfalfa, apparently is more acceptable, being in the range of 20% of the mean (28). Because sampling error varies with degree of aggregation, population density, field size, and sample size (26), further research is needed to determine the range of sampling precision that is attainable with nematodes. This endeavor should include repeated sampling of several fields over a number of sample times. The very small k values (0.27 to 2.0) obtained for *Meloidogyne* spp. (based largely on single soil cores) thus result in unmanageable sample numbers (Table 3) for acceptable levels of precision for some purposes. This problem can be overcome partially for IPM purposes by use of large composite samples of 1.0-1.5 kg, as suggested by Seinhorst (63) for *Globodera rostochiensis*. The precision in sampling low population density fields and in describing the associated distributions may be improved by the use of bioassays (46,53,54). This approach is especially important for virulent *Meloidogyne* spp.

The direction of row-placement often has a major influence on the distribution pattern of those nematodes (49). Similarly, edaphic factors such as soil texture and other environmental factors, may be related to the distribution of nematodes and should be considered in sampling (55,56). On the basis of available information, more intensive sampling is needed for row crops (Barker et al., unpublished) as compared to solidly planted crops (28).

Another source of variation in sampling involves the mechanics of the collection of specific cores by given individuals. This problem can be minimized by standardization of the distance from plants, depth, angles, and inclusion of roots by sampling only when moisture is near optimum. The loci of sample collection should be correlated or closely related to the location of maximum fibrous root growth (7,39,45). On annual crops, this usually is within 10-20 cm of the base of the plants, whereas the collection of soil cores on most perennial plants should be in the area of the dripline of the foliage (7,45) (Fig. 6-C, D). The depth of sample collection is dependent on the host, nematode species, and type of soil. Variation also is introduced among counts made by different observers in the laboratories. Use of duplicate samples minimizes these errors, but is impractical in commercial-scale laboratories.

CARE AND STORAGE OF SAMPLES

Plant-parasitic nematodes are thin-walled, short-lived microorganisms. Improper handling, such as exposure to the sun, temperature extremes or improper storage results in high mortality rates and erroneous information.

Table 3. Examples of numbers of samples needed based on model below[a].

Value of k[b]	Predetermined precision (E)[c]	Number of samples needed per mean number of nematodes[d] 10	100	1000
0.3	0.05	5276	5138	5124
	0.10	1319	1284	1281
	0.20	330	321	320
	0.30	147	143	142
1.0	0.05	1690	1552	1538
	0.10	423	388	385
	0.20	106	97	96
	0.30	47	43	43
3.0	0.05	666	528	514
	0.10	167	132	128
	0.20	42	33	32
	0.50	19	15	14

[a] Model is: $n = (Z\alpha/2)^2 \dfrac{1/\bar{x} + 1/k}{E^2}$, Karandinos (36).

[b] k = aggregation parameter, based on negative binomial distribution.

[c] E = precision parameter (as % of mean).

[d] Refers to numbers of soil cores needed per k and $\bar{x}$ combination per level of precision.

These problems are particularly important where extraction methods are dependent on nematode motility or where bioassays are utilized. The handling of soil and root samples during collection and transit to the laboratory warrant special care. After collection, each sample should be sealed in a plastic bag and placed in an insulated container or kept out of the sun at a moderate temperature, preferably less than 30°C (59). The effects of shipping samples on nematode survival have received only limited study. In one comparison, samples collected and mailed to the laboratory yielded fewer numbers of *Belonolaimus* spp. and *Paratrichodorus minor* than samples that were transported directly to the lab (R. E. Motsinger, personal communication). Other species such as *Hoplolaimus columbus* and *Rotylenchulus reniformis* were not adversely affected by transit through the mail. Another study showed that the use of styrofoam containers gave 1.4 to 3 times more specimens of *Helicotylenchus dihystera, P. minor* and *Xiphinema americanum* upon extraction by elutriation and centrifugation than did samples shipped (during 14 days) in non-insulated containers (59). Soil samples from quarantined areas may be treated before shipment by saturation with the fixative FAA and picric acid or 5% formalin. These nematodes must be extracted by flotation methods.

Laboratory storage of soil samples may facilitate hatching and increased extractable numbers of nematodes, or improper storage may lead to nematode death, which would result in misleading assay results. The two most important parameters in sample storage are temperature and moisture. Adequate moisture and a temperature of 10-13°C result in relatively uniform recovery rates of *Meloidogyne incognita* over a period of weeks, as compared to numbers in non-stored samples (10) (Fig. 7). However, excessive moisture or extreme temperatures may result in rapid declines of nematode populations in storage. An exception to this was observed with *Criconemella xenoplax*: samples stored with 16-24% moisture (V/W) for 1-7 days before extraction increased the recovery of this nematode (38).

METHODS OF CHARACTERIZING NEMATODE COMMUNITIES

The precision of estimates of nematode population densities is dependent on the relative density, degree of aggregation, type of sample and reliability of the extraction methods utilized. Errors among samples of *Globodera rostochiensis* are about 2 times greater than that expected for Poisson distributions (64,65). Estimates of the errors encountered for *Meloidogyne* spp. vary more than this.

Numerous additional methods of analyzing

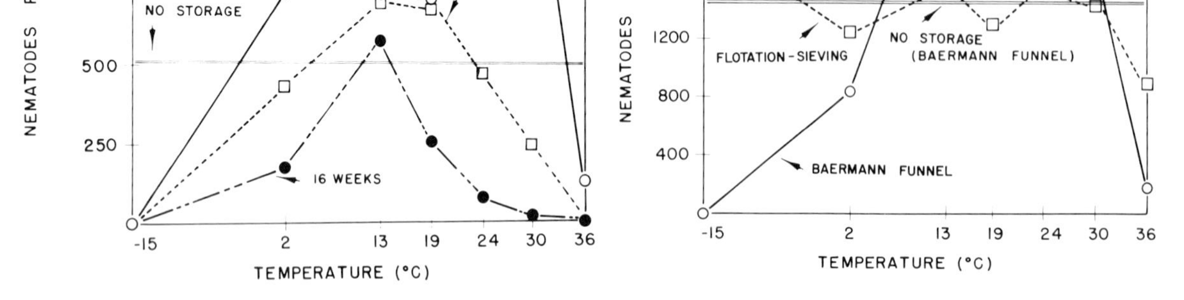

Fig. 7. Survival of *Meloidogyne incognita* as affected by temperature and storage time. A) Short-term effects; B) Mean effects of temperature on yields with two extraction methods over 16 weeks storage [latter after Barker et al. (10)].

nematode communities are available (20,49,50, 66). These range from frequencies of occurrence (50) to complex spectral analysis (49). Only some of the most frequently used procedures are included herein. The application of these and other ecological principles should improve the reliability and usefulness of sampling for nematodes.

Most nematode assays are based on numbers or density per unit volume or weight of soil. In addition to this absolute measure of density, one may determine how many specimens/species are in a sample relative to the total mean number of nematodes for all species. This relative density is calculated as a percentage (50):

$$\textit{Relative density} = \frac{\textit{no. of individuals of a species in a sample}}{\textit{total no. of individuals in a sample}} \times 100.$$

The sum of all relative densities for a community equals 100%.

Models are available for determining relative biomass such as that developed by Andrassy (3):

$$G = \frac{a^2 b}{(1.6)(1{,}000{,}000)}$$

where G = biomass in μg; a = the greatest body width (μm); b = body length (μm); 1.6 = constant for correcting volume of nematode; and 1,000,000 is a factor for converting μm^3 to μg. Typical biomass values for given nematodes are available (3).

Two types of frequency may be calculated. Frequency is a measure of distribution uniformity or rate of occurrence (50). *Absolute frequency* is given as a percentage:

$$\frac{\textit{no. of samples containing a species}}{\textit{no. of samples collected}} \times 100.$$

Relative frequency is calculated as:

$$\frac{\textit{absolute frequency of species}}{\textit{sum of frequency of all species}} \times 100.$$

Absolute frequency is an independent calculation for rate of occurrence for each species. In contrast, relative frequency gives the total frequency of all species on the basis of 100% (50).

The relationship of population density and frequency can be expressed as a *prominence value* (50):

$$\textit{Prominence value} = \textit{density}\ \sqrt{\textit{frequency}}\ .$$

These determinations are based on the absolute density and frequency per species.

The importance of a nematode species in a community may be estimated by calculation of the *importance value* (50):

$$\textit{Importance value} = \textit{relative frequency} + \textit{relative density} + \textit{relative biomass}.$$

The sum of all relative biomass values for a community also equals 100 as with relative frequency and density (50). A discussion of the potential biological value of this concept in comparison of nematodes of different trophic groups is available (50).

An "index of similarity" may be determined for comparison of nematode infestations of two areas or test sites. The following is an example for the models summarized by Norton (50):

$$IS_s = \frac{2c}{a + b}$$

where IS_s is the index of similarity (after Sorenson); c = no. of species that the two samples or areas have in common; and a and b = no. of species in samples of area a and b, respectively.

As indicated earlier, advances in computer science greatly facilitate data management and mapping techniques needed in nematode population studies. Simulation models are becoming available for evaluation of various sampling procedures. A number of computer programs also are available for fitting nematode population data to standard distributions, as well as mapping the nematode distributions over space and time. Three-dimensional distribution maps can be developed from nematode population data by computer (29). These programs are helpful in the study of nematode communities, improvement of sample methods, and in relation of nematode spatial patterns to various edaphic factors such as soil texture (28).

NEEDS FOR ADDITIONAL RESEARCH

Although much progress has been made in characterizing nematode communities and improving sampling procedures, much additional research is still needed in this area. First, the inherent errors associated with various sampling schemes are still open to question. Estimated errors obtained by some investigators are quite low (28,44), in the range of 10 to 20% standard error, whereas those of others are quite high (49,63). Another problem area that has received limited attention concerns the integration of sampling and experimental errors in relation to expected damage caused by nematodes (24). The contagious spatial patterns pose problems in modeling nematode population dynamics as well as plant response to these pests (52). Variation in nematode distribution patterns from field to field and over time as well as the impact of management practices need further clarification. Basic sampling methodology still needs further development. There is especially a need for automated sampling.

Sampling nematode communities does provide data that are valuable in many integrated pest management purposes. For example, in spite of the variation encountered, sufficiently reliable information can be derived that enables growers to manage serious nematode problems through assay programs. These assays are quite useful in diagnosing the causes of crop production problems, including those caused primarily by nematodes, as well as the involvement of nematodes in disease complexes. In contrast, the development of highly reliable quantitative data still warrants extensive research.

Literature Cited

1. Alby, T., J. M. Ferris, and V. R. Ferris. 1983. Dispersion and distribution of *Pratylenchus scribneri* and *Hoplolaimus galeatus* in soybean fields. J. Nematol. 15:418-426.
2. Allen, M. W. 1965. Ecology of cotton root-knot nematode in California. Nematologica 11:33 (Abstr.).
3. Andrassy, I. 1956. The determination of volume and weight of nematodes, pp. 73-84 *In* English translation of selected East European papers in nematology, eds. B. M. Zuckerman, M. W. Brzeski, and K. H. Deubert. Univ. Mass. Cranberry Exp. Stn., East Wareham, Massachusetts.
4. Anscombe, F. J. 1950. Soil sampling for potato root eelworm cysts. Ann. Appl. Biol. 37:286-295.
5. Atkinson, H. J. and A. J. Ballantyne. 1977. The adenosine triphosphate content of cysts and *Globodera pallida* and *G. rostochiensis* as a possible quantitative estimate of field populations. Ann. Appl. Biol. 87: 407-414.
6. Atkinson, H. J. and G. B. Sykes. 1981. An analysis from aerial photography of the spread of seed-borne *Ditylenchus dipsaci* in Lucerne. Nematologica 27:235-241.
7. Barker, K. R. (Chairman). 1978. Determining nematode population responses to control agents, pp. 114-125 *In* Methods for evaluating plant fungicides, nematicides and bactericides, ed. E. I. Zehr. Am. Phytopathol. Soc., St. Paul, Minn.
8. Barker, K. R. and C. L. Campbell. 1981. Sampling nematode populations, pp. 451-473 *In* Plant parasitic nematodes, Vol. III, ed., B. M. Zuckerman and R. A. Rohde. Academic Press, New York. 508 pp.
9. Barker, K. R., C. J. Nusbaum, and L. A. Nelson. 1969. Seasonal population dynamics of selected plant-parasitic nematodes as measured by three extraction procedures. J. Nematol. 1:232-239.
10. Barker, K. R., C. J. Nusbaum, and L. A. Nelson. 1969. Effects of storage temperature and extraction procedure on recovery of plant-parasitic nematodes from field soils. J. Nematol. 1:240-247.
11. Barker, K. R., P. B. Shoemaker, and L. A. Nelson. 1976. Relationships of initial densities of *Meloidogyne incognita* and *M. hapla* to yield of tomato. J. Nematol. 8: 232-239.
12. Barker, K. R., F. A. Todd, W. W. Shane, and L. A. Nelson. 1981. Interrelationships of *Meloidogyne* species with flue-cured tobacco. J. Nematol. 13:67-79.
13. Boag, B. 1981. Observations on the population dynamics and vertical distribution of Trichodorid nematodes in a Scottish forest nursery. Ann. Appl. Biol. 98:463-469.
14. Brodie, B. B. 1976. Vertical distribution of three nematode species in relation to certain soil properties. J. Nematol. 8:243-247.
15. Caubel, G., D. Mugniery, and R. Rivoal.

1972. [Distribution of the bulb and stem nematode, *Ditylenchus dipsaci* (Kuhn) Filipjev, in the soil of a field of red clover under attack.] (In French) Ann. Zool. Ecol. Anim. 4:385-393.
16. Church, B. M., H. C. Gough, and J. F. Southey. 1959. Soil sampling procedures for potato root eelworm cysts. Plant Pathol. 8:146-151.
17. Cobb, N. A. 1918. Estimating the nema population of soil. U.S. Dep. Agric. Agric. Tech. Circ. 1. 48 pp.
18. Cockran, W. G. 1977. Sampling techniques, 3rd ed. Wiley, New York. 428 pp.
19. Cooke, D. A., H. E. McKinney, and I. J. Thomason. 1979. A rapid method of sampling surface soil. J. Nematol. 11:202-204.
20. Elliott, J. M. 1977. Some methods for statistical analysis of samples of benthic invertebrates. Freshwater Biol. Assoc. Sci. Publ. 25. 144 pp.
21. Esser, R. P., J. B. MacGowan, and H. M. Van Pelt. 1965. Two new nematode subsampling tools. Plant Dis. Rep. 49:265-267.
22. Fernandez, L. G. 1978. Distribución horizontal de algunos géneros de nemátodos fitoparasitos en terrenos agricolas de Costa Rica. Turrialba 28:67-69.
23. Ferris, H. 1969. The vertical distribution of the root-knot nematode (*Meloidogyne javanica*) in a Rhodesian sandy soil. Rhod. J. Agric. 7:15-24.
24. Ferris, H. 1984. Nematode damage functions: The problems of experimental and sampling error. J. Nematol. 16:1-9.
25. Gates, C. E. and F. G. Ethridge. 1972. A generalized set of discrete frequency distributions with FORTRAN program. Math. Geol. 4:1-24.
26. Gérard, G. and P. Berthet. 1971. Sampling strategy in censusing patchy populations, pp. 59-67 *In* Statistical ecology, Vol. I: Spatial patterns and statistical distributions, ed. G. P. Patil, E. C. Pielou, and W. E. Walters. Penn State Univ. Press, University Park, Penn. 582 pp.
27. Godfrey, G. H. 1934. Indicator plants for measuring soil populations of the root-knot nematode, *Heterodera marioni* (Cornu) Godfrey. Soil Sci. 38:3-37.
28. Goodell, P. and H. Ferris. 1980. Plant-parasitic nematode distributions in an alfalfa field. J. Nematol. 12:136-141.
29. Goodell, P. and H. Ferris. 1980. Graphic techniques for illustrating distribution data. J. Nematol. 12:151-152.
30. Goodell, P. B. and H. Ferris. 1981. Sample optimization for five plant-parasitic nematodes in an alfalfa field. J. Nematol. 13: 304-313.
31. James, P. E. 1974. A vacuum nematode cyst sampler. Agric. Eng. 55:27.
32. Johnson, A. W., C. C. Dowler, and E. W. Hauser. 1974. Seasonal population dynamics of selected plant-parasitic nematodes on four monocultured crops. J. Nematol. 6:187-190.
33. Jones, F. G. W. 1975. Accumulated temperature and rainfall as measures of nematode development and activity. Nematologica 21:62-70.
34. Jones, F. G. W. and R. A. Kempton. 1978. Population dynamics, population models and integrated control, pp. 333-361 *In* Plant nematology, ed. J. F. Southey. Minist. Agric., Fish. and Food Technol. Bull. 7. Her Majesty's Stationery Office, London. 440 pp.
35. Jutras, P. J. and A. C. Tarjan. 1965. A novel hydralic soil auger-screen shaker unit for the collection of root samples. Proc. Soil and Crop Sci. Soc. Fla. 24:154-158.
36. Karandinos, M. C. 1976. Optimum sample size and comments on some published formulae. Bull. Entomol. Soc. Am. 22:417-421.
37. Kleijburg, P. 1960. Soil sample examination as a base for advisory work about stem nematodes. Nematologica 5:22-27.
38. Lawrence, E. G. and E. I. Zehr. 1978. Improvement of techniques for determining populations of *Macroposthonia xenoplax* in dry soil. Phytopathology 68:1102-1105.
39. Lopez, C. R. 1981. Observaciones sobre la distribución espacial de *Meloidogyne incognita* despues de la cosecha en dos plantios de tabaco burley. Turrialba 31:11-14.
40. Lownsbery, B. F., J. T. Mitchell, and E. E. Head. 1959. An electric auger for nematological soil sampling in orchards. Plant Dis. Rep. 43:918-919.
41. Marks, C. F., J. M. Elliot, J. R. Rainforth, M. C. Watson, and B. Back. 1974. Aerial photography—an aid in surveying for damage by root-lesion nematode in flue-cured tobacco. Can. Plant Dis. Surv. 54: 105-107.
42. McBurney, T. 1981. Field assessment of stem nematode (*Ditylenchus dipsaci*) from visual symptoms in peas. Nematologica

27:115-117.
43. McSorley, R. 1982. Simulated sampling strategies for nematodes distributed according to a negative binomial model. J. Nematol. 14:517-522.
44. McSorley, R. and J. L. Parrado. 1982. Estimating relative error in nematode numbers from single soil samples composed of multiple cores. J. Nematol. 14:522-529.
45. McSorley, R. and J. L. Parrado. 1982. Spatial arrangement of nematodes around four species of tropical fruit trees. Nematropica 12:247-255.
46. McSorley, R. and J. L. Parrado. 1982. Plans for the collection of nematode soil samples from fruit groves. Nematropica 12:257-267.
47. McSorley, R. and J. L. Parrado. 1983. A bioassay sampling plan for *Meloidogyne incognita*. Plant Dis. 67:182-184.
48. Nirula, K. K. and B. T. Raj. 1971. Vertical distribution of *Meloidogyne incognita* in potato fields. Indian Phytopathol. 24:375-378.
49. Noe, J. P. 1984. Spatial population dynamics of plant-parasitic nematodes. Ph.D. thesis, North Carolina State Univ., Raleigh, N.C. 96 pp.
50. Norton, D. C. 1978. Ecology of plant-parasitic nematodes. Wiley, New York. 268 pp.
51. Orion, D., B. Nissim-Bistritsky, and R. Hochberg. 1982. Using color infrared aerial photography to study cotton fields infested with *Meloidogyne incognita*. Plant Dis. 66: 105-108.
52. Perry, J. N. 1983. Effects of spatial heterogeneity on Jones' model for cyst-nematode population dynamics and crop root damage. J. Appl. Ecol. 20:849-856.
53. Prot, J. C. and C. Netscher. 1978. Improved detection of low population densities of *Meloidogyne*. Nematologica 24:129-132.
54. Prot, J. C. and C. Netscher. 1979. Influence of movement of juveniles on detection of fields infected with *Meloidogyne*, pp. 193-203 *In* Root-knot nematodes (*Meloidogyne* species) systematics, biology and control, ed. F. Lamberti and C. E. Taylor. Academic Press, New York. 477 pp.
55. Prot, J. C. and S. D. Van Gundy. 1981. Effect of soil texture and the clay component on migration of *Meloidogyne incognita* second-stage juveniles. J. Nematol. 13:213-217.
56. Prot, J. C. and S. D. Van Gundy. 1981. Influence of photo-period and temperatures on migrations of *Meloidogyne* juveniles. J. Nematol. 13:217-220.
57. Proctor, J. R. and C. F. Marks. 1975. The determination of normalizing transformations for nematode count data from soil samples and of efficient sampling schemes. Nematologica 20:395-406.
58. Raeuber, A. and H. Stelter. 1963. Die Sicherheit bei der Bestimmung der Verseuchung eines Feldes mit Kartoffelnematoden (*Heterodera rostochiensis* Woll.). Biol. Zentralbl. 82:455-464.
59. Rickard, D. A. and K. R. Barker. 1982. Nematode assays and advisory services, pp. 8-20 *In* Nematology in the southeastern region of the United States, ed. R. D. Riggs. South. Coop. Series Bull. 276, Arkansas Agric. Exp. Stn., Fayetteville, Ark. 206 pp.
60. Sanwald, E. F. 1978. Some basic considerations on surveying and inventorying by remote sensing damage caused by *Heterodera schachtii* Sch. in sugar beets. Ann. Phytopathol. 10:273-274.
61. Sanwald, E. F. 1979. Mapping of nematode damage to sugar-beet crops by aerial infra-red photography. Zuckerindustrie 104: 125-129.
62. Sasser, J. N. 1980. Root-knot nematodes: a global menace to crop production. Plant Dis. 64:36-41.
63. Seinhorst, J. W. 1982. The distribution of cysts of *Globodera rostochiensis* in small plots and the resulting sample errors. Nematologica 28:285-297.
64. Southey, J. F. 1970. Principles of sampling for nematodes, pp. 1-4 *In* Laboratory methods for work with plant and soil nematodes, 5th ed., ed. J. F. Southey. Minist. Agric., Fish. and Food Tech. Bull. 2. Her Majesty's Stationery Office, London.
65. Southey, J. F. 1974. Methods for detection of potato cyst nematodes. EPPO (Eur. Mediterr. Plant Prot. Organ.) Bull. 4:463-473.
66. Southwood, T. R. E. 1978. Ecological methods with particular reference to the study of insect populations, 2nd ed. University Printing House, Cambridge, England. 524 pp.
67. Taylor, L. R. 1961. Aggregation, variance and the mean. Nature 189:732-735.
68. Thomason, I. J., E. P. Caswell, and H. E. McKinney. 1983. Utility of tare soil sam-

pling for detection and management of *Heterodera schachtii*. Nematropica 13:122 (Abstr.).
69. Turner, G. O. 1974. A device for obtaining uncontaminated and undisturbed soil samples. Down to Earth 30:18-21.
70. Weingartner, D. P., J. R. Schumaker, and G. C. Smart, Jr. 1983. Why soil fumigation fails to control potato corky ringspot disease in Florida. Plant Dis. 67:130-134.

CHAPTER 2

NEMATODE EXTRACTION AND BIOASSAYS

K. R. Barker

Plant Pathology Department
North Carolina State University
Raleigh, North Carolina, USA

Comtemporary nematologists with an interest in population dynamics are fortunate in that numerous assay methods have been developed. Although methods for routine monitoring of many soilborne microflora and microfauna are still unavailable, nematologists have made major advancements with these techniques (16,21,31,41,53,64). Nevertheless, the selection of the best assay procedure for given nematode species continues to pose problems. For example, the importance of ectoparasitic nematodes was overlooked until the 1950's (18,19) because scientists failed to associate these external feeders with host damage.

This chapter enumerates many of the factors that should be considered in nematode assays, especially with regard to *Meloidogyne* spp. The polyspecific nature of nematode communities, however, requires that we be concerned with several genera of plant-parasitic nematodes. Thus, methods for extracting most major nematode taxa are included herein. The advantages and deficiencies of these procedures also are discussed. Topics emphasized include the biological and ecological considerations in selection of assay methods, care and preparation of samples, relative efficacy of extraction procedures, use of vital stains, practical application of bioassays, special considerations in using certain assay methods, outlines of the selected procedures or methods, and types of data that should be reported. Special methods such as use of hydrolytic enzymes to extract females of *Meloidogyne* spp. (28) are not included. Extraction methods not described herein are outlined in other reviews (4,25,31,35,41,54,58). Since sample size has been discussed in the previous chapter, it will not be covered here.

BIOLOGICAL AND ECOLOGICAL CONSIDERATIONS IN SELECTION OF ASSAY METHODS

Nematode-assay methods should be chosen that yield the highest or most reliable percentage of the target nematode species from given soil and/or plant samples. Two or three extraction procedures often are required for samples that contain divergent species, such as *Meloidogyne*, *Heterodera* and/or *Pratylenchus*. As will be emphasized in this chapter, extraction of eggs for many species is equally important as that of juveniles and/or adults (5,6,14,26). The life stages that are expected in a given soil sample affect the selection process. These populations may consist of relatively motile forms to quiescent forms, such as cysts, individual eggs, eggs in masses, or various combinations of these. High proportions of endoparasitic nematodes are in roots, root fragments, or other plant parts during certain periods of the year, whereas ectoparasites occur primarily in the soil. The actual selection of assay methods depends on the kinds and numbers of nematodes present, the host plant, equipment available, the nature and condition of the samples, including soil types, time of collection, and other management practices such as nematicide treatments.

The various biophysical characters that influence the suitability of different assay methods have been enumerated (42,47,68). Relative nematode size and differential staining are the basis for some qualitative methods. Mechanical resistance of nematodes is critical to efficient blender preparations; the activity of various nematode species is important in the Baermann methods; and both the size and shape of nematodes are relevant to the use of sieving. Knowledge of specific weights, related sedimentation rates, and sizes of nematode types is helpful in the application of flotation, elutriation, centrifugation, and decanting techniques (69,70). Edaphic factors cannot be overlooked in this selection process. Soil texture is important particularly for samples containing high percentages of organic matter or clay (69). Other factors that must be considered include population dynamics, sample storage conditions, the expected population levels as influenced by management practices, and time elapsed after the previous crop. Bioassays may be required in instances where populations are barely detectable (6,38,46,57).

CARE OF SAMPLES

Soil samples for nematode assays must be considered perishable and handled accordingly. Exposure to temperatures above 40°C for a short time kills some species. All soil and root samples should be placed in plastic bags to prevent drying, kept out of the sun, and transported in insulated containers.

Ideally, nematodes should be extracted within a week after soil samples are collected. If delay is necessary, sample storage at 10-15°C prolongs the recoverability of nematodes (see Chapter 1). Lower storage temperatures may be desirable in northern regions. High temperatures (25-40°C) allow hatching of eggs and rapid aging of motile forms, but low temperatures (<10°C) may cause chilling injury (50,63,71). Changes in numbers of most species are minimal at intermediate temperatures (10-15°C) (9,63,67).

Adequate mixing of composite soil samples before removal of an aliquant (usually 100 to 1000 cm^3) for nematode extraction is also important when the entire sample cannot be processed. Such samples should be screened first and then mixed thoroughly by a suitable procedure, such as by being run through soil sample splitters (Fig. 1) two to four times. Also, soil samples may be mixed effectively by coning and quartering. Certain elutriators that handle large samples automatically mix the soil (12; Fig. 2). Mixing the composite sample avoids undue variation associated with nematode aggregation. Although mixing soil is necessary, special care, including gentle mixing or use of flotation procedures, must be given to samples containing *Xiphinema*, *Longidorus*, and *Paratrichodorus* spp. These taxa are often injured and lose their motility by mechanical mixing of soil. Roots or other plant tissues should be cut and mixed when an aliquant is selected for extraction. Macerating tissues in blenders may be satisfactory in some instances, but is not generally recommended due to the release of toxic substances in plants such as peach.

Fig. 1. Multiple rifflers for mixing and splitting soil samples.

All nematode samples must be handled so that the specimens can be easily identified and counted. The addition of chemical fixatives may be necessary when soil is collected from quarantined fields, but the difficulties in counting nematodes are increased (7). Microscopes equipped with mechanical stages greatly facilitate the counting (Fig. 3). Various counting dishes are available for this purpose. An improved counting dish that eliminates the border effect was developed by DeGrisse (23). Simple plastic boxes (7.5 × 3.3 cm) with grids etched on the bottom are satisfactory for most purposes. Bias in counting can be avoided by use

Fig. 2. Semiautomatic elutriator for extracting nematodes and soil fragments from soil (After Byrd, et al., 1976).

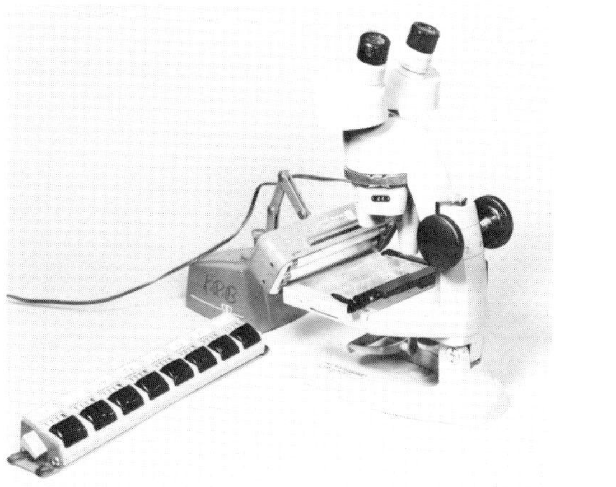

Fig. 3. Stereoscopic microscope with mechanical stage for counting nematodes.

of sample numbers rather than identified treatments (39). A few studies have been directed toward partial automation of nematode counts (10,43).

Special systems must be developed for disposal of soil residue and nematodes from fields under quarantine or from other countries. Disposal systems that handle large numbers of soil samples have been developed in Florida (62). In North Carolina, a series of large screens of differently sized openings are used to minimize the chances of major soil-inhabiting pests being spread (K.R. Barker, unpublished).

RELATIVE EFFICIENCY OF EXTRACTION METHODS

Much effort has been invested toward the development and improvement of various procedures for the assay of soil and plant samples for nematodes. Only limited research has been directed toward the characterization of the efficiency of most extraction techniques (Table 1). The sedimentation rates for various species of nematodes have been determined recently (68). The finding that these rates vary for different species poses problems in the quantitative extraction of several species.

Rates of recovery, as estimated by various researchers, have ranged from 10-100%. Extraction efficiency depends on several factors, including the method, soil type, nematode species and size, and the laboratory. For example, density-flotation curves for different nematode species are governed by solute density, solute osmotic activity, and the physiological properties of the target species (70). Nematode viability and function may be adversely affected by improper selection of solute for density extraction. In contrast, the recovery of particular nematode species may be enhanced by the selection of certain solutes. Use of the lowest effective density for flotation extraction minimizes damage caused by high osmotic activity. Polymers with low osmotic activity such as Ficoll and Percoll (Pharmacia Fine Chemicals) may have potential in the extraction of nematodes (70). Examples of actual ranges of efficiency of extraction by selected methods are summarized in Table 1.

Extraction efficiency can be enhanced by repeated sieving with an appropriately sized sieve. Seinhorst (52) suggested that nematodes can be separated, essentially quantitatively, from a suspension by relatively few runs through a sieve with openings not greater than 1/10 of the nematode length; sieves that are 1/5 of the nematode's length are too coarse for quantitative extraction (52). Increasing the volume of suspension lowers extraction efficiency, especially when small sieves are used. The theoretical recovery from a suspension may be estimated by the following formula:

$$\textit{percentage recovery} = 100\ (1-a^{x}),$$

where a = proportion of total number of nematodes present of a given length which pass through the sieve, and x = number of times sieved (52,53). Banks of sieves for giving maximum efficiency have been used extensively in some European laboratories (53), but rarely in the U.S.A. Differences in these factors undoubtedly are partially responsible for the discrepancies in efficiency reported for given extraction methods.

Several parameters, in addition to relative efficiency, must be considered in the selection and evaluation of nematode assay procedures. Time requirements and personnel availability, as well as other variables, such as the influence of soil type and numbers and kinds of species present, must be given careful attention. For example, centrifugal-flotation, in which the standard sugar solution is used, has been listed to have five major advantages over other methods. It is simple, rapid, and the nematodes are available for study immediately after the extraction. Centrifugal flotation also yields a greater proportion of nematodes, especially the slowly moving forms such as *Criconemella* and *Hemicriconemoides*, and it recovers eggs, especially those of ectoparasites (3). Still, this method may yield less than 50% of many species present (69,72). The efficiency of centrifugal-flotation may be enhanced by supplemental pre-extraction treatments, including the use of sodium-containing detergents (73) and a pre-extraction incubation after the addition of water (36).

The sugar-flotation-sieving technique developed by Byrd et al. (13) is a simple and rapid extraction method useful in working with nematodes such as the juveniles of *Meloidogyne* spp. This procedure clears colloids from suspension with a flocculating agent "Separan" (the Dow Chemical Company) and leaves the nematodes in a 0.7 to 1.0M sucrose solution. It compares favorably with centrifugal flotation for *Meloidogyne* juveniles and most nematode taxa, except the Criconemoids. If cost is a factor, blackstrap molasses may be substituted for the sucrose (48,49). Greater nematode recovery rates can be achieved with the molasses, even through the same specific gravities of both flotation solutions are utilized, probably because the molasses has a slightly higher viscosity as compared to sucrose solutions. Sugar-flotation sieving also extracts anhydrobiotic forms of nematodes from dry, desert soil (27).

A step toward the mechanization and better standardization of nematode extraction was made by Byrd et al. (12) in the development of two semi-automatic elutriators (Fig. 3). These apparatus include some of the features of the Oostenbrink funnel elutriator (41) and the air-stirred homogenizers of Coolen and D'Herde (22) and of Gooris and D'Herde (29). Key features of these machines include a mechanism for collection of root fragments, a sample-splitter that divides the water-nematode suspension from large samples of 500-1000 cm^3 into manageable portions, and the provision for collection of dual samples including possible assays of different nematode forms and/or reproductive structures of certain fungi (20,44). Collected root fragments may be utilized in the sodium-hypochlorite methods for extraction of eggs of *Meloidogyne* spp. (14), or incubated in a Seinhorst mist (51) for extraction of migratory endoparasites, such as *Pratylenchus* spp. Cysts of *Heterodera* or *Globodera* may be caught on 60-mesh (250-μm) sieves and subsequently counted, and they may also be exposed to the

Table 1. Relative efficiency of various methods of extracting nematodes.

Nematode	Extraction method	Percent efficiency (range)[a]	Time required	References
Meloidogyne spp.				
Juveniles from soil				
	Baermann trays/funnels (BF)[b]	30-45	2-5 days	69
	Centrifugal flotation (CF)[b]	17-20	5-10 min	69,72
	Elutriation + CF	10-35	5-15 min	12
	Sieving + BF	30-54	2-5 days	12,69
Juveniles and egg fractions				
	Baermann trays/funnels	20	2-5 days	—
	Centrifugal flotation	10	5-10 min	—
	Elutriation + NaOCl			
	Egg extraction	20-60	10-20 min	14
	Sieving + BF-Seinhorst mist	13-31		69
Criconemella spp.				
	Centrifugal flotation	25-62	5-10 min	12,69,72
	Elutriation + CF	70-79	5-15 min	12
	Sieving + BF	1-15	2-5 days	69
Pratylenchus spp.				
Soil fraction				
	Baermann trays/funnels	35-45	2-5 days	69
	Centrifugal flotation	17-29	5-10 min	69
	Sieving + BF	22-36	2-5 days	69
Soil and root fractions				
	Centrifugal flotation (soil only)	10-30	5-10 min	69
	Elutriation + Seinhorst mist	20-80	2-14 days	41
	Sieving + mist	32-40	2-5 days	69
Xiphinema spp.				
	Baermann trays/funnels	25-30	2-5 days	69
	Elutriation + CF	2-60	5-10 days	69
	Sieving + BF	27-35	2-5 days	69

[a]Data do not reflect actual variation which may be 100% or more (69). Rates vary with investigator and soil types.

[b]Combined with sieving.

sodium-hypochlorite (14) or the glass tissue homogenizer for the release of eggs to be counted.

Nematodes such as the second-stage juveniles of *Meloidogyne* spp. are collected on a 400-mesh (38-μm) sieve and further cleaned by a modified version of Jenkins' centrifugal-flotation method (34) which includes two steps to minimize the loss of these small forms. In the first step, the nematodes and fine soil particles are washed into a beaker with approximately 150 ml water and allowed to settle for at least 1 hour. The water is then decanted and the remainder, including the nematodes in the debris, is poured into 50-ml centrifuge tubes. The settling-decanting method of concentration retains nematodes usually lost by sieving. In the second step, the nematodes are centrifuged in a sucrose solution (454 g sucrose in 1-liter solution). The suspended nematodes are collected on a 500-mesh (26-μm) or a 635-mesh (20-μm) sieve.

Only a few techniques are available for extraction of nematodes from plant tissues. As was indicated previously, the Seinhorst mist technique (51) is best for this purpose. Other techniques such as the flask-shaker (11) are satisfactory, but do not approach the efficiency of the Seinhorst-mist. This basic principle of spraying a mist of water onto infected plant tissues on Baermann funnel-like supports may be modified for large-scale applications. Advantages of the mist-chamber technique include its ease of automation and yield of large numbers of relatively vigorous, clean specimens. Disadvantages include the extended time of 3 or more days for extraction, and the data obtained are more qualitative than quantitative. Some nematodes continue to develop in tissues in the mist chamber with the result that many nematodes are recovered after 2 or more weeks of exposure.

Other techniques that have limited use for extraction of nematodes from plant tissues include Young's Mason jar incubation (74), and the use of an airstream to enhance the activity and recovery of nematodes from plant tissues (17). The shaker-incubation system greatly enhances nematode recovery from plant tissues (11) because it provides aeration as well as suppression of microbial activity. Nevertheless, we have found it to be much less efficient than the mist system. About 90% of the recoverable *Pratylenchus brachyurus* may be extracted in 4 days from cotton roots incubated on a shaker in a mixture of 10 μg/ml ethoxyethyl mercuric chloride (Aretan) and 50μg/ml dihydrostreptomycin sulfate (11). Two weeks are required to give the same nematode yield in water alone.

The addition of other chemicals such as hydrogen peroxide may also enhance recovery of nematodes from plant tissues. Incubation of 5-10 g of citrus root infected with *Radopholus similis* or *Tylenchulus semipenetrans* in plastic bags with 10 ml of 1-3% H_2O_2 for 2 days may yield numbers of nematodes sufficient for determination of the relative magnitude of infection (59,60,61). A dilute solution of H_2O_2 also gives greater yields of *R. similis* and *Helicotylenchus* spp. from banana roots in Baermann pans (30).

VITAL STAINS

It is difficult to determine readily the physiological state of many nematodes after extraction, especially of sedentary ectoparasites, such as *Criconemella* spp. Thus, vital stains have considerable potential, but are not completely reliable as a supplemental tool in nematode assays. Materials including meldola blue and nile blue A and other stains have been used with varying degrees of success to differentiate dead from living specimens of several genera, including *Helicotylenchus*, *Pratylenchus*, *Tylenchorhynchus*, and *Xiphinema*. Chrysoidin also may give acceptable results where high magnification is used (40). New blue R and potassium permanganate also have promise as vital stains (33,56). Unfortunately none of these materials give satisfactory results in all laboratories. Thus, improved methods for distinguishing living from dead nematodes, especially sedentary ectoparasites, still are needed for studies of nematicide efficacy. Other approaches, such as the use of fluorescent microscopes, should be developed and evaluated for this purpose.

BIOASSAYS

The incidence of *Meloidogyne* spp. may be indicated by the relative extent to which the roots are galled and necrotic (75). Galling indices on a scale of 0-100 (8,75) obtained after mid-season are useful in evaluation of nematicide efficacy or estimation of actual yield losses. Yield losses increase linearly as root-gall or necrosis indices increase (8). The slope of such regression lines may be relatively sharp in areas where severe damage is caused by these nematodes, such as in coarse-textured soils. The actual relationship of gall indices to nematode numbers and yield would vary with different climatic and geographic locations.

Root-gall indices should be especially useful in determination of the magnitude of nematode infection of perennial crops such as peach. This approach is also very practical in states and countries where Nematode Advisory Programs are not available. Growers can be taught to use this type of assessment based on root symptoms in the field. Detailed considerations of bioassays are given later in this chapter.

ASSAY PROCEDURES

Selection of assay procedures depends on the kind and number of nematodes present, the host, the nature and condition of the samples (including soil texture), the time of collection, and the purpose of the nematode assay. Actively moving juveniles or adults (in roots or soil) can be extracted by (I) Baermann trays or modifications thereof; (II) a combination of flotation, sieving or both and Baermann methods; or (III) flotation (or elutriation) and cotton-wool methods. Methods adapted to both motile and nonmotile forms include (I) sieving, (II) centrifugal flotation, (III) sugar-flotation sieving, and (IV) Seinhorst's two-Erlenmeyer flask, sedimentation apparatus. The flotation methods are of limited use for endoparasites when most are inside the roots or other plant parts, but they are superior for ectoparasites. As indicated earlier, the mist system is ideal for extracting nematodes from roots, but incubation on a gyratory shaker is also satisfactory (11). Methods for extracting free eggs and egg masses are now being developed (1,14,26). Extraction of cysts of *Heterodera* and *Globodera* spp. also requires special techniques, such as the Fenwick or flotation can, or "heavy sugar" (50% by wt.) centrifugation (24). If population densities are below a detectable level, an appropriate bioassay may be used (e.g., 'Rutgers' tomato for most *Meloidogyne* spp.; 'Lee' soybean for *Heterodera glycines*).

Specific outlines for selected methods of extracting nematodes follow. Several methods that require special equipment or excessive labor inputs or both are not described. As indicated earlier, many nematodes are lost during sieving in any procedure. Such losses often can be minimized if the nematodes are allowed to settle out of suspension and excess water is decanted instead of sieved. A second passage of the suspension through the sieves may be necessary when population densities are low. Required equipment and chemicals (5,6) are listed for each of the following methods.

Assays For *Meloidogyne* Species and Other Nematodes In Soil

I. BAERMANN TRAYS (65,66) This procedure or the Baermann funnel is useful for extracting nematodes from small soil samples and root fragments containing certain eggs and juveniles of *Meloidogyne* or *Heterodera* (in spring only for cyst nematodes) and migratory nematodes in debris extracted by elutriators.

A. Equipment: 17.5-cm-diameter (20-mesh; 850 μm), plastic screen (sieve-type) support with 4-mm-high support legs or similar metal unit; wet-strength tissues; 20-cm-diameter, 500-mesh (26-μm) sieve; epoxy-resin-coated, aluminum pie pans.

B. Procedure:

1. Mix the soil.
2. Place 100 cm^3 of soil uniformly over the paper tissue, which is superimposed on a plastic or stainless steel screen. *Note:* Do not use copper screens.
3. Place the screen with soil in the pie pan and add water just to cover the soil.
4. Incubate the samples at 21-24°C (cover or stack to reduce evaporation). Add water as needed.
5. Collect the nematodes from the pans after 3 days. If dirty, clean by pouring through a 500-mesh sieve. For maximum recoveries, nematodes should be collected over a 3- to 14-day period.

II. COMBINATION OF BAERMANN TRAYS WITH ELUTRIATION OR SIEVING The combination of Baermann trays with elutriation (12,25) or sieving (64), a modification of the Christie-Perry (19) procedure, is still invaluable for extracting nematodes from soil and root fragments.

A. Equipment: 10-, 14-, and 400-mesh (2000-, 1400-, and 38-μm) sieves (U.S. Series); elutriator; screen supports; 15- to 20-cm-diameter pie pans (stainless steel or resin treated); wet-strength facial tissue, milk filter, or muslin filter.

B. Procedure:

1. Add a nonmixed 500-cm^3 soil sample to the elutriator.
2. Collect the root and soil fractions from 10- and 400-mesh sieves, respectively (see semiautomatic elutriator), or use the modified Cobb's decanting-sieving method.
3. Place these fractions on wet-strength facial tissues of filter supported on screen.
4. Carefully add sufficient water just to cover the residue. Add more water as needed with time.
5. Collect the nematodes after incubation at 21-24°C for 3 days. For maximum recoveries, make collections over a 10- to 14-day period.

III. CENTRIFUGAL FLOTATION (modified) This is an excellent method (34) for routine assays for extracting juveniles of *Meloidogyne* from soil. It is one of the best procedures when *Criconemella* is the target genus.

A. Equipment: Centrifuge with horizontal (swinging bucket) head with 50-ml or larger tubes; operation to 420 *g*; mechanical stirrers; 35-, 60-, 400-, and 500-mesh (500-, 250-, 38- and 26-μm) sieves; 100-, 150-, and 1000-ml beakers; and soil-sample splitters (W. S. Tyler Co., Mentor, OH 44060) or coarse sieves for mixing.

B. Chemicals: Sucrose solution, 454 g, in

sufficient water to make 1 liter of solution (specific gravity, 1.18, or 38.5 % by weight). *Note*: Commercial detergents (73) and moisture treatments (36) may enhance recovery by this method.

C. Procedure:

1. Mix the soil.
2. Place 100-cm^3 of soil in a 1000-ml beaker and add sufficient water to bring the total volume to 600 ml.
3. Stir for 20 sec and allow the soil to settle for 60 sec. A maximum settling time of 20-30 sec is best for *Criconemella* spp.
4. Decant onto a 35-mesh sieve over a 400-mesh sieve. Hold sieves at an angle of 35-40° during all decanting processes to minimize the chance of small nematodes passing directly through the sieve.
5. Using a wash bottle, rinse the 35-mesh sieve while still over the 400-mesh sieve (excessive rinsing washes small nematodes through both sieves). For *Heterodera* or *Globodera* spp., add a 60-mesh sieve between the 35- and 400-mesh sieves to collect cysts.
6. Wash the debris and nematodes from the 400-mesh sieve into a 150-ml beaker.
7. Pour the washings (step 6) into 50-ml centrifuge tubes.
8. Place the tubes in the centrifuge. *Be sure to balance the tubes.*
9. Centrifuge at 420 *g* for 5 min.
10. Decant water from the tubes; nematodes are in the soil pellet in the bottom of the tubes.
11. Refill centrifuge tubes with sucrose solution and mix with motorized stirring rod or vibrator mixer.
12. Centrifuge for 30 sec at 420 *g*. The nematodes will remain suspended in sucrose solution. *Do not* use the brake on certain centrifuges, as it may cause sufficient vibration to dislodge the pellet.
13. Decant the sucrose solution nematode suspension onto a 500-mesh sieve. Pour slowly when recovering small nematodes.
14. Rinse the residue and nematodes from the 500-mesh sieve into a 150-ml beaker. About 20 ml of water is suitable for making counts.

IV. FLOTATION-SIEVING METHOD (13) This method is well suited for extracting nematodes of a large size, such as *Xiphinema* spp., as well as juveniles of *Meloidogyne* spp. from sandy soils. It generally is not as efficient as centrifugal flotation, but is acceptable if centrifuges are not available.

A. Equipment: Mechanical stirrers; 35-, 400-, and 500-mesh (500-, 38- and 26-μm) sieves; 100-, 150-, and 1000-ml beakers; wash bottles or mist hoses; and soil sample splitter or sieves (for mixing).

B. Chemicals: Sucrose solution (22% by weight); Separan NP10® (Dow Chemical Company, Midland, MI 48640), final concentration of 12.5 μg/ml. *Note:* Molasses (48,49) can be used in lieu of sucrose (specific gravity, 1.1).

C. Procedure:

1. Mix soil.
2. Place a 100-cm^3 subsample in a 1000-ml beaker and add sufficient 0.7M sucrose and Separan (12.5 μg/ml) solution to bring total volume to 500 ml.
3. Stir with a motorized stirrer for 20 sec.
4. Allow the soil to settle for about 2 min.
5. Decant the liquid onto a 35-mesh sieve over a 400-mesh sieve.
6. Rinse the 35-mesh sieve while it is still over the 400-mesh sieve.
7. With a wash bottle, rinse the nematodes and debris from the 400-mesh sieve into a 150-ml beaker using about 50 ml of water.
8. Swirl the 150-ml beaker and allow the contents to settle for 5-10 sec.
9. Decant the nematode suspension onto a 500-mesh sieve and rinse the residue into a 150-ml beaker using about 20 ml of water.

V. SEMIAUTOMATIC ELUTRIATOR (12) This approach to nematode extraction includes an elutriator similar to Oostenbrink's (41), plus a sample splitter and sieve shaker. It may be used in combination with Baermann trays or centrifuge. The major advantage of this elutriator is the added option for root extraction which makes egg assays for *Meloidogyne* spp. possible. Other available elutriators (25,41,52, 53) do not provide this option.

A. Equipment: Simple elutriator, Oostenbrink type; aqueous sample splitter; water and air supplies; motorized sieve shaker; 10-, 20-, 35-, 60-, 400-, and 500-mesh (2000-, 850-, 500-250-, 38- and 26-μm) sieves. These may be semiautomated with time clocks, etc. (12).

B. Procedure:

1. Add 500 cm^3 of nonmixed soil to the elutriator, with air and water flowing at desired

rates.

2. Run the elutriator for 3-4 min, catching roots on the 35-mesh sieve over a sample splitter and "free" nematodes on the 400-mesh sieve on the motorized shaker. A 10- or 20-mesh sieve should be used in lieu of the 35-mesh sieve for large nematodes such as *Xiphinema* as these may be trapped on a 35-mesh sieve.

3. Rinse the sieves.

— For eggs of *Meloidogyne* spp., roots are processed from the 35-mesh sieve by NaOCl method (14). See Extraction of eggs.

— For *Pratylenchus* and other migratory endoparasites, roots are trapped on the 35-mesh sieve and incubated in a mist chamber.

— Cysts of *Heterodera* or *Globodera* may be collected on the 60-mesh sieve under 10- or 20-mesh sieves. The eggs from cysts may be extracted by the NaOCl methods (14) or with a glass tissue grinder.

— Any fraction (1/15, 1/5 or all) of nematodes in the soil is collected on sieves on the shaker. *Criconemella* and related genera may be cleaned by centrifugation (first method described), and other species by Baermann tray (or funnel), sugar-flotation sieving, or centrifugal flotation, with 500-mesh sieves.

VI. COBB'S DECANTING AND SIEVING The simple, modified version of Cobb's (4,21,64) sifting and gravity method is useful in extracting nematodes for inoculation purposes and routine assays when combined with Baermann Trays. Addition of Separan eliminates the need for numerous sievings. Thorne (64) has described the original method; see Townshend (65,66) for illustrations.

A. Equipment: 10-liter pails; 10-, 25-, 35-, 60-, 100-, 200-, 325-, and 500-mesh (2000-, 710-, 500-, 250-, 150-, 75-, 45-, and 26-μm) sieves; 50-, 150-, 250-, and 500-ml beakers.

B. Chemicals: Separan with sucrose can be used to reduce the steps involved. A stirrer and 600-ml suspension with 100 cm^3 of soil should be used as under "sugar-flotation-sieving."

C. Procedure:

1. Place 500 cm^3 of soil in a large pan or pail and cover well with water containing Separan (12.5 μg/ml). Thoroughly break all lumps and allow them to settle about 2 min.

2. Decant onto a 35-mesh sieve over a 325-mesh sieve.

3. Resuspend the original soil in water and repeat step 2 if maximum recovery is desired.

4. Combine the washings from steps 2 and 3, stir, and allow them to settle for 10 sec.

5. Decant through a 500-mesh sieve and rinse the residue containing nematodes into a clean beaker for counting [may place on Baermann tray (or funnel)for cleaner samples].

Extraction of Eggs of *Meloidogyne* spp. and Those of Other Taxa

Methods of extracting free eggs from soil are available (26), but hatching must be induced before identification is possible. Assays of eggs produced in "masses" or in cysts are more practical (14).

I. ELUTRIATION-DISSOLUTION OF GELATINOUS MATRICES OF EGG MASSES AND STAINING FOR POPULATION STUDIES (14) Variation of egg numbers in the field sometimes causes problems with this assay method. It is useful, however, for midseason to late-season assays of *Meloidogyne* spp. and other nematodes that form egg masses and/or cysts.

A. Equipment: Sample splitter or semiautomatic elutriator; 150- and 600-ml beakers; stirrers; 15-cm household sieve; 35-, 200-, and 500-mesh (500-, 75- and 26-μm) sieves; exhaust hood; 5-ml dipper; compressed air (for cleaning sample splitter and use with elutriator).

B. Chemicals: Sodium hypochlorite (NaOCl), antifoam spray, acid fuchsin, and lactic acid.

C. Procedure:

1. Mix the soil.

2. With water flow adjusted to 350-ml/sec, or 60-80 ml/sec if air-water mixture is used, place a 500-cm^3 sample of soil in the elutriator. Turn the water on for 2-3 min, trapping root fragments on a 35-mesh sieve.

3. With a spray nozzle, wash the residue off the sieve into a 600-ml beaker and add water to 200 ml.

4. Add 20 ml of 5.25% NaOCl and spray with an antifoam agent.

5. Stir the mixture in an exhaust hood for 10 min.

6. Using a household sieve to retain debris, take a 5-ml sample with a dipper and rinse it into a 150-ml beaker.

7. Pour this subsample onto a 500-mesh sieve and wash the eggs from the sieve into a clean 150-ml beaker. At this point, there should be about 20 to 25 ml of suspension.

8. Add two drops of 0.35% acid fuchsin in 25% lactic acid and boil for 1 min under the exhaust hood. A microwave oven is useful for this purpose.

9. Allow the mixture to cool before counting.

II. NaOCl-EXTRACTION OF *MELOIDOGYNE* SPP. EGGS FOR INOCULUM (32) The procedure using sodium hypochlorite for dissolving egg masses of *Meloidogyne* or cysts of *Globodera* or *Heterodera* spp. has been modified for extracting eggs to be used as inoculum (32). The procedure as follows is very helpful for this purpose, but special efforts are required to minimize the exposure to the NaOCl. Even with normal precautions, only about 20% of the eggs extracted with NaOCl produce infective juveniles (32).

A. Equipment and chemicals: same as previously described, excluding the stain.

B. Procedure:

1. For *Meloidogyne* spp., collect and cut 6- to 12-week-old infected tomato roots into 1-2 cm segments.

2. Shake root segments vigorously (manually) in 200 ml of a 0.5% to 1.0% (a.i.) NaOCl solution for 1 to 4 min.

3. Quickly pass NaOCl solution through a 200-mesh (75-μm) sieve, nested over a 500-mesh sieve to collect freed eggs.

4. Then quickly place the 500-mesh sieve with eggs under a stream of cold water to remove residual NaOCl (rinse for several minutes).

5. Rinse remaining roots with water to remove additional eggs and then collect them by sieving.

6. Standardize numbers of eggs per unit volume, and inoculate plants directly with eggs.

7. For maximum precision in experiments, the eggs should be placed on 500-mesh nylon hatching sieves in 1- to 2-cm-deep chlorine-free water. Tap water allowed to sit in the laboratory for 2-4 days before use is practically chlorine free due to evaporation. The hatching juveniles are then collected and used as inoculum.

This same procedure may be used for freeing eggs of *Heterodera* or *Globodera* from cysts, but the time of exposure to NaOCl will need to be increased up to 8-10 minutes.

III. EXTRACTION OF EGGS BY CENTRIFUGATION This topic has been covered by various investigators (1,26,37,55).

Separation of Nematodes From Plant Tissues

I. MODIFIED SEINHORST MIST APPARATUS (MIST CHAMBER This is the most widely used method for obtaining nematodes from plant tissues (51).

A. Equipment: Time clock, water mixer-warmer, water regulator and filter, solenoid switch, cover (fiberglass, Plexiglas, stainless steel or other suitable material) with doors to funnel racks or supports, superfine nozzles, plastic petri dishes, 10-cm diameter glass funnels with rubber tubing, 1-liter plastic cups with holes in the bottom, clamps, 500-mesh (26-μm) sieve (specifications available from K. R. Barker).

B. Procedure:

1. Place a representative sample of roots or other plant tissues in plastic cups superimposed over an open Baermann funnel. This funnel is supported over a second Baermann funnel by a plastic petri dish with a 2.5-cm hole in the center. Use wet-strength facial tissues in the cups to reduce debris.

2. Set the time clock to regulate desired mist (on 1 min, off 2 min). Adjust the water mixer to the flow rate that gives a temperature of 24°C.

3. Collect the nematodes from the funnels every 3 to 5 days through a 14-day period (nematodes such as *Pratylenchus* spp. will continue to emerge for weeks).

4. Concentrate and clean the nematode suspensions with a 500-mesh sieve, if necessary.

II. SHAKER (11) Several researchers have found this shaker procedure to yield numbers of *Pratylenchus* and other species from roots similar to those obtained from mist chambers.

A. Equipment: Gyratory shaker, 400- and 500-mesh (38- and 26-μm) sieves, 125-ml flasks, 150-ml beakers.

B. Chemicals: Ethoxyethyl mercuric chloride (Aretan) and streptomycin sulfate or other suitable antibiotics.

C. Procedure:

1. Wash roots and cut into 1- to 2-cm segments.

2. Place a representative sample of root tissue (0.5 to 5.0 g) in a 125-ml flask.

3. Cover the tissue with a solution of 10 μg/ml of ethoxyethyl mercuric chloride and 50 μg/ml of streptomycin sulfate.

4. Incubate the mixture at 100 rpm for 48 hr.

5. Collect nematodes on a 400- or 500-

mesh sieve; rinse them into a 150-ml beaker and count.

III. BLENDER—BAERMANN TRAY (31) Although many problems are encountered with this procedure, it is useful for limited situations.

A. Equipment: Blender, beakers, 325-mesh (45-μm) sieve, and Baermann funnels or pans.

B. Chemicals: Ethoxyethyl mercuric chloride (Aretan) in a 10 μg/ml solution; other materials such as antibiotics or certain fungicides (captan) may be used instead of Aretan.

C. Procedure:

1. Rinse the plant tissues until they are free of soil.
2. Weigh the tissues to be processed and place them in the blender. Use 50 g of tissue or less (fresh weight basis) with 200 ml of water in 1.9-liter (2-qt) blender.
3. Homogenize the mixture for 15 sec.
4. Decant the suspension from the blender, including the rinse water, over a 325-mesh sieve.
5. Using a wash bottle containing antibiotic solution (see Shaker method), wash the debris from the sieve into a beaker.
6. Pour the liquid that passed through the sieve over the sieve again.
7. Combine the material collected on the sieve in the second passage with that from the first passage.
8. Gently decant the suspension of material collected on the sieve over the filter of the Baermann apparatus.
9. Fill the Baermann apparatus with sufficient antibiotic solution so that the debris on the filter is just barely submerged.
10. Replenish the incubation solution with water as needed.
11. Collect the nematodes after incubation at 21-24°C for 2 to 3 days.

In addition to the combined procedure just described, blending followed by wet sieving is useful for extraction of certain nematodes such as *Radopholus similis* on banana.

Bioassays of Nematode Populations (38,46)

With low natural populations of nematodes such as *Meloidogyne, Heterodera* and *Ditylenchus* spp., bioassays are the most reliable procedures (6,8,38,57). The following is a typical bioassay for *Meloidogyne.*

I. EQUIPMENT Fumigated sandy loam soil, 10-cm clay pots, 15-cm plastic pots, pot labels, 3-week-old 'Rutgers' or other susceptible tomato cultivar seedlings, nutrient solution, and greenhouse space. Other recommended tomato cultivars are 'Person Al,' 'Heinz 1350,' 'Marglobe,' 'Bonny Best,' and 'Manapal.'

II. PROCEDURE

A. Fill the bottom 5-cm layer of a 10-cm clay pot with sterile soil.

B. Add 250 cm^3 of test soil to a given clay pot (above). The field soil used for a bioassay should represent an entire plot or a portion of a given field. Large fields should be marked off into units of about 1 ha with two composite 20-core (2.5-cm diam each) samples used for bioassays from each (38). This sampling-bioassay scheme will give a standard-error to mean ratio of about 25%. A greater number of bioassays will be needed if increased precision is desired (38).

C. Transplant a tomato seedling to a pot, and fill the remainder of the pot with sterile soil. Place the 10-cm pot in a 15-cm plastic pot to minimize contamination.

D. Grow plants for 5 to 6 weeks at 24-28°C, providing nutrients and water as needed. *Do not overwater.*

E. Harvest the plants by washing the roots carefully. Rate nematode development by use of the gall index (Table 2). The numbers of eggs also may be determined (see NaOCl method).

The third root-knot index (1-6) in Table 2 (nomograph) is recommended when the effects of nematicides on *Meloidogyne* spp. treated in the field or greenhouse are being evaluated. The other schemes in the nomograph (Table 2) are acceptable.

Some confusion may be encountered when galling caused by different *Meloidogyne* spp. is rated, since the number and size of galls incited by a given number of juveniles varies with nematode species and host plant. This problem can be minimized by concentrating primarily on the proportion of roots galled rather than the size of galls. When the cause of galls is uncertain, roots should be stained with 0.05% acid fuchsin in lactophenol and cleared in lactophenol (15). The contents of galls (numbers of juveniles, eggs, and adults) can then be determined.

Gall indices for *Meloidogyne* spp. should be limited to root galling. If the rating of necrosis is desired, a separate rating system such as that of Powell *et al.* (45) is suitable for rating necrosis associated with root-knot. This classification system is: 0, no necrosis; 1, less than 10% of the root system necrotic; 2, 11-25% necrotic; 3, 25-50% necrotic; 4, 51-75% necrotic; and 5, 76-100% necrotic. Based on all root systems per treatment, the disease or necrosis index is as follows:

$$\textit{Disease index} = \frac{\begin{bmatrix}\textit{Number of}\\ \textit{plants in}\\ \textit{class } 1 \times 1\end{bmatrix} + \begin{bmatrix}\textit{Number of}\\ \textit{plants in}\\ \textit{class } 2 \times 2\end{bmatrix} + \ldots \begin{bmatrix}\textit{Number of}\\ \textit{plants in}\\ \textit{class } 5 \times 5\end{bmatrix}}{\textit{Number of plants in treatment} \times 5} \times 100$$

A similar lesion or necrosis index is also useful for rating banana roots for infection by *Radopholus similis*.

Numbers of cysts may be counted on bioassay plants for *Heterodera* spp. [cereal cyst—*H. avenae* on wheat (57) or soybean cyst—*H. glycines* on soybean].

Table 2. Nomograph of root-knot galling indices for *Meloidogyne* spp.

Galling index systems[a] 0-4	0-5	1-6[b]	0-10	Percentage of total root system galled
0	0	1	0	0
	1	2	1	10
	2	3	2	20
1				
			3	30
			4	40
2		4	5	50
	3			
			6	60
			7	70
3				
	4	5	8	80
			9	90
4	5	6	10	100

[a]Roots are scored for degree of galling using one of several root-galling indexes, all of which are comparable and relatively interchangeable. Minimum of ten root systems should be evaluated per treatment.

[b]This scheme is recommended for use in evaluating nematicides [after Barker (5)].

Extraction of Cysts

Meaningful assays for *Globodera* and *Heterodera* species must include numbers of cysts, eggs and juveniles. The use of a reliable method for determination of numbers of eggs in cysts, and in egg masses if present, is essential for evaluation of the efficacy of nematicides and other management practices on *Heterodera* or *Globodera* spp. A glass tissue homogenizer is satisfactory for this purpose. The NaOCl (sodium hypochlorite) method described earlier also gives good results in dissolving cysts and freeing the eggs for counts. The NaOCl concentration may need to be 2-3 times that indicated for egg masses of *Meloidogyne* spp.

I. FLOTATION-MODIFIED FENWICK CAN (58) This procedure is useful for extracting cysts from dry soil, but centrifugation with 50% sucrose (by wt.) and other methods are becoming more widely used.

A. Equipment and Chemicals: Modified Fenwick can, 60- to 80-mesh (250- to 180-μm) sieve with bowl, camel's hair brush (No. 00, 0 or 1), glass tissue homogenizer (Ten-Broeck), plain glass slide, 100-ml bottle, pipette, aquarium pump or air supply, 400-ml Erlenmeyer flask, 250-ml volumetric flask, funnel, petri dish, 18.5-cm diameter filter papers, acetone or acetone-carbon tetrachloride 3:1 (2,6,58). Ethanol-glycerin [9:1] may be used in lieu of acetone-carbon tetrachloride (6).

B. Procedure:

1. Mix the soil thoroughly.

2. Fill the modified Fenwick can with water. Place the sample of 100 cm^3 of well-mixed soil in the top sieve (20.5-cm diameter, 18- or 24-mesh sieve).

3. Wash the sample into the apparatus via the funnel. The coarse material is retained on the top sieve, heavy soil particles such as sand sink to the bottom of the apparatus, and the floating cysts are carried off over the overflow collar.

4. Cysts, root debris, and other particles are collected on a 20-cm diameter sieve (60- to 80-mesh). Particles of 175 μm or smaller pass with water through the sieve.

5. After washing, dry the debris at room temperature. Transfer the somewhat dried debris retained on the sieve to a 250-ml-capacity flask.

6. Pour technical acetone, or a mixture of 3 parts acetone and 1 part carbon tetrachloride, into a volumetric flask up to level 1 (neck of flask). Shake the flask and fill it completely. *Caution:* Use an exhaust hood.

7. After 1 min, decant the floating cysts and debris through a filter paper (18.5-cm

diameter) in a glass funnel into a volumetric or Erlenmeyer flask while rotating the original flask. The acetone passes through the filter.

8. Place the filter in a petri dish and view it through a stereoscopic microscope (magnification, 50 ×) with overhead light. Pick up the cysts with a camel's hair brush (no. 00, 0, or 1) and transfer them to a small watch glass containing moist filter paper. Identify the cysts under the dissection microscope using an overhead light. Use a camel's hair brush to transfer the cysts of the desired species into a small drop of water in the glass tube of the homogenizer. Place the piston in the tube and *carefully* rotate it by hand. Pour the eggs and juveniles that were released from the cysts into a bottle. Fill the bottle with water, up to 100 ml. Mix the suspension carefully using compressed air. Pipette out two 100-ml aliquants and place in dishes for counting. As indicated earlier, eggs also may be freed by NaOCl and numbers determined.

II. CENTRIFUGATION WITH HEAVY SUGAR (24), MODIFIED This procedure is useful for isolating cysts and juveniles of *Heterodera* and *Globodera* spp., but problems may be encountered with some fine clay soils.

A. Equipment: Same as for centrifugal flotation for routine use, except 25-, 100- and 400-mesh (710-, 150, and 38-μm) sieves are needed.

B. Chemicals: 1.8 M (50% by weight) sucrose solution (615 g of sucrose) dissolved in sufficient warm water to make 1 liter of solution (1.23, specific gravity).

C. Procedure:

1. Wash 100 cm^3 of soil through a 25-mesh sieve and collect it in a beaker (use about 1.0 liter of water).
2. Mix the suspension thoroughly and allow it to settle for 5 sec.
3. Pour the supernatant through a 100-mesh screen. Add a 400-mesh sieve for juveniles.
4. Wash any residue from the screen into a centrifuge tube or tubes with 1.8M sucrose solution.
5. Centrifuge at 420 *g* for 2.5 min.
6. Collect the supernatant on a 100-mesh screen. Add a 400-mesh sieve for juveniles.
7. Rinse thoroughly.
8. Wash the sample into a beaker, using about 20 ml of water.
9. Crush the cysts with a Ten-Broeck homogenizer or dissolve them with NaOCl, as described under Flotation Can (section I), and count the eggs and juveniles.

III. SEMIAUTOMATIC ELUTRIATOR This method (12) is useful for recovering cysts of *Globodera* or *Heterodera* (see previous description Assays for *Meloidogyne* V. Semiautomatic Elutriator).

IV. SEINHORST'S EXTRACTION PROCEDURE FOR *HETERODERA* OR *GLOBODERA* CYSTS FROM MOIST SOIL (see reference 54).

REPORTING RESULTS

The type of data reported is dependent on the purpose of an assay. For experiments, replicate and treatment means should be determined for the following:

1. Mean numbers of nematodes per 100-cm^3 or greater volume of soil.
2. Numbers of nematodes per gram of root or other plant parts (fresh or dry weight) when endoparasites are extracted. The total nematodes per plant for greenhouse experiments should be included.
3. For *Globodera* or *Heterodera* spp., numbers of juveniles and cysts as well as numbers of eggs per cyst.
4. For *Meloidogyne* spp., numbers of eggs and juveniles per unit volume of soil and/or weight of roots (fresh or dry weight); also record root-knot indices based on the index given under bioassay procedure. Where possible, the number of eggs and juveniles of *Meloidogyne* and *Heterodera* or *Globodera* should be determined for advisory purposes.
5. Plant response. Precision in determination of nematode population responses to resistant cultivars or nematicides depends on how the samples are collected, handled, stored, extracted, and counted. With proper execution in each of these activities, variation usually ranges from 10-30%. These errors vary, however, with nematode species and population densities. When egg numbers for *Meloidogyne* spp. are estimated, the coefficient of variation is considerably greater than with other assays. Thus, proper statistical analyses of all data must be used when the variance for a given experiment, including coefficient of variation and standard error, is estimated. Since some zero readings and considerable variation are normally encountered, the $\log_{10}(P+1)$ or another transformation may be necessary for nematode

population data. Standardization of the reporting of nematode population data is needed to facilitate communications concerning nematicide efficacy. This issue is especially important since most assay procedures vary in their efficiencies (69). A notation giving the extraction technique and rates of recovery for the target species for the soils involved should be provided with nematode population data.

Literature Cited

1. Acedo, J. R. and V. H. Dropkin. 1982. Technique for obtaining eggs and juveniles of *Heterodera glycines*. J. Nematol. 14:418-420.
2. Anderson, S. 1970. A method for the separation of *Heterodera* cysts from organic debris. Nematologica 16:222-226.
3. Ayala, A., J. Roman, and A. C. Tarjan. 1963. Comparisons of four methods for isolating nematodes from soil samples. J. Agric. Univ. Puerto Rico 47:219-225.
4. Ayoub, S. M. 1977. Plant nematology, an agricultural training aid. NemaAid Publ., Sacramento, Calif. 195 pp.
5. Barker, K. R. (Chairman). 1978. Determining nematode population responses to control agents, pp. 114-125 *In* Methods for evaluating plant fungicides, nematicides, and bactericides, ed. E. I. Zehr. Am. Phytopathol. Soc., St. Paul, Minn. 141 pp.
6. Barker, K. R. (Chairman). 1985. Determining nematode population responses to control agents *In* Methods for evaluating plant fungicides, nematicides and bactericides (revised ed.), ed K. D. Hickey. Am. Phytopathol. Soc., St. Paul, Minn. (In press).
7. Barker, K. R., G. V. Gooding, Jr., A. S. Elder, and R. E. Eplee. 1972. Killing and preserving nematodes in soil samples with chemicals and microwave energy. J. Nematol. 4:75-79.
8. Barker, K. R., F. A. Todd, W. W. Shane, and L. A. Nelson. 1981. Interrelationships of *Meloidogyne* species with flue-cured tobacco. J. Nematol. 13:67-79.
9. Barker, K. R., C. J. Nusbaum, and L. A. Nelson. 1969. Effects of storage temperature and extraction procedure on recovery of plant-parasitic nematodes from field soils. J. Nematol. 1:240-247.
10. Baxter, R. I. 1969. An intermittent or continuous counter for use in nematology. Nematologica 14:599.
11. Bird, G. W. 1971. Influence of incubation solution on the rate of recovery of *Pratylenchus brachyurus* from cotton roots. J. Nematol. 3:378-385.
12. Byrd, D. W., Jr., K. R. Barker, H. Ferris, C. J. Nusbaum, W. E. Griffin, R. H. Small, and C. A. Stone. 1976. Two semi-automatic elutriators for extracting nematodes and certain fungi from soil. J. Nematol. 8:206-212.
13. Byrd, D. W., Jr., C. J. Nusbaum, and K. R. Barker. 1966. A rapid flotation-sieving technique for extracting nematodes from soil. Plant Dis. Rep. 50:954-957.
14. Byrd, D. W., Jr., H. Ferris, and C. J. Nusbaum. 1972. A method for estimating numbers of eggs of *Meloidogyne* spp. in soil. J. Nematol. 4:266-269.
15. Byrd, D. W., Jr., T. Kirkpatrick, and K. R. Barker. 1983. An improved technique for clearing and staining plant tissues for detection of nematodes. J. Nematol. 15:142-143.
16. Cairns, E. J. 1960. Methods in nematology—A review, pp. 33-84 *In* Nematology, ed. J. N. Sasser and W. R. Jenkins. Univ. North Carolina Press, Chapel Hill, N.C. 480 pp.
17. Chapman, R. A. 1957. The effects of aeration and temperature on the emergence of species of *Pratylenchus* from roots. Plant Dis. Rep. 38:794-795.
18. Christie, J. R. 1959. Plant nematodes—their bionomics and control. Fla. Agric. Exp. Stn., H. and W. B. Drew Co., Gainesville, Fla. 256 pp.
19. Christie, J. R. and V. G. Perry. 1951. Removing nematodes from the soil. Proc. Helminthol. Soc. Wash. 18:106-108.
20. Clark, C. A., J. N. Sasser, and K. R. Barker. 1978. Elutriation procedures for quantitative assay of soils for *Rhizoctonia solani*. Phytopathology 58:1234-1236.
21. Cobb, N. A. 1918. Estimating the nema population of soil. U. S. Dep. Agric. Agric. Circ. No. 1. 48 pp.
22. Coolen, W. A. and C. J. D'Herde. 1972. A

method for the quantitative extraction of nematodes from plant tissue. Minist. Agric. Res. Admins. State Agric. Res. Centre, Ghent, Belgium. 77 pp.
23. deGrisse, A. 1963. A counting dish for nematodes excluding border effect. Nematologica 9:162.
24. Dunn, R. A. 1969. Extraction of cysts of *Heterodera* species from soils by centrifugation in high density solutions. J. Nematol. 1:7 (Abstr.).
25. Flegg, J. J. M. and D. J. Hopper. 1970. Extraction of free-living stages from soil, pp. 5-22 *In* Laboratory methods for work with plant and soil nematodes, ed. J. F. Southey. Minist. Agric. Fish. and Food Tech Bull. 2. Her Majesty's Stationery Office, London.
26. Flegg, J. J. M. and D. G. McNamara. 1968. A direct sugar-centrifugation method for the recovery of eggs of *Xiphinema, Longidorus* and *Trichodorus* from soil. Nematologica 14:156-157.
27. Freckman, D. W., D. T. Kaplan, and S. D. VanGundy. 1977. A comparison of techniques for extraction and study of anhydrobiotic nematodes from dry soils. J. Nematol. 9:176-181.
28. Goday, G. and R. Rodriguez-Kabana. 1983. An enzymatic technique for obtaining *Meloidogyne* females for biological control studies. Nematropica 13:75-78.
29. Gooris, J. and C. J. D'Herde. 1972. A method for the quantitative extraction of eggs and second stage juveniles of *Meloidogyne* spp. from soil. Minist. Agric., Ghent, Belgium. 36 pp.
30. Gowen, S. R. and J. E. Edmunds. 1973. An evaluation of some simple extraction techniques and the use of hydrogen peroxide for estimating nematode populations in banana roots. Plant Dis. Rep. 57:678-681.
31. Hooper, D. J. 1970. Extraction of nematodes from plant material, pp. 34-38 *In* Laboratory methods for work with plant and soil nematodes, ed. J. F. Southey. Minist. Agric. Fish. and Food Tech. Bull. 2, Her Majesty's Stationery Office, London.
32. Hussey, R. S. and K. R. Barker. 1973. A comparison of methods of collecting inocula for *Meloidogyne* spp., including a new technique. Plant Dis. Rep. 57:1025-1028.
33. Jatala, P. 1975. Efficiency of potassium permanganate in differentiating between live and dead nematodes. Ann. Appl. Biol. 80:109-113.
34. Jenkins, W. R. 1964. A rapid centrifugal-flotation technique for extracting nematodes from soil. Plant Dis. Rep. 48:692.
35. Jones, F. G. W. 1955. Quantitative methods in nematology. Ann. Appl. Biol. 42:372-381.
36. Lawrence, E. G. and E. I. Zehr. 1978. Improvement of techniques for determining populations of *Macroposthonia xenoplax* in dry soil. Phytopathology 68:1102-1105.
37. McClure, M. A., T. H. Kruk, and I. Misaghi. 1973. A method for obtaining quantities of clean *Meloidogyne* eggs. J. Nematol. 5:230.
38. McSorley, R. and J. L. Parradd. 1983. A bioassay sampling plant for *Meloidogyne incognita*. Plant Dis. 67:182-184.
39. Moriarty, F. 1960. Laboratory errors associated with the estimation of the population density of *Heterodera* species in soil. Ann. Appl. Biol. 48:665-680.
40. Ogiga, I. R. and R. H. Estey. 1974. The use of meldola blue and nile blue A, for distinguishing dead from living nematodes. Nematologica 20:271-276.
41. Oostenbrink, M. 1960. Estimating nematode populations by some selected methods, pp. 85-102 *In* Nematology, ed. J. N. Sasser and W. R. Jenkins. Univ. North Carolina Press, Chapel Hill, N.C. 480 pp.
42. Oostenbrink, M. 1970. Comparison of techniques for population estimates of soil and plant nematodes, pp. 249-255 *In* Methods of study in soil ecology, ed. J. Phillipson. Proc. Paris Symp. UNESCO, Paris. 303 pp.
43. Patel, T. R. and B. A. McFadden. 1976. A simple spectrophotometric method for measurement of nematode populations. Anal. Biochem. 70:447-453.
44. Phipps, P. M., M. K. Beute, and K. R. Barker. 1976. An elutriation method for quantitative isolation of *Cylindrocladium crotalariae* microsclerotia from peanut field soil. Phytopathology 66:1255-1259.
45. Powell, N. T., P. L. Melendez, and C. K. Batten. 1971. Disease complexes in tobacco involving *Meloidogyne incognita* and certain soil-borne fungi. Phytopathology 61:1332-1337.
46. Prot, J. C. and C. Netscher. 1978. Improved detection of low population densities of *Meloidogyne*. Nematologica 24:129-132.
47. Rickard, D. A. and K. R. Barker. 1982. Nematode assays and advisory services, pp. 8-20 *In* Nematology in the southeastern region of the United States, ed. R. D. Riggs.

South. Coop. Ser. Bull. 276, Arkansas Agric. Exp. Stn., Fayetteville, Ark. 206 pp.
48. Rodriguez-Kabana, R. and P. S. King. 1972. The use of sugarcane molasses as an economical substitute in the extraction of nematodes from soil by the flotation-sieving technique. Plant Dis. Rep. 56:1093-1096.
49. Rodriguez-Kabana, R. and P. S. King. 1975. Efficiency of extraction of nematodes by flotation-sieving using molasses and sugar and by elutriation. J. Nematol. 7:54-59.
50. Sayre, R. M. 1964. Cold-hardiness of nematodes. I. Effects of rapid freezing on the eggs and larvae of *Meloidogyne incognita* and *M. hapla*. Nematologica 10:168-179.
51. Seinhorst, J. W. 1950. De betekenis van de toestand van de grond voor het optreden van aantasting door het stengelaaltje (*Ditylenchus dipsaci* (Kuhn) Filipjev). Tijdschr. Plantenziekten 56:289-348.
52. Seinhorst, J. W. 1956. The quantitative extraction of nematodes from soil. Nematologica 1:249-267.
53. Seinhorst, J. W. 1962. Extraction methods for nematodes inhabiting soil, pp. 243-256 *In* Progress in soil zoology, ed. P. W. Murphy. Buttersworth, London. 398 pp.
54. Seinhorst, J. W. 1964. Methods for the extraction of *Heterodera* cysts from not previously dried soil samples. Nematologica 10:87-94.
55. Shaad, N. W. and J. T. Walker. 1975. The use of density-gradient centrifugation for the purification of eggs of *Meloidogyne* spp. J. Nematol. 7:203-204.
56. Shepherd, A. M. 1962. New blue R, a stain that differentiates between living and dead nematodes. Nematologica 8:201-208.
57. Simon, A. 1980. A plant assay of soil to assess potential damage to wheat by *Heterodera avenae*. Plant Dis. 64:917-919.
58. Southey, J. F., ed. 1970. Laboratory methods for work with plant and soil nematodes. Minist. Agric. Fish. and Food Tech. Bull. 2. Her Majesty's Stationery Office, London. 148 pp.
59. Tarjan, A. C. 1960. A comparison of polyethylene plastic bags and glass jars as incubation chambers for obtaining nematodes from roots. Plant Dis. Rep. 44:574-577.
60. Tarjan, A. C. 1967. Influence of temperature and hydrogen peroxide on the extraction of burrowing nematodes from citrus roots. Plant Dis. Rep. 51:1024-1028.
61. Tarjan, A. C. 1972. Observations on extracting citrus nematodes, *Tylenchulus semipenetrans*, from citrus roots. Plant Dis. Rep. 56:186-188.
62. Taylor, A. L. 1974. Disposal of nematode-infested soil and plant debris from a nematology laboratory. Soil and Crop Sci. Soc. Fla. Proc. 33:97-100.
63. Thomason, I. J., S. D. VanGundy, and J. D. Kirkpatrick. 1964. Motility and infectivity of *Meloidogyne javanica* as affected by storage time and temperature in water. Phytopathology 54:192-195.
64. Thorne, G. 1961. Principles of nematology. McGraw-Hill Inc., N.Y. 553 pp.
65. Townshend, J. L. 1962. An examination of the efficiency of the Cobb decanting and sieving method. Nematologica 8:293-300.
66. Townshend, J. L. 1963. A modification and evaluation of the apparatus for the Oostenbrink direct cottonwood filter extraction methods. Nematologica 9:106-110.
67. VanGundy, S. D., A. F. Bird, and H. R. Wallace. 1967. Aging and starvation in larvae of *Meloidogyne javanica* and *Tylenchulus semipenetrans*. Phytopathology 57:559-571.
68. Viglierchio, D. R. and R. V. Schmitt. 1983. On the methodology of nematode extraction from field samples: comparison of methods for soil extraction. J. Nematol. 15:438-444.
69. Viglierchio, D. R. and R. V. Schmitt. 1983. On the methodology of nematode extraction from field samples: comparison of methods for soil extraction. J. Nematol. 15: 450-454.
70. Viglierchio, D. R. and T. T. Yamashita. 1983. On the methodology of nematode extraction from field samples: density flotation techniques. J. Nematol. 15:444-449.
71. Vrain, T. C., K. R. Barker, and G. I. Holtzmann. 1978. Influence of low temperature on the rate of development of *Meloidogyne incognita* and *M. hapla* larvae. J. Nematol. 10:166-176.
72. Weber, D. E. and A. S. Williams. 1968. Evaluation of the centrifugal-flotation technique for the quantitative recovery of three types of ectoparasitic nematodes. Nematologica 14:18 (Abstr.).
73. Wehunt, E. J. 1973. Sodium-containing detergents enhance the extraction of nematodes. J. Nematol. 5:79-80.
74. Young, T. W. 1954. An incubation method for collecting migratory endoparasitic

nematodes. Plant Dis. Rep. 38:794-795.
75. Zeck, W. M. 1971. A rating scheme for field evaluation of root-knot nematode infestations. Pflanzenschutz-Nachr. Bayer 24: 141-144.

Glossary of descriptive terms

Bioassay—use of appropriate hosts to determine relative population levels. Bioassays are used primarily for endoparasites such as root-knot and cyst nematodes, but can be used for any nematode under suitable propagation conditions. They are usually done in greenhouse.

Detection level—minimum population density of a nematode species that can be identified routinely from a soil or plant tissue sample. This level varies with each sampling and extraction technique.

Horizontal head—for centrifuge, head is so structured that individual tubes (swinging bucket) are in horizontal position when centrifuge is operating.

Molar solution—1 g mol wt in sufficient water to make 1 liter of solution.

Population density—number of nematodes per unit volume of soil or weight of plant tissue.

Relative *g* force (relative centrifugal force)—for centrifuge, force (minimum, average, maximum) times that of gravity: Rel. g force = $0.00001118 \times r \times RPM^2$ (r, radius in centimeters).

Sieve-mesh number—number of wires (filaments running in each direction) per inch.

Part Two

Staining and Culture

STAINING AND HISTOPATHOLOGICAL TECHNIQUES IN NEMATOLOGY

M. E. Daykin
Plant Pathology Department
North Carolina State University
Raleigh, North Carolina, USA

and

R. S. Hussey
Plant Pathology Department
University of Georgia
Athens, Georgia, USA

Investigation of host-parasite relationships involving endoparasitic *Meloidogyne* species requires use of methods that facilitate observations of nematodes inside roots. Limited information may be obtained simply by the staining of nematodes so they may be observed in the interior tissues of intact roots. For more detailed studies of host-parasite relationships, infected roots may be prepared by histological methods and sectioned before examination. Numerous microtechniques have been developed for studying nematode-host relationships (30). Methods that are reliable and pertinent to experiments involving *Meloidogyne* species will be discussed herein.

STAINING NEMATODES IN PLANT TISSUE

Research on *Meloidogyne* species often involves study of the penetration and development of specimens within intact roots. Many procedures for staining and clearing nematode-infected plant tissues have been developed, and most of these are satisfactory for various kinds of tissues (10,11,12, 19,30). The acid-fuchsin-lactophenol method (21), developed in 1941, has been the most widely used procedure. However, two other procedures which are noteworthy as being very reliable, relatively simple to use, and superior to the acid-fuchsin-lactophenol method are discussed in the following paragraphs.

Sodium-Hypochlorite-Acid-Fuchsin Method (5)

This new method for clearing and staining nematode-infected root tissues has several advantages over other procedures. First, it eliminates exposure of personnel to toxic substances (e.g. phenol) which are utilized in other methods (21). Also, since root tissue is cleared with sodium hypochlorite (NaOCl) prior to staining with acid fuchsin, it does not become heavily stained. As a result, the time required for destaining is shortened, and frequently destaining is unnecessary. This method is excellent for staining endoparasitic nematodes in soybean and cotton roots. The tissue-clearing step (treatment with NaOCl) may need to be modified slightly for roots of other plant species.

PROCEDURE

1. Wash infected roots and place them in a 150-ml beaker. They may be cut into segments or stained intact; however, large root systems should be cut into sections for staining.
2. Clear the roots by adding 50 ml of tap water plus an appropriate amount of chlorine bleach (5.25% NaOCl). The amount of bleach needed depends on the age of the

root tissue. The following suggestions are guidelines for the quantity of bleach to be added:

a. Young roots - add 10 ml 5.25% NaOCl,
b. Moderate-age roots - add 20 ml 5.25% NaOCl,
c. Older or more ligneous roots - add 30 ml 5.25% NaOCl.

3. Soak roots for 4 min in the NaOCl solution and agitate occasionally.
4. Rinse roots for approximately 45 sec in running tap water then soak them in tap water for 15 min to remove any residual NaOCl which may affect staining with acid fuchsin.
5. Drain the water and transfer roots to a beaker with 30-50 ml of tap water.
6. Add one ml of stock acid-fuchsin-stain solution to the water. Stock solution is prepared by dissolving 3.5 g acid fuchsin in 250 ml acetic acid and 750 ml distilled water.
7. Boil the solution for about 30 sec on a hot plate or in a microwave oven.
8. Cool the solution to room temperature, drain it from the roots, and rinse the roots in running tap water.
9. Place roots in 20-30 ml of glycerin acidified with a few drops of 5N HCl and heat to boiling for destaining.
10. Following the destaining process, roots may be either stored in acidified glycerin with little change in contrast between nematodes and root tissue or examined as described in the following.

McBryde Method (22,29)

This method requires no heating and destaining, is fairly rapid, and is easily regulated. A disadvantage of this procedure is that destaining is done with chloral hydrate which is a controlled substance in the United States.

PROCEDURE

1. Fix and stain infected roots for 6 to 24 hr in a solution of equal parts of 95% ethanol and glacial acetic acid, containing 0.0175 mg/ml acid fuchsin (17.5 mg/liter).
2. Destain root tissue by soaking it in a saturated solution of chloral hydrate (4.5 g/ml of H_2O) for 12 to 24 hr.
3. Rinse roots with stained nematodes in tap water and store them in acidified glycerin. If nematode color fades with time, the roots may be stained again by this method after being rinsed in tap water.

Lactophenol Method (10,21)

This method has been the one most widely used in the staining of nematodes in plant tissue. Infected roots of young plants, or small roots of older plants, respond best to this technique. This staining technique is apparently not suitable for roots with a high fat content, such as most perennials, since their roots will retain too much of the stain. The disadvantage of this procedure is that the destaining process is not easily regulated and often takes several days. The process is accelerated when stained roots are placed in clear lactophenol and autoclaved for 10 min at 15 lb pressure (103 kiloPascals)(3).

PROCEDURE

1. Prepare lactophenol by mixing liquid phenol (500 ml), lactic acid (500 ml), glycerin (100 ml) and distilled water (500 ml).
2. Prepare stain solution by adding 5 ml of a 1% stock solution of acid fuchsin or cotton blue per 100 ml of lactophenol. The concentration of stain may need to be varied depending on the age of tissues.
3. Add stain to a beaker and bring to a boil on a hot plate. Immerse infected roots in boiling stain for about 1 min, rinse them in tap water, and destain in clear lactophenol solution until maximum contrast between the nematodes and root tissue is obtained. The solution may be heated in a water bath (100°C) or an oven (70°C) for about 90 to 120-min in lieu of direct boiling of the stain. Destaining usually requires from a few hours up to several days.

COUNTING NEMATODES IN ROOT TISSUE

Stained endoparasitic nematodes may be counted directly in intact root tissue or following root maceration.

Direct Counting

Roots are easily examined when distributed in a small amount of glycerin on a petri dish cover (glass or plastic) and pressed against the cover with the petri dish bottom. Marking a grid on the petri dish aids in counting the nematodes under a stereoscopic microscope at 40× magnification.

Extracting Nematodes

Considerable time is required for direct counting of nematodes inside large root systems. Nematodes, however, can be freed from the roots by maceration, and a subsample can be removed for counting (20). Roots may be macerated in a Waring blender or a Silverson laboratory mixer (Silverson Machines, Ltd., Buckinghamshire, England). Nematodes can be separated from the root tissue by sieving. However, care must be taken to ensure that nematodes are not ruptured or destained during maceration.

STAINING MELOIDOGYNE EGG MASSES

Nematode reproduction should be assessed when the resistance of cultivars or breeding lines to *Meloidogyne* species is evaluated or when host ranges of different populations are compared. Reproduction can be estimated by the number of egg masses produced per plant. Staining egg masses red with phloxine B facilitates counting (9,14). Galled roots are placed in an aqueous solution of phloxine B (0.15 g/liter tap water) for 15-20 min. However, the exact concentration of phloxine B is not critical. Root systems are rinsed in tap water to remove residual stain on the roots. Phloxine B primarily stains the gelatinous egg sac and not viable eggs. Staining egg masses, however, does not provide information on the actual number of eggs produced per plant. Eggs must be extracted from a root system with NaOCl as outlined by Hussey and Barker (15), and processing is enhanced by the use of laboratory stirrers (6). A subsample of eggs to be stained should be transferred to 30 ml of water in a small beaker, mixed with 2 drops of stain (3.5 g acid fuschin, 250 ml lactic acid, and 750 ml water), and heated to a boil (6). Eggs then are counted under a stereoscopic microscope.

DISTINGUISHING LIVING FROM DEAD NEMATODES

Certain nematode studies require distinction between live and dead nematodes. Mortality may be based on nematode motility assayed by the Moje technique (23,25) or by direct observation of nematode activity (25). Several stains have been reported as being useful for differentiating living from dead nematodes. Live nematodes normally do not stain. New blue R (0.05% aqueous solution) (27), potassium permanganate (0.062-0.50% $KMnO_4$ in aqueous solution) (16), and chrysoidin (24) have been used to stain dead nematodes. However, results with these stains are not always definitive; some nematodes may only partially stain or stain very lightly (24).

METHODS OF MICROTECHNIQUE IN NEMATOLOGY

Histological studies involving *Meloidogyne* species require the capacity to prepare thin sections of infected root tissue with a minimum of distortion. The examination of such sections with the microscope can then be used to study and compare the reactions of various tissues to nematodes and treatments. Tissues first embedded in a matrix of either paraffin or plastic will be easily sectioned and handled. The sectioning of paraffin-embedded tissue requires less sophisticated equipment, and a greater quantity of tissue may be more rapidly examined. However, thinner sections can be cut from plastic-embedded tissue and greater cytologic detail may be observed. The following steps are required in the preparation of material for histopathological study according to the paraffin method.

Tissue Selection and Preparation

Samples of healthy and infected root tissue should be included in all histopathological studies. It is also often advisable to collect infected tissue at different stages of nematode development. Roots must be dug, not pulled, out of the soil and washed gently, but thoroughly, under running tap water to remove all soil particles. Even a small particle of soil adhering to the tissue will dull the microtome knife during sectioning. Tissue should then be cut into smaller pieces and placed into a container of fixative as soon as possible. Small roots should be cut with a sharp razor blade into lengths of about 1 cm while immersed in a drop of water. Large roots and stems (greater than 1 cm diameter) should also be cut longitudinally as well for a better exchange of chemicals during processing. When tissues are prepared for fixation, care must always be taken that the tissue is not crushed or allowed to dessicate during handling.

Fixation

Fixation kills and hardens the tissue and preserves the cellular structure. One of the best and most common fixatives is formalin-aceto-alcohol (FAA) (17): 90 ml of 50% ethanol, 5 ml of glacial acetic acid, and 5 ml of 37% formaldehyde. Another fixative that works well is formalin-propiono-propanol (FPP): 90 ml of 50% isopropyl alcohol, 5 ml of proprionic acid, and 5 ml of 37% formaldehyde. The proportions of reagents given for FAA and FPP are satisfactory for most tissues, but the concentrations of acids and formaldehyde may be varied for certain types of material if poor results are obtained with the standard concentrations.

In fixation, the tissue should be submerged in a volume of fixative at least 10 times greater than that of the volume of the tissue to ensure that the fixative does not become overly diluted by water from the tissue. If pieces do not sink rapidly after they are put into the fixative, the container of fixative may be placed under a mild vacuum to draw air out of the tissue more quickly. Tissue must remain in the fixative for a minimum of 24 hr to several days, depending on its thickness. Material may also be stored indefinitely in the fixative.

Dehydration

Dehydration removes water from the tissue. Water must be removed gradually if plasmolysis is to be avoided. Therefore, dehydration is accomplished by moving the tissue stepwise through increasingly higher concentrations of alcohols. When FAA is used as the fixative, the tertiary-butyl-alcohol (TBA) dehydration schedule should be followed (17) (Table 1). If the material has been fixed in FPP, the isopropyl-alcohol (IPA) dehydration schedule should be followed (Table 2).

When solutions in the dehydration schedule are changed, the liquid is drained from the container holding the tissue, and then the tissue is covered immediately with the next solution. Care must be taken that the material is never allowed to dessicate. The time the material is allowed to remain in the various dehydrating solutions depends on its thickness. For example, fine roots may need the minimum length of time whereas thick woody material will require the maximum length of time. Material allowed to remain too long in the higher concentrations of alcohols, however, may become brittle.

Infiltration

In this step, alcohols in the tissue are replaced by paraffin so that the tissue is saturated with a pure solution of paraffin. When the TBA dehydration schedule (17) has been followed, the 100% TBA solution in step 8 is first replaced with a 1:1 mixture of 100% TBA and paraffin oil. The tissue is allowed to remain in this solution for 1 hr or more, depending on its thickness. Shortly before the next step, another container is 3/4 filled with melted paraffin, and the paraffin is allowed to solidify slightly. The tissue in the TBA-paraffin oil mixture is then placed on top of the solidified paraffin and is covered with a layer of the TBA-

Table 1. Tertiary butyl alcohol dehydration schedule (17).

			Quantity (ml) needed for solution			
Step	**% Alcohol**	**Time**	**Distilled water**	**95% ethanol**	**100% ethanol**	**100% TBA**
1	50	2 hr or more	50	40	0	10
2	70	overnight	30	50	0	20
3	85	1-2 hr	15	50	0	35
4	95	1-2 hr	0	45	0	55
5	100	1-3 hr	0	0	25	75
6[a]	100	1-3 hr	0	0	0	100
7	100	1-3 hr	0	0	0	100
8	100	overnight	0	0	0	100

[a]It is important to keep the tertiary butyl alcohol changes in a warm place as the chemical solidifies at 25.5°C.

Table 2. Isopropyl alcohol dehydration schedule.

Step	% Alcohol	Time	Quantity (ml) needed for solution	
			Distilled water	100% IPA
1	70	1 day-1 wk	30	70
2	90	1 day-1 wk	10	90
3	100	1 day-1 wk	0	100
4	100	1 day-1 wk	0	100

paraffin oil solution. This container is placed uncovered in an oven that is set slightly above the melting point of the paraffin. The tissue sinks to the bottom of the container as the paraffin melts. After 1 to 3 hr, the TBA-paraffin oil-paraffin mixture is poured off and replaced with pure melted paraffin. The uncovered container then is placed back in the oven for about 3 hr. This step should be repeated at least once more. Next, the melted paraffin is replaced by a specialized type of melted paraffin, such as Paraplast (Sherwood Medical Industries, St. Louis, MO 63103) or TissuePrep (Fisher Scientific Co., Fair Lawn, NJ 07410), which is made especially for histological use. Tissue should remain in the oven in this paraffin overnight. After one more exchange with fresh melted Paraplast or TissuePrep for about 1 hr, the tissue is ready for embedding.

The infiltration procedure to be followed when the IPA dehydration schedule has been used is much simpler but requires a greater length of time. First, part of the 100% IPA in the last dehydration step is poured off so that the tissue in the bottom of the container remains covered. The container is then filled with chips of Paraplast or TissuePrep and is placed uncovered in an oven set at 59-60°C. Once the chips have melted, the paraffin-IPA mixture is poured off and replaced with pure melted Paraplast or TissuePrep. This paraffin should be exchanged for freshly melted Paraplast or TissuePrep at least twice more at 3- to 4-day intervals. Small, fine roots should be ready to embed 1 wk after beginning the infiltration procedure while thicker tissues may take 1 to 2 wks longer.

Embedding

In embedding, the tissue is positioned in cooling paraffin so that it can be sectioned after hardening. Molds for embedding may either be constructed in the lab out of folded paper, or metal base molds and embedding rings designed especially for histology may be purchased. Molds should first be coated with a thin layer of glycerin. The tissue sample is then poured or carefully lifted into the mold with heated forceps and additional melted paraffin is added to fill the mold. This step may be done on a hot plate set at 60°C. However, an embedding table, consisting of a rectangular metal plate which has a heat source at one end and which becomes progressively cooler toward the other end, gives increased control of the embedding procedure. The filled mold is next moved to a cooler surface either on the laboratory bench or on the embedding table. As soon as the paraffin begins to solidify on the bottom of the mold, the tissue is rapidly oriented in the desired fashion with a heated dissecting needle. Once the paraffin begins to solidify over the top of the mold, the mold is plunged into ice water and left there until the paraffin is completely solidified. After hardening, the paraffin is removed from the mold and may be cut into smaller blocks which can either be mounted on wooden blocks with melted paraffin or inserted directly into the microtome. Samples of tissue may be stored in these blocks indefinitely if kept in a cool place.

Sectioning

Sectioning of paraffin blocks is done on a rotary microtome equipped with a knife or disposable razor blade. A knife must be used when the tissue is tough or woody. However, for other types of tissue, razor blades will cut sections equal in quality to the knife-cut sections. Because razor blades can be frequently replaced with fresh blades, they have the advantage of not requiring sharpening and therefore are useful when working with root tissue, which often carries soil particles that can rapidly dull a cutting edge.

Excess paraffin surrounding the tissue should be trimmed away before sectioning, leaving at least 1 mm around the tissue. Care should be taken so that opposite edges of the trimmed block face are parallel. The block is then cooled in ice water for at least 5 minutes, inserted into the microtome clamp, and one edge of its face aligned parallel to the knife edge. Sections 8-12 μm in thickness are usually cut for histological studies. Tissue that is especially tough or woody will section more easily if the trimmed block is first soaked overnight in the refrigerator in a softening solution consisting of 90 ml of 1% sodium lauryl sulfate and 10 ml of glycerin (2). The excess paraffin on the face of the block must first be trimmed away, exposing the tissue so that the softening solution can penetrate.

As the sections are cut, the edge of each section should adhere to the previous section to form a ribbon. A sharp knife edge and proper knife angle are most important in obtaining a ribbon. The back of the knife edge should also be checked frequently and cleaned if necessary as paraffin build-up there will adversely affect ribbon formation. As the ribbon increases in length, it should be held away from the microtome with a dissecting needle or brush. It is then removed from the knife edge with a second needle and transferred, shiny side down, to a clean, flat surface. The ribbon should never be touched with one's hands. The ribbons can either be mounted immediately or stored in a cool, dust-free place for several weeks if necessary.

Ribbon Mounting

Ribbon mounting adheres sections to glass microscope slides so that they can be stained. Ribbons must first be cut into shorter lengths so that they can be fit onto the slides. Slides may be labeled with a diamond pencil or, if the glass on one end is frosted, with a lead pencil. The surface of the slides is then coated with a small amount of Haupt's adhesive (17) which consists of 1 g powdered high-grade gelatin, 100 ml distilled water, 2 g phenol crystals, and 15 ml glycerin. For preparation of the adhesive, the gelatin is dissolved in the water at 30°C, the phenol and glycerin are added, and the solution is filtered. Before the adhesive dries on the slides, the slides are flooded with a 2-3% formalin solution, which should be made fresh each day. The flooded slides are then placed on a warming tray held at 35 to 40°C, and segments of the ribbon are floated on the slides. As the slides warm up, the ribbon will flatten out and the liquid will evaporate. After several hours, when the slides are completely dry, they can be removed and stored indefinitely.

Staining

The process of staining removes the paraffin from the sections and increases the contrast in the tissues. Three staining procedures that have been used for nematode-infected root material are Johansen's Quadruple Stain (17) (Table 3), Sass Safranin and Fast Green Stain (4,26) (Table 4), and Triarch Quadruple Stain (13) (Table 5).

Solutions made from dry, powdered stains should always be filtered before they are first used. If only a few slides are being stained, the alcohols and staining solutions may be kept in Coplin jars and the slides moved individually after each time period. However, larger containers and racks that hold 25 or 50 slides are much more convenient when staining larger quantities. Stains and alcohols that are not being used over long periods should be stored in tightly capped bottles.

In these three staining schedules (Tables 3,4,5), containers holding water rinses should be emptied and refilled with fresh water after each group of slides moves through. Rinses containing alcohols should be changed when the liquid becomes heavily stained. Staining solutions and xylene require replacement less often.

After completion of the staining procedure, coverslips are mounted with a few drops of either balsam or Permount (Fisher Scientific Co., Fair Lawn, NJ 07410). Slides are first removed from the xylene, which is always the final step in a staining procedure, and laid on a flat, absorbent surface. The mounting medium is then applied to the surface of the slide before the xylene evaporates, and a coverslip is lowered gradually over the slide. A minimum of mounting medium should be used, as any excess will run out over the surface of the coverslip. Finished slides should be left flat to dry for at least 24 hr at room temperature. However, the medium will harden better if the slides are held on a 60°C warming tray overnight.

The safranin in the previously discussed

Table 3. Johansen's quadruple stain (17).

Step	Solution	Time
1	xylene	5 min
2	xylene-absolute ethanol (1:1)	5 min
3	95% ethanol	5 min
4	70% ethanol	5 min
5	safranin O solution[a]	6-24 hr
6	rinse in tap water	
7	1% aqueous methyl violet 2B	10-15 min
8	rinse in tap water	
9	95% ethanol-methyl cellosolve-tertiary butyl alcohol (1:1:1)	15 sec
10	fast green FCF solution[b]	10-15 min
11	95% ethanol-tertiary butyl alcohol (1:1) plus 0.5% glacial acetic acid	15 sec
12	orange G solution[c]	3 min
13	clove oil-methyl cellosolve-95% ethanol (1:1:1)	15 sec
14	clove oil-absolute ethanol-xylene (1:1:1)	15 sec
15	xylene	5 min
16	xylene	5 min or longer

[a]The safranin O solution is prepared by dissolving 4 g safranin O in 200 ml methyl cellosolve. When the safranin is dissolved, add 100 ml 95% ethanol and 100 ml distilled water. Finally, add 4 g sodium acetate and 8 ml formalin.

[b]The fast green FCF solution is prepared by adding 0.25 g fast green FCF to 50 ml of a solution composed of methyl cellosolve and clove oil (1:1). After the fast green has dissolved, 150 ml 95% ethanol, 150 ml tertiary butyl alcohol, and 3.5 ml glacial acetic acid are added.

[c]The orange G solution is prepared by dissolving 1 g orange G in 200 ml methyl cellosolve and then adding 100 ml 95% ethanol.

Table 4. Sass safranin and fast green stain (4,26).

Step	Solution	Time
1	xylene	5 min
2	absolute ethanol	5 min
3	95% ethanol	5 min
4	70% ethanol	5 min
5	50% ethanol	5 min
6	30% ethanol	5 min
7	1% aqueous safranin O	1-12 hr
8	rinse in tap water	
9	30% ethanol	3 min
10	50% ethanol	3 min
11	70% ethanol	3 min
12	95% ethanol	3 min
13	0.1% fast green FCF in 95% ethanol	5-30 sec
14	absolute ethanol	15 sec
15	absolute ethanol	3 min
16	xylene-absolute ethanol (1:1)	5 min
17	xylene	5 min
18	xylene	5 min or longer

Table 5. Triarch quadruple stain (13).

Step	Solution	Time
1	xylene	5 min
2	xylene	5 min
3	xylene-absolute ethanol (1:1)	5 min
4	95% ethanol	5 min
5	70% ethanol	5 min
6	1% safranin O in 50% ethanol	5-15 min
7	rinse in distilled water	
8	1% aqueous crystal violet	1-2 min
9	rinse in distilled water	
10	absolute ethanol	30 sec
11	absolute ethanol	30 sec
12	orange G[a]-fast green[b] (135 ml-15 ml)	3 min
13	orange G-fast green (145 ml-5 ml)	2 min
14	orange G-fast green (148 ml-2 ml)	2 min
15	orange G	2 min
16	absolute ethanol	1 min
17	xylene	5 min
18	xylene	5 min or longer

[a]Orange G is prepared by dissolving 0.4 g orange G in 100 ml clove oil.

[b]Fast green is prepared by dissolving 1 g fast green FCF in 100 ml absolute ethanol.

stains turns lignified or cutinized cell walls red whereas the fast green generally turns cellulose walls greenish. Starch grains stain purple by methyl violet and crystal violet. Nematodes in tissue vary from brownish to red. Triarch Quadruple Stain gives the best contrast consistently and takes the least amount of time. Also, a wide variety of tissue types may be stained without altering the staining time. However, since it requires a large amount of clove oil, it is the most expensive stain to prepare. Staining times in the Johansen's Quadruple Stain and Sass Safranin and Fast Green Stain may need to be adjusted for the best contrast between cell types to be obtained.

PLASTIC EMBEDDING OF PLANT TISSUE FOR LIGHT MICROSCOPY

The previous sections presented information on preparation of histological sections by the traditional embedding method in which paraffin is used. Plastic-embedded material, however, provides superior preservation of cell structure (8). Plastic-embedded material will not entirely replace paraffin-embedded material, which still can be used for initial examination of infected tissue. However, critical tissue evaluation is enhanced with plastic-embedded material due to the added resolution obtainable with thin sections. The cytologic detail seen in 3-μm-thick sections of plastic-embedded material is superior to that which can be obtained in paraffin-embedded sections.

Several procedures for plastic embedding are available (7,8). Feder and O'Brian (8) discuss some new methods that they have thoroughly tested. One general procedure involves the fixation of small pieces of tissue in 4% glutaraldehyde in 0.1M phosphate buffer at pH 7.2; rinsing twice in buffer; dehydration in a graded, acetone series; and infiltration with and embedding in soft Spurr's resin (28). Sections 3-μm thick are cut from a carefully trimmed block on a Sorvall MT-2 ultramicrotome on a glass knife and floated on distilled water on a gelatin-coated slide at 60°C. Sections are stained with toluidine-blue and permanently mounted with Polymount resin (7). Excellent results have been obtained with plastic-embedded tissue that is infected with nematodes (1,18,31).

Literature Cited

1. Acedo, J. R., V. H. Dropkin, and V. D. Luedders. 1984. Nematode population attrition and histopathology of *Heterodera glycines*-soybean associations. J. Nematol. 16:48-57.
2. Alcorn, S. M. and P. A. Ark. 1953. Softening paraffin-embedded plant tissues. Stain Technol. 28:55-56.
3. Alves, L. M. and G. B. Bergeson. 1966. A quick destaining procedure for showing contrast between nematodes and root tissue. Plant Dis. Rep. 51:511.
4. Berlyn, G. P. and J. M. Miksche. 1976. Botanical microtechnique and cytochemistry. Iowa State Univ. Press, Ames, Iowa.
5. Byrd, D. W., Jr., T. Kirkpatrick, and K. R. Barker. 1983. An improved technique for clearing and staining plant tissue for detection of nematodes. J. Nematol. 15:142-143.
6. Byrd, D. W., Jr., H. Ferris, and C. J. Nusbaum. 1972. A method for estimating numbers of eggs of *Meloidogyne* sp. in soil. J. Nematol. 3:378-385.
7. Dawes, C. L. 1971. Biological techniques in electron microscopy. Harper and Row, Publ., Inc. 193 pp.
8. Feder, N. and T. P. O'Brien. 1968. Plant microtechnique: some principles and new methods. Am. J. Bot. 55:123-142.
9. Fenner, L. M. 1962. Determination of nematode mortality. Plant Dis. Rep. 46: 383.
10. Franklin, M. T. and J. B. Goodey. 1959. A cotton-blue lactophenol technique for mounting plant parasitic nematodes. J. Helminthol. 23:175-178.
11. Godfrey, G. H. 1935. The demonstration of plant parasitic nematodes in host tissues. Phytopathogy 25:1026-1030.
12. Goodey, T. 1937. Two methods for staining nematodes in plant tissues. J. Helminthol. 15:137-144.
13. Hagquist, C. W. 1974. Preparation and care of microscope slides. Am. Biol. Teacher 36:414-417.
14. Holbrook, C. C., D. A. Knauft, and D. W. Dickson. 1983. A technique for screening peanut for resistance to *Meloidogyne arenaria*. Plant Dis. 57:957-958.
15. Hussey, R. S. and K. R. Barker. 1973. A comparison of methods of collecting inocula of *Meloidogyne* spp., including a new technique. Plant Dis. Rep. 57:1025-1028.
16. Jatala, P. 1969. A new staining technique for differentiating live and dead nematodes. Phytopathologica 59:1033.
17. Johansen, D. A. 1940. Plant microtechnique. McGraw-Hill Book Co., New York. 523 pp.
18. Jones, M. G. K. and H. L. Payne. 1978. Early stages of nematode-induced giant-cell formation in roots of *Impatiens balsamina*. J. Nematol. 10:70-85.
19. Kirkpatrick, J. D. and W. F. Mai. 1957. A new staining technique for in situ observation of *Pratylenchus penetrans* and other endoparasitic nematodes. Phytopathology 47:526 (Abstr.).
20. Marks, R. T. and L. A. McKenna. 1981. A single procedure for the efficient estimation of potato cyst nematode larvae in lactophenol on treated roots. Ann. Appl. Biol. 97:323-324.
21. McBeth, C. W., A. L. Taylor, and A. L. Smith. 1941. Note on staining nematodes in root tissue. Proc. Helminthol. Soc. Wash. 8:26.
22. McBryde, M. C. 1936. A method of demonstrating rust hyphae and haustoria in unsectioned leaf tissue. Am. J. Bot. 23:686-689.
23. Moje, W. 1959. Structure and nematicidal activity of allylic and acetylenic halides. J. Agric. and Food Chem. 7:702-707.
24. Moriarity, F. 1964. The efficacy of chrysoidin, new blue R and phloxine B for determining the viability of beet eelworm, *Heterodera schachtii* Schm. Nematologica 10:644-646.
25. Rich, J. R., N. T. Keen, and I. J. Thomason. 1977. Association of coumestans with the hypersensitivity of lima bean roots to *Pratylenchus scribneri*. Physiol. Plant Pathol. 10:105-116.
26. Sass, J. E. 1951. Botanical microtechnique. Iowa State College Press, Ames, Iowa. 228 pp.
27. Shepherd, A. M. 1962. New blue R, a stain that differentiates between living and dead nematodes. Nematologica 6:181-200.
28. Spurr, A. R. 1969. A low-viscosity epoxy

resin embedding medium for electron microscopy. J. Ultrast. Res. 26:31-43.

29. Southards, C. J. 1965. Host-parasite relations of the lesion nematodes, *Pratylenchus brachyurus, P. zeae,* and *P. scribneri*, and flue-cured tobacco. Ph.D. thesis, North Carolina State University, Raleigh, N.C. 101 pp.
30. Southey, J. F., ed. 1970. Laboratory methods for work with plant and soil nematodes. Minist. Agric. Fish. and Food Tech. Bull. 2. Her Majesty's Stationery Office, London. 148 pp.
31. Wyss, U. 1975. A routine method for investigations on the histology of feeding sites of plant-parasitic nematodes. Nematologica 21:110-111.

CHAPTER 4

GNOTOBIOTIC TECHNIQUES FOR PLANT-PARASITIC NEMATODES

S. R. Koenning and K. R. Barker

Department of Plant Pathology
North Carolina State University
Raleigh, North Carolina, USA

The purpose of this chapter is to provide the reader with a basic understanding of gnotobiology as it applies to research with plant-parasitic nematodes. The authors have attempted to summarize the current status, future applications and the potential pitfalls of gnotobiological research. The scope of this discussion has been limited, as much as possible, to myceliophagous and phytophagous nematodes. More detailed information is available in reviews by Zuckermann (65,66) and Vanfleteren (61) and in selected methods texts (17,24). Because of the space limitations of this chapter, some literature citations are limited to more general treatments.

Gnotobiology is the study of a single species in the absence of other organisms or in the presence of known species. Mycologists and microbiologists have employed these techniques to great advantage for nearly 100 years. Many nematologists, however, have been reluctant to pursue this area of endeavor. This hesitancy is attributable to several factors: gnotobiology is tedious and time consuming, requires expertise and specialized equipment; nematodes are rarely found associated with only one or a few other organisms; and phytonematologists often must bring non-sterile soil and/or plant tissues into their laboratories, a practice incompatible with gnotobiological technique. Lastly, plant disease caused by plant-parasitic nematodes is modified, to a greater or lesser extent, by associated organisms.

Root-knot nematodes, *Meloidogyne* spp., were successfully cultured monoxenically by Byers in 1914 (4). He utilized these techniques to describe the entire life cycle of *Heterodera marioni* (*Meloidogyne* spp.) (4). Tyler (58) used monoxenic cultures in 1933 to demonstrate parthenogenesis in root-knot nematodes. Further progress in phytonematode gnotobiology, however, was delayed until the 1950's (65,66).

The discovery of the etiology of cauliflower disease of strawberry by Crosse & Pitcher (9) and the application of Koch's postulates to phytonematology by Mountain (37,38) led to renewed interest in gnotobiology by nematologists and plant pathologists. Further impetus in this area of biology was gained with the use of excised roots (37,56) and the discovery that alfalfa callus tissue (27) could be used to rear large quantities of certain plant parasites. The axenic culture of the stylet-bearing nematodes, *Aphelenchoides rutgersi* (42,43), *Aphelenchus avenae* (19) and *Bursaphelenchus xylophilus* (3), has provided insights into nematode physiology which are likely to have a major impact on phytonematology. The axenic culture of nematode species strictly parasitic on higher plants would represent a milestone in nematology.

Although gnotobiology proved to be a useful research tool in the study of many stylet-bearing nematodes, similar progress with the more fastidious *Meloidogyne* and *Heterodera/Globodera* spp. has been limited. Notable exceptions to this statement include: the use of monoxenic culture to demonstrate the influence of host nutrition on the development and sex ratios of *Meloidogyne incognita* (34,35); the use of excised root systems to evaluate resistance to *Meloidogyne* spp. (15); and the separation of nematode species by symptoms (55). More recently, a number of researchers have used excised root systems in monoxenic studies of various aspects of the biology and host-parasite relationships of certain cyst nematodes and the reniform nematode (28,29,30,31).

The difficulties inherent in gnotobiological research must be overcome if progress is to be made in several areas of nematological research. Nematode cultures free of associated microorganisms are essential in biochemical and disease-complex studies. Small, sealed, monoxenic nematode cultures are useful when the effects of various physical or chemical factors on the population dynamics of nematodes are to be quantified because confounding factors are minimized. For example, it is extremely difficult to quantify temperature effects accurately on nematodes in pot cultures because soil moisture will vary with temperature. Genetic studies are facilitated by monoxenic culture since cross contamination of cultures can be prevented.

TERMINOLOGY

Terminology to define the level of gnotobiological culture was developed by Dougherty (12,13) and has been universally accepted (Table 1). Confusion may be avoided if one remembers that the prefix (mono, di or tri) refers to the number of associated species, not the total number of species present.

Media are defined as being holidic, meridic or oligidic (12,13). A holidic medium is one with chemically defined constituents, other than purified inert materials, before compounding. Meridic refers to those media composed of a holidic base to which is added at least one substance or preparation of unknown structure, such as a protein, regardless of purity. Oligidic media are those containing crude materials which supply most of the dietary requirements, such as chick embryo or liver extract.

METHODOLOGY

Aseptic Techniques, Steps in Cultivation and Nematode Nutrition

Some general considerations are in order before specific techniques are discussed. A course in microbiology or a techniques course in plant pathology should be considered as a prerequisite to gnotobiological research. Experience and knowledge gained in such courses will aid the investigator in system design and problem solving.

Utmost cleanliness and good microbiological technique are essential in monoxenic culture of nematodes. Media on which plant tissues are grown are rich in a number of organic constituents. In addition to supporting plant growth, they are excellent media for microbial growth. Thus, any lapse in technique can result in contamination of cultures.

A sealed room with restricted air movement will provide the best atmosphere for culture work. A microvoid (Fig. 1) fitted with a microscope or laminar flow hood (Fig. 2) is necessary in most instances. Working areas should be sterilized with suitable phenolic sterilants (NaOCl or ethanol solutions also may be used). Non-sterile plant or animal tissues, fungal cultures, and soil should be excluded from this environment. Tools used in handling nematodes and plant tissue must be sterilized after each operation with flame or a strong sterilant such as $HgCl_2$ (1000 ppm).

Techniques for the establishment of organisms in laboratory culture for the first time vary greatly for different target species. Success depends upon the provision of a manageably restricted environment that meets all physical and chemical needs. At the present time,

Table 1. Terminology used to describe the level of gnotobiological culture (12).

Term	Number of associated species	Total number of species
Gnotobiotic	Known	Indeterminate
Axenic	None	One
Synxenic	One or more	Two or more
Monoxenic	One	Two
Dixenic	Two	Three
Trixenic	Three	Four
Polyxenic	Several to many	Several to many
Xenic or Agnotobiotic	Unknown	Unknown

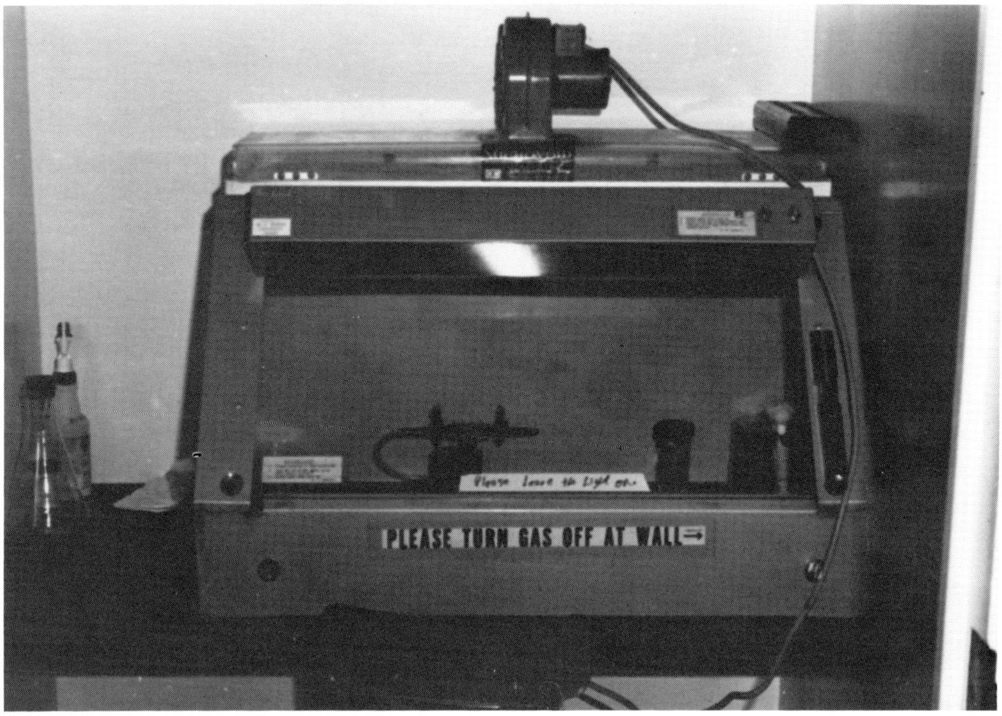

Fig. 1. A microvoid for aseptic techniques. Air is forced through a filter by the fan on top creating a positive air pressure which keeps contaminants outside. Microvoids may be altered to accommodate a microscope. Caution should be employed when work is conducted with toxic chemicals since vapors will also be forced out.

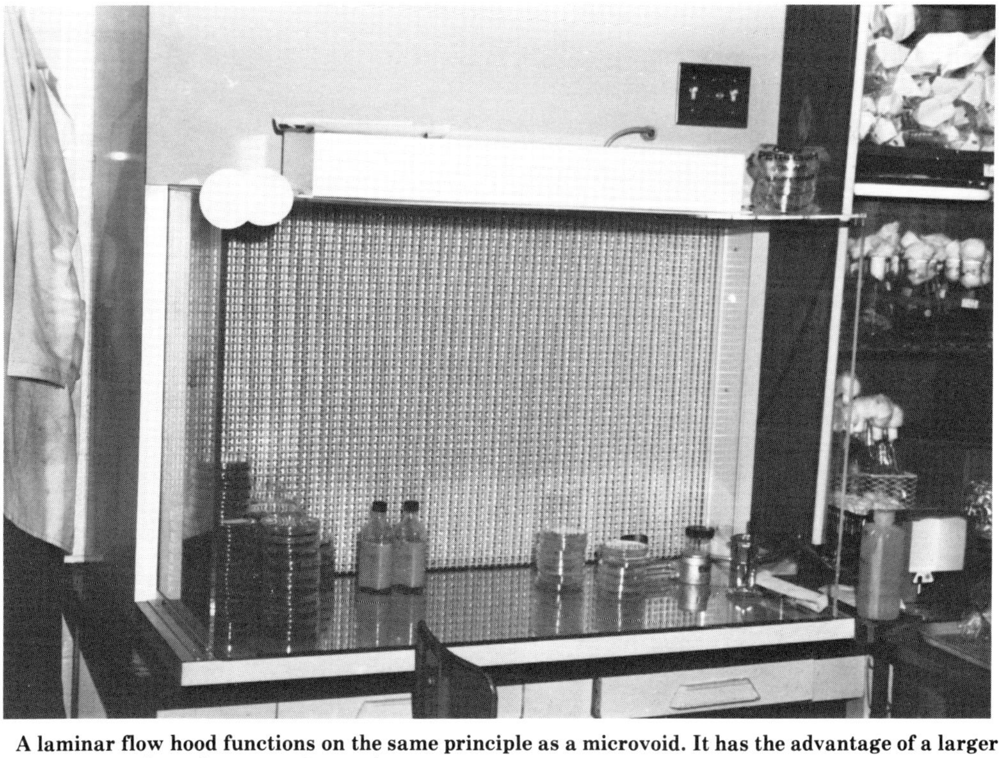

Fig. 2. A laminar flow hood functions on the same principle as a microvoid. It has the advantage of a larger working space and ready access for a microscope.

empirical methodology still dominates most initial efforts at rearing organisms in the test-tube or Petri dish. Frequently, there is a long period of trial and error before a species is successfully established. One must therefore seek a more natural environment that yields a plentiful supply of nematodes or run the risk of investing time and effort only to exhaust that supply before succeeding (Table 2).

Agnotobiotic or Xenic Culture

Most nematodes inhabit an environment in which they are in intimate contact with microbial flora and fauna. Many plant parasites, especially the migratory parasites, have such an environment though they are inside the host tissue. Nematodes such as *Ditylenchus destructor* may actually feed more on the associated fungal population than on the host tissue (potato) in many cases. Xenic cultivation should first be achieved before a more highly refined method can be approached. Many parasitic and potentially parasitic nematodes as yet have not been carried through this step. *Xiphinema americanum* is one of the most widespread plant parasites, but often it is difficult to grow under greenhouse conditions (18). Thus, some environmental condition, such as temperature, moisture or an additional food or host requirement, is missing. When one finds a high population of this species, the common saprophagous nematodes and other plant parasites are usually absent (18). Thus, it is possible that it may also be a predator as well as a plant parasite or that it requires a different flora from that needed by the saprophagous nematodes.

Agnotobiotic culture of nematodes will continue to be utilized for many research purposes. The simple requirements of agnotobiotic culture make it the only feasible approach for many nematologists at the present time.

Gnotobiotic Culture

SYNXENIC CULTURE One can appreciate the need for the identification of the organisms normally occurring with a nematode. This knowledge would refine the cultivation one step to *synxenic* culture. The effects of the associated species on the nematode in question may then be considered. Under field conditions, numerous fungi and nematodes, as well as bacteria, probably contribute to root rots and other plant diseases. Thus, regardless of which group of soil organisms are under study in relation to a root problem, one must be aware of whether an axenic or monoxenic culture is in use. The next step in cultivating a given nematode would be to determine which single organism will provide an adequate diet for the species in question. Thus, numerous parasites have been grown on various tissue cultures or in some instances on fungi. Several free-living nematodes have been grown monoxenically on bacteria such as *Escherichia coli*. Species from some of the important genera of plant-parasitic nematodes have been reared monoxenically.

The more specialized (fastidious) parasites such as *Meloidogyne* spp. require special consideration in monoxenic culture. Sayre (53) demonstrated that these nematodes grow or reproduce only on tissues that contain vascular elements; the females are always associated with the vascular tissues. By means of tissue-culture techniques, the mechanism by which nematodes induce the polynucleate giant cells of the host can be determined.

Little is known about the nutritional requirements of plant-parasitic nematodes. Generally, nutritional levels that support host tissue growth also favor nematode reproduction. However, there are a number of exceptions to this statement. Levels of N, P, K may affect

Table 2. Steps in gnotobiology.

Culture level	Procedure	Major constraints
Xenic	Increase nematodes in greenhouse or field	Interactions undetermined, mixed cultures
Monoxenic	Axenize plants or plant tissues, and nematodes	Equipment, unnatural environment
Axenic	Determine nutritional requirements	Additional equipment

nematodes and their damage differently. Dolliver et al. (11) found that low calcium levels inhibited *Aphelenchoides ritzemabosi* in tissue culture, whereas reduced levels of Zn, Mn, Fe, Mg, or K had no effect. Chelated Fe (EDTA) greatly suppressed nematode reproduction.

Host nutrition has a major influence on most aspects of the biology of *Meloidogyne* spp. The sex ratios, rate of development and penetration of excised cucumber roots by *M. incognita* were strongly influenced by the concentrations of sucrose, iron chelate and macronutrients used in the media (34,35). Increased concentrations generally enhanced penetration, development and high female/male sex ratio, but these trends were reversed at the higher concentrations (34,35). Low sucrose concentrations resulted in 100% males. Later research (45), in which excised tomato was used on two chemically defined media [a formulation (MS) by Murashige and Skoog (40) and a formulation (STW) by Skoog et al. (24), Appendix A], showed that limited development of this nematode on the MS media was related to the poor development of giant cells. The poor nematode and giant-cell development was attributed to the high concentration of ammonium nitrate in the MS media (45).

There are a few other publications which suggest that altered host physiology and condition might favor nematode reproduction. Dolliver (10) observed an increase in the population of *Pratylenchus penetrans* in garden pea associated with foliage detachment that detrimentally affected plant growth. He suggested that changes in the pattern of metabolic products of the root may be correlated with changes in nematode numbers. A similar situation was observed with this nematode by Mountain & McKeen (39) when they added *Verticillium dahliae* to soil in which tomato or eggplant was planted. Nematodes reproduced two times as fast in *V. dahliae*-infected plants. This increase was attributed to possible changes in host physiology that rendered nutritional factors essential to nematode reproduction more available. Certain fungicides such as Terraclor increased *P. penetrans* on strawberry (48).

AXENIC CULTURE Much information on the growth requirements of parasitic nematodes is needed before most of them can be grown axenically. The majority of these pests then must be cultured monoxenically before axenic culture can be considered. The physiological and physical requirements must be elucidated before axenic culture can be achieved. The more specialized plant parasites in the superfamily Heteroderoidea may require feedback from their hosts. Axenic culture of these parasites may, therefore, depend on devising some type of dynamic system that mimics the host. In contrast, axenic culture of several free-living nematodes has been achieved. A number of others have been grown monoxenically or xenically.

Myers (42,43) first reported the axenic culture of a stylet-bearing nematode (*Aphelenchoides rutgersi*). *Aphelenchus avenae* was cultured axenically by Hansen et al. (19) and *Bursaphelenchus xylophilus* (the pine wilt nematode) has been so cultured (3). All of these nematodes apparently feed primarily on fungi.

Dougherty and his research group in California were pioneers in axenically culturing nematodes and did much in working out some of their nutritional requirements. One or more of the following undefined substrates were added to the early developed media in order to obtain axenic cultures (14):

1) Liver extract — ML,
2) Ammonium sulfate-precipitated liver protein fraction — LPF,
3) Chick embryo extract — CEE.

Chick embryo extract, as the only supplement in a nutrient medium, would not support growth of *Caenorhabditis briggsae*; liver extract was also required for reproducible results.

The following are considered essential amino acids: arginine, histidine, isoleucine, leucine, lysine, methionine, phenylalanine, threonine, tryptophane, and valine (2). Other research showed that the amino acids alanine, aspargine, cysteine, glutamine, glycine, proline, serine and tyrosine can be synthesized by nematodes (59). By using a basal medium consisting of minerals, amino acids and a number of vitamins plus the chick embryo extract, Nicholas et al. (44) were able to demonstrate requirements for certain vitamins. By varying the concentration of a number of vitamins and omitting each one in turn, they found that the following were deleterious if omitted (44,62): thiamine, riboflavin, folic acid, calcium panthothenate, niacinamide and pyridoxine. Biotin and cyanocobalamin were not shown to be required, but may well be provided by the CEE as they are required only in low amounts.

Even though *C. briggsae* can now be cultivated continuously on a chemically defined (*holidic*) medium, the research until recently resulted in its being cultured in partly defined (*meridic*) media (47), each consisting of a solution of known compounds (the basal medium) supplemented with a small amount of *crude extract*. These are important advances, since only a few nematodes have been grown or maintained in crude (*oligidic*) media made from cell-free extracts of animals or bacteria.

The most recent research showed that *C. elegans* and *C. briggsae* can be cultured on a holidic medium (60,62). The constituents supplied by the "growth factor" amendments were 1) a sterol (19,61,62) and 2) a heme moiety (21,61). Nevertheless, the addition of a protein from soy peptone enhanced nematode reproduction (62).

The pinewood nematode, *B. xylophilus*, responds similarly on such media (3). One medium supporting the reproduction of this nematode consisted of 4% soy peptone and 1% Difco yeast extract supplemented with 50 μg/ml of cholesterol and 50 μg/ml bovine hemoglobin. The second medium consisted of a modified *Caenorhabditis* medium buffered to pH 7.4 and supplemented with the cholesterol and bovine hemoglobin. Better growth and development occurred on the undefined soy peptone medium, but the *Caenorhabiditis* medium has the advantage of being chemically defined (2).

In summary, certain amino acids (59), a dietary sterol (20,61,62), a heme moiety (21,61), and a protein "growth factor" (62) are dietary essentials for *C. briggsae*. Croll & Matthews indicates that the growth factor may be taken up in the intestine by phagocytosis (8). The protein "growth factor," although not absolutely required, enhances the uptake of heme by the nematodes and may improve the amino acid balance (62).

System Design

Whether working with pot cultures in a greenhouse or axenic culture in the laboratory, the investigator is concerned with providing a suitable environment for the experimental organism. Since these environments are small ecosystems of greater or lesser complexity, a systems approach is helpful in the design and/or selection of the experimental apparatus (Fig. 3).

Once the nematodes' requirements for nutrition have been fulfilled, primary considerations in system design include experimental objectives, equipment and space available. These factors place certain constraints upon the ecosystem that can be constructed and utilized successfully (Fig. 3). These constraints, once appreciated, should deter the nematologist from planning overly ambitious experiments requiring gnotobiotic techniques.

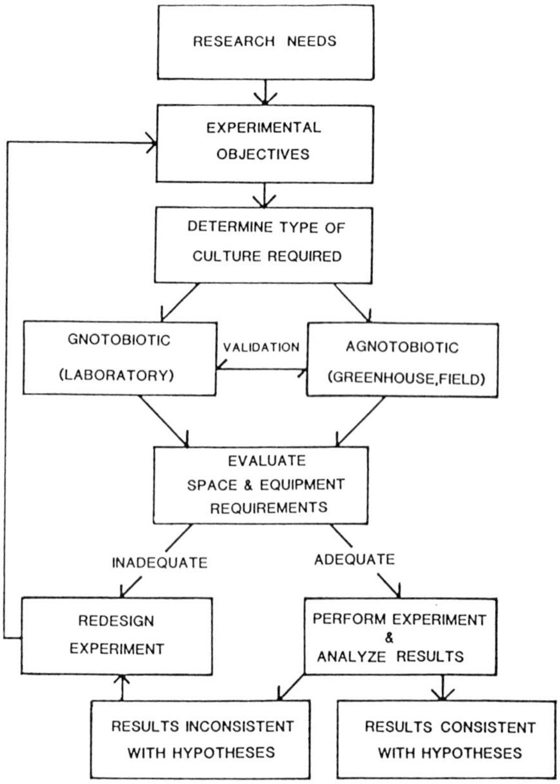

Fig. 3. Conceptual processes by which the type of culture (agnotobiotic vs. gnotobiotic) for use in nematological research may be chosen. Agnotobiotic techniques are the only feasible approaches for many researchers because of constraints on equipment etc. Where possible, results from gnotobiotic experiments should be verified by agnotobiotic techniques and vice versa.

In-vitro systems are the most commonly used technique. Advantages include availability, ease of observation and the limited space requirements. Several disadvantages, however, are inherent to these systems. They include the necessity of storage in an area protected against environmental extremes because of small size and limited gas exchange in tightly sealed containers. Small container size also places certain limitations on experiment duration. Test tubes, flasks and petri plates have been used successfully. Important considerations in selecting the type of container include

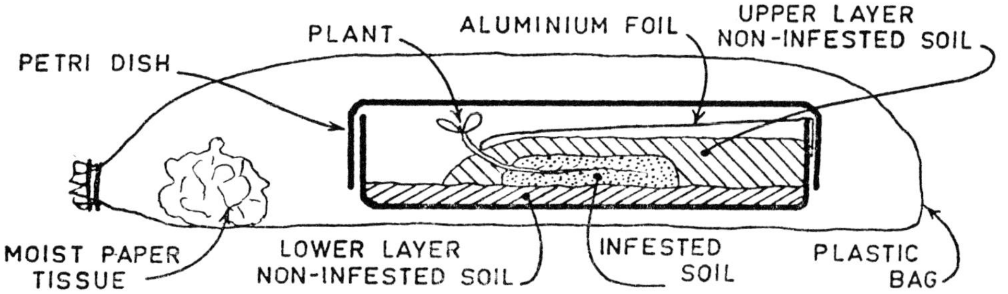

Fig. 4. A simple isolator used by Kable and Mai (25) to maintain constant moisture (Courtesy of E. J. Brill, publishers).

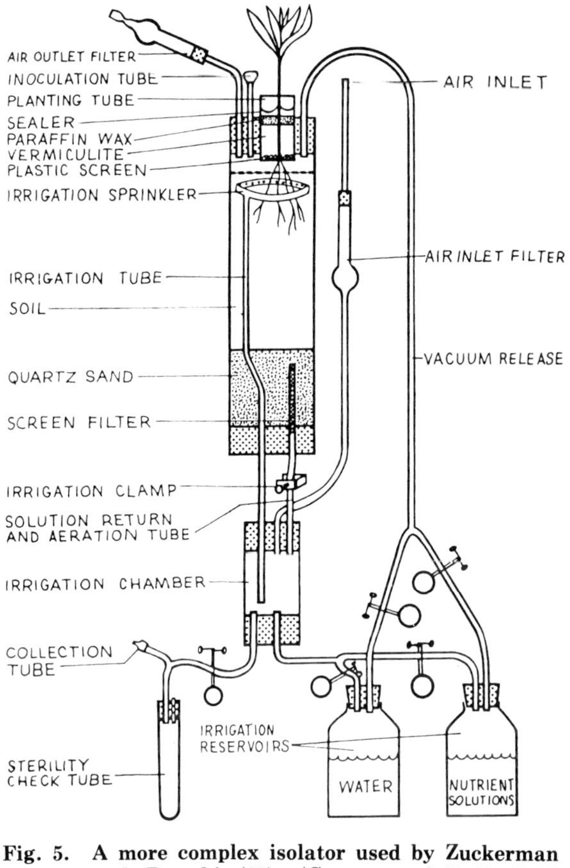

Fig. 5. A more complex isolator used by Zuckerman and Brzeski (67). (Courtesy of E. J. Brill publishers).

ease of access, storage space, number required and volume. The system must either be maintained in a germ-free atmosphere or sealed to prevent contamination. Since a germ-free atmosphere is difficult to maintain, tightly sealed containers that prevent dessication and contamination may restrict gas exchange. Thus, unless provisions are made to supply air to the system aseptically, the experimenter must either ignore the effects of respiration, limit experiment duration or open each container periodically and risk contamination.

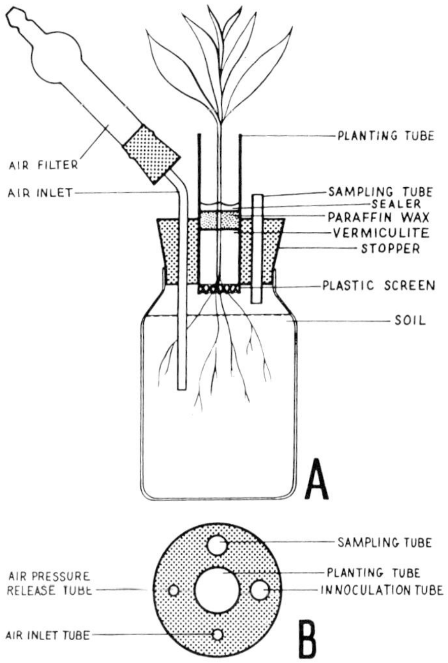

Fig. 6. A smaller isolator used by Zuckerman and Brzeski (67). A) Lateral view. B) Top view. (Courtesy of E. J. Brill publishers).

Various types of isolators have been designed to maintain plants under germ-free conditions (Figs. 4,5,6) (25,67). Zuckerman (66) outlines the advantages and disadvantages of the various types of isolators. The larger, more complex, isolators allow for aseptic monitoring of gas exchange and root exudates (66). The ultimate in isolators for plant pathogens can be found at the Plant Disease Research Laboratory, USDA-ARS, Frederick, Maryland, where an entire complex is devoted to research on exotic plant pathogens (36). Although this facility is designed to contain foreign pests, the technology employed could be applied to gnotobiological research on a large scale.

Complex isolators, however, may serve experimental needs inefficiently and be wasteful in terms of time and equipment. Simple, but effective, isolators for intact plants, with the aerial portions exposed to the external environment, have also been designed (67). Major advantages of the simpler isolators include low cost and the ability to replicate experiments for statistical analyses. Excised plant parts (generally roots) are more popular than the whole-plant systems for monoxenic nematode culture because they are amenable to the smaller types of isolators.

Production of Sterile Media for Axenic Plant Growth

Obligate plant-parasitic nematodes still require a host plant for growth and reproduction. Vigorous plants or plant parts must first be established in axenic culture before monoxenic culture may be attained. The problem is twofold: first, provision of a suitable medium for plant growth, and second, axenization of plants. Axenization is the process by which a species is separated from all microbial associates. The term *axenization* is preferred to the term *sterilization* because the latter can imply inability to germinate or reproduce.

SOIL Plants and their associated nematodes are normally found in soil. Soil systems offer certain advantages to the investigator working with intact plants. A sandy-loam potting mixture has moisture and nutrient-holding properties which make it a good substrate for plant growth. Such a potting mix will generally be adequate for nematodes. The most common methods for soil sterilization employ physical (heat, steam) or chemical [fumigation with methyl bromide or propylene oxide (1)] means. Sterilized soil must be stored and aerated for 6 to 8 weeks if phytotoxicity (due to heavy metals and/or ammonia when heat is used, or due to residual fumigants and/or ammonia if chemical means are employed) is to be avoided. Aseptic storage of sterilized soil requires additional space and a delay in the onset of experiments. Furthermore, stored soil should be assayed for microbial activity before and after use in an experiment.

AQUACULTURE - AGAR CULTURE Plants grown in aquaculture are suitable hosts for plant-parasitic nematodes. However, provisions must be made for aeration of the solution to supply oxygen. Such a system is advantageous in experiments requiring timed inoculation and permits easy access, observation, and roots free from debris or soil.

Agar culture (a form of aquaculture) is the most common method used to rear axenic plants or plant tissues. The ease of observation and detection of contamination in agar cultures are responsible for its popularity and wide usage. The ease in detection of contamination in agar cultures is unfortunately related to the ease with which they become contaminated. Other problems associated with agar culture are related to supplying adequate nutrition for the plant. Intact plants will require only a simple nutrient solution (modified Hoagland's solution) which can be autoclaved with the agar. Excised roots will require vitamins and carbohydrates in addition to mineral nutrients. A medium developed by Skoog et al. (24) (STW medium, Appendix A) has been used to grow excised roots of tomato, corn and soybean for nematological research (28,29,30,31,45). The investigator may regulate root growth to some extent by altering the concentration of nutrients and vitamins in the media. The addition of plant-growth regulators to a medium used for excised roots will result in the formation of callus-undifferentiated plant tissue. Callus tissue is useful for rearing nematodes which require no vascular elements.

The selection of media used in agar culture is somewhat dependent on the plant species used. Alfalfa callus will grow well on nutrient agar with minor additions [Riedel's medium (50) Appendix C], whereas soybean may require a more complex medium for growth (26) [Shenk and Hildebrandt's medium (54) Appendix B].

Axenization of Plants

Procedures for the axenization of plants have been derived empirically and vary with laboratory and plant species. Plant seeds are most easily separated from associated microflora and are the logical starting point. A first step should involve selection of high quality seed and removal of discolored or diseased specimens. The second stage usually involves treatment with broad-spectrum biocidal chemicals (mercuric chloride, hydrogen peroxide, sodium hypochlorite, etc.). The efficacy of these chemicals can be increased by sonication and/or detergent treatments. The plant species and seed contaminants will determine the chemical concentration and duration of exposure required for axenization. Rigorous chemical treatments are required for hard-seeded plants [e.g., concentrated H_2SO_4 for alfalfa (27)] and will kill seeds of other plants. Finally, the seeds must be rinsed thoroughly with sterile distilled water for removal of these chemicals, and seeds should be placed on nutrient agar or PDA for several days so contamination can be detected.

Compounding holidic or meridic media requires some special equipment. Many complex organic molecules are heat labile. They must be filter sterilized before addition to the autoclaved medium.

Axenization of Plant Seeds

Several methods for the axenization of plant seeds are available. The following methods should be used as a guide, but modification may be necessary.

A. Concentrated sulphuric acid (H_2SO_4) for alfalfa (27): place seeds (alfalfa) in concentrated H_2SO_4 for 30 min and rinse several times in distilled water. Place on 2% water agar for germination, and transfer seedlings, when about 2.6 cm long, to nutrient-agar slants.

B. Concentrated hydrogen peroxide (H_2O_2) for alfalfa seeds (R. S. Hussey, personal communication):
   1. Soak seeds in concentrated H_2O_2 for 25 min;
   2. Rinse with sterile H_2O;
   3. Place seeds in tubes containing medium for growing callus tissues.

C. Sodium hypochlorite (NaOCl) for alfalfa or tomato seeds:
   1. *Sterilize* a 30- or 50-ml beaker,
      1 petri dish with #9 filter paper (3 to 4 sheets),
      2 flasks 1/3 filled with distilled H_2O,
      2 empty flasks — (1 for NaOCl, 1 for ethanol).
   2. Pour 100 to 200 seeds into the sterile beaker. Fill beaker with a 1:1 mixture of 95% ethanol and 5.25% NaOCl. Stir gently —allow to soak 8 min.
   3. Drain alcohol-NaOCl mixture and rinse 4 to 6 times with the sterile H_2O.
   4. Drain on the sterile paper in the petri dish.
   5. Place 8-12 seeds per dish on the germination plates.
   6. Allow seeds to grow for 3 days before use.

D. Mercuric chloride for soybean seeds (S. R. Koenning & M. P. Ko, unpublished):
   1. Visually select whole, apparently disease-free seed.
   2. Soak seeds for 5-7 min in 70-95% ETOH.
   3. Soak seeds in 2500-5000 ppm $HgCl_2$ for 5-7 min (2.5-5.0 g $HgCl_2$ in one liter H_2O + 7 ml concentrated HCl).
   4. Rinse 5-7 times in sterile distilled H_2O.
   5. Place 6-12 seeds on petri plates containing PDA or nutrient agar.
   6. Use only contamination-free plates after 5-10 days.
   7. Excise roots and grow on SH media for callus tissue culture or grow excised roots on SH media minus growth regulators, or modified STW medium for root system cultures. (Note: remaining cotyledons limit tissue discoloration.)

E. Wash seeds (cucumber) in running tap water in small bag of cheesecloth. Transfer to 20% solution of chloramine-T for 1 hr, then rinse three times with sterile distilled water. Transfer to 1.5% water agar and germinate at 20° C in dark. Excise radicals when they are 80-90 mm long (3 days) and transfer to culture medium (34).

Production of Axenic Nematodes

The method for axenization of a particular nematode species is affected by the life history, trophic level, habitat and available facilities. Phytonematodes are usually easier to axenize than free-living nematodes because the stylet acts as microbial filter. Nematode eggs may be treated with substances which would kill vermiform stages. Similarly, methods used to surface sterilize various members of the Secernentea apparently were fatal to trichodorids (67,68).

An extraction method which does not injure the nematodes must be chosen, and this varies with the target species. Methods which rely on nematode motility allow one to start with viable nematodes. Egg extraction of *Meloidogyne* or *Heterodera* spp. by NaOCl (22,63) will yield large numbers of eggs and second-stage juveniles. Juveniles which migrate through tissue in the first several hours after NaOCl extraction, however, should be discarded since they may have been injured during extraction. The NaOCl treatment will have removed many contaminants, but other treatments will be required. Allowing the juveniles to migrate into a disinfectant [Aretan (methoxyethyl mercuric chloride) or Malachite green] may suffice for final surface sterilization.

Chemical and/or physical means are usually employed to axenize nematodes. Nematode movement, size, density and osmotic tolerances can be used to reduce the numbers of associated organisms. A chemical treatment, though, is usually required to free nematodes from microbes completely. Important factors to be considered in choice of a chemical treatment are biocidal spectrum, exposure duration and temperature. Cool temperatures may slow uptake of the toxicant by the nematode. Nematodes will not tolerate prolonged exposure to strong all-purpose biocides, such as mercuric chloride, and exposure time must be limited. Antibiotics have been used with general biocides to reduce the biocidal concentration or exposure time, thus increasing success. Antibiotics for axenization of nematodes are preferred by many investigators because they are relatively non-toxic to nematodes (and workers). Disadvantages of antibiotic treatments include: 1) the narrow spectrum of most antibiotics which necessitates the use of several compounds, 2) expense, 3) storage requirements, and 4) the need for filter sterilization of the solutions.

Plant-parasitic nematodes which are virus vectors may require special axenization techniques. For culture of virus-free nematodes, specimens may need to be transferred after molting or reared on a host which is not susceptible to the virus. Nematodes infected with rickettsia-like organisms may require special chemotherapy (32).

The following methods for extraction and axenization of nematodes have been used successfully by various investigators. This list is not complete, and other methods are available (17,24,65,66).

Methods of Aseptically Collecting Inocula of *Meloidogyne* spp.

A. Extraction of eggs with sodium hypochlorite (NaOCl). (Caution—use exhaust hood when working with NaOCl) (23).

1. Collect infected tomato roots with light brown egg masses (about 7-10 weeks after inoculation) and wash gently.
2. Cut roots into 2- to 5-cm lengths.
3. Place roots in 0.5% sodium hypochlorite (a.i.) and shake or stir with motorized stirrer for 4 min (use ±20 g roots in 200 ml NaOCl solution in a 0.5-liter container).
4. To stop action of NaOCl, pour the NaOCl-egg suspension and roots over a 425-μm (40-mesh) sieve into 8000-ml bucket, and rinse roots on a 425-μm sieve to collect freed eggs (building up volume to 6-8 liters).
5. Pour egg suspension in bucket onto 26-μm (500-mesh) sieve, and *immediately rinse* the eggs on the 26-μm sieve with stream of *cold tap water to remove residual NaOCl.*
6. Rinse eggs off sieve and add water to give desired concentration.
7. Determine volume of suspension to be added/experimental unit and dispense 1/3 of this for each of 3 "runs" into beakers, using flask-type pipetters ("repeater pipetter," Fisher Scientific Co., Pittsburg, PA).

Notes: About 20% of these eggs will give rise to infective juveniles. This same method may be used for extraction of eggs from *Heterodera glycines* except that the NaOCl concentration must be increased to 1% (Byrd & Barker, unpublished).

Additional Note: Survival of eggs on screens and other equipment increases the chance of culture contamination. Other methods should be employed for stock cultures.

B. Hatching and collection of *Meloidogyne* spp. juveniles (23,63):

1. Extract eggs by NaOCl method.
2. Place eggs on 26-μm sieve in a closed chamber with sufficient water to cover eggs.
3. Collect juveniles daily.
4. Inoculate plants immediately *or* store at 10-15°C and aerate until ready to use juveniles as inoculum.
5. Dispense with flask-type pipetter as outlined for eggs.

Notes: About 20% of the juveniles hatching will infect tomato at 24°C. Use of tap water which has been boiled in an open container and allowed to remain uncovered overnight will increase hatch and infectivity (A.C. Triantaphyllou, personal communication).

Surface Sterilization of Nematodes

(Sterilize tools with $HgCl_2$, 1000 ppm)

A. Hibitane (for *Meloidogyne* spp.):

1. Eggs — expose for 4 to 8 min in 0.5% hibitane.
2. Juveniles — expose for 2 to 3 min in 0.5% hibitane *or* 5 to 10 min in 0.1% hibitane (A.C. Triantaphyllou, personal communication).

B. NaOCl (for eggs of *Meloidogyne* spp.): Expose for 2 to 3 min in 0.525% (a.i.) sodium hypochlorite (10% Clorox).

C. Malachite green + Streptomycin sulfate: Pass nematodes singly with a needle through 5 or 6 baths of a 20 ppm solution of malachite green + 1000 ppm streptomycin sulfate (27).

D. Aretan/centrifugation: Wash nematodes 3 times in 133 ppm Aretan by centrifugation. After final washing and centrifugation, add Aretan to bring volume to 30 ml and leave suspension overnight at 15°C (34).

E. Combination of A & B.

F. Combination of C & D.

G. Hydrogen peroxide (H_2O_2) for eggs *Heterodera* spp. (M.P. Ko, personal communication):

1. Obtain eggs by crushing cysts in a Ten Broeck homogenizer.
2. Layer approximately 5-10 ml of nematode suspension over a 40-ml sucrose solution (20% w/w), and centrifuge at 400*g* for 3 minutes.
3. Collect eggs at the sucrose/water interface (use suction), wash onto a 26-μm sieve and rinse with water.
4. Sterilize eggs in 4% H_2O_2 for 15 min and rinse 5 times with sterile distilled water.

H. Streptomycin sulfate, penicillin-G and Nystatin [*Bursaphelenchus xylophilus* (3)]. Wash through 5 to 7 changes of a solution of sterile 0.9% NaCl containing 10 mg/ml streptomycin sulfate, 5 mg/ml Nystatin (freshly prepared) and 16,000 units/ml of penicillin-G. Incubate for 5 min in each wash.

Transfer and Culture Maintenance

Transfer of axenic nematodes to plant tissue is best performed under a microvoid, laminar flow hood or in a sterile room. Nematodes are commonly transferred in a water droplet with a pipette or individually with a mercuric-chloride-sterilized pulp canal file. Egg masses of root-knot or cyst nematodes can easily be transferred aseptically from one monoxenic culture to the next. Similarly, pieces of infected plant tissue can be transferred to fresh callus to establish new cultures of *Pratylenchus* or *Ditylenchus* spp.

Harvesting of axenic inocula for experiments is more difficult. A Tiner trap (57) is an efficient but costly means to accomplish this goal. Autoclaved petri plates, containing screens covered with tissue and sterile distilled water or a disinfectant, are probably the simplest means. Once established, monoxenic cultures are relatively easy to maintain. The optimal interval for transfer should be determined empirically. The more virulent pathogens will require more frequent transfer, but this can be delayed if cultures are maintained at temperatures which will slow nematode development. Since the rate of population increase is related to the amount of host tissue and P_i, the transfer of a few nematodes to a large amount of host tissue will result in large populations (Fig. 7). As more information is gained about factors which influence survival stages in nematodes, the inducing of nematodes to enter dormant stages may make it possible for axenic cultures to be maintained for long periods (26).

Preventing contamination of gnotobiotic cultures is an exercise in pest control. Because a single contaminant cannot be tolerated, exclusion of unwanted organisms is a necessity. A properly designed system will help circumvent this problem, but additional steps may be required. Such steps might include storage of sealed cultures in oil, periodic fumigation of chambers and the use of insecticidal paper on shelves.

The addition of antibiotics or antifungal compounds to agar culture may help alleviate some problems with contamination. These compounds may, however, have subtle effects on plant tissues and/or nematodes which may alter experimental results. Furthermore, many require filter sterilization before addition to autoclaved media.

IMPORTANCE AND APPLICATIONS

Gnotobiotic techniques can be used in many aspects of nematology. The extensive use of *Caenorhabditis elegans* as an experimental animal is due in part to its adaptability to monoxenic or axenic culture. The use of nematodes as model biological systems has yielded much information about the aging process, genetics, embryology and cell lineages (8,49,68,69).

Development of axenic culturing techniques for nematodes offers many benefits. The physiology of these animals could be studied thoroughly if they could be grown in an holidic medium as well as providing ideal tools for studying a wide range of biological problems. Nematode mitochondria have been isolated and characterized from axenic culture (41,52). Investigators have had some success in culturing nematode ovarial explants (22,64), and this

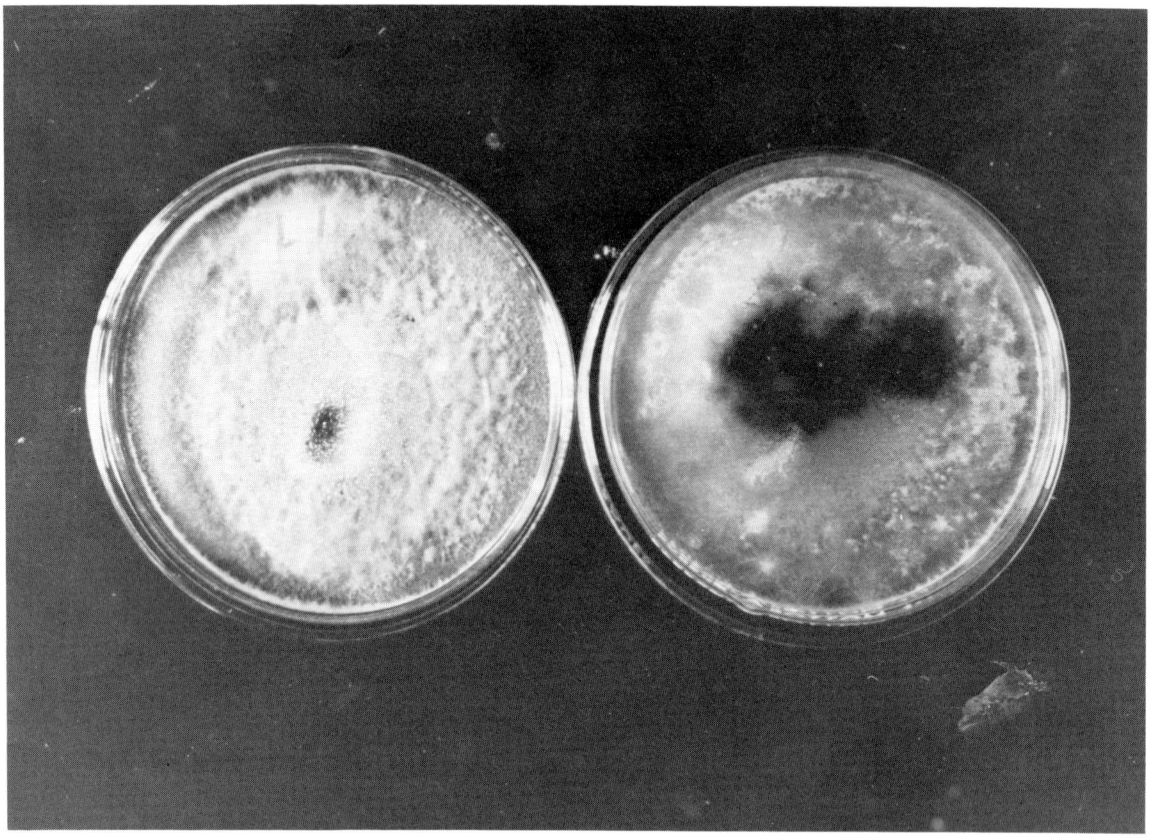

Fig. 7. Monoxenic cultures of *Bursaphelenchus xylophilus* reared on fungal mycelium. Plate on the right was inoculated by simultaneous transfer of nematodes and mycelium (note the dark areas where the fungus was killed). The petri plate on the left was inoculated with the nematode after the fungus was allowed to cover the plate and will ultimately yield five to ten times as many nematodes as the other plate.

may ultimately lead to nematode cell-tissue culture. It is not absolutely necessary to advance to axenic culture on a holidic medium to investigate the physiology of a nematode. However, it is mandatory that they be cultured under monoxenic conditions if physiological investigations are to progress. In this manner, contaminating metabolites of the host may be largely removed. If the subject is grown polyxenically, or in the presence of bacteria, the metabolites of these contaminating organisms become a problem.

A major advantage of gnotobiotic culture (agar cultures) is ease of observation. Such cultures may, therefore, be important as educational as well as research tools. Observation is essential to most behavioral studies, and observations of nematode feeding may permit us to interfere in this process.

Monoxenic culture of plant-parasites facilitates studies of nematode mating. Hybridization studies require rigorous controls which are difficult to maintain in pot cultures. The degree to which different nematodes hybridize is important from both a taxonomic and an agricultural point of view.

Monoxenic cultivation of nematodes such as *Meloidogyne* spp. may offer an additional tool in the study of their speciation as well as physiological differences. Schuster and Sullivan (55) found that they could distinguish certain species of nematodes by the response of excised roots in agar culture: e.g. *Meloidogyne incognita*—prevented root-hair formation on the developing galls; *M. hapla* induced root-hair formation on the galls; and *Nacobbus batatiformis* prevented root-hair formation on the galls, with most epidermal cells about the gall becoming spherical. A similar procedure could be adapted to the host differential test used to distinguish *Meloidogyne* spp.

The genetics of host-parasite interactions can be investigated in detail through the use of monoxenic cultures. Soybean cultivars have been evaluated for their resistance to *Heterodera glycines* in excised root cultures (30). The resistance of tomato to *Meloidogyne* spp. is maintained in excised roots (15). One person can perform 300-400 inoculations/week and roots can be scored for galling in 3 to 9 days in this system (15). The maintenance of large numbers of different species of plant-parasitic nematodes in one location can only be accomplished through monoxenic culture. Such culture collections (51) can be used in the screening for plant resistance to many different nematode species. Although pest resistance must ultimately be tested in the field, the genetics and physiology of resistance will most probably be determined in the laboratory. Mathematical models to predict the length of time required for the appearance of resistance-breaking biotypes can be constructed, tested and validated within a few years' time from laboratory data. Similar field research could require a decade or longer.

The precision of microplot experiments can be increased through the use of monoxenic cultures of nematodes. Martin et al. (33) developed a quantitative technique by which microplots could be inoculated with monoxenic lesion nematode cultures. NaOCl extraction of eggs of *Meloidogyne* and *Heterodera* spp. (23) will remove many microbial associates from greenhouse-reared inoculum. Some contaminants remain, but the NaOCl treatment reduces both beneficial and pathogenic organisms to low levels and minimizes their confounding effects. Open pot and microplot experiments will not remain monoxenic for long, but the initial minimization of background biological influences is necessary for repeatable and credible results.

The effects of physical factors on nematode population dynamics are more easily measured in the laboratory setting. Quantification of temperature effects on nematodes in excised roots (46) or hanging drops (16) is more accurate because the confounding influence of moisture is reduced. Such experiments may reveal how much variance is inherent within nematode biology and ecology.

Microcosm experiments have been utilized to establish the importance of nematodes in soil communities and how they influence nutrient turnover in soil ecosystems (1,5,6,7). Basic ecological research in microcosms may demonstrate beneficial effects of free-living and plant-parasitic nematodes.

Various areas of nematode control can benefit from gnotobiotic technique. Monoxenic cultures have been used in the evaluation of nematicidal chemicals (43). Biocontrol agents may require cultivation in monoxenic or dixenic cultures. *Seinura* spp. (a nematode predator) must be raised on nematodes and thus a dixenic system could be applied. Most obligate parasites of nematodes have not been reared in pure culture (32). *Bacillus penetrans* or other parasites of nematodes might be more easily reared in monoxenic or dixenic cultures. The economic feasibility of nematode biocontrol agents may ultimately depend upon our ability to rear them in axenic or monoxenic culture.

Literature Cited

1. Anderson, R. V., E. T. Elliott, J. F. McClellan, D. C. Coleman, C. V. Cole, and H. W. Hunt. 1978. Trophic interactions in soils as they affect energy and nutrient dynamics. III. Biotic interactions of bacteria, ameobae, and nematodes. Microb. Ecol. 4:361-371.
2. Balasubramanian, M. and R. F. Myers. 1971. Nutrient media for plant-parasitic nematodes—II. Exp. Parasitol. 29:330-336.
3. Bolla, R. I. and W. Jordan. 1982. Cultivation of the pine wilt nematode, *Bursaphelenchus xylophilus,* in axenic culture media. J. Nematol. 14:377-381.
4. Byars, L. P. 1914. Preliminary notes on the cultivation of the plant parasitic nematode, *Heterodera radicicola*. Phytopathology 4:323-326.
5. Cole, C. V., E. T. Elliot, H. W. Hunt, and D. C. Coleman. 1978. Trophic interactions in soils as they affect energy and nutrient dynamics. V. Phosphorous transformation. Microb. Ecol. 4:381-387.
6. Coleman, D. C., C. V. Cole, H. W. Hunt, and D. A. Klein. 1978. Trophic interactions in soils as they affect energy and nutrient dynamics. I. Introduction. Microb. Ecol. 4:345-349.
7. Coleman, D. C., R. V. Anderson, C. V. Cole, E. T. Elliot, L. Woods, and M. K. Campion.

1978. Trophic interactions in soils as they affect energy and nutrient dynamics. IV. Flows of metabolic and biomass carbon. Microb. Ecol. 4:373-380.

8. Croll, N. A. and B. E. Matthews. 1977. Biology of nematodes. Halsted Press, John Wiley & Son, New York. 201 pp.
9. Crosse, J. E. and R. S. Pitcher. 1952. Studies in the relationship of eelworms and bacteria to certain plant diseases. I. The etiology of strawberry cauliflower disease. Ann. Appl. Biol. 39:475-486.
10. Dolliver, J. S. 1961. Population levels of *Pratylenchus penetrans* as influenced by treatments affecting dry weight of wando pea plants. Phytopathology 51:364-367.
11. Dolliver, J. S., A. C. Hildebrandt, and A. J. Riker. 1962. Studies on reproduction of *Aphelenchoides ritzemabosi* (Schwartz) on plant tissues in culture. Nematologica 7: 294-300.
12. Dougherty, E. C. 1959. Introduction to axenic culture of invertebrate metazoa: A goal. Ann. N. Y. Acad. Sci. 27:27-54.
13. Doughtery, E. C. 1960. Cultivation of aschelminths, especially rhabditid nematodes, pp. 297-317 *In* Nematology—fundamentals and recent advances with emphasis on plant parasitic and soil forms, ed. J. N. Sasser and W. R. Jenkins. Univ. North Carolina Press, Chapel Hill, N.C. 480 pp.
14. Dougherty, E. C., E. H. Hansen, W. L. Nicholas, J. A. Mollett, and E. A. Yarwood. 1959. Axenic cultivation of *Caenorhabditis briggsae* (Nematoda: Rhabditidae) with unsupplemented and supplemented chemically defined media. Ann. N.Y. Acad. Sci. 77: 176-217.
15. Dropkin, V. H. and W. R. Boone. 1966. Analysis of host-parasite relationships of root-knot nematodes by single-larva inoculations of excised tobacco roots. Nematologica 12:225-230.
16. Ferris, H., H. S. DuVernay, and R. H. Small. 1978. Development of a soil-temperature data base on *Meloidogyne arenaria* for a simulation model. J. Nematol. 10:39-42.
17. Goodey, J. B. 1963. Laboratory methods for work with plant and soil nematodes. Minist. Agric., Fish. and Food Tech. Bull. No. 2, Her Majesty's Stationery Office, London. 72 pp.
18. Griffin, G. D. and K. R. Barker. 1966. Effects of soil temperature and moisture on the survival and activity of *Xiphinema americanum*. Proc. Helminthol. Soc. Wash. 33:126-130.
19. Hansen, E., E. J. Buecher, and A. A. F. Evans. 1970. Axenic culture of *Aphelenchus avenae*. Nematologica 16:330.
20. Hieb, W. F. and M. Rothstein. 1968. Sterol requirements for reproduction of a free-living nematode. Science 160:778-780.
21. Hieb, W. F., S. L. R. Stokstad, and M. Rothstein. 1970. Heme requirement for reproduction of a free-living nematode. Science 168:143-144.
22. Hirumi, H., T. A. Chen, and K. Maramorosch. 1968. Intra-uteral development of the root-knot nematode in organ culture. Second Int. Colloquium on invertebrate tissue culture, Instituto Lombardo: Fondazione Baselli 147-152.
23. Hussey, R. S. and K. R. Barker. 1973. A comparison of methods of collecting inocula on *Meloidogyne* spp., including a new technique. Plant Dis. Rep. 57:1025-1028.
24. Jacob, J. Js' and J. Van Bezooijen. 1971. A manual for practical work in nematology (revised edition). Int. Postgraduate Nematology Course, Int. Agric. Center, Wageningen, The Netherlands.
25. Kable, P. F. and W. F. Mai. 1968. Influence of soil moisture on *Pratylenchus penetrans*. Nematologica 14:101-122.
26. Koenning, S. R. 1984. The ecology and damage potential of *Pratylenchus brachyurus* on soybean. Ph.D. thesis, North Carolina State Univ., Raleigh, N.C. 128 pp.
27. Krusberg, L. R. 1961. Studies on the culturing and parasitism of plant-parasitic nematodes, in particular *Ditylenchus dipsaci* and *Aphelenchoides ritzemabosi* on alfalfa tissues. Nematologica 6:181-200.
28. Lauritis, J. A., R. V. Rebois, and L. S. Graney. 1983. Development of *Heterodera glycines* Ichinohe on soybean, *Glycine max* (L.) Merr., under gnotobiotic conditions. J. Nematol. 15:272-281.
29. Lauritis, J. A., R. V. Rebois, and L. S. Graney. 1982. Technique for gnotobiotic cultivation of *Heterodera glycines* Ichinohe on *Glycine max* (L.) Merr. J. Nematol. 14:422-424.
30. Lauritis, J. A., R. V. Rebois, and L. S. Graney. 1982. Screening soybean for resistance to *Heterodera glycines* Ichinohe using monoxenic cultures. J. Nematol. 14:593-594.
31. Lauritis, J. A., R. V. Rebois, and L. S. Graney. 1983. Life cycle of *Heterodera zeae*

Koshy, Swarup, and Sethi on *Zea mays* L. axenic root explants. J. Nematol. 15:115-119.

32. Mankau, R. 1981. Microbial control of nematodes, pp. 475-494 *In* Plant parasitic nematodes, Vol. III, ed. B. M. Zuckerman and R. A. Rohde. Academic Press, New York. 508 pp.
33. Martin, M. J. and M. Riedel. 1983. A technique for quantitative use of nematodes from monoxenic tissue culture as soil inoculum. Plant Dis. 67:275-277.
34. McClure, M. A. and D. R. Viglierchio. 1966. Penetration of *Meloidogyne incognita* in relation to growth and nutrition of sterile, excised cucumber roots. Nematologica 12: 237-247.
35. McClure, M. A. and D. R. Viglierchio. 1966. The influence of host nutrition and intensity of infection on the sex ratio of *Meloidogyne incognita* in sterile agar culture of excised cucumber roots. Nematologica 12:248-258.
36. Melching, J. S., K. R. Bromfield, and C. H. Kingsolver. 1983. The plant pathogen containment facility at Frederick, Maryland. Plant Dis. 67:717-722.
37. Mountain, W. B. 1955. A method of culturing plant parasitic nematodes under sterile conditions. Proc. Helminthol. Soc. Wash. 21:49-52.
38. Mountain, W. B. 1960. Acceptable standards of proof and approaches for evaluating nematode relationships, pp. 422-425 *In* Nematology—fundamentals and recent advances with emphasis on plant parasitic and soil forms, ed. J. N. Sasser and W. R. Jenkins. Univ. North Carolina Press, Chapel Hill, N.C. 480 pp.
39. Mountain, W. B. and C. D. McKeen. 1962. Effect of *Verticillium dahliae* on the population of *Pratylenchus penetrans*. Nematologica 7:261-266.
40. Murashige, T. and F. Skoog. 1962. A revised medium for rapid growth and bioassays with tobacco callus. Physiol. Plant. 15:473-497.
41. Murfitt, R. R., K. Vogel, and D. Rao Sanadi. 1976. Characterization of the mitochondria of *Caenorhabditis elegans*. Comp. Biochem. Physiol. 53:423-430.
42. Myers, R. F. 1967. Axenic cultivation of *Aphelenchoides sacchari* Hooper. Proc. Helminthol. Soc. Wash. 34:251-255.
43. Myers, R. F. 1972. Assay of nematicidal chemicals using axenic cultures of *Aphelenchoides rutgersi*. Nematologica 18: 447-457.
44. Nicholas, W. L., E. Hansen, and E. C. Dougherty. 1962. The B-vitamins required by *Caenorhabditis briggsae*. Nematologica 8:129-135.
45. Orion, P., W. P. Wergin, and B. Y. Endo. 1980. Inhibition of syncytia formation and root-knot nematode development on cultures of excised tomato roots. J. Nematol. 12:196-203.
46. Olowe, T. and D. C. M. Corbett. 1976. Aspects of the biology of *Pratylenchus brachyurus* and *P. zeae*. Nematologica 22: 202-211.
47. Pertel, R. 1964. Axenic cultivation of two species of Rhabditid nematodes on a commercial medium. Nematologica 10:343.
48. Rich, S. and P. M. Miller. 1964. Verticillium wilt of strawberries made worse by soil fungicide that stimulates meadow nematode populations. Plant Dis. Rep. 48: 246-248.
49. Riddle, D. L. 1978. The genetics of development and behavior in *Caenorhabditis elegans*. J. Nematol. 10:1-16.
50. Riedel, R. M., J. G. Foster, and W. F. Mai. 1973. A simplified medium for monoxenic culture of *Pratylenchus penetrans* and *Ditylenchus dipsaci*. J. Nematol. 5:71-77.
51. Riedel, R. M., M. Alves del Lima, and M. Martin. 1983. Establishment of nematode germplasm banks, pp. 880-903 *In* Handbook of plant cell culture, Vol. 1: Techniques for propagation and breeding, ed. D. A. Evans, W. R. Sharp, P. V. Ammirato, and Y. Yamada. Macmillan Publ., New York. 950 pp.
52. Rothstein, M., F. Nichols, and P. Nichols. 1970. Nematode biochemistry — XII. Mitochondria from the free-living nematode *Turbatrix aceti*. Int. J. Biochem. 1:695-705.
53. Sayre, R. M. 1958. Plant tissue culture as a tool in the study of the physiology of the root-knot nematode, *Meloidogyne incognita* Chit. Ph.D. thesis, Univ. Nebraska, Lincoln, Nebr. 61 pp.
54. Schenk, U. and A. C. Hildebrandt. 1972. Medium and techniques for induction and growth of monocotyledenous and dicotyledenous plant cell cultures. Can. J. Bot. 50:199-204.
55. Schuster, M. L. and T. Sullivan. 1960. Species differentiation of nematodes through host reaction in tissue culture. I. Comparisons of *Meloidogyne hapla*, *Meloi-*

dogyne incognita, and *Nacobbus batatiformis.* Phytopathology 50:874-875.

56. Tiner, J. D. 1960. Culture of the plant parasitic nematode genus *Pratylenchus* on sterile excised roots. I. Their establishment and maintenance. J. Parasitol. 9:121-126.
57. Tiner, J. D. 1961. Cultures of the plant parasitic nematode genus *Pratylenchus* on sterile excised roots. II. A trap for collection of axenic nematodes and quantitative initiation of experiments. Exp. Parasitol. 11:231-240.
58. Tyler, J. 1933. Reproduction without males in aseptic cultures of the root-knot nematode. Hilgardia 7:373-415.
59. Vanfleteren, J. R. 1973. Amino acid requirements of the free-living nematode *Caenorhabditis briggsae.* Nematologica 19: 93-99.
60. Vanfleteren, J. R. 1974. Nematode growth factor. Nature 148:255-257.
61. Vanfleteren, J. R. 1978. Axenic culture of free-living, plant parasitic and insect parasitic nematodes. Annu. Rev. Phytopathol. 16:131-157.
62. Vanfleteren, J. R. 1980. Nematodes as nutritional models, pp. 47-79 *In* Nematodes as biological models, Vol. 2, ed. B. M. Zuckerman. Academic Press, New York. 312 pp.
63. Vrain, T. C. 1977. A technique for the collection of larvae of *Meloidogyne* spp. and comparison of eggs and larvae as inocula. J. Nematol. 9:249-251.
64. Viglierchio, D. R., A. R. Maggenti, and R. N. Johnson. 1969. Axenic ovarial explants from the marine nematode *Deontostoma californicum* on culture media. J. Nematol. 1:70-83.
65. Zuckerman, B. M. 1969. Gnotobiotic culture and its significance to plant nematology. Helminthol. Abstr. 38:1-15.
66. Zuckerman, B. M. 1971. Gnotobiology, pp. 159-184 *In* Plant parasitic nematodes, Vol. II, ed. B. M. Zuckerman, W. F. Mai, and R. A. Rohde. Academic Press, New York. 347 pp.
67. Zuckerman, B. M. and M. W. Brzeski. 1965. Methods for the study of plant parasitic nematodes in gnotobiotic root culture. Nematologica 11:453-466.
68. Zuckerman, B. M. 1980. Nematodes as biological models, Vol. 1. Academic Press, New York. 312 pp.
69. Zuckerman, B. M. 1980. Nematodes as biological models, Vol. 2. Academic Press, New York. 306 pp.

APPENDIX A.

An agar medium for plant growth [STW medium; from Lauritis et al. (24,28)].

Components

Macronutrient salts (Stock A):

Solution 1

$Ca(NO_3)_2 \cdot 4H_2O$, 1.44 g
KNO_3, 0.80 g
KCl, 0.65 g
KH_2PO_4, 0.38 g
NH_4NO_3, 0.40 g
Distilled H_2O q.s., 800 ml

Solution II

$MgSO_4 \cdot 7H_2O$, 0.72 g
Distilled H_2O q.s., 200 ml

Combine solutions I and II to make 1 liter of Stock A, store @ 5°C.

Iron (Stock B):

$FeSO_4 \cdot 7H_2O$, 0.56 g
Na_2EDTA, 0.74 g
Distilled H_2O q.s., 100 ml

Aerate Stock B for 24 hrs, store in dark bottle @ 5°C.

Micronutrient salts (Stock C):

$ZnSO_4 \cdot 7H_2O$, 0.027 g
$MnSO_4 \cdot H_2O$, 0.049 g
H_3BO_3, 0.016 g
KI, 0.0075 g
Distilled H_2O q.s., 100 ml

Store @ 5°C.

Vitamins and amino acids (Stock D):

Nicotinic acid, 0.050 g
Pyridoxine HCl, 0.075 g
Thiamine HCl, 0.010 g
Sodium glycinate, 0.262 g
Distilled H_2O q.s., 100 ml

Filter sterilized Stock D, store 1.0 ml aliquots @ −20°C.

To make 1 liter of nutrient agar medium, combine 100 ml of Stock A, 5 ml of Stock B, 10 ml of Stock C, 20 g of sucrose, 15 g of Difco Noble agar and distilled H_2O q.s. 1000 ml. Autoclave medium @ 44.4 kg/cm^2 for 15 min. Add 1 ml of Stock D and adjust pH with acid or base to 5.6-5.8 after medium has cooled to about 60°C.

APPENDIX B.

Shenk and Hildebrant's Plant Tissue Culture Medium [for callus tissue (54)].

Preparation of stock solutions

Major elements stock solution

KNO_3, 50.0 g
$MgSO_4 \cdot 7H_2O$, 8.0 g
$NH_4H_2PO_4$, 6.0 g
$CaCl_2 \cdot 2H_2O$, 4.0 g
DDH_2O, 1000.0 ml

Use 50 ml/liter of the final medium.

Minor elements stock solution

$MnSO_4 \cdot H_2O$, 200.0 mg
H_3BO_3, 100.0 mg
$ZnSO_4 \cdot 7H_2O$, 20.0 mg
KI, 20.0 mg
$CuSO_4 \cdot 5H_2O$, 4.0 mg
$Na_2MoO_4 \cdot 2H_2O$, 2.0 mg
$CoCl_2 \cdot 6H_2O$, 2.0 mg
DDH_2O, 1000.0 ml

Use 50 ml/liter of the final medium.

Iron compounds stock solution

$FeSO_4 \cdot 7H_2O$, 1.5 g
Na_2 EDTA, 2.0 g
DDH_2O, 100.0 ml

Heat the ingredients in DDH_2O. Use 1 ml/liter of final medium.

Growth regulators stock solution

2,4-D, 50.0 mg
p-acid chlorophenoxyacetic, 200.0 mg
Kinetin, 10.0 mg
DDH_2O, 100.0 ml

Use 1 ml/liter of the final medium.

Organic stock solution

Thiamine HCl, 500.0 mg
DDH_2O, 33.0 ml
Nicotinic acid, 500.0 mg
DDH_2O, 33.0 ml (add KOH to dissolve if necessary)
Pyridoxine HCl, 50.0 mg
DDH_2O, 33.0 ml

Mix after each ingredient is dissolved completely. Use 1 ml of the mixture/liter of the final medium.

Preparation of SH tissue culture medium

To Flask No. 1 (2-liter capacity), add 12 g purified agar and 1 liter distilled water, stir with a magnetic bar and heat on a hot plate.

To Flask No. 2 (1-liter capacity), add

Major elements stock solution, 100 ml
Minor elements stock solution, 100 ml
Iron compounds stock solution, 2 ml
Growth regulators stock solution, 2 ml
Organic stock solution, 2 ml
Inositol, 2 g
Sucrose, 60 g
Distilled water, 800 ml

Stir with a magnetic bar. Check pH with a pH meter. The reading will be 4.0 ± 0.1. Bring the pH to 5.8 to 5.9 with 1N NaOH. After the agar is dissolved completely, add the contents of Flask No. 2 to Flask No. 1. Autoclave at 121°C and 44.4 kg/cm^2 for 15 min.

APPENDIX C.

A simplified media for plant tissue callus.

Reidel et al. (50) simplified medium

Sucrose, 20 g
Yeast extract (Difco), 5 g
2,4-D (from a stock solution with 1 mg 2,4-D per ml H_2O), 2 mg
Difco-Bacto agar, 10 g
H_2O (distilled), 1000 ml

Part Three

Identification

IDENTIFICATION OF *MELOIDOGYNE* SPECIES ON THE BASIS OF DIFFERENTIAL HOST TEST AND PERINEAL-PATTERN MORPHOLOGY

K.M. Hartman and J.N. Sasser

Department of Plant Pathology
North Carolina State University
Raleigh, North Carolina, USA

The extensive morphological variation among and within root-knot nematode species makes their identification difficult. Because many *Meloidogyne* species infect a wide range of plants, inferences about species identity cannot be based solely on the host, even though some species do have very narrow host ranges. For example, although hundreds of plant species are susceptible to *M. incognita* (Kofoid & White) Chitwood, blueberry (*Vaccinium* sp.) and azalea (*Rhododendron* sp.) are the only known hosts of *Meloidogyne carolinensis* Eisenback (9).

Accurate nematode identification cannot be overemphasized, especially in the initiation of research programs and in the development of control strategies. Major control tactics such as crop rotation and resistant cultivars are generally species or race specific. Therefore, in the development of a root-knot-resistant cultivar, the exact identity (that is, species and host race) of the nematode population being tested must be known. Likewise, useful crop rotation recommendations cannot be made unless the identities of the predominant nematode species and races are known. If the identity of a population is unknown or only assumed, the crop selected for use in a rotation scheme may cause dramatic increases in nematode populations.

DEVELOPMENT OF THE DIFFERENTIAL HOST TEST—HISTORICAL BACKGROUND

Following the first written account of the root-knot nematodes by Berkeley (1) in 1855, host lists began to be compiled. The number of known hosts eventually exceeded 2000 plant species (2,11,15,33). These compilations were developed when root-knot nematodes were considered to be a single species, *Heterodera marioni* (Cornu) Goodey.

Certain inconsistencies appeared in host range reports before the complexity of *Meloidogyne* species had been elucidated. For example, in 1889, Neal (17) concluded that peanut (*Arachis hypogaea* L.) was "badly affected" in Florida, whereas later, Bessey (2) reported growing peanut in heavily infested fields without any infection. Godfrey (11) indicated that peanut was immune or slightly susceptible to this nematode and suggested that Spanish peanut could be used in crop rotation schemes. Sherbakoff (23), in 1939, reported considerable root-knot injury to cotton (*Gossypium hirsutum* L.) grown on land previously planted to cotton, but observed no injury to cotton grown on land previously planted to tomato, (*Lycopersicon esculentum* Mill.), even though tomato had been severely injured by this disease. He suggested as a working hypothesis that "the host specialization of the nemas [*sic*] is due primarily to the segregation and survival of favored genetic races."

Influenced by Sherbakoff, Christie and Albin (5) tested the differential responses of populations of *Heterodera marioni* to Persian clover (*Trifolium resupinatum*) to demonstrate the diversity within this pest. They were able to distinguish at least 5 "races" from 14 populations based upon host response and galling characteristics. Christie (4) and Christie and Havis (6) further documented the diversity of *H. marioni* by using other plant species.

Chitwood, in 1949, (3) removed the root-knot nematodes from the genus *Heterodera* and reassigned them to *Meloidogyne*, a genus described earlier by Göldi (12) for a root-knot infecting coffee in Brazil. Chitwood conducted a detailed morphological study of many characters of males, juveniles and females, including perineal patterns, and described five species and one subspecies: namely, *Meloidogyne incognita; M. incognita acrita* Chitwood, 1949; *M. javanica* (Treub, 1885) Chitwood, 1949; *M. arenaria* (Neal, 1889) Chitwood, 1949; *M. hapla* Chitwood, 1949; and *M. exigua* Göldi, 1887. Even though Chitwood emphasized morphological characters in his description, he also noted some important host differences. For example, *M. hapla* was the only root-knot nematode known to reproduce on strawberry (*Fragaria* sp.), and *M. incognita* failed to infect peanut but reproduced readily on cotton.

Early evidence of variation within the species of *M. incognita* was reported by Martin in 1954 (16). He noted that parasitism of cotton among isolates of *M. incognita* ranged from none to severe. About the same time, Sasser (19) tested 50 plant species and varieties commonly used in rotation programs in the southeastern United States for susceptibility to the most common species of *Meloidogyne*. Many were resistant to some species of *Meloidogyne* and susceptible to others. One half of the plant species tested were either uniformly susceptible or uniformly resistant to all five root-knot species.

Sasser (19) proposed a simple and practical method for identification of the four major species of *Meloidogyne* based on their responses to a series of differential hosts and the type of galling induced. Accordingly, an unknown population of a single common species of root-knot nematode could often be identified by its reaction to four plant species: peanut, watermelon (*Citrullus vulgaris* Schrader), pepper (*Capsicum frutescens* L.) and *Lycopersicon peruvianum*. *M. hapla* and *M. arenaria* were separated from *M. javanica* and *M. incognita* by their ability to reproduce on peanut. Furthermore, *M. arenaria* was separated from *M. hapla* by its response to watermelon, wheat (*Triticum vulgare* Villars), barley (*Hordeum vulgare* L.) or corn (*Zea mays* L.). *M. incognita* was differentiated from *M. javanica* by its reproduction on pepper, and *M. incognita acrita* Chitwood was separated by its reproduction on *Lycopersicon peruvianum*. *M. incognita acrita,* however, was eventually synonymized with *M. incognita* by Triantaphyllou and Sasser (30,31, 32).

This early differential host test became less reliable as more root-knot nematode populations were obtained from widely separated geographical regions. Several populations of *M. incognita* were found to infect the resistant tobacco cultivar 'NC95' (20). This variety was considered to be practically immune to the United States populations of *M. incognita*. Populations of *M. arenaria* were also found to vary in the ability to parasitize peanut (20); some populations produced severe galling on peanut roots and pods, whereas others did not.

A new differential host test for identification of unmixed populations of the four common species was proposed in 1977: *M. incognita, M. arenaria, M. javanica* and *M. hapla* (22). The six differential hosts used were cotton 'Deltapine 16,' tobacco 'NC95,' pepper 'California Wonder,' watermelon 'Charleston Gray,' peanut 'Florunner,' and tomato 'Rutgers' (25). Based on the susceptibility or resistance of these hosts, *M. incognita* populations were classified into four host races. Populations of *M. arenaria* were placed into two host races according to their ability to parasitize peanut. To date, neither a genetic nor a biochemical basis has been found for identification of host races within species. Many inconsistencies in host responses reported by plant breeders and nematologists may be accounted for by the discovery of host races.

THE DIFFERENTIAL HOST TEST—PURPOSE AND LIMITATIONS

The differential host test, in conjunction with perineal pattern morphology, has several objectives: 1) to give a preliminary indication of the four common species based on the "usual" host response (Table 1), 2) to detect mixed populations, 3) to detect new and undescribed species, and 4) to detect host races and other pathogenic variation within the common species as evidenced by a different or unexpected host response. Without backup by morphological characters, this test is not reliable for the identification of *Meloidogyne* populations. If several species or an uncommon species were present, they could not be discerned by a "usual" host response. The results of hundreds of tests have proven it to be fairly reliable for species and race identification of single populations of the four common species when identifications are verified with morphological characters, such as perineal patterns.

Any deviation from the responses expressed in Table 1 indicates pathogenic variation of the population. For example, some populations of *M. javanica* and *M. arenaria* race 2 have been detected that do not have the usual response. The discovery of such variants may eventually lead to the establishment of additional physiological races. Hundreds of differential host tests involving populations from many parts of the world have given some insight into the stability of the reaction of *Meloidogyne* species. Variation in pathogenicity to the differential hosts among populations of the major species of *Meloidogyne* is relatively uncommon. The International *Meloidogyne* Project (IMP) has summarized the responses of nearly 1000 populations of the four most common species of *Meloidogyne* and their races to the differential hosts (Table 2).

Physiological (or "host") races of *M. incognita* reflect the population's ability to reproduce on cotton 'Deltapine 61' ('Deltapine 16' is no longer available) and tobacco 'NC 95' (*Nicotiana tabacum* L.) (Table 1). These races are generally distinct and intermediate responses are not common. *M. incognita* with its four races is the most commonly encountered species from agricultural areas with 71, 12, 15, and 2% of the populations of this species encountered being race 1, 2, 3, and 4, respectively. The races of *M. incognita* have not been distinguished from each other morphologically and do not correlate with the two existing cytological races described by Triantaphyllou (29,30).

The two physiological races of *M. arenaria* are defined by their ability to parasitize peanut. Race 1, which infects peanut, is least common and accounts for only 20% of the *M. arenaria* populations collected by the IMP. As with the physiological races of *M. incognita*, those of *M. arenaria* cannot be identified on the basis of morphology (7) and show no correlation with *M. arenaria's* cytological races (27).

Physiological races of *M. javanica* or *M. hapla* have not been named, although about 7% of the *M. javanica* populations collected by the IMP parasitize pepper. Host response differences between the two cytological races of *M. hapla* have not been discovered (28).

The four common species and their host races generally can be identified on the basis of their response to the differential host test. When more than one species occurs in a population, interpretation of host responses becomes more complex. For example, population E933 from the Canary Islands gave a host response similar to *M. incognita* race 4. However, after examination of perineal patterns

Table 1. Usual response of the four common *Meloidogyne* species and their races to the North Carolina Differential Host Test.

Meloidogyne species and physiological races	Differential Host Plants[a]					
	Cotton	Tobacco	Pepper	Watermelon	Peanut	Tomato
M. incognita						
Race 1	−	−	+	+	−	+
Race 2	−	+	+	+	−	+
Race 3	+	−	+	+	−	+
Race 4	+	+	+	+	−	+
M. arenaria						
Race 1	−	+	+	+	+	+
Race 2	−	+	−	+	−	+
M. javanica	+	+	−	+	−	+
M. hapla	−	+	+	−	+	+

[a]Cotton, Deltapine 61; tobacco, NC 95; pepper, Early California Wonder; watermelon, Charleston Gray; peanut, Florunner; tomato, Rutgers; (−) indicates a resistant host; (+), a susceptible host.

Table 2. Reaction of populations of *Meloidogyne* species and races on the North Carolina Differential Host Test[a].

Species and physiological race	Number of populations studied	Number of countries from which samples were received	Percentage of populations pathogenic to the respective differentials[b]					
			Cotton	Tobacco	Pepper	Watermelon	Peanut	Tomato
M. incognita								
Race 1	372	54	0	0	98	100	0	100
Race 2	62	27	0	98	90	100	0	100
Race 3	79	20	100	0	100	100	0	100
Race 4	10	6	100	100	100	100	0	100
M. javanica	311	59	0	100	7	98	1	100
M. hapla	88	23	0	87	88	0	95	100
M. arenaria								
Race 1	13	5	0	62	85	100	100	100
Race 2	51	18	0	100	35	100	0	100

[a]Revised from Sasser and Carter (21).

[b]Cotton cv. 'Deltapine 61;' Tobacco cv. 'NC 95;' Pepper cv. 'California Wonder;' Watermelon cv. 'Charleston Gray;' Peanut cv. 'Florunner;' Tomato cv. 'Rutgers.'

from each infected host, two species were discovered. *M. javanica* was parasitizing tobacco, watermelon and tomato, whereas cotton, pepper, and tomato were parasitized by *M. incognita*. Because of the widespread occurrence of mixed populations, individuals from each infected differential host must be examined for perineal pattern, male head shape, or other morphological characters, if all of the species present are to be identified.

GUIDELINES FOR CONDUCTING THE DIFFERENTIAL HOST TEST—METHODOLOGY

Collection and Propagation of *Meloidogyne* Populations

Field populations are best established from egg masses collected from infected plant roots rather than from root fragments or field soil. If necessary, roots or field soil can be used to establish the culture, although there is an increased chance that other plant-parasitic nematodes and pathogens will be introduced.

Egg masses (20 to 40) should be selected randomly from several plants from the same field and placed in a 1% saline solution or in tap water. This inoculum is placed beneath the roots of a suitable susceptible host, such as a tomato seedling (cv. 'Rutgers'), as it is transplanted into a 15-cm-diameter pot filled with sterile soil. For optimal nematode egg production, the tomato seedlings should be grown for 45 to 60 days. However, nematode cultures can be maintained for several months if plants are cared for properly. For long-term maintenance, the culture must be transferred periodically to new tomato seedlings.

Common Sources of Contamination By Other *Meloidogyne* Populations

The possibility of contamination by other root-knot populations should be considered. For example, non-sterile soil, water splashing or draining from one pot to another, and unclean hands or equipment are potential sources of contamination.

CONTAMINATED SOIL Soil should be steam sterilized for 2 hrs. before use. Afterwards, periodic checks for contamination will indicate whether the steam sterilization has been effective. A 'Rutgers' tomato planted into a pot of soil and examined 45-60 days later will have a noticeably galled root system if the soil has become contaminated. Elutriation or Baermann funnel techniques can also be used to check for presence of other plant-parasitic nematodes.

WATER SPLASHED FROM ADJACENT POTS Contamination by splashing can be avoided if splash guards are placed between pots. These devices can be made out of plastic-coated screen, plastic, fiberglass, or other suitable materials. Plants should be spaced widely apart and watered as gently as possible if splash guards are not used.

CONTAMINATED GREENHOUSE BENCHES Culture pots can be elevated above the bench surface on inverted drainage saucers. This arrangement will prevent drainage water from coming in contact with adjacent pots. If culture pots are placed on a screen-mesh surface, excess water will drain from pots without coming in contact with other pots. Accumulated soil and algae should be cleaned periodically from bench surfaces with sodium hypochlorite (NaOCl).

CONTAMINATED UTENSILS Hands, utensils and laboratory surfaces should be carefully washed so adhering egg masses will be removed. A dilute solution (0.52% a.i.) of household bleach [sodium hypochlorite (NaOCl)] applied to laboratory surfaces will destroy nematode eggs and juveniles.

Procedure for Collection, Extraction and Calibration of Root-Knot Nematode Eggs For Inoculum (14)

1. Cut away the above-ground portion of the plant, and gently wash soil from the roots with running tap water.
   Note: To verify the identity of the nematode inoculum, remove 10 to 20 small galls containing single mature females from the roots. The galls can be stored in 1% saline solution and kept refrigerated for several hours until perineal patterns can be cut.
2. Prepare 200 ml of 0.52% NaOCl solution using commercial-strength bleach (5.25% NaOCl).
3. Cut root system into 2-4 cm pieces and place in the 1-liter flask.
4. Add the 200 ml of 0.52% NaOCl, stopper the flask, and shake vigorously for 3 minutes.
5. After shaking the flask, quickly pour the solution through a 200-mesh (75-μm) sieve nested on a 500-mesh (26-μm) sieve. Rinsing with tap water washes eggs through the upper sieve. After removal of the 200-mesh sieve, gentle and thorough rinsing of the eggs collected on the 500-mesh sieve removes all residual bleach.

   Note: If nematode eggs are exposed to NaOCl for more than 5 minutes, a large percentage of them may lose infectivity (see Chapter 2). The sieving process should be completed within 2 minutes. If it is not, stop sieving and begin rinsing eggs that have been collected up to that point.
6. Wash the eggs collected on the 500-mesh sieve into a graduated cylinder and bring the volume to 100 ml with tap water.
7. Pour a known volume (100 ml) of the egg suspension into a 250-ml flask. There are generally too many eggs to be easily counted in a 1-ml aliquant of egg suspension, so a dilution series is necessary.
8. Take a 0.5-ml sample of eggs and transfer by pipette to another flask containing 4.5 ml tap water.

   Note: When an aliquant of the egg suspension is transferred, it is very important that the suspension be well agitated so that eggs are evenly dispersed throughout the known volume. The flask is vigorously swirled, air is blown forcefully into the suspension with a pipette, and the eggs are quickly transferred before they settle.
9. Pipette a 0.5-ml aliquant of the diluted suspension to the counting dish and immediately disperse eggs with tap water so they fall randomly on the counting surface. Count the eggs covering 1/4 of the counting surface and multiply that number by 80 to obtain the number of eggs/ml in the original solution.
   Note: Counting dishes can be made from small, rectangular, transparent, plastic containers that can be used under the dissecting microscope. The surface area can be subdivided into fractions with lines scratched on the surface with a needle.
10. Repeat the dilution series and average the results of the two replicates.
11. Multiply the average number of eggs/ml by the total volume (100 ml) to determine the total number of available eggs. At least 90,000 eggs are usually needed for the test since the three replications of the six differential hosts must each be inoculated with 5000 eggs. (5000 eggs/plant $\times$ 18 plants = 90,000 eggs.)
12. Divide the total number of eggs needed (90,000) by the average number of eggs/ml calculated in step 4. The result of this calculation will be the number of milliliters of the inoculum suspension containing the desired number of eggs. Measure this amount from the original suspension and bring the volume up to 90 ml (5 ml of inoculum/plant).

Initiation of the Differential Host Test

1. Transplant three healthy seedlings (2- to 4-true-leaf stage) of each of the following cultivars into 10-cm pots with a sterile soil mixture: tomato 'Rutgers,' tobacco 'NC 95,' pepper 'Early California Wonder,' peanut 'Florunner,' watermelon 'Charleston Gray,' and cotton 'Deltapine 61.'
2. Pipette 5 ml of well-stirred inoculum near the base of each plant.
3. Water plants very lightly to wash eggs into the soil. Overwatering may wash the inoculum out of the pot.
4. Allow the plants and nematodes to develop about 60 days.

   Note: The test should be conducted at a temperature range of 24° to 30°C. The resistant reaction of cotton and tobacco is temperature dependent. Therefore, at high temperatures (>30°C), populations of *M. incognita* may be induced to infect one or both of these hosts, and thus the race may be erroneously identified.

Termination and Evaluation of the Differential Host Test

1. Remove plant stem and leaves and gently wash soil free from the root system.
2. If egg masses are quite small or few in number, stain them with a solution of Phloxine B (15 mg/liter of water or 0.0015%, C.I. No. 45410) for 15 to 20 minutes to enhance visibility. The stain is absorbed by the gelatinous matrix surrounding the eggs and gives it a pink to red color. The roots themselves remain unstained or only very lightly stained. Eggs do not absorb the stain and remain viable (8,13).
3. Count the number of galls and the number of egg masses separately for each root system and assign a rating index number according to the scale: 0 = no galls or eggs masses, 1 = 1-2 galls or egg masses, 2 = 3-10, 3 = 11-30, 4 = 31-100, and 5 = over 100 galls or egg masses. Host-plant types that have an average gall and egg mass index of 2 or less are designated resistant (−), provided that the average rating of the 'Rutgers' tomato (susceptible control) is greater than 4. Host plants on which nematode reproduction is moderate to high (that is, those with an average gall and egg mass index greater than 2) are considered to be susceptible (+) (see Tables 1 and 2).

   Note: Females often develop in close proximity to each other, thus forming large compound galls. When galls are counted, the number of females present in large compound galls should be estimated so the actual number of infections will be portrayed more accurately.
4. Remove 15 to 20 separate galls containing single mature females from each infected host type for preparation of perineal patterns. Make one microscope slide of females from each host.

   Note: Unless a single egg mass was used to establish the culture from a field population, a given culture may contain more than one species.

PREPARATION OF PERINEAL PATTERNS FOR OBSERVATION

Several methods have been described for preparation of perineal patterns (10, 18, 24). Some techniques involve fixation and staining of root material prior to removal of females, but Taylor & Netscher's procedure (26) utilizes fresh root material, thereby eliminating exposure of eggs to toxic fixatives. Staining is unnecessary since it does not enhance the clarity of the diagnostic details of perineal patterns. Perineal patterns can be prepared from fresh root material that has been stored up to several weeks in a refrigerator. A method outlined is modified from Taylor and Netscher (26).

The following materials are useful in the preparation of perineal patterns: a small, surgical eye knife or single-edged razor blade, forceps, half spear or teasing needle, pulp canal file, 45% lactic acid, glycerin, disposable plastic petri dish, syracuse dish, clean microscope slides and coverslips, and sealing compound.

1. Select galls with mature females. Place in syracuse dish with tap water. Single galls are preferable to compound galls.
2. Tease the root tissue apart with forceps and half spear to remove adult females (Fig. 1).
3. Rupture the cuticle of the female near the neck and gently push the body tissues out (Fig. 2).
4. Place the cuticle in a drop of 45% lactic acid on a plastic petri dish. The lactic acid facilitates the removal of body tissues that adhere to the cuticle after trimming. Collect 10 to 20 cuticles in the drop and let

them stand for 30 minutes to several hours. All cuts with the eye knife are done on the surface of a plastic petri dish to minimize the dulling of the knife. The knife can be re-sharpened with a fine-grit sharpening stone.

5. Cut the cuticle in half (equatorially) with cataract knife. Remove the cuticle with the perineal pattern from the drop of 45% lactic acid. Place it next to the drop and trim the perineal pattern to a square (Fig. 3). Place the trimmed perineal pattern back in the 45% lactic acid. Cut 5 to 10 perineal patterns per sample.
6. Thoroughly clean debris from the perineal pattern with the pulp canal file.
7. Transfer the perineal patterns to a drop of glycerin on a clean glass microscope slide. Align the perineal patterns so that they are in a straight line and the anus is oriented down. The interior surface of the cuticle must be placed against the glass. Press the perineal pattern gently against the glass with the pulp canal file.
8. Gently place the coverslip on the glycerin drop. Excess glycerin can be absorbed by a piece of filter paper. If there is insufficient glycerin under the coverslip, a small drop can be placed on the edge of the coverslip.
9. Seal the coverslip and label the slide (Fig. 4).

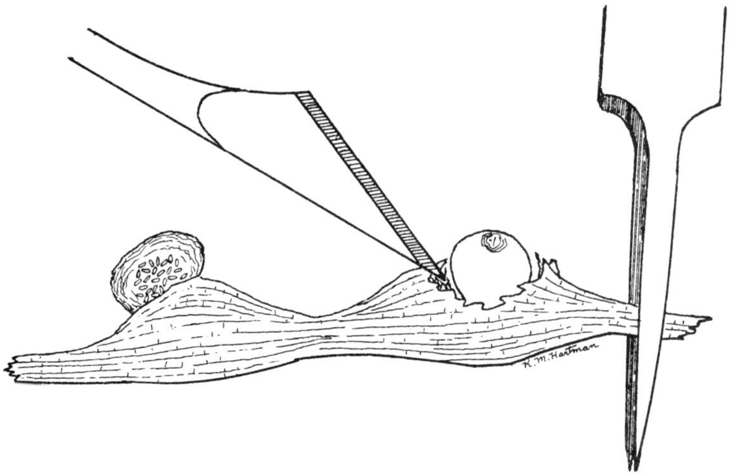

Fig. 1. Female being removed from root fragment.

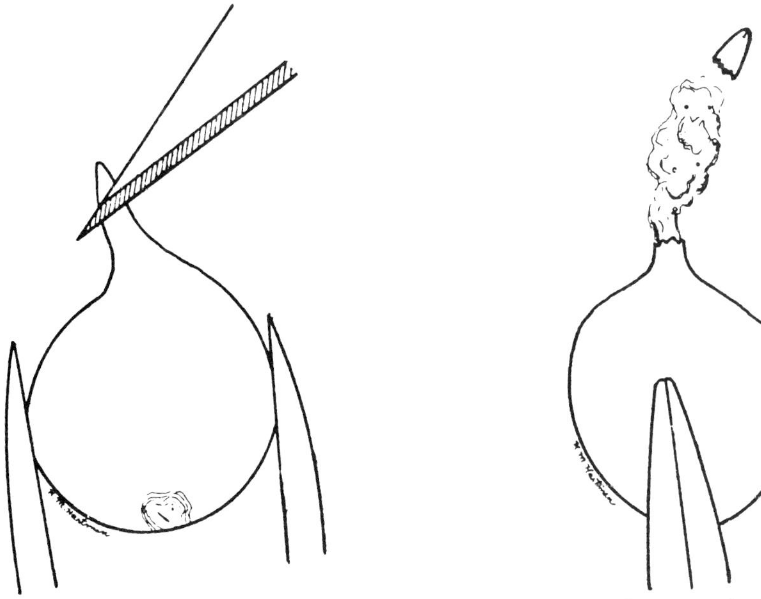

Fig. 2. Excised female being ruptured and body tissues gently removed.

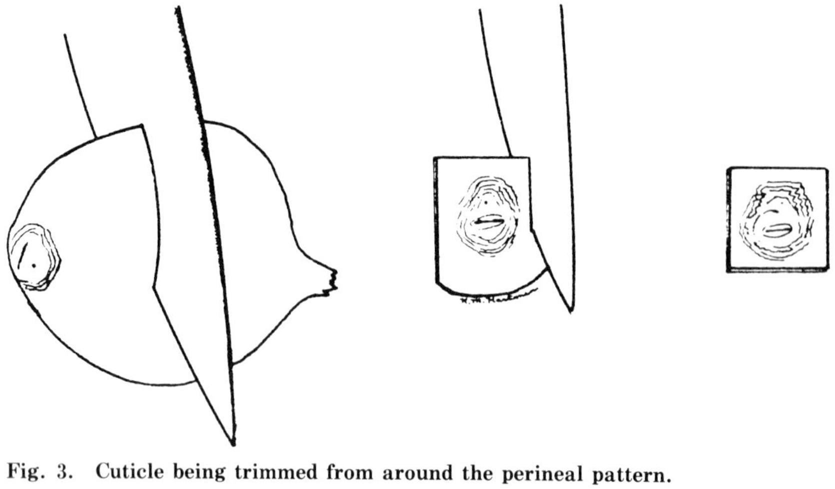

Fig. 3. Cuticle being trimmed from around the perineal pattern.

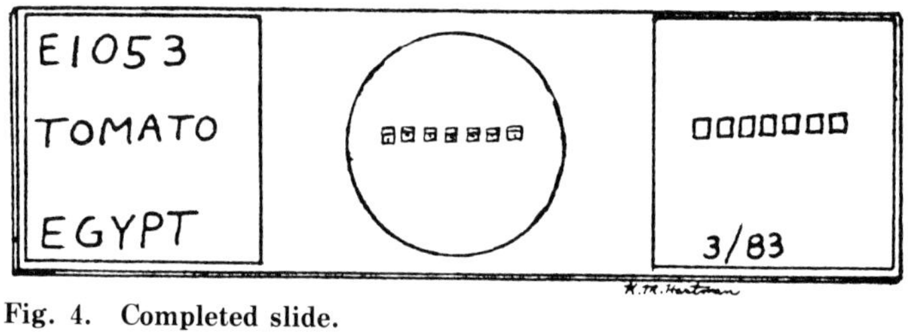

Fig. 4. Completed slide.

Literature Cited

1. Berkeley, M. J. 1855. Vibrio forming cysts on cucumber. Gardeners' Chronicle 14:220.
2. Bessey, E. A. 1911. Root-knot and its control. U.S. Dep. Agric. Bur. Plant Ind. Bull. 217. 89 pp.
3. Chitwood, B. G. 1949. Root-knot nematodes. I. A revision of the genus *Meloidogyne* Goeldi 1887. Proc. Helminthol. Soc. Wash. 16:90-104.
4. Christie, J. R. 1946. Host parasite relationships of the root-knot nematode, *Heterodera marioni*. II. Some effects of the parasite on the host. Phytopathology 36:340-352.
5. Christie, J. R. and F. E. Albin. 1944. Host-parasite relationships of the root-knot nematode, *Heterodera marioni*. I. The question of races. Proc. Helminthol. Soc. Wash. 11:31-37.
6. Christie, J. R. and L. Havis. 1948. Relative susceptibility of certain peach stocks to races of the root-knot nematode. Plant Dis. Rep. 32:510-514.
7. Cliff, G. M. 1983. Morphological and serological comparisons of members of the *Meloidogyne arenaria* species complex, with characterization of two new forms. Ph.D. thesis, North Carolina State Univ., Raleigh, N.C. 153 pp.
8. Dickson, D. W. and F. B. Struble. 1965. A sieving-staining technique for extraction of egg masses of *Meloidogyne incognita* from soil. Phytopathology 55:497 (Abstr.).
9. Eisenback, J. D. 1982. Description of the blueberry root-knot nematode, *Meloidogyne carolinensis* n. sp. J. Nematol. 14:303-317.
10. Esser, R. P., V. G. Perry and A. L. Taylor. 1976. A diagnostic compendium of the genus *Meloidogyne* (Nematoda: Heteroderidae). Proc. Helminthol. Soc. Wash. 43:138-150.
11. Godfrey, G. H. 1923. Root-knot: its cause and control. U.S. Dep. Agric. Farmers' Bull. 1345. 26 pp.
12. Göldi, E. A. 1887. Relatoria sobre a molestia do cafeeiro na provincia do Rio de Janeiro. Arch. Mus. Nac. Rio de Janeiro 8:7-121.

13. Hartman, K. 1982. Enhancement technique for egg masses of the root-knot nematode with Phloxine B, pp. 129-130 *In* Proc. Third IMP Res. & Plan. Conf. on Root-knot Nematodes, *Meloidogyne* spp., Region II, March 22-26, 1982. Centro Internacional de la Papa (CIP), Lima, Peru. 233 pp.
14. Hussey, R. S. and K. R. Barker. 1973. A comparison of methods of collecting inocula of *Meloidogyne* spp. including a new technique. Plant Dis. Rep. 57:1025-1028.
15. Marcinowki, K. 1919. Parasitisch und semiparasitisch an pflangen lebende nematoden. Arbeiten aus der Kaiserlichen Biologischen Anstalt fur Land und Forstwirtschaft, Berlin 7(1):1-192.
16. Martin, W. J. 1954. Parasitic races of *Meloidogyne incognita* and *M. incognita* var. *acrita*. Plant Dis. Rep. 227:86-88.
17. Neal, J. C. 1889. The root-knot disease of the peach, orange, and other plants in Florida, due to the work of the Anguillula. U.S. Dep. Agric. Div. Entomol. Bull. No. 20. 31 pp.
18. Netscher, C. 1978. Morphological and physiological variability of species of *Meloidogyne* in West Africa and implication for their control. Meded. Landbouwhogesch., Wageningen, 78-3. 46 pp.
19. Sasser, J.N. 1954. Identification and host-parasite relationships of certain root-knot nematodes (*Meloidogyne* spp.). Md. Agric. Exp. Stn. Bull. A-77. 31 pp.
20. Sasser, J. N. 1972. Physiological variation in the genus *Meloidogyne* as determined by differential hosts. Eur. Mediterr. Plant Prot. Organ. (OEPP/EPPO) Bull. 6:41-48.
21. Sasser, J. N. and C. C. Carter. 1982. Root-knot nematodes (*Meloidogyne* spp.): Identification, morphological and physiological variation, host range, ecology, and control, pp. 21-32 *In* Nematology in the southern region of the United States, ed. R. D. Riggs. South. Coop. Ser. Bull. 276, Arkansas Agric. Exp. Stn., Fayetteville, Ark., 206 pp.
22. Sasser, J. N. and A. C. Triantaphyllou. 1977. Identification of *Meloidogyne* species and races. J. Nematol. 9:283 (Abstr.)
23 Sherbakoff, C. D. 1939. Root-knot nematodes on cotton and tomatoes in Tennessee. Phytopathology 29:751-752.
24. Taylor, A. L., V. H. Dropkin and G. C. Martin. 1955. Perineal patterns of root-knot nematodes. Phytopathology 45:26-34.
25. Taylor, A. L. and J. N. Sasser, 1978. Biology, identification and control of root-knot nematodes (*Meloidogyne* spp.). Coop. Publ. Dep. Plant Pathol., North Carolina State Univ., and U.S. Agency Int. Dev., Raleigh, N.C. 111 pp.
26. Taylor, D.P. and C. Netscher. 1974. An improved technique for preparing perineal patterns of *Meloidogyne* spp. Nematologica 20:268-269.
27. Triantaphyllou, A. C. 1963. Polyploidy and parthenogenesis in the root-knot nematode *Meloidogyne arenaria*. J. Morphol. 113: 489-499.
28. Triantaphyllou, A. C. 1966. Polyploidy and reproductive patterns in the root-knot nematode *Meloidogyne hapla*. J. Morphol. 118:403-414.
29. Triantaphyllou, A. C. 1979. Cytogenetics of root-knot nematodes, pp. 85-109 *In* Root-knot nematodes (*Meloidogyne* species) systematics, biology and control, ed. F. Lamberti and C. E. Taylor. Academic Press, New York. 477 pp.
30. Triantaphyllou, A. C. 1981. Oogenesis and the chromosomes of the parthenogenetic root-knot nematodes *Meloidogyne incognita*. J. Nematol. 13:95-104.
31. Triantaphyllou, A. C. and J. N. Sasser. 1959. Morphological and physiological variation in *Meloidogyne incognita* and *M. incognita acrita*. Phytopathology 49:553 (Abstr.).
32. Triantaphyllou, A. C. and J. N. Sasser, 1960. Variation in perineal patterns and host specificity of *Meloidogyne incognita*. Phytopathology 50:724-735.
33. Tyler, J. 1944. The root-knot nematode. Univ. Calif. (Berkeley) Agric. Exp. Stn. Circ. 330. 30 pp.

TECHNIQUES FOR PREPARING NEMATODES FOR SCANNING ELECTRON MICROSCOPY

J. D. Eisenback

Department of Plant Pathology
North Carolina State University
Raleigh, North Carolina, USA

The recent applications of scanning electron microscopy (SEM) to the study of plant-parasitic nematodes demonstrate the increasing importance of these techniques to nematologists (45,91,92). The SEM is significantly different from the light microscope, and thus successful observation of nematodes requires specialized preparative procedures. In the initial SEM investigations of plant-parasitic nematodes, several different processing techniques were evaluted (37,38,39). The early SEM micrographs of nematodes revealed new information when compared with light micrographs and illustrated the value of SEM in the study of the external morphology of nematodes even though many early preparations contained artifacts. Preparation techniques now available generally produce specimens free of artifacts and quite stable under the high vacuum and intense electron beam of the microscope.

In SEM, the image is formed by electrons emitted from the surface of the specimen; therefore, external structures that are translucent and difficult to see in the light microscope are readily visible with SEM. Furthermore, imaging is enhanced in SEM by the extremely large depth of focus when compared with the light microscope. The depth of focus and the opacity of the specimen combine to form an image that appears three-dimensional and is generally easy to interpret (Fig. 1).

SEM has greatly aided the clarification of the surface morphology of many nematodes (14,45,74,84,92), and has become an important tool in their morphology and taxonomy. Several taxonomic clarifications have been made on the basis of SEM observations (3,15,24,45,46, 68,92,95), and species descriptions increasingly include SEM studies (4,26,33,34,35,79,80,86, 92,95,96).

Other subdisciplines of nematology have utilized SEM, and its importance to the science of nematology undoubtedly will increase as existing techniques are improved and new ones developed (65,66). Significant contributions have already been made in nematode ecology (19,20,81), host-parasite relationships (54,55, 60,61,69,70,92), and biological control (22,92).

Reviews of SEM techniques have been published recently (91,92,94), and numerous texts on theory and operation of the SEM and preparation of biological material for SEM are available (5,8,31,36,40,41,42,64,65,75,76,88, 90). The purpose of this chapter is to provide some insight into and details of specific techniques that have been shown to be useful in the preparation of nematodes for SEM.

HANDLING NEMATODES

Nematodes and structures dissected from them are microscopic and thus require special handling in preparation for SEM. Careful handling procedures are absolutely necessary for successful specimen processing. Such procedures require initial selection of specimens and subsequent transfer of these specimens through several preparative solutions and processes.

Specific individuals must be selected from bulk quantities of material extracted by appropriate means from soil or host tissues (2,32, 49,50). Harsh extraction techniques should not be used. The method used in the initial selection of individuals depends upon the size,

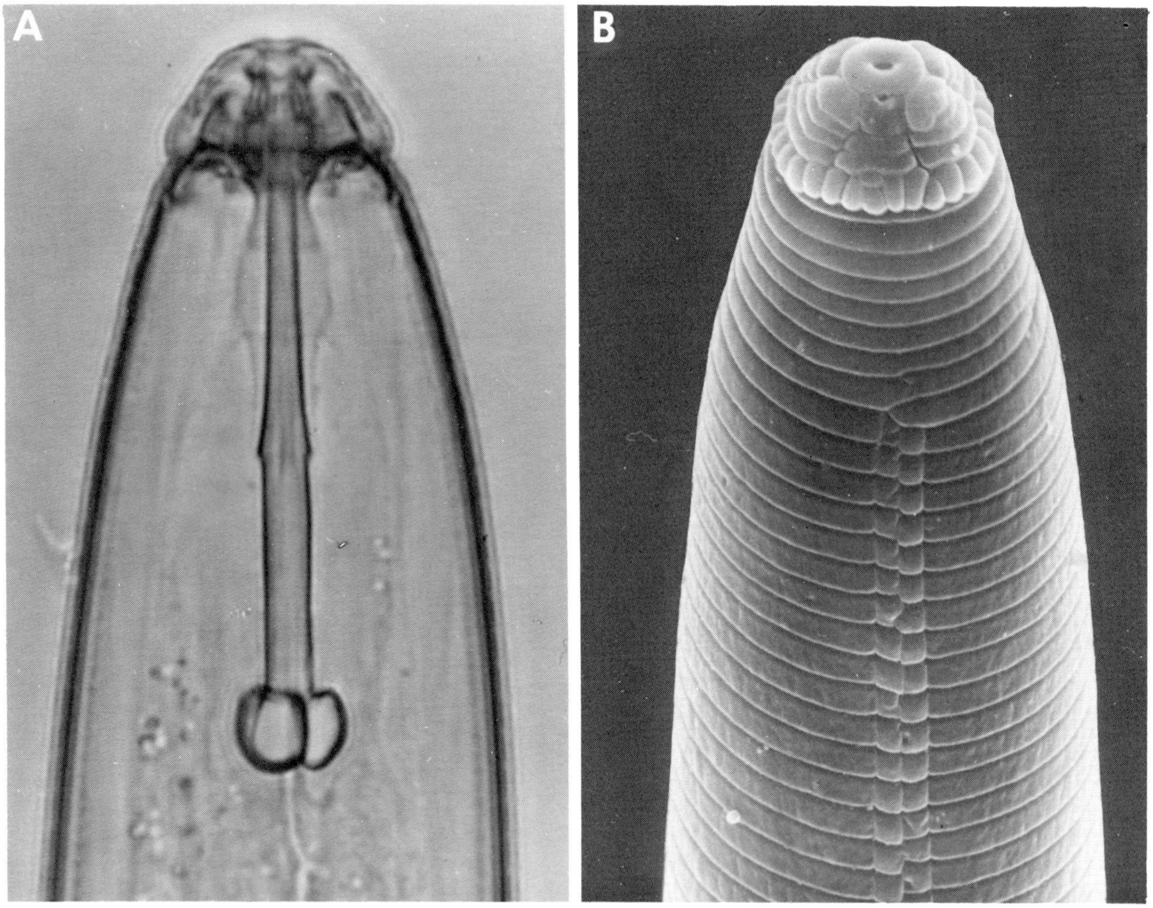

Fig. 1. A) Light and B) scanning electron micrograph, respectively, of the anterior end of a female of *Hoplolaimus galeatus* (Cobb) Filipjev and Shuurmans Stekhoven. In the light microscope the specimen is transparent, contrast of surface details is often lacking, the depth of focus is narrow, and resolution is limited. The scanning electron microscope has a large depth of focus, increased contrast, high resolving power, and the specimen appears opaque.

quantity, and quality of specimens available for preparation. Generally, hand picking with a dental file or similar tool is best in most cases because only quality individuals of the same species and in the same developmental stage are selected. Hand picking usually insures a minimal loss of material, a concentration of specimens into a small amount of solution, and significant reduction of contaminating debris. However, if the specimens are numerous, in good physiological condition, and in a relatively clean monospecific solution, then the initial selection can be made with a micropipette. Specimens are gently drawn into the pipette, briefly allowed to settle, and then transferred in a small amount of solution to a processing container. Micropipetting is less stressful and much faster than hand picking, particularly with large, saccate female forms.

NEMATODE PROCESSING CONTAINERS

After initial selection, the nematodes are placed in a processing container. Such containers should allow for a rapid exchange of fluids, minimize loss of specimens, reduce shock to tissues, restrict contamination, and in some instances provide a smooth substrate for viewing specimens in SEM.

Containers used to process nematodes for SEM functionally work by filtration. Diffusion may also be involved, particularly in containers that are capped by membranes with very small pore sizes (41). The specimens are retained in these containers as the solutions are exchanged. The exchange is facilitated either by a pipette, absorbent paper (18), syringe (67), or vacuum (47).

Although a few commercially manufactured

containers can be used or adapted for processing nematodes, most are user-made from a variety of materials. Materials used, however, should be tested for adverse reactions with the preparative solutions.

Beem Capsule Container

Perhaps the most widely used and highly recommended chamber is easily made from two polyethelene BEEM capsules (Fig. 2A) (16,43) which are frequently used in thin-sectioning electron microscopy and are available from many suppliers. A versatile specimen holder (Fig. 2) can be fashioned from the severed caps of two BEEM capsules. With a hole-puncher 4 mm in diameter, two holes are made in one cap, which is used for the bottom of the chamber; and 2-6 holes are punched in the other cap, which is used for the top. The two holes in the bottom insure proper fluid exchange, whereas the holes in the top code for different samples. One of the capsules is discarded, and the tapered end of the second capsule is cut off and discarded. The hollow cylinder is saved. Two small, 20-mm square or circular pieces of nylon screen or other appropriate filter (Fig. 2B) are cut, and one piece is placed over an end of the cylinder, and the bottom cap is snapped in place. This chamber should be placed in a Stendor dish or similar glass container that has a close-fitting lid (Figs. 2C-D).

Specimens can be hand picked or pipetted in the BEEM capsule container after the glass container has been filled with an appropriate solution. As exchange of fluids become necessary, most of one solution is withdrawn from the glass dish and replaced with the next solution. Between exchanges, particularly during dehydration and infiltration with a highly volatile intermediate fluid prior to critical point drying, the glass container should be tightly covered. The polyethelene capsules are not affected by the majority of solutions including many organic solvents. Osmium tetroxide blackens not only the nematodes, but also the chamber and nylon cloth. Prior to critical point drying, the chamber is closed with a filter and cap to prevent the loss of specimens (Fig. 2D). After drying, the bottom of the chamber is removed, and the specimens are easily accessible for mounting on a specimen stub.

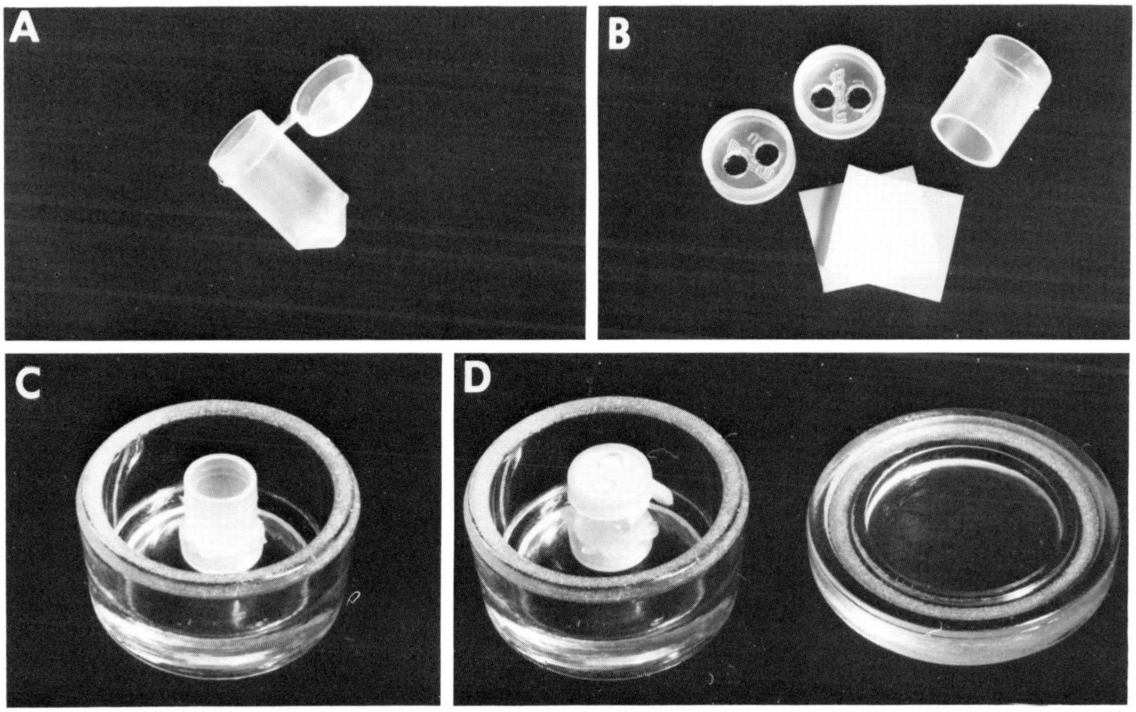

Fig. 2. BEEM capsule container. A) A BEEM capsule. B) Component parts of the container: two capsule caps with holes punched in them, capsule with conical portion cut off and discarded, and two pieces of nylon filter. C) Container without top in a Stendor dish ready for the exchange of solutions. D) Container with cap, ready for critical-point drying.

Porous Polyurethane Container

The porous polyurethane container, commercially manufactured and originally designed for storage and treatment of contact lenses, is suitable for processing nematodes from fixation through critical point drying (CPD) (91) (Fig. 3). The container measures 12 mm×12 mm, with an inside diameter of 6 mm and a pore size between 10-15 μm. The tight-fitting lid is made of the same material. The polyurethane is extremely porous and fluids rapidly drain through the container. The porosity may be a disadvantage, however, if the specimens become trapped in the pores, or if the specimens inadvertently dry out between steps in the procedures.

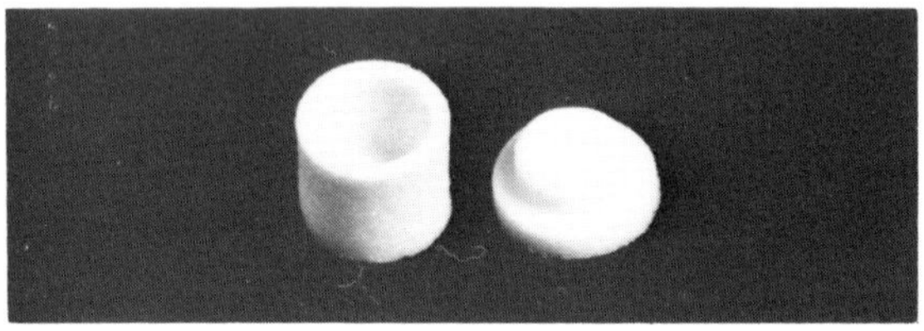

Fig. 3. Porous polyurethane container and lid.

Micro-Tissue Capsule Assembly

The micro-tissue capsule assembly (Fig. 4), primarily intended for the infiltration of tissue with plastic for thin-sectioning electron microscopy, is a manufactured version of the mini-sieve chamber (18). It is made from a glass tube that is 12 mm long with a 4-mm outside diameter and has ends covered with an appropriate filter held in place with teflon caps.

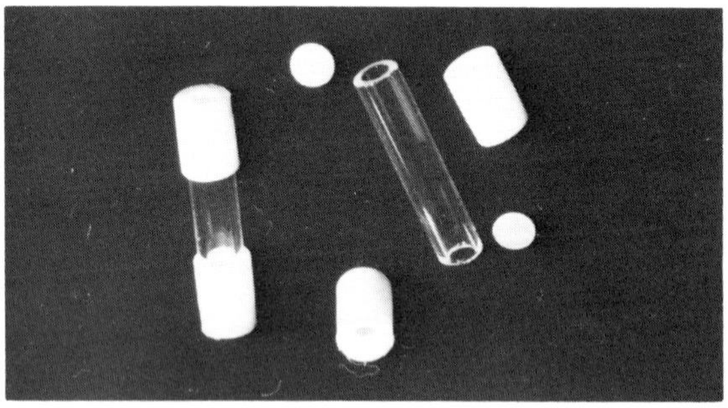

Fig. 4. Micro-tissue capsule assembly.

Coverslip Dissecting Chamber

Hard, cuticularized parts, as well as soft tissues, can be dissected from nematodes for observation with the SEM. Because the dissected tissues are so small, there are special problems in their preparation. The tissues must be observed in the same chamber in which they are removed because they cannot be readily transferred from one chamber to another. Therefore, all solution changes and other preparative procedures should take place in the dissection chamber. This chamber has to be transparent and small enough to be observed with a dissecting microscope. Specimens must adhere tightly to the surface of the container to avoid being washed away by the rushing currents that sometimes occur between exchange solutions due to a violent interaction. Contaminating debris must be avoided as it may obscure small dissected tissues. Finally, the chamber must be small enough to fit the SEM mounting stub; the surface of the container should provide a smooth substrate for viewing of the specimen; and dissected specimens should be marked on the chamber for ease of location in the microscope.

A simple chamber (28), made from a small round glass coverslip (12.7 mm in diameter) and a glass microscope slide, has been used successfully (28,29) (Fig. 5). The coverslip is attached to the center of the slide with a small dot of nail polish. The nail polish holds the coverslip onto the slide during the processing of the specimens, but it allows the coverslip to be removed easily for placement on the microscope mounting stub. The glass slide facilitates handling of the coverslip. A thick ring of nail polish, 7 mm in diameter, is centered on the coverslip. A large drop of fluid is placed in the ring on the coverslip. Whole nematodes are hand picked and placed into this solution, and the desired tissues are dissected in the same solution. After a successful dissection, the additional fixing solutions can be pipetted on top of each other, or the entire coverslip can be removed from the slide and placed in separate containers of the solutions. Several coverslips can be processed simultaneously if different types of holders are used (85). When the specimens are dry, their location on the coverslip is marked with a small triangle of tape.

BASICS OF NEMATODE FIXATION

Preparation of nematodes for SEM often requires specialized fixation, dehydration, drying, and mounting procedures that conform

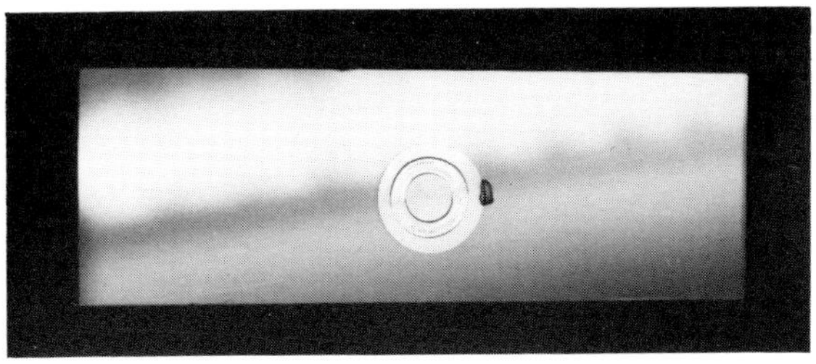

Fig. 5. Coverslip dissecting chamber.

to certain physical requirements and limitations of the microscope. Nematodes must be sufficiently rigid to prevent collapse and deformation of surface topology when they are exposed to the instruments' high energy beam and high vacuum. They also must be electrically conductive to prevent the image-destroying buildup of electrons within tissues.

Living specimens are preferable for preparation, but sometimes only previously fixed or glycerin-infiltrated nematodes are available. Each type of material must be prepared in a different manner, although previously fixed specimens and even nematodes in glycerin can be infiltrated with ethanol and critical point dried. Living nematodes, in contrast, should not be infiltrated with glycerin for viewing with the SEM. In special situations, however, it may be necessary to view glycerin-infiltrated material without further processing.

Overview of the Fixation Process

Fixation is the first and most important step in preparing nematodes for SEM. Specimens previously prepared for light microscopy probably are not suitable for SEM. Formaldehyde-based fixatives and, particularly, heat killing techniques do not adequately preserve tissues for electron microscopy.

Two different approaches to the fixation of nematodes have been used. One approach emphasizes stopping cellular activity quickly to prevent tissue destruction through autolytic activities. In rapid fixation methods, nematodes are immersed in a full-strength solution of fixative at room temperature.

The alternative approach emphasizes the effect of fixative concentration on the tonicity of the buffer. With a sequential fixation method, tissues are slowly fixed in increasing concentrations of the fixative.

Qualities of Fixatives and Buffers

The ideal fixative should stop all cellular activity immediately without any subsequent change in cell volume (78). The final appearance of properly fixed specimens should be similar to that of living specimens observed by light microscopy. There should be no apparent structural damage revealed as cracks or wrinkles, and specific structures should have similar appearances following the use of several different fixatives. Unfortunately, there is no ideal fixative, and generally two or more fixatives are used in combination to obtain the best results (78).

For SEM of nematodes, chemical fixation is usually a two-step process. An aldehyde, or mixture of aldehydes, is incorporated into the tissues for the primary fixation of protein molecules (6,51,52), and secondarily, the unsaturated fats are stabilized by post-fixing with osmium tetroxide (77,78).

The most common aldehyde used to process nematodes is glutaraldehyde, although acrolein and mixtures of glutaraldehyde and formaldehyde, or acrolein and glutaraldehyde, may give similar or better results. Because glutaraldehyde is a dialdehyde, it is most effective for fixing proteins. Each molecule has two reactive groups that are both capable of cross-linking proteins. The stability of fixation in general is also dependent upon the concentration, duration, temperature, and pH of the specific aldehyde used.

Aldehydes adequately fix proteins, but other molecules are unaffected and subject to subsequent extraction. Excess extraction of unfixed molecules is detrimental to tissue preservation and may cause distortion and/or shrinkage. Secondary fixation with osmium tetroxide stabilizes unsaturated fats and prevents their extraction.

Fixatives in an aqueous solution drastically upset the physiological pH of the tissues. The resulting osmotic effects cause large changes in cell volume, and either shrinkage or swelling occurs (82). Fixative molecules, therefore, must be transported to the cells by a buffering system (83). The most commonly used buffers for nematode preparations are the sodium-cacodylate and the phosphate buffers. Both give similar results, and are compatible with the aldehydes and osmium tetroxide. Sodium-cacodylate contains arsenic and, therefore, should be handled with caution (53). The toxicity of this buffer keeps it relatively free of contaminating organisms. The phosphate buffers are non-toxic, although some microorganisms may grow in the solution. They also cause some specimens to become brittle.

The most important aspect of the buffering system is its tonicity. When the nematode is immersed in the solution, there should be no volumetric change within the cells. The tonicity of the buffer is directly related to the concentration of the fixative and the composition of the buffers, and can be changed by the addition of nonelectrolytes or electrolytes (51,78).

Fixatives that are strong often cause specimen shrinkage and/or distortion because they are hypertonic. Weak solutions or fixatives initially preserve morphology without changing the cell volume. Unfortunately, weak concentrations do not harden the tissue enough to withstand deleterious effects of other preparative solutions. One answer to the question of fixative concentration is sequential fixation. Sequential fixation begins with a weak fixative, and the concentration is slowly increased until it is quite strong (1-3%). The concentration can be increased by gradually adding a double-strength fixative to the nematodes in a known volume of solution or by transferring the nematodes through a series of fixatives of increasing concentration.

Preparation of Fixatives and Buffers

SODIUM CACODYLATE BUFFER A working solution of 0.1 M sodium cacodylate buffer is made by mixing 21.4 g of sodium cacodylate, or 15.5 g of anhydrous sodium cacodylate with enough distilled water to make 500 ml. The pH of the buffer can be adjusted with concentrated hydrochloric acid.

This working solution is used to buffer all of the fixatives and as a wash after primary and secondary fixation. Sodium cacodylate contains arsenic and *should be handled carefully*. The working solution should be kept refrigerated and can be stored with little contamination by microorganisms. The pH will drop during storage, but can be readjusted with concentrated hydrochloric acid.

SODIUM PHOSPHATE BUFFER A stock solution of 0.2 M sodium phosphate buffer can be made with a combination of monobasic and dibasic sodium phosphate. Monobasic sodium phosphate (26.85 g) is added to distilled water to make 500 ml. In a separate solution, 13.80 g of dibasic sodium phosphate are added to distilled water to make 500 ml. The working solution of 0.1 M phosphate buffer can be adjusted to the desired pH by combining different volumes of monobasic and dibasic sodium phosphate buffer and adding distilled water to make 100 ml. All stored solutions should be refrigerated and discarded when contaminating organisms are evident.

0.2 M sodium phosphate buffer

pH	Monobasic	Dibasic
6.8	25.5 ml	24.5 ml
7.0	19.5 ml	30.5 ml
7.2	14.0 ml	36.0 ml

GLUTARALDEHYDE Concentrated glutaraldehyde, 70%, is commercially available in 2-ml ampules. Under a fume hood, the concentrated glutaraldehyde is added to the necessary quantity of buffer to achieve the desired concentration. Glutaraldehyde should be handled with caution, always with rubber gloves and under a fume hood.

Concentration	Buffer	70% Glutaraldehyde
2%	68 ml	2 ml
3%	44.7 ml	2 ml
4%	33.0 ml	2 ml

FORMALDEHYDE An aqueous solution, 10%, of formaldehyde is made from paraformaldehyde powder to ensure a menthol-free working solution. Under a fume hood, 10 g of paraformaldehyde powder are dissolved into 100 ml of water by heating to approximately 60°C. Sodium hydroxide (1.0 N) is added one drop at a time until the solution clears. This solution is cooled before use, and stored and refrigerated in tightly stoppered glass bottles. When a precipitate forms, the solution is discarded.

FORMALDEHYDE / GLUTARALDEHYDE A mixture of 2% formaldehyde/4% glutaraldehyde is made by adding 3.5 ml of 10% aqueous paraformaldehyde and 2.0 ml of 70% glutaraldehyde to 33.0 ml of 0.1 M buffer solution.

OSMIUM TETROXIDE, 2% Pure osmium is available in crystalline form. Because the crystals dissolve rather slowly, the solution should be prepared the day before it is needed. Osmium readily reacts with any organic material. All glassware should be thoroughly cleaned with soap and water, acid cleaned, and air dried. Under a fume hood and with rubber gloves and goggles, 1 g of osmium tetroxide and 50 ml of buffer are combined in a bottle.

Dissolution of the osmium can be hastened by sonication of the solution. Osmium can be stored in a refrigerator in a stoppered dark glass bottle, but leakage of vapors may occur.

PREPARATION OF LIVING SPECIMENS

A large selection of preparative techniques is available for nematodes from diverse habitats, but no single technique is useful for all specimens. Each genus or species may require a different procedure for the optimum preservation of morphological detail; however, certain generalized processes are common to most preparations. Basic preparative procedures include fixation, dehydration, drying, mounting, and enhancement of electrical conductivity of the specimens. Each step in the process is important to the success of subsequent steps. Distortions that may occur in the initial stages of preparation are usually magnified by the remaining steps.

As mentioned previously, two generalized methods of fixing nematodes for SEM have been widely utilized. In one scheme, the nematodes are relaxed at cold temperatures and then gradually fixed by a slow increase in the concentration of a cold fixative. The other scheme emphasizes the speed with which specimens are plunged and fixed in a strong concentration of fixative at room temperature. There are many variables in each scheme that can be adjusted to optimize the preparation of specific materials.

Following primary fixation, the nematodes should be rinsed with cold buffer for removal of any excess fixative. The post-fixative may react with the primary fixative and contaminate the surface of the nematodes with dense precipitate. The rinsing should be brief to avoid the extraction of unfixed lipids. Three or four changes within a 15-min period is adequate provided the specimens in solution are agitated between exchanges of fluids.

Primary fixation is greatly enhanced by post-fixation with osmium tetroxide, which stains the nematodes black and gives them an electrical charge. Osmium effectively stabilizes the unsaturated fats and increases the electron density of the specimens. Since the nematodes are stained black, dried specimens can be easily found and mounted. In addition, the electrical charge facilitates subsequent handling with a fine metal wire.

Osmium is prepared with the same buffer as the primary fixative. A two-percent concentration yields adequate fixation within 4-8 hr in the cold (0-4°C), but longer fixation does not seriously affect the tissue. Osmium is extremely dangerous and should be used only under a fume hood, and rubber gloves and goggles should be worn. The specimens and the osmium are placed in an ice-cooled water bath while in the fume hood.

Post-fixation is followed by rinsing with several changes of cold buffer within a 15-min period. The unreduced osmium must be removed to prevent formation of surface precipitates with subsequent solutions.

Sequential Fixation Technique

Sodium cacodylate buffer (0.1 M) at pH 7.0, 2% formaldehyde/4% glutaraldehyde, and 2% osmium tetroxide solutions are prepared and chilled at 0-4°C. The nematodes are picked into 20 drops of tap water in a small Bureau-of-Plant-Industry (BPI) dish. The BPI dish is placed in a petri dish lined with moist filter paper. The nematodes are chilled in a refrigerator at 0-4°C for 15-30 min and gradually fixed by adding one drop of formaldehyde/glutaraldehyde every 10-15 min until 20 drops of fixative have been added. The specimens should be kept in the fixative at 0-4°C for an additional 24-48 hr.

After primary fixation, the nematodes are transferred in a small quantity of fixative into a BEEM capsule container that is placed in a Stendor dish and cooled in an ice bath. The excess fixative is washed from the specimens with 3-4 changes of cold buffer within a 15-min period. Too much washing extracts unfixed molecules. When solutions are exchanged, most of the old solution is withdrawn with a pipette from the Stendor dish, and immediately the new solution is added. Gentle, periodic agitation of the chamber increases fluid exchange.

The specimens are post-fixed with 2% osmium tetroxide for 8-16 hr or longer in the cold, 0-4°C. This can be done overnight. After post-fixation, the nematodes are washed with several changes of cold buffer within a 15-min period. The specimens are allowed to return slowly to room temperature during the last wash. The specimens are now ready for dehydration and drying.

Rapid Fixation Technique (91)

Phosphate buffer (0.05M) at pH 6.8, 3%

glutaraldehyde, and 2% osmium tetroxide solutions are prepared. At room temperature, the live nematodes are transferred directly into the fixative in a vial or the specimens are gently centrifuged to the bottom of a tube, the water is withdrawn and the fixative is added immediately. For best results, the nematodes must be quickly placed in the fixative and left for at least 1.5 hr at room temperature. If necessary, the specimens can be stored in glutaraldehyde in the cold, 0-4°C, after the initial fixation at room temperature. The fixed specimens are transferred into a chamber within a Stendor dish and washed with 5-6 complete changes of buffer for 10-15 min each. They are post-fixed with 2% osmium tetroxide for 2 hr at room temperature. The excess osmium is washed from the tissues with 4-6 complete changes of buffer for 10-15 min each. The nematodes are now ready for dehydration and drying.

Dehydration and Intermediate Fluids

After fixation, the nematodes must be dehydrated and possibly infiltrated with a fluid that is miscible with the CPD solution. Generally, ethanol is recommended as the major dehydration solvent. If carbon dioxide is used for critical point drying, an intermediate fluid is not necessary; however, if Freon 13 is used, then the ethanol has to be exchanged with miscible liquid freon.

Dehydration should be gradual to avoid osmotic shock and to ensure complete removal of water. A six- or seven-graded series of ethanol in water is adequate if the steps are 15-20 min long. The last step of 100% ethanol should be repeated three times with a freshly opened bottle of ethanol to eliminate hygroscopic absorption.

If Freon 13 is used for CPD, the ethanol should be replaced with an intermediate fluid, generally Freon 113. A five- or six-graded series of freon in absolute ethanol of 15-20 min each is required to exchange the ethanol with freon. The last step of 100% freon should be repeated three times. During the exchange of the highly volatile alcohols and freons, it is extremely important to prevent accidental drying of the specimens.

Drying

Nematodes that have been fixed and dehydrated are still in a liquid environment. Successful observation with the scanning electron microscope, however, requires that the specimens be completely dry. Severe distortions often result when nematodes are dried because evaporating liquids exert pressure on the surface of the tissue. Three different strategies have been developed that avoid the damaging effects of surface-tension phenomena: critical point drying, freeze drying, and air drying from a volatile, low-surface-tension liquid.

CRITICAL POINT DRYING The most widely used technique for drying nematodes is CPD. It is the only drying strategy that completely eliminates the harmful forces of surface tension (1,7,12,13). At a certain temperature and pressure, a solution reaches a "critical point" where the liquid and gaseous phases are indistinguishable. Therefore, the liquid in a specimen, if heated to the critical point, can be converted into a gas and removed. In this way, the damaging effects of surface tension can be avoided. Unfortunately, water is not an ideal transitional fluid because its critical temperature and pressure are extremely high. The most widely used transitional fluids, freon and carbon dioxide, have much lower critical points but are not miscible with water; consequently, the water within the specimens must be replaced with a compatible intermediate fluid. The transitional fluid selected also affects the CPD procedure.

CRITICAL POINT DRYING WITH FREON Specimens are fixed, dehydrated, and infiltrated with an intermediate fluid as previously discussed. A critical point dryer should be used with caution; its condition should be carefully checked before each use. The chamber is cooled to 10-15°C below room temperature, and the entry and exit ports are closed. The valve on the freon cylinder is opened. The cover of the critical-point-drying chamber is removed, and the capped container with nematodes is placed inside. The nematode specimen holder should retain a small quantity of the intermediate fluid to prevent the specimens from drying out in the air. Also, the transfer must be rapid, and the lid of the critical-point-drying chamber should be replaced quickly. The chamber is sealed and Freon 13 slowly introduced. Only freon that is oil-free should be used. As the pressure rises, a liquid meniscus can be seen through the viewing port as the chamber slowly fills. When the chamber is 75-85% full, the entry port and the valve on the cylinder of Freon 13 is closed.

The chamber is slowly heated, approximately 4°C per min, to at least 10°C above the critical temperature of 28.9°C. The pressure is allowed to rise to 300-400 psi above the critical pressure of 561 psi. The increase in temperature and pressure should be gradual to avoid excess shock to the tissues. Excess pressure above 900-1000 psi can be vented by slightly opening the exit port. The temperature is maintained at least 10°C higher than the critical temperature, and the exit port is slowly opened. Gradually, the pressure is reduced by a rate of approximately 100 psi/min or slower. When the chamber reaches atmospheric pressure, the chamber is opened and the specimens are removed and stored in a desiccator. The nematodes can be safely stored for several weeks in a desiccator before they are mounted.

CRITICAL POINT DRYING WITH CARBON DIOXIDE The nematodes are fixed and dehydrated as previously discussed. Ethanol is miscible with carbon dioxide and serves as both the dehydration and intermediate fluid. The critical point dryer should be carefully examined and cooled to 10-15°C below room temperature. The entry and exit valves on the chamber are closed and the cylinder of carbon dioxide is opened. Carbon dioxide is commonly contaminated with water and oil and should be filtered with a commercially manufactured filter before use. The cover of the drying chamber is removed, and the nematodes are quickly transferred in a tightly capped container with a small amount of ethanol. The cover is immediately replaced and securely sealed. Compressed carbon dioxide is gradually admitted, and as the pressure rises, a liquid meniscus can be seen through the viewing port as the chamber fills. When the chamber is 80% full, the exit port is opened, and the liquid not covering the specimens is allowed to escape. The exit port is closed, and the chamber is refilled. The exchange of solutions, repeated six to eight times, eliminates any traces of ethanol. When the nematodes are in a pure solution of carbon dioxide, the chamber is gently heated to approximately 10°C above the critical temperature of 31°C, and the pressure is allowed to rise to approximately 1400 psi. After the chamber reaches the critical point, the temperature is maintained and the pressure is gradually reduced by opening the exit port slightly. The maximum rate of pressure reduction should be 100 psi/min. The specimens are removed from the chamber when the pressure is zero and stored in a desiccator.

FREEZE DRYING Freeze drying involves two steps: 1) rapid freezing of the cellular fluids, and 2) sublimation of the frozen liquid under a high vacuum. Harsh surface tensions are avoided because there is a direct transition from a solid to a gaseous phase (63). When fresh, unfixed specimens are frozen in water, the formation of large ice crystals often damages the specimens. Tissues that are adequately fixed, or fixed and dehydrated with a nonpolar solvent such as amyl acetate or ethanol, are less susceptible to damage (74).

Only a few nematologists have utilized freeze drying for preparing nematodes for SEM (38, 39,73,89), as this technique apparently causes many artifacts. However, in earlier studies with freeze-dried nematode preparations, specimens were killed with a hot fixative. Perhaps artifacts resulted from the method of fixation rather than the freeze drying. Very good results have been obtained with some free-living nematodes (73), and freeze drying may be considered as an alternative technique for drying certain plant-parasitic nematodes.

AIR DRYING Stylets and spicules are the only structures of nematodes that are stable enough to withstand the disruptive effects of surface tension caused by air drying from water (Fig. 6) (27,28,29,30; J. D. Eisenback, unpublished). The damaging force of surface tension can be greatly reduced by air drying from a highly volatile liquid such as Freon 113, ethanol, or amyl acetate (58,62). Whole nematodes dried in this manner undergo considerable shrinkage and collapse; however, satisfactory results have been obtained with isolated cells of nematodes (Fig. 7) (J. D. Eisenback, unpublished). Details of the technique are presented in the section of this chapter on the preparation of nematode sperm for observation with the SEM.

Alternatives to Drying

Infiltration of nematodes with an epoxy resin that is subsequently polymerized (11,17,18,38) or substitution of cellular water with a liquid of extremely low volatility (37,84,91) are two alternatives to drying. Resin infiltration is very difficult, and the results are quite inconsistent. Substitution of water with a liquid of low volatility is also unsatisfactory, except under special circumstances in which the material has already been infiltrated with glycerin. Details

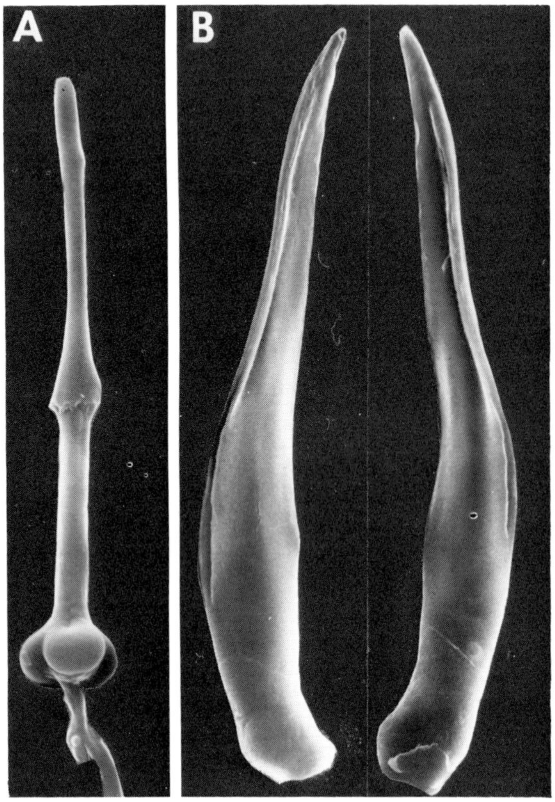

Fig. 6. Scanning electron micrographs of A) stylet and B) spicules, respectively, dissected from a *Meloidogyne incognita* (Kofoid and White) Chitwood male prepared by air drying.

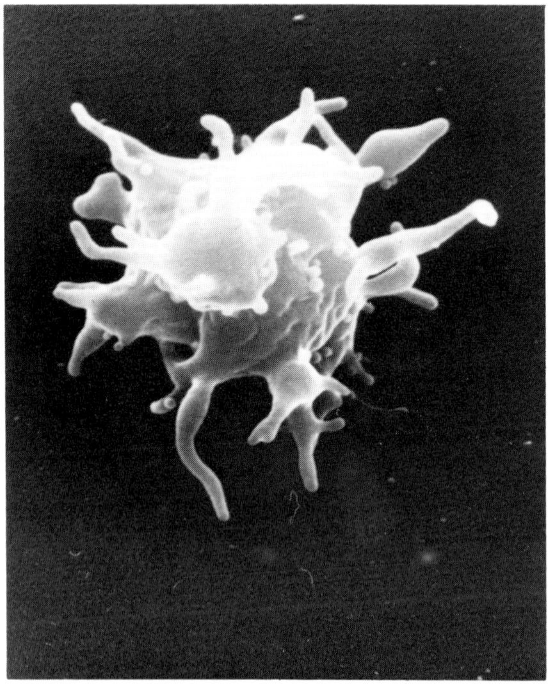

Fig. 7. A sperm of a *Heterodera glycines* Ichinohe male dissected and dried from a highly volatile liquid.

of processing and examining previously fixed nematodes are presented in a separate section in this chapter.

Mounting Dry Nematodes (72)

The whole dried nematodes are mounted on an SEM mounting stub with the aid of a dissecting microscope. A mounting block made from a clean piece of acrylic plastic approximately 6 cm×9 cm×1.2 cm (Fig. 8A) greatly facilitates manipulation of the specimens. The stub is mounted in a hole of appropriate diameter in the center of the mounting block. A small, 10-mm square piece of double-coated adhesive tape is attached on top of the stub, and a short length (12 mm) of hair is placed on the tape (Fig. 8B). The nematodes are transferred from the specimen container to the stub with a dental pulp canal file by placement of the file alongside the posterior portion of the nematode (Fig. 8C). Extreme caution has to be exercised since specimens are easily damaged. The nematodes are propped up against a hair to ensure adequate viewing angles (Fig. 8D). The specimen should be oriented perpendicular to the hair, with the anterior portion of the nematode slightly projecting above the hair, and at a 45° angle to the surface of the stub. Observation of many specimens is greatly simplified if they are all mounted at the same level and are oriented in the same direction.

Enhancement of Electrical Conductivity

Dehydrated nematodes are very poor conductors of electricity. Conductivity is necessary for high quality images to be obtained in the SEM because image formation is achieved by an interaction of the specimen with a high energy electron beam. If the specimen is a poor conductor, electrons accumulate within the specimen and charging artifacts obscure the image with bright streaks of white across the viewing screen and micrograph. Specimen conductivity also affects the yield of secondary electrons discharged by the interaction of the beam and the specimen (21).

METAL COATING Generally, a metal coating is applied to the surface to render the nematodes electrically conductive. The metal, usually a gold/palladium alloy, can be applied to the surface either by evaporation or sputtering. The method generally depends upon the type of coater available, but in general the sputter coat produces a more even coating (21,25). Usually, coating is applied at a thickness of 100-300Å.

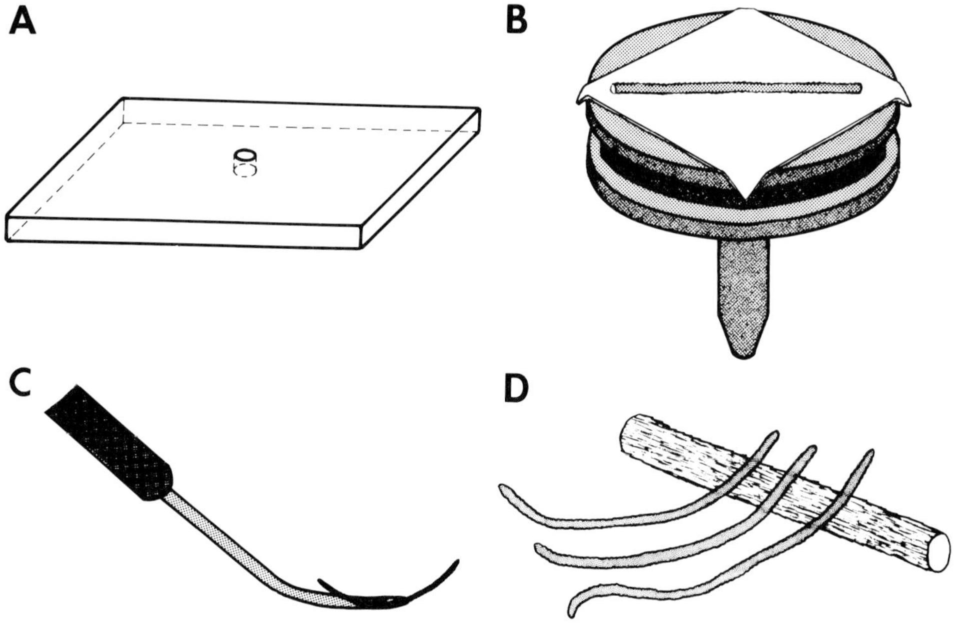

Fig. 8. Mounting technique for dried whole nematodes. A) Special mounting block made out of clear acrylic. B) Stub with square of double-coated tape and 12 mm length of human hair. C) Dental file with dried nematode. D) Nematodes placed against a hair so that the anterior ends are aligned at the same angle as close to the stub as possible.

Too much coating obscures surface details, whereas too little coating may cause charging artifacts. Fortunately, specimens can be coated, viewed, and recoated if necessary; therefore, it is better to coat too thinly the first time to avoid the risk of overcoating and losing details.

ALTERNATIVES TO COATING The alternative to coating with metal is the impregnation of the specimens with a conductive material (71). The most commonly used non-coating technique is the osmium-thiocarbohydrazide-osmium method. There are no reported uses of this method for nematodes, but the technique certainly seems worthy of consideration.

Thiocarbohydrazide ligates more molecules of osmium to the tissues, and thus the electrical conductivity of the specimen is increased. Only thiocarbohydrazide of very high quality should be used. After the first exposure to osmium in the fixation scheme, the excess osmium is removed with a thorough rinsing. The specimens are then exposed to an aqueous solution of thiocarbohydrazide, thoroughly washed again, and reexposed to osmium. After treatment with osmium, the nematodes are dehydrated and dried as previously discussed. If the specimens still lack electrical conductivity, they can be treated a second time in thiocarbohydrazide, washed, and exposed to osmium a third time. Generally, specimens treated with thiocarbohydrazide and osmium are harder, less fragile, and better preserved (21,71).

PREPARATION OF PREVIOUSLY FIXED MATERIAL

It is most desirable to use live nematodes in good condition for SEM preparation, but sometimes only previously fixed material, either stored in a formaldehyde-based fixative or infiltrated with glycerin, are available. Previously fixed nematodes can be prepared with the same techniques used for live ones. Glycerin-infiltrated nematodes can be directly examined or the glycerin can be removed and the specimens rehydrated, post-fixed in osmium, dehydrated, and dried.

Observation of Glycerin-Infiltrated Nematodes

Glycerin-infiltrated specimens are placed on a piece of filter paper and the excess glycerin is drained. Several specimens are mounted on a small glass chip with the anterior portion protruding slightly beyond the edge of the chip (Fig. 9A). The nematodes are attached to the glass with a small amount of nail polish or thinned Zut, and the glass chip is mounted onto a microscope stub with adhesive (Fig. 9B) (84). The specimens should be mounted per-

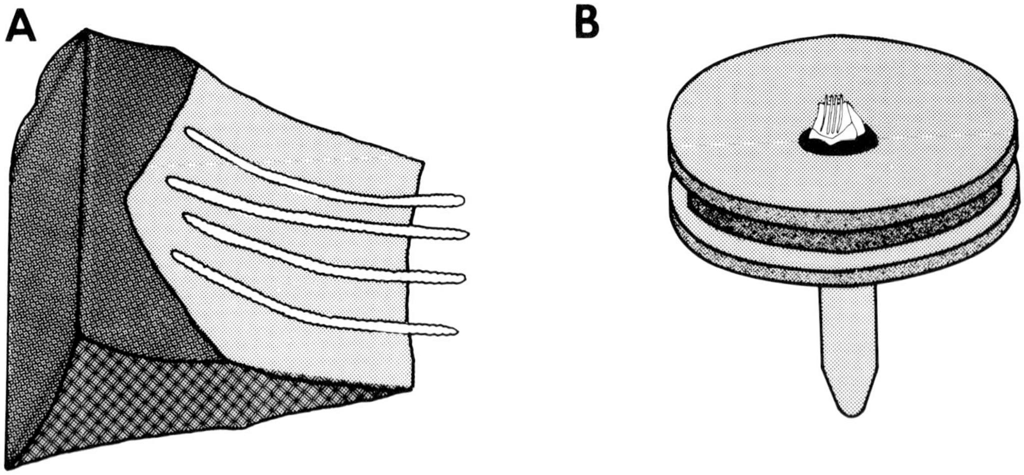

Fig. 9. Mounting technique for glycerin-infiltrated specimens. A) Nematodes mounted on a glass chip. B) Glass chip with nematode mounted on a microscope stub.

pendicular to the surface of the stub, coated with 100-300 Å of gold/palladium alloy, and viewed at relatively low accelerating voltages of 5 to 10 Kv. Specimens are sensitive to beam damage and cannot be viewed for long periods at high magnifications.

Drying Glycerin-Infiltrated Specimens

A graded series of glycerin/dilutent is prepared. The dilutent is 30% absolute ethanol in 70% distilled water. Percent glycerin in dilutent in the graded series is 85, 65, 45, 25, 5 and 0. The first concentration of glycerin/dilutent (85%) is added to a nematode-processing container in a Stendor dish. The nematodes are hand picked from glycerin into the container. At 15-30 min intervals, one half of the old solution is withdrawn and replaced with the next solution in the graded series. The last step is repeated three times to ensure that the glycerin is completely exchanged with the alcohol/water dilutent. The ethanol is removed from the specimens with 15 min exchanges of 15% ethanol, 5% ethanol, and three 15-min exchanges of 100% water. The material can then be post-fixed in osmium, washed, dehydrated, and dried as previously discussed for live specimens.

PREPARATION OF DISSECTED SPECIMENS

Because the SEM can only image details of surface morphology, internal structures must be dissected out of the nematode to be viewed. Also, certain cuticular structures, perineal patterns and vulval cyst cones are more easily examined and visualized if they are excised from the specimen. The pieces of cuticle and cuticularized stylets and spicules are processed with solutions and techniques very similar to those previously described. Because these structures are not hydrated, they can withstand severe treatment and still maintain their structural integrity. Generally, they do not need to be fixed and can be dissected in harsh liquids such as lactic acid and glycerin, and can be successfully air dried.

Stylets of *Meloidogyne* Females

Several coverslip dissecting chambers (Fig. 5) are prepared. The center of the ringed coverslip is marked on the bottom of the glass slide with a permanent marking pen, and the ring on the coverslip is filled with a large drop of 45% lactic acid.

A young, egg-laying female is dissected from root tissue and placed into a large drop of 45% lactic acid in a clear plastic petri dish. The head of the female is cut off just posterior to the median bulb with a sharp eyeknife or similar tool (Fig. 10A). With a micropipette or forceps, the excised female head is transferred from the plastic dish to the drop of lactic acid on the ringed coverslip (Fig. 10B). With the aid of a dissecting microscope at 60-80 magnifications, the stylet and attached cuticular esophageal lining is pushed out through the cut opening of the head with a finely sharpened dental file (Fig. 10C). The attached muscles and debris are cleaned from the stylet and lining by being swished in the lactic acid with the dental file. Gentle pressure with the file on the lining of the median bulb attaches the stylet to the coverslip.

One drop of 2% formalin, dropped onto the lactic acid every minute for several minutes until the coverslip is flooded, fixes the specimen and makes it stick to the coverslip. The lactic acid is replaced by the formalin (Fig. 10D). The stylet is allowed to remain in the formalin undisturbed for 5 to 10 min. Visual inspection ensures that the stylet is still attached to the coverslip. If it is not attached, it is carefully reattached with the file. The formalin is drained from the coverslip (Fig. 10E) and then refilled with fresh formalin. The lactic acid must be completely replaced by the formalin or the stylet will be obscured by the residual lactic acid. Lactic acid will not evaporate, whereas formalin will. The excess formalin is blotted from the slide, and the coverslip is finally drained. The slide and coverslip are allowed to air dry overnight in a desiccator. The location of the stylet is marked on the coverslip with the point of a small triangular piece of tape (Fig. 10F). The coverslip is re-

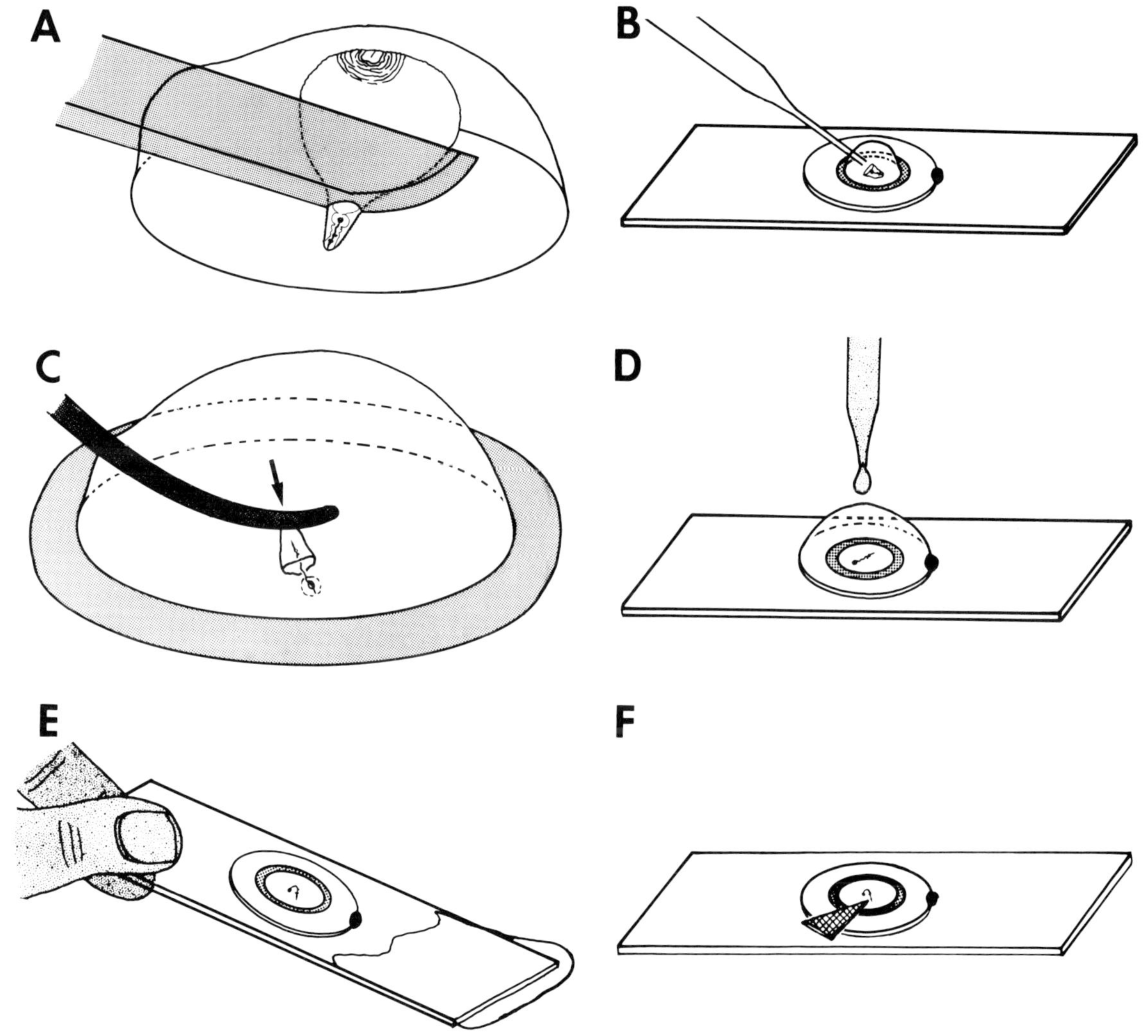

Fig. 10. Diagrams illustrating the steps in the dissection of stylets of root-knot nematode females. A) Live specimen is transferred into a large drop of 45% lactic acid on a clear plastic dish, and the head is cut off posterior to the median bulb with a sharp eyeknife. B) The head is pipetted into another drop of lactic acid on a specially prepared coverslip dissecting chamber. C) The stylet and attached cuticular lining of the esophagus are pushed out through the cut opening. D) Formalin is pipetted onto the lactic acid one drop at a time until the lactic acid is completely removed. E) The formalin is then drained and the stylet allowed to air dry. F) The location of the stylet is marked with a triangular piece of tape.

moved from the glass slide and attached to a microscope stub with double-coated tape or other suitable substance. An electrical connection is made between the top surface of the coverslip and the stub with a small dot of colloidal graphite in isopropyl alcohol or silver paint and allowed to dry. The specimens are coated with approximately 200 Å of gold in a sputter coater or vacuum evaporator and examined in the scanning electron microscope. Viewing the specimen perpendicular to the beam avoids foreshortening of the image. The stylet can be located on the stub by a search of the area immediately around the point of the tape marker. Locating the stylet is easiest if the scope is operated in television mode.

Stylets of *Meloidogyne* Males and Second-stage Juveniles

Stylets of root-knot males and second-stage juveniles are prepared similarly to those of females. The vermiform stages are placed directly into lactic acid on a ringed coverslip, and the head is cut off posterior to the median bulb. The stylet is then pushed out through the cut opening and processed as described for the female.

Stylets of Other Nematode Genera

Stylets of other genera can be dissected and viewed by SEM using the same basic procedures developed for *Meloidogyne* (Fig. 11). The shaft and knobs of some plant-parasitic nematode genera, however, dissolve in lactic acid, and other dissecting solutions must be employed. Stylets from *Criconemella* spp. (10) and *Heterodera* second-stage juveniles have been dissected successfully in sodium hypochlorite dilutions of various concentrations (Fig. 12) (J. D. Eisenback, unpublished).

Spicules

Nematode spicules can be dissected with a technique similar to that of stylet extraction. Whole specimens are picked into lactic acid on a ringed coverslip. The tail is cut off slightly anterior to the spicules and allowed to soak for 20-30 min. The spicules are forced out of the spicular pouch by pushing on them with a dental root canal file, and they are pushed out through the cut opening. Spicules are strongly attached to the body and are very difficult to remove. Also, they do not easily stick to the coverslip. Finally, when they are successfully mounted and observed, orientation on the coverslip cannot be determined easily and comparisons among specimens is difficult (Fig. 6).

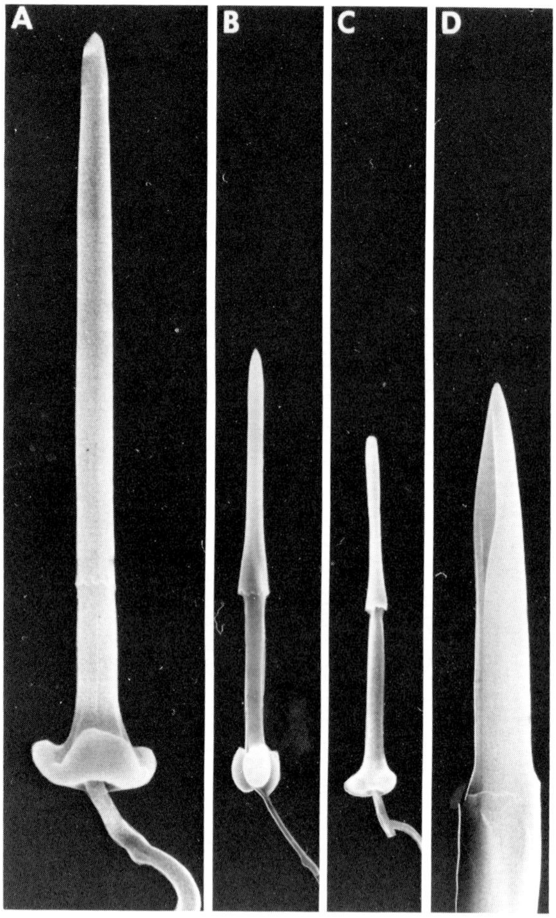

Fig. 11. Dissected stylets of different nematodes. A) *Criconemella* spp., female; B) *Hoplolaimus galeatus* (Cobb) Filipjev and Shuurmans Stekhoven, female; C) *Heterodera glycines* Ichinohe, second-stage juvenile; D) Odontostyle of a dorylaim.

Perineal Patterns (29) (Fig. 13)

Live, young, egg-laying females are dissected from root tissue, and placed into a large drop of 45% lactic acid on a clear plastic dish. The head of the female is removed with a sharp eyeknife or similar tool, and the body contents are pushed out through the cut opening (Fig. 14 A,B). The posterior one third of the remaining cuticle is cut off and trimmed to a small square containing the perineal pattern (Fig. 14 C,D). The remaining body tissues are gently cleaned from the pattern with a dental file (Fig. 14E) and transferred to a drop of anhydrous glycerin in a deep-well slide or small glass dish (Fig. 14F). Ten to thirty patterns are pre-

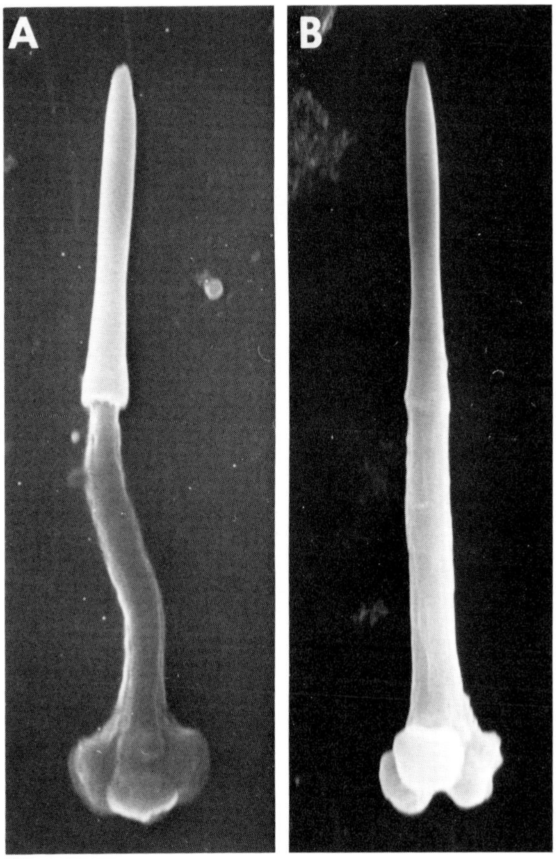

Fig. 12. Dissected stylets of *Heterodera glycines* Ichinohe second-stage juveniles. A) Stylet dissected in 45% lactic acid showing dissolution of shaft and knobs. B) Stylet dissected in 0.05% sodium hypochlorite.

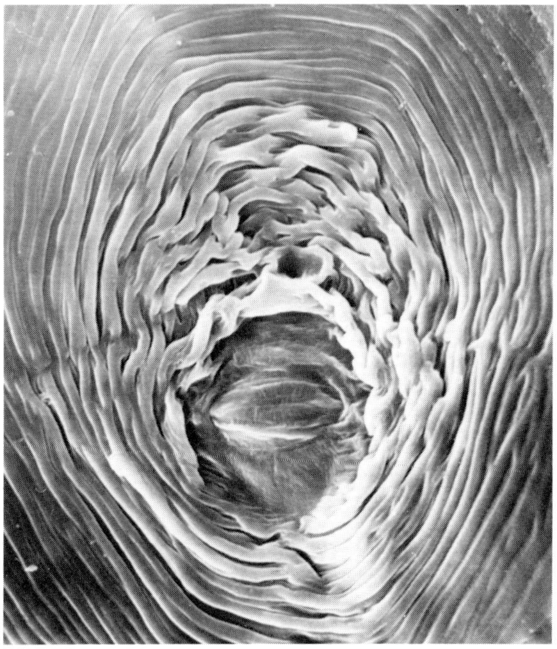

Fig. 13. Perineal pattern of *Meloidogyne chitwoodi* Golden *et al.*

pared in this way and stored in glycerin overnight in a desiccator.

Each perineal pattern is transferred with a dental file to a narrow strip of filter paper marked at 2-mm intervals (Fig. 14G). The pattern should be placed right side up on the filter paper because after draining it will be impossible to distinguish the inner and outer sides. The patterns are allowed to drain for 2 to 14 days in a desiccator. They are removed from the desiccator and transferred with a dental file onto a stub covered with double-coated tape. The patterns are arranged in closely spaced rows with the pattern side up (Fig. 14H) and coated with approximately 200 Å of gold in a sputter coater or vacuum evaporator. The specimens are viewed with the stub perpendicular to the beam so the image will not be foreshortened. The quality of the preparation can be assessed by comparisons with perineal patterns viewed in the light microscope.

Vulval Cyst Cones

Brown cysts are collected, sonicated in tap water, and fixed for at least 48 hr in 5% formalin (Fig. 15). They are transferred into 45% lactic acid, and the cyst cones are cut from the perineal area with a sharp eyeknife. The dissected cyst cones are placed in anhydrous glycerin and stored in a desiccator overnight. The cones are removed from glycerin and placed on a narrow strip of filter paper. The cones are transferred to an oven at 40°C and left for 1 hr. They can be stored in a desiccator for 1 wk. The well-drained specimens are mounted with double-coated tape on microscope stubs and coated with approximately 200 Å of gold by means of a sputter coater or vacuum evaporator (46).

PREPARATION OF TISSUES AND CELLS

Some internal organs and structures such as sperm and esophageal glands have been dissected from live nematodes and processed for SEM. Such live tissue cannot be air dried, but must be fixed, dehydrated, and dried in a manner similar to whole nematode preparations (87). Because of their small size, however, some tissues can be fixed, dehydrated,

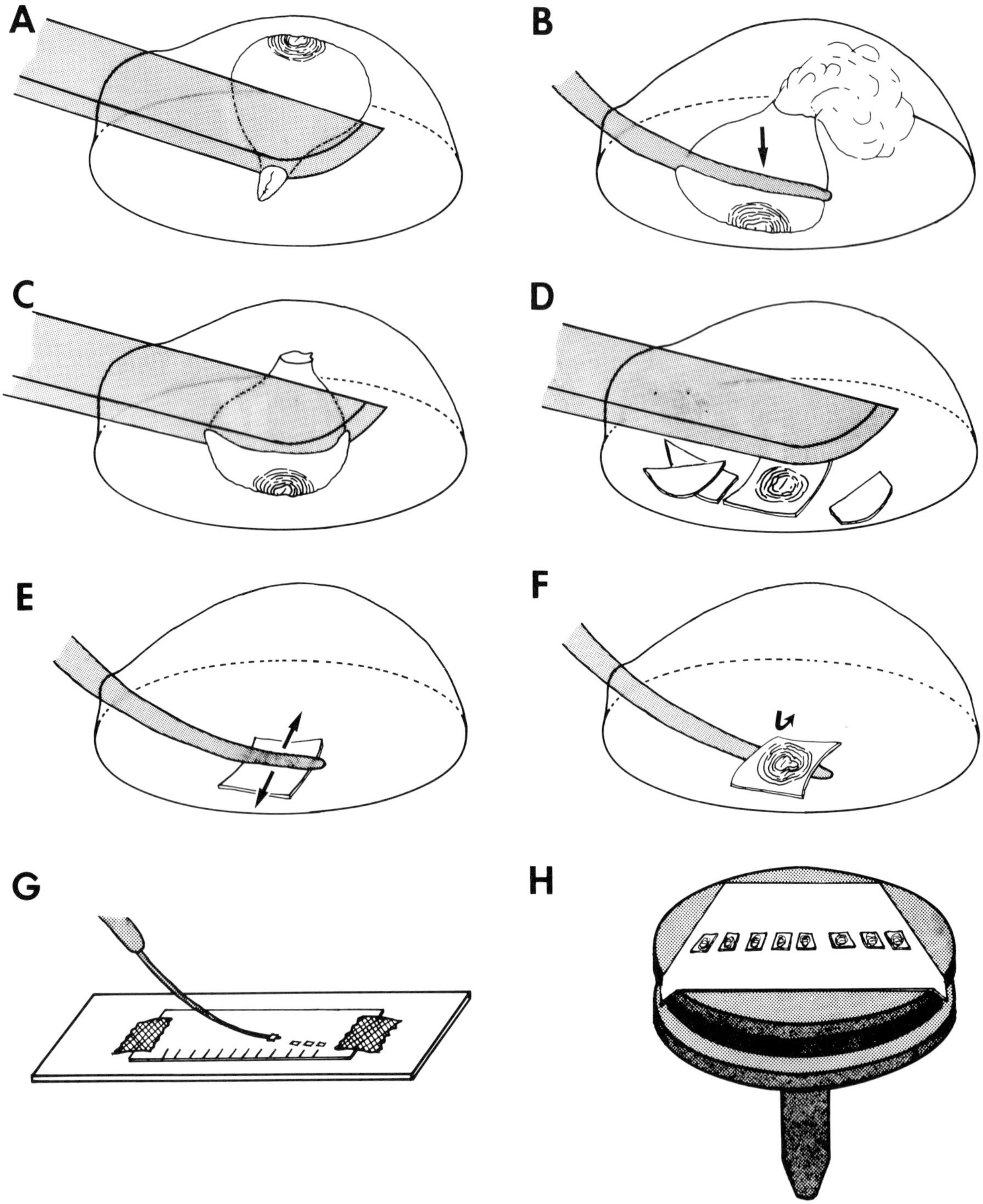

Fig. 14. Diagrams illustrating steps in the preparation of perineal patterns for scanning electron microscopy. A) Female is placed into a drop of 45% lactic acid and the head is cut off with a sharp eyeknife. B) The body contents are pushed out through the cut opening using a dental file. C) The posterior one-third of the cuticle is cut off. D) The perineal pattern is trimmed to a small square piece of cuticle. E) The cut pattern is turned upside down and thoroughly, but gently, cleaned with a dental file. F) The pattern is transfered and stored in anhydrous glycerin in a dessicator overnight. G) The patterns are placed right side up on a small piece of filter paper and stored in a desiccator for several days. H) The patterns are mounted in closely spaced rows on a stub with double-coated tape.

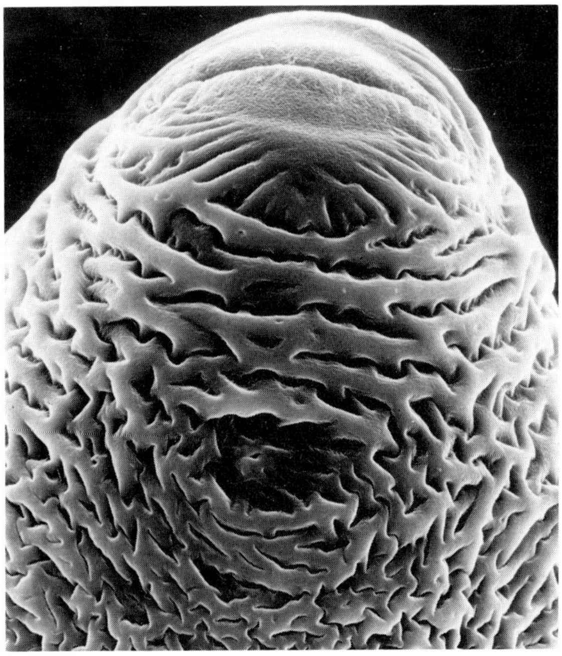

Fig. 15. Vulval cyst cone of *Heterodera trifolii* Goffart. (Courtesy H. Hirschmann.)

and air dried from a volatile liquid such as freon or acetone. Such a procedure is not only less expensive and faster than critical point drying, but also gives comparable results.

Sperm From Seminal Vesicle of Male

A large drop of 0.9% saline solution is placed within the ringed area of a coverslip dissecting chamber (Fig. 16). A young, active male is handpicked into the saline (Fig. 16A), and the tail end of the nematode is cut off anterior to the head of the spicules with an eyeknife (Fig. 16B). Sperm, viewed under a dissecting microscope at 60× magnification or greater, can be pushed out of the epithelial lining of the seminal vesicle and vas deferens by gentle pressure with a dental file on the anterior portion of the testis (Fig. 16C). Some of the sperm will attach to the coverslip. The sperm can then be fixed in 2-4 drops of 4% glutaraldehyde in a 0.1 M sodium-cacodylate buffer (pH 7.0) for 15-25 min at room temperature (Fig. 16D). The coverslip is then removed from the slide and placed in a Stendor dish. The sperm are washed with several changes of buffer: most of the old solution is withdrawn and immediately new buffer is added. The coverslip should remain covered with liquid at all times.

The sperm are dehydrated with a graded series of 10, 30, 50, 70, 95, and 100% ethanol for 5 min at each step (Fig. 16E). Three 5-min changes of absolute ethanol are necessary for complete dehydration. The specimens are infiltrated with the highly volatile Freon 113 in a 5-step graded series of 5 min each. The last step of absolute freon is repeated three times, and the coverslip is removed from the freon with forceps and allowed to air dry (Fig. 16F). The coverslip is stuck to a glass slide, and the group of sperm are located and marked with a triangular piece of tape.

Sperm From Inseminated Females

A live, young, egg-laying female is placed in a large drop of 0.9% saline in a plastic petri dish. Cutting the head of the female with a sharp eyeknife relieves body pressure. Then the posterior body portion is cut off at the perineal region, and the reproductive tract is pulled from the body cavity with a dental file. The entire gonad is uncoiled and, together with a small amount of solution, is gently picked up with a small loop of fine wire and placed in a drop of saline on a coverslip dissecting chamber. The gonad is cut posterior to the spermatheca, and sperm are released through the cut opening by gentle pressure with a dental file on the walls of the ovary. The sperm that attach to the coverslip are fixed, dehydrated, and dried as was discussed for sperm dissected from males.

SPECIMENS IN HOST TISSUE

SEM facilitates detailed observation of nematode morphology, as well as detailed examination of the nematode-host relationship. If the interaction occurs at the surface, such as is the case with nematode penetration, standard preparative procedures for whole nematodes can be used. Generally, however, the interaction occurs deep within the tissues at the feeding site. Since the SEM can visualize only surface structures, the host must be dissected to reveal the internal morphology. Fracturing a frozen specimen or cutting the specimen with a sharp blade are two possible methods of dissection.

Cryofracturing

Small pieces of nematode-infected tissues are fixed, washed, post-fixed, and dehydrated as previously discussed for whole nematodes. The tissues in 100% ethanol are transferred

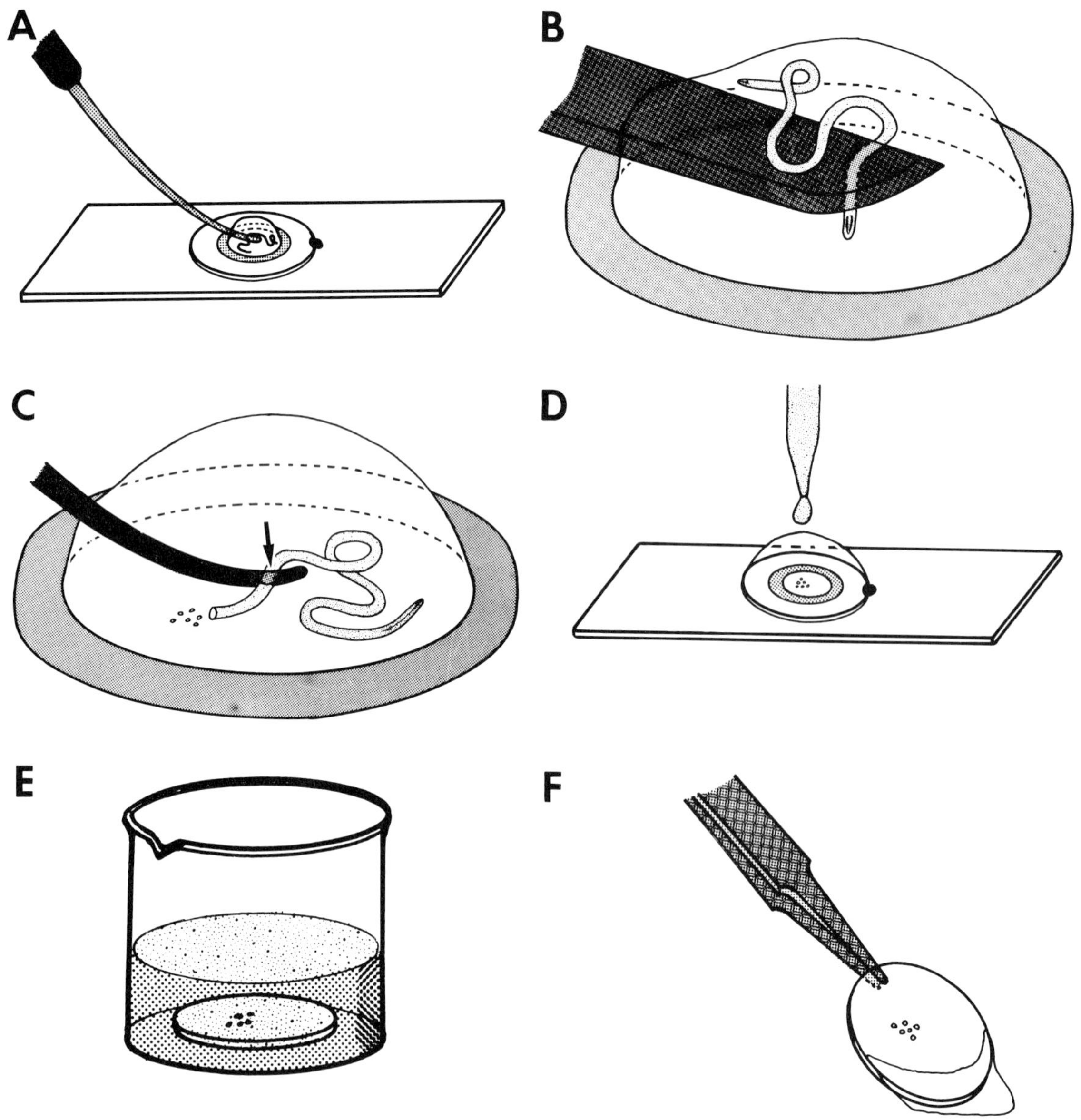

Fig. 16. Diagrams illustrating the steps in dissecting sperm from male nematodes. A) Transfer live, young males into a drop of 0.9% saline. B) Cut off tail just anterior to the spicules. C) Force sperm out of the vas deferens by gently pressing on the anterior portion of the testis with a dental file. D) Fix sperm with drops of glutaraldehyde. E) Dehydrate and infiltrate sperm with a highly volatile liquid. F) Remove sperm from the final solution and allow to air dry.

with tweezers or forceps into liquid nitrogen. A razor blade or eyeknife is precooled in liquid nitrogen, and the frozen specimens are "cut" while still submerged in the nitrogen. Actually, the tissue cannot be cut but instead will fracture posterior to the "cutting" point or randomly shatter into many pieces. The pieces are transferred back into absolute ethanol and allowed to thaw at room temperature. The specimens are critical point dried, mounted, and coated as previously discussed. Whole nematodes can also be cryofractured in a similar fashion. Because the specimens are so small, however, specialized handling procedures must be utilized (48,91).

Dissection and Digestion (55,70)

Small pieces of nematode-infected material are fixed, washed, post-fixed, and dehydrated as was discussed for whole nematodes. The plant tissues are viewed under a dissecting scope and cut in longitudinal and cross sections with a clean, sharp razor blade. As a result, the nematodes and feeding sites are exposed. Then the specimens are critical point dried, coated, and viewed.

Treating the specimens dissolves away the cytoplasm and reveals the modified structure of the inner cell wall. After post-fixation in osmium and washing in distilled water, the cytoplasm is digested. The tissues are cut in longitudinal and cross sections, treated in 1% aqueous periodic acid for 2 min, and rinsed with five changes of distilled water of 2 min each. They are transferred into 4% potassium hydroxide for 30 min at 55°C, washed in distilled water, transferred to 1% acetic acid for 5 min, and washed with distilled water for 10 min. The tissues are hardened in 2% osmium for 3 hr, washed, dehydrated, and critical point dried.

OPTIMAL USE OF THE MICROSCOPE

While specimen preparation is a vital part of successful imaging of biological materials, equally important is the manipulation of many variables controlling the production of a high quality image on the microscope (40). The image obtained is not permanent and has relatively poor resolution. The ultimate goal of scanning electron microscopy is, therefore, the production of a permanent record of the image, a scanning electron micrograph.

Viewing

Because many parameters control image formation, determining the cause of poor picture quality is sometimes difficult. Separation of faulty microscope function from inadequate specimen preparation can be achieved by the use of a non-biological standard specimen. If a high quality picture can be obtained with the standard, then the preparation may be at fault. Several different types of standard specimens are available from suppliers of products for electron microscopy.

Nematodes that are prepared for scanning electron microscopy should be handled very carefully because they are quite delicate. Coated specimens viewed shortly after drying have the best image quality. When immediate viewing is not possible, dried nematodes should be stored in a desiccator. Mounting the specimens as close to the stub as possible shortens the electrical pathway. Unnecessary exposure to the beam should be avoided; focus and exposure of the photograph should be rapid; and the lowest magnification possible should be used when the specimen is not being examined.

Specimen orientation to the beam is important for a good photograph, and comparisons of images are easiest if the viewing angles are similar. A slight change in the orientation can dramatically affect the appearance of the specimen. Morphological comparisons should be made from face and lateral views. Major external taxonomic characters of nematodes are present mainly on the head. The form of the labial disc, medial lips, head region, and sensory structures are valuable features and best seen in face views. Lateral views are important because they can be correlated with light microscope observations. The two different orientations complement each other because the change in position affects the lighting, and sometimes minor details that are obscured in one view are visible in the other. In addition, other viewing angles can supplement certain morphological details that may not be seen in the standard face and lateral views.

The extremely large depth of focus in the SEM can cause a foreshortening of the image that can lead to misinterpretation of the image. For this reason, some dissected parts of nematodes such as stylets and perineal patterns should be viewed only in an angle perpendicular to the electron beam.

Photography

Photographs provide a highly resolved, permanent record that can be analyzed and reviewed. Most SEMs are equipped with a camera that utilizes 10.2 × 12.7-cm (4×5-in) Polaroid Land film. The exposure is automatic, and the developing and printing are instant.

Several adjustments of the microscope greatly affect the quality of the photograph. The most important variables are contrast/brightness, focus, astigmatism correction, scan rate, spot size, and working distance. Generally, a good micrograph has some pure white areas, some pure black areas, and a range of gray tones. Contrast is the ratio of light to dark areas. It is an aesthetic quality that is sometimes difficult to evaluate. In the SEM, contrast is adjusted by two dependent controls—contrast and brightness. Both controls should be set at the recommended levels. Experience with certain specimens may require a setting different from the recommended levels.

Critical focus and astigmatism correction must be made prior to the exposure of the film. Both of these steps are best evaluated at a one- or two-step increase in magnification above the final desired magnification. The highest resolution is obtained with the slowest scan rate. Because a slowed scan rate means a long exposure (60-90 sec), vibrations around the microscope should be eliminated during the exposure. The spot size should be kept small, and the working distance should be adjusted according to the magnification: the higher the magnification, the shorter the working distance.

FILM Two types of Polaroid Land film are commonly used. Type-52 film gives a positive print, has an ISO value of 400, and a resolution of 35-40 lines/mm. Type-55 film yields both a positive print and a highly resolved negative. It has an ISO value of 50. The print has a resolution of 22-25 lines/mm, but the negative has a resolution of 150-160 lines/mm. The increased resolution of the negative gives a superior print compared to the original. Because of the high resolution and increased flexibility that a negative has over a print alone, Type-55 film is the preferred film for recording SEM images.

PRINTING NEGATIVES Conventional black- and white-photographic techniques are used to make enlargements or prints from Type-55-film negatives. Many texts are available for consultation (9,23,44,57,59). Because the negative has a much greater resolution than the original print, the final print done in the darkroom will have a much improved image. The Polaroid print and negative should be considered as an intermediate step to the finished print. To get the most out of the recorded image, the microscopist must be proficient in the darkroom.

Biological materials may exhibit much contrast because of the interaction of the beam with the edges of the specimen. For nematodes, especially in face view, this phenomenon results in a negative that is difficult to print with good range of gray tones. The darkroom techniques of dodging or burning-in usually produce the highest quality prints (Fig. 17).

AIRBRUSH TECHNIQUES Sometimes the specimen cannot be oriented properly, and a photograph with a good format is not obtained. However, with airbrush techniques (56), the background can be painted and the format changed (Fig. 18). Airbrushing can also clean the background in certain instances making the photograph more aesthetically pleasing (Fig. 19).

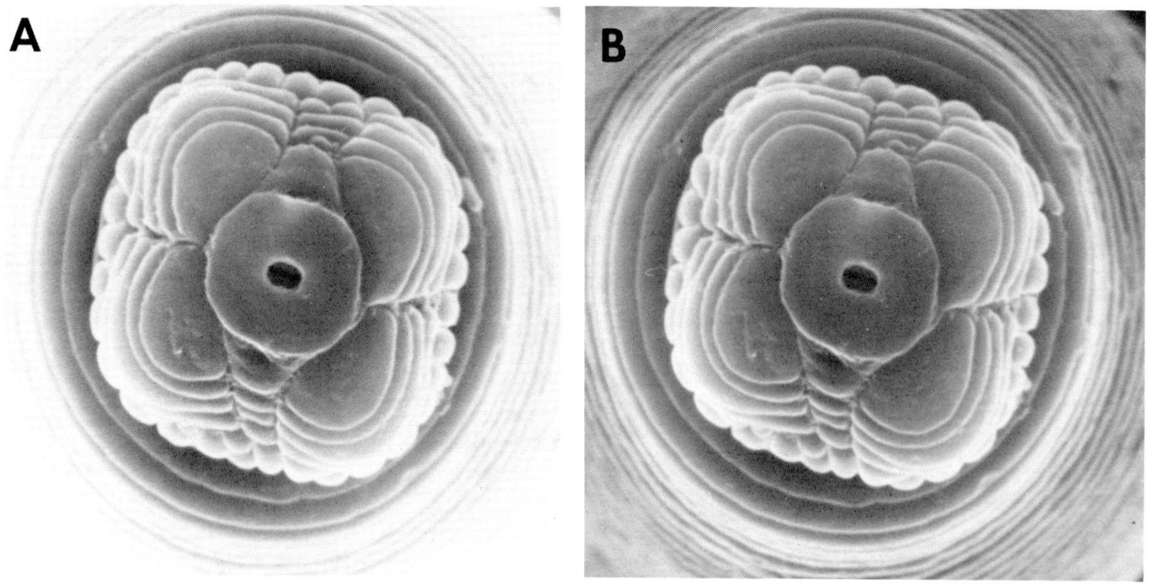

Fig. 17. Two prints of the same negative, illustrating the effects of dodging and burning in on the quality of the finished photograph. A) A standard print of a male of *Hoplolaimus galeatus* (Cobb) Filipjev and Shuurmans Stekhoven. B) A darkroom manipulated print.

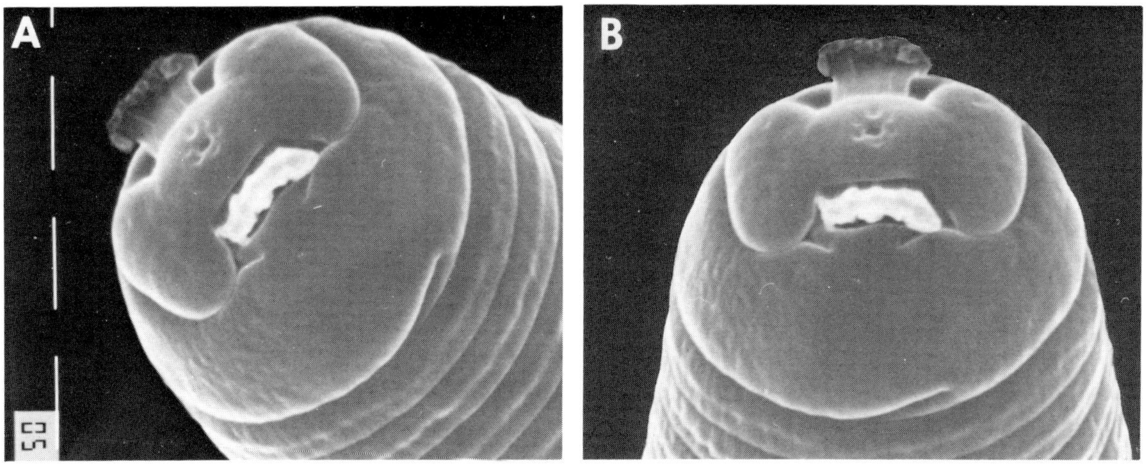

Fig. 18. Two prints of the same negative illustrating the effects of airbrush technique in altering the format of the orginal print. A) Improperly oriented specimen. B) Format changed by airbrush techniques.

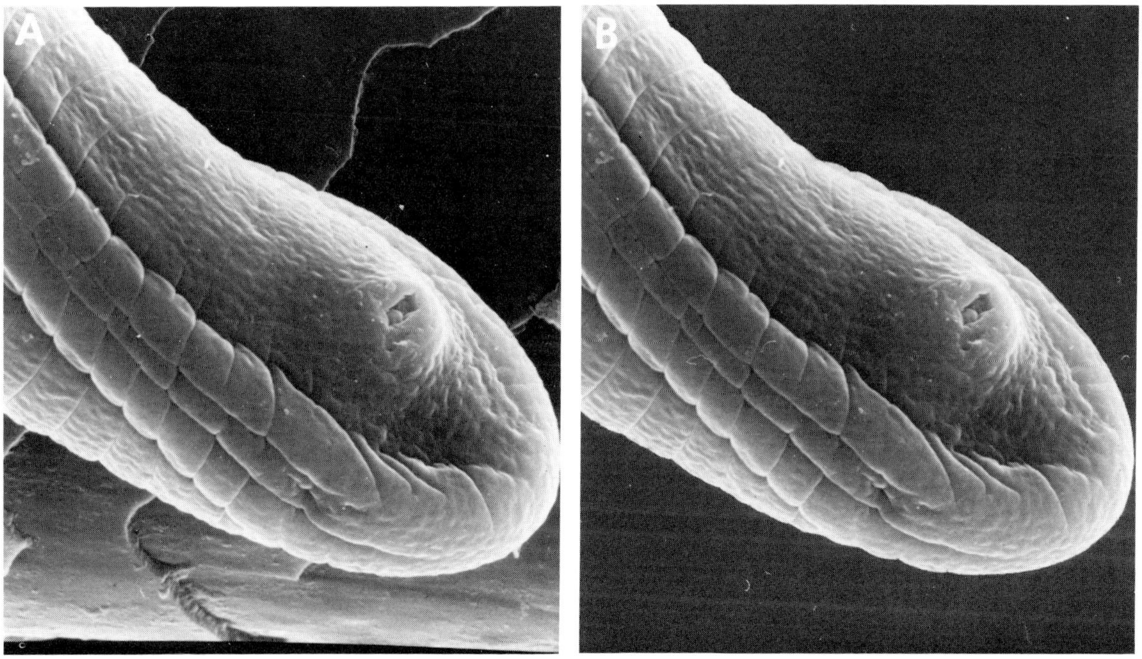

Fig. 19. Two prints of the same negative illustrating the effects of airbrush technique in making a more aesthetically pleasing background. A) A specimen with a distracting background. B) Distracting background removed by airbrushing.

Airbrushing the background of a photograph is recommended for prints in which the specimen has the proper contrast but an unacceptable background (Figs. 20A,B). The specimen must be covered with a layer of frisket (Fig. 20 B) that adheres tightly to the emulsion of the photograph. The specimen is traced with a sharp knife which cuts through the layer of frisket (Fig. 20C). The frisket is then removed from the background, and the background is gradually airbrushed with medium to dark gray designer's gouache. After the background is completely covered, the paint is allowed to dry thoroughly (Fig. 20D). Finally, the frisket that protected the specimen from the paint is removed without damaging the photographic emulsion (Figs. 20E,F).

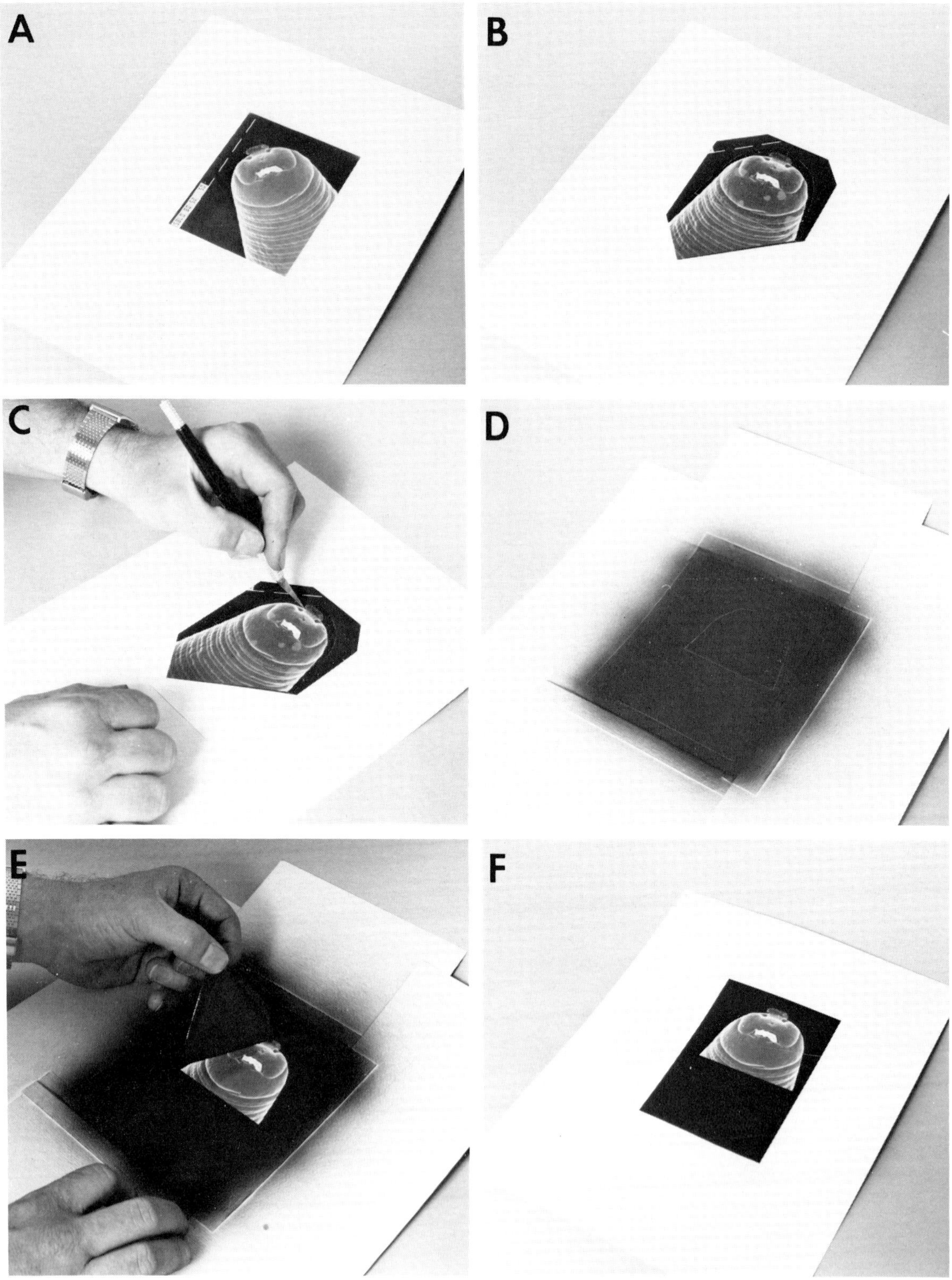

Fig. 20. Photographs illustrating steps in airbrushing prints to alter the format of the original photograph or to remove a distracting background. A) The original print with the wrong orientation. B) A good print with the proper orientation covered with a layer of frisket. C) Trace around the edge of the specimen with a sharp knife cutting through the frisket but not the photograph, and remove the frisket from the background. D) Gradually airbrush the background with a medium or dark gray designer's gouache, and allow the paint to dry. E) Carefully remove the frisket to expose the specimen. F) The finished product.

Stereo-pairs

The use of two pictures of the same specimen taken from different angles can greatly enhance the perception of the image by making the specimen appear to be three-dimensional (Fig. 21). Although several methods can be used to make stereo pairs (93), the specimen-tilt technique is the simplest and easiest to use.

When a stereo-pair of micrographs is made by means of the specimen-tilt technique (91), one photograph of the specimen is made by standard methods. An outline of the specimen is marked on the screen of the display tube with a wax pencil, and the specimen is tilted 4-10° with the stage tilt control. Tilting the nematode changes many factors such as focus, contrast/ brightness, and placement of the specimen on the screen. The specimen is realigned on the screen to match the pencil outline, and the image is refocused with the Z control on the stage. Contrast and brightness levels are readjusted, and a second photograph of the specimen is made. The stereo pair can be viewed, projected or published by means of several different techniques (40,93).

CONCLUSIONS

Although the light microscope is likely to remain the most important instrument used in the study of nematodes, SEM provides new information that cannot be obtained with any other present technique. It has given us a better view of the organism itself, of the relationship between the nematode and the host, and of the interaction of nematodes and other organisms. Certainly SEM will become even more important in nematology as new techniques and procedures are developed.

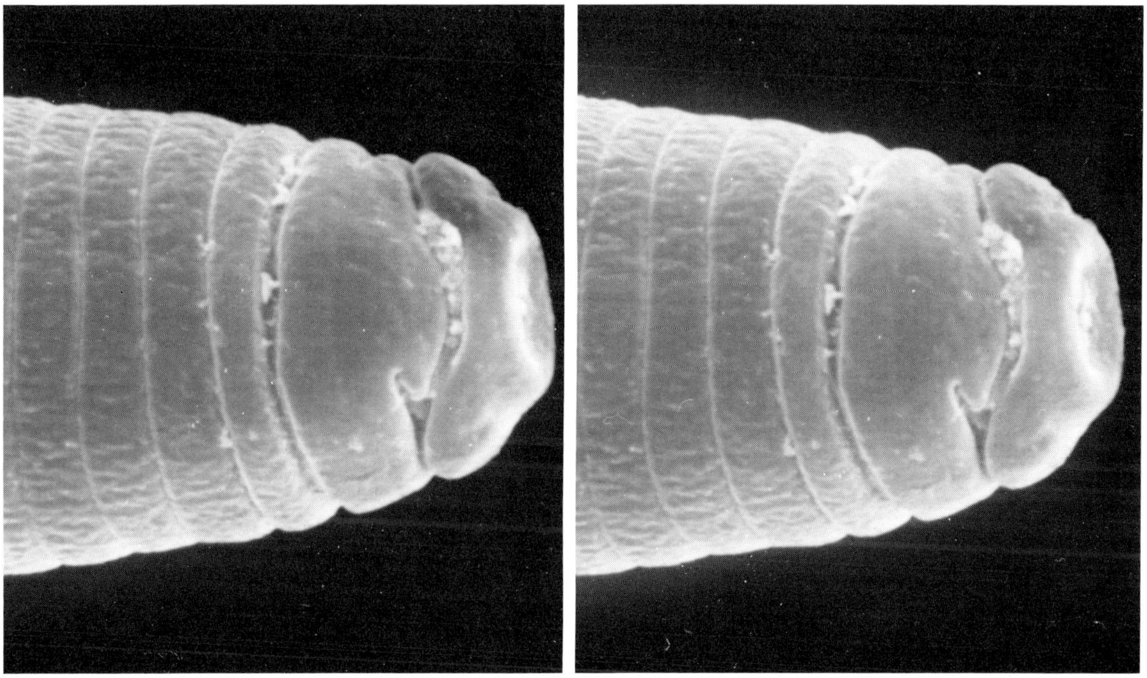

Fig. 21. Stereo pair of a male of *Meloidogyne incognita* (Kofoid & White) Chitwood.

Literature Cited

1. Anderson, T. F. 1951. Techniques for the preservation of three dimensional structure in preparing specimens for the electron microscope. Trans. N.Y. Acad. Sci. 13:130-134.
2. Ayoub, S. M. 1980. Plant nematology, an agricultural training aid. NemaAid Publ., Sacramento, Calif. 195 pp.
3. Baldwin, J. G., M. Luc and A. H. Bell. 1983. Contribution to the study of the genus *Pratylenchoides* Winslow (Nematoda: Tylenchida). Rev. Nematol. 6:111-125.
4. Baldwin, J. G., M. Mundo-Ocampo, and A. A. Othman. 1983. *Cryphodera utahensis* n. sp. (Heteroderidae), a new species from wild rose in Utah. J. Nematol. 15:182-191.
5. Black, J. T. 1974. The scanning electron microscope: operating principles, pp. 1-43 *In* Principles and techniques of scanning electron microscopy, Vol. 1, ed. M. A. Hayat. Van Nostrand Reinhold, New York. 273 pp.
6. Bowes, J. H. and C. W. Cater. 1966. The reaction of glutaraldehyde with proteins and biological materials. J. R. Microsc. Soc. 85:193-200.
7. Boyde, A. 1978. Pros and cons of critical point drying and freeze drying for SEM. Scanning Electron Microsc. II:303-314.
8. Boyde, A. and C. Wood. 1969. Preparation of annimal tissues for surface-scanning electron microscopy. J. Microsc. (Oxf.) 90:221-249.
9. Carrol, J. S. 1979. Photographic lab handbook, 5th ed. Amphoto Book, Garden City, N.Y. 720 pp.
10. Chen, T. A. and G. Y. Wen. 1980. Electron microscopy of the stomastylet and esophagus of *Criconemoides curvatum*. J. Nematol. 12:72-83.
11. Clark, S. A. and A. R. Stone. 1975. A simple method for preparing nematodes for scanning electron microscopy, using Spurr's low-viscosity epoxy resin. Nematologica 21:256-266.
12. Cohen, A. L. 1974. Critical point drying, pp. 44-112 *In* Principles and techniques of scanning electron microscopy, Vol. 1, ed. M. A. Hayat. Van Nostrand Reinhold, New York. 270 pp.
13. Cohen, A. L. 1979. Critical point drying—principles and procedures. Scanning Electron Micros. II:303-324.
14. Cohn, E. and M. Mordechai. 1983. Wall structure of mature female cystoid body of *Meloidoderita kirjanovae* (Nematoda: Criconematoidea). J. Nematol. 15:325-327.
15. Corbett, D. C. M. and S. A. Clark. 1983. Surface features in the taxonomy of *Pratylenchus* species. Rev. Nematol. 6:85-98.
16. Day, J. W. 1974. A beem capsule chamber-pipette for handling small specimens for electron microscopy. Stain Technol. 49: 408-410.
17. DeGrisse, A. T. 1974. A method for preparing nematodes and other soft tissues for scanning electron microscopy. Meded. Fac. Landbouwwet. Rijksuniv. Gent. 38:1685-1695.
18. DeGrisse, A. T. 1979. Modification of the mini-sieve and prepolymerized plate techniques for use in electron microscopy. J. Nematol. 11:196-199.
19. Demeure, Y., D. W. Freckman, and S. D. Van Gundy. 1979. Anhydrobiotic coiling of nematodes in soil. J. Nematol. 11:189-195.
20. Demeure, Y., D. W. Freckman, and S. D. Van Gundy. 1979. *In vitro* response of four species of nematodes to desiccation and discussion of this and related phenomena. Rev. Nematol. 2:203-210.
21. DeNee, P. B. and E. R. Walker. 1975. Specimen coating technique for the SEM. A comparison study. Scanning Electron Microsc. I:225-232.
22. Dunn, M. T., R. M. Sayre, A. Carrell, and W. P. Wergin. 1982. The use of the SEM to evaluate the fungal parasites of pest nematodes. Scanning Electron Microsc. III:1351-1357.
23. Eastman-Kodak Co. 1973. Electron microscopy and photography. Kodak Publ. No. P-236, Eastman Kodak Co., Rochester, N.Y. 48 pp.
24. Ebsary, B. A. 1981. Generic revision of Criconematidae (Nematoda): *Crossonema* and related genera with a proposal for *Neocrossonema* n. gen. Can. J. Zool. 59: 103-114.
25. Echlin, P. Coating techniques for scanning electron microscopy and x-ray microanalysis. Scanning Electron Microsc. I:109-132.
26. Eisenback, J.D. 1982. Description of the

blueberry root-knot nematode, *Meloidogyne carolinensis* n. sp. J. Nematol. 14:303-317.
27. Eisenback, J. D. 1982. Morphological comparison of head shape and stylet morphology of second-stage juveniles of *Meloidogyne* species. J. Nematol. 14:339-343.
28. Eisenback, J. D. and H. Hirschmann. 1982. Morphological comparison of stylets of male root-knot nematodes (*Meloidogyne* spp.). Scanning Electron Miscrosc. III: 837-843.
29. Eisenback, J. D., H. Hirschmann, and A. C. Triantaphyllou. 1980. Morphological comparison of *Meloidogyne* female head structures, perineal patterns, and stylets. J. Nematol. 12:300-313.
30. Ellenby, C. and E. M. Wilson. 1969. Scanning electron microscope observations on the spear of the second-stage larva of *Heterodera rostochiensis*. Nematologica 15: 290-291.
31. Everhart, T. E. and T. L. Haynes. 1972. The scanning electron microscope. Sci. Am. 226:54-69.
32. Flegg, J. J. and D. J. Hooper. 1970. Extraction of free-living stages from soil, pp. 5-22 *In* Laboratory methods for work with plant and soil nematodes, ed. J. F. Southey. Minist. Agric., Fish. and Food Tech. Bull. 2. Her Majesty's Stationery Office, London. 440 pp.
33. Garcia-Martinez, R., A. L. Taylor, and G. C. Smart, Jr. 1982. *Meloidogyne cruciani* n. sp., a root-knot nematode from St. Croix (U.S. Virgin Islands) with observations on morphology of this and two other species of the genus. J. Nematol. 14:292-303.
34. Giblin, R. M. and H. K. Kaya. 1983. *Bursaphelenchus seani* n. sp. (Nematoda: Aphelenchoididae), a phoretic associate of *Anthophora bomboides stanfordiana* Cockerall, 1904 (Hymenoptera: Anthophoridae). Rev. Nematol. 6:39-50.
35. Golden, A. M. and R. H. Mulvey. 1983. Redescription of *Heterodera zeae*, the corn cyst nematode, with SEM observations. J. Nematol. 15:60-70.
36. Goldstein, J. I. and H. Yakowitz, eds. 1975. Practical scanning electron microscopy. Plenum Press, New York. 582 pp.
37. Green, C. D. 1967. Preparation of nematodes for examination under the stereoscan electron microscope. Nematologica 13:279-282.
38. Green, C. D., A. R. Stone, R. H. Turner, and S. A. Clark. 1975. Preparation of nematodes for scanning electron microscopy. J. Microsc. (Oxf.) 103:89-99.
39. Hammon, R. A. 1969. The preparation of nematode material for scanning electron microscopy. J. Microsc. 90:273-274.
40. Hayes, T. L. 1973. Scanning electron microscope techniques in biology, pp. 153-214 *In* Advanced techniques in biological electron microscopy, ed. J. K. Koehler. Springer-Verlag, New York. 304 pp.
41. Hayes, T. L. and J. B. Pawley. 1975. Very small biological specimens, pp. 45-81 *In* Principles and techniques of scanning electron microscopy, Vol. 3, ed. M. A. Hayat. Van Nostrand Reinhold, New York. 226 pp.
42. Hayes, T. L., R. F. W. Pease, and L. W. MacDonald. 1968. The scanning electron microscope: principles and applications in biology and medicine. Adv. Biol. Med. Phys. 12:85-137.
43. Hayunga, E. G. 1977. A specimen holder for dehydrating and processing very small tissue samples. Trans. Am. Microsc. Soc. 96:156-158.
44. Hedgecoe, J. 1979. The book of photography. Alfred A. Knopf Inc., New York. 256 pp.
45. Hirschmann, H. 1983. Scanning electron microscopy as a tool in nematode taxonomy, pp. 95-111 *In* Concepts in nematode systematics, ed. A. R. Stone, H. M. Platt, and L. F. Kahlil. Academic Press, New York. 388 pp.
46. Hirschmann, H. and A. C. Triantaphyllou. 1979. Morphological comparison of members of the *Heterodera trifolii* species complex. Nematologica 25:458-481.
47. Hogger, C. H. and R. H. Estey. 1976. Chamber for critical-point drying of nematodes and other biological specimens. J. Nematol. 8:357-358.
48. Hogger, C. H. and R. H. Estey. 1977. Cryofracturing for scanning electron microscope observations of internal structures of nematodes. J. Nematol. 9:334-337.
49. Hooper, D. J. 1969. Extraction and handling of plant and soil nematodes, pp. 20-36 *In* Nematodes of tropical crops, ed. J. E. Peachey. Commonw. Bur. Helminthol. (St. Albans, Herts, England) Tech. Commun. No. 40.
50. Hooper, D. J. 1970. Extraction of nematodes from plant material, pp. 34-38 *In*

Laboratory methods for work with plant and soil nematodes, ed. J. F. Southey. Minist. Agric., Fish., and Food Tech. Bull. 2. Her Majesty's Stationery Office, London. 440 pp.

51. Hopwood, D. 1969. Fixatives and fixation: A review. Histochem. J. 1:323-360.
52. Hopwood, D. 1972. Theoretical and practical aspects of glutaraldehyde fixation. Histochem. J. 4:267-303.
53. Humphreys, W. J. 1977. Health and safety hazards in the SEM laboratory. Scanning Electron Microsc. I:537-544.
54. Inserra, R. N., N. Vovlas, G. D. Griffin and J. L. Anderson. 1983. Development of the false root-knot nematode, *Nacobbus aberrans,* on sugarbeet. J. Nematol. 15:288-296.
55. Jones, M. G. K. and V. H. Dropkin. 1976. Scanning electron microscopy of nematode-induced giant transfer cells. Cytobios 15: 149-161.
56. King, W. S. and A. L. Slade. 1954. The airbrush technique of photographic retouching. The Macmillan Co., New York. 115 pp.
57. Kirkpatrick, K. 1975. Basic darkroom. Peterson's how-to photographic library. Peterson Publishing Co., Los Angeles, Calif.
58. Lamb, J. C., IV, and P. Ingram. 1979. Drying of biological specimens for SEM directly from ethanol. Scanning Electron Microsc. II:459-464.
59. Langford, M. 1981. The darkroom handbook. Alfred A. Knopf Inc., New York. 352 pp.
60. Lauritis, J. A., R. V. Rebois, and L. S. Graney. 1983. Development of *Heterodera glycines* Ichinohe on soybean, *Glycines max* (L.) Merr., under gnotobiotic conditions. J. Nematol. 15:272-281.
61. Lauritis, J. A., R. V. Rebois, and L. S. Graney. 1983. Life cycle of *Heterodera zeae* Koshy, Swarup, and Sethi on *Zea mays* L. axenic root explants. J. Nematol. 15:115-119.
62. Liepens, A. and E. de Harven. 1978. A rapid method for cell drying for SEM. Scanning Electron Microsc. II:37-44.
63. MacKenzie, A. P. 1976. Principles of freeze-drying. Transplant Proc. 8(2) Suppl. 1: 181-188.
64. Maugel, T. K., D. B. Bonar, W. J. Creegan, and E. B. Small. 1980. Specimen preparation techniques for acquatic organisms. Scanning Electron Microsc. II:57-77.
65. McClure, M. A. 1981. Electron microscope histochemistry, pp. 89-124 *In* Plant parasitic nematodes Vol. III, ed. B. M. Zuckerman and R. A. Rhode. Academic Press, New York. 508 pp.
66. McClure, M. A. and B. M. Zuckerman. 1982. Localization of cuticular binding sites of concanavalin A on *Caenorhabditis elegans* and *Meloidogyne incognita.* J. Nematol. 14:39-44.
67. McClure, M. A. and L. J. Stowell. 1978. A simple method of processing nematodes for electron microscopy. J. Nematol. 10: 376-377.
68. Mulvey, R. H. and A. M. Golden. 1983. An illustrated key to the cyst-forming genera and species of Heteroderidae in the western hemisphere with species morphometrics and distribution. J. Nematol. 15: 1-59.
69. Mundo-Ocampo, M. and J. G. Baldwin. 1983. Host-parasite relationships of *Atalodera* spp. (Heteroderidae). J. Nematol. 15:234-243.
70. Mundo-Ocampo, M. and J. G. Baldwin. 1983. Host response to *Sarisodera hydrophila* Wouts and Sher, 1971. J. Nematol. 15:259-268.
71. Murphy, J. A. 1978. Noncoating techniques to render biological specimens conductive. Scanning Electron Microsc. II:175-193.
72. Murphy, J. A. 1982. Considerations, materials, and procedures for specimen mounting prior to scanning electron microscopic examination. Scanning Electron Microsc. II:657-696.
73. Nicholas, W. L. 1975. The biology of free-living nematodes. Clarendon Press, Oxford. 219 pp.
74. Noel, G. R. and B. A. Stanger. 1983. Observations on the inner labial sensilla of second-stage larvae of *Heterodera glycines.* J. Nematol. 15:146-148.
75. Nowell, J. A. and J. B. Pawley. 1980. Preparation of experimental animal tissue for SEM. Scanning Electron Microsc. II: 1-19.
76. Oatley, C. 1972. The scanning electron miscroscope. Cambridge University Press, Cambridge. 192 pp.
77. Palade, G. E. 1952. A study of fixation for electron microscopy. J. Exp. Med. 95:285-298.
78. Pentilla, A., H. Kalimo, and B. F. Trump.

1974. Influence of glutaraldehyde and/or osmium tetroxide on cell volume, ion content, mechanical stability, and membrane permeability of Ehrlich ascites tumor cells. J. Cell Biol. 63:197-214.

79. Robbins, R. T. 1982. Description of *Dolichodorus grandaspicatus* n. sp. (Nematoda: Dolichodoredae). J. Nematol. 14:507-511.
80. Robbins, R. T. 1982. Description of *Hoplolaimus magnistylus* n. sp. (Nematoda: Hoplolaimidae). J. Nematol. 14:500-506.
81. Rossner, J. and R. N. Perry. 1975. Water loss and associated surface changes after desiccation in *Rotylenchus robustus*. Nematologica 21:438-442.
82. Sabatini, D. D., K. G. Bensch, and R. J. Barnett. 1962. New means of fixation for electron microscopy and histochemistry. Anat. Rec. 142:274.
83. Schiff, R. and J. F. Gennaro. 1979. The role of the buffer in the fixation of biological specimens for transmission and scanning electron microscopy. Scanning 2:135-148.
84. Sher, S. A. and A. H. Bell. 1975. Scanning electron micrographs of the anterior region of some species of Tylenchoidea (Tylenchida: Nematoda). J. Nematol. 7:69-83.
85. Sweney, L. R., N. J. Laible, and B. L. Shapiro. 1976. Some methods for more efficient processing of SEM specimens. J. Microsc. (Oxf.) 108:335-338.
86. Volvas, N. 1983. Observations on the morphology and histopathology of *Quinisulcius acti* on corn. Rev. Nematol. 6:79-83.
87. Ward, S., Y. Argon, and G. A. Nelson. 1981. Sperm morphogenesis in wild-type and fertilization-defective mutants of *Caenorhabditis elegans*. J. Cell Biol. 91: 26-44.
88. Watson, L. P., A. E. McKee, and B. R. Merrell. 1980. Preparation of microbiological specimens for scanning electron microscopy. Scanning Electron Microsc. II: 45-46.
89. Weise, R. W. 1973. Preparation of helminths for scanning electron microscopy, pp. 438-439 *In* Proc. 31st Electron Microsc. Soc. Am., ed. C. J. Arceneaux. Claitor's Publishing Division, Baton Rouge, La.
90. Wells, O. C. ed. 1974. Scanning electron microscopy. McGraw Hill, New York. 421 pp.
91. Wergin, W. P. 1981. Scanning electron miscroscopic techniques and applications for use in nematology, pp. 175-204 *In* Plant parasitic nematodes, Vol. III, ed. B. M. Zuckerman and R. A. Rhode. Academic Press, New York. 508 pp.
92. Wergin, W. P. 1982. Bibliography on the use of the scanning electron microscope (SEM) in studies of plant parasitic nematodes. Scanning Electron Microsc. III:831-836.
93. Wergin, W. P. and J. B. Pawley. 1980. Recording and projecting stereo pairs of scanning electron micrographs. Scanning Electron Microsc. I:239-249.
94. Wergin, W. P. and A. R. Stone. 1981. Techniques for preparation and examination of plant parasitic nematodes in the scanning electron microscope. Scanning Electron Microsc. III:169-176.
95. Williams, K. J. O. 1981. Revision of the genus *Discocriconemella* de Grisse and Loff, 1965 and the erection of the new genus *Acrozostron* (Nematoda: Criconematidae). Syst. Parasitol. 2:133-138.
96. Yang, B. and J. D. Eisenback. 1983. *Meloidogyne enterolobii* n. sp. (Meloidogynidae), a root-knot nematode parasitizing pacara earpod tree in China. J. Nematol. 15:381-391.

CHAPTER 7

CYTOLOGICAL METHODS FOR THE STUDY OF OOGENESIS AND REPRODUCTION OF ROOT-KNOT NEMATODES

A. C. Triantaphyllou
Department of Genetics
North Carolina State University
Raleigh, North Carolina, USA

Various methods have been employed in cytological and cytogenetic studies of root-knot nematodes. Nevertheless, only the "Feulgen" and the "propionic-orcein" methods have given relatively satisfactory results. Presumably, the Feulgen method is the most reliable because it specifically stains only nucleic acids and none of the other cell components. This method has been used by Nigon (4) in the cytological study of several free-living nematodes and more recently by Dalmasso and his associates (1) in the study of several plant-parasitic species, including those of *Meloidogyne*. A description of the Feulgen method can be found in many general books on cytology (2,3,5).

I used the Feulgen method in my early cytological studies on *Meloidogyne* with some degree of success, but later discontinued its use in favor of the propionic-orcein method which gave superior and equally reliable results. In comparison to the propionic-orcein method, the Feulgen method involves a more complicated procedure, especially if the commercially produced Schiff's reagent is not available, and also the chromosomes stained by this method are smaller, less deeply stained and less discrete because they are spread over a much smaller area.

The propionic-orcein method, when properly employed, gives very satisfactory results. The chromosomes stain very heavily; they contrast sharply against the practically unstained cytoplasm; and they appear relatively large because of the swelling effect of the propionic acid. Furthermore, the cytoplasm remains in a semifluid state for several days after staining, and this feature allows further spreading of the chromosomes by application of gentle pressure on the coverslip.

The propionic-orcein method, as employed in our laboratory for the study of *Meloidogyne* chromosomes, has been described in some detail earlier (6). Attempts by a small number of nematologists to use this method have not been universally successful. Apparently, there are several steps in the procedure that are critical if good results are to be obtained. In the following sections, the method is redescribed in more detail, in the hope that it will become more widely used not only with *Meloidogyne* spp., but also with many other types of nematodes.

Selecting the appropriate nematode material for cytological study and using properly prepared stain are two aspects of utmost importance which are often overlooked by prospective users of the proprionic-orcein method. These aspects will be discussed first.

1. PREPARATION OF PROPIONIC-ORCEIN STAIN

The best quality of orcein stain can be prepared according to the method described previously (6). However, comparable results are obtained when 2.2 g orcein stain (natural or synthetic) is added to 100 ml of glacial propionic acid and boiled gently for at least 20 minutes (warning: boiling can suddenly become violent). The solution is cooled and then diluted by the addition of 100 ml of distilled water. The solution is passed through a fine filter paper for removal of any undissolved stain particles and is stored in glass dropping bottles (pipette bottles with ground glass stopper - Fig. 1A), so that evaporation of the acid and precipitation

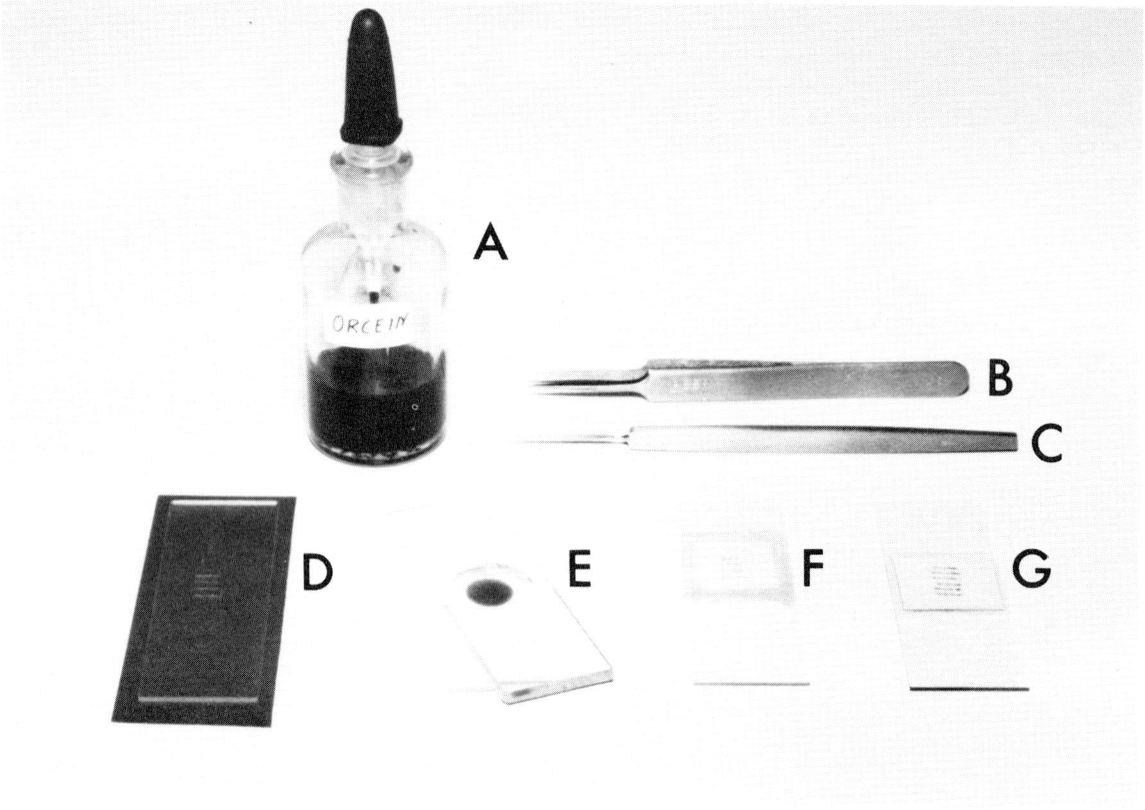

Fig. 1. Staining *Meloidogyne* chromosomes: A) glass dropping bottle containing orcein stain; B) fine forceps; C) eyeknife; D) microscope slide with smears of four females; E) microscope slide with the smears and a cavity slide over it covering the drop of stain to prevent evaporation; F) temporary mount—a slide with four stained smears and a coverslip sealed with paraffin-lanolin mixture; G) permanent mount—a slide with six stained smears embedded in euparal.

of stain particles will be minimized. Such diluted stain will be good for several months but may require periodic filtration.

2. SELECTION OF NEMATODE MATERIAL FOR CYTOLOGICAL STUDY

The best material for studying oogenesis of root-knot nematodes consists of live, relatively young, egg-producing females, developing in the roots of very favorable host plants that are lightly infected and grown under the most favorable environmental conditions. Greenhouse cultures maintained at 23 to 28°C on favorable host plants, properly watered and fertilized, are the best source of nematode material. Infected roots of plants brought into the laboratory from the field usually are not very satisfactory for cytological studies. Females should be at the peak of their reproductive period, i.e. produce a large number of eggs per day. Usually, young females with white egg masses, containing about 100 to 300 eggs, are in the proper developmental stage. Very young and very old females produce eggs at a low rate and have only a few oocytes in the proper stage for cytological study. Very young females, however, are often favorable for the study of oogonial divisions, and very old females can give valuable information on advanced stages of maturation of oocytes (stages beyond telophase I) and cleaving eggs.

A. Procedure for Obtaining Root-knot Nematode Females

Wash the roots of infected greenhouse plants shortly before you plan to use them for extraction of females. If for any reason you have to delay extraction of the females for a few hours or overnight, keep the roots wet in plastic bags at room temperature. Do not refrigerate them.

Extract females from the galled roots using forceps and a dissecting needle, or a surgical eyeknife, and a stereoscope at 10-15× magnification. Females can be extracted directly from the wet roots and transferred immediately into a BPI dish containing 0.9% NaCl or, preferably, KCl (Fig. 2E). Alternatively, small pieces of galled roots can be placed into the same solution in a Syracuse watch glass (Fig. 2D), and the females can be extracted directly into the solution. Never attempt to extract females in distilled water; they usually burst in a few minutes because of high turgor pressure. Females thus extracted should be used for cytological preparations immediately, or within an hour after extraction from the roots; otherwise the process of oogenesis is affected unfavorably and so is the quality of the preparations.

B. Procedure for Obtaining Root-knot Nematode Eggs

The young egg-producing females as described in the previous section, provide information about oogonial divisions and the early stages of maturation of the oocytes, usually up to metaphase I and, rarely, up to telophase I. The second maturation division, fusion of sperm and egg pronuclei and cleavage divisions cannot be studied from such nematode material. Usually, even before metaphase I, or as soon as a spermatozoon enters an oocyte, the egg membranes that will eventually form the egg shell become progressively less permeable to the fixatives and stains. For the study of such advanced stages of maturation, material consisting of advanced oocytes and young mature eggs are obtained as follows.

Select individual root galls containing egg-laying females of different ages from a good, lightly infected root system. Transfer a gall to a clean Syracuse watch glass containing 0.9% KCl solution. Using forceps and a needle, tease off the egg mass of each female on the gall. Subsequently, extract each female from the gall, and transfer it to a BPI dish containing KCl solution. Discard the root gall, and repeat this procedure with the next gall until you have collected 15 to 30 females in the BPI dish and an equal number of egg masses in the Syracuse watch glass. A small number of eggs will have separated from the egg masses in the Syracuse watch glass and be free in the KCl solution. These are the last of the eggs laid by the females, and they are free because they were not embedded firmly into the hardened gelatinous matrix. These eggs are usually in an advanced stage of maturation or in an early cleavage division and are the ones needed for investigation of these advanced stages. Agitate the egg masses slightly with a needle to make certain that all the unattached eggs are separated and are spread individually in the solution. Using a fine pipette, collect and transfer these free eggs into another BPI dish. The egg masses can be discarded. Each of the two BPI dishes, one containing the eggs and the other containing the females, will be processed separately (proceed with Section 5).

3. PREPARATION OF SMEARS FOR THE STUDY OF OOGONIAL DIVISIONS AND EARLY STAGES OF MATURATION OF OOCYTES

Using a fine pipette, draw four of the females extracted earlier from the galled roots (Section 2A), and transfer them with a small drop of water on a clean slide (rinsing with alcohol and wiping with tissue paper is sufficient). With very fine forceps, remove one female from the drop, and deposit it in the center of the same slide with as little water as possible. Wait for a few seconds until all the water that surrounds the female has evaporated. Then, holding the female with the forceps, make a cut at the neck region with a sharp, surgical eyeknife. Immediately draw the female body with the forceps along the slide applying slight pressure so that the body contents are smeared uniformly along a strip about 0.5 to 1 cm long. Slight tapping of the female body with the point of the eyeknife can help to make a more uniform smear. Discard the cuticle of the female. The smear dries in a few seconds and becomes clearly visible as a white strip. Smear the other three females in parallel strips, spaced about 2 mm from each other (Fig. 1D). Proceed then immediately with the hydrolysis of the material.

A. Hydrolysis

In practically every cytological procedure described for other animals and plants, hydrolysis in 1N HCl, usually carried out at 60°C, follows fixation of the material (2,3,5). When this procedure is employed for root-knot nematodes, it results in unsatisfactory preparations. The oocytes appear filled with large globules, and the cytoplasm absorbs much stain, thus obscuring observation of the chromosomes. However, I have found that slight hydrolysis, carried out at room temperature before fixation, improves the quality of the preparations considerably. This treatment af-

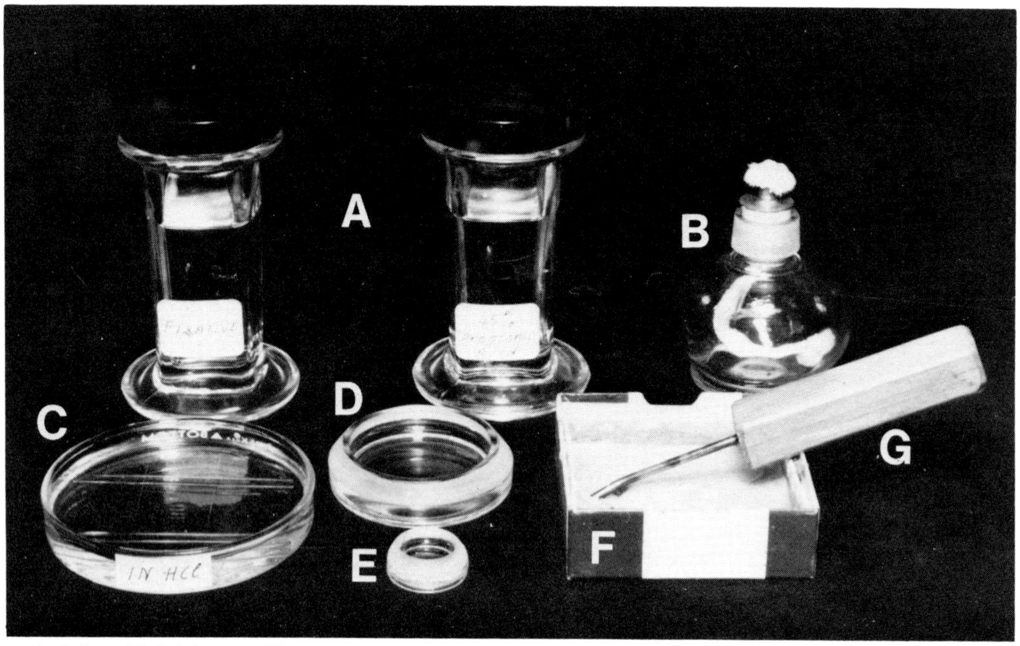

Fig. 2. Staining *Meloidogyne* Chromosomes: A) Coplin staining dishes with fixative and 45% propionic acid, respectively; B) glass alcohol burner; C) petri dish with 1N HCl. The inverted slide with the smears rests on two glass rods; D) Syracuse watch glass; E) BPI dish; F) cardboard box with sealing medium, paraffin-lanolin mixture; G) applicator used in sealing the coverslip of temporary slides with sealing medium.

fects in some way (change of the pH?) the cytoplasm of the oogonia and oocytes which remains optically homogeneous, without any lipid or protein globules, and quite clear, practically unstained.

Proceed as follows for the hydrolysis. Immerse the slide with the smears in a 1N HCl solution prepared by mixing about 10 ml of reagent HCl in 100 ml distilled water. This should be done immediately after you have made the smears (Section 3), i.e. before the nematode material has dried excessively.

Note: Some material may be lost from the surface of the slides during hydrolysis. To minimize the loss of material, invert the slide and then lower it into the dish containing the HCl solution (Fig. 2C). The nematode material will thus be pressed against the slide as it touches the surface of the HCl solution and will not be washed sideways. If, in spite of these precautions, much material is still lost in a later step, allow the smears to dry for a longer period, i.e. 3 to 5 minutes, before immersing the slide into the HCl. In fact, the smears can be prepared and kept for several days, and possibly months, in a dry environment before they are processed for staining, with only slight reduction in quality of the final preparations. Still, for best results hydrolysis should be done immediately, without drying the material excessively.

Leave the slide in the HCl solution for about 5 to 10 minutes at room temperature, then remove it and wipe dry with tissue paper, leaving wet only the smears and a small area surrounding them.

Note: You should keep track of the side of the slide that carries the smears to avoid wiping them out during this step. The smears at this stage are transparent and only faintly visible.

The slide is now ready for fixation and should be immersed in the fixative before the smears dry out completely. It is important that most of the water is removed from the slide before immersion into the fixative because the fixative itself contains no water and should be kept free from water contamination. The presence of water also will create turbulence when mixed with the fixative, and this may cause further loss of material.

B. Fixation

Immerse the slide in a Coplin staining dish (Fig. 2A) or other glass container filled with freshly prepared fixative consisting of three parts of absolute ethyl alcohol and one part glacial acetic acid. About 20 to 30 minutes of fixation are sufficient for oogonia and young oocytes at metaphase I, but 40 to 60 minutes may be required for oocytes at anaphase I, telophase I, or more advanced stages of maturation. During fixation, the nematode material turns white, and the smears are easily visible on the slide.

C. Staining (20-40 min)

Remove the slide from the fixative and wipe it dry with tissue paper leaving wet only the smears and a small area surrounding them. Place the slide on a perfectly level surface and wait a few seconds, until much of the fixative has evaporated from the area surrounding the smears. The smears themselves should be still wet. Apply one or two drops of orcein stain on the smears. Cover the stain with a deep-well (cavity) slide to prevent evaporation of the acid and precipitation of stain particles (Fig. 1E).

Note: For best results, place the slide on a piece of plate glass previously leveled perfectly on a working table. If the slide is not perfectly level, the stain will have the tendency to move away from the smears and be sucked between the two slides. The same will happen if too much stain has been applied on the smears because the drop of stain spreads over a larger area as soon as it is covered with the cavity slide.

D. Mounting

Remove the cavity slide, and then hold the slide with the smears vertically on a piece of tissue paper to drain the excess stain. The smears are still soaked with stain. To remove the rest of the stain, immerse the slide for 3 to 5 seconds in a glass staining dish containing 45% propionic acid (Fig. 2A).

Note: Some material may be lost during this step, especially if a lower concentration of propionic acid is used or if the washing period is extended.

Remove the slide from the 45% propionic acid and place it on absorbent paper on the working table. The upper side of the slide that carries the smears is still wet. Take with the forceps a No-1, 22-mm-square coverslip, dip it momentarily in the staining dish with the 45% propionic acid to wet it, and then apply it on the wet smears. Absorb the excess acid solution with tissue paper until the coverslip settles firmly over the smears and does not move sideways. Any side movement of the coverslip should be avoided because the soft and delicate oocytes are easily destroyed. Allow the slide to dry slowly for 5 to 10 minutes, and seal the mount as soon as some air starts moving under the edges of the coverslip.

A sealing medium can be prepared as follows. Add equal weights of paraffin and lanolin in a beaker, and place it in a hot water bath until the paraffin melts. Put the liquefied mixture into small flat cardboard containers, e.g. slide boxes, and allow to cool and solidify (Fig. 2F). A wire that is approximately 2 to 3 mm thick, 10 mm long, and attached to a wooden holder makes a suitable sealing applicator (Fig. 2G). Heat the wire end of the applicator over the flame from an alcohol burner (Fig. 2B), and then touch for a few seconds the paraffin-lanolin mixture to produce a small well of melted sealing material. Subsequently, draw the heated applicator along the edges of the coverslip, touching simultaneously the surface of the slide and the coverslip. The applicator may have to be reheated several times to ensure that the sealing medium is hot when deposited on the slide. If the medium is not sufficiently hot, the bonding with the glass surfaces will not be strong, and air may move under the coverslip within a few days. When application is performed properly, the mount may not dry out even after 6 months. The slide thus prepared is a temporary mount, ready for microscopic examination.

4. EXAMINATION OF TEMPORARY PREPARATIONS

Slides prepared in Sections 3 to 3D constitute temporary preparations that should be examined microscopically within 3 or 4 days. During this period the oogonia and oocytes, as well as the rest of the nematode material surrounding them, remain soft. If further spreading of the chromosomes is desired, gentle pressure can be applied locally on the coverslip with a needle under low (100×) magnification. Within 4 to 5 days, the material hardens. Pressure is no longer effective in spreading the chromosomes and can only cause cracking, thus damaging the preparation. Still, such older preparations can be in satisfactory condition for cytological observations for several weeks, and even 2 to 3 months, if the mount is sealed properly so that it will not dry out.

The smears can be surveyed through a 10× objective on a compound research microscope.

Advanced oocytes which have attained the shape of an egg, young oocytes which are still spherical or ovoid, the tips of the ovaries which usually remain intact, the oviduct-spermatotheca region of each gonad, and possibly other tissues of the nematode body can be observed (see Vol. I, Chapter 9, Fig. 2).

The most clearly visible eggs in the entire preparation are those that are filled with globules and often stained reddish. These eggs represent the advanced oocytes that have already developed beyond anaphase I and have a thick egg shell that is not easily permeable to the fixative and the stain. They are not useful for cytological observations. *Note:* To study such advanced oocytes, one needs to follow a special cytological procedure (see Section 5).

Some eggs are transparent or semitransparent, slightly tinted or completely unstained. These are young oocytes at prometaphase or metaphase I and are the most useful for cytological observations. They usually are located close to the spermatothecae, although in many preparations, they may be found in any part of the smear. Observe and study such eggs at a high magnification, preferably with a 90 or 100× objective and 10 to 15× eyepieces. To increase the contrast of the reddish-purple-stained chromosomes, use a green filter, such as Kodak #58 (546 nm). With the green filter, the chromosomes are seen as black, whereas the surrounding cytoplasm or nucleoplasm is greenish and transparent. You may use a camera lucida or a drawing tube to make sketches of the chromosomal figures. Similarly you can photograph interesting figures using a medium or high-contrast, black-and-white film.

For the study of oogonial divisions, observe in high magnification the tips of the ovaries. At the apex there is always an apical cell (cap cell) with a large nucleus, which never divides. Further posteriorly, there are many spherical cells, the oogonia, most of which are in a resting stage, but some of which may be dividing. Dividing oogonia can be examined at late prophase or at metaphase when the chromosomes are usually well stained and discrete.

5. PREPARATION OF SLIDES FOR THE STUDY OF ADVANCED STAGES OF MATURATION, CLEAVAGE DIVISIONS, AND THE PROCESS OF FERTILIZATION

For this step, use the nematode material obtained according to the instructions given in section 2B.

Proceed first with the BPI dish that contains the eggs. Add 2 to 3 drops of a mixture of 0.1N KOH in 0.4% NaOCl. Observe with a stereoscope the egg shells becoming thinner as they dissolve through the action of the bleaching solution. Some eggs burst and their contents dissolve and disappear. Allow the reaction to proceed for 6 minutes, and then using a pipette, fill the BPI dish with distilled water to stop the reaction. Within 2 to 3 minutes, the eggs should settle to the bottom of the dish.

Note: If an insufficient amount of water is added to the NaOCl solution, the eggs will remain in suspension, and the excess water cannot be removed without loss of many eggs.

Using a fine pipette, remove most of the bleaching solution leaving the eggs undisturbed in the bottom of the dish. Refill the dish with distilled water. Allow a few minutes for the eggs to settle to the bottom, and remove the water again. Repeat the water change another time, but at this time fill the dish with 0.9 KCl rather than distilled water to avoid bursting the eggs which have very thin membranes. The total washing period should be at least 10 minutes. Finally, concentrate the eggs by again removing the water, and add 2 or 3 drops of a 1% gelatin solution.

Note: Prepare fresh gelatin solution the same day you need it. In a small glass vial, add 5 to 10 ml of distilled water and the appropriate amount of gelatin. Heat over a small flame until the gelatin dissolves. Allow to cool. The solution now should be thin and very fluid.

Mix the eggs with the gelatin solution and then transfer a drop of egg suspension to the center of a clean microscope slide. Spread the drop over a slightly larger area (about 1 cm square), and allow the gelatin solution to dry completely but not excessively (10 to 15 minutes). The slide can now be processed for staining in the same manner as the smears of females, starting with hydrolysis, fixation, etc. (Section 3A-D).

Note: The success of this method depends largely on the degree of dissolution of the egg membranes during NaOCl treatment. Prolonged treatment will destroy all the very young eggs, whereas too short a treatment will not increase sufficiently the permeability of the shells of the old eggs. For best results, the time of NaOCl treatment should be varied and also the time of fixation extended to 1 or 2 hours.

Treat the females, contained in the other BPI dish (Section 2B), in the same manner as the eggs. However, puncture the females with a needle or an eyeknife just before the 2-3 drops of the KOH-NaOCl mixture are added. In about 1 minute, remove and discard the cuticles of the females, using stainless steel forceps. Observe in the stereoscope the egg suspension, and stop the reaction as soon as you see some of the young eggs with thin membranes burst and dissolve. Usually, a treatment period of 3 to 4 minutes is sufficient. By the end of this treatment, all the components of the nematode body will have been dissolved completely, and only the eggs at telophase I, or more advanced stages, will still be present. Wash and transfer the eggs into the gelatin solution in the same manner as described for the eggs obtained from the egg masses.

6. PREPARATION OF PERMANENT MOUNTS ACCORDING TO THE QUICK-FREEZE METHOD

Temporary mounts containing smears of females (Section 3D) or NaOCl-treated eggs (Section 5) can be made permanent by the method described below. However, some of the quality of the temporary preparations is lost in permanent mounts. Similarly, some valuable material on the slides, especially oocytes which have been previously squashed in order for their chromosomes to be spread, may be lost from the slides during the process. For this reason, the cytological study, including preparation of camera-lucida drawings and photographs, should be completed before a mount is made permanent. Temporary mounts should be made permanent as soon as possible, preferably within a few weeks or a month.

The procedure for making a slide permanent is as follows. Place the slide on a flat surface of a piece of dry ice (solidified CO_2). With the eraser ends of two lead pencils, or any other means, press the slide firmly against the dry ice to ensure quick and complete contact. The slide freezes instantaneously. Leave the slide on the dry ice for 3 to 10 minutes, and then using a scalpel, remove the sealing medium (1:1 paraffin-lanolin, Section 3D) from around the coverslip. Insert a razor blade under one corner of the coverslip, and flip it free. Transfer the slide and the coverslip successively through the following alcohol baths (petri dishes can be used as containers):

a) 95% ethyl alcohol for 2 minutes,
b) 100% ethyl alcohol for 1 minute,
c) 100% ethyl alcohol for 1 minute.

Place the slide on absorbent paper on the working table. Do not wipe the slide. Add one drop of euparal on the smears or the eggs and cover with the same coverslip. The alcohol will evaporate in a few minutes. Allow about 1 hr until the euparal hardens slightly around the edges of the coverslip and then apply a small weight (e.g., a small glass vial filled with sand) to press the coverslip against the smears. The euparal will solidify slowly, and you can safely remove the weight and examine the slide microscopically within about 2 or 3 days.

7. SUMMARY OF PROCEDURES FOR STAINING *MELOIDOGYNE* CHROMOSOMES WITH PROPIONIC ORCEIN

A. Have available the following supplies and small equipment:
   1. Fine forceps (Fig. 1B),
   2. Eyeknife (Fig. 1C),
   3. Two Coplin staining dishes or equivalent glass dishes (Fig. 2A),
   4. Syracuse watch glasses (Fig. 2D),
   5. BPI dishes (small Syracuse watch glasses - Fig. 2E),
   6. Microscope slides and #1 coverslips,
   7. Deep cavity slides,
   8. Pipettes (drawn to a fine point).

B. Prepare the following solutions:
   1. Propionic-orcein stain (see text, Section 1)
   2. 1N HCl (10 ml of reagent HCl + 100 ml distilled water),
   3. 0.9% KCl or NaCl,
   4. Fixative (three parts of 100% ethyl alcohol and one part glacial acetic acid),
   5. 45% propionic acid.

C. Procedure:
   1. Smear female nematodes (see text, Section 3) or spread eggs (Section 5) on slides (Fig. 1D),
   2. Hydrolysis (Section 3A - Fig. 2C),
   3. Fixation (Section 3B - Fig. 2A),
   4. Staining (Section 3C - Fig. 1E),
   5. Mounting (Section 3D - Fig. 1F, 2F-G).

Literature Cited

1. Dalmasso, A. and T. Younes. 1970. Etude de la gametogénèse chez *Xiphinema mediterraneum*. Nematologica 16: 51-54.
2. Darlington, C. D. and L. F. La Cour. 1975. The handling of chromosomes, 6th ed. John Wiley & Sons, New York. 201 pp.
3. McGregor, H. and J. Varley. 1983. Working with animal chromosomes. John Wiley & Sons, New York. 250 pp.
4. Nigon, V. 1949. Modalités de la reproduction et le déterminisme du sexe chez quelques nématodes libres. Fac. Sci., Univ. Paris. 132 pp.
5. Sharma, A. K. and A. Sharma. 1980. Chromosome techniques, theory and practice, 3rd ed. Butterworths & Co., London. 711 pp.
6. Triantaphyllou, A. C. 1979. Cytogenetics of root-knot nematodes, pp. 85-108 *In* Root-knot nematodes (*Meloidogyne* species) systematics, biology and control, ed. F. Lamberti and C. E. Taylor. Academic Press, London. 477 pp.

CHAPTER 8

ELECTROPHORETIC METHODS FOR THE STUDY OF ROOT-KNOT NEMATODE ENZYMES

P. R. Esbenshade and A. C. Triantaphyllou

Departments of Plant Pathology and Genetics, respectively
North Carolina State University
Raleigh, North Carolina, USA

The use of electrophoretic techniques in the study of soluble proteins and specific enzymes in many organisms has become very popular since the early 1960s. The theory and application of disc electrophoresis with polyacrylamide gels was first described in detail by Ornstein (13) and Davis (2) in 1964. The first biochemical taxonomic research on root-knot nematodes, conducted by Dickson et al. (3) in 1971 and Hussey et al. (7) in 1971, employed disc electrophoresis for the comparison of proteins and enzymes of various *Meloidogyne* species.

In 1978, Dalmasso and Berge (1) adapted a miniature extraction and separation procedure (5,8,11,14) for the enzymatic analysis of extracts obtained from a small number, or even single females, of *Meloidogyne*. This microdisc system utilized separate gels for each sample. It was further improved by the use of thin gel slabs in which many samples can be processed simultaneously (9).

In this chapter, the thin-slab polyacrylamide gel procedure that we have used in the investigation of various enzymes of root-knot nematodes is described in some detail (4). Except for a few slight modifications, this procedure is the same one employed earlier in the collaborative work between our laboratory and the INRA, Station de Recherches sur les Nematodes, Antibes, France (9).

This method applies specifically to the study of adult females of *Meloidogyne*. Certain modifications of the extraction procedure will be required before it can be applied to other developmental stages of *Meloidogyne* or to other types of nematodes.

SELECTION OF NEMATODE MATERIAL

Young, egg-laying females developing on a favorable host provide the best material for the study of most enzymes. Such material can easily be obtained from greenhouse cultures or from naturally infected host plants brought from the field.

Care must be given to the preparation of nematode material. The infected roots should be gently but thoroughly washed free of soil and accompanying organic matter. Females with medium-sized, white egg masses can be teased out of the roots with forceps, a dissecting needle or an eyeknife, and the aid of a stereoscope at 10X magnification. The nematodes are then transferred to chilled "extraction medium" (see next Section) in a solid watch glass or a cavity plate (Fig. 1A) resting on a bed of ice.

EXTRACTION MEDIA

The extraction medium in which the females will be macerated can have a great effect on the solubility and activity of a particular enzyme. For this reason, different formulations are used in the study of different enzymes. Extraction media used for the enzymes included in this chapter are listed in Table 1. Media prepared with sucrose as the densifier should be used within one week after preparation because of the risk of fungal contamination. Media containing glycerol have longer storage life and remain unfrozen at −10°C if the glycerol content is increased from the recommended 20%

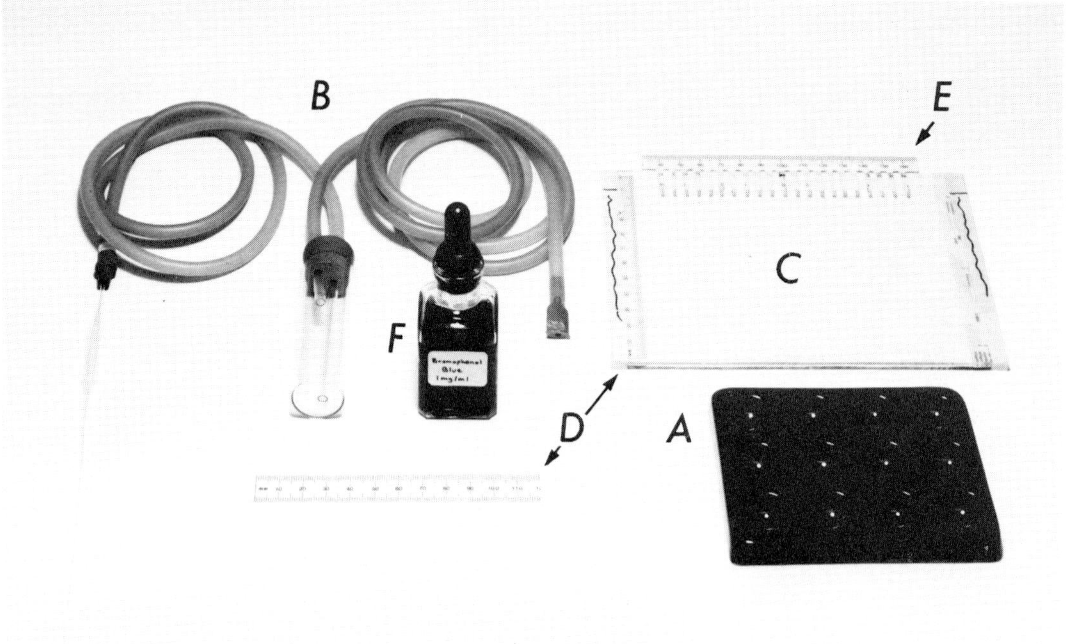

Fig. 1. **Gel cassette and associated equipment. A) Cavity plate to hold nematode females prior to maceration; B) Mouth-operated pipette apparatus, mouthpiece (right), trap (center), and transfer pipette (left); C) Cassette in which the polyacrylamide gel will be cast; D) Spacer; E) Sample comb with the teeth inserted between the glass plates of a cassette; F) Dropping bottle with bromophenol blue dye solution.**

Table 1. Extraction media[a].

Sucrose-Triton (Cat, Est, Got, Gpd, Sod):
20% (w/v) sucrose or glycerol
2% (w/v) Triton X-100
distilled water

TRIS/HCl [Trudgill + Carpenter (16)] (Mdh/Sod):
20% (w/v) sucrose or glycerol
0.1% TRIS
0.1% ascorbic acid
0.1% cysteine/HCl
distilled water
adjust to pH 8.0 with HCl

[a]Solutions are filtered after preparation and stored in the refrigerator between use.

(Table 1) to approximately 35%. The higher concentration reduces the denaturing effect of freezing and thawing on proteins and better preserves enzymatic activity.

EXTRACTION TUBES

Nematodes are macerated in small tubes with the help of a glass pestle. The tubes are constructed from commercially available micro-hematocrit capillary tubes, 75 mm long x 1.2 mm ID. These tubes are thoroughly washed with several changes of acetone for removal of coatings (heparin and dye) applied for hematology use and air dried overnight (Fig. 2A). Each tube is broken in the middle after being knicked with a file (Fig. 2B). Each half-tube is then heat-sealed at one end when the rough end is melted over a bunsen flame (Fig. 2C). Only the tip of the tube should be placed into the flame, while the tube is rotated between the thumb and finger. Some practice will be required in the production of usable tubes. The tubes should have a flat bottom and perfectly parallel sides that allow the pestle to be inserted and rotate freely against the bottom of the tube. If the walls of the extraction tube are tapered, the pestle becomes wedged against the walls before it reaches the bottom and will not be effective in macerating the nematodes.

PESTLES

The pestles to be used for macerating the nematodes are prepared from glass rods (soda-lime glass). The center of a 3- to 4-mm-thick and about 120-mm-long glass rod is heated over a bunsen flame until it becomes soft. The rod is then removed from the flame, and the ends are pulled in opposite directions so the melted portion is drawn out in the center to approxi-

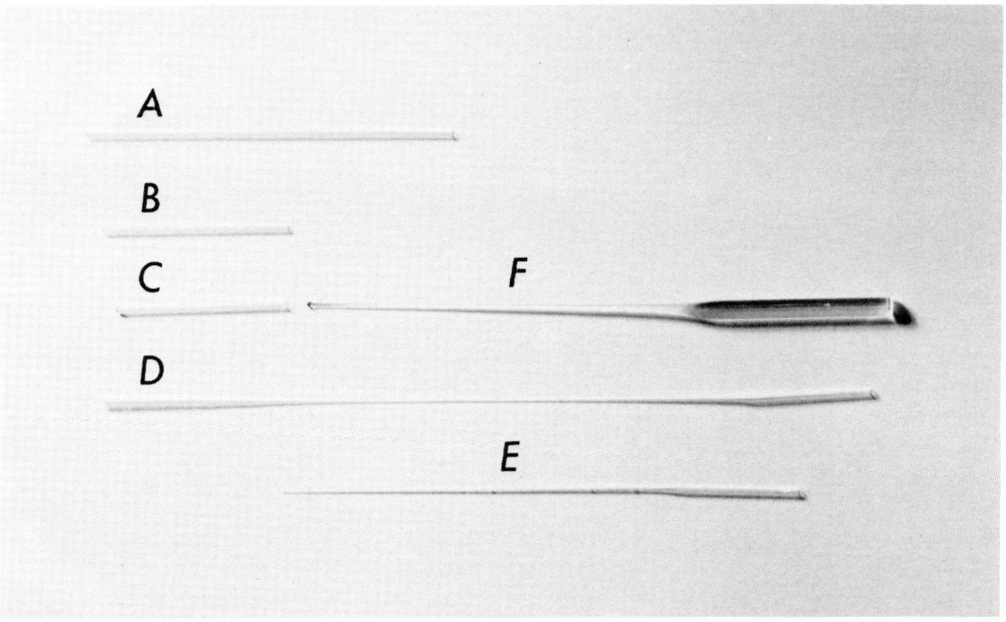

Fig. 2. Miniature glassware used for maceration of nematode females and for transfer of the aqueous nematode samples. A) Microhematocrit tube after acetone wash; B) and C) Microhematocrit tube broken into two halves, the lower half having been heat sealed at one end to form an extraction tube; D) Microhematocrit tube that has been heated and drawn out to form a sample pipette; E) Sample pipette; F) Pestle used for maceration of nematode females inside the extraction tube (C).

mately 0.5 to 0.7 mm in thickness. They are held in this position until the glass in the center cools off and becomes rigid. The rod is broken in the center at the thinnest region, thus producing two pestles. The thin end of each pestle is reinserted into the flame and rotated until it melts to produce a small ball, slightly larger than the inside diameter of the extraction tube (about 1.3 mm in diameter). This ball is then carefully sanded with fine emery cloth on all sides and the end so the pestle will fit into the extraction tube with close allowance. The final steps of sanding and checking of the fit of the ball into the bottom of the tube is done with the help of a stereoscope at 10× magnification. The sanding also produces abrasive surfaces required to macerate the nematodes. The pestle should be able to reach the bottom of the extraction tube and rotate freely making good contact with the bottom of the tube (Fig. 2F).

PIPETTES

Two types of small pipettes are needed. The first, a transfer pipette is used to transfer female nematodes into the extraction tubes. It is a pasteur pipette (soda-lime glass) that has been heated and drawn out to a thickness slightly smaller than the internal diameter of the extraction tube (Fig. 1B). The second type, a sample pipette, is used to transfer the aqueous fraction of the nematode extract (Fig. 2E). This type is a microhematocrit tube that has been heated and drawn out to a very small size that will fit between the glass plates of the cassette and into the sample well of the stacking gel (see the Section **Electrophoresis and Staining of the Gels**). Both types of pipettes can be controlled by means of a mouth-operated apparatus consisting of a series of thin rubber hoses and a trap (Fig. 1B) (17). The mouth-operated apparatus enables the operator to transfer the nematodes and the extracted samples quickly and with exceptional precision. The pipette can be thoroughly washed between samples if the end is placed into a beaker containing distilled water and filled and emptied four to five times.

EXTRACTION PROCEDURE

1. Female nematodes are harvested and placed in chilled extraction medium.
2. One to four nematodes with a minimal amount of extraction medium are transferred into an extraction tube by means of a transfer pipette.
3. The females are macerated by rotation of the pestle against the bottom and sides of the extraction tube.

4. Additional extraction medium is added to increase the total volume of the crude sample to about 10 μl.

5. The crude sample is frozen at $-10°C$ until electrophoresis. The samples are stored in labelled bottles or vials with air-tight lids for long periods (3-4 months) without noticeable loss of activity of most enzymes.

SAMPLE COMB

Wells are created in the stacking gel into which the nematode extract will be loaded by means of the sample comb. Although sample combs of different types are available commercially, they can preferably be constructed in the laboratory from the same plastic material that is used for the spacers separating the glass plates of the cassette (Fig. 1C). Alternatively, plastic rulers, about 0.6 or 0.7 mm thick, usually available at bookstores, can be employed as material to construct the spacers (Fig. 1D) and comb (Fig. 1E). Combs are designed with about 10 to 25 teeth, depending upon the size of the cassette. Teeth [1.5 (bottom) x 20 x 3.0 mm (top)] should be spaced at least 3 mm apart so that fusion of the bands from neighboring samples is avoided. The material between the teeth of the comb is removed with a sharp knife, and then the teeth are rounded off with the edge of the knife.

ELECTROPHORESIS APPARATUS

There are many thin-slab (0.6- to 1.5 mm-thick gels) electrophoresis apparatus available commercially that can be used in electrophoretic work with nematodes (Fig. 3, Table 2). Each apparatus requires a slightly different technique by which the polyacrylamide gel is cast. The principle, however, is the same. The acrylamide gel is cast in a cassette (Fig. 1C) made up of two glass plates separated by two spacers (Fig. 1D) of certain thickness. The cassette is held together by various devices (clamps, tape, etc.) according to the instructions provided by the manufacturer of the electrophoretic apparatus. See Maurer (10) for a general description of polyacrylamide gel methods.

CASSETTES

The glass plates used for construction of cassettes are washed carefully with soap and water, rinsed in distilled water and dried after each use. Immediately before construction of the cassettes, the glass plates are washed with 95% ethanol and air dried at room temperature.

The cassettes are constructed according to the manufacturer's procedures, and then rinsed in a 1% solution of Kodak Photo Flow 200. The Photo Flow coating prevents the gel from adhering to the glass and facilitates its removal from the cassette later on. The cassettes are placed vertically to drain out the excess Photo Flow, allowed to air dry for about 1 hr, and then are placed into an oven to dry at 70°C for 3 hr.

Table 2. Laboratory equipment list[a].

Refrigerator/Freezer
Microhematocrit tubes (Fisher # 02-668-68)
Microhematocrit centrifuge (Fisher # 05-040)
Power Supply 0-300 volts DC (Bio-Rad # 165-4700)
(Fisher # 09-528-111)
pH Meter
Glass rods - 4 mm flint glass
Pasteur pipettes - flint glass (Fisher # 13-678-6A)
Electrophoresis apparatus (Bio-Rad # 165-1422)
(Pharmacia GE-2/4)
(Bethesda Research Labs # 1070 HE, # 1072 HE)
Bunsen burner
Rubber gloves
Face mask - particulate (Fisher # 13-995-15)
Glass dishes - 2 x 15 x 20 cm
Vacuum pump (Fisher # 01-182)
Vacuum dessicator or bell jar
Incubator - temperature controlled
Magnetic stir plate and stir bars
Cavity plate (Fisher # 13-748B)

[a]Manufacturers and products mentioned here are only a small portion of the potential suppliers of equipment and chemicals needed for electrophoresis. The mention of a particular product is for descriptive purposes only and does not imply that other similar products are not suitable.

CASTING POLYACRYLAMIDE GELS

The stock solutions used in the preparation of the separation and stacking acrylamide gels are given in Table 3. The gel systems described here (Tables 3, 4) have been used successfully in our laboratory, but there are many other systems that may also give satisfactory results (12,18).

The separation gel is prepared first. The chemical stock solutions (Table 3) are mixed in the proper ratio, according to Table 4, but without the ammonium persulfate solution (Table 3, G). The mixture is placed in a vacuum

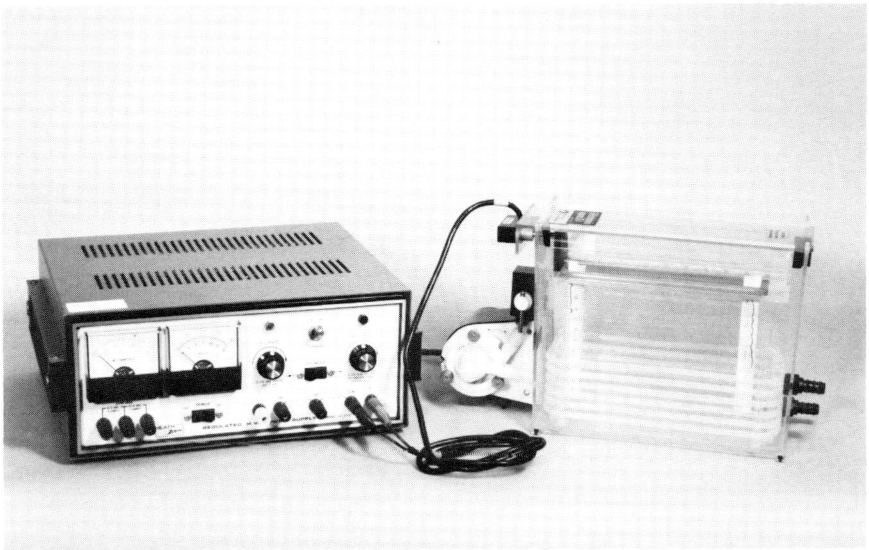

Fig. 3. **Assembled electrophoresis apparatus (right) and power supply (left).**

Table 3. Stock solutions for acrylamide gel preparation [Davis (2)][a].

Solution A 1 N HCl, ~35 ml TRIS, 36.6 g TEMED, 0.23 ml Adjust to pH 8.9 with 1 N HCl Water to 100 ml	**Solution A2 - esterase gels** 1 N HCl, ~20 ml TRIS, 18.3 g TEMED, 0.23 ml Adjust to pH 8.6 with 1 N HCl Water to 100 ml
Solution B 1 N HCl, ~48 ml TRIS, 5.98 g TEMED, 0.46 ml Adjust to pH 6.7 with 1 N HCl Water to 100 ml	**Solution C** Acrylamide, 28.0 g BIS, 0.735 g Water to 100 ml
Solution D Acrylamide, 10.0 g BIS, 2.5 g Water to 100 ml	**Solution E** Riboflavin, 4 mg Water to 100 ml
Solution F Sucrose, 40.0 g Water to 100 ml	**Solution G** Ammonium persulfate, 0.14 mg Water to 100 ml
Bridge-buffer-stock solution TRIS, 6.0 g Glycine, 28.8 g Adjust to pH 8.3 Water to 1 liter Dilute 1 part of stock solution with 9 parts of water before use.	**Bromophenol-blue dye** Bromophenol blue, 1 mg Bridge buffer, 1 ml

[a]All stock solutions are filtered after preparation and stored in the refrigerator.

Table 4. Acrylamide gel preparation [Davis (2)][a].

Separation Gel - small pore gel	
1 part stock solution A or A2 (esterase)	
2 parts stock solution C	
1 part water	
4 parts stock solution G	
pH 8.9 (8.8 - 9.0)	
Stacking Gel - large pore gel	
1 part stock solution B	
2 parts stock solution D	
1 part stock solution E	
4 parts stock solution F	
pH 6.7 (6.6 - 6.8)	

[a]Stock solutions are warmed to room temperature prior to gel preparation.

(5 micrometers Hg) for 10 to 15 minutes for removal of much of the air (oxygen will inhibit the chemical polymerization of the acrylamide gel). The proper amount of ammonium persulfate solution is measured in a separate beaker. After the acrylamide gel solution is removed from the vacuum, the ammonium persulfate solution is carefully poured down the side of the beaker containing the acrylamide and slowly mixed. When completely mixed, the solution is poured into the cassette following the specific procedure recommended by the manufacturer. The cassette is filled to a level 15 mm from the top. This amount leaves sufficient space for the stacking gel to be applied later. The cassette is secured on a stand in a vertical position. A thin layer of water is then applied evenly on the top of the separation gel, by means of the mouth-operated sample pipette. This layer of water flattens the top surface of the separation gel and prevents oxygen in the air from inhibiting polymerization. Even distribution of the water, which produces a flat surface on the top of the separation gel, is very essential and requires some practice (2,11). The cassette should be left undisturbed approximately 30 minutes for polymerization of the gel. Routinely, we prepare the separation gels one day prior to their use.

After gel polymerization, the water layer on top of the separation gel is removed with a sample pipette, until no water is visible on the surface. The stacking gel can then be poured on top. The stacking gel is prepared from the stock solutions in Table 3, as described in Table 4. While the stacking gel solution is mixing, the comb is inserted between the glass plates at the top of the cassette. The comb is positioned so that the teeth are about 5 mm from the surface of the separation gel. The stacking gel solution is added by means of a pipette and flows around the teeth of the comb, filling the entire space to the top of the cassette. The stacking gel is polymerized by a reaction activated by ultra-violet light (2). A fluorescent lamp without a diffusing cover is placed very close to the cassette containing the unpolymerized stacking gel. After 15 minutes, additional stacking gel solution may be added to replace any that has leaked during polymerization and to bring the gel back to the top of the cassette. Polymerization may take up to 30 minutes.

The wells can be numbered on the outside of the glass plate of the cassette with an indelible marking pen. After polymerization, the comb is carefully removed by being slowly pulled upward. The empty wells left in the stacking gel are cleaned by means of a sample pipette which removes any loose pieces of gel. The cassette is placed in the electrophoresis apparatus, and the upper and lower tanks are filled with bridge buffer (Table 3). A sample pipette is used to remove any air trapped in the sample wells so that the wells become filled with bridge buffer.

ELECTROPHORESIS AND STAINING OF THE GELS

Sample tubes containing the nematode extracts (see Section **Extraction Procedure**) are removed from the freezer and are orderly arranged in a numbered rack. The numerical sequence of the samples is maintained through the entire electrophoretic procedure, so that each sample can be properly identified later on the stained gel. The sample tubes are placed into the microhematocrit centrifuge in numerical order and spun at 13,000 *g* for 15 minutes. Centrifugation and electrophoresis are done at low (~5°C) temperature. The samples are removed from the centrifuge and placed back into the sample rack in the same numerical sequence. As a result of the centrifugation, the nematode extract is separated into three fractions: the insoluble pellet at the bottom of the tube, the clear aqueous phase in the middle, and the lipid-rich layer at the top. The tube is knicked with a file just below the lipid layer and broken so that the lipid fraction can be discarded. The clear aqueous phase is drawn into a sample pipette and transferred carefully into a sample well in the stacking gel.

The sample is viscous and more dense than the bridge buffer, because it contains sucrose or glycerol, and will remain in the well without mixing with the buffer. After all the samples are loaded in the sample wells of the stacking gel, a drop of bromophenol-blue dye solution (Table 3) is added to the bridge buffer in the upper tank, and the electric current is turned on. The electric potential is set at 80 volts for the first 30 minutes and allows stratification of the proteins in the sample according to their relative charge (13). After 30 minutes it is increased to 200 (or 300) volts and kept at this level for the remainder of the separation period. Approximately 15 minutes after the electricity is turned on, a definite band of bromophenol dye should be visible in the gel.

Approximately 30 minutes before the bromophenol dye band is expected to reach a point marking 100 mm of migration, preparation of the enzyme-stain solution (Table 5) is initiated. Generally, the buffer, substrates, chelators, and coenzymes are mixed together first on a stir plate. The dye and other photosensitive chemicals are added to the buffer solution just prior to use. It is a good practice to cover the stain solution with an opaque cover to prevent light from affecting the chemicals. When the band of bromophenol-blue dye has migrated approximately 100 mm, the power supply is turned off, and the cassette is removed from the electrophoretic apparatus. The clamps or tape holding the cassette together are removed, and the glass plates are pried apart. The adherence of the gel to one of the plates allows the gel to be handled easily while cutting off the unused portion. The bottom of the gel is cut along the band of bromophenol dye, thus marking the bottom of the gel. Then the lower corner of the gel that corresponds to the position of the last sample loaded in the gel is cut off diagonally. The physical marking of the gel at this point is essential for later identification of the samples, especially after the gel is stained and stored. After the gel is properly marked, it is washed off the glass plate into a glass dish using a stream of water from a wash bottle. The plate with the gel is held just above and with some slope toward the dish, so that the gel will slide easily into the dish, without tearing. The water is poured out of the dish, and the staining solution is poured over to cover the gel thoroughly. The dish is covered with a plastic sheet and placed into the incubator in the dark at 37°C. The staining solution should be agitated every 10 minutes by slow movement of the dish from side to side, to ensure even staining.

The dish is removed from the incubator, and the stain solution is poured off after about 30 to 60 minutes of incubation, i.e. when the enzyme bands are sufficiently stained. Tap water is added to wash the gel from any remaining staining solution, and then the gel is stored in 7% acetic acid solution, in an air-tight container. The enzyme phenotypes of the nematodes under study are now recorded and can be compared to those of the controls in the same gel.

Table 5. Staining solutions for enzyme detection.

Catalase (Cat) [Harris + Hopkinson (6)]:

Place gel in dish resting on a bed of ice, chill 10 minutes.

Cover the gel with a 0.01% solution of hydrogen peroxide, for 20 minutes.

Rinse gel with two changes of distilled water.

Cover the gel with stain solution, agitate slowly for 1-2 minutes.

Rinse gel with water, store gel in 7% acetic acid.

Regions of Cat activity appear as clear areas on a blue-green background.

Catalase stain solution

Potassium ferricyanide, 0.5 g
Ferric chloride, 0.5 g
Water, 100 ml

Esterase (Est) [Harris + Hopkinson (6)]:

Cover gel with staining solution, and incubate at 37°C for approximately 1 hr.

Regions of Est activity appear as dark areas on a clear background.

Esterase stain solution

Potassium phosphate buffer 01.M pH 7.2, 100 ml
EDTA, 30 mg
Fast Blue RR salt, 60 mg
α-Napthylacetate 40 mg dissolved in 2 ml acetone add dropwise to the buffer while stirring. Vacuum filter to remove insoluble material.

Glutamate-oxaloacetate transaminase (Got) [Harris + Hopkinson (6)]:

Cover gel with stain solution, and incubate at 37°C for approximately 1 hr.

Got bands appear as blue regions on an orange background.

continued on next page

Table 5 (continued)

Got stain solution
α-Ketoglutaric acid, 37 mg
L-Aspartic acid, 133 mg
PVP-40, 0.5 g
EDTA, 50 mg
Sodium phosphate (dibasic), 1.42 g
Fast Blue BB salt, 100 mg
Distilled water, 100 ml

Glycerol-3-phosphate dehydrogenase (Gpd) α-glycerophosphate dehydrogenase [Harris + Hopkinson (6)]:

Cover gel with staining solution, and incubate at 37°C for approximately 1 hr.
Regions of Gpd activity appear as dark bands on a clear background.

Gpd stain solution
TRIS, 0.29 g
Water, 80 ml
Adjust to pH 8.0 with HCl
DL-Glycerol-3-phosphate, 1.3 g
Sodium pyruvate, 400 mg
NAD, 40 mg
MTT, 10 mg
PMS, 10 mg in 10 ml water

Malate dehydrogense/superoxide dismutase (Mdh/Sod) [Harris + Hopkinson (6)]:

Cover gel with stain solution, and incubate at 37°C for approximately 1 hr.
Mdh bands will appear as dark bands in the gel.
Sod bands will appear as clear bands in the gel. The Sod bands will fade in a few hours, so results must be recorded as soon as possible.

Mdh/Sod stock solutions
(Stock solutions can be stored at 5°C)

Stock solution A
Sodium carbonate, 10.6 g
L-Malic acid, 1.34 g
Water up to 100 ml

Stock solution B
TRIS, 6.06 g
Adjust to pH 7.1 with 2N HCl
Water up to 100 ml

Mdh/Sod Stain solution
Stock solution A, 10 ml
Stock solution B, 15 ml
NAD, 50 mg
NBT or MTT, 30 mg
Water, 50 ml
PMS, 2 mg in 26 ml water

Superoxide dismutase (Sod) [Ravindranath + Fridovich (15)]:

Cover gel with stain solution, incubate at 37°C for 20 minutes, in the dark.
Remove from the incubator, and place under light. Bands of Sod activity will resolve as clear areas in the gel.

Sod stain solution
TRIS, 0.61 g
Water, 100 ml
Adjust to pH 8.2 with HCl
Sodium EDTA, 7.5 mg
Riboflavin, 4 mg
NBT, 10 mg

Literature Cited

1. Dalmasso, A. and J. B. Bergé. 1978. Molecular polymorphism and phylogenetic relationships in some *Meloidogyne* spp.: application to the taxonomy of *Meloidogyne*. J. Nematol. 10:323-332.
2. Davis, B. J. 1964. Disc electrophoresis. II. Method and application to human serum proteins. Ann. N. Y. Acad. Sci. 121:404-427.
3. Dickson, D. W., D. Huisingh, and J. N. Sasser. 1971. Dehydrogenases, acid and alkaline phosphatases and esterases for chemotaxonomy of selected *Meloidogyne, Ditylenchus, Heterodera,* and *Aphelenchus* spp. J. Nematol. 1:1-16.
4. Esbenshade, P. R. and A. C. Triantaphyllou. 1985. Use of enzyme phenotypes for identification of *Meloidogyne* species. J. Nematol. 17:6-20.
5. Grossbach, U. 1965. Acrylamide gel electrophoresis in capillary columns. Biochim. Biophys. Acta 107:180-182.
6. Harris, H. and D. A. Hopkinson. 1976. Handbook of enzyme electrophoresis in human genetics. North-Holland Publ. Co., New York/Oxford. 356 pp.
7. Hussey, R. S., J. N. Sasser, and D. Huisingh. 1972. Disc-electrophoretic studies of soluble proteins and enzymes of *Meloidogyne incognita* and *M. arenaria*. J. Nematol. 4:183-189.
8. Hyden, H., K. Bjurstam, and B. McEwen.

1966. Protein separation at the cellular level by micro disc electrophoresis. Anal. Biochem. 17:1-15.

9. Janati, A., J. B. Berge, A. C. Triantaphyllou, and A. Dalmasso. 1982. Nouvelles donnees sur l'utilisation des isoesterases pour l'identification des *Meloidogyne*. Rev. Nematol. 5:147-154.

10. Maurer, H. R. 1971. Disc electrophoresis and related techniques of polyacrylamide gel electrophoresis. Walter de Gruyter, Berlin/New York. 222 pp.

11. Neuhoff, V. 1973. Micromethods in molecular biology. Springer-Verlag, New York/Berlin. 428 pp.

12. Ogita, Z. and C. L. Markert. 1979. A miniaturized system for electrophoresis on polyacrylamide gels. Anal. Biochem. 99:233-241.

13. Ornstein, L. 1964. Disc electrophoresis. I. Background and theory. Ann. N.Y. Acad. Sci. 121:321-349.

14. Pun, J. Y. and L. Lombrozo. 1964. Microelectrophoresis of brain and pineal proteins in polyacrylamide gel. Anal. Biochem. 9:9-20.

15. Ravindranath, S. D. and I. Fridovich. 1975. Isolation and characterization of a manganese-containing super-oxide dismutase from yeast. J. Biol. Chem. 250:6107.

16. Trudgill, D. L. and J. M. Carpenter. 1971. Disc electrophoresis of proteins from *Heterodera* species and pathotypes of *Heterodera rostochiensis*. Ann. Appl. Biol. 69:35-41.

17. Varma, M. K. 1977. A new device for rapid picking of nematodes from aqueous suspension. Indian J. Nematol. 7:94-95.

18. Wendel, J. F. and C. W. Stuber. 1984. Plant isozymes: enzymes studied and buffer systems for their electrophoretic resolution in starch gels. Isozyme Bull. 17:4-11.

Part Four

Assessment of Economic Importance

CHAPTER PAGE

CHAPTER 9

THE APPLICATION OF MICROPLOT TECHNIQUES IN NEMATOLOGICAL RESEARCH

K.R. Barker

Plant Pathology Department
North Carolina State University
Raleigh, North Carolina, USA

Nematologists have been concerned with the basic host-parasite interactions of nematodes and their hosts, including damage potential and population dynamics, during the last 30 years. There is an increasing need for host responses to nematodes, including growth and yield, to be quantified so that basic information can be provided for integrated pest management purposes. The many confounding factors often implicated in nematode effects on plants must be identified and characterized systematically for meaningful data to be developed. Some of the components of these interactions, such as host suitability and efficiency, can be determined in greenhouse experiments. In contrast, the damage potential of single species of nematodes can be determined with reasonable confidence only under field conditions where normal environmental stress is imposed upon the host plants. Because of the polyspecific nature of nematode communities and the contribution of associated microflora in natural field infestations, many problems are often encountered in differentiating damage caused by single species of nematodes as well as the related population dynamics under such conditions. Small field plots with barriers to retain the nematodes, often called microplots, have much potential for these purposes.

Nematode population experiments in field microplots, specifically designed to overcome many of the difficulties associated with field plots, were first conducted by Jones in the early 1950's (9). By plotting initial population levels versus final numbers of *Heterodera schachtii* on sugar beet, and then comparing this plot to that for nonhost crops and fallow, he developed the notion of a ceiling level, which evolved into the concept of an equilibrium density. The use of microplots, which gives more control of the experimental variables than do typical field experients, has been employed by many nematologists, plant pathologists, and agriculturalists.

The application of microplot techniques in nematology is an important component to be considered in the decision to use microplots. Types of barriers that may be constructed, methods of installation, and procedures used in infestations with various nematodes can then be determined. Some examples of results that can be derived from such experiments will be discussed. Some advantages of these techniques will be presented, and selected problems and shortcomings will be enumerated.

APPLICATIONS

Nematologists have used microplots extensively for many reasons. Advantages inherent in microplot techniques include several facets. They are utilized in field conditions, but minimize variation normally encountered in the field. Many plots may be placed in a small area. Most importantly, they offer a means of differentiating the effects of various biotic and abiotic factors on the population dynamics of nematodes. This methodology is becoming increasingly important in investigations of the quantitative relationships between nematodes and plants. They have much promise in the characterization of disease complexes, including those with plant-parasitic nematodes.

Microplot methodology also has a number of inherent problems that must be considered. As with greenhouse and phytotron experiments, results obtained in microplots should be validated in parallel full-scale field tests. Data obtained may apply to limited geographic and climatic areas. This criticism, however, is also true with most field experiments. Multiple species contamination can be a serious problem in microplots, especially with nematodes such as *Paratrichodorus* spp. which often are difficult to eradicate with fumigant nematicides. Similarly, special efforts are required for eradicating nematodes on deep-rooted plants prior to establishing experiments in such niches. One must be aware that methyl bromide treatments may suppress plant growth which can be overcome with appropriate waiting periods. Other shortcomings include the fact that these plots require much hand labor and frequent use of expensive nematicide treatments. The greatest disadvantage of microplot research concerns the fact that data interpretation and conclusions usually cannot be extrapolated directly to the field situation. This is particularly important with regard to the relationship of initial nematode densities to crop growth and yield.

TYPES OF MICROPLOTS

Most microplots employed for nematode research range in size from about 25 × 25 cm to about 1 m^2 (soil surface area) [Figs. 1,2,3]. The smaller plots often pose problems for moderate- to large-sized plants, since available soil moisture becomes limited rather quickly. This problem is minimized when automated irrigation systems are used. Perennial plants such as peach may require that these plots be in the range of approximately 5 × 5 m. Plots of approximately one-half square meter are satisfactory for most annual plants as well as smaller perennial plants.

Various types of materials have been utilized in the construction of microplots. Jones (9) used slotted concrete posts for corners and sided each plot with paving stones. These posts and paving stones were jointed with "Bitumastic." The surface area of these plots was 0.65 m^2 or 1/6000 of an acre. Similar concrete paving stones have been used by other investigators (14). The depth of these and other barriers usually ranges from 30 to about 60 cm. Although unnecessary in most soils, drainage tiles may be installed in poorly drained locations (14).

High quality fiberglass (flat, clear, 15% acrylic fortified, about 2mm thick) is an excellent material for constructing rectangular microplots of almost any size (2,18). This rather heavy fiberglass is cut to the desired lengths and the corners cemented together. The basic procedure for assembling these materials is described as follows. The corners to be cemented together must be sanded with a belt sander. Both sides of the ends to be joined should be sanded.

Several precautions are necessary. Some type of mask should be worn to prevent inhalation of fiberglass dust. Protective clothing should be worn and a shower taken immediately after working with the sander. All work with these materials should be done in open areas. Immediately prior to assembly, the sanded areas should be cleaned with acetone. Small strips of duct tape can be used to hold the four pieces of fiberglass together before attaching the fiberglass tape. The precut strips of fiberglass tape (10 cm wide × 70 cm) are dipped into a polyester resin (Aropol; Asheland Chemical Co., Ashland, Kentucky) with activator (Lupersol DDM; 60% methyl ethyl ketone peroxide in dimethyl phathalate; Lucidol Division, Pennwalt Corp., Buffalo, NY) and then placed on one side of each joint. The pieces of fiberglass tape are simply pulled through a pan filled with the resin plus hardener before being attached to the fiberglass. A second coat of resin should be applied with a paint brush after the first coat has dried. Attaching the fiberglass tape to both sides of each joint increases the durability of the fiberglass barriers. The materials for constructing such microplot barriers may be purchased at most plastic supply houses.

A lighter fiberglass (flat, clear, 15% acrylic fortified, 5 oz/ft^2 (0.16 g/cm^2), 0.9 mm thick [Lascolite Crystalite, Lasco Industries, Florence, Kentucky] is ideal for making circular fiberglass plots of any size. This material can be cut, sanded, and assembled into the desired size with resin and fiberglass tape, as described for the rectangular plots. Similar materials manufactured by other companies should be satisfactory. A much improved, simpler and more efficient method of construction of the circular fiberglass barriers was developed by Rodriguez and Beute (Pers. Comm.). They drill paired 3/16" diam holes in the ends to be connected and rivet them together with pop-gun rivets of

Fig. 1. General view of microplots with various species of plants and nematodes [after Barker et al. (2)].

the same size as the drilled holes. Washers of the appropriate size should be used with these rivets to strengthen the seams. These seams then may be sealed with any high-quality silicone caulking material. Some researchers indicate that the latter step is unnecessary, but I suggest that it be followed.

Glazed or unglazed tiles of various sizes as small as 25 × 25 cm have been used successfully by a number of investigators (3,11,12,17, 18). Unglazed tiles, 25 cm diam by 30 cm long, have proven to be satisfactory for investigation of disease complexes on potato (11,12).

TYPES OF INSTALLATION

The rectangular type of microplot barrier must be installed by the opening of trenches with a tractor-mounted backhoe or a "ditch witch." As indicated earlier, a provision for drainage by the use of tile or gravel may be necessary in poorly drained areas. One must be concerned with damage to the fiberglass or other materials during the back-filling operations. This problem is minimized when the soil is replaced by hand with a shovel. A major disadvantage of these types of plots is the disruption of the soil profile for the deeper barriers.

Circular fiberglass barriers can be installed with the aid of a device (M-cutter) that is mounted on a tractor-powered gear box and is normally used for soil augers (2). This apparatus removes the soil around a 70-to-80-cm-diam ring (about 10 cm wide) to depths of 50-55 cm, thereby facilitating establishment of permanent fiberglass barriers for microplots with minimal disturbance of the soil profile (Fig. 4). A very different method of barrier installation was developed by Johnson et al. (8). Their system utilized high water pressure to remove the soil in a 4-cm band on the perimeter of the 76-cm-diam circle to a depth of 50 cm. Unfortunately, both of these methods are better suited to sandy soils. It may be necessary to use a backhoe in very heavy clay or rocky soils. The clay-tile barriers may be installed on a permanent basis with a backhoe (17,18), or the smaller ones may be placed in deep furrows on an annual basis (11,12).

Fig. 2. **Comparison of rectangular (A) and circular (B) fiberglass microplot barriers. Note controls (C,D) vs. *Meloidogyne* species-infected plants (A,B).**

PREPARATION AND INFESTATION OF MICROPLOTS

Considerable effort and expense are required in preparation of microplots for infestation by the desired numbers and kinds of nematodes and/or other soil organisms. The soil must be treated with methyl bromide at a rate of 50-150 g/m^2 to kill indigenous nematodes and other soil pests. The soil in each plot should be tilled to a depth of 20-30 cm, and 3 or more holes at least 40 cm deep should be made in each plot to facilitate penetration by the methyl bromide. The entire plot area should be covered with plastic prior to treatment with methyl bromide. Nematodes or plants should not be introduced into the plots for 1-2 months after treatment. Tillage of the soil two or three times after removal of the plastic cover will enhance the dissipation of the toxic gas.

Greenhouse- or monoxenically grown inocula may be used for infesting microplots to the desired levels (3,4,8,11,12,14,19,21,22). *Meloidogyne* spp. can be increased on tomato in the greenhouse. Infected plants should grow for approximately 10-12 weeks before the harvesting of inocula. Standardized, chopped, infected roots or NaOCl-extracted eggs may be utilized to infest the plots to desired levels (3). NaOCl-extracted eggs should be employed for more critical experiments (4). One should expect about 20% of the eggs to give rise to juveniles that will be successful in penetrating host roots when this technique is used. Monoxenic culture of nematodes for this and other purposes is described in the chapter by Koenning and Barker (10). This type of monoxenic inocula has been used successfully for *Pratylenchus* species (12).

Appropriate numbers of mycorrhizal fungi, such as *Glomus* spp. and other beneficial organisms (*Rhizobium* spp. for legumes), also should be introduced into each plot at the time of nematode infestation (4,19). These organisms

Fig. 3. Installation of unglazed clay tiles in the study of the early dying complex of potato in Ohio. A) View of tiles used as single-plant plots; B) Use of "ditch-witch" to open furrows for placement of tiles (Courtesy of M. J. Martin, R. M. Riedel and R. C. Rowe).

should be applied at the same rates to all plots including the "no-nematode" controls. Where infected root material is used to establish the desired levels of nematodes, appropriate amounts of healthy root material should be incorporated into the control plots and lower density plots to give an equal amount of root material for each plot.

The quality of data obtained from such microplots is greatly dependent on preparation, precise infestation, and thorough but not excessive mixing of the inoculum into each plot. Microplot experiments can give useful quantitative data for most crops. Because of minimal competition, an abnormal result sometimes encountered is an annual crop yield two- to three-fold greater than that which occurs with normally spaced plant populations. This problem can be overcome with continuous normal spacing of plants within and between the microplots (11,12).

The production and preparation of inocula for other nematode species and genera pose varied problems. If possible, monoxenic inocula should be utilized. Our limited experience with *Pratylenchus* spp. has shown that greenhouse-reared inoculum of these nematodes may result in enhanced growth of soybean. This response apparently was due to the introduction of beneficial organisms along with the nematodes in the infected root material. We overcame much of this by infesting all plots with spores of *Glomus* spp. and *Rhizobium* spp. Thus, tissue culture as outlined by Martin et al. (12) should be used as a medium on which to increase species of this genus.

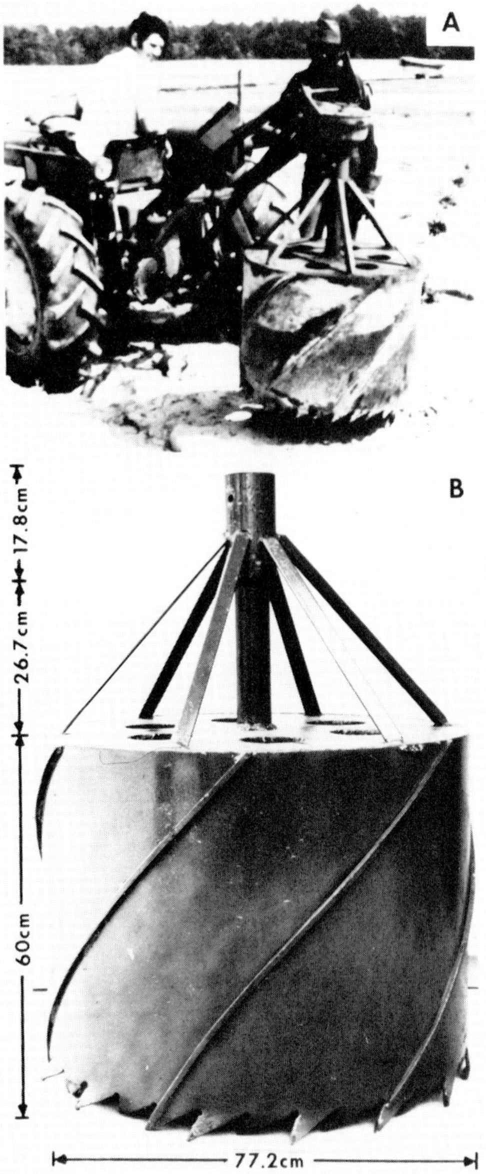

Fig. 4. Microplots. A) Tractor-mounted apparatus for digging ring-like holes for circular fiberglass barriers. B) Details of construction of M-cutter [after Barker et al. (2)].

EXAMPLES OF RESULTS OBTAINED FROM VARIOUS RESEARCH PROGRAMS

During the last ten years, estimates of the effects of various nematode species on yield potential of given crops have been developed through microplot techniques (2,4,7,11,16,19, 22). In addition to the actual effects of given nematodes on plant growth and yield, the relative importance has been determined, in some instances, for various physical and environmental parameters, including soil type, temperature, and soil moisture (3,5,19). Similarly, differential responses of various cultivars to specific species have been elucidated (3,19,21, 23).

The population dynamics of many nematodes have been characterized with varying degrees of success in microplots. Reproductive factors, equilibrium densities and rates of decline can be determined reasonably well with single species under these conditions. Nevertheless, population responses of nematodes under the semi-sterile conditions may not reflect those of naturally-infested fields. Still, this approach is one of the best solutions available for characterization of nematode-host interactions (5,9, 14,21,22,23). Although limited information on these types of interactions is available, information obtained to date concurs with that from greenhouse experiments. Generally, multiple nematode-species infestations antagonize nematode build-up and, frequently, damage host plants (3, 17).

Only limited progress has been made in identification and differentiation of the roles of certain soil microfauna and microflora associated with various disease complexes in this microcosm (3,4,11,13,15,20). An exàmple of one of the successful studies is that of the organisms involved in the early-dying complex of potato caused by *Verticillium dahliae* and *Pratylenchus penetrans* (11,12). Extensive results also have been obtained with *Heterodera schachtii* and various fungi on sugar beet (14). This type of investigation is much more difficult when disease complexes involve *Meloidogyne* spp. Since these nematodes predispose many plants to attack by fungi and bacteria that are normally non-pathogenic, the precise role of given microflora cannot be easily determined under these experimental conditions (3,4).

Microplots can be used in experiments designed for development and evaluation of nematode control tactics. They have been useful for evaluation of the relative resistance of various tobacco cultivars and sweet potato to the major *Meloidogyne* spp. (Gwynn, Moyer & Barker, unpub.). The differential efficacy of various nematicides can be determined on single and multiple species of nematodes in microplots (4,6,7). These plots are also useful for the study of biocontrol agents (1). The impact of rotation regimes and other management practices on nematode populations can also be determined under these limited conditions (5,14,

21,23).

Data generated through microplot techniques not only have limited application to other geographic areas, but also may vary from year to year (3,4,19). As previously indicated, these limitations are also inherent with most field experiments. One should be cognizant of the potential inherent limitations of data resulting from either approach when applications for research or applied purposes are considered. These problems are particularly important in reference to the effects of nematodes on yield potential of various crops, as discussed in detail in other papers (3,4,19). Thorough knowledge of the characteristics of all types of fields and microplots should enable nematologists to be more effective in elucidating the population dynamics and damage potential of plant-parasitic nematodes.

Literature Cited

1. Al-Hazmi, A.S., D.P. Schmitt, and J.N. Sasser. 1982. Population dynamics of *Meloidogyne incognita* on corn grown in soil infested with *Arthrobotrys conoides*. J. Nematol. 14:44-50.
2. Barker, K.R., B.I. Daughtry, and D.W. Corbett. 1979. Equipment and techniques for establishing field microplots for the study of soilborne pathogens. J. Nematol. 11:106-108.
3. Barker, K.R. and T.H.A. Olthof. 1976. Relations between nematode population densities and crop responses. Annu. Rev. Phytopathol. 14:327-353.
4. Barker, K.R., F.A. Todd, W.W. Shane, and L.A. Nelson. 1981. Interrelationships of *Meloidogyne* species with flue-cured tobacco. J. Nematol. 13:67-79.
5. Cotton, J. 1977. Effect of annual and perennial cropping regimes in microplots on the population density and vertical distribution of *Xiphinema diversicaudatum* and on soil porosity. Ann. Appl. Biol. 86:397-404.
6. DiSanzo, C.P. 1982. Effect of foliar applications of carbofuran and a related compound on plant parasitic nematodes under microplot and field conditions. J. Nematol. 14:208-212.
7. Jakobsen, J. 1980. The relationship between population density of *Heterodera schachtii*, soil temperature, treatment with aldicarb and sugar beet yields. Tidsskr. Planteavl. 84:537-546 (English summary).
8. Johnson, J.T., J.R. Rich, and A.W. Boatright. 1981. A technique for establishing microplots in the field. J. Nematol. 13:233-235.
9. Jones, F.G.W. 1956. Soil populations of beet eelworms (*Heterodera schachtii* Schm.) in relation to cropping. II. Microplot and field plot results. Ann. Appl. Biol. 44:25-56.
10. Koenning, S.R. and K. R. Barker. 1985. Gnotobiotic techniques and plant-parasitic nematodes, pp. 49-66 *In* An advanced treatise on *Meloidogyne*, Vol. II, Methodology, ed, K.R. Barker, C.C. Carter, and J.N. Sasser. University Graphics, Raleigh, N.C.
11. Martin, M.J., R.M. Riedel, and R.C. Rowe. 1982. *Verticillium dahliae* and *Pratylenchus penetrans*: Interactions in the early dying complex of potato in Ohio. Phytopathology 72:640-644.
12. Martin, M.J., R.M. Riedel, and R.C. Rowe. 1983. A technique for quantitative use of nematodes from monoxenic tissue culture as soil inoculum. Plant Dis. 67:275-277.
13. McKinley, R.T. and P.W. Talboys. 1979. Effects of *Pratylenchus penetrans* on development of strawberry wilt caused by *Verticillium dahliae*. Ann. Appl. Biol. 92: 347-357.
14. Meagher, J.W. and R.H. Brown. 1974. Microplot experiments on the effect of plant hosts on populations of the cereal cyst nematode (*Heterodera avenae*) and on the subsequent yield of wheat. Nematologica 20:337-346.
15. Mueller, J. 1980. Interactions between *Heterodera schachtii* and soil fungi on sugarbeet. Phytopathol. Z. 97:357-363 (English summary).
16. Olthof, T.H.A. 1978. Influence of population densities of *Heterodera schachtii* on sugarbeets grown in microplots. J. Nematol. 10:255-258.

17. Ross, J.P. 1964. Interaction of *Heterodera glycines* and *Meloidogyne incognita* on soybeans. Phytopathology 54:304-307.
18. Ross, J.P. and J.A. Harper. 1970. Effect of *Endogone* mycorrhiza on soybean yields. Phytopathology 60:1552-1556.
19. Schmitt, D.P. and K.R. Barker. 1981. Damage and reproductive potentials of *Pratylenchus brachyurus* and *P. penetrans* on soybean. J. Nematol. 13:327-332.
20. Shoemaker, P.B. and K.R. Barker. 1979. Influence of *Meloidogyne incognita* and *Verticillium dahliae* on tomato at different inoculum densities. Phytopathology 69:532 (Abstr.).
21. Townshend, J.L. and J.W. Potter. 1980. Population behavior of *Meloidogyne hapla* under four forage legumes in microplots. Can. J. Plant Sci. 60:293-294.
22. Townshend, J.L. and J.W. Potter. 1982. Forage legume yields and nematode population behavior in microplots infested with *Paratylenchus projectus*. Can. J. Plant Sci. 62:95-100.
23. Wilson, W.R. and N. French. 1975. Population studies of stem eelworm (*Ditylenchus dipsaci*) in microplots, field plots and fields in the north of England under different crop rotations. Plant Pathol. 24:176-183.

CHAPTER 10

NEMATODE POPULATION DYNAMICS WITH EMPHASIS ON DETERMINING DAMAGE POTENTIAL TO CROPS

K. R. Barker, D. P. Schmitt, and J. L. Imbriani

Department of Plant Pathology
North Carolina State University
Raleigh, North Carolina, USA

The challenge of determining the damage potential of nematodes to given crops has received much study in recent years. Most of the research relating to tolerance limits or damage and economic thresholds has been limited to single species-host combinations. Experiments concerning concomitant nematode species have been limited primarily to descriptions of their antagonistic or synergistic interactions (see Chapter 17, Vol. 1).

Because of the voluminous treatment of population dynamics in the literature, our discussion will be limited largely to examples concerning *Meloidogyne* spp. Topics receiving major emphasis include the increased need for threshold information on nematodes as a result of the development of integrated pest management (IPM) programs, related aspects of population dynamics of nematodes, characterization of host status, methodology for determining damage potential, and the types of damage thresholds and hazard indices that are useful in applied nematode control programs.

Few scientists attempted to quantify plant responses to nematodes prior to 1960. Gerald Thorne was one of the first to use soil sampling and soil assay as a basis for assessing potential nematode damage (*Heterodera schachtii* on sugar beet) in the 1920's. He later suggested that the large errors inherent in estimates of nematode populations practically precluded the possibility of sampling entire fields, predicting populations, or informing growers of the advisability of planting certain crops (69). In spite of these difficulties, he pointed out that such nematode-assay data are of value in gaining insight for effective use of control tactics. His guidelines for sampling, which include taking more than one large sample, collecting soil from different depths (particularly on deep-rooted perennials), and being cognizant of the life-cycle of the target nematodes, apply today. These problems in relating crop yields to nematode population densities are largely responsible for the limited number and size of ongoing advisory programs (49). Still, there is considerable optimism that the challenges in relating numbers and kinds of nematodes to crop productivity can be met if we are willing to invest the necessary resources.

Even with the contributions of early and current nematologists, prediction of crop losses due to nematodes is as much an art as a science. Like Thorne (69), many current nematologists feel that the many factors influencing the relationships between population densities of nematodes and crop responses prohibit an accurate characterization of these interactions. Nevertheless, considerable progress has been made in development of methods by which field populations of nematodes and their relationships to crop growth and yield can be characterized quantitatively (19,20,31,45,55,59,61).

A general negative relationship existing between preplant population densities of nematodes and crop growth and yield (6,45,55) may be modified by environmental factors and crop management practices. Seinhorst (60), however, concludes there is little evidence that external factors influence the tolerance of a plant to nematodes. In spite of this disagreement, information generated on nematode-host relationships for one climatic area should not be considered as valid for other areas. Host responses are more predictable in geographic areas with a less variable climate, than in areas such as North Carolina where seasons and soils vary greatly. For example, tobacco can tolerate a relatively high number of *Meloidogyne incognita* during an ideal growing season, whereas this crop may be severely damaged by moderate numbers of this pest under stress conditions such as lack of rainfall. The contagious distribution of nematodes may result in overestimates of damage potential (40,46). Characterization of host responses to nematodes can be further confounded by the erratic occurrence of other organisms such as fungi or bacteria that interact synergistically with nematodes (47).

There are at least three approaches to practical nematode management: 1) leave the control of these pests entirely to the grower (75); 2) teach growers how to utilize simple bioassay techniques for root-knot (36) and cyst nematodes (see Chapter 2); or 3) generate the best estimates of population levels of various pathogenic nematodes that would be expected to cause disease at economic levels on specific crops. Much of the information would be qualitative; as research improves the precision or predictability of crop responses to these pests, quantitative data could be used.

The need for reliable threshold information is becoming more urgent for developed and developing countries. Direct crop losses due to nematodes still are estimated to be in the range of 5-10% of the crop value annually. In addition, the costs of corrective actions, particularly the use of nematicides and rotation are increasing each year. The utility of resistant cultivars has been lost in many cases due to excessive use or because disease complexes reduce their effectiveness. Thus, nematologists have an unusual opportunity to redirect control tactics toward reduction of serious losses through nematode and crop management approaches rather than the traditional preventive control program. A recent treatment on integrated nematode control is available (10).

The basic strategy used in managing nematodes is the reduction of initial populations, although the use of resistant cultivars involves VanderPlank's (73) second strategy for pest control, the reduction of the rate of population increase. These facts are related to the low motility of nematodes which enhances the potential of preplant population densities as a quantitative predictive tool. Nematode management problems may be based on their population densities at the time a crop is established or on previously associated symptoms. Unfortunately, damage caused by nematodes, although sometimes obvious, typically is not associated by the grower with nematode feeding. Foliage and root symptoms of nematodes often are similar to those caused by other pathogens or by adverse management practices or weather conditions. The field appearance of plants affected by nematodes usually involves stunting in given foci, resulting in a spotty appearance with less vigorous plants that fail to respond to proper crop management practices. The root galls and other abnormal growths (Fig. 1) caused by *Meloidogyne* spp. are easily recognized by the trained observer but still may be overlooked by many growers.

The major problems in attempting to relate nematodes to agronomic problems have been summarized by Norton (42). These include: 1) determination of nematode species which are pathogenic; 2) correlation of nematode densities with crop loss; 3) elucidation of how the environment affects nematode activity; and 4) development of management strategies for these pests. The relationships of nematode population numbers to the performance of perennial crops involve additional difficulties. Nondetectable or subsequently introduced nematode populations may increase over time and cause serious damage. Because these pests are obligate parasites, their numbers are not always correlated with induced damage on given plants. Once severe injury occurs on a plant, the nematode population may decline sharply below that on healthy plants. This fact causes much confusion in diagnosing nematode-induced problems, especially on perennial plants.

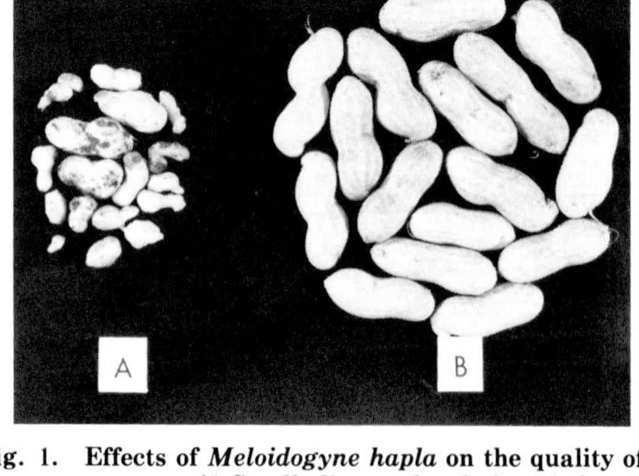

Fig. 1. **Effects of *Meloidogyne hapla* on the quality of peanuts. A) Small, distorted pods from infected plants grown in microplots; B) Pods from healthy plants.**

POPULATION DYNAMICS

Knowledge of the population changes of nematodes is essential in the collection of meaningful soil samples for nematode assay. Typical seasonal population fluctuations of nematodes were discussed in Chapter 1. Management practices greatly influence rates of activity and survival of *Meloidogyne* spp. In addition, edaphic factors including soil-water potential have a striking effect on the survival of species such as *Meloidogyne javanica* (70). The time required for a 50% reduction in soil infestation at −0.16 (wet) or −92 bars (dry) was about 2½ to 3 days as compared to 110 days at −1.1 bars (moist). Temperature has an equally important impact on these pests (20, 24,71).

Management practices (e.g. subsoiling) may shift the spatial and temporal distribution of nematodes. Without subsoiling, confined roots of plants may be more susceptible to nematode damage (29). In contrast, subsoiling may result in decreased populations of nematodes in the upper soil layer and favor deeper penetration of the soil by these pests (29). The effect of subsoiling may reduce the need for a nematicide, but this benefit may be temporary (29,38, 39). Competition among polyspecific communities of nematodes may result in one species becoming dominant and could result in the nematodes having greater or smaller effects on yield. *Meloidogyne incognita* has been shown to suppress the reproduction of *Rotylenchulus reniformis* on sweet potato, but both still adversely affected the quality of this crop (68) [see Chapter 17, Vol. I by J. D. Eisenback].

The importance of geographic differences in the population dynamics of nematodes, including *M. incognita*, can be appreciated by comparing the survival rates of this nematode as given in the data on sampling for North Carolina (see Chapter 1) compared to the survival rates encountered in Florida (77). Relatively large population densities of *Meloidogyne* spp. can develop on potato or other crops in Florida even though no nematodes were detected in preseason or early season soil samples (77). In contrast, detection of these nematodes in North Carolina is rarely a problem in the spring of the year.

The relationship of edaphic factors to the reproduction and survival of nematodes has been of interest for many years (64,65), but varies by taxa (72). For example, only slight gall development caused by *M. javanica* may occur in a fine-textured soil compared to moderate to severe gall development in a coarse-textured soil (64). *Ditylenchus dipsaci*, in contrast, is favored by fine-textured soils (72).

Knowledge of the biological and ecological requirements of nematodes can be utilized to minimize or limit associated losses. A delay in planting wheat until the temperature has dropped below an activity threshold of 18°C for *M. incognita* limits the penetration of juveniles in wheat roots and maximizes the utility of this crop in rotation as a tactic for the control of this pest (50). Although not fully understood, *M. incognita* apparently has a diapause that may be involved in its persistence during climatic stress and/or when host plants are absent (15). Limited evidence suggests that some nematodes have the capacity to adapt to lower temperatures when exposed to that condition over a period of years. The annual practice of early planting and harvesting of potatoes to limit losses caused by *Globodera rostochiensis* apparently resulted in this pest adapting to lower temperatures (28).

REPRODUCTIVE POTENTIAL

Unfortunately, life tables are unavailable for most nematodes, but general information is available on the reproductive potential of *Meloidogyne* spp. Tyler (71) found that a *Meloidogyne* species produced as many as 2882 eggs per female. Precise information on the fecundity (number of potential eggs produced) and fertility (number of viable eggs produced) is still needed for most nematode species. Differential fecundity may be related to the overall virulence of different species. For example, the average number of eggs in egg masses produced by *M. hapla, M. incognita* and *M. javanica* were about 300, 700, and 1000 respectively (K. R. Barker, unpublished). The relative virulence of these three species on tobacco follows this same order (8).

A number of concepts and terms are useful in the characterization of nematode population changes. The ratio of the final population (P_f) to initial population (P_i) is described as the reproduction factor (45). The precise model for calculating this ratio varies with the type of reproductive curve for given nematodes (45). The equilibrium density is the population level at which multiplication or reproduction of a nematode just suffices to maintain the population, i.e. birth rate = death rate. Another useful concept developed by Seinhorst (56) for nematode population studies is that of the maximum rate of reproduction: $a = e^{rt}$, where a is the maximum rate of reproduction; e = the base of the natural log (2.3); r = rate of reproduction; and t = time. The term underpopulation refers to the population densities at which multiplication rates decline with declining densities, a situation which is usually encountered with amphimictic species (57). Surplus population refers to excessively high populations at which very low ratios of nematodes reach maturity due to overcrowding and mass root invasion causing root death (12).

An excellent treatment of population change of nematodes is provided by Norton (41). Additional population parameters included in his treatment are natality which is the rate of production of new individuals and mortality. He also provides models for characterizing growth patterns, including that for exponential growth or logistic growth. Two of the parameters of the logistic model or curve have been used to categorize organisms that have differential reproductive characteristics. These are the intrinsic growth rate (r) and the asymptote (K) (41). The "r strategists" theoretically have a high reproductive rate and short life span. Norton suggests that the unstable habitat of cultivated fields, plowed annually, could favor the r selection of nematodes such as *Pratylenchus*. The K selection species are better competitors and usually live in a more stable habitat than the r selection species (41). *Meloidogyne* spp. that maintain relatively stable populations on perennial plants may fit this category. The r and K terminology is relative and changes from one nematode-plant community to another.

CHARACTERIZING HOST STATUS

Several useful treatments of the concepts of host suitability, efficiency and sensitivity in plant-nematode relationships are available (6, 13,16,30). Basically, host efficiency relates to how well a plant supports parasite reproduction. Host sensitivity is the plant's response to the parasite. One scheme utilizes the terms resistant and susceptible to refer to host efficiency, whereas tolerance and intolerance were used to refer to host sensitivity (13). A similar scheme was recently suggested in relation to virus infection (14). This usage has caused some debate in virology (66), as well as nematology because of the fact that resistance and susceptibility traditionally have been used as a reference of the host's response to infection in plant pathology and in the medical field. The concept of tolerance particularly is a focal point of much debate, including whether or not such a phenomenon exists in reference to plant disease (26,66). A resolution of this issue will require further research and will not be attempted herein.

The scheme proposed by Seinhorst (55,56) for quantitative characterization of host sensitivity and efficiency is one of the most useful. It involves the concept of tolerance limit (T), the population density below which no loss in growth and yield occurs. The term threshold density, which refers to the same point on the curve, also has been used to indicate the minimal population density that would suppress plant growth. Thus, sensitive hosts would have low tolerance limits whereas resistant hosts would have very high tolerance limits. The status of a plant as a host for a given nematode species is summarized by the maximum rate of reproduction (a) and the equilibrium density (E) (56). Although these two terms are not always closely correlated, both of these constants would be high for a good host and low for a poor host. The maximum rate of reproduction as well as the equilibrium density for a species depends on inherent characters of the nematode and the plant as well as environmental conditions. The general model developed by Seinhorst (55,58,59) for determination of tolerance limit and yield responses to nematodes is:

$$Y = m + (1 - m)Z^{P-T}$$

where Y = a ratio between yield at nematode density P and that in the absence of nematodes; m = the yield at nematode densities in which all available space is occupied by the nematodes (minimal yield); Z = a constant less than 1 (= to the proportion of plants not attacked at population density $P = 1$); P = the population density, which is usually $\geq T$ ($Y = 1$ where P is equal to or less than T); and T = the nematode density below which no detectable loss in yield occurs. The Z value depends on nematode species, host species, and external conditions. Z^{-T} usually ranges from 1.05 to 1.15 (55,58). In addition to relating initial nematode numbers to yield in this model, it is useful in

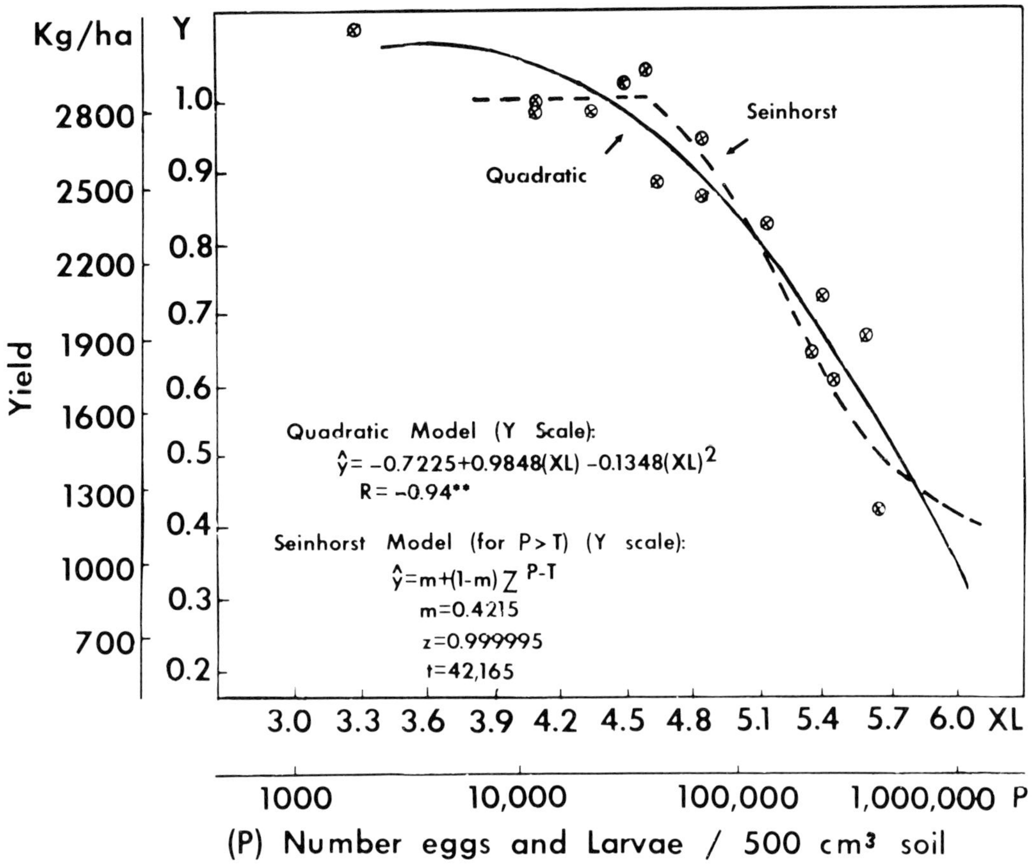

Fig. 2. **Comparison of quadratic and Seinhorst regression models for relating midseason numbers of *Meloidogyne* spp. to yield of tobacco [range of nematodes established with various chemical soil treatments; y = relative yield, XL = $\log_{10}$ of nematode numbers; after Barker et. al. (8)].**

depicting plant responses to differential population densities affected by nematicide treatments (Fig. 2). An algorithm for fitting data to the Seinhorst curve and examples of validation of this model are available (17,25). A more complex version of this basic model is available for characterization of plant nematode interactions that might involve stimulation as well as growth suppression (61).

METHODOLOGY FOR DETERMINING DAMAGE POTENTIAL

Several empirical experimental methods are useful in elucidating nematode-crop relationships. Characteristics such as relative host efficiency can be determined in greenhouse or phytotron experiments; however, host sensitivity or tolerance, as determined under these types of experiments, may not coincide with those determined in field experiments. Most of the data have been generated on the effects of single species of nematodes. Such information is necessary and useful but must be evaluated in the context of agricultural ecosystems.

A number of approaches may be used in the establishment of a range of nematode population densities by which associated crop losses may be evaluated under field conditions. A relatively simple and effective, but laborious, method is the use of sequential sampling to delineate plots that have a wide range of initial population densities. Kinloch (32) used a range of natural initial population densities of *Meloidogyne incognita* as plots to generate regression models relating yield to numbers of nematodes as determined at various sampling times. Rotation regimes or a series of nematicides with differential efficacy also are useful in establishing ranges of target species of nematodes for evaluation of damage potential. Oostenbrink (45) and Nusbaum (43) have used crop rotation effectively in determining the relative virulence of various nematodes on crops. Differences in soil fertility and structure must be considered carefully in such experiments.

Nematicides have been used more widely for these purposes (8,38,39,51,52,53), but this technique has a number of inherent problems. Many nonfumigant nematicides give inadequate control and thereby may not reflect the true potential crop yield; others may be phytotoxic or even enhance plant growth directly (1,48,53).

The utilization of microplots in the establishment of damage thresholds is covered by K. R. Barker (see Chapter 9). As emphasized by a number of researchers, this approach should not be used as a sole method for determination of crop responses to nematodes. Instead, it should be used with parallel naturally-infested field experiments for determination of damage potential. Techniques utilized for infestation of microplots may be modified for infestation of field sites (2); a 0.125% agar solution reduces the settling of *Meloidogyne* eggs and keeps them suspended for several hours. Although initial results were somewhat variable, this approach may be useful in the infestation of relatively large field plots with varying numbers of *Meloidogyne* spp. for experimental purposes. A machine dispenser has also been developed for rapid delivery of standard volumes of *Globodera rostochiensis* cysts (35).

Damage thresholds for *Meloidogyne* spp. generated under one set of conditions may be of limited value and sometimes misleading. The importance of soil texture and other edaphic factors cannot be overemphasized. Care and maintenance of the experiment can produce results different from those which would occur under natural conditions. For example, supplemental irrigation enables some plants parasitized by *Meloidogyne* spp. to compensate in varying degrees for nematode damage (3,37). The age of plants when infected by nematodes also affects their responses (18,27,44,62,63,78).

The impact of *Meloidogyne* spp. on the quality of plants such as carrot (74) and on peanut and other crops must be evaluated as well as their effects on yield. The influence of *M. hapla* on the quality and size of peanut is depicted in Figure 1. Thus, the many interactions of the environment, host, and nematode must be considered in the development of meaningful damage potential information.

Modeling (Chapter 16) undoubtedly will allow us to understand better these many interactions involving *Meloidogyne* spp. (19,20,22, 31). For example, interaction of *Meloidogyne incognita* and *Rhizoctonia solani* on cotton is dependent on initial levels of nematode as well as soil temperature (11), but few efforts involve quantification and modeling of such disease complexes. Thus the modeling approach must have the flexibility of including the multiple interactions among nematode species (17).

Newly developed computer-video image systems have much potential in relating nematode crop performance to nematode numbers. Foliage-disease symptoms of plants have been quantified by use of microcomputer-digitized video-image analyses (33). This system may have application for characterization of root- and shoot-growth responses to various levels of *Meloidogyne* spp. Simple, direct photographic techniques and other approaches also provide non-destructive estimates of growth responses of plants to nematodes (9,67).

A wide range in problems of statistical analyses of nematode-plant environmental interactions has been discussed by Wallace (76). He emphasizes that we must distinguish between qualitative and quantitative interactions. Furthermore, he stresses that we must utilize multifactorial experiments and analyze and interpret these correctly in order to identify the various determinants associated with diseases involving nematodes. Many of the multiple comparisons of statistical analyses used for such experiments do not provide for a measure of interactions (34,76). Greater emphasis on appropriate statistical design and analyses is needed for characterization of the many interactions of nematodes with hosts and the environment (see Chapters 13-14).

DAMAGE THRESHOLDS AND HAZARD INDICES

Many of the tolerance limits or damage thresholds developed prior to 1976 for *Meloidogyne* spp. and other nematodes have been summarized (6). A few examples of damage thresholds of *Meloidogyne* spp. are given in Table 1. Only wide ranges can be given in many cases because responses vary greatly or insufficient research has been conducted to establish precise levels. These relationships between crop yield and nematode population data usually are too variable to allow an estimate of the tolerance limit with confidence (21). As more research is conducted, precision will increase.

A thorough treatment on the derivation, requirements, and various considerations concerning the use of economic thresholds in

Table 1. Comparison of damage thresholds (DT) and nematode hazard indices (NHI) for *Meloidogyne* species on certain crops[a]

	Damage thresholds (no. nematodes/100 cm^3 soil) and hazard indices (0-100) per nematode species per crop (Spring samples)							
	M. arenaria		*M. hapla*		*M. incognita*		*M. javanica*	
Crop	**DT[a,b]**	**NHI[a,b]**	**DT**	**NHI**	**DT**	**NHI**	**DT**	**NHI**
Cabbage	—	—	+	0	150-1000	10-50	—	—
Corn	200	20-60	+	0	50-1000	10-50	—	—
Cotton	+	0	+	0	10-100++	0-90	+	0
Peanut	1-40++	0-100++	10-200	10-80	+	0	—	—
Peach	2	50-100	2	—	2	50-100	2	50-100
Snap bean	100-600	30-80	—	—	100-600	30-80	—	—
Soybean	6-70	30-100	22-250	5-70	10-250	10-90	6-50	30-100
Tobacco	1-2	90-100	20-100	10-30	1-40	20-90	1-2	90-100
Tomato	2-100	50-100	20-200++	10-100++	2-100++	50-100	—	—
Watermelon	2-50++	70-100	—	—	2-50++	70-100++	—	—

[a]Damage thresholds (DT) for nematode numbers per 100 cm^3 of soil [partially after data (Fall) for North Carolina, Alabama, Georgia and Florida as given by Rickard and Barker (49)]. Nematode hazard indices (NHI) given on a percentage basis of risk (actual risk dependent on several factors, including nematode numbers and races, relative virulence of specific nematode, previous crop, edaphic factors and temperature). 0 indicates no problem, 100 indicates a high (maximum) hazard.

[b]+ indicates nonhost; ++ indicates host status depends on nematode race; — indicates no information available.

nematology is available (21). Economic threshold refers to the population density of a pest at which the value of the damage caused is equal to the cost of the control. At population densities up to the economic threshold, there is no advantage to pest control since the cost would be greater than the increased return of the crop. Unfortunately, the variability in these relationships also has limited the utilization of economic thresholds of nematodes. Ferris (21) indicates that this concept has been ignored in nematology for four primary reasons: 1) limited information on the relationship between nematode densities and plant-damage functions; 2) difficulties in the assay of nematode population densities in a field; 3) effort required for arrival at the decision; and 4) the availability of low-cost pesticides. Still, we should emphasize that the need for this type of information is much greater today than just a few years ago because the more effective nematicides are expensive or being withdrawn from the market.

An example of one approach to the derivation of economic thresholds is given in Figure 3 for *Heterodera glycines* on soybean. In this work (D. P. Schmitt, unpublished), the damage function curve in the loamy sand soil was linear with increasing inoculum. The application of arithmetic rates of aldicarb to high populations of the nematode effected a linear increase in yield up to 70% yield potential without nematodes. This relationship varied with soil type.

Although the variability associated with characterization of plant-growth responses to nematodes prohibits the determination of precise tolerance limits, sufficiently reliable information can be obtained to form the basis for a meaningful advisory service to growers (4,5,49). For example, the regression curves for tomato losses due to *M. incognita* in the Coastal Plain of North Carolina over 2 years were relatively similar (7). The marketable yield loss for the first year per 10-fold increase in numbers of juveniles of *M. incognita* was 24% compared to 19% for the second year (Fig. 4). The importance of edaphic factors in these relationships is depicted in Figure 5 for two microplot soils infested with *Pratylenchus brachyurus* (54) and for *Meloidogyne* spp. at different locations on tobacco in Figure 6. The damage from both nematodes was limited by a finer soil texture and/or cooler soil temperature.

Challenges and prospects of utilizing damage threshold information in advisory programs have been reviewed recently (4). Much of the advice offered to growers today is based on qualitative or at best semi-quantitative assays because of the low to moderate level of preci-

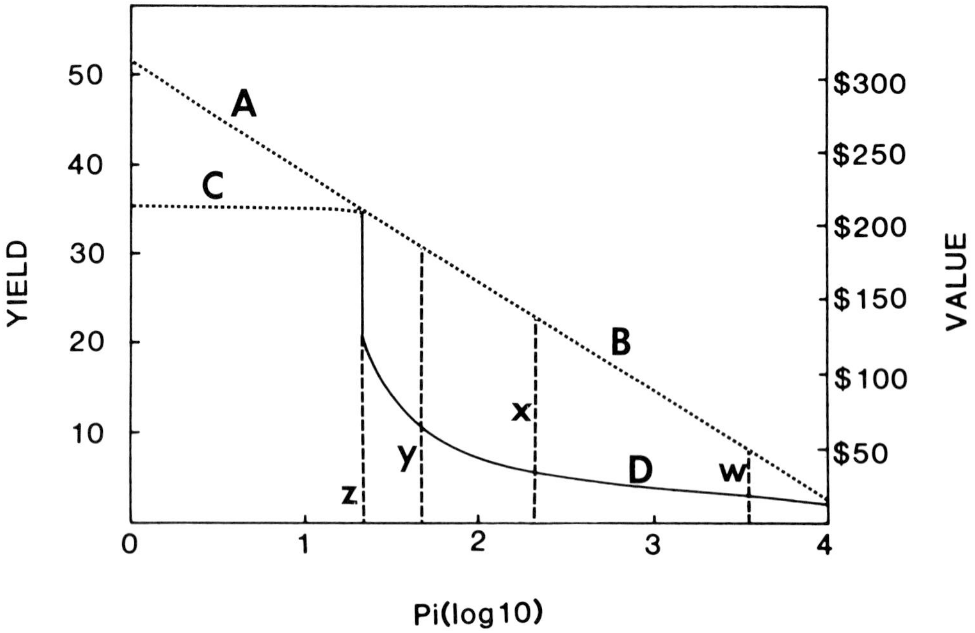

Fig. 3. Summary of components useful in determining economic threshold for *Heterodera glycines*: Line AB—linear damage function of *H. glycines* on soybean grown in a loamy sand with no nematicide treatments; curve BC—damage function of *H. glycines* in same soil treated with various rates of aldicarb [note that yields (bushels/acre) could not be increased with aldicarb beyond 70% of that in plots without nematodes; w = 1.0, x = 2.0, y = 4.0 and z = 8.0 lb (a.i.) aldicarb/acre (broadcast)]. Curve D = control-cost function for aldicarb.

sion in present sampling schemes and the limited available information on nematode-damage functions as influenced by host cultivar, nematode species and races, and the environment.

Because of these difficulties in developing precise damage threshold information, some type of qualification should accompany any recommendations to growers based on most available information. One approach would be to combine sampling error as an indication of the precision of any threshold parameter (23). Unfortunately, this type of data is unavailable for most nematode-crop combinations and would make the information more complex for grower interpretation. Another interim approach for use of nematode hazard indices has been suggested (4). This scheme is based on the relative chance of an economic problem, depicted by an indices series of 0-100 with 100 representing the maximum hazard. Actual indices would be based on the relative virulence of the nematodes in question as well as the precision of the threshold data available. For example, *Meloidogyne incognita* is highly virulent on okra and would have a hazard index of approximately 90-100. In contrast, this species would be given an index of 10-30 for most soybean cultivars in North Carolina. The lack of precise information on a given combination such as *M. hapla* on Ladino clover might be indicated by use of broad hazard indices of approximately 20-70% risk. As more precise damage-potential data become available for that particular combination, hazard indices could be narrowed. Examples of currently used damage thresholds and proposed hazard indices are offered in Table 1 for selected crops involving *Meloidogyne* spp.

Use of hazard indices, as proposed, is less desirable than use of economic thresholds. However, this approach does offer information that would greatly facilitate the development of management decisions by the grower. Although such data are of more limited quality than desired, advising growers of potential problems and the relative risk involved should enable them to make reasonably sound management decisions. Without monitoring, nematode management is very difficult.

For *Meloidogyne* spp., growers can develop their own monitoring system by utilizing bioassay plants. These bioassays may be based on relative gall indices as determined on the actual crop plants, or soil samples may be collected and bioassay plants such as tomato

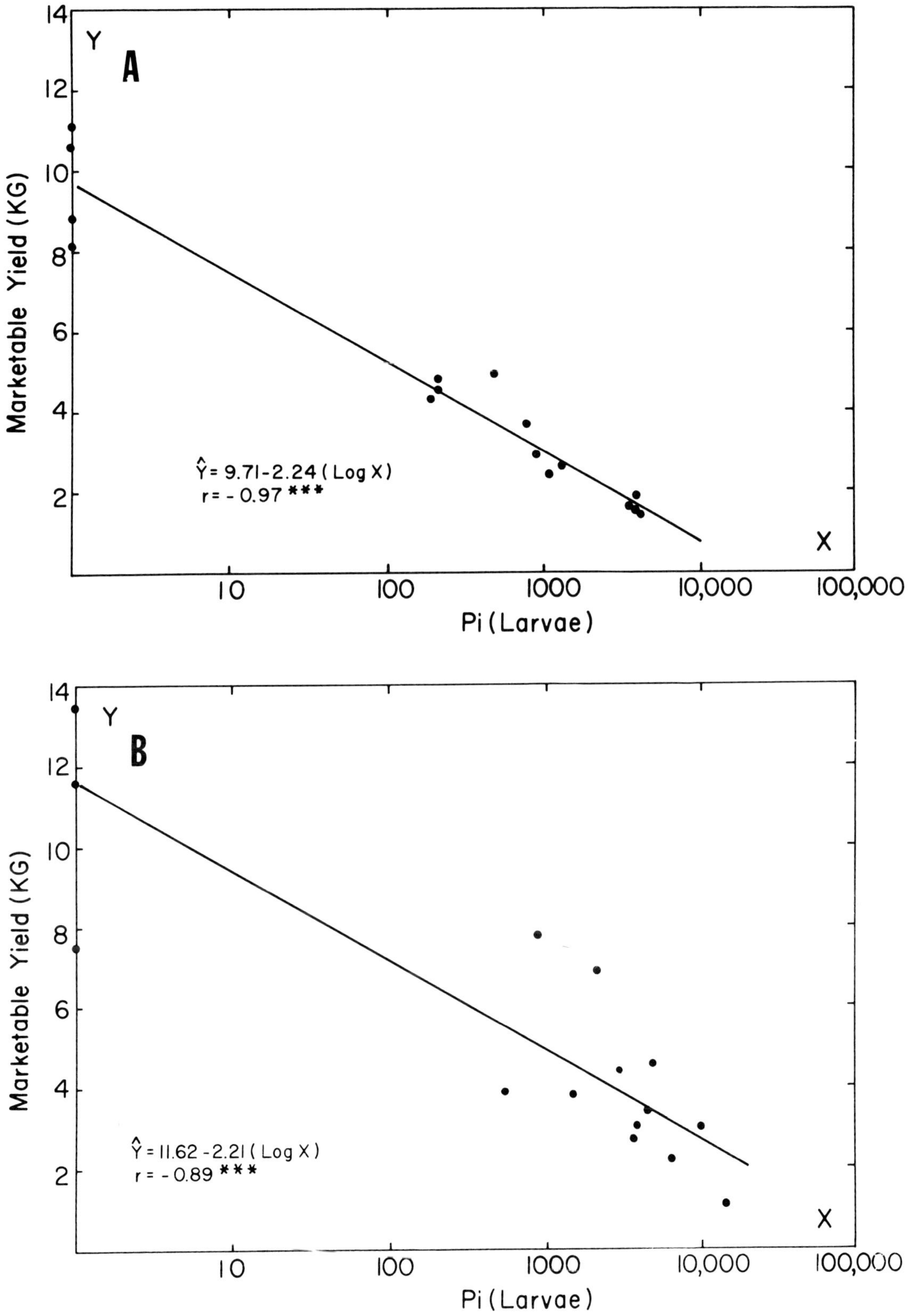

Fig. 4. Relationship of initial population densities of juveniles (larvae) of *Meloidogyne incognita* to the marketable yield of tomato as determined in microplots in the coastal plain of North Carolina: A) yield response after initial infestation; B) effects of carry-over population on yield for following year. [after Barker, et al. (7)].

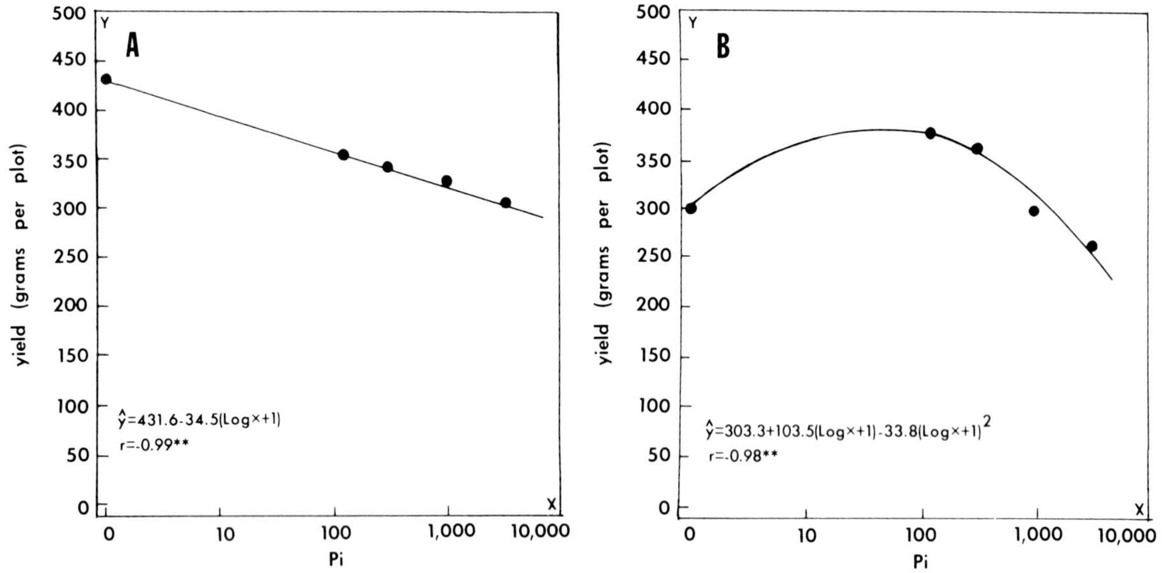

Fig. 5. Relationship of initial densities of *Pratylenchus brachyurus* and soil type to yield of 'Forrest' soybean: A) Norfolk sandy loam; B) Appling sandy clay loam. Y = predicted yield in g/plot, X = initial numbers of nematodes/500 cm^3, * = significant at P = 0.05, ** = significant at P = 0.01 [After Schmitt and Barker (54)].

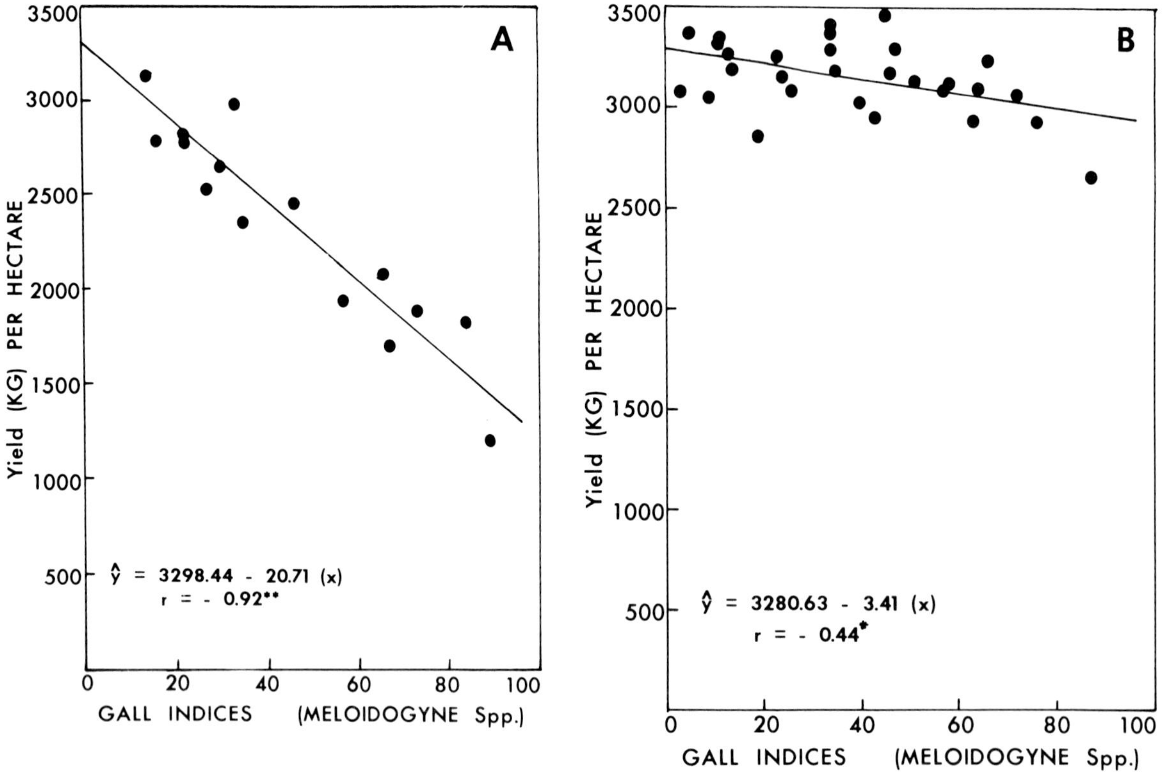

Fig. 6. Effects of location on impact of root-knot galling on tobacco yields in North Carolina. A) Coastal Plain (Johnston County); B) Piedmont (Surry County). After Barker et al. (8).

grown in them (4,36). This approach has particular utility in providing information on which management tactics may be based. It would be especially helpful where no threshold information or advisory programs are available. Gall indices determined at mid-season may be more useful than those done at harvest in tropical locations. This is especially true where disease complexes develop since associated root decay may destroy the galled roots.

Thus, considerable progress has been made in developing nematode-damage thresholds. Nevertheless, our ability to predict actual crop response and yield, as based on natural infestations of nematodes, remains somewhat limited. Development of more reliable sampling procedures and nematode-damage functions poses a major challenge. We would like to encourage more nematologists to accept this challenge rather than leaving the responsibility of nematode management largely to the individual grower.

Literature Cited

1. Altman, J. and K. M. Tsue. 1965. Changes in plant growth with chemicals used as soil fumigants. Plant Dis. Rep. 49:600-602.
2. Ball, D. A. and H. Ferris. 1982. A technique for inoculating field sites with *Meloidogyne* eggs. J. Nematol. 14:420-422.
3. Barker, K. R. 1982. Influence of soil moisture, cultivar, and population density of *Meloidogyne incognita* on soybean yield in microplots. J. Nematol. 14:429 (Abstr.).
4. Barker, K. R. and J. L. Imbriani. 1984. Nematode advisory programs—status and prospects. Plant Dis. 68:735-741.
5. Barker, K. R. and C. J. Nusbaum. 1971. Diagnostic and advisory programs, pp. 281-301 *In* Plant parasitic nematodes, Vol. I, ed. B. M. Zuckerman, W. F. Mai, and R. A. Rohde. Academic Press, New York. 345 pp.
6. Barker, K. R. and T. H. A. Olthof. 1976. Relationships between nematode population densities and crop responses. Annu. Rev. Phytopathol. 14:327-353.
7. Barker, K. R., P. B. Shoemaker, and L. A. Nelson. 1976. Relationships of initial densities of *Meloidogyne incognita* and *M. hapla* to yield of tomato. J. Nematol. 8:232-239.
8. Barker, K. R., F. A. Todd, W. W. Shane, and L. A. Nelson. 1981. Interrelationships of *Meloidogyne* species with flue-cured tobacco. J. Nematol. 13:67-79.
9. Benson, D. M. 1978. A direct versus a photographic technique as non-destructive estimates of growth response on perennial ornamentals affected with nematode decline. Plant Dis. Rep. 62:68-70.
10. Bird, G. W. 1981. Integrated nematode management for plant protection, pp. 355-375 *In* Plant parasitic nematodes, Vol. III, ed. B. M. Zuckerman and R. A. Rohde. Academic Press, New York. 508 pp.
11. Carter, W. W. 1975. Effects of soil temperatures and inoculum levels of *Meloidogyne incognita* and *Rhizoctonia solani* on seedling disease of cotton. J. Nematol. 7:229-233.
12. Chitwood, B. G. and J. Feldmesser. 1948. Golden nematode population studies. Proc. Helminthol. Soc. Wash. 15:43-55.
13. Cooke, R. 1974. Nature and inheritance of nematode resistance in cereals. J. Nematol. 6:165-174.
14. Cooper, J. I. and A. T. Jones. 1983. Responses of plants to viruses: Proposals for the use of terms. Phytopathology 73:127-128.
15. DeGuiran, G. 1979. A necessary diapause in root-knot nematodes. Observations on its distribution and inheritance in *Meloidogyne incognita*. Rev. Nematol. 2:223-231.
16. Dropkin, V. H. and P. E. Nelson. 1960. The histopathology of root-knot nematode infections in soybean. Phytopathology 50: 442-447.
17. Duncan, L. W. and H. Ferris. 1983. Validation of a model for prediction of host damage by two nematode species. J. Nematol. 15:227-234.
18. Fawole, B. and W. F. Mai. 1979. Influence of plant age, light intensity, nematode inoculum level, and their interactions on tomato growth and reproduction of *Meloidogyne hapla*. J. Nematol. 11:199-201.
19. Ferris, H. 1976. Development of a computer-simulation model for a plant-nematode system. J. Nematol. 8:255-263.

20. Ferris, H. 1978. Modification of a computer simulation model for a plant-nematode system. J. Nematol. 10:198-201.
21. Ferris, H. 1978. Nematode economic thresholds: derivation, requirements, and theoretical considerations. J. Nematol. 10: 341-350.
22. Ferris, H. 1981. Dynamic action thresholds for diseases induced by nematodes. Annu. Rev. Phytopathol. 19:427-436.
23. Ferris, H. 1984. Nematode damage functions: the problems of experimental and sampling error. J. Nematol. 16:1-9.
24. Ferris, H., S. M. Schneider, and M. C. Stuth. 1982. Probability of penetration and infection by root-knot nematode, *Meloidogyne arenaria* in grape cultivars. Am. J. Enol. Vitic. 33:31-35.
25. Ferris, H., W. D. Turner, and L. W. Duncan. 1981. An algorithm for fitting Seinhorst curves to the relationship between plant growth and preplant nematode densities. J. Nematol. 13:300-304.
26. Gaunt, R. E. 1981. Disease tolerance—an indicator of thresholds? Phytopathology 71:915-916.
27. Griffin, G. D. 1981. The relationship of plant age, soil temperature, and population density of *Heterodera schachtii* on the growth of sugarbeet. J. Nematol. 13:184-190.
28. Hominick, W. M. 1982. Selection of a rapidly maturing population of *Globodera rostochiensis* by continuous cultivation of early potatoes in Ayrshire, Scotland. Ann. App. Biol. 100:345-351.
29. Hussey, R. S. 1977. Effects of subsoiling and nematicides on *Hoplolaimus columbus* populations and cotton yields. J. Nematol. 9:83-86.
30. Hussey, R. S. and R. W. Roncadori. 1982. Vesicular-arbuscular mycorrhizae may limit nematode activity and improve plant growth. Plant Dis. 66:9-14.
31. Jones, F. G. W. and R. A. Kempton. 1978. Population dynamics, population models and integrated control, pp. 333-361 *In* Plant nematology, ed. J. F. Southey. Minist. Agric., Fish. and Food Tech. Bull. 7. Her Majesty's Stationery Office, London. 440 pp.
32. Kinloch, R. A. 1982. The relationship between soil populations of *Meloidogyne incognita* and yield reduction of soybean in the coastal plain. J. Nematol. 14:162-167.
33. Lendou, S. E. and R. R. Webb. 1983. Quantification of foliar plant disease symptoms by microcomputer-digitized video image analysis. Phytopathology 73:520-524.
34. Madden, L. V., J. K. Knoke, and R. Louie. 1982. Considerations for the use of multiple comparison procedures in phytopathological investigations. Phytopathology 72: 1015-1017.
35. Marks, R. J. and L. A. McKenna. 1981. A dispenser for the rapid delivery of standard inocula of the cyst nematode *Globodera rostochiensis*. Ann. Appl. Biol. 97:119-122.
36. McSorley, R. and J. L. Parrado. 1981. A bioassay sampling plan for *Meloidogyne incognita*. Plant Dis. 67:182-184.
37. Mein, S., H. R. Wallace and J. M. Fisher. 1978. Water relations of tomato (*Lycopersicum esculentum* Mill. cv. early Dwarf red) infected with *Meloidogyne javanica* (Treub) Chitwood. Physiol. Plant Pathol. 13:275-281.
38. Minton, N. A., M. B. Parker, and B. G. Mullinix, Jr. 1978. Effects of cultivars, subsoiling and fumigation on soybean yields and *Meloidogyne incognita* populations. J. Nematol. 10:43-47.
39. Minton, N. A., M. B. Parker, and C. E. Perry. 1980. Effects of one and two applications of nematicides on nematode populations and soybean yields. J. Nematol. 12:294-299.
40. Noe, J. P. 1984. Spatial population dynamics of plant-parasitic nematodes. Ph.D. thesis, N. C. State Univ., Raleigh, N.C. 96 pp.
41. Norton, D. C. 1978. Ecology of plant-parasitic nematodes. Wiley & Sons, New York. 268 pp.
42. Norton, D. C. 1983. Maize nematode problems. Plant Dis. 67:253-256.
43. Nusbaum, C. J. and H. Ferris. 1973. The role of cropping systems in nematode population management. Annu. Rev. Phytopathol. 11:423-440.
44. Olthof, T. H. A. 1983. Effect of plant age and transplanting damage on sugarbeets infected by *Heterodera schachtii*. J. Nematol. 15:555-559.
45. Oostenbrink, M. 1966. Major characteristics of the relation between nematode and plants. Meded. Landbouwhogesch. Wageningen 66-4. 46 pp.
46. Perry, J. N. 1983. Effects of spatial heterogeneity on Jones' model for cyst-nematode

population dynamics and crop root damage. J. Appl. Ecol. 20:849-856.

47. Powell, N. T. 1979. Internal synergisms among organisms inducing disease, pp. 113-133 *In* Plant disease an advance treatise, Vol. IV, ed. J. G. Horsfall and E. B. Cowling. Academic Press, New York. 466 pp.
48. Ragab, S. M. 1981. Cotton growth and nutrient uptake following temik aldicarb application. J. Agric. Sci., Cambridge, UK 97:731-737.
49. Rickard, D. A. and K. R. Barker. 1982. Nematode assays and advisory services, pp. 8-20 *In* Nematology in the southern region of the United States, ed. R. D. Riggs. South. Coop. Ser. Bull. 276, Arkansas Agric. Exp. Stn., Fayetteville, Arkansas. 206 pp.
50. Roberts, P. A., S. D. Van Gundy and H. E. McKinney. 1981. Effects of soil temperature and planting date of wheat on *Meloidogyne incognita* reproduction, soil populations, and grain yield. J. Nematol. 13: 338-345.
51. Rodriguez-Kabana, R. and J. C. Williams. 1981. Assessment of soybean yield losses caused by *Meloidogyne arenaria*. Nematropica 11:105-113.
52. Rodriguez-Kabana, R., J. C. Williams, and R. A. Shelby. 1982. Assessment of peanut yield losses caused by *Meloidogyne arenaria*. Nematropica 12:279-288.
53. Sasser, J. N., K. R. Barker, and L. A. Nelson. 1975. Correlations of field populations of nematodes with crop growth responses for determining relative involvement of species. J. Nematol. 7:193-198.
54. Schmitt, D. P. and K. R. Barker. 1981. Damage and reproductive potentials of *Pratylenchus brachyurus* and *P. penetrans* on soybean. J. Nematol. 13:327-332.
55. Seinhorst, J. W. 1965. The relation between nematode density and damage to plants. Nematologica 11:137-154.
56. Seinhorst, J. W. 1967. The relationships between population increase and population density in plant parasitic nematodes. III. Definition of the terms host, host status and resistance. IV. The influence of external conditions on the regulation of population density. Nematologica 13:429-442.
57. Seinhorst, J. W. 1968. Underpopulation in plant parasitic nematodes. Nematologica 14:549-553.
58. Seinhorst, J. W. 1970. Dynamics of populations of plant parasitic nematodes. Annu. Rev. Phytopathol. 8:131-156.
59. Seinhorst, J. W. 1979. Nematodes and growth of plants: Formalization of the nematode-plant system, pp. 231-256 *In* Root-knot nematodes (*Meloidogyne* species) systematics, biology and control, ed. F. Lamberti and C. E. Taylor. Academic Press, New York. 477 pp.
60. Seinhorst, J. W. 1981. Growth and yield of oats at a range of *Heterodera avenae* densities and under different watering regimes. Nematologica 27:52-71.
61. Seinhorst, J. W. and H. den Ouden. 1971. The relation between density of *Heterodera rostochiensis* and growth and yield of two potato varieties. Nematologica 17:347-369.
62. Seinhorst, J. W. and J. Kozlowska. 1977. Damage to carrots by *Rotylenchus uniformis,* with a discussion on the cause of increase of tolerance during the development of the plant. Nematologica 23:1-23.
63. Singh, N. D. 1975. Effect of inoculum level and plant age on pathogenicity of *Meloidogyne incognita* and *Rotylenchulus reniformis* on tomato and lettuce. Plant Dis. Rep. 59:905-908.
64. Sleeth, B. and H. W. Reynolds. 1955. Root-knot nematode infestation as influenced by soil texture. Soil Sci. 80:459-461.
65. Steiner, G. 1952. The soil in its relationship to plant nematodes. Soil Sci. Soc. Fla. Proc. 12:24-29.
66. Tavantzis, S. M. 1984. The use of terms of responses of plants to viruses: A reply to recent proposals. Phytopathology 84:379-380.
67. Taylor, P. L., J. M. Ferris, and V. R. Ferris. 1973. Technique for quantifying injury to seedling soybeans by *Pratylenchus penetrans* without sacrificing the plant. J. Nematol. 5:68-69.
68. Thomas, R. J. and C. A. Clark. 1983. Population dynamics of *Meloidogyne incognita* and *Rotylenchulus reniformis* alone and in combination, and their effects on sweet potato. J. Nematol. 15:204-211.
69. Thorne, G. 1961. Principles of nematology. McGraw-Hill, new York. 553 pp.
70. Towson, A. J. and W. J. Apt. 1983. Effect of soil water potential on survival of *Meloidogyne javanica* in fallow soil. J. Nematol. 15:110-115.
71. Tyler, J. 1938. Egg output of the root-knot nematode. Proc. Helminthol. Soc. Wash.

5:49-54.
72. Upadhyay, R. S., M. Oostenbrink, and A. M. Khan. 1972. The effect of different soil types on the density of nematode populations. Indian J. Nematol. 2:42-53.
73. VanderPlank, J. E. 1975. Principles of plant infection. Academic Press, New York. 216 pp.
74. Vrain, T. C. 1982. Relationship between *Meloidogyne hapla* density and damage to carrots in organic soils. J. Nematol. 14:50-57.
75. Wallace, H. R. 1973. Nematode ecology and plant diseases. Edward Arnold, London; Crane, Russak, New York. 228 pp.
76. Wallace, H. R. 1983. Interactions between nematodes and other factors on plants. J. Nematol. 15:221-227.
77. Weingartner, D. P. and J. R. Schumaker. 1981. Comparisons of nematode control and potato yields in large plots treated with different combinations during 1977-1981. Proc. Fla. State Hortic. Soc. 94:141-144 (pub. 1982).
78. Wong, T. K. and W. F. Mai. 1973. Pathogenicity of *Meloidogyne hapla* to lettuce as affected by inoculum level, plant age at time of inoculation and temperature. J. Nematol. 126-129.

CROP LOSS ASSESSMENT METHODS: CURRENT SITUATION AND NEEDS

P. S. Teng

Department of Plant Pathology

University of Minnesota

St. Paul, Minnesota, USA

Pest-induced crop losses are a feature of both primitive and modern agricultural systems involved in food or fiber production. It is common knowledge among agricultural scientists that "yield gaps" exist among countries and even among farms in the same locality with comparable management and environment (46). Some data also exist to show that the level of pest-induced losses has been relatively constant in spite of advances made in chemical control, breeding and management technology (16,20). Because crop losses continue to be encountered while agriculture generally has become more input and energy intensive, much research has been directed toward identifying the causes of loss, quantifying losses, and forecasting losses for disease management (19).

International concern about crop losses led to the Food and Agriculture Organization (FAO), United Nations, convening an international symposium in 1967 (9). One conclusion of the symposium was that methodology for crop-loss assessment was inadequate and deserved high priority for research. As a result of the symposium, FAO initiated a loose-leaf series titled *Crop Loss Assessment Methods* which appeared in three supplements (5,6). Further, FAO conducted regional crop-loss workshops through its International Collaborative Program, and methodologies discussed at the workshops in India (12) and in New Zealand (7) have been published. A second international symposium on crop-loss assessment was held in Minnesota, U.S.A., in 1980 (43) and involved scientists interested in abiotic causes of loss as well as pest-induced losses.

A general introduction to the subject of crop-loss assessment is the book by Ordish (28). Aspects of the methodology for loss assessment were reviewed in 1974 (17), in 1979 (19), and in 1984 (46,47). The American Phytopathological Society convened a symposium in 1982 to discuss loss estimation, with modeling (24) and survey (39) being two of the topics addressed. The mathematics of loss forecasting is also documented by Teng (40), whereas Teng & Oshima (45) have elaborated on the conceptual and practical limitations of current assessment methodology. It is therefore not the intention of this chapter to present another exhaustive review but rather to adopt a "project" approach to discussing the various components of a practical program.

RATIONALE AND CONCEPTS

Several objectives may be identified for any disease-loss appraisal program, and these relate to whether yield loss or crop loss is under consideration. MacKenzie (23) identified yield loss as the loss that may be mainly attributed to a single causal factor and measured in relatively small areas such as experimental plots. He considered crop loss to be the loss estimated over large cropping areas, i.e. an integrated value which is not equal to the sum of individual yield losses. This distinction enables the division of objectives for disease-loss appraisal into those concerned with macro-level phenomenon versus the micro-level phenomenon. For example, crop-loss estimates are urgently needed for evaluation of the economic benefits from plant-protection programs, whether they be national or state programs, developing country or developed country programs (16). Furthermore, regional crop-loss programs will provide the data needed for allocating resources in response to identified priorities for research or extension. The same statistics, collected over many seasons, will also provide a basis for evaluating the impact of changing technology in agriculture and enable future plant protection needs to be identified. The value of regional crop loss programs cannot be underestimated for their contribution to improving our understanding of the field status of crops and pests. Indeed, the surveillance data on disease severity resulting from a crop-loss program is of value itself in defining the extent of endemic pest infestations, as is shown by the development in the U.S.A. of a Cooperative National Plant Pest Survey and Detection System (41).

Plant diseases were originally studied because of the losses they caused, yet today it is paradoxical that there are only a few reliable estimates of loss (17). Reliable loss data are requisite to the development of rational disease-management programs, whether these involve the use of resistant cultivars, pesticides, cultural methods, or their combination. James (16) argued that it is not sufficient to state that disease causes loss. The magnitude of the loss must be quantified so that it can be related to the gain resulting from control. With yield-loss assessment, we see a potential for doing just this. Coupled with public concern over environmental pollution, yield-loss and crop-loss assessments are needed to rationalize the use of pesticides in agriculture and to evaluate objectively the returns from pesticide usage through programs such as the National Pesticide Impact Assessment Program (NAPIAP) in the U.S.A. On farmers' fields, yield-loss models provide a means of deciding on whether any spray application will be profitable (41,45).

STRATEGY FOR LOSS ASSESSMENT

Crop-loss data may be obtained using different approaches, dependent on the resolution of the data required for decision-making and the definition of loss used. The FAO considers crop loss to be the difference between attainable yield and actual yield (5). Attainable yield is that achieved when crops are grown under optimal conditions using fully the available technology in an area, whereas actual yield is what farmers get under their prevailing husbandry (56). Loss is the reduction in quantity and/or quality of the economic yield of a crop. Although some workers have suggested that it is necessary to obtain loss data in quantities such as kg/ha, others have argued that knowing the magnitude of loss (e.g. 1 = none, 5 = very severe) may be acceptable for setting policy on crop protection (17,39). The following approaches have been used to collect regional crop-loss data (38):

1. Expert testimony, where knowledgeable scientists are asked to make a "statement of authority" on the extent of loss based on their experience with the crop and diseases in an area.
2. Enquiries, where estimates are solicited from a broad range of people concerned with the production of a crop in an area, and a consensus is developed on the extent of loss. This approach resembles the "delphi" procedure used in research management.
3. Literature reviews, where published work not specifically designed for loss assessments is evaluated for its value in giving estimates. Examples are multi-location fungicide and cultivar evaluation trials.
4. Remote sensing, where satellite imagery is used in estimating crop area, crop yield and crop losses. This technique has worked well for diseases that result in total plant loss, as in some nematode-caused diseases.
5. Experiment/survey approach, where yield and crop loss models are developed through field experiments and the models coupled with survey data to provide regional loss estimates. James & Teng (19) are of the opinion that this is the most objective method of deriving crop loss data for decision-making at all levels.

There is currently no single approach for collecting regional crop loss data that has been generally accepted. The experiment/survey approach is perhaps the most direct and empirical approach. Van der Graaff (50) has suggested that the indirect methods (approaches 1-3) be used to improve the reliability of direct methods (approaches 4-5), providing data for cross-validation. The experiment/survey approach is to be favored as it contains many elements of the scientific method as we know it today. This approach is discussed here. The general strategy for using this approach requires the following:

—Quantification of pathogen/disease,
—Collection of data to measure the disease-loss relationship,
—Modeling of the disease-loss relationship,

and

—Development and use of regional crop loss data bases.

Each of these will be discussed in the following sections, with the emphasis on the current situation and future needs.

Quantification of Pathogen and Disease

Although fungal, bacterial and viral pathogens commonly cause identifiable symptoms on plants which may be used for designing a disease measurement method, nematodes have generally been quantified in terms of their numbers per soil or plant unit. In the context of loss assessment, quantification is essential for providing the pathogen/disease "descriptor" used in estimating the relationship between the pathogen/disease and yield or loss. Pathogen numbers may be related to the intensity of plant symptoms, which is in turn a visual indication of the stress imposed on the plant, resulting in measurable loss. The choice of a descriptor, therefore, should be made with some knowledge of yield physiology, to reflect the interaction between pathogen and yield (39). It is interesting that with nematodes, preplant nematode density is the most common descriptor used in quantitative relationships of pathogen loss (1).

Disease incidence is the proportion of plants infected in a population, commonly expressed as a percentage, whereas disease severity is the proportion of plant tissue infected (19). The FAO has used disease intensity to mean either disease incidence or disease severity (5). Disease assessment is the process of determining disease intensity in a population of plants using an accepted method. Methods in common use by plant pathologists for field disease assessment are disease keys, standard area diagrams, remote sensing, and population counts. Disease keys and standard area diagrams rely on the determination of severity in comparison with a predefined key or series of diagrams depicting different degrees of severity (15). The severity assessed for a plant part like a leaf includes the infected area as well as any accompanying chlorosis or necrosis. Remote sensing has been successfully used for assessment of losses due to pests that cause total plant loss but has been only marginally successful with pests that affect only plant parts (4,19). Pathogen counts, with the exception of nematodes, are not widely used in quantifying plant diseases. Methodology for making nematode counts is generally well-established (36,53). However, with other types of pathogens, methods for pathogen/disease quantification have not achieved the same degree of standardization when compared with methods for diagnosis.

A problem in loss assessment is the determination of a representative mean value of the pathogen or disease in a cropping unit using the designated method of assessment. Sampling for diseased populations is a relatively under-researched area in comparison with insect sampling, and it was only recently that this has been recognized (39). The distribution of a pathogen or disease in any spatial unit may be mathematically described as a frequency distribution with estimated parameters, e.g. normal or negative binomial. Preliminary indications of the type of distribution are obtained by examining the mean:variance ratio of the sample mean of disease intensity. Knowledge of the type of distribution in a field enables sampling protocol to be designed to obtain a representative mean in an economical manner (21,26). Nematode populations commonly occur as clusters, suggesting that the pattern of taking samples from a field is important (1,11).

Some recent uses of microprocessor technology for disease measurement suggest that we may see more reliable methods in field use in the future. Pederson & Fiechtner (29) have designed a portable, low-cost data acquisition system for measuring canopy reflectances, which may be used for determining the mean effect of a pest on a crop in terms of reduced crop vigor. A laboratory-based video image analysis unit was used by Lindow & Webb (22) to measure the area of infected leaf tissue and proportion of infection. These workers have also begun testing the taping of images of diseased leaves in the field with portable videocameras and analyzing the images in the laboratory. Image analysis is routinely used for measuring the root area of plants, and it is conceivable that there will be developments allowing its use for measuring nematode number in a sample.

Collection of Data to Measure The Disease-Loss Relationship

This represents the experimental phase of a disease-loss program, in which field data are collected either from fields with natural epidemics or from experimental treatment plots with different disease intensities.

THE SINGLE TILLER/PLANT METHOD Fields in any cropping area can commonly be found in one season with varying disease, leading Richardson et al (31) to use a single tiller method for collecting data to model the disease-loss relationship. In this method, hundreds of tillers (shoots) are tagged, with care being made to select tillers reflecting a wide range of severities, including zero and maximum disease. Fields are visited using a predefined survey procedure, and in each field, disease intensity is assessed and tillers harvested. Each tiller is then a single datum point for regression analysis. The method is a derivation of the paired-plant method (3), in which pairs of healthy and diseased plants are tagged and observations made on them through the growing season. The advantages of the single tiller or paired plant method are that natural epidemics are used and there is economy of labor, space and time (19,31). However, the method has mathematical limitations, and models developed have only been able to explain a small proportion of the variation in yield due to disease (19). Inter-plant differences in yield are a major source of variation in single tiller/plant studies. Hau et al. (13) improved on the method by using measurements of plant parts not affected by disease, but related to potential yield, to correct for differences in observed yield, and were able to reduce some of the variation.

THE SYNOPTIC METHOD In an attempt to determine the effect on wheat yield from multiple factors, Stynes (37) developed a "synoptic" procedure in which parts of farmers' fields were intensively sampled throughout the season. Variables measured included disease, insects, nematodes, soil and water properties. Models were developed which explained a significant proportion of the yield variation caused by several factors. A simpler procedure has also been used in developing countries to determine production constraints on farmers' fields by the International Rice Research Institute (IRRI) and in Colombia (30). The synoptic method allows crop-loss profiles (6) to be developed, showing the contribution of each constraint in reducing attainable yield to actual yield. In the U.S.A., Wiese (54) has modified the procedure for field peas but found that the models developed were not stable over seasons. A limitation common to both the single tiller and synoptic methods is that the range of disease severities for each disease may not be wide enough in any season, leading to an underestimation of its importance as a yield constraint.

FIELD PLOT TECHNIQUES Plots arranged using an experimental design such as a randomized complete block are common in crop-loss work. The plots are either paired-treatment or multiple-treatment (9), where treatments are desired levels of disease or pathogen population. In crop loss assessment, the aim of treatment is to ensure that epidemics with different characteristics are generated using methods that may not necessarily be economical (17). In the paired-plot approach, healthy (protected) and diseased (unprotected or inoculated) plots are situated near each other to constitute a replicate and the pairs repeated over many locations (19). With multiple-treatment experiments, treatment extremes range from healthy (no disease) to maximum disease, with intervening levels of disease as the other treatments. Levels of disease have been generated on cultivars with different susceptibilities, but comparable potential yield, by variation in planting date, by use of fungicides, by use of isogenic lines and by differential inoculation (17,19,52). Nematode populations can be varied by growing hosts or non-hosts prior to experimentation, by inoculation or by mixing infested with non-infested soil (35).

An important consideration in plot techniques is the plot size. Although nematologists have generally used microplots, pathologists working with airborne pathogens have had to use larger plots because of the problem of interplot interference (17). In practice, there is a trade off between reducing inherent yield variation by increased plot size and increasing the variation due to soil factors when plot size is increased. In general, small plot size, as opposed to large plot size, results in higher between-plot variation and requires a larger number of replicates for the same difference between two treatments to be detected (18).

Recently, there has been recognition of the inability of standard, experimental designs with replication to provide data that can explain the full range of interaction between crop yield, disease intensity and crop development stage (45-47). The relationship between yield loss and disease at different growth stages was conceptualized as a three-dimension response surface (42), where at each growth stage, disease-loss may be represented by curves like the inverse of the Seinhorst model (1). To derive a holistic model of the three-dimension surface would require data from a wide range of epidemics, more than can be obtained with replicated experiments. Teng & Oshima (45) have argued for the use of response surface methodology, with its emphasis on treatment number rather than replication number, to provide the data needed for modeling disease loss. Further, the statistical techniques for modeling disease-loss relationships are variance-reduction techniques like least-squares regression, which assumes that there is no or minimum variation in the independent variable (disease). With standard experimental designs, this assumption is violated when averaging across replicates while with response surface designs, treatment values may be obtained without variance.

Experiments to model nematode population-loss relationships commonly focus on the initial nematode population. This approach is in marked contrast to experiments involving fungal pathogens where repeated assessments during the season are needed to determine the crop growth state(s) most sensitive to disease. It may be that the precision of nematode models could be greatly improved if data were available on population levels during the cropping season.

Modeling of the Disease-Loss Relationship

A mathematical model is a concise way of representing any system. In crop-loss experiments, the usefulness of the data generated would be limited if the data were not reduced into a simple form. Because of the many forms of the disease-loss relationship, there is no universal mathematical model to fit all these forms. The forms of the relationships range from linear to sigmoid, and Teng (40) has postulated that there are nine possible shapes of the disease-loss curve. Further, the mathematical description of the relationship depends on the disease descriptor (independent variable) used, such as disease severity at one growth stage or area-under-the disease-progress-curve (19,52). With nematodes, the log of nematode density is commonly used as the independent variable (1). The majority of mathematical models describing the disease-loss relationship have been derived using least squares regression techniques, although recently, simulation modeling has been attempted (10,48). With regression models, some workers have suggested that several statistical criteria be used to evaluate each model: F, r, s and t (19,42). Furthermore, assumptions in the data collected for modeling need to be recognized and tested. For example, regression assumes that the variables show a normal distribution yet this assumption is often violated in samples of disease data (19).

The empirical disease-loss models reported in the literature may be grouped into single-point models, multiple-point models, integral models, response-surface models, non-linear models and synoptic models (40)

SINGLE-POINT MODELS Single-point models relate loss to disease intensity at a specific time in the life of a crop, either a critical growth stage or a predetermined number of days into the growing season. With wheat stem rust, loss may be estimated from the percent stem rust severity at the 3/4-berry stage (X) using the model:

$$\% \text{ Loss} = -25.53 + 27.17 \ln X \qquad (33).$$

In this example, the 3/4-berry stage was identified as a time when the crop was most sensitive to rust. James & Teng (19) have cautioned that fitting a single-point model to a data set does not imply that no other growth stages respond to diseases but rather that a particular stage only shows good statistical correlation. Furthermore, it is often necessary to incorporate some physiological knowledge into regression models, to ensure that the models are biologically meaningful (42). Another form of single-point models is that for estimating losses in potato due to late blight from the number of blight-free days (X), using the model:

$$\text{Yield, t/ha} = 234.0 - 1.706\, X \qquad (40).$$

Single-point models are the most common type of disease-loss model in the literature, mainly because they require relatively less data to develop. However their application appears restricted to short-duration, late epidemics with stable infection rates. This type of model assumes that disease dynamics before and after the single point in fields resembles that encountered in the original experiments.

MULTIPLE-POINT MODELS The multiple-point models relate yield loss to several disease assessments during a crop's life. The disease descriptors used in models have been either disease increments during a defined period or disease intensities at identified growth stages. An example is the model for estimating wheat-yield loss due to leaf rust from three growth stages, $X2$ (percent rust/tiller at boot stage), $X5$ (percent rust on flag leaf at early berry stage), and $X7$ (percent rust on flag leaf at early dough stage), using the model (2):

$$\% \text{ Loss} = 5.3788 + 5.5260X2 - 0.3308X5 + 0.5019X7.$$

The workers who developed this model found that even though they could determine several single-point models from the same data, the multiple-point model explained the most variation in yield loss due to rust. Multiple-point models are particularly suited for epidemics that are long in duration, have unstable infection rates, and affect more than one yield component.

INTEGRAL MODELS Integral models relate loss to a disease descriptor derived from summing disease intensities over a specified period of crop growth. The idea may be attributed to VanderPlank (51), who proposed the area-under-disease-progress-curve (AUDPC) as a method for analysis of wheat stem rust data. An example is the model for estimating loss in cowpea due to Cercospora leafspot:

$$\% \text{ Loss} = 0.43 \text{ AUDPC} - 14.95 \quad (34).$$

In general, AUDPC models cannot distinguish between late or early epidemics since two progress curves with very different onset times and infection rates could give the same area under the curve. AUDPC models have been successfully applied for short-duration, late epidemics (17,19). Some workers have improved the predictive ability of AUDPC models by assigning weighting factors to the disease assessments made at different growth stages which are used to calculate the AUDPC (14).

OTHER MODELS The relationship among disease, crop growth stage and loss was conceptualized as a three-dimension response surface by Teng & Gaunt (42) and may be generalized as % Loss = f (Disease, Crop stage), thereby enabling loss estimation if the disease intensity and growth stage are known. This response surface model may also be considered an integrated series of single-point models, and several workers have developed models fitting the concept (40). Response surface models require substantially more data to develop than the other models discussed previously and have led to research on alternative ways of experimentation to collect data, as discussed in a previous section.

The majority of disease-loss models have assumed a linear relationship, while it is generally recognized that biological relationships may be non-linear. Madden et al. (25) have modified the Weibull Distribution Function, with very flexible curve-fitting capabilities, for use on disease-loss relationships, and have demonstrated that good fits to data may be obtained.

More than one model can commonly be found to fit a set of experimental data on any disease-loss system. Although it is generally advisable to collect more data than is necessary for modeling because of the lack of prior knowledge on the form of the model, with some diseases and crops, enough is known of yield physiology to enable postulation of potential relationships (42). This approach can do much to guide the design of experiments and pinpoint growth stages where it may be useful to have more treatments. The intended use of a model is another consideration in determining the form of the model, whether it be single point or multiple point. In surveys, where fields may be visited only once, several different single-point models would be needed. To forecast potential yield loss may require a multiple-point or integral model that can account for fluctuating rates of disease progress in response to factors like fungicide application.

Development and Use of Regional Crop-Loss Data Bases

Regional crop-loss data bases have been developed for several countries through the use of sample surveys and the indirect methods previously discussed. This endeavor represents the next phase in most crop-loss programs (17,19). A regional data base is considered here to be a set of crop-loss data collected over many definable spatial units such as counties. Although the method of procuring the crop-loss data may be different, it is common that such data bases provide information on the prevalence (distribution), incidence and severity of diseases and/or the related losses in percentages or tonnage (41).

In developing regional loss data bases, James and Teng (19) have recommended that sound statistical sampling methods be used to determine field averages and to select representative fields. Some of the within-field sampling methods were previously discussed. Selection of fields to represent a region has been based on simple random sampling, stratified sampling or multistage sampling. Stratification based on county (or similar) lines, crop reporting districts, or state boundaries appears to be most common in the literature, although stratification according to soil types, agroecological zones or production systems has more interpretative value to agricultural scientists (41). One of the longest ongoing survey programs is that in England and Wales which started in 1967 on wheat and barley (20). The program has demonstrated its usefulness in keeping administrators, researchers and farmers aware of the status and importance of cereal diseases and regularly publishes loss figures. In this program, loss is estimated by determining field disease severity, which is then used with a disease-loss model. Another long-term survey program is that of the U.S.D.A. Cereal Rust Laboratory in St. Paul, which monitors changes in rust virulence as well as estimates losses through expert opinion and actual field data from major wheat states. The literature on *ad hoc* disease surveys is plentiful, and although not allowing historical trend analysis, these surveys have been useful in identifying diseases for further research and extension (19). Some states in the U.S.A., such as North Carolina, collate crop-loss information each year from varied sources (27).

With developments in computer technology, Rijsdijk & Zadoks (32) have plotted isoloss lines for Western Europe showing the mean annual loss in cereals due to each key pest. A combination of expert opinion, mail surveys and literature reviews was used to provide the data. These workers also plotted risk maps to show parts of the region where a certain disease has historically caused significant loss. A new program in the U.S.A. sponsored by the U.S.D.A., Animal and Plant Health Inspection Service, began in 1982 to develop a national pest survey data base for use in export certification (41). This program, the Cooperative National Plant Pest Survey & Detection Program, relies heavily on computers and telecommunications for exchange of weekly pest data among states and a central computer and, in the future, will be capable of providing real-time pest and crop data during the growing season. Individual state systems for computerized, on-line pest management are in place, e.g. in Michigan and Minnesota (41). In Europe, the Dutch "EPIPRE" system of national pest management has been extended to other countries (55). This system projects increases in pest populations and the potential economic loss, thereby providing a basis for control. In the U.S.A., research is also being directed at linking regional crop loss data bases to decision-making at the farm level. The "RUSTMAN" model for timing fungicide sprays on sweet corn makes recommendations based on rust severity in the field, estimated field loss, and rust severity in surrounding fields (44). With increasing miniaturization of electronic equipment, matched by increasing computer power, it may soon be possible to deploy highly sophisticated models of nematode-loss systems in farmers' fields for pest monitoring and management.

CONCLUDING REMARKS

A crop-loss program should have a flexible structure to enable it to respond to changing needs and objectives. The different approaches reviewed in this chapter, and the specific methods discussed, allow for this flexibility. In line with the "project" approach of this chapter, it is conceivable that the first year of a new crop-loss program would have field plot experiments, sample survey to determine the distribution and intensity of pests, and use of the single tiller/plant method as part of the survey. Concurrently, development of the methods of data analysis and information storage and delivery would have to be conducted. At the end of the first season, some preliminary disease-loss data would already be available for use by a varied clientele. The second and subsequent seasons serve to increase the reliability of the models and procedures developed during the first. There is a dynamic cycling process, with feedback on performance of the models to the research such that each succeeding season should see improvement in the overall crop loss program.

Plant pathology may be regarded as a practical exercise in problem-solving, aimed at the economic control of plant diseases (19). Two elements are evident in the resolution of a problem—first the definition and then the solution. In the context of plant pathology, crop-loss assessment represents the problem-definition element and is therefore a vital component in any plant-protection program since it is illogical to attempt a solution of a problem that has not been defined. With increasing resource awareness in agricultural research worldwide, the technology of crop-loss assessment is likely to play an important role in both developing and developed countries in the rational allocation of scarce resources.

The current authoritative publication on global crop loss statistics is the 1967 book by Cramer (8), which drew heavily from older sources such as the 1965 U.S.D.A. handbook, "Losses in Agriculture" (49). These statistics are being used to set and justify many plant-protection programs, with figures like 35% average preharvest crop loss per year due to all pests being commonly cited (45). There is an urgent need to update global statistics on crop losses, especially in the developing countries. One way by which this may be done is through increased exchange of information such as the workshop for which this chapter was written.

Literature Cited

1. Barker, K. R. and T. H. A. Olthof. 1976. Relationships between nematode population densities and crop responses. Annu. Rev. Phytopathol. 14:327-353.
2. Burleigh, J. R., A. P. Roelfs, and M. G. Eversmeyer. 1972. Estimating damage to wheat caused by *Puccinia recondita tritici*. Phytopathology 62:944-946.
3. Chester, K. S. 1950. Plant disease losses: their appraisal and interpretation. Plant Dis. Rep. Suppl. 193:189-362.
4. Chiang, H. C. and V. R. Wallen. 1971. Detection and assessment of crop diseases and insect infestations by aerial photography, section 3.1.7 *In* Crop loss assessment methods, ed. L. Chiarappa. FAO/CAB, Slough, England. 123 pp.
5. Chiarappa, L., ed. 1981. Crop loss assessment methods. FAO/CAB, Slough, England.
6. Chiarappa, L., ed. 1971. Crop loss assessment methods—supplement 3. FAO/CAB, Slough, England. 123 pp.
7. Close, R. C, R. E. Gaunt, I. C. Harvey, A. J. McCully, F. R. Sanderson, and P. S. Teng. 1978. Epidemiology and crop loss assessment. Lincoln College, Canterbury, New Zealand. 280 pp.
8. Cramer, H. H. 1967. Plant protection and world crop production. Pflanzenschutz-Nachr. Bayer 20:1-524.
9. FAO, UN. 1967. Background papers presented at the FAO Symposium on Crop Losses. FAO, Rome. 330 pp.
10. Ferris, H. 1978. Modification of a computer simulation model for a plant-nematode system. J. Nematol. 10:198-201.
11. Goodell, P. B. and H. Ferris. 1981. Sample optimization for five plant-parasitic nematodes in an alfalfa field. J. Nematol. 12: 304-313.
12. Govindu, H. C., G. K. Veeresh, P. T. Walker, and J. F. Jenkyn. 1980. Assessment of crop losses due to pests and diseases. Univ. Agric. Sci., Bangalore, India. 300 pp.
13. Hau, B., J. Kranz, H. J. Dengel, and J. Hamelink. 1980. On the development of loss assessment methods in the tropics, pp. 254-263 *In* Crop loss assessment. Minn. Agric. Exp. Stn. Misc. Publ. No. 7. 326 pp.
14. Hills, F. J., L. Chiarappa, and S. Geng. 1980. Powdery mildew of sugarbeet: disease and crop loss assessment. Phytopathology 70:680-682.
15. James, W. C. 1971. An illustrated series of assessment keys for plant diseases, their preparation and usage. Can. Plant Dis. Surv. 51:39-65.
16. James, W. C. 1978. Rationale and concepts of disease loss appraisal, pp. 1-1 to 1-6 *In* Epidemiology and crop loss assessment, ed. R. C. Close, et al. Lincoln College, New Zealand.
17. James, W. C. 1974. Assessment of plant diseases and losses. Annu. Rev. Phytopathol. 12:27-48.
18. James, W. C. and C. S. Shih. 1973. Size and shape of plots for estimating yield losses

from cereal foliage diseases. Exp. Agric. 9:63-71.
19. James, W. C. and P. S. Teng. 1979. The quantification of production constraints associated with plant diseases. Appl. Biol. 4:201-267.
20. King, J. E. 1980. Cereal survey methodology in England and Wales, pp. 124-133 *In* Crop loss assessment. Minn. Agric. Exp. Stn. Misc. Publ. No. 7. 326 pp.
21. Lin, C. S., G. Poushinsky, and M. Mauer. 1979. An examination of five sampling methods under random and clustered disease distribution using simulation. Can. J. Plant Sci. 59:121-130.
22. Lindow, S. E. and R. R. Webb. 1983. Quantification of foliar plant disease symptoms by microcomputer-digitized video image analysis. Phytopathology 73:520-524.
23. MacKenzie, D. R. 1983. Toward the management of crop losses, pp. 82-92 *In* Challenging problems in plant health, ed. T. Kommedahl and P. H. Williams. Am. Phytopathol. Soc., St. Paul, Minn.
24. Madden, L. V. 1983. Measuring and modeling crop losses at the field level. Phytopathology 73:1591-1599.
25. Madden, L. V., S. P. Pennypacker, C. E. Antle, and C. H. Kingsolver. 1981. A loss model for crops. Phytopathology 71:685-689.
26. Main, C. E. and C. H. Proctor. 1980. Developing optimal strategies for disease-loss sample survey, pp. 118-123 *In* Crop loss assessment. Minn. Agric. Exp. Stn. Misc. Publ. No. 7. 326 pp.
27. Main, C. E., B. W. Essex, and A. W. Bragg. 1980. Crop losses in North Carolina due to plant diseases and nematodes. N. C. State Univ. Dep. Plant Pathol. Spec. Publ. No. 1. 106 pp.
28. Ordish, G. 1952. Untaken harvest: man's loss of crops from pest, weed and disease. Constable, London. 171 pp.
29. Pederson, V. D. and G. Fiechtner. 1980. A low-cost, compact data acquisition system for recording visible and infrared reflection from barley crop canopies, pp. 71-75 *In* Crop Loss Assessment. Minn. Agric. Exp. Stn. Misc. Publ. No. 7. 326 pp.
30. Pinstrup-Andersen, P., N. de Londono, and M. Infante. 1976. A suggested procedure for estimating yield and production losses in crops. PANS (Pest Artic. and News Summ.) 22:359-365.
31. Richardson, M. J., M. Jacks, and S. Smith. 1975. Assessment of loss caused by barley mildew using single tillers. Plant Pathol. 24:21-26.
32. Rijsdijk, F. H. and J. C. Zadoks. 1979. A data bank on crop losses. Eur. Mediterr. Plant Prot. Organ. (OEPP/EPPO) Bull. 9:297-303.
33. Romig, R. W. and L. Calpouzos. 1970. The relationship between stem rust and loss in yield of spring wheat. Phytopathology 60:1801-1805.
34. Schneider, R. W., R. J. Williams, and J. B. Sinclair. 1976. Cercospora leaf spot of cowpea: models for estimating yield loss. Phytopathology 66:384-388.
35. Seinhorst, J. W. 1981. Methods for generating different pest/yield relationships (Nematodes), pp. 85-89 *In* Crop loss assessment methods-supplement 3, ed. L. Chiarappa. FAO/CAB.
36. Southey, J. F. 1978. Plant nematology. Minist. Agric., Fish. and Food Tech. Bull. 7. Her Majesty's Stationery Office, London. 440 pp.
37. Stynes, B. A. 1975. A synoptic study of wheat. Ph.D. thesis, University of Adelaide, South Australia.
38. Teng, P. S. 1981. Data recording and processing for crop loss models, pp. 105-109 *In* Crop loss assessment methods-supplement 3, ed. L. Chiarappa. FAO/CAB, Slough, England.
39. Teng, P. S. 1983. Estimating and interpreting disease intensity and loss in commercial fields. Phytopathology 73:1587-1590.
40. Teng, P. S. 1985. Construction of predictive models. II. Forecasting crop losses *In* Advances in Plant Pathology, Vol. 4, ed. C. A. Gilligan. Academic Press, London. In press.
41. Teng, P. S. 1984. Surveillance systems in disease management. FAO Plant Prot. Bull. In press.
42. Teng, P. S. and R. E. Gaunt. 1981. Modeling systems of disease and yield loss in cereals. Agric. Syst. 6:131-154.
43. Teng, P. S. and S. V. Krupa, eds. 1980. Crop loss assessment. Minn. Agric. Exp. Stn. Misc. Publ. No. 7. 326 pp.
44. Teng, P. S. and P. R. Montgomery. 1982. RUSTMAN: a portable, microcomputer-based economic decision model for sweet corn common rust control. Phytopathology

72:1140 (Abstr.).
45. Teng, P. S. and R. J. Oshima. 1983. Identification and assessment of losses, pp. 69-81 *In* Challenging problems in plant health, ed. T. Kommedahl and P. H. Williams. Am. Phytopathol. Soc., St. Paul, Minn.
46. Teng, P. S. and W. W. Shane. 1983. Crop losses due to plant pathogens, pp. 81-89 *In* Proc. Tenth Int. Congr. Plant Prot., Brighton, England, Vol. I.
47. Teng, P. S. and W. W. Shane. 1984. Assessment of crop losses caused by plant pathogens. Reviews in Plant Sciences. CRC Press, Cleveland, Ohio. In press.
48. Teng, P. S., M. J. Blackie, and R. C. Close. 1978. Simulation modelling of plant diseases to rationalize fungicide use. Outlook Agric. 9:273-277.
49. U.S. Dep. Agric./Agric. Res. Service. 1965. Losses in agriculture. U.S. Dep. Agric., Agric. Handbook No. 291. 120 pp.
50. Van der Graaff, N. 1981. Increasing reliability of crop loss information: the use of indirect data, pp. 65-69 *In* Crop loss assessment methods-supplement 3, ed. L. Chiarappa. FAO/CAB, Slough, England. 123 pp.
51. VanderPlank, J. E. 1963. Plant diseases: epidemics and control. Academic Press, London. 349 pp.
52. Walker, P. T. 1983. Crop losses: the need to quantify the effects of pests, diseases and weeds on agricultural production. Agric., Ecosystems and Environ. 9:119-158.
53. Webster, J. M., ed. 1972. Economic nematology. Academic Press, London. 563 pp.
54. Wiese, M. V. 1980. Comprehensive and systematic assessment of crop yield determinants, pp. 262-269 *In* Crop loss assessment. Minn. Agric. Exp. Stn. Misc. Publ. No. 7. 326 pp.
55. Zadoks, J. C. 1983. An integrated disease and pest management scheme, EPIRE, for wheat, pp. 116-129 *In* Better crops for food, Ciba Foundation Symposium 97. Pitman Books, London
56. Zadoks, J. C. and R. D. Schein. 1979. Epidemiology and plant disease management. Oxford Univ. Press, New York. 427 pp.

THE ROLE OF SYSTEMS SCIENCE AND DATA MANAGEMENT IN NEMATOLOGY

G. W. Bird[1], *R. L. Tummala*[2], *S. H. Gage*[3]

Michigan State University
East Lansing, Michigan, U.S.A.

A modern era of the Science of Nematology evolved immediately after the discovery of the nematicidal properties of 1,3-dichloropropene-propanes by Carter in 1943 (6). During the past 40 years, our understanding of nematodes has increased greatly (13, 20). Today, *Caenorhabditis elegans* is considered as a model metazoan system for biotechnology research (19). Although nematology is still a relatively young discipline with inadequate personnel and fiscal resources, significant progress has been made in making agricultural production system managers aware of the importance of phytopathogenic nematodes and related management procedures designed to minimize crop losses (2).

Nematology is currently in a state of transition. This change is dictated by environmental and human-health concerns related to both fumigant and non-fumigant nematicides. These challenges are similar to those facing other pestology disciplines. As this transition proceeds, the role of timely and high quality information about nematodes and their relationships to specific agricultural production systems will become significantly more important than in the past. In our current "age of information," the use of a systems approach and data-base management technology is imperative (5, 7, 8, 12, 15, 16). The objective of this contribution is to provide nematologists with an introduction to these topics. General references and examples related to nematology are included as a basis for further study. Because of the nature of recent research by the senior author, nematodes other than *Meloidogyne* spp. will be used as examples. The principles, however, should be of value in research, extension and teaching programs related to *Meloidogyne* spp. The chapter is divided into sections on systems science, role of computers in nematology, and data-base management technology.

SYSTEMS SCIENCE

The word science can be defined as a body of knowledge and the requisite methodology. Systems science is a discipline that provides a framework for the study of interactions among related entities (5, 7, 17). Systems science consists of a set of basic procedures that can be used in the description, modeling, evaluation; design and optimization of systems. This discipline evolved out of necessity over the past 40 years as a way of dealing with complex military operations and space exploration development, and has now become of significant value in the evaluation and management of agricultural and natural resource production systems.

Conceptual Models

Systems have real or conceptual boundaries. The first step in a systems approach is identification of these boundaries. This process consists of the dichotomization of the system (universe of concern) into the object of control and external environment (Fig. 1). If the objective of a model is to summarize existing nematology literature about *Heterodera schachtii* (sugarbeet-cyst nematode) and *Brassica capitata* (cabbage) in a quantitative manner with predictive capabilities, the system boundaries can be defined so the host-parasite relationship is

[1]Nematologist, Department of Entomology.
[2]Systems Scientist, Department of Electrical Engineering and Systems Science.
[3]Environmental Entomologist, Department of Entomology.

the object of control and all other factors influencing this interaction are considered as the external environment (Fig. 2). This separation does not imply that the system and the environment are mutually independent. On the contrary, it provides a framework for quantitative evaluation of the interdependence between the object of control and its surrounding environment.

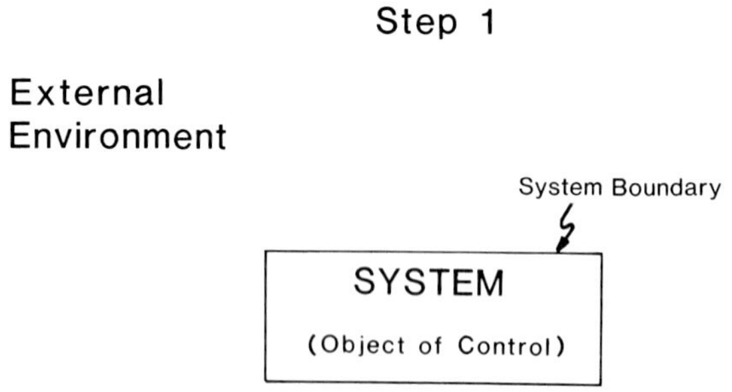

Fig. 1. A conceptual model illustrating the dichotomization of a system (universe of concern) into the object of control and external environment.

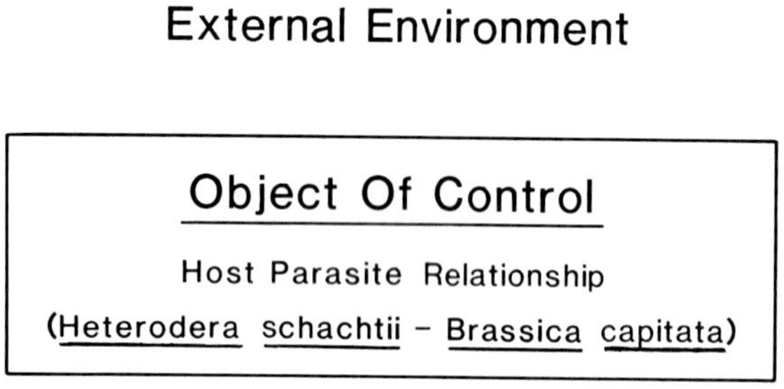

Fig. 2. A conceptual model illustrating the dichotomization of the *Heterodera schachtii-Brassica capitata* host-parasitic relationship into the object of control and external environment.

The second step consists of identification of the system stimuli and responses. For example, application of the nonfumigant nematicide fenamiphos minimizes damage to cabbage by *H. schachtii* (1). This system can be illustrated as a "Black Box" model, with the nematicide represented as a stimulus (s_1) of the object of control (Fig. 3). The improved cabbage yield is represented as a response (r_1) of the system. Although this very simple conceptual model is of limited value, it can be used as the beginning of a process for evaluation of complex phenomena, such as nematode behavior or the impact of plant-parasitic nematodes on an agricultural production system. Although the use of a systems approach in nematology and other biological sciences has usually been limited to ecological and population dynamics studies, the procedure is also of value for physiologic and socio-economic components of science.

In the implementation of scientific methodology (scientific method), it is very important to define the problem, research objectives and research hypotheses clearly and concisely. When these procedures are incorporated into a systems approach, they can be represented, in part, by definition of the desired system responses ($r_1 \ldots r_n$). This part of systems methodology must be done with great care. As with all other aspects of science and scientific method, the procedure is objective dependent.

The third step in the systems approach is the dichotomization of the external environment into controllable and noncontrollable factors. Fenamiphos and NPK fertilizer applications are examples of controllable system inputs. Rainfall and air temperature, however, are usually noncontrollable (Fig. 4).

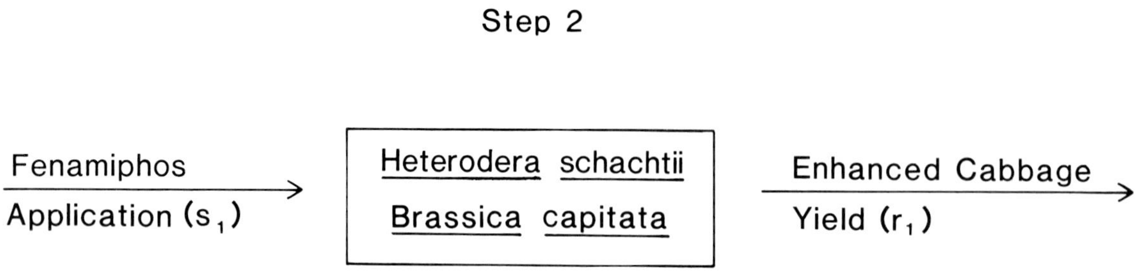

Fig. 3. A "Black Box" model of the *Heterodera schachtii-Brassica capitata* system illustrating the increased cabbage yield (r_1) resulting from the application of fenamiphos (s_1).

STEP 3

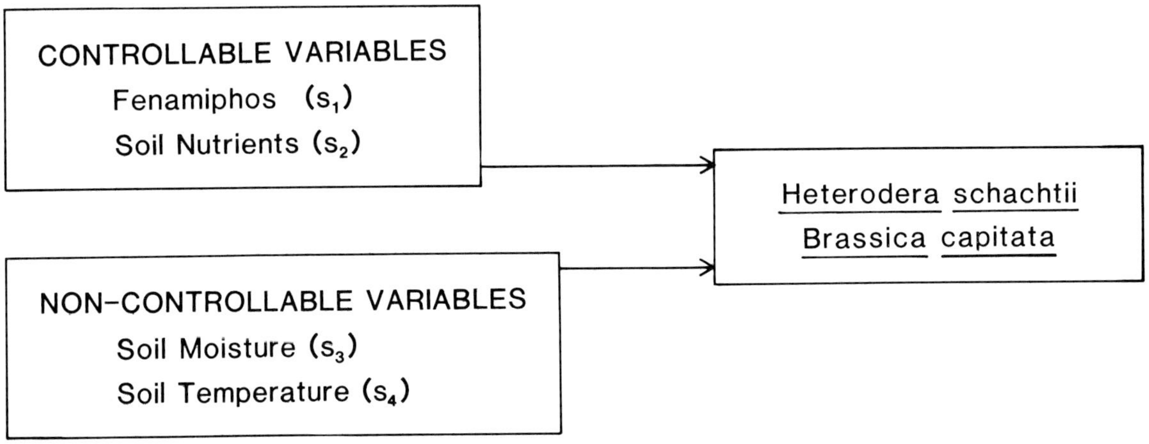

Fig. 4. A conceptual model illustrating the dichotomization of the external environment into controllable and non-controllable factors.

Steps 4&5

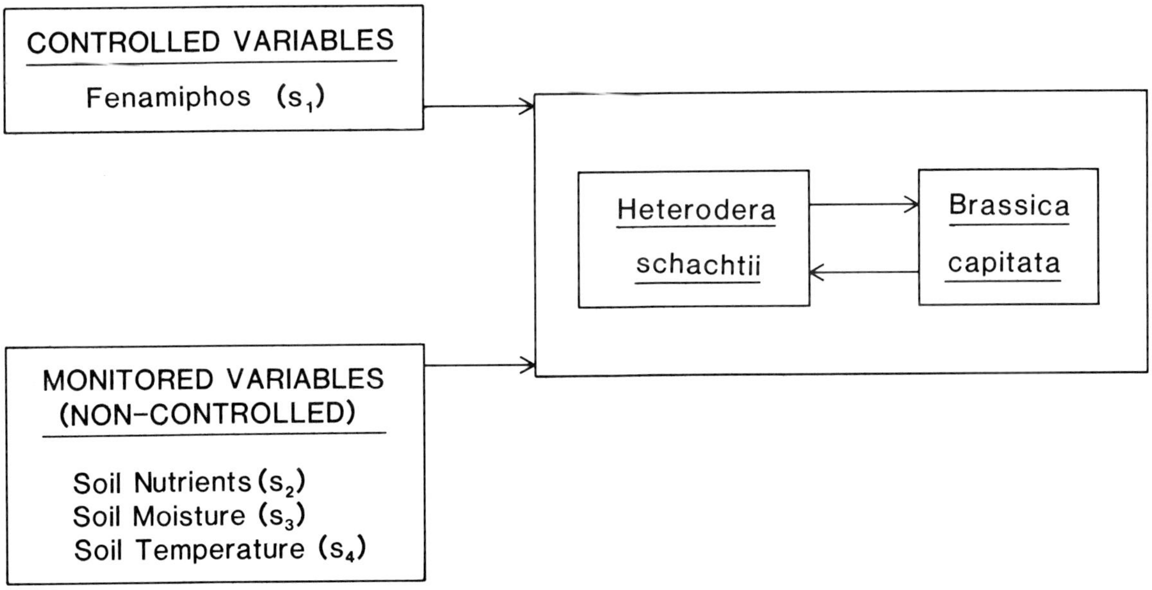

Fig. 5. A conceptual model illustrating the dichotomization of the external environment into controlled and non-controlled (monitored) factors.

In the fourth step, you determine which of the controllable stimuli are to be controlled, and which will be included as noncontrolled variables. In the *H. schachtii-B. capitata* example, it may not be necessary to control soil nutrients (Fig. 5). Therefore, soil nutrients would be considered with air temperature and rainfall as a noncontrolled variable. If specific noncontrolled variables have significant impacts on the object of control, they must be monitored and perhaps should be classified as monitored variables.

As currently described, the *H. schachtii-B. capitata* model is still of relatively limited value. One reason is that the object of control is defined in inadequate detail. In the fifth step, additional resolution is obtained when the object of control is divided into two subsystems: the sugarbeet cyst nematode and the cabbage plant (Fig. 5). One of the most powerful aspects of systems science is the discrete component approach. This approach is used for detailed evaluation of specific components of the system and identifying their interactions. The components can be removed from the system, studied as free-body models, and then reinserted into the model (Fig. 6), thus allowing complexity to be incorporated into the system response (Fig. 7). Although the sugarbeet-cyst nematode-cabbage example is not a complete biological representation of the life cycle of this nematode, these components are all that is necessary for achieving the desired objectives of the model. The structure of a model should never be more complex than what is needed to complete the objective.

After a conceptual model is completed, various quantitative procedures can be used to make the model suitable for specific purposes such as prediction of the population density (*P*) of *H. schachtii* at some future time, $\Delta t + 1$ (Fig. 8). A simple representation of this is described by the following state equation:

$$P(t + \Delta t) = P(t) - m(N,T)N(t) + C(N)N(t) + E(N,T)N(t) + J(t)$$

where $m(N,T)$ = percent natural nematode mortality during the interval $(t, t + \Delta t)$; $C(N)$ = percent nematode population control from application of the nematicide fenamiphos during the interval $(t, t + \Delta t)$; $E(N,T)$ = nematode eggs produced during the interval $(t, t + \Delta t)$; and $J(t)$ = 2nd stage juveniles or eggs introduced or migrating into the system during the interval $(t, t + \Delta t)$.

Models of biological systems can be static or dynamic. In a static model the desired responses are obtained by manipulation of current stimuli without any historical information about the system (Fig. 9). A dynamic model, however, requires an understanding of the current state of the system, which considers the historical development of the system resulting in its current state (Fig. 10).

Creating a model system to substitute for a real system can be a primary objective of

Step 6

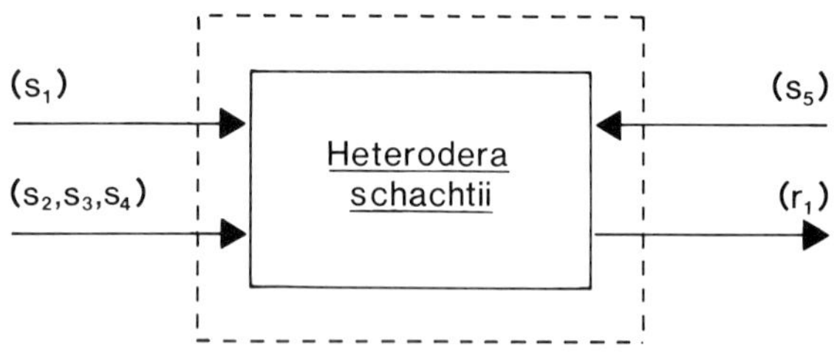

Fig. 6. Example of a "Black Box" free body model of *Heterodera schachtii* with the associated system constraints ($s_1 \ldots s_4$ and r_1).

Step 7

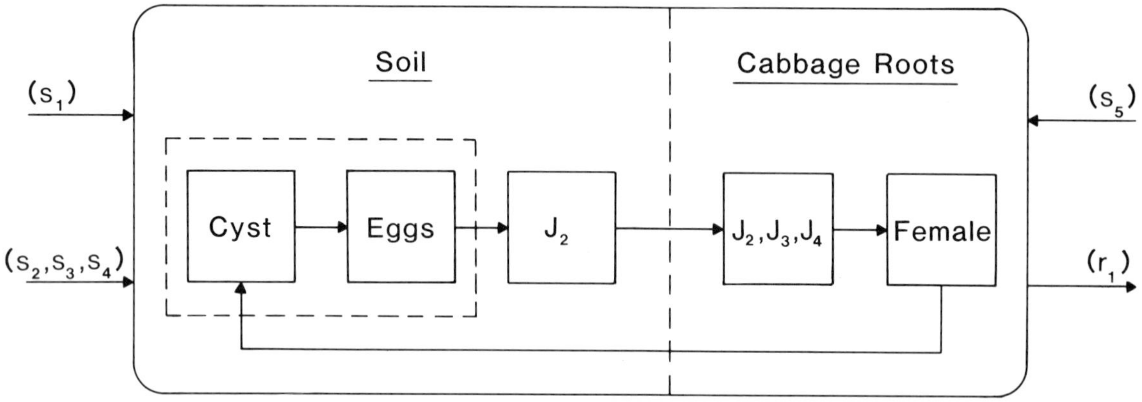

Fig. 7. A conceptual model of selected components of the life cycle of *Heterodera schachtii*.

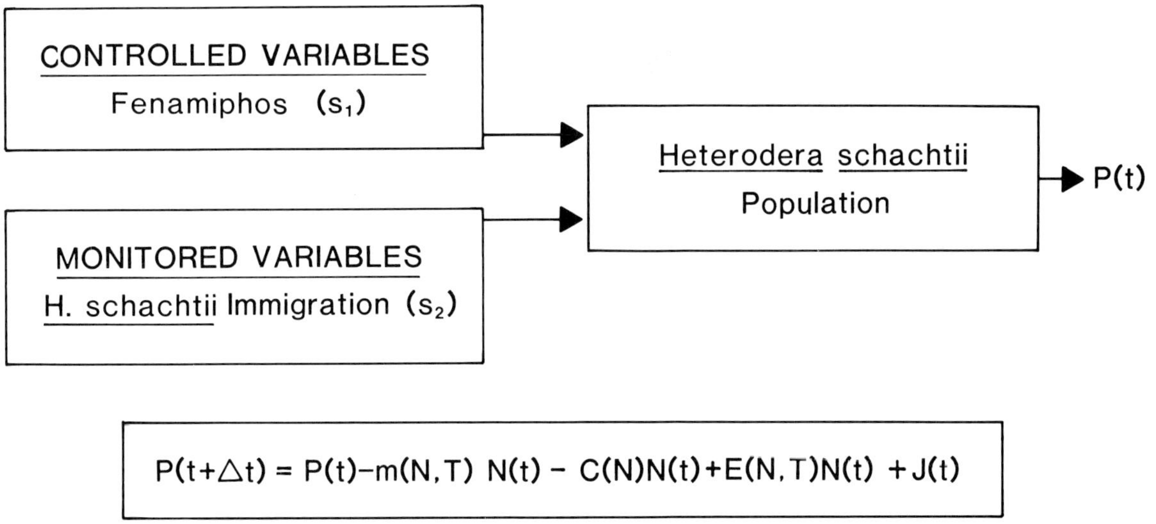

Fig. 8. Model of the population dynamics of *Heterodera schachtii*, with a state equation of $P(t+\Delta t) = P(t) - m(N,T)\ N(t) - C(N)\ N(t) + E(N,T)\ N(t) + J(t)$.

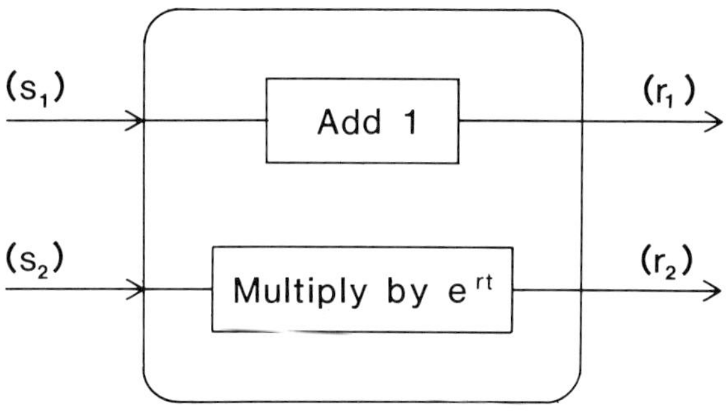

Fig. 9. A conceptual illustration of a static model.
$r_1 = s_1 + 1$
$r_2 = s_2e^{rt}$

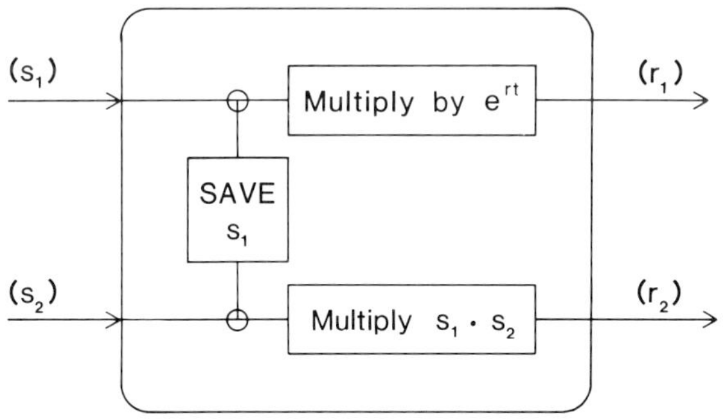

Fig. 10. A conceptual illustration of a dynamic model.

systems analysis. This type of modeling may be done to obtain information concerning the consequences of proposed actions or policies on the system. A model, however, is only an abstraction of a real system—not the system. The model should reflect the dominant features of the system, which in turn should depend on the given objective. Thus, if the objective is to describe the population dynamics of a nematode species, it should be possible to include only features associated with the population as a whole, rather than individuals within the population.

A model can be a quantitative representation of a system. Although a conceptual diagram can be called a model, the definition can be restricted to models that are quantitative representations. For example, the increase of a nematode population is exponential in an essentially unlimited, favorable environment (Fig. 11). As time passes, the population density increases until the number of individuals approaches infinity, as indicated in the equation:

$$P_t = P_o e^{rt};\ t > 0$$

where P_t = nematode population density at some t; P_o = nematode population density at some arbitrarily set time; and t = time (units of time such as hours, days, weeks, years) set according to the objectives of the model.

The model includes the relevant attributes within the context of the defined objectives and is a logical framework for illustration of the interactions among the components of the system.

A model may be a hypothesis, dealing with the behavior of the system as described by the best information available. A model is usually

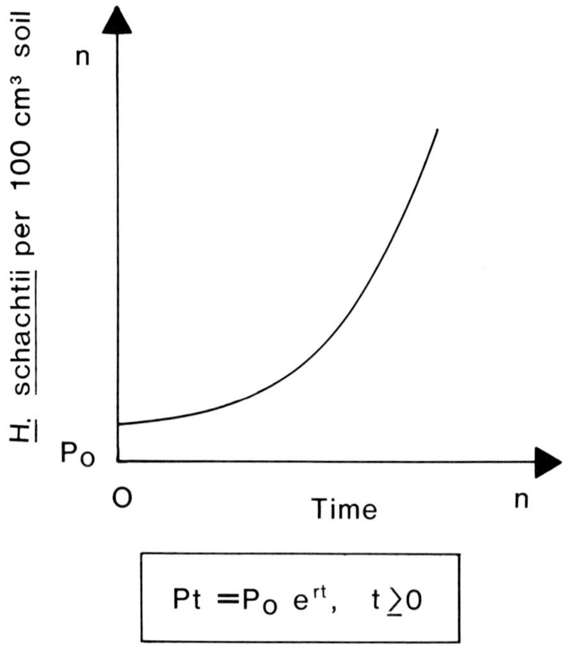

Fig. 11. Fate of the population density of *Heterodera schachtii* in an unlimited favorable environment.

the end result of an iterative process, continuously updated with new information, thus providing a concise summary of the past information of the system and a description of its current state. One of the greatest advantages is that users can determine the effects of manipulations in the real systems through the use of its imitation, the model. Models can be used to enhance understanding of behavior of the system under various hypothetical conditions at very low cost. They can be used to design and test integrated nematode management strategies.

Models must be realistic, including the ability to follow the general form of the system being modeled. Assuming the model hypotheses are realistic, the model can predict the time course of the model variables within a given degree of precision. A model also must be general enough to include all possible situations.

Resolution is the opposite of generality and relates to the number of attributes of a system that a model attempts to reflect. Time scale is an example of resolution. A model designed to show yearly trends in nematode populations would most likely lack the resolution necessary to predict changes in the system at 1-min. phase intervals.

No modeling effort is complete without validation. This phase is the most critical and difficult task. Validation is easier if the parameters of the model can be measured and verified in the field or under controlled conditions through experimentation. The system structure is then verified when the overall system behavior is compared with that of the model behavior. Validation is usually accomplished by comparison of the trends, such as nematode population density, obtained in an independent set of data with those from the model. If the predicted population densities agree with the observed data at some satisfactory level of probability, the real world system is considered as identified, and the model accepted for use in systems analysis and design. If the degree of agreement between the predicted nematode population density and the measured data is inadequate, then the model should be modified. This refinement may involve changes in the perceived interactions within the system (system structure), equation forms, or values of the constants. One important point in the validation process is that the data used to construct the model must always be independent of the data used to validate the model.

Once the model is validated or accepted as a representation of the system under study, then it can be used for various types of analyses. A sensitivity analysis of a specific parameter (i.e., soil temperature) can be useful at this stage to evaluate the impact of this parameter on the system. One can check for possible errors in the estimated values and their potential for control of the system by artificial modification of these parameter values (17).

Computer Simulations

If a model is used for a computer simulation, it is usually necessary for the known biology of the system to be incorporated both as graphic and mathematical functions (Fig. 12). In 1982, Dr. Annette Klieneke-Borchers (University of Hannover, Federal Republic of Germany) and the senior authors of this chapter constructed such a computer simulator for the cabbage-sugarbeet cyst nematode (NEMACAB). The simulation was designed to predict cabbage yield and final nematode population densities in the presence or absence of the organocarbamate nematicide, aldicarb (Table 1). Although the conceptual models used for the development of computer simulations frequently appear to be complex, it is imperative that the output be user oriented. A corn-nematode crop loss/nematicide cost-benefit model (NEMACORN, programmed in 1981 by D. Miller and G. W. Bird) is an example of a model that has been useful for decision-making related to nematode control programs in Michigan corn production. Output from a simulation of a 100-acre corn field with three areas with nematode population densities above the damage threshold is presented as an example of NEMACORN (Table 2).

Although both NEMACORN and NEMACAB are useful for assisting with management decisions, they also can be used for research and education. A third nematode/crop computer simulation, POTATOPEST, has been used for identification of nematology research priorities and experimental simulations (3, 4). A hypothesis for research on the joint action of the root-lesion nematode, *Pratylenchus penetrans*, and Colorado potato beetle on potato was derived from an experimental simulation with POTATOPEST (14).

NEMACORN, NEMACAB and POTATOPEST have been used successfully for educational purposes. POTATOPEST is used to assist in the teaching of the principles of nematode population dynamics in an academic course in nematology for undergraduate students. POTATOPEST is also used in an introductory course in integrated pest management (IPM) to teach students how to calculate economic thresholds, the optimization point where marginal revenue from nematode control is equal to the marginal cost of control. Both NEMACORN and POTATOPEST also have been used in extension education programs, particularly in extension field staff in-service training and comprehensive IPM scout training schools. When used as an integral component of a teaching program, computer simulations are powerful educational tools for which there are currently no substitutes.

Table 1. Output from NEMACAB (a computer simulation of the interactions among *Heterodera schachtii* and *Brassica capitata*) for an initial population density of 2,000 eggs and juveniles per 100 cm^3 of soil in the presence and absence of the organocarbamate nematicide, aldicarb.

Weeks after feeding (accumulative degree days)	*H. schachtii* eggs and juveniles per 100 cm^3 soil		Plant weight (g) Root		Plant weight (g) Shoot		Head diameter (CM)	
	+aldicarb	−aldicarb	+aldicarb	−aldicarb	+aldicarb	−aldicarb	+aldicarb	−aldicarb
0 (0)	2000	2000	0	0	0	0	0.0	0.0
2 (218)	1627	1627	22	22	51	39	0.0	0.0
4 (424)	928	960	43	40	102	45	0.0	0.0
6 (712)	1367	3184	60	55	335	194	0.0	0.0
8 (964)	2186	6311	79	74	660	448	10.2	9.3
10 (1204)	2713	6986	90	80	1444	987	13.5	11.6
12 (1416)	3427	7624	97	80	2317	1588	17.3	14.2
14 (1554)	4083	8932	100	81	2796	1934	19.4	15.7
16 (1641)	4707	10283	101	82	3150	2199	20.9	16.8

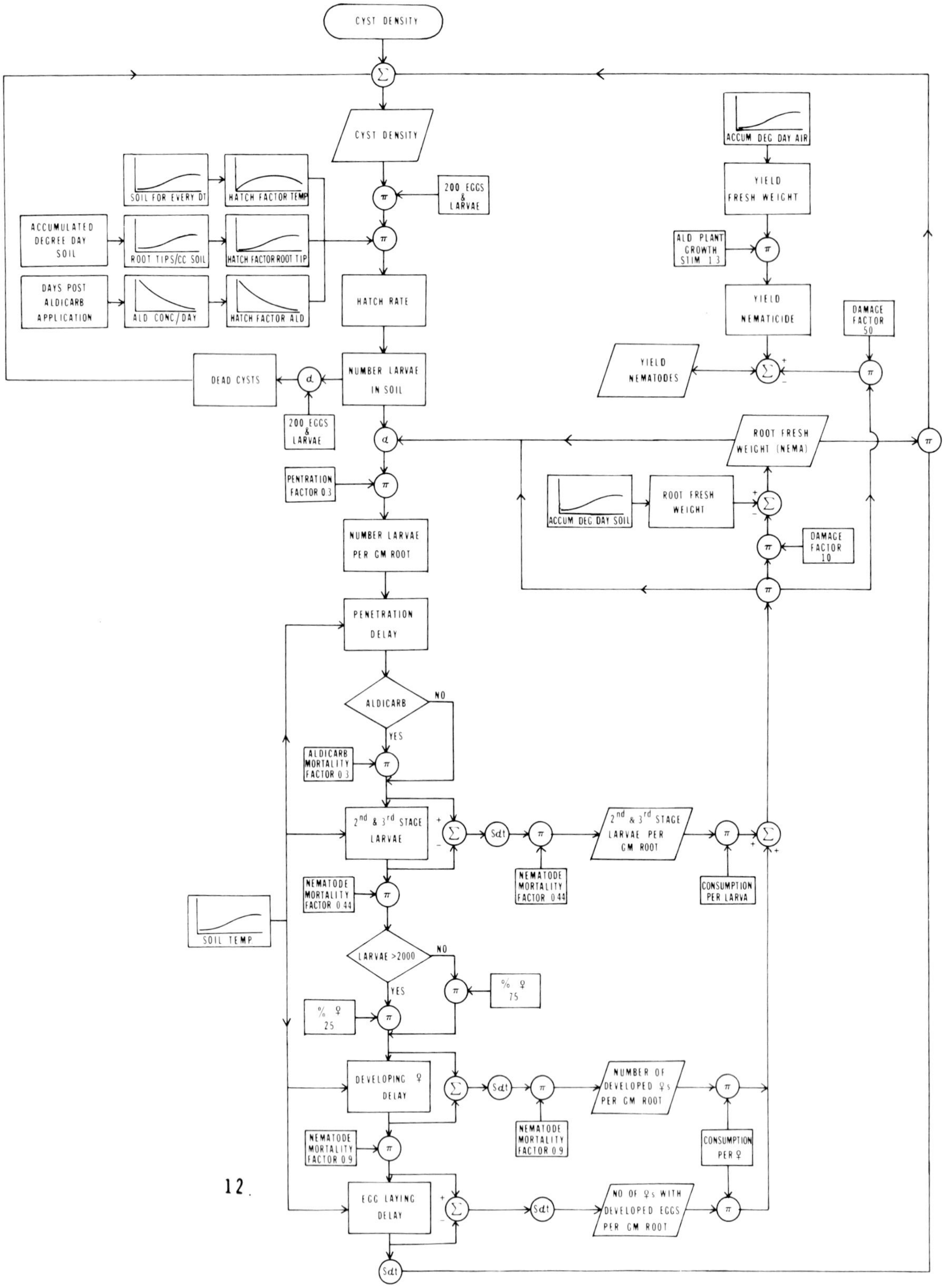

Fig. 12. A conceptual model of a computer simulation of *Heterodera schachtii* associated with *Brassica capitata*.

Table 2. Computer program output from NEMACORN, a microcomputer program for analysis of the benefits of nematicide application in corn production.

NEMACORN Input		
	Field Size	100 acres
	Potential Yield	125 bu/A
	Estimated Price	$3.50/bu
	Nematicide Cost	$15,000/A
	Nematode Problem Description	
	Area #1 (circle, 400 ft in diameter)	
	25 ft radius	50% loss
	50 ft radius	25% loss
	100 ft radius	10% loss
	200 ft radius	5% loss
	Area #2 (ellipse)	
	10 × 100 ft	30% loss
	Area #3 (circle)	
	50 ft radius	50% loss
NEMACORN Output		
	Infested Area	3.1 acres
	Crop Loss	32.3 bu
	Estimated Gross Without Nematode Problem	$43,750.00
	Estimated Gross Without Nematicide	$43,637.00
	Estimated Net With Nematicide	
	Entire Field Treated	$42,250.00
	Infested Areas Treated	$43,702.90
	Treatment Benefits (Losses)	
	Entire Field Treated	($1,387.00)
	Infested Areas Treated	65.94
	Nematicide Costs	
	Entire Field	$1,500.00
	Infested Areas	47.06

ROLE OF COMPUTERS IN NEMATOLOGY

Nematologists have used computers for data analysis for many years. Today, numerous other attributes of computers are readily available for use in nematode research, teaching, extension, and regulatory programs. The availability of microcomputers and user-friendly software have made word processing a common feature of many nematology programs. Computer graphics is another area related to data analysis in which there have been numerous changes in the past few years. Both regular computer graphics and three-dimensional systems are now readily available in black and white or color through commercial software packages. Numerous advances have also been made in data collection directly from computerized laboratory instrumentation and environmental monitoring devices. In this presentation, the role of computers in nematology will be limited to computerized information communication and data-base management systems.

DATA-BASE MANAGEMENT SYSTEMS

The science of data-base management has evolved rapidly during the past 20 years (15). Several specific data-base management systems (DBMS) have been designed for pest-management purposes, and DBMS-software packages are currently commercially available for many computer hardware systems. Modern computer and communications technology make it possible to integrate and manage data bases on an interactive basis. In an "age of information," it is essential that appropriate high quality data be readily available for diverse uses, including integrated nematode management decision-making (2). Information may be needed for making short-term decisions on a real-time basis. It may also be required for long-term decisions based on historical information. PMEX (Pest Management Executive System) and CCMS (Cooperative Crop Monitoring System) are examples of DBMS that have been used in the Michigan State University IPM program during the past 5-10 years (10, 17,18).

Pest Management Executive System (PMEX)

PMEX is an on-line computerized communications system that has been in use since 1975 (10, 17, 18). It operates on the MSU mainframe computer (CYBER 750) and uses telecommunications to interact with remote teletype terminals, minicomputers, and microcomputers throughout Michigan. The system allows for bidirectional communications among a variety of users, including nematologists, growers, extension agents and agribusiness representatives. The information available through PMEX can be classified as biological monitoring data, environmental monitoring data, predictive models and agroecosystem management recommendations. Because PMEX is an executive system, no knowledge of computers is required for users. A new user can be taught the basics of the system in less than 20 minutes. A PMEX User's Guide provides additional details (10). PMEX information is provided to users through automatic scheduling, which is an elementary form of artificial intelligence (AI). Automatic scheduling is based on a profile designed to provide each user with the information most likely to be of the greatest value. Electronic mail (CONTACT) and individual programs selected by the user are also available. PMEX access is free to MSU personnel. It is available on a fee basis to the private sector and other public sector users.

CONTACT is used by growers, extension agents and scouts to provide extension specialists and other users with biological monitoring information on a close to real-time basis. Extension specialists use CONTACT to disseminate pest, crop or biocontrol ALERTS to agricultural clientele. From 1976 through 1982 information from the MSU Nematode Diagnostic Service Laboratory could be accessed through a PMEX program named NEMALAB. This DBMS allowed users to request information about nematodes in numerous formats including state, county, crop, nematode, or grower summaries. NEMALAB ouput for the samples from Montcalm County which were processed for nematodes in 1980 is given as an example (Table 3).

The environmental monitoring component of PMEX consists of four programs, AG-ADVISORY, FORECAST, TEMPCAST, and WEATHER. AGADVISORY (Agricultural Advisory) is a narrative interpretation of how past, present and forecasted weather conditions will affect current farm operations (11). FORECAST is a weather forecast for Michigan and includes such items as predicted temperature, and the probability of precipitation for 10

Table 3. Output from the PMEX NEMALAB program associated with the MSU Nematode Diagnostic Service Laboratory.

NEMALAB REPORT

Calculated for Montcalm County by Nematode Genus

Summary for all crops

Date range is Jan. 1 to Dec. 31, 1980

28 SAMPLES PROCESSED

	Soil[a]					Root Tissue				
	1	2	3	4	x̄/100 cm^3	1	2	3	4	x̄/1.0 g
Pratylenchus	5	4	17	2	30	18	2	7	1	47
Meloidogyne	26	0	2	0	46	28	0	0	0	0
Heterodera-Not found										
Trichodorus-Not found										
Xiphinema	25	3	0	0	4	28	0	0	0	0
Tylenchorhynchus	22	2	4	0	23	28	0	0	0	0
Helicotylenchus	27	0	1	0	32	28	0	0	0	0
Hoplolaimus-Not found										
Criconemoides	25	0	3	0	29	28	0	0	0	0
Hemicycliophora-Not found										
Ditylenchus-Not found										
Paratylenchus	21	4	3	0	11	28	0	0	0	0
Longidorus-Not found										
Aphelenchoides-Not found										
Pratylenchoides-Not found										

[a] Nematode popultion density index: 1) 0, 2) 1-10, 3) 11-100, 4) >100/100 cm^3 or 1.0 g root.

weather districts and 22 zones. TEMPCAST provides a 5-day temperature forecast for the 10 weather districts. WEATHER summarizes weekly and seasonal weather conditions, daily weather, soil temperature data, and various degree-day accumulations for any one or combination of the 65 weather stations in the Michigan Agricultural Weather Network.

The predictive models available through PMEX include programs such as BIOSCHED. This program provides current weather and biological monitoring data from various geographic regions and advises IPM scouts when and where they should be looking for specific pests, crop development stages or biocontrol agent. EVAPOTRANS predicts moisture loss due to evapotranspiration. The output of other programs such as WEEVILCOST, FERTILIZER and IRRIGATE include production system management recommendations. WEEVILCOST is an alfalfa weevil control cost-benefit assessment model. FERTILIZER provides recommendations for fertilizer applications based on soil nutrient laboratory analysis results. IRRIGATE is designed to assist with irrigation scheduling decisions. Many of the pest-management recommendations made through PMEX, however, are disseminated through CONTACT.

PMEX currently has about 200 users. Although it provides specific on-line information in as close to real-time as possible, it was not designed as a flexible DMBS. In the late 1970's, CCMS (Cooperative Crop Monitoring Service) was developed to overcome this deficiency.

Cooperative Crop Monitoring Service (CCMS)

CCMS is a DMBS designed at Michigan State University. Since monitoring crops for evaluation of performance with respect to pest occurrence, development, abundance and crop damage is an essential part of IPM, and various agencies and individuals allocate considerable resources toward the inspection of crops for specific problems, the availability of a high quality DMBS is imperative. Unfortunately, most of this type of information has not been standardized, even though the same or similar information may have been collected by a number of different institutions. This problem frequently precludes the use of these data in many types of evaluations. CCMS provides users with systematic and semi-quantitative information about current and potential problems in an array of crops throughout Michigan. In addition, the system produces an historical data base which enables individuals to examine trends in crop-pest dynamics and other features useful in forecasting damage and loss potential caused by pests. CCMS was designed as a DBMS for personnel already collecting this type of information. It assists them in the standardization of information so that it will be useful to as many individuals and agencies as appropriate. CCMS offers a standard, preserved format for information on crops, pests, beneficial organisms, control, economics and the environment.

CCMS is computer-based for rapid data processing and summarization. A TI990 minicomputer (Texas Instruments) was designed for CCMS, and appropriate DBMS software written. Data for CCMS are recorded and submitted via mail on computer mark-sense forms. Individual forms have been developed for each crop, based on the needs of specific clientele. Cooperators are taught how to use CCMS. Electronic data-recording instruments will most likely replace the mark-sense forms in the relatively near future.

The most popular summaries generated from CCMS data are year-end reports. These are prepared for individual fields in IPM programs and are made available to growers through private IPM organizations. MSU currently prepares these summaries as payment for the data submitted by growers and scouts. All CCMS data are maintained confidentially. The year-end reports are used by growers for long-range planning and decision-making related to crop rotation, IPM strategies and pesticide purchases. The field (firm-level) data in the year-end reports are frequently compared with other summaries, such as those prepared for analysis of local, regional, or state situations.

In 1983, the MSU Nematode Diagnostic Service Laboratory terminated the NEMALAB program on PMEX and initiated NEMALAB II on CCMS. This substitution has provided a much more powerful and flexible DBMS. However, a number of changes had to be made in the information submitted in conjunction with soil and root tissue for nematode analysis (Table 4). It will be used in 1985 to generate computerized nematode reports that will be returned to clientele (Table 5).

In 1984, CCMS began a transition from the hardware-software DBMS designed at MSU to a VAX minicomputer using INGRESS as the DBMS software. This change was successful and is a good example of how rapid developments in the computer industry have made DBMS technology widely available. The transition, however, still required the assistance of computer and DBMS specialists.

CONCLUSIONS

Both systems science and DBMS are very likely to play important roles in the future of nematology. Computer technology will continue to develop rapidly. For example, al-

Table 4. Michigan State University Nematode Diagnostic Service Laboratory sample submission form.

Grower ______ Present Crop ______

Address ______ Variety ______

______ zip ______ Planting Date ______

County ______ Township ______ Future Crop 19 ____: ______

Section ______ Quarter Section ______ Past Crops 19 ____: ______

19 ____: ______

CCMS Manual # ______ Site # ______

Field I.D. ______ Sample Date ______ Nematicide Treatment ______ (date)

Soil Type ______ ______ (chemical)

SAMPLE METHOD (check): As described in Ext. Bull. E-800 ( ) MSU IPM Manual ( ) Other ( ) describe

REASON FOR SAMPLE (check): Problem Avoidance/IPM ( ) Problem Diagnosis ( )

DISEASE SYMPTOMS

Above Ground Symptoms	% of Plants in Field With Symptoms
Yellowing	
Necrosis	
Stunting	
Wilting	
Other	

Distribution (check): Entire Planting ( ) Localized ( ) Scattered ( )

Below Ground Symptoms (check): Galls ( ) Excessive branching ( )

Rot ( ) Reduced Growth ( )

COMMENTS:

SAMPLE SUBMITTED BY: ______

ADDRESS: ______

______ COUNTY ______

DO NOT WRITE BELOW THIS LINE

NDSL Sample Number: ______ Date Received ______

though "Fifth Generation" computer hardware and software are not currently available to biologists, it is important to mention the future potential of artificial intelligence (AI) to the science of nematology (9). When AI becomes a reality, it should have a significant beneficial impact on nematology. This developing resource will be particularly important for a field with limited human and financial resources. It is imperative for nematologists to work closely with professional systems scientists, computer programmers, and the knowledge engineers of the "Fifth Generation" (9). Through these interactions, future nematology research

Table 5. Computerized client output from the MSU Nematode Diagnostic Service Laboratory.

NEMATODE DIAGNOSTIC SERVICE LABORATORY REPORT

Grower: DONALD SMITH
County or State: VAN BUREN
Sample date: 2/10/83
Present crop: POTATO, Planted on 0/0/83
Field I.D.: NW
Future crop: POTATO, Planting date: 0/0/84
Sample submitted by: TOM JONES

NDSL sample no.: 483
Received on: 2/15/83
Analysis method:
Centrifuge, Shaker

Symptom distribution: SCATTERED
Above ground symptoms: YELLOWING (10%); WILTING (10%)
Below ground symptoms: REDUCED GROWTH; GALLS

NEMATODES RECOVERED FROM SAMPLE

Nematode	# per 100 cm^3 soil	# per 1.0 g tissue	Total	Threshold
ROOT LESION NEMATODE	50	55	105	+
NORTH. ROOT-KNOT NEMATODE	10	100	110	+
STUNT NEMATODE	6		6	-
DAGGER NEMATODE	3		3	-

Threshold + = at or above; - = below; +/- = border line
? = level unknown

Occurrence of beneficial nematodes and fungi:
SAPROPHAGOUS NEMATODES - FEW
PREDACEOUS NEMATODES - FEW
ENDOMYCORRHIZAL FUNGI - COMMON
NEMATODE TRAPPING FUNGI - NONE

DIAGNOSIS: Nematode problem.
GENERAL RECOMMENDATION: Action advisable.
REFERENCES: Control of Insects, Diseases, and Nematodes on Commercial Vegetables. MSU Ext. Bul. No. E-312, see pages 44 and 45.
SPECIFIC RECOMMENDATION: Apply fumigant or non-fumigant nematicide. Plant a variety that is tolerant to the root-lesion nematode.
COMMENTS: Root-lesion nematode is most likely serving as a predisposition agent for wilt fungi. This results in the disease complex known as "early-die" of potatoes. Superior is the most susceptible variety to this disease complex.

G. W. Bird, Extension Nematologist

and education programs will remain at the cutting-edge of science. Our society cannot afford to have well-trained nematologists investing significant amounts of time working on modeling, hardware, and software systems in isolation from the experts in systems science and computer programming. A highly significant number of young nematologists already have been introduced to systems science and modern computer technology. This progress should provide a strong base for the interdisciplinary interaction necessary for nematology to provide scientific leadership and answers to the important challenges facing our agriculture and natural resource production systems.

Literature Cited

1. Abawi, G. S. and W. F. Mai. 1983. Increase in cabbage yield by fenamiphos treatment of uninfested and *Heterodera schachtii*-infested field soils. Plant Disease 67:1343-1346.
2. Bird, G. W. 1981. Integrated nematode management for plant protection, pp. 355-375 *In* Plant parasitic nematodes, Vol. III, ed. B. M. Zuckerman and R. A. Rohde. Academic Press, New York. 508 pp.
3. Bird, G. W. and I. J. Thomason. 1980. Integrated pest management: the role of nematology. BioScience 30:670-674.
4. Bird, G. W., C. Coley, M. Sorette, and E. Meister. 1975. Potential of heuristic models for development of nematode research priorities. Nematropica 5:19 (Abstr.).
5. Brooks, Jr., F. P. 1975. The mythical man month: essays on software engineering. Addison-Wesley Pub. Co., Reading, Mass. 195 pp.
6. Carter, W. 1943. A promising new soil amendment and disinfectant. Science 97: 383-384.
7. Churchman, C. W. 1968. The systems approach. Dell Pub. Co. Inc., New York. 243 pp.
8. Dubos, R. 1968. So human an animal. Charles Scribner's Sons, New York. 267 pp.
9. Feigenbaum, E. A. and P. McCorduck. 1983. The fifth generation. Addison-Wesley Pub. Co., Reading, Mass. 275 pp.
10. Gage, S. H., D. J. Miller, R. A. Antosiak, and R. Rosenbaum. 1984. PEST Management Executive System (PMEX) user's guide. Pest Management Tech. Rep. No. 28. Dep. Entomol., Michigan State Univ., E. Lansing, Mich. 89 pp.
11. Haynes, D. L., R. K. Brandenburg, and P. D. Fisher. 1973. Environmental monitoring network for pest management systems. Environ. Entomol. 2:884-899.
12. Naisbitt, J. 1984. Megatrends: Ten new directions transforming our lives. Warner Books, Inc., New York. 333 pp.
13. Nickle, W. R, ed. 1984. Plant and insect nematodes. Marcel Dekker Inc. Pub., New York. 925 pp.
14. Noling, J. W., G. W. Bird, and E. J. Grafius. 1984. Joint influence of *Pratylenchus penetrans* (Nematoda) and *Leptinotarsa decemlineata* (Insecta) on *Solanum tuberosum* productivity and pest population dynamics. J. Nematol. 16:230-234.
15. Rogers, E. M. 1982. Diffusion of innovations (3rd ed.). The Free Press, New York. 453 pp.
16. Rullo, T. A. 1980. Advances in data base management, Vol. I. Heyden, Philadelphia, Penn. 207 pp.
17. Tummala, R. L. and D. L. Haynes. 1977. On-line pest management systems. Environ. Entomol. 6:339-349.
18. Welch, S. M. 1984. Developments in computer-based IPM extension delivery systems. Annu. Rev. Entomol. 29:359-381.
19. Zuckerman, B. M., ed. 1980. Nematodes as biological models, Vols. I & II. Academic Press, New York. 312 & 306 pp.
20. Zuckerman, B. M. and R. A. Rohde, eds. 1981. Plant parasitic nematodes, Vol. III. Academic Press, New York. 508 pp.

Part Five

Statistics and Modeling

DESIGN OF GREENHOUSE AND FIELD EXPERIMENTS FOR NEMATODE INVESTIGATIONS

L. A. Nelson

Department of Statistics
North Carolina State University
Raleigh, North Carolina, USA

There is a great need for those charged with the responsibility of planning and conducting experiments to be educated in some basic statistical principles. The purpose of this chapter is to introduce some of these experimental concepts. In addition to being familiar with these fundamental concepts, the researcher should seek advice from a competent statistician on the many details of planning an experiment, especially if the statistician is to be involved in the analysis and interpretation of data from the experiment.

VARIATION

Most biological data are quite variable, reflecting the natural variability in soils, plants, and animals. Variability should not be considered to be an undesirable property of experimental material, and thus of data. To the contrary, variability should be anticipated prior to an experiment and managed properly so that it does not mask effects in the data which are of considerable interest and importance. There are accepted ways by which natural variation of experimental plots may be controlled. One of these, a device called *grouping* or *blocking*, has been incorporated into many of the commonly used experimental designs.

IMPORTANCE OF GOOD EXPERIMENTAL DESIGN IN CONTROLLING VARIATION

The planning stage is an important phase in experimentation. Without good planning, it is possible that the results obtained from an experiment will be invalid or, even if they are valid, that maximum efficiency from the labor and money will not be attained. Good planning, especially choice of a good experimental design, is like insurance against built-in deficiencies in design which can later haunt the researcher. Experiments in nematology and plant pathology are often subjected to an extremely high degree of variability. Coefficients of variation often exceed 40%, and occasionally they are greater than 100%. Good experimental design is one way of coping with this variability.

For any experimental situation, there are usually one or two experimental designs which are obvious choices for adoption. Two important criteria used are degree of error control which the design provides and complexity with regard to lay-out and data analysis.

Some designs work well for situations in which the natural plot size is not uniform for the various factors within an experiment (e.g., the split-plot design). Some designs are useful for controlling two sources of extraneous variation (e.g., the latin square design). Still other designs will accommodate large numbers of treatments (e.g., the lattice designs). A researcher will first need to determine the special requirements of a given experimental situation and then choose an appropriate design which will have the characteristics needed. A good rule to be followed in the choice of a design is selection of the simplest design available which will provide the degree of error control required.

ELEMENTS OF EXPERIMENTAL DESIGN

Three elements of experimental design are replication, randomization, and local control.

Replication

Replication assures the availability of an estimate of experimental error and also provides a more stable estimate of means. There are several ways of estimating the number of replications required in an experiment. Table 2.1 in Cochran & Cox (4) may be used as a guide if the coefficient of variation from previous experiments is available. Other information necessary for use of this table is the degree of precision required, the magnitude of differences to be measured, and the significance level to be used. This table is accompanied by examples of its use. Regardless of the results obtained by use of this table, the amount of land, material, time, and money available for the study place an upper limit on the number of replications to be used. There is also a lower limit (usually three replications per site for field experiments) below which the experiment will not be precise enough to produce conclusive results.

Field experiments usually are conducted in series over space and/or time in order to provide a sampling of environments over which recommendations are to be made. This repetition over space and/or time provides another form of replication and broadens the base of the inferences. Also, interaction of treatments with environmental effects can be detected through replication of this kind. A sizeable interaction might imply that separate recommendations need to be made for two or more subsets of the environments.

Randomization

Randomization assures that both the estimates of error and the treatment effects will be unbiased. Thus, no treatment has a higher probability than any other of being assigned to a plot which has unusual conditions. Randomization is an objective process which involves use of random number tables or random number generators for digital computers. Some judgment must be exercised in implementing the randomization process. A randomization outcome which has a systematic appearance would usually call for a re-randomization.

Some workers have systematically arranged the treatments in one replication but randomized them in other replications. The first-mentioned replication is then used for demonstrational purposes. It would seem to be good practice to exclude the demonstrational replication from the analysis of variance which combines the results from the other replications. The utility of that replication then would be only demonstrational.

Separate randomization should be carried out for each site and/or year in a series of experiments. The exception to this is the perennial experiment, in which case a single randomization for all growing seasons is carried out.

Local Control

Local control (also called grouping or blocking) is a method of grouping together units which are similar with respect to certain environmental variables (e.g., natural soil fertility). The commonly used experimental designs differ from one another with respect to the number of restrictions on randomization, the restrictions being necessary because of the blocking. The randomized complete block design has one restriction on randomization: a complete set of treatments is randomized within each of the blocks. The latin square design has two restrictions on randomization: each row contains a complete set of the treatments and each column does also.

The alternative to randomized designs is systematic experiments. Situations involving completely systematic arrangement of treatments are usually to be avoided. They often cause problems in analysis and interpretation of the resulting data.

Most experimental designs currently being used in nematode investigations incorporate all three of the elements just mentioned.

SPECIFIC DESIGNS TO BE USED FOR NEMATODE INVESTIGATIONS IN THE FIELD

No attempt will be made here to describe the various experimental designs available for nematode investigations in the field. Such descriptions are available in several books on experimental design: Cochran & Cox (4), Cox (5), Federer (6), Snedecor & Cochran (19), Steel & Torrie (20). The randomized complete block, latin square, and split plot are three designs which are logical choices for field experiments. The first two involve two different degrees of the use of local control. The third is more of a nested situation because levels of one or more factors are nested within the plots for another factor. This design lends itself well to the case where one factor requires larger plots than the other factor(s).

One of the most commonly used designs is the randomized complete block design because it is a reasonably precise design, and yet it is simple to use. The blocks have administrative utility in addition to providing the error control. Ordinarily, the block size should be held below 15 treatments when this design is used. Otherwise, incomplete block designs may be more appropriate. In this connection, the writer has found lattice designs to be quite useful for plant pathological experiments where the number of treatments is large (e.g., 25-100).

The latin square design is more appropriate for situations in which error control in two dimensions is required and the number of treatments is ten or less. The latin square is a precise design, although slightly more complicated than the randomized complete block design.

EXPERIMENTAL MATERIAL

The experimental unit is the item(s) to which a treatment is applied. Some examples of experimental units are trees, petri dishes, and pots of soil in a greenhouse. In the field, the experimental units are plots of land.

Some researchers have become confused as to the identity of the experimental unit in their experiment and have laid out an experiment in such a way that no estimate of experimental error was available. The estimate of sampling error which was available could not be used for drawing inferences about treatments.

In some experiments, there is both an experimental unit and a sampling unit (i.e., sampling of individuals is carried out within the experimental unit). In this case, it is important to keep a clear-cut perspective of the distinction between the two types of units and the error term associated with each.

SELECTION OF TREATMENTS FOR AN EXPERIMENT

Careful selection of the treatments to be used in an experiment assures that an experiment will accomplish its purposes and avoids confusion at the data analysis stage. An important criterion in treatment selection is the purpose of the experiment. In some cases, it is to spot a winner. In other cases, it is to estimate a response curve or surface. Equal spacing of rates in this latter situation may or may not be desirable. More important is the need to obtain precise information in the region of the optimum.

If contrasts among treatments are to be made in the data analysis, it is important to select treatments which will permit the contrasts of interest to be made in a straightforward manner at the analysis stage. Confusion often results at the analysis stage if there was lack of a clear-cut set of contrasts in the mind of the researcher when the experiment was designed.

Factorial Arrangement of Treatments

One common pattern of treatments which has desirable properties is called a factorial arrangement. The treatments consist of all combinations of two or more factors, each of which has two or more levels. The data resulting from such an arrangement permit the estimation of main effects and interaction effects. The main effects are stable estimates because for each level of a factor, averages are taken over all levels of the other factors. This type of averaging is known as "hidden replication." Interaction effects reflect the fact that the pattern of response to one factor depends upon the particular level of the other factor(s).

In addition to hidden replication and the possibility of estimating interaction effects, a factorial arrangement has the advantage that one factor broadens the scope of inferences for the other factors.

One disadvantage of the factorial is that the number of treatments becomes quite large as the number of factors and the number of levels of these factors increase. For this reason, some have used incomplete factorials. Although this design saves space, labor, and experimental material, the disadvantage is that the analysis is somewhat more complicated in that the effects for each factor must be adjusted for those of the other factor(s). Care should be taken when treatments in incomplete factorials are selected so that important comparisons are possible.

It is quite common to include in the set of treatments those treatments which are factorially arranged and, in addition, certain treatments which are not factorially arranged. All treatments are assigned to the plots according to the randomization scheme without distinction as to which are factorially arranged and which are not. The factorial partitioning of the treatment sum of squares is then made at the data analysis stage.

As an example, suppose there are two nematicides A and B, each applied to the soil at two different rates, r_1 and r_2. In addition, an untreated control, C, is included in the experiment. The five treatments are as follows: Ar_1, Ar_2, Br_1, Br_2, and C. The first four treatments are factorially arranged, and the last one is outside the factorial arrangement. Contrasts which would seem logical to be obtained as part of an analysis of variance are:

	degrees of freedom
Nematicides (N)	1
Rates (R)	1
$N \times R$	1
C vs other treatments	1

LEVELS OF FACTORS IN A FACTORIAL Specific classes of qualitative factors and levels of quantitative factors must be chosen when a factorial is being planned. In addition, there is a choice of either equal or non-equal spacing of the levels for the quantitative factor(s). The total range in rates is the first decision to make in the case of the quantitative factor. It is possible to set the range in such a way that it will cause a misrepresentation of the response relationship. Previous experiments may serve as a basis for establishing a reasonable range. Once a range is established, the next decision is how many levels to use for a factor and whether or not they should be equally spaced.

If a straight-line relationship exists between the response and the rate, two levels will be adequate for that factor. If curvilinearity is suspected, at least three levels are called for in order to have a check for curvature. For factors with more complicated response curves, more levels are needed for these factors in order for the parameters of the model to be estimated.

Use of Controls in Plant Disease Studies

Controls are included as a basis for comparison or benchmark treatment in plant-disease experiments. It is important to include them, especially if the degree of effectiveness of the other treatments is unknown. As mentioned in the discussion on plot interference, the inclusion of a control may be detrimental in some plant-disease studies because it may cause plot-interference problems. In such cases, use of a treatment with known effects for the standard in lieu of a control is recommended.

In many studies, more than one type of control is needed in order to pinpoint the mechanisms of other treatment effects. For materials which consist of mixtures of several chemicals which are used to control more than one pathogen, each check should be a factorial treatment combination of presence or absence of the various components. In this situation, there could be a number of different checks in an experiment. As another example, the treatment material may be facilitated in its action by a wetting agent. In this case, a check with the wetting agent only, one with the treatment material only without the wetting agent, and one with neither treatment chemical nor wetting agent are needed.

FIELD PLOT TECHNIQUE

Experimental plot technique can have a pronounced effect upon the results obtained from an experiment. Two important aspects of plot technique are plot size and shape.

Plot Size

Plot size has been shown by Smith (18) to affect the precision of an experiment. He developed quantitative techniques for estimation of plot size from uniformity trial data. Koch & Rigney (11) adapted Smith's techniques to utilize data from designed experiments in the estimation of optimal plot size.

Smith's original method involved calculation of variances for plots of various sizes from the uniformity trial data and then regression of the logarithms of these variances on the logarithms of the plot sizes. The resulting regression coefficient was termed the coefficient of soil heterogeneity. Smith used cost data (both the cost proportional to plot size and the cost which is independent of plot size) along with this coefficient of soil heterogeneity to estimate optimal plot size for a given crop. The relationship which he developed for unguarded plots was

$$X_{opt} = \frac{b \quad K_1}{(1-b)\,K_2}$$

where X_{opt} = optimal plot size, b = coefficient of soil heterogeneity, K_1 = cost per plot which is independent of plot size, and K_2 = cost which is proportional to plot size.

For guarded plots, if a block consists of a single row of plots side by side, the end guards (area A) may be considered as lying outside the block. The area occupied by the side guards is related to the test area, X, by the ratio $B = (W - w)/w$, where w is the width of the test area and W is the width of the test area plus guards. If Kg is the cost per unit area for handling the guard areas, the cost per plot is given by

$$K_1 + K_2X + Kg(A + BX) .$$

The optimal plot size is then calculated to be

$$X_{opt} = \frac{b(K_1 + KgA)}{(1-b)(K_2+KgB)}$$

Pearce (16) confirmed that Smith's (18) approach is a good approximation to the relationship between variance and plot size, although it lacks a firm theoretical basis. This relationship appears to hold over the limited range of plot sizes which is used in this method.

Hatheway & Williams (8) refined the mathematical aspects of the Koch & Rigney (11) work, taking into consideration the fact that yields of plots of different sizes formed from the same basic plots are correlated and that the yields of these plots have different variances. They claimed that their approach should also be used when dealing with data from uniformity trials.

Although Smith's procedure (18) and related techniques are useful as guides, plot size and shape are often dictated by many factors such as the dimensions of the equipment to be used and the labor available. The ultimate choice of plot dimensions will be obtained only after a number of factors are considered carefully, and it will usually be a compromise. Smith found that the loss in precision is not great over a range of soil heterogeneity values if the plot size is one-quarter to four times the optimum size. Also, slight reductions in precision due to non-optimal plot size may be offset by an increased number of replications in an experiment.

Plot Shape

Most examples of the use of Smith's procedure (18) deal with crop yield studies rather than plant disease. The general recommendation resulting from such experiments has been to use long, narrow plots. These recommendations are not necessarily to be followed for studies of plant diseases, especially those spread by airborne inoculum. Long, narrow plots have too much exposure to contaminants from neighboring plots, which leads to interplot interference (probably negligible with nematodes). Van Der Plank (21) used the term "representational error" to refer to the effect on the response reading of a plot caused by contaminants of a neighboring plot. Square plots are in order for many plant disease experiments. The size of the square plot will be dependent upon the amount of protection from neighboring plots which is needed.

Plot Interference

The existence of interference among plots has been established by several investigators (1,2,3,10). In many cases, the use of buffer areas around the harvest area of a plot (e.g., border rows of some crops) as traditionally done in yield tests, will not be adequate to handle the sizeable interferences which occur in some plant-disease research. Fortunately, this problem is less severe with most nematode-induced diseases than with those caused by many fungi which infect plant foliage.

There has been more research designed to establish that interferences among neighboring plots exist than studies to determine what techniques should be used to reduce this interference. Paysour & Fry (15) found that the magnitude of inoculum loss from a plot and exchange with other plots was a function of plot size, shape, spacing between plots, and steepness of the disease gradient. Neighboring plots that differed markedly in disease severity were more influenced by interplot interferences than were plots with similar disease severities. Plot sizes and spacings that limited interplot interference to acceptable levels were identified. James & Shih (9) did not recommend a specific plot size or shape for plant-disease assessment studies. They indicated that only by conducting uniformity trials using disease records rather than yield data could such information on plot size and shape be obtained. There are other considerations in plant-disease plot technique. Large plots (a minimum of 90 m^2) are required to create certain types of epidemics with characteristics similar to a large field. One method by which the plot interference problem could be handled was suggested by James & Shih (9) for cereal-crop disease experiments. They used areas between plots, the buffer crop being a cereal variety not susceptible to the disease under study. The problem with such an approach is that wherever there are gaps in the field (different heights of crops), there will be edge effects on the perimeters of the treatment plots due to the availability of more moisture, sunlight, etc., than in the center of the treatment plot. The total size of the experiment becomes quite large also if such a procedure is adopted. As a result, this problem usually produces a tendency towards more variability in the experimental results. Some of this plot interference can be minimized in nematode experiments by the use of 4-row plots and by collection of all data from the two center rows of each plot.

The selection of treatments to be used in nematode and plant-disease research can often have a pronounced effect on the interference among plots. The omission of untreated checks often lessens the degree of interplot interference. When chemicals are used for eradication, interference problems can occur when the range of effectiveness of the materials is great. This problem should receive careful consideration when nematicide experiments are designed, especially in locations where high rainfall or irrigation often results in movement of a compound from one plot to adjacent ones. Shoemaker (17) described three different plot techniques to be used in experiments involving airborne pathogens, depending upon the goal of the experiment and the stage of testing of chemicals.

EXPERIMENTAL TECHNIQUE

It is not enough to use a suitable design for an experiment and to select the experimental material in an appropriate way. Good experimental technique is also necessary for the success of an experiment. Although it is not strictly a statistical matter, statisticians can make an important contribution to a research project, pointing out areas where the researcher can improve technique and thereby increase the precision.

Nelson (14) pointed out that some ways of improving experimental technique are to:

1. Write out procedures for conducting various phases of the experiment and a time schedule for their execution;
2. Make all personnel dealing with the treatments, plots, and data aware of the various sources of error and the need for good technique;
3. Apply the treatments uniformly;
4. Exercise sufficient control over external influences so that every treatment produces its effect under controlled, comparable conditions;
5. Devise suitable measures of the effects of treatments;
6. Prevent gross errors.

Data from previous years may give clues as to how experimental technique might be improved. Statisticians may design new exploratory experiments to investigate ways of improving technique in a particular experiment. Additional information often may be obtained from existing experiments by superimposing sampling and measurement studies on existing plots. This approach allows estimation of variance components for plots, samples in plots, and measurements on samples in plots.

USE OF SUPPLEMENTARY VARIABLES FOR STATISTICAL CONTROL

Readings on supplementary variables should be made when it is clear that sizeable variation in the response variation due to exogenous variables exists and it cannot be controlled by the experimental design. Such information may be used later in an analysis of covariance to adjust treatments for the information on those exogenous variables and to improve the estimate of experimental error. One commonly used covariable is number of plants per plot.

ON-FARM EXPERIMENTS

There has been considerable interest recently in designing experiments to be conducted on farms where disease problems actually occur. Such trials are usually carried out on farmer-owned land using as much as practical those resources available to the farmer. The treatments to be included can be carefully selected from more extensive experiments on research stations. The farmer can also have some input as to which practices he would like to have compared. Usually, the current farmer practice is one of the treatments.

On-farm experiments not only have a research utility, but also they are of value for extension purposes. The research is usually planned by a multidisciplinary team, which could mean that the results would be of multidisciplinary utility. The close working relationship developed between the disciplinary extension specialists and clients makes possible the feed-back of information in two different directions.

There are a number of types of experiments

ranging from those conducted on the experiment stations to those conducted strictly on farms. In this section, we are describing primarily the experiments on farms either by a trained technician or by the farmer under the supervision of the technician.

Difficulties in Conducting On-farm Experiments

There are many reasons why results of on-farm research may be unreliable. First, the farmer is not usually trained for field plot research work and therefore does not know the importance of good experimental plot technique. Second, it usually takes more replications than are feasible for the farmer to use in order to have the necessary precision. Third, the farmer does not often have the equipment necessary for small plot work, so this results in large plots being used. Large plots usually produce stable results, but at the expense of large areas of land being used. For a fixed area of land, large plots imply a smaller number of replications. In developing countries, one may need to combine several adjacent small farms in order to obtain the land area necessary for one experiment.

Experiments on nematodes often should be done as a network over a number of locations. Replication in time (years) is also desirable. It would be very difficult to standardize the size of plot, variety, experimental technique, etc., over an entire series of locations if different farmers were involved at the various locations. For this reason, each on-farm experiment would need to stand more or less on its own from an interpretational point of view. The benefit of the stability of results from averaging over locations would be lost if it were necessary to conduct a self-contained experiment at each location and the experiments did not have common design, levels of auxiliary variables and management. The possibility of estimating interaction would also not exist.

If precision is a problem on an on-farm experiment, it means that non-response to new technologies may occur even when the new technology is effective. This problem will send the wrong signal to the farmer who might be anxiously awaiting the results of the experiment conducted on his land.

This discussion should not discourage those who wish to conduct on-farm experiments. It is more to inform those individuals that the characteristics of these experiments (less replications, large plots, lack of experienced technicians to carry out all phases) suggest that their precision will not be high. This type of test results in less information for the money spent than for a research station experiment, but there are compensating factors such as the extension utility. There is no doubt but that the problems which the farmer is facing can be identified better through this approach.

The design of on-farm experiments needs considerably more study before this research practice is adopted on a widespread basis. There ought to be some innovative approaches in adapting traditional experimental designs to those new on-farm applications.

GREENHOUSE EXPERIMENTS

The principles of experimental design apply to greenhouse experiments just as they do to other types of experiments. Again, the goal is to design an experiment that is as simple as possible and which will provide the necessary information on the problems being studied.

Gomez & Gomez (7) reported that greenhouse experiments with rice are not as precise as one might expect because such few plants are involved per experimental unit, which causes much variation among experimental units. Also, since space is often limited, border rows are seldom used to take care of the competition effects.

Some of the sources of variation in the greenhouse are as follows:

1. Temperature gradients due to steam pipes under the bench, doors, and proximity to ventilators;
2. Shading effects from outside caused by structure of the greenhouse, adjacent buildings or trees;
3. Moisture differentials caused by air currents near door openings, higher evapotranspiration by plants on the outside of the bench and by non-uniformity in methods of water application;
4. Light intensity (especially in northern latitudes during the winter months).

A careful study of the conditions in the greenhouse should be made when planning a greenhouse experiment. Visual inspection may suffice in some cases, but devices for measuring light, temperature, and relative humidity will provide the needed objective data to identify variability patterns in the greenhouse environment. Uniformity trials (blank experiments) may be useful in some cases for assessing the variability patterns. LeClerg (12) reported the results of such a uniformity trial for sugar beets grown in a greenhouse.

The experimental designs used in the greenhouse will often be the same as those used in field experiments. The randomized complete block design is often a good choice because the blocking which it provides will control some of the environmental variation mentioned above. The latin square design will be used in greenhouse experiments in which two-way error control is required. The completely randomized design is appropriate if it is known that the environment is quite uniform within the greenhouse. This design has the advantage that equal replication is not required from one treatment to the next.

GROWTH CHAMBER EXPERIMENTS

Experiments conducted in growth chambers are often deficient with respect to one or more aspects of experimental design, primarily in the use of replication and randomization. Work of Lee & Rawlings (13) which was conducted in the North Carolina State University phytotron confirmed that the basic principles of experimental design are as important in growth chamber experiments as in the greenhouse or field. They pointed out that growth chambers are not as uniform within as many have assumed.

It is essential that replications of growth chambers either in time or space be provided if temperature, relative humidity, or day length are the controlled treatment factors. The chamber then is the experimental unit and needs to be replicated either over space or, better yet, over time (multiple runs). Plant-to-plant variation within the chamber is never a proper estimate of experimental error for the temperature, relative humidity or day length factors.

For within-chamber experiments, it is usually best to have multiple plant experimental units and to assign the treatments randomly to the units according to an appropriate design, such as the completely randomized design or, if blocking is needed, according to the randomized block design. The experimental unit in a phytotron often consists of an individual plant. Even in the case of "within-chamber" experiments, repetition of the experiment over chambers and/or time is desirable to assure that an individual chamber effect has not biased the results.

It is also possible to combine the "among-chamber" and "within-chamber" experiments into one split-plot-like experiment that provides information on variation due to the temperature, relative humidity or day length, as well as on a factor or several factors which are randomized within the chamber, and also the interaction(s) between the factor(s). The design considerations, as discussed for the among-chamber and within-chamber experiments, apply for the whole-plot and sub-plot portions, respectively, of this combined experiment.

CONCLUSIONS

The planning phase of experimentation is important. There are many decisions to be made at that time which might ultimately affect the reliability and the outcome of an experiment. One or two experimental designs often emerge as logical choices based on the criteria of simplicity and also precision. The treatments which will provide the information needed on the questions under study need to be chosen carefully. The comparisons to be made at the data analysis stage will need to be determined at the planning stage in order that the appropriate treatments will be included, that the analysis will be straightforward, and that it will provide the necessary discrimination among important comparisons.

One must be clear about what is the appropriate experimental unit in an experiment and then use randomization for assigning the treatments to the experimental units. Replication is needed for an estimate of experimental error and for stabilization of estimates of means. Replication in time (growing seasons) and space (different locations) is also desirable. Local control, if carried out effectively, improves the precision of experiments. Several important experimental designs incorporate this device. In addition to the experimental design, an appropriate experimental technique is necessary. Several ways of improving technique were suggested. For variation due to exogenous variables which the experimental design fails to control, statistical control (covariance) is often useful.

Plot technique is quite important in field experiments. Size and shape of plot to employ will often be determined by a number of practical considerations regardless of the estimates of optimal sizes obtained by quantitative procedures such as that of Smith (18). One such consideration in disease studies is plot interference. This problem can be controlled most readily by choice of treatments to include in the experiment, but there are other techniques such as use of large square plots, separation of plots, and use of systematic designs. This author does not recommend use of systematic designs because they often cause statistical and interpretational problems.

On-farm research has several important advantages, but often precision is lacking. It is often difficult to design the experiments as a network in order that the treatment effects may be averaged over a number of environments. More work needs to be done to develop innovative techniques for designing on-farm experiments.

Greenhouse experiments require the same design principles as field experiments. Local control is often needed due to drafts, shading effects, temperature variation, etc. Greenhouse experiments do not have the advantage of plot stability that field experiments have because fewer plants are usually involved.

Growth chamber experiments often have serious design defects. There often is not a good estimate of experimental error for the temperature, humidity or daylight treatment (the whole-plot treatment).

Literature Cited

1. Bainbridge, A. and J. F. Jenkyn. 1976. Mildew reinfection in adjacent and separated plots of sprayed barley. Ann. Appl. Biol. 82:477-484.
2. Cammack, R. H. 1958. Factors affecting infection gradients from a point source of *Puccinia polysora* in a plot of *Zea mays*. Ann. App. Biol. 46:186-197.
3. Christ, R. A. 1957. Control plots in experiments with fungicides. Commonwealth Phytopathological News 3:54,62.
4. Cochran, W. G. and G. M. Cox. 1962. Experimental designs, 2nd ed. Wiley, New York. 611 pp.
5. Cox, D. R. 1958. Planning of experiments. Wiley, New York. 308 pp.
6. Federer, W. T. 1955. Experimental design: theory and application. MacMillan and Co., New York. 544 pp.
7. Gomez, K. A. and A. A. Gomez. 1984. Statistical procedures for agricultural research with emphasis on rice, 2nd ed. Wiley, New York. 680 pp.
8. Hatheway, W. H. and E. J. Williams. 1958. Efficient estimation of the relationship between plot size and the variability of crop yields. Biometrics 14:207-222.
9. James, W. C. and C. S. Shih. 1973. Size and shape of plots for estimating yield losses from cereal foliage diseases. Exp. Agric. 9:63-71.
10. James, W. C., C. S. Shih, L. C. Callbeck, and W. A. Hodgson. 1973. Interplot interference in field experiments with late blight of potato (*Phytophthora infestans*). Phytopathology 63:1269-1275.
11. Koch, E. J. and J. A. Rigney. 1951. A method of estimating optimum plot size from experimental data. Agron. J. 43:17-21.
12. LeClerg, E. L. 1935. Factors affecting experimental error in greenhouse pot tests with sugar beets. Phytopathology 11:1019-1025.
13. Lee, C. and J. O. Rawlings. 1982. Design of experiments in growth chambers—uniformity trials in the North Carolina State University Phytotron. Crop Sci. 22:551-558.
14. Nelson, L. A. 1978. Use of statistics in planning, data analysis, and interpretation of fungicide and nematicide tests, pp. 2-14 *In* Methods for evaluating plant fungicides, nematicides and bactericides, ed. E. I. Zehr. Amer. Phytopathol. Soc., St. Paul, Minn. 141 pp.
15. Paysour, R. E. and W. E. Fry. 1983. Interplot interference: A model for planning field experiments with aerially disseminated pathogens. Phytopathology 73:1014-1020.
16. Pearce, S. C. 1976. An examination of Fairfield Smith's law of environmental variation. J. Agric. Sci. 87:21-24.
17. Shoemaker, P. B. 1974. Fungicide testing: Some epidemiological and statistical considerations. Fungicide Nematicide Tests 29:1-3.
18. Smith, H. F. 1938. An empirical law de-

scribing heterogeneity in the yield of agricultural crops. J. Agric. Sci. 28:1-23.
19. Snedecor, G. W. and W. G. Cochran. 1980. Statistical methods, 7th ed. Iowa State Univ. Press, Ames, Iowa. 507 pp.
20. Steel, R. G. D. and J. H. Torrie. 1980. Principles and procedures of statistics: a biometrical approach, 2nd ed. McGraw-Hill, New York. 633 pp.
21. Van Der Plank, J. E. 1963. Plant diseases: Epidemics and control. Academic Press, New York. 349 pp.

CHAPTER 14

ANALYSIS AND INTERPRETATION OF DATA FROM NEMATOLOGICAL EXPERIMENTS

J.P. Noe

Department of Plant Pathology
North Carolina State University
Raleigh, North Carolina, USA

Analysis of data from nematological experiments should follow general guidelines and principles that apply to any experimental data. Guidelines for the correct application of statistical methods are well established and detailed in a number of comprehensive textbooks. Several texts are particularly oriented toward applications in the biological sciences (5,6,14, 16). Consultation with a statistician, along with reference to a comprehensive text on methodology, is essential to ensure correct analysis of data. The type of data analysis to be used should be considered as early as possible in the design of experiments, as better designs may be available if considered prior to the experimental work. This chapter is not meant to be a complete treatment of all topics related to analysis of nematological data. Rather, it is an introduction to basic steps of data analysis, a description of necessary terms, and an explanation of specific problems in the analysis of data from nematological experiments.

PRELIMINARY STEPS IN DATA ANALYSIS

Examination of Data

An examination of the nature and structure of recorded information is the first step in data analysis. One may discover obvious errors in the recording of data and determine the suitability of certain types of statistical procedures. During visual inspection of the data, especially when plotted or arranged into frequency tables, trends may become apparent that will lead to the identification of appropriate methods of analysis.

Type of Data

Data may be qualitative or quantitative. The former represents a system of classification, the latter an experimental measurement. Quantitative data may be continuous, assuming all possible values within a given range of numbers, or discrete, assuming only certain values. Measurements are usually considered to be continuous, whereas count data are discrete because they may assume only integer values.

Plotting

Data including quantitative variables should be plotted prior to more involved analyses. Where data have been measured as a response to another quantitative variable (independent variable), they should be plotted versus that variable. Where data have been measured in response to qualitative or classification treatments, the data should be arranged by treatment and plotted on the same axis. Plotting data will indicate trends, which may suggest appropriate methods of analysis, such as linear versus curvilinear regression. It is possible to have an apparently good linear regression, as measured by an r-squared value, but in fact to have a curvilinear relationship in the data (Fig. 1).

Plotting will also indicate the occurrence of outliers in a data set. Outliers are data with extreme values, completely removed from data in similar treatments. A few erroneous data points can significantly alter the parameter estimates of a regression analysis, resulting in an incorrect model (Fig. 2). Outliers are usually caused by an accident or unusual situation in the experiment. There should be a good bio-

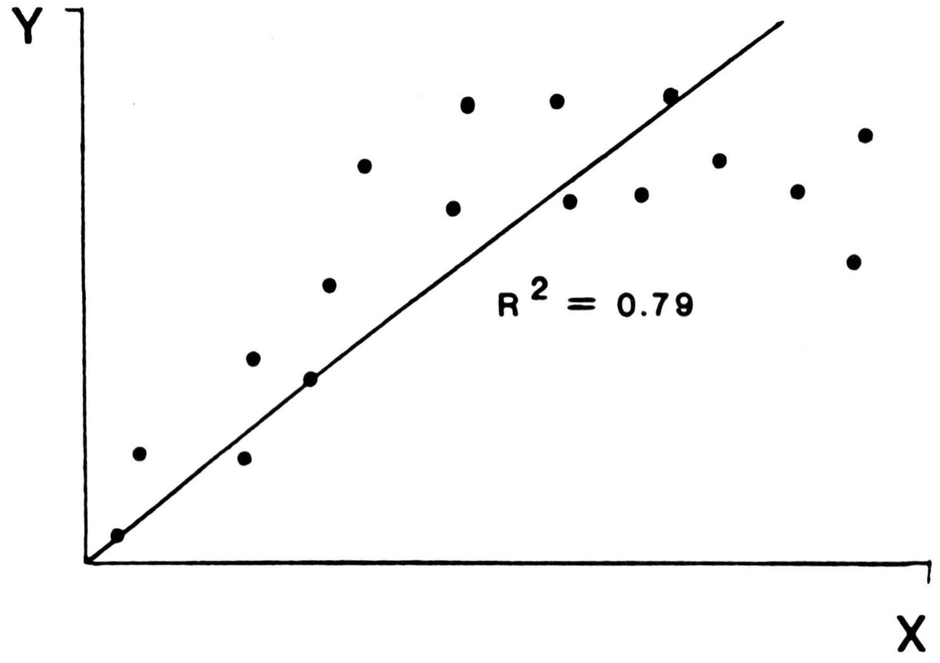

Fig. 1. Plot showing a good-fit linear regression, according to the R^2 value. However, the wrong model was selected. The true relationship is probably quadratic, not linear. The R^2 value cannot be relied on as an indication of goodness-of-fit. Data should be plotted, as shown. A plot of the residuals, after regression analysis, would also indicate poor model selection.

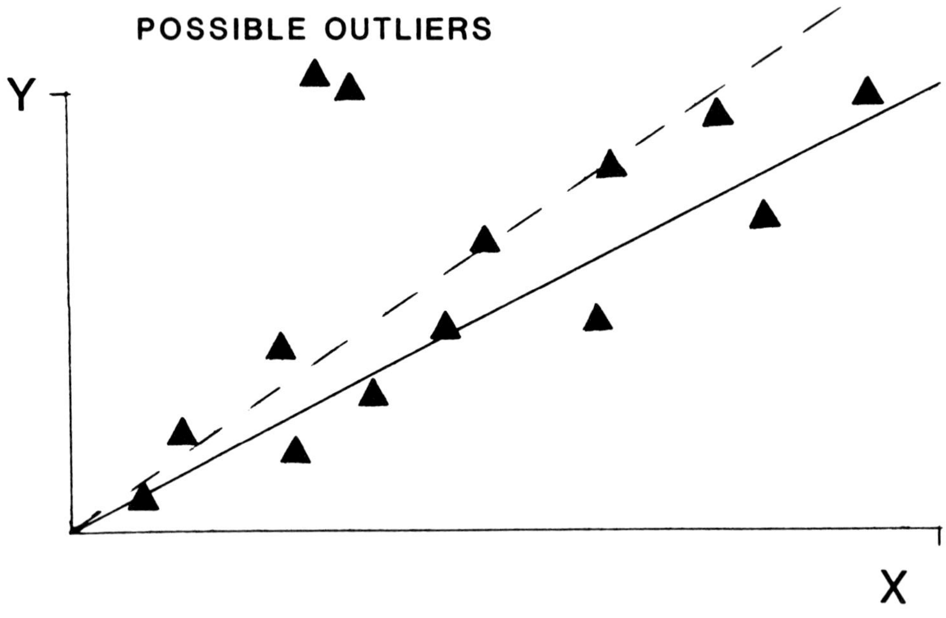

Fig. 2. Plot of regression lines fitted to experimental data, with outliers included (----), and with outliers removed (———). Including erroneous data can alter the estimated response model.

logical or operational reason to believe that a data point is an outlier. Known outliers should be removed from the data set before any further analyses are attempted.

Missing Data

Accidents may result in loss of data. Crops may be damaged by a factor not related to the treatments, inoculum may not be viable, or samples may be lost in processing. Missing data may not be critical in some types of analysis, such as regression analysis. Missing data are important, however, when a factorial design experiment is analyzed. Data with missing values may be analyzed in an unbalanced design, but this type of analysis is complex and difficult to interpret. It is recommended that the missing data be replaced with a value computed by an appropriate missing-value formula (16; pp. 209,227,389,427) and analyzed as a balanced design.

Standardization

Standardization is a method by which data measured on widely differing scales can be converted to a single, common scale with the same mean and variance. The need for standardization often arises when ecological data have been collected. Temperature, humidity, rainfall, and soil parameters such as texture, fertility, and pH may all be measured for a given experiment. There is no common unit of measurement in these variables, and scales (magnitude of numbers) are largely arbitrary. Standardization is not necessary in the application of certain methods, such as regression analysis, because regression coefficients compensate for differing scales of measurement. Standardization is usually required for multivariate statistical techniques, such as principal component, factor, and canonical variate analysis.

Converting data to percentages is a common method of standardization. Yield or nematode count data from different growing seasons or locations often are converted to percentages before analysis, so that treatments may be compared across years or locations. This technique should be used with caution, since it may lead to errors and misinterpretation of results. Dividing data from differing years by differing denominators will change the variance structure and invalidate an analysis of variance. Ideally, analyses should be done on actual data, with an appropriate partitioning of the sums-of-squares to account for effects of different years and locations. Data may then be reported as percentages, with analyses and interpretations based on the actual data.

Data Transformation

In order to meet assumptions necessary for the application of statistical methods, it is often necessary to transform data from one scale to another by applying a mathematical function to the data prior to analysis. Partial transformation of a variable or transformation of data from different treatments with differing transformations is not valid.

Data transformation is particularly important when the variance among treatments is not homogeneous or when variance is a function of the mean. A transformation which corrects this problem is referred to as a variance-stabilizing transformation. A common transformation for this purpose is $log(x+1)$. The $(+1)$ allows for the transformation of data with zero values. As an illustration of the effects of this function, the log-transformation was applied to count data of *Meloidogyne incognita* (Kofoid and White) Chitwood, producing a more symetrical, normal-appearing frequency distribution (Fig. 3). Log-transformation is also useful to remove the dependence of variance on the mean. In count data for *M. incognita*, there is a strong linear relationship between the variance, and mean, but after transformation the relationship appears random (Fig. 4). Removing the dependence of variance on the mean is essential to the correct application of most statistical methods.

Other transformations have been proposed and may be appropriate for certain data. The square-root transformation $\sqrt{x}$ may be appropriate for count data with very low values (less than 10 individuals per sample). Two transformations have been proposed specifically for data with a negative binomial distribution (1,6; p. 33). The selection and use of any of these transformations is strictly up to the judgement of the researcher, aided by a statistician. There are no guidelines for determining the need for a transformation or for selecting a specific transformation which is appropriate for a given data set. A report on the use of transformations with nematological count data (13) concluded that the only way to select a transformation was to try several and empirically determine which

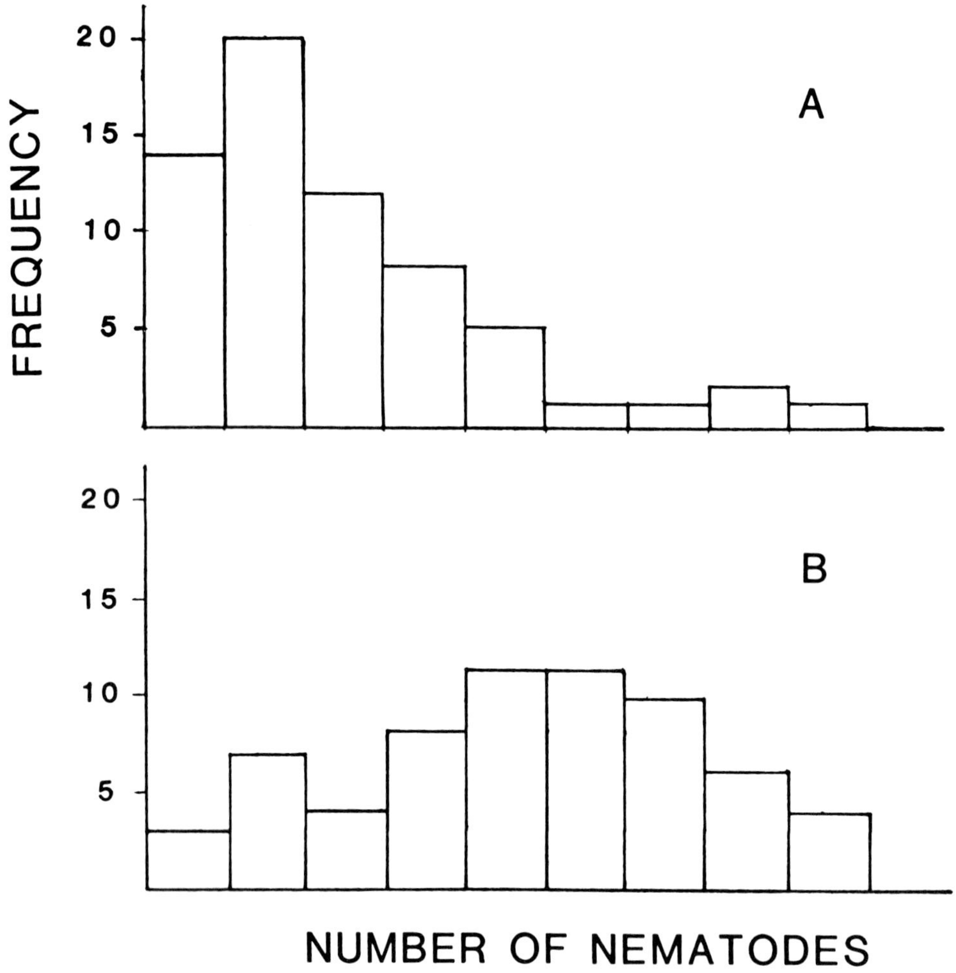

Fig. 3. Frequency distributions of *Meloidogyne incognita* count data, showing effect of log-transformation. Untransformed data, A, are highly skewed, whereas after transformation, B, the distribution is more symmetrical, and likely to be more suitable for normal-based statistical methods.

was most effective. Tests are available to assess whether or not a transformation has successfully altered the variance structure of data (6; p. 109).

Frequency Distributions

Count data should be arranged into frequency distributions for descriptive analysis. Frequency distributions will provide information on the underlying structure of the population being measured and help to determine its suitability for analysis by methods based on an assumption of the normal distribution. Frequency distributions are formed by the arrangement of the data into classes according to their rank. A class represents a range of values which occurs in the data. Classes are selected at equal intervals to cover the entire range of values in the data. Usually the data are arranged into 10-20 classes, depending on the range of values in the data, and the total number of samples. The number of samples that are contained in a given class are plotted versus the class intervals to form a frequency distribution, or histogram.

The class frequencies can be expressed as probabilities, which leads to the analysis of statistical probability distributions. The normal distribution, upon which most statistical methods are based, is a symmetrical, unimodal

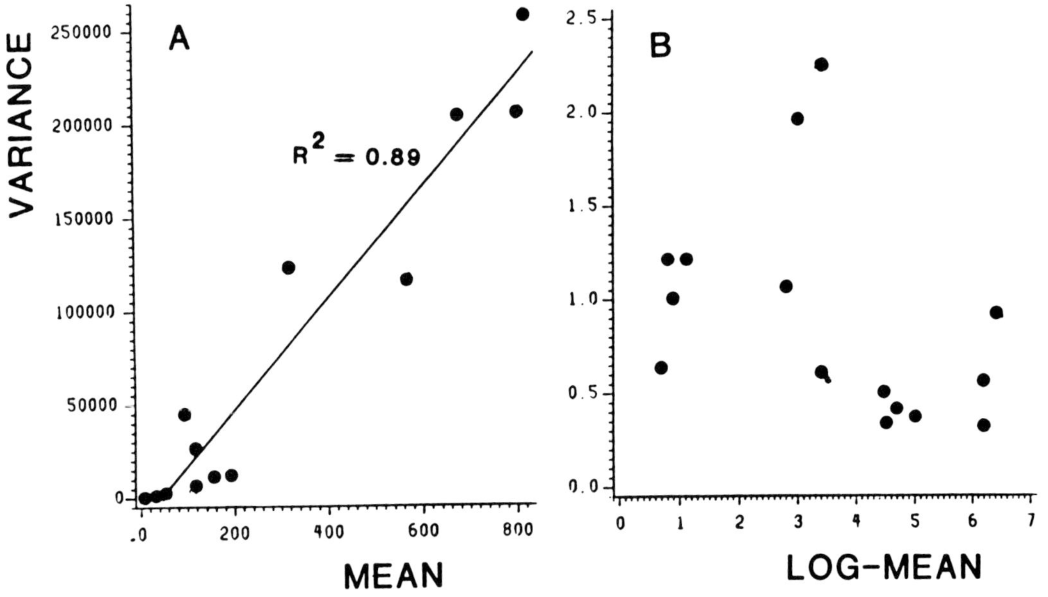

Fig. 4. Plots of mean density (number of nematodes/500 cm^3 soil) and variance for *Meloidogyne incognita* counts from 15 fields. Untransformed data, A, show a strong relationship between the mean and variance, which is unacceptable for the application of analysis of variance techniques. After transformation, B, there is no apparent relationship, and these data could be used in analysis of variance.

(one peak) distribution. An examination of the frequency distribution of experimental data should give an indication as to its symmetry, or approximate normality. Data that have very asymmetric, or skewed, frequency distributions may require transformation before the application of parametric statistical procedures. Data that are measurements of continuous variables may be expected to have a normal distribution. Yield data and environmental variables are examples of continuous measurement data.

Nematode population counts usually do not have normal frequency distributions (2,9,12). Nematode counts are highly skewed or asymmetric and typically have large variance-to-mean ratios. Egg counts, egg-mass indices, and gall ratings are usually even more skewed than juvenile and adult counts.

Many nematode population counts can be described by a negative-binomial distribution (2,7,9,12). The negative binomial is a discrete probability distribution which has been fitted to many kinds of biological count data (3,17). The negative binomial is a positively skewed distribution, which has a variance much greater than the mean. The negative binomial is decribed by two parameters: the mean, or average of the count data, and an exponential parameter k. The relationship between a frequency histogram for counts of *M. incognita* and a negative-binomial curve is shown in Fig. 5. The negative-binomial curve is fitted to the data by an iterative selection of the parameter k. After a negative-binomial distribution has been fitted to nematode counts, k may provide information about the population being sampled. A smaller value of k indicates that the data is less suitable for analysis with parametric methods. A smaller value of k may also indicate that the population is spatially clustered (15; p. 40), which means that most of the nematodes are occurring in relatively small clusters in the area sampled.

Visual gall and necrosis ratings on plant roots are representative of a logarithmic scale. The human eye is more sensitive to differences at the lower and higher ends of the scales, so that visual intervals in the middle of the scales are larger than intervals at either end. Where a rating scheme of zero (no galls) to ten (maximum galling) is used, fewer galls would be required to progress from a rating of 0 to 1 (usually only 2 or 3 galls) than from a rating of four to five, where hundreds of galls may be involved. Usually, rating intervals are larger in

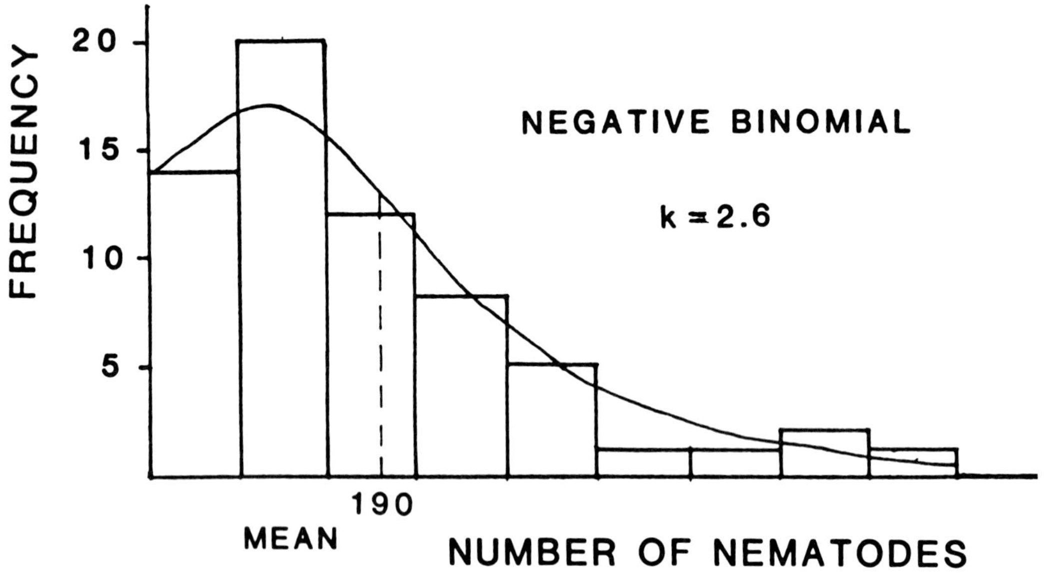

Fig. 5. Frequency distribution of *Meloidogyne incognita* counts showing a negative binomial probability model fitted to the data, with mean of 190 nematodes/500 cm^3 soil, and parameter *k* of 2.6.

the middle of visual scales, where differences are more difficult to discern, than at either end. These uneven rating intervals lead to data which are not suitable for normal statistical analyses. There are special statistical methods available for the analysis of data from rating schemes (14; p. 132).

ANALYSIS AND INTERPRETATION

Analysis of Variance and Regression (Least-Squares Estimation)

A large portion of nematological research has been concerned primarily with hypothesis testing. Analysis of variance leading to a test for differences among means has been the norm. Current trends in statistics, however, are toward the estimation of effects, as opposed to simple hypothesis testing (10). Where appropriately done, estimation of parameters with regression analysis, which leads to the development of a response model, will yield more accurate information than analysis of variance; however, the type of data analyzed will determine appropriate procedures to be employed. Where treatments are qualitative in nature, such as different nematicides or cultivars, analysis of variance and hypothesis testing are the only alternatives. Where treatments are quantitative levels of a continuous variable, such as different rates of the same nematicide or differing temperatures, regression analysis is appropriate.

Consider the analysis of yield versus differing rates of a nematicide. Hypothesis testing may indicate that one rate performed better than another rate, or that all nematicide treatments performed better than the check. These conclusions may not be all that can be gained from such an experiment. In regression analysis, a response model is selected and the parameters of that model are estimated (Fig. 6). If the model is suitable for the data, additional information is available (10). The expected yield without a nematicide (intercept), the response to increases or decreases in the rate of nematicide (slope), and the optimum rate of nematicide (maximum) can be obtained from parameter estimates. The information gained from analysis of variance and regression analysis are compared in Table 1. The selection of a response model will affect experimental design. Usually, more data points, or more levels of quantitative treatments, are required to completely describe the relationships among variables. This means more treatments in the experimental design, which brings economics and feasibility into consideration. The goals of the experiment will determine whether the additional effort is worthwhile.

In an experimental design requiring analysis of variance, the selection of orthogonal comparisons may increase the information gained from hypothesis testing. Orthogonal compari-

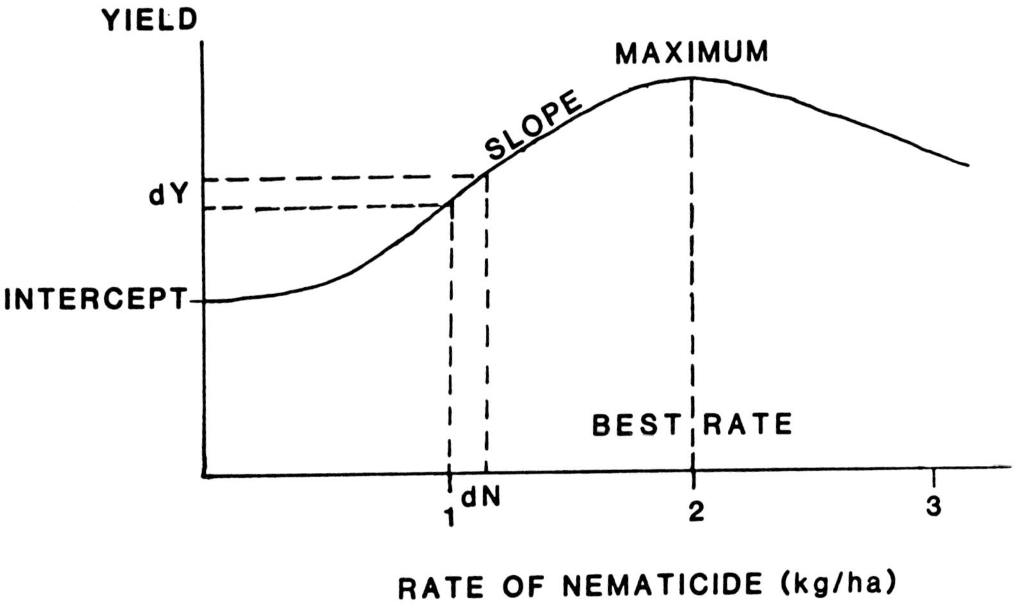

Fig. 6. Theoretical response model for host yield on rate of nematicide. The intercept will show the expected response with no nematicide. The slope will indicate expected yield response for a given increase in the rate of nematicide (dY/dN). The maximum indicates an optimal rate of nematicide, within the range of values used in the experiment. The response curve should never be extended beyond the range of experimental data.

Table 1. Comparison of information gained from hypothesis testing and parameter estimation.

Hypothesis testing (ANOVA, LSD'S, tests of means)

1. Means of treatments are/are not different.
2. A particular treatment mean differs from another.
3. A particular group of treatment means differs from other groups.

Parameter estimation (regression, least-squares analysis)

1. Mathematical model for expected response of dependent variable (yield, number of nematodes, etc.) to differing levels of treatment.
2. Expected response of dependent variable in the absence of treatment (intercept of regression line).
3. Rate of response to increase or decrease in treatment levels (slope).
4. Optimum rate of treatment (maximum).

son allows the researcher to contrast the effect of one group of treatments with the effect of another group. All cultivars with a certain parent or gene for resistance can be compared to susceptible cultivars. Nematicides with a given mode of action could be compared to those with a different mode of activity. It is essential that planned comparisons make biological sense to the researcher. The selection of appropriate comparisons should be made with the assistance of a statistician.

In any application of variance analysis, it is essential that correct error terms be specified for each hypothesis to be tested. The use of inappropriate error terms has been a common mistake in agricultural applications of these procedures (11). One cannot assume that a general-use computer program will provide a correct partitioning of the sums of squares for all models which may be involved in an analysis. In complex factorial experiments, with high levels of sampling error, as often arise in nematological research, this problem may be critical. Failure to remove sampling and experimental error from other effects may mean the difference between success and failure in an experiment. An excellent discussion of the applications of analysis of variance and hypothesis testing is given by Chew (4).

Comparison of Means

Analysis of variance may lead to the conclusion that there are significant differences among the means of treatments. A next logical question is concerned with determining which mean is the largest, or which treatment is the best. Many tests for the separation of means have been developed for this purpose. Tests commonly available include Fisher's LSD, Student-Newman-Keuls' (SNK), Tukey's HSD, Sheffe's, and the Waller-Duncan K-ratio test. One or more of these tests would be available with any general-use statistical computer program.

There is continuing controversy among statisticians over which of these procedures is the best for mean comparisons, probably because none of the tests is 'best' in every possible way. They all have particular strengths and weaknesses in their power and error rates. The power of a test refers to its ability to detect significant differences, when they exist. The type-I error rate, on the otherhand, is the likelihood that a test will show significant differences, when there are none. Usually, these opposing characteristics are balanced in the selection of a mean separation test. Two tests that are currently in favor, Fisher's LSD and the SNK, can be contrasted, in that the LSD has more power than the SNK, but it is also more likely to suffer from type-I errors (5; p. 269). The SNK may be the choice of reseachers who are conservative and who demand more protection from reporting erroneous significant differences, while the LSD may be more useful in detecting all differences among experimental treatments. Another consideration, however, is that Fisher's LSD should be protected, meaning that it should be applied only where the analysis of variance F-test was significant. The SNK does not specifically require a significant F-test, as it has other means of protecting the overall error rate. Of the other tests listed, Scheffe's and Tukey's HSD are generally more conservative than the LSD or SNK (5; p. 269). The Waller-Duncan has a completely different approach to error control and cannot be compared to other mean-separation tests. The selection of an appropriate test and significance testing levels should be made in consultation with a statistician in the design phase of experimentation. Comprehensive descriptions of the tests listed are available (4, pp. 15-24; 5, pp. 262-271).

Assumptions Necessary for the Application of Parametric Statistics

The use of standard normal-based statistical methods generally requires that (6, p. 98)

1. The data represent samples drawn from normally distributed populations, and
2. The populations to be compared all have the same variance and, therefore, that the mean and variance be independent.

Of the two assumptions, the first may be violated, and usually is, without serious consequence (16, p. 168). Violation of the second assumption is a serious problem and commonly occurs in the analysis of nematode count data. Heterogeneity of variance and dependence of variance on the mean may lead to the statistical inference of significant differences, when there is in fact no difference among treatments (14, p. 276). This error is not desirable for scientific work. Tests are available to determine whether or not the assumptions are met for a particular set of experimental data (5, p. 294).

Analysis of variance is considered a very robust statistical procedure. It is relatively insensitive to violations of its assumptions and will give reliable results in most cases. Unfortunately, it is impossible to quantify this degree of robustness. There is no predetermined method to specify the degree to which your data can vary from the necessary assumptions and still provide reliable statistical inferences.

When there is doubt about normality, homogeneity of variance, or the dependence of variance on the mean, it is recommended that analyses be done on actual and transformed data. A comparison of the results may show differences due to a failure to meet the assumptions of a given analysis. In extreme cases, it is recommended that nonparametric procedures be used.

Nonparametric Statistics

Nonparametric methods for comparisons of samples do not require the assumptions underlying normal-based statistical methods. Specifically, there are not requirements for a normal distribution or for homogeneity of variance. The correct application of nonparametric statistics does require, however, that the samples be drawn from populations with the same probability distributions. This requirement is another important reason for looking at the frequency distributions of data before analysis.

Where the assumptions for parametric methods have not been met, nonparametric methods are more powerful than their normal-based counterparts (14, p. 132; 16, p. 534). There is a smaller likelihood of making incorrect inferences about the data if the appropriate techniques are used. Most available nonparametric techniques are for hypothesis testing, but there are some simple procedures for determination of the relationship between two variables.

There are drawbacks to the use of nonparametric methods. One problem is that usually medians are being compared as opposed to means. This approach may not be a drawback in a practical sense because the mean is not a particularly worthwhile statistic in a highly skewed distribution. Another drawback is that nonparametric procedures are not available for the analysis of complex factorial designs. The experimental design of many common nematological experiments would preclude the application of nonparametric methods. Some available methods and their normal-based counterparts are listed in Table 2. While the current selection among these methods is limited, the development of robust procedures is receiving much attention in statistical research, and more suitable methods should soon be available to biological researchers (8).

Interpretation of Results

Interpretation depends largely on the experimental design selected. A complete understanding of the design and of the methods used in analysis is critical. A statistician should be consulted in the interpretive process. Results must also be interpreted with regard to the biology of an organism. In the collection of experimental data related to nematode population indices, time of observation is extremely important since the nematode is a living, moving, changing organism. Usually, when we assess a population, we are taking a thin slice out of time and freezing it. We then assume that this frozen moment in the population dynamics of the nematode will provide an accurate assessment of treatment effects. In nematicide tests, the difference in population levels may only be apparent at mid-season. Readings taken too early or too late will give ambiguous results. Population counts should be taken several times during a growing season to ensure an accurate representation of treatment effects. Time, life-cycle stages, and environmental conditions must all be considered in the interpretation of nematological research data.

Spatial patterns of the nematodes must be considered during interpretation, as well as during the experimental design phase. Since plant-parasitic nematodes usually occur in clusters in a field, it is likely that some experimental plots will have high numbers of nematodes, whereas others will have very few. This large difference in natural inoculum rates can cause many problems in experiments for the evaluation of nematicides or resistant cultivars. Control plots with few nematodes may perform as well as treated plots, and some treated plots may perform poorly due to very large nematode counts. Appropriate randomization in an experimental design and the use of a suitable number of replications is critical to account for these differences. During interpretation of results, the researcher should be aware of plots which have very high or very low nematode counts.

Meaningful interpretation of results is also dependent on the selection of correct error terms in analysis of variance and hypothesis testing. High sampling errors often occur in

Table 2. Nonparametric methods, and their parametric equivalents.

Experimental design	Nonparametric	Parametric
Comparison of two samples	Mann-Whitney	T-Test
Relationship of two variables	Spearman-rank correlation	Simple correlation
One factor, more than two levels	Kruskal-Wallace	One-way ANOVA
Two factors, more than two levels	Freidman rank-test (no interactions)	Two-way ANOVA

nematological field experiments. These error terms must be partitioned appropriately for testing of treatment effects. Inspired selection of treatment comparisons for orthogonal contrasts can add to the total information gained from an experiment. Selection of contrasts should be made carefully, however, and all comparisons should be planned in advance.

Data transformations may help to stabilize variances and provide a more accurate assessment of treatments. Data transformation may be a liability, however, in the interpretation of regression analyses. Working with transformed data may disguise the true relationships among variables (i.e., change a curvilinear relationship to a linear type), thereby masking something of biological significance (the occurrence of maximum, minimum, or optimum levels). In any case where interpretation of results does not make good biological sense, a careful examination of every phase of the experiment and subsequent analyses is necessary.

Literature Cited

1. Anscombe, F. J. 1948. The transformation of Poisson, binomial, and negative binomial data. Biometrika 35:246-254.
2. Barker, K. R. and C. L. Campbell. 1981. Sampling nematode populations, pp. 451-472 *In* Plant parasitic nematodes, Vol. 3, ed. B. M. Zuckerman and R. A. Rohde. Academic Press, New York. 508 pp.
3. Bliss, C. I. and R. A. Fisher. 1953. Fitting the negative binomial distribution to biological data. Biometrics 9:176-200.
4. Chew, V. 1977. Comparisons among treatment means in an analysis of variance. ARS/H/6. Agriculture Research Service, U.S. Dep. Agric., Hyattsville, Maryland. 64 pp.
5. Dowdy, S. M. and S. Wearden. 1983. Statistics for research. John Wiley & Sons, New York. 537 pp.
6. Elliott, J. M. 1977. Methods for the statistical analysis of samples of benthic invertebrates. Freshwater Biol. Assoc. Sci. Publ. 25. 156 pp.
7. Goodell, P. B. and H. Ferris. 1980. Plant parasitic nematode distributions in an alfalfa field. J. Nematol. 12:136-141.
8. Huber, P. J. 1981. Robust statistics. John Wiley & Sons, New York. 308 pp.
9. McSorley, R. and J. L. Parrado. 1982. Estimating relative error in nematode numbers from single soil samples composed of multiple cores. J. Nematol. 14:522-529.
10. Nelson, L. A. 1978. Use of statistics in planning, data analysis, and interpretation of fungicide and nematicide tests, pp. 2-14 *In* Methods for evaluating plant fungicides, nematicides, and bactericides. Am. Phytopathol. Soc. & Soc. Nematol., St. Paul, Minnestoa. 141 pp.
11. Nelson, L. A. and J. O. Rawlings. 1983. Ten common misuses of statistics in agronomic research and reporting. J. Agron. Educ. 12:100-105.
12. Noe, J. P., K. R. Barker, and D. P. Schmitt. 1981. Comparison of four soil sampling methods for estimating field populations of plant parasitic nematodes. J. Nematol. 13:452 (Abstr.).
13. Proctor, J. R. and C. F. Marks. 1974. The determination of normalizing transformations for nematode count data from soil samples and of efficient sampling schemes. Nematologica 20:395-406.
14. Snedecor, G. W. and W. G. Cochran. 1967. Statistical methods. Iowa State University Press, Ames, Iowa. 593 pp.
15. Southwood, T. R. E. 1978. Ecological methods. Chapman and Hall, London. 524 pp.
16. Steel, R. G. D. and J. H. Torrie. 1980. Principles and procedures of statistics. McGraw-Hill, New York. 633 pp.
17. Taylor, L. R., I. B. Woiwod, and J. N. Perry. 1978. The density dependence of special behavior and the rarity of randomness. J. Anim. Ecol. 47:383-406.

MULTIVARIATE ANALYSIS IN NEMATODE TAXONOMY

Ted H. Emigh and Major M. Goodman

Department of Genetics and Crop Science, respectively
North Carolina State University
Raleigh, North Carolina, USA

CHOOSING CHARACTERS

Characters that vary greatly among populations in relation to their variability within populations are clearly more useful for systematic purposes than are those that show lesser variability among populations in proportion to their variances within populations. Similarly, characters that are less affected by the environment in which the organisms occur usually are more useful than those that are highly subject to environmental effects. Analysis of variance can be used to select characters which have such desirable properties. Standard computer programs are available that perform such analyses and provide estimates of the proportion of total variation attributable to differences among populations, differences among collection sites, and differences among years or seasons of the collections. Such programs are found in the UCLA Biomed and the SAS software packages, at least one of which should be available at most major computing centers. In simple cases, it often is unnecessary to do more than carefully examine your data. For example, for only a few populations with data collected in only a few years or locations, the data (either raw data or means) might look like that in Table 1. Clearly, despite being affected by year-to-year (and/or season-to-season and/or location-to-location) sampling, charac-

Table 1. Sample data.

	Character X			
	Taxon A	**Taxon B**	**Taxon C**	**Average**
Year (or Season or Location) 1	5	10	15	10
Year (or Season or Location) 2	10	15	20	15
Year (or Season or Location) 3	7	13	16	12
Year (or Season or Location) 4	8	13	18	13
Average	7.5	12.75	17.25	12.5

	Character Y			
	Taxon A	**Taxon B**	**Taxon C**	**Average**
Year (or Season or Location) 1	5	10	15	10
Year (or Season or Location) 2	12	10	9	10.3
Year (or Season or Location) 3	4	10	15	9.7
Average	7	10	13	10

ter X is still a useful character effectively separating taxa A, B, and C. However, character Y is quite unreliable. It does not consistently separate the taxa: a comparison of taxon A collected in year 2 compared to taxon D in year 1 is more similar than taxon A is to itself in years 1 and 2.

Since the amounts of time, effort, and money available for any study are limited, the number of characters that can be studied is also limited. There is little reason to study both members of a pair of highly correlated characters when a third character with lower correlations is available and has an equivalent proportion of its variation attributable to differences among populations. In most studies, the number of available mutually non- (or slightly) correlated characters with high proportions of their variability attributable to differences among populations is large in relation to the amount of time available for measurement. Under such circumstances, a choice of which characters to use arises. Common sense suggests that the characters should as much as possible depict the entire organism, not some particular feature (Table 2). For best discrimination among populations, we would like σ_p^2 to be large in relation to $\sigma_s^2 + \sigma_y^2 + \sigma_{py}^2 + \sigma^2$. It is particularly important that σ_p^2 be large with respect to σ_{py}^2 or we will have little confidence in the relative accuracy of the observed population's means.

MATHEMATICS NEEDED BY TAXONOMISTS

In addition to making possible comparisons among large numbers of individuals and populations, computers make possible the intelligent use of correlated characters. The understanding of many of the procedures used for correlated characters does require some knowledge of basic undergraduate mathematics.

The use of mathematical techniques in systematics has been limited by the general lack of mathematical training received by taxonomists. It seems reasonable to expect that in the future almost all students expecting to pursue graduate work in any field of science will complete at least a year of calculus and a semester of matrix algebra. At present, however, most biologists possess little more mathematics than somewhat rusty high school algebra. Even that is quite sufficient for most currently used procedures in numerical taxonomy. This rather fortuitous circumstance results from the fact that the more complex mathematical procedures usually arise due to correlations among the characters being utilized. Choice of non-correlated characters, makes the mathematics much less difficult. Thus, a small amount of preliminary field work designed to determine which characters are essentially uncorrelated with other characters can often make up for a gap in one's knowledge of higher mathematics. Such preliminary studies can often also be used to select characters that not only differ from population to population, but that also yield consistent results from year to year and from region to region.

GROUPING POPULATIONS

Once the characters have been chosen, the populations sampled, and the specimens measured, then data are available to use in grouping the populations. This process usually has two steps. First, some measure of similarity or dissimilarity between members of each possible pair of populations studied is chosen. The number of possible similarity measures is very large indeed (1), but two general kinds have

Table 2. Example ANOVA.

Source	d.f.	Mean square	Expected mean square
Populations	9	20.2	$\sigma^2 + 5\sigma_{py}^2 + 3\sigma_s^2 + 15\sigma_p^2$
Sampling Sites Within Populations	40	5.2	$\sigma^2 + 3\sigma_s^2$
Years	2	10	$\sigma^2 + 5\sigma_{py}^2 + 50\sigma_y^2$
Populations x Years	18	5	$\sigma^2 + 5\sigma_{py}^2$
(Sites/Populations) x Years	80	4	σ^2

been used widely.

Matching and Correlation Coefficients

In addition to the familiar product-moment correlation coefficient, other measurements of agreement between characters are sometimes used. These measurements include matching coefficients and various types of nonparametric procedures, usually resulting in some form of nonparametric correlation coefficient. These procedures usually result in values ranging from 0 for no agreement to +1 for complete agreement (and sometimes −1 for perfect disagreement). As an example, consider two characters, say knob length of a nematode and knob width of the same nematode, each divided into three classes: large, medium, and small (Table 3).

A total of 53 nematodes are examined. Suppose, for each, we score it as a one if knob length and knob width fall in the same class, say small, and zero if they fall in different classes, such as small knob length but medium knob width. The average score for all 53 nematodes would then be a matching coefficient whose possible range would be from zero (for no matches) to one (for all 53 matches).

It is possible to use matching and correlation coefficients to measure the similarity between pairs of taxa as well as between pairs of characters. The data for the first taxon are treated as the X variable, and the data from the second taxon are treated as the Y variable. Then the matches between X and Y are tallied, or the correlation coefficient between X and Y is calculated. Details are given by Sneath and Sokal (1), but it should be noted that the characters are often standardized by subtracting the character mean and subsequently dividing each observation by the standard deviation among population means for that character before calculating the correlation or matching coefficient.

Distance Measures

The simplest measure of the square of the distance between two taxa i and j is:

$$D_{ij}^2 = [(X_{1i}-X_{1j})^2 + (X_{2i}-X_{2j})^2 + \ldots + (X_{ni}-X_{nj})^2] = \sum_{k=1} (X_{ki}-X_{kj})^2,$$

Where X_{kj} is character k of taxon j. Usually the characters are standardized so that various scales of measurement such as millimeters, meters, counts, etc. will not be confounded. The standardization most commonly used is the division of each character by its among-taxa standard error. Other standardizations also can be employed.

Such a simple measure of distance works quite well for characters that are reasonably non-correlated. However, when the value of one character can be reasonably predicted on the basis of the knowledge of the values of some of the other characters, the equation results in distances which differ more amongst themselves than is justified on the basis of the data. Such a situation can be avoided if the characters are adjusted or transformed so that they are not correlated before the distances are calculated. The resulting distances are called generalized distances. The procedures involved require a knowledge of matrix algebra and will not be discussed at this point.

For six samples and two characters (X and Y), suppose we have the measurements given in Table 4. To standardize these scores, we can subtract 30 from each value of X and then divide each value by $\sqrt{1220} = 34.9$. For Y, we subtract 50 and divide by $\sqrt{1110} = 36.2$. Thus, our standardized scores appear as in Table 5. (Actually, due to small rounding errors, the means and variances of the standardized scores do differ trivially from 0 and 1, respectively.) Since the correlation between the sample means of X and Y is large (about 0.96), a decision should be made as to whether to use just X, just Y, both X and Y, or neither X nor Y. In practice, we should probably choose either X

Table 3. Numbers of nematodes with small, medium and large knob widths and knob lengths.

	Small	Medium	Large	Total
Knob Length	15	25	13	53
Knob Width	17	23	13	53

or Y or a new 'character' Z that combines X and Y in an 'optimal' manner. Any one of the three methods should give about the same results. For now, we will ignore this problem and calculate the distances from both characters:

$$D^2_{I,II} = \{-.72 - (0.57)\}^2 + \{-0.83 - (0.41)\}^2 = (0.15)^2 + (0.42)^2 = 0.0225 + 0.1764 = 0.20$$

Thus $D_{I,II} = \sqrt{0.20} = 0.45$.

The complete distance table appears in Table 6. Clearly samples I, II, III, and IV differ little, while V and VI differ from each other as well as differing greatly from the other samples. The same conclusions are evident from either X or Y alone, but the technique can be used for many characters at once, where simple inspection of the data may not be sufficient for reaching conclusions.

CLUSTERING PROCEDURES

Once measures between taxa similarity or distance have been obtained, several techniques can be used to group the taxa. The most widely used one first groups those taxa that are mutually most similar or closest together to form groups, then proceeds to group the groups together on the same basis, treating remaining individual taxa as groups. This process results in a tree-like diagram called a dendrogram, where the taxa are located at the branch tips and where a pair of branches join together at the average distance (or average similarity) between all possible pairs of taxa located on the two branches; the first member of a pair being from one branch, and the second member from the other. In the example, the first step would be to group samples I and III, which have the smallest distance, 0.14. Second, samples

Table 4. Sample data for six samples and two characters.

Sample	I	II	III	IV	V	VI	Mean	Variance
X	5	10	5	15	55	90	30	1220
Y	20	35	15	40	90	100	50	1310

Table 5. Standardized data

Sample	I	II	III	IV	V	VI	Mean	Variance
X_s	−0.72	−0.57	−0.72	−0.43	0.72	1.72	0	1
Y_s	−0.83	−0.41	−0.97	−0.28	1.10	1.38	0	1

Table 6. Distance table for sample data.

	I	II	III	IV	V	VI
I	—	0.45	0.14	0.62	2.41	3.29
II	0.45	—	0.58	0.19	1.99	2.91
III	0.14	0.58	—	0.75	2.52	3.39
IV	0.62	0.19	0.75	—	1.38	2.72
V	2.41	1.99	2.52	1.38	—	1.34
VI	3.29	2.91	3.39	2.72	1.34	—

II and IV would be grouped with their distance of 0.19. Then samples V and VI are grouped at a distance of 1.34, since they are closer to each other than to any other sample. The average distance from {I, III} to {II, IV} is (0.45 + 0.62 + 0.58 + 0.75)/4 = 0.60, which is much less than the distance of either to {V, VI}, so {I, III} and {II, IV} are grouped at 0.60. Finally, the average distance from {I, II, III, IV} to {V, VI} is (2.41 + 1.99 + 2.52 + 1.38 + 3.29 + 2.91 + 3.39 + 2.72)/8 = 2.58 which completes the dendrogram (Fig. 1). A quick inspection of the data indicates that either the use of X alone or Y alone would have resulted in a dendrogram of essentially the same form.

Instead of average distances, smallest or largest distances between the members of one group and those of another can be used for clustering. However, the agreement between the original table of distances and the distances read from the dendrogram obtained is much less than is obtained using average distances. In this example, using smallest distances between members of different groups would change the grouping of {I, III} with {II, IV} to 0.45 rather than 0.60, and of {I, II, III, IV} with {V, VI} to 1.38 rather than 2.58, but would leave the form of the dendrogram unchanged. Using largest distances would result in a dendrogram of the same form also, but with {I, III} grouping with {II, IV} at 0.75 and with {I, II, III, IV} grouping with {V, VI} at 3.39.

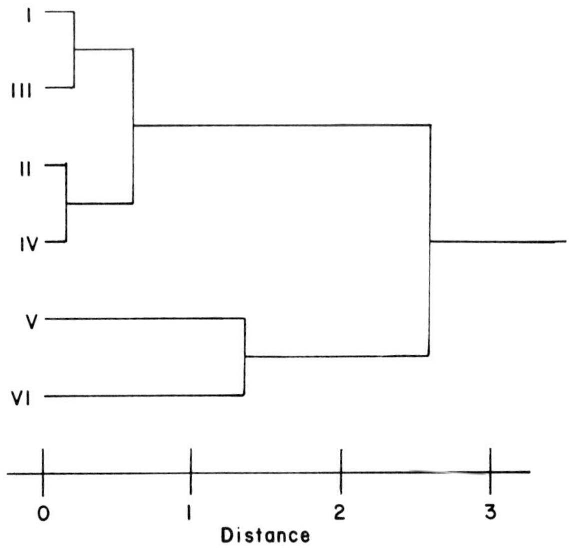

Fig. 1. Example of a dendrogram.

Rather than distances being averaged, the means themselves can be averaged after each clustering and the distances recomputed. The latter technique is called centroid clustering. Results generally differ trivially from distance averaging.

Several theoretically appealing alternatives to these procedures have been proposed. One is to form k groups in such a manner that the variation among the within-group distances is as small as possible in relation to the among-group variation. A second is to find the dendrogram for which the dendrogram distances fit closest to the actual distances among the taxa. Several criteria of closeness of fit could be used, but the usual one is least squares

$$Min \sum (D_{ij} - d_{ij})^2,$$

which is the minimum sum of squares of differences between the distances, D_{ij}, and the dendrogram distances, d_{ij}. Neither procedure has proven to be practical in the case of more than a few taxa. Good clustering procedures are found in the UCLA Biomed and NTS software packages, but *not* in SAS.

PRINCIPAL COMPONENTS AND THE BIPLOT

When it is necessary to work with sets of highly correlated characters, the simple application of distance equations is often unsatisfactory. Calculating distances from several highly correlated characters is essentially equivalent to calculating them from a single one of the correlated characters and then multiplying the distances obtained by an unknown constant. When several sets of highly correlated characters plus a few relatively uncorrelated ones must be dealt with, then the distance equations are clearly inappropriate.

The biplot is a graphical technique that to a large extent is based on principal components. In Table 5, we have two measurements of each of six populations. If we take each measurement as representing an axis on a graph, then we may use the observations to plot the populations (Fig. 2). As a result, we may cluster by physical distances. In this case, since the distances on this plot are the same as the distances calculated earlier, the clustering will be the same. However, when more than two characters are present, then we use principal components to come

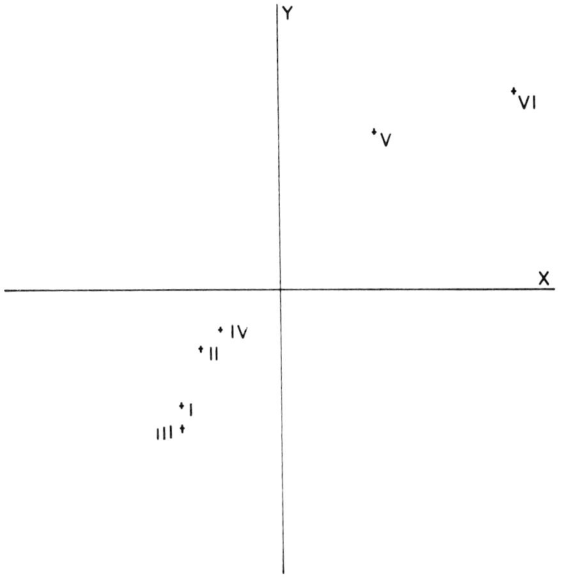

Fig. 2. Sample biplot.

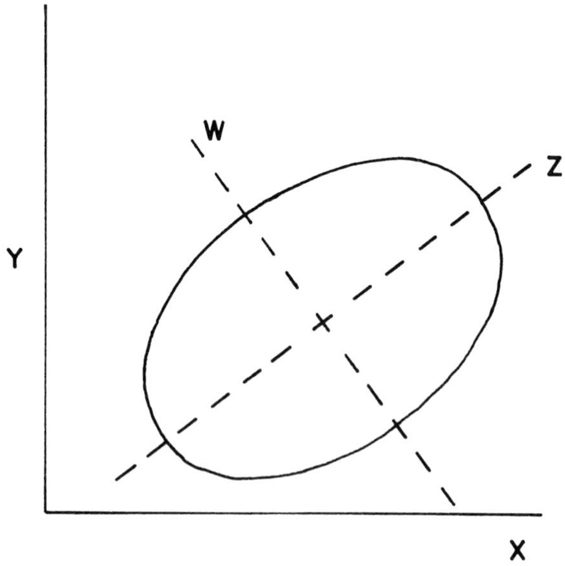

Fig. 3. Sample principal component plot.

up with functions of the characters that may act to cluster populations. The principal components simply are the new characters obtained by transforming the original characters to an uncorrelated set. If X and Y are two correlated characters, then the W-Z axes define the principal components (Fig. 3). An individual nematode or a taxon mean has values of W and Z that correspond to its values of X and Y. Theoretically, these values could be read from a graph or calculated arithmetically, but in practice computer programs provide them directly. The computations involved require a knowledge of matrix algebra, but an understanding and use of the technique does not.

The biplot can be described independently of principal components, even though its roots are in principal components. We may consider the data as a set of n characters (morphological or genetic) on m individuals or populations. A typical value is represented as y_{ij}, $i=1, \ldots n$ and $j=1, \ldots, m$, and are organized in a $n \times m$ matrix Y. The rows of Y represent characters and the columns, individuals or populations, as in the example in Table 4. Any matrix, Y, can be written in terms of its singular value decomposition as

$$Y = \Sigma \lambda_\alpha p_\alpha q'_\alpha$$

The value is called the singular value, and the p and q vectors are the singular column (character) vector, respectively. The biplot method involves using the two largest sets of singular values and singular vectors to approximate the Y matrix. We use $(\lambda_1 p_1, \lambda_x p_2)$ as the coordinates for the populations and (q_1, q_2) as the coordinates for the characters, plotting both sets of coordinates on the same graph. Populations that are near each other have greater similarity to each other than to populations that are further apart. The interaction of the plot of the populations and the plot of the characters gives some information on the nature of the differences among the populations.

EXAMPLES WITH NEMATODE DATA

In the following dendrograms and biplots (Figs. 4, 5), several taxa of nematodes were evaluated. Two sets of data were obtained: morphometric and classification. The morphometric measurements were: stylet length, knob width, knob length, DGO, and excretory pore. The classifications were: chromosome numbers, banding patterns for various enzymes, race, and survival on various media. In the interests of space, intermediate steps are not given.

Literature Cited

1. Sneath, P.H.A. and R.R. Sokal. 1973. Numerical taxonomy; the principles and practice of numerical classification. W.H. Freeman, San Francisco, Calif. 573 pp.

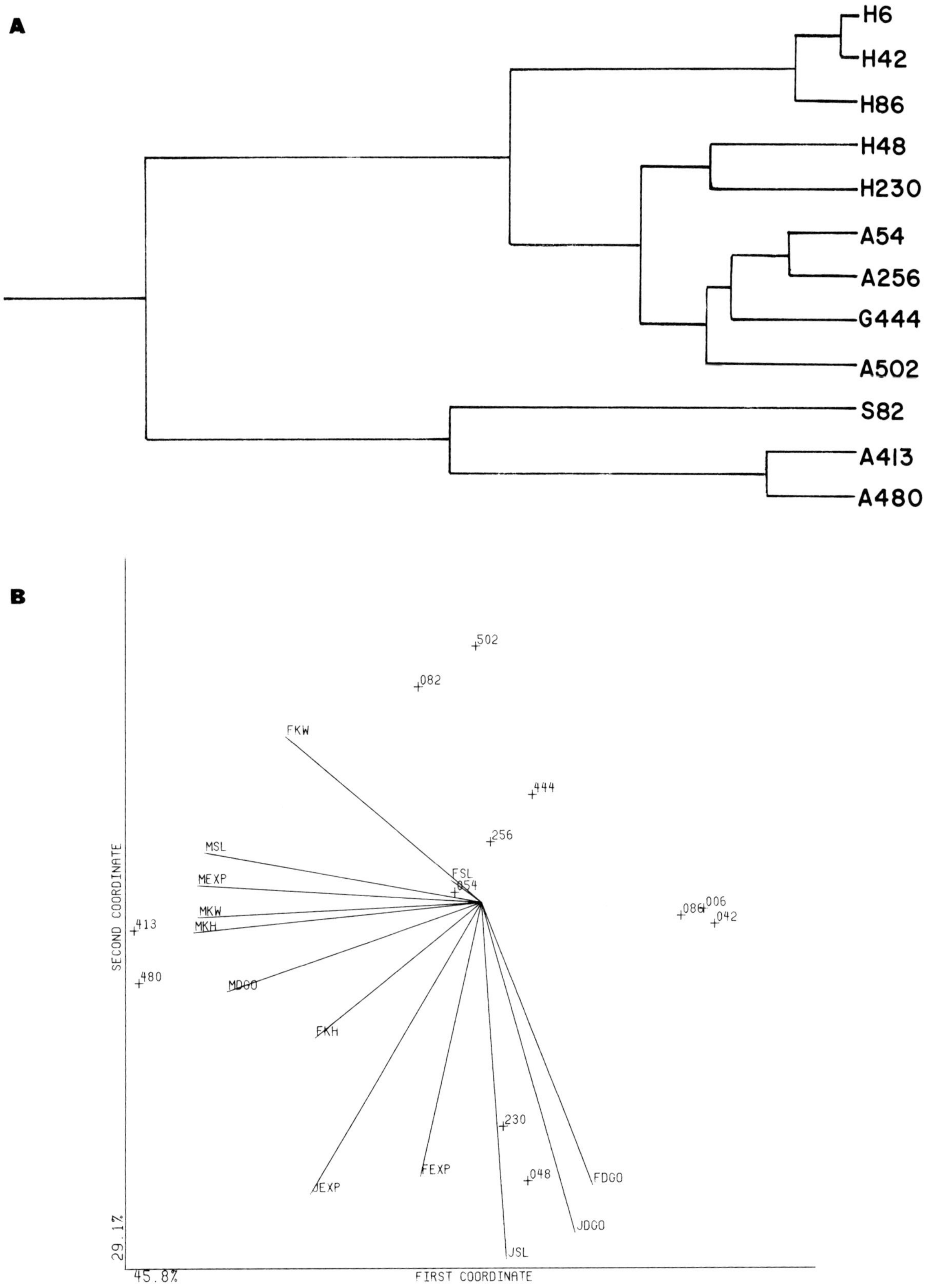

Fig. 4. Analysis of nematode populations. A) Dendrogram. B) Biplot.

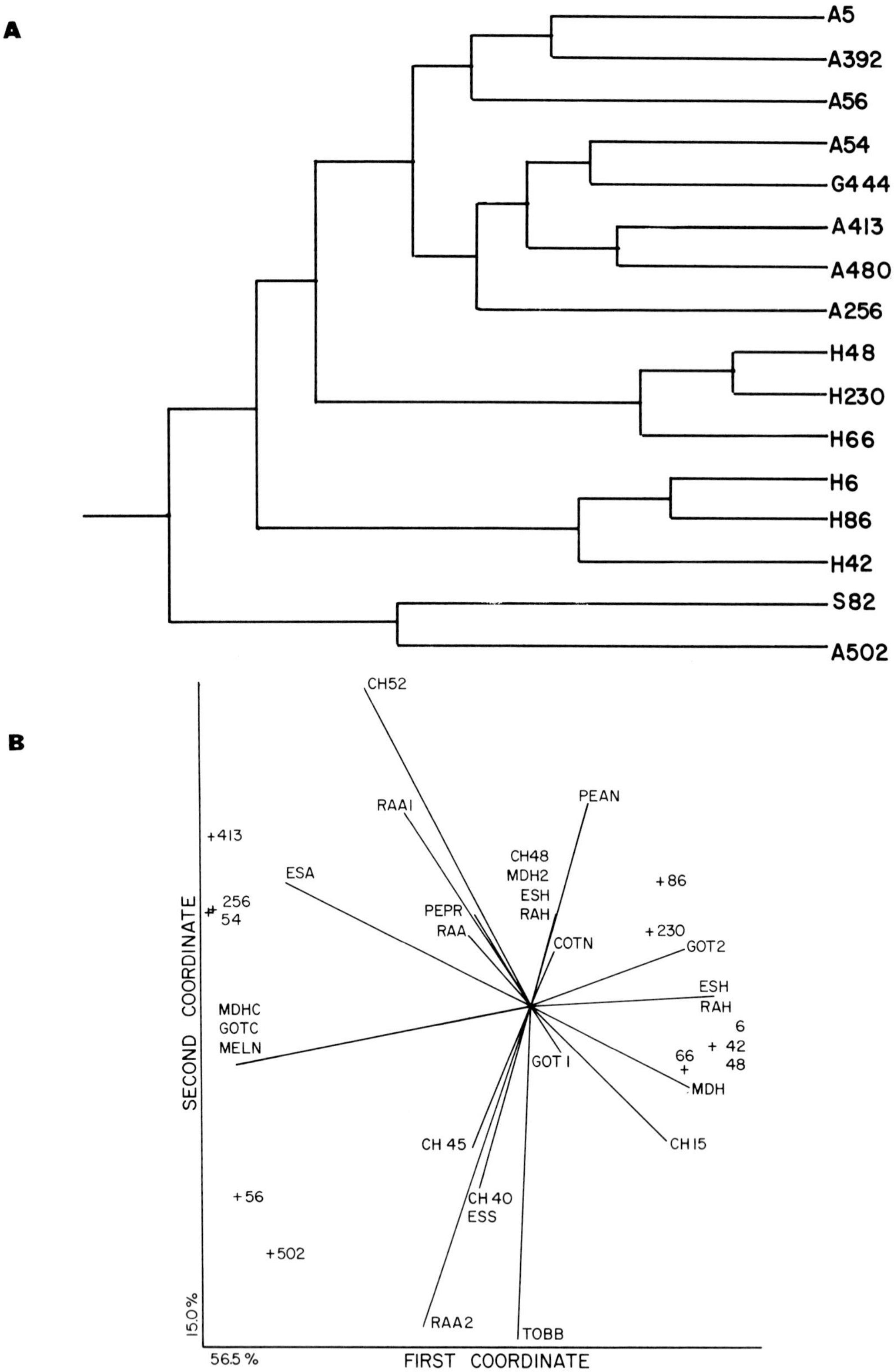

Fig. 5. Analysis of nematode population without male measurements. A) Dendrogram. B) Biplot using classification variables.

BASIC MODELING STRATEGIES FOR NEMATODE MANAGEMENT

H. Ferris

Department of Nematology
University of California
Riverside, California, USA

Models, in any sense of the word, are simplifications of larger, more complex systems, structures, or events. They promote understanding of the form and function of the system they represent. The model is a formal statement of hypothesis of system function or structure at a specified level of resolution. The hypothesis is usually developed as equations describing the processes and interactions within the system. Such formally stated hypotheses applied to agricultural pests are testable at the field level and can be reevaluated and refined based on field observations. Ultimately, they can be used as a rational basis for management decisions.

Several types of models are in use in integrated pest management. Critical point models predict crop yield, crop loss or pest densities from a single observation of the state of the system at a particular time. Multiple-point models also make such predictions, but are based on monitoring the state of the system at a series of times. Such models frequently fall into the category of analytical models since they consist of one or a group of equations for which a closed-form solution is possible. They are often correlative rather than explanatory since they cannot be sufficiently complex to describe biological phenomena realistically.

Simulation models, whose solution is determined by numerical integration techniques, are more complex. These models are generally mechanistically explanatory at a level of resolution compatible with the objectives of the modeling process. Other models of importance in integrated pest management fall into the category of management decision models. They may be extensions of analytical or simulation models. They might, for example, predict the relationship between the intensity of a monitoring or sampling process and the associated precision of the pest-population estimate.

Integrated pest management is, in practice, effective management of the crop and its associated community. The complexity of this process involves the understanding of how individual pest populations interact with the crop and with each other. Such understanding would allow prediction of repercussions throughout the system resulting from perturbation of one population or of one part of the system. The interaction among the individual populations is integrated essentially by the plant. The effects of this integration from a practical standpoint are reflected in yield and quality of the crop. A plant model is essential to the development of a multiple-pest mechanistic model. However, it may not be necessary for the simplistic analytical models, based on population assessment prior to planting, which are frequently used for nematode management.

TYPES OF MODELS

Descriptors of Spatial Distribution

Like many organisms, plant-parasitic nematodes generally exhibit an aggregated distribution pattern reflective of such aspects of their biology and ecology as feeding habits, food sources, and reproductive patterns (Fig. 1). Frequently, the distribution pattern is adequately described by the negative binomial model, a function of the current population mean and its index of dispersion (k) (5). Since the index of dispersion is a function of the mean and vari-

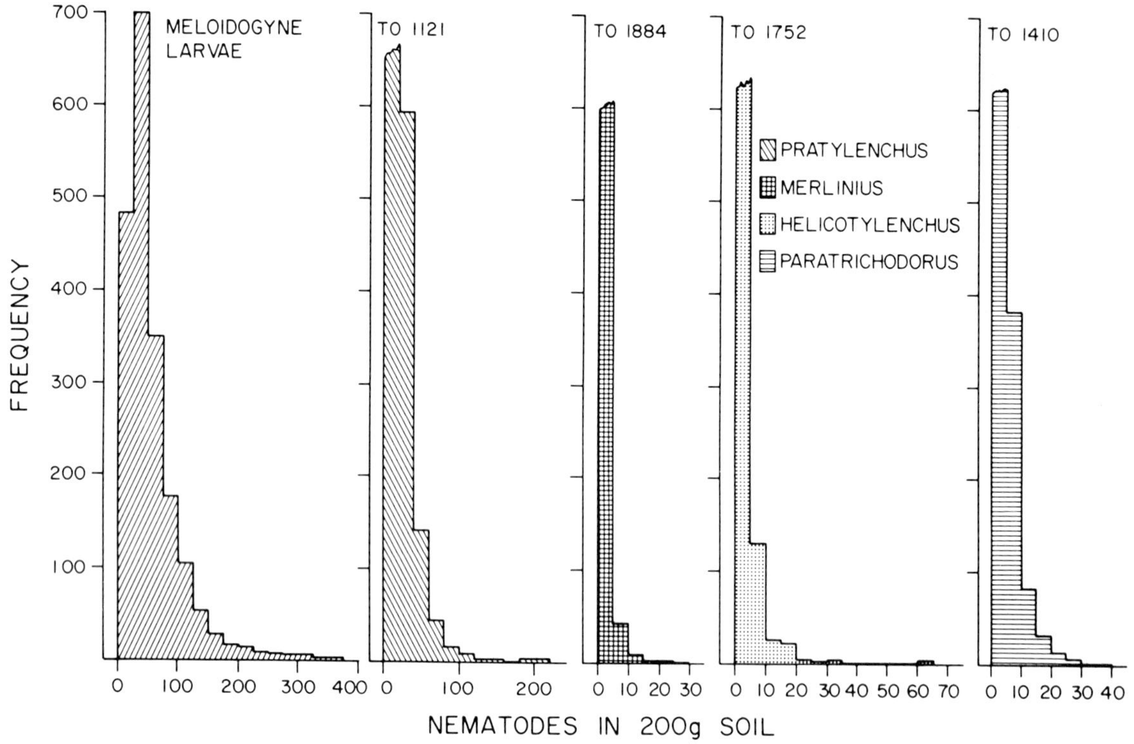

Fig. 1. Frequency distribution of nematode densities in soil samples from an alfalfa field. [from Goodell and Ferris (15)].

ance, it varies with time and population density. Means ($\bar{x}$) and variances (s^2) of sample counts can be empirically related by a power function such that $s^2 = a\bar{x}^b$, where b is an index of aggregation for the species (29,30). Assuming the stability of such indices, or knowledge of the environmental factors which influence their magnitude, it is possible to calculate the probability of obtaining a specific sample population estimate from a population of known mean and variance (5).

Knowledge of population mean and variance relationships allows generation of formulae indicating the number of samples (n) required to measure a population with a certain level of precision, given a specific population size and distributional condition (20,28). The general form of such relationships ($n = C^2(s^2/\bar{x}^2)$), where C is a probability constant provided by the standard normal deviate (Z or its t estimator) and the acceptable proportional deviation (d) from the true mean, such that $C = t/d$. The relationship can be used for determining required sampling intensity for crop management decisions of various costs (10), or the probable deviation associated with the population mean estimate given a specific sampling intensity (Fig. 2).

Spatial descriptors provide the basis for risk assessment associated with population monitoring activities and subsequent strategies in integrated pest management.

Phenology and Population Models

Both analytical and simulation forms of phenology and population models fall into this category. The systems are generally considered to be temperature driven, and the rate determining or limiting constraints include host presence, host quality, moisture availability and age-specific responses of the organisms.

Perhaps the simplest of these mechanistic models applicable to pest management are phenology models, which predict the timing of developmental events. Such events might include the appearance of various life stages of pest populations, or stages in the growth cycle of a crop relative to prevailing environmental conditions. Models at this level are useful for timing crop protection activities, or for the timing of monitoring and assessment efforts for the detection of pest species. If the expected age structure of the population is known relative to some starting point such as planting

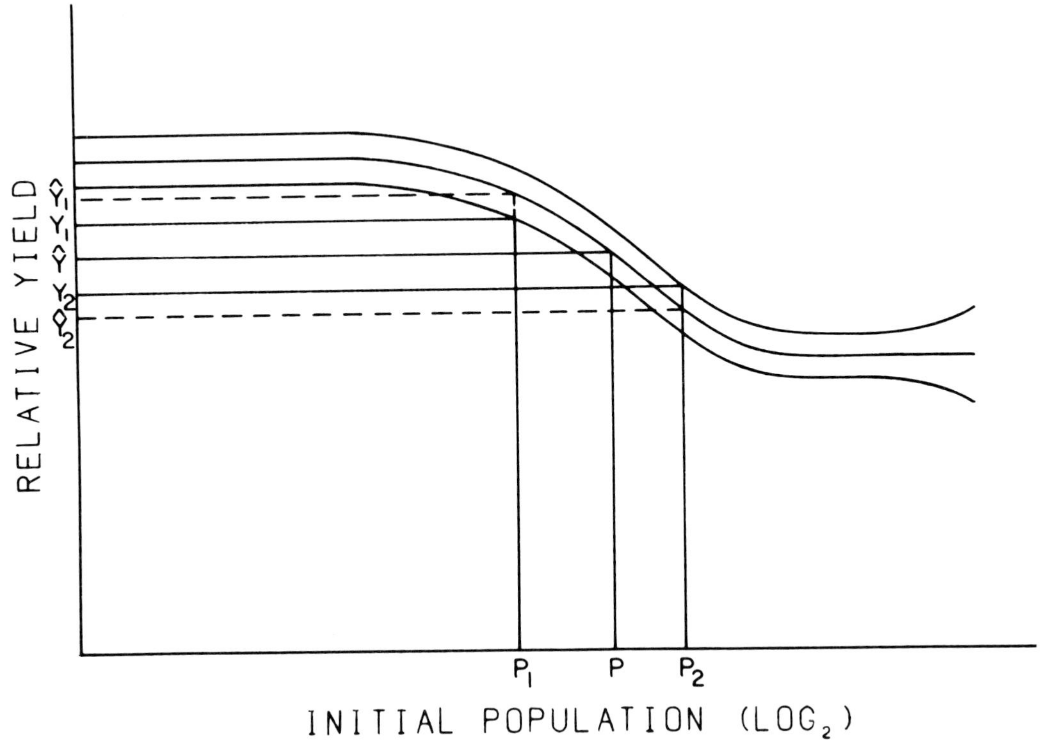

Fig. 2. Relationship between yield and nematode density indicating confidence bands associated with the damage function, economic decision value (y), economic threshold population (P), predictive population interval (P_1 to P_2), and economic decision interval (y_1 to y_2). [from Ferris (10)].

date, and only one or a few life stages of the population are being measured by the assessment technique, the probable true magnitude of the whole population can be predicted from the sample. Similarly, if the population age- and stage-structure can be determined by accumulating heat units (11,12,13,14,25,31), it may be possible to determine when overwintering populations of eggs will hatch, or when a population of hatched juveniles in the absence of a host will have depleted their food reserves sufficiently to be non-infective to a crop (Fig. 3).

Population models are more complex than phenology models. They predict or explain not only the timing of events, but also the magnitude. A pest population model would predict not only occurrence, but also the magnitude of the pest outbreak and, if linked with a plant model, the magnitude of the expected crop damage. Such models can form a basis for economic threshold considerations with greater flexibility and responsiveness to environmental perturbations than can critical or multiple point analytical models.

The population models, which predict the magnitude of population changes (either growth or decline), are potentially useful from a management standpoint in assessing the expected growth of a population under a specific crop or sequence of crops. The rate of population decline given overwintering environmental conditions may be similarly predictable, allowing projection of preplant nematode densities from samples taken at various times during the overwintering period and also allowing analysis of population changes under specific cropping sequences.

Population models may be of the analytical form frequently used in nematology: for example, the relationship between final population and initial population (27). They may also be simulation models at various levels of complexity (1,6,7,24). Models for root-knot nematodes, for example, take into account variability associated with single age cohort developmental rates as a function of the variation in the nematode and the variability of the plant-host system (12,13,14). Thus, the mean and variance for the number of heat units required for egg development of *Meloidogyne arenaria* (11), the number of heat units over which a freshly hatched, second-

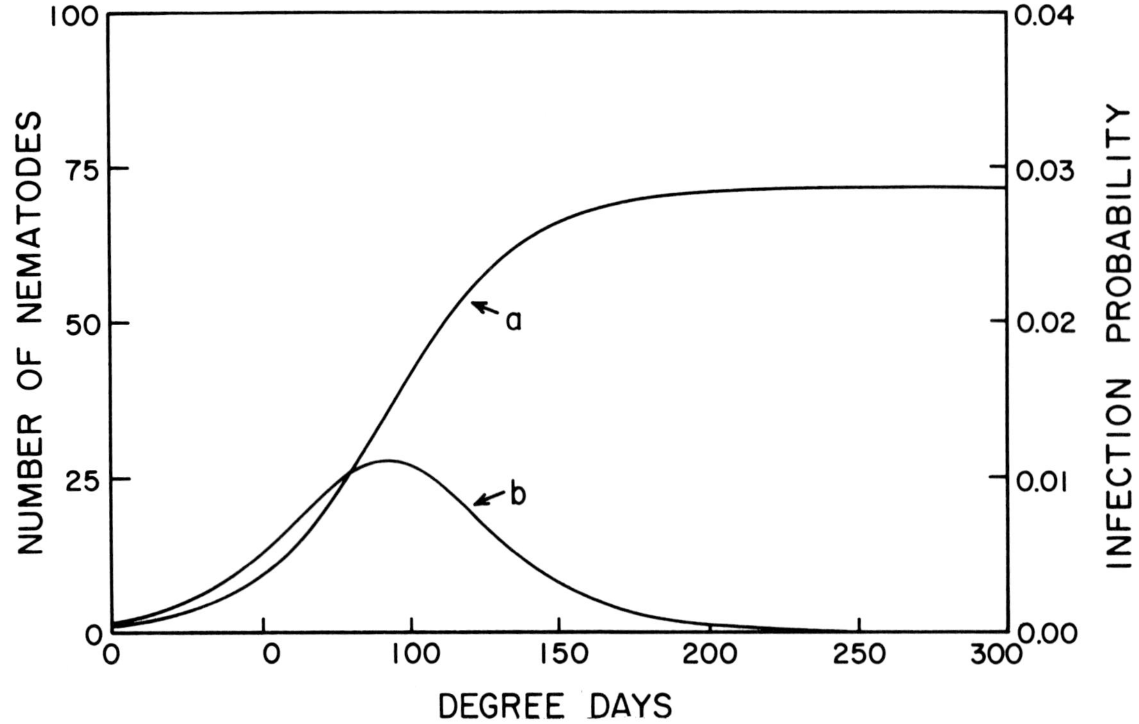

Fig. 3. Cumulative number of nematodes (a) and probability of penetration (b) in the vascular region of grape roots cv. 'French Colombard' relative to length of exposure (DD_{10}) to a single age cohort of *Meloidogyne arenaria* juveniles. [from Ferris et al., (13)].

stage juvenile is able to penetrate different varieties of grape roots, the variance associated with this penetration period, and the probability that penetration will occur have been determined (13). Similarly, the host dependence of the developmental period and its variance (12), and the variability in the fecundity period and rate of egg production/female (14) have been calculated. Survivorship values for each developmental stage have been determined (Table 1). The development of simulation models of this form requires an interactive plant-host model, the appropriate coupling structures for the effects of the nematode on the host supply/demand system (33), and the effects of the physiological stress level on the nematode biology (14) (Fig. 4 & 5).

Yield Loss Models

Damage-function predictions from population densities are critical to management decisions (6,18,34,35). Such models may be of the typical analytical form, either critical or multiple point (2,4,19,27), or they may be simulation models (1,6,7,24). Simulation models generally require a plant model in tandem with the pest models in predicting crop growth or yield loss. They are at a greater level of complexity than the analytical models and tend to have greater variability in their predictiveness. They have the advantage, however, of allowing multiple pest linkages in a biologically rational manner rather than relying on multiple regression approaches only valid for the existing data set.

Utilization of yield loss models forms the basis for economic threshold determinations. The economic threshold for nematodes can be defined as that population of nematodes which will reduce crop value by an amount equivalent to the cost of controlling or managing the nematode population. At another level of resolution, it is that level to which the population must be reduced to maximize the difference between crop value and management costs (8). These determinations usually involve critical-point analytical models; however, similar application can be made of simulation models. Forward predictions in time are necessary, involving projections of expected crop value and projections of regional weather conditions. The threshold determinations provide a rational basis for management decision.

Table 1. Stage-specific development, fecundity, and survivorship data for *Meloidogyne arenaria* on three grape varieties (from 11,12,13,14).

	Nematode life stage						
	Egg	Infective J_2		Parasititic J_3/J_4		Adult	
Variety	$\bar{x}^a$	$\bar{x}^b$	p^c	$\bar{x}^d$	p^e	$\bar{x}^f$	E^g
French Colombard	176 ± 7.5	115 ± 58	0.36	532 ± 102	1.0	580 ± 70	0.81 ± 0.07
Carignane	176 ± 7.5	113 ± 56	0.22	532 ± 90	1.0	660 ± 60	0.98 ± 0.02
Thompson Seedless	176 ± 7.5	73 ± 36	0.11	655 ± 104	1.0	550 ± 60	0.53 ± 0.03

[a] Mean and standard error of egg development period (DD_{10}).
[b] Mean and standard error of probable penetration period (DD_{10}).
[c] Proportional survivorship of penetration under unstressed conditions.
[d] Mean and standard error of development to maturity (DD_{10}).
[e] Proportional survivorship of development period under unstressed conditions.
[f] Mean and standard error of egg production period (DD_{10}).
[g] Mean and standard error of egg production/female/DD_{10}.

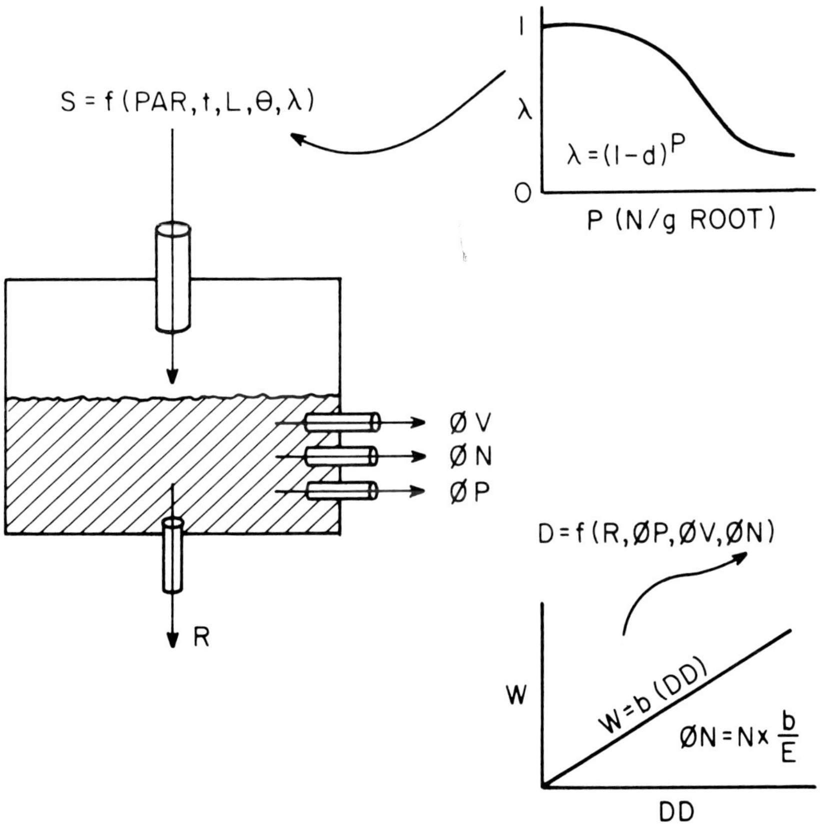

Fig. 4. Coupling structures between plant and nematode based on a metabolic pool concept of plant growth. Supply is a function of PAR, temperature, leaf area, photosynthetic efficiency of current leaf age structure, and physiological efficiency as influenced by current nematode population density. Demand is a function of maintenance respiration and the cost of new tissue formation (vegetative or propagative) including growth, respiration, and nematode demand, as a function of nematode-growth rate [adapted from Wang et al. (33) and Ferris et al. (14)].

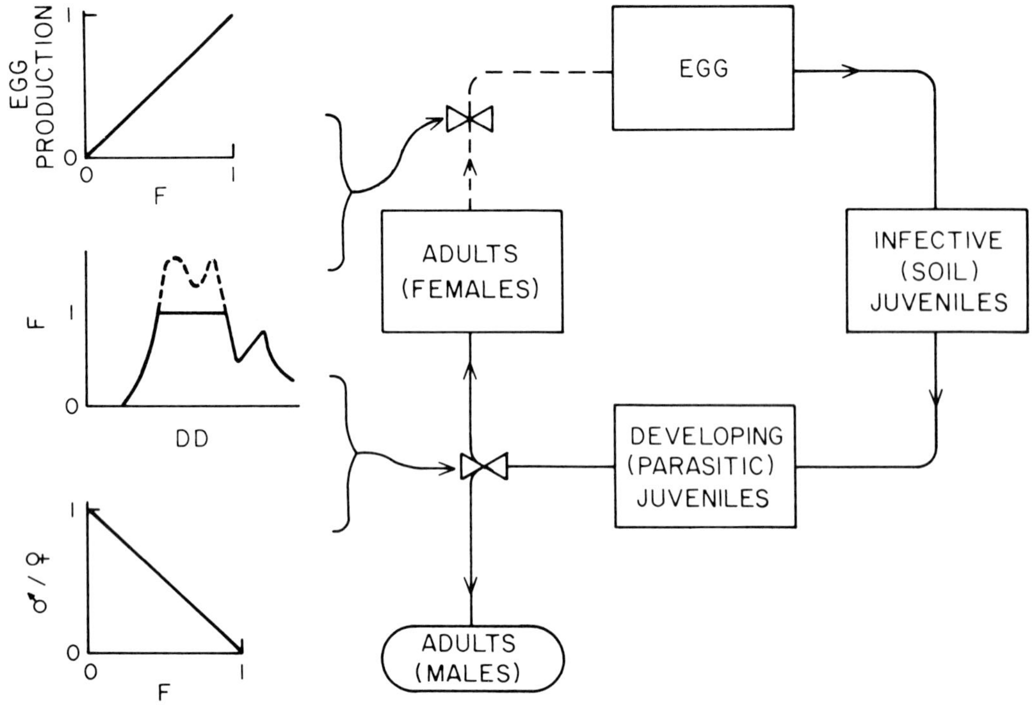

Fig. 5. Coupling structures between plant and nematode. Effect of plant physiological stress (*F*) as influenced by phenology, metabolite demand priorities and physiological stress on nematode population development (sex ratio and egg production rate). [from Ferris et al., (14)].

Management Optimization Models

Models of this type represent the next level of organization. Their accuracy and predictive reliability is constrained by the resolution and error in lower level models on which they are based. They provide a framework for optimization of the management process. Among the simplest forms are management models predicting control efficacy for a particular management approach based upon environmental conditions. Thus, the expected efficacy of a fumigant nematicide is influenced by the soil texture and pore spaces, and soil moisture and temperature, resulting in predictability in the amount of nematicide required to obtain an acceptable level of control (21,22,23). A higher level of organization is to superimpose the control-cost model on the nematode-damage function so that the amount of control to be invested, in order to reduce the nematode population to a point at which the difference between the crop value and cost of control is a maximum, can be determined. This method is an optimization approach to the economic threshold, in fact, the true economic threshold (8,17), but one difficult to achieve with available information due to lack of data on control cost functions (Fig. 6). Similar control-cost functions can be developed for crop-rotation programs where the cost of reducing the nematode population to specified levels is predictable relative to the reduced value obtained by the nonhost crop (3,8,26).

Management and optimization models can be considered in the short and in the long term. In the short term, the returns for single growing seasons may be optimized by reduction of the nematode population to a specific level, but the problem may be exacerbated for subsequent crops (2,3,8). However, data generated for analytical models of population increase during a single growing season on a given host crop (2,3) allow prediction from a preplant population estimate of the expected crop value (or crop loss) and the expected final nematode population. Further, data on nematode overwintering survival (2,16,32) allow prediction of the expected nematode-population densities at the start of the following cropping season. These relationships allow prediction of expected crop yield or value on a series of possible subsequent crops. Calculations of this type generate the possibility of optimizing cropping sequences for nematode management over multiple years (3). Such approaches are particularly appropriate with plant-parasitic nema-

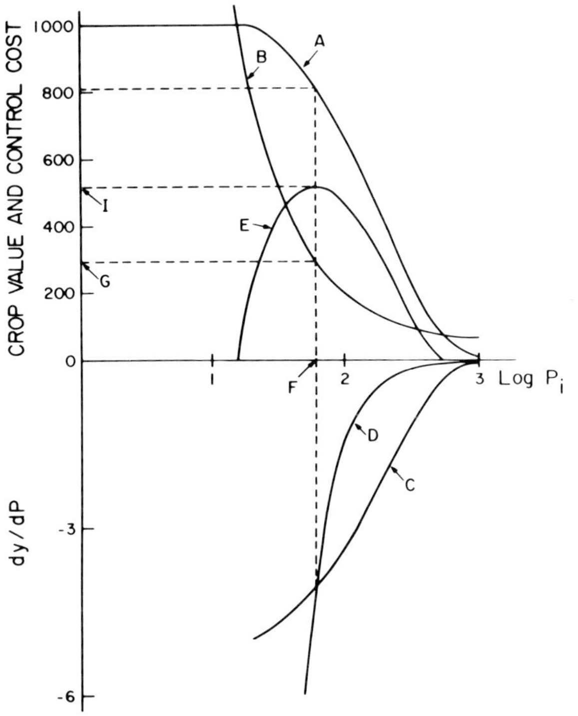

Fig. 6. Determination of the economic threshold by maximizing the difference (curve E) between the nematode damage function (curve A) and the control cost function (curve B). The economic threshold is the population level at which the derivatives of the damage function (curve C) and the control-cost function (curve D) interact. [from Ferris (8)].

todes, and perhaps with other soilborne pathogens, because of relatively slow reproductive and dispersal rates and yield loss predictability from simple analytical models.

The selection of a cropping sequence, of course, involves some forecasting of crop values over a series of years and an expectation that average weather conditions will prevail. However, the commencement of a particular cropping sequence does not lock the user into that program should conditions change over the period for which the predictions were made. The optimal solution may involve reassessment of the system through biological or economic monitoring at the beginning of each growing season and selection of the most profitable route from that point forward. Linear programming techniques are adaptable to this approach.

SUMMARY AND CONCLUSIONS

The status of modeling as a basis for integrated pest management strategies for nematodes is conceptually well advanced. The structures of models of nematode distribution, population phenology and development and those for yield loss due to nematodes have been developed and tested in various crops. Primarily these have been analytical models, simple to develop and generally useful. Attempts at simulation models have been useful in the investigation of nematode and plant systems in demanding clear hypothesis statements and in delineating efficient research approaches. They have not yet been implemented in pest management programs.

Current studies on the required intensity of nematode-population assessment relative to crop value and to the cost of the anticipated management approach (10) indicate that the basic approach of measuring a nematode population at the beginning of a growing season to determine the required management is viable. The resolution required is crop and nematode specific as well as management-cost specific. Further, since the expected damage per nematode decreases with increased nematode densities (3,4,9,19,27), and since fairly large increases in nematode density are required to achieve proportionally much lower increases in crop loss, the required precision of population assessment for management decisions may not be cost-prohibitive (10). With root-knot nematodes, where the population density can frequently be estimated by a gall rating from the previous crop, the use of modeling strategies for prediction in integrated pest management for nematodes is especially promising. The basic utility of simulation models is in hypothesis statement and testing and in the definition of priorities. Critical-point analytical models of crop loss, nematode increase, and overwintering survivorship that have rational and understandable parameters can be developed readily and are easily implemented. The biology, motility and reproductive patterns of nematodes make them an attractive biological system for the implementation of quantitative strategies.

Literature Cited

1. Bird, G.W. and I.J. Thomason. 1980. Integrated pest management: the role of nematology. Bioscience 30:670-674.
2. Duncan, L.W. 1983. Predicting effects of plant-parasitic nematode communities on crop growth. Ph.D. thesis. Univ. Calif., Riverside. 160 pp.
3. Duncan, L.W. and H. Ferris. 1983. Effects of *Meloidogyne incognita* on cotton and cowpeas in rotation, pp. 22-26 *In* Proc. Beltwide Cotton Prod. Res. Conf. 347 pp.
4. Duncan, L.W. and H. Ferris. 1983. Validation of a model for prediction of host damage by two nematode species. J. Nematol. 15:228-234.
5. Elliott, J.M. 1977. Some methods for the statistical analysis of samples of benthic invertebrates. Freshwater Biol. Assoc. Sci. Publ. 25. 160 pp.
6. Ferris, H. 1976. Development of a computer-simulation model for a plant-nematode system. J. Nematol. 8:255-263.
7. Ferris, H. 1978. Modification of a computer simulation model for a plant-nematode system. J. Nematol. 10:198-201.
8. Ferris, H. 1978. Nematode economic thresholds: derivation, requirements and theoretical considerations. J. Nematol. 10:341-350.
9. Ferris, H. 1982. Approaches to the assessment of crop losses due to nematodes, pp. 16-19 *In* Proc. Beltwide Cotton Prod. Res. Conf. 302 pp.
10. Ferris, H. 1984. Probability range in damage predictions as related to sampling decisions. J. Nematol. 16:1-9.
11. Ferris, H., H.S. DuVernay, and R.H. Small. 1978. Development of a soil-temperature data base on *Meloidogyne arenaria* for a simulation model. J. Nematol. 10:39-42.
12. Ferris, H. and W.A. Hunt. 1979. Quantitative aspects of the development of *Meloidogyne arenaria* larvae in grapevine varieties and rootstocks. J. Nematol. 11:168-174.
13. Ferris, H., S.M. Schneider, and M.C. Stuth. 1981. Probability of penetration and infection by *Meloidogyne arenaria* in grape cultivars. Am. J. Enol. & Vitic. 33:31-35.
14. Ferris, H., S.M. Schneider, and M.C. Semenoff. 1984. Distributed egg production functions for *Meloidogyne arenaria* in grape varieties, and consideration of the mechanistic relationship between plant and parasite. J. Nematol. 16:178-183.
15. Goodell, P.B. and H. Ferris. 1980. Plant-parasitic distributions in an alfalfa field. J. Nematol. 12:136-141.
16. Goodell, P.B., H. Ferris, and N.C. Goodell. 1983. Overwintering population dynamics of *Meloidogyne incognita* in cotton. J. Nematol. 15:480 (Abstr.).
17. Headley, J.C. 1972. Defining the economic threshold, pp. 100-108 *In* Pest control strategies for the future. Nat. Acad. Sci., Washington, D.C. 376 pp.
18. James, W.C. 1980. Economic, social and political implications of crop losses; a holistic framework for loss assessment in agricultural systems, pp. 10-16 *In* Crop loss assessment, eds. P.S. Teng and S.V. Krupa. Univ. Minn. Agric. Exp. Stn. Misc. Publ. 7-1980. 327 pp.
19. Jones, F.G.W. and R.A. Kempton. 1978. Population dynamics, population models and integrated control, pp. 333-361 *In* Plant nematology, ed. J.F. Southey. Minist. Agric., Fisheries and Food Tech. Bull. 7. Her Majesty's Stationery Office, London. 440 pp.
20. Karandinos, M.G. 1976. Optimum sample size and comments on some published formulae. Bull. Entomol. Soc. Am. 22:417-421.
21. McKenry, M.V. 1978. Selection of preplant fumigation. Calif. Agric. 32:15-16.
22. McKenry, M.V. 1981. The nature, mode of action, and biological activity of nematicides, pp. 59-73 *In* CRC Handbook of pest management in agriculture, Vol. 3, ed. D. Pimentel. CRC Press, Inc., Boca Raton, Fla. 656 pp.
23. McKenry, M.V. and I.J. Thomason. 1974. 1,3-Dichloropropene and 1,2-dibromoethene compounds. II. Organism dosage-response studies in the laboratory with several nematode species. Hilgardia 42:422-438.
24. McSorley, R. and J.M. Ferris. 1979. PHEX: A simulation of lesion nematodes in corn roots. Indiana Agric. Exp. Stn. (Purdue) Res. Bull. 959. 22 pp.
25. Milne, D.L. and D.P. DuPlessis. 1964. Development of *Meloidogyne javanica* (Treub) Chit., on tobacco under fluctuating soil

temperatures. S. Afr. J. Agric. Sci. 7:673-680.

26. Roberts, P.A., I.J. Thomason, and H.E. McKinney. 1981. Influence of nonhosts, crucifers, and fungal parasites on field populations of *Heterodera schachtii*. J. Nematol. 13:164-170.
27. Seinhorst, J.W. 1965. The relation between nematode density and damage to plants. Nematologica 11:137-154.
28. Southwood, T.R.E. 1971. Ecological methods with particular reference to the study of insects. Chapman and Hall, London. 391 pp.
29. Taylor, L.R. 1961. Aggregation, variance and the mean. Nature 189:732-735.
30. Taylor, L.R. 1971. Aggregation as a species characteristic, pp. 357-372 *In* Statistical ecology, Vol. I: Spatial patterns and statistical distributions, ed. G.P. Patil, E.C. Pielou, and W.E. Waters. Penn. State Univ. Press, University Park, Penn. 582 pp.
31. Tyler, J. 1933. Development of the root-knot nematode as affected by temperature. Hilgardia 7:391-415.
32. Vrain, T.C. 1978. Influence of chilling and freezing temperatures on infectivity of *Meloidogyne incognita* and *M. hapla*. J. Nematol. 10:177-180.
33. Wang, Y.H., A.P. Gutierrez, G. Oster, and R. Daxl. 1977. A population model for plant growth and development coupling cotton-herbivore interaction. Can. Entomol. 109: 1359-1374.
34. Zadoks, J.C. 1980. Yields, losses, and costs of crop production—three views, with special reference to wheat growing in the Netherlands, pp. 17-22 *In* Crop loss assessment, ed. P.S. Teng and S.V. Krupa. Univ. Minn. Agric. Exp. Stn. Misc. Publ. 7-1980. 327 pp.
35. Zadoks, J.C. and R.D. Schein. 1979. Epidemiology and plant disease management. Oxford Univ. Press, New York. 427 pp.

Index

INDEX

Subject and Taxonomic Index